Yong Wang (Ed.)
ESSE 2017
De Gruyter Proceedings

De Gruyter Proceedings

Volume 2

ESSE 2017

Proceedings of the International Conference on Environmental Science and Sustainable Energy

Edited by Yong Wang

DE GRUYTER

Editor
Prof. Dr. Young Wang
East Carolina University
Department of Geography, Planning, and Environment
Greenville, NC 27858
USA
wangy@ecu.edu

ISBN 978-3-11-053913-4
e-ISBN (PDF) 978-3-11-054004-8
e-ISBN (EPUB) 978-3-11-053914-1

Library of Congress Cataloging-in-Publication Data
A CIP catalog record for this book has been applied for at the Library of Congress.

Bibliographic information published by the Deutsche Nationalbibliothek
The Deutsche Nationalbibliothek lists this publication in the Deutsche Nationalbibliografie; detailed bibliographic data are available on the Internet at http://dnb.dnb.de.

Printing and binding: CPI books GmbH, Leck
♾ Printed on acid-free paper
Printed in Germany

www.degruyter.com

Preface

Environmental science is interdisciplinary integrating physical, biological and information sciences (including ecology, biology, physics, chemistry, zoology, mineralogy, oceanology, limnology, soil science, geology, atmospheric science, and geodesy) to study our environments, and to address environmental problems and issues. The Sustainable Energy is energy that is consumed at insignificant rates compared to its supply and with manageable collateral effects on environments. The 2017 International Conference on Environmental Science and Sustainable Energy [ESSE2017] was held 23-25 June 2017 in Suzhou, China. The annual conference was focused on the environmental science and sustainable energy, and provides a venue for professionals, researchers, and students to stimulate, share, and articulate ideas with each other. We received a total of 200 submissions worldwide. The Technical Program Committee (TPC) worked diligently to complete the review before the deadline. The final technical program consisted of 84 papers that are grouped into ten categories. The categories were Environmental Science and Ecology, Strategy of Sustainable Development, Energy Science and Technology, Control Theory and Application, Materials, Artificial Intelligence and Computer Science, Sensors, Control System and Automation, Advanced Design and Manufacturing, Emerging Fields and Other Related Fields. There were 4 keynote speeches. The keynote speakers coming from China, Hongkong of China, Singapore, and USA were nationally and internationally recognized experts in their fields. The proceedings were published as a volume in De Gruyter Publishing Company. Some excellent and well-written papers were identified and recommended to the journals that belong to Engineering Index and Science Citation Index Expanded databases.

We greatly appreciates the research community for contributing a wide range of original research manuscripts, and the publisher, De Gruyter Publishing Company for allocating resources making this special volume possible. We would like also to express my sincere gratitude to TPC members, organizers, staffs, and volunteers for their expertise, enthusiasm, time, and effort. Finally, we would like to thank all authors, presenters, and participants for making the ESSE2017 successfully.

ESSE2017 Organizing Committee

DOI 10.1515/9783110540048-202,

Committees

General Chair

Prof. NEAL Tai-Shung Chung, National University of Singapore, Singapore

Co-Chair

Prof. Teik-Thye Lim, Nanyang Technological University, Singapore

Prof. ZhaoYang Dong, The University of Sydney, Australia

Prof. Chung-Shin Jonathan Yuan, National Sun Yat-Sen University, Taiwan

Prof. Pierluigi Siano, University of Salerno, Italy

Editor

Prof. Yong WANG, East Carolina University, USA

Prof. ZhaoYang Dong, The University of Sydney, Australia

Technical Program Committee

Dr. Tseung-Yuen Tseng, National Chiao Tung Universi, Taiwan

Prof. GOH HUI HWANG, Universiti Teknologi Malaysia, Malaysia

Dr. MOHD YUHAZRI BIN YAAKOB,;Universiti Teknikal Malaysia Melaka, Malaysia

Dr. CHU Lee Man, The Chinese University of Hong Kong, China

Dr. Peng-Sheng Wei, National Sun Yat-Sen University, Taiwan

Dr. Hong-Fei LIU, Institute of Physics, Chinese Academy of Sciences, Beijing, Singapore

Prof. I-Shyan Hwang, Yuan Ze University, Taiwan

Prof. Mona Fouad Mohamed Kaiser, Suez Canal University, Egypt

Prof. Yushan Zhao, University of Wisconsin, USA

Prof. Nilay COSGUN, Gebze Technical University, TURKEY

Dr. Gilbert Y S CHAN, The Hong Kong Polytechnic University, Hong Kong

Dr. Mahmoud Ghofrani, University of Washington Bothell, USA

Dr. Soner GOKTEN, Baskent University, TURKEY

Prof. TZU-CHEN HUNG, National Taipei University of Technology, Taiwan

Dr. Marcin Dębowski, University of Warmia and Mazury in Olsztyn, Polish

Dr. Małgorzata Aleksandrzak, West Pomeranian University of Technology, Polish

Dr. HEFA CHENG, Peking University, China

Dr. Shaimaa Omran, National Research Centre, Egypt.

Dr. MEGAT AHMAD KAMAL MEGAT HANAFIAH, Universiti Teknologi MARA Pahang, Malaysia

Prof. Candida Milone, University of Messina, Italy

Prof. BAKHTIER FAROUK, University of Delaware, USA

Dr. Zai-Xing HUANG, University of Wyoming, China

Dr. Hui-Ming Wee, Chung Yuan Christian University, Taiwan

Dr. JUAN ANTONIO ZAPIEN, University of Hong Kong, Mexican

Dr. Jamshid Aghaei, Iran University of Science and Technology, Iranian

Prof. Guo-Qian CHEN, Peking University, China

DOI 10.1515/9783110540048-204,

Dr. Rong-Jong Wai, National Taiwan University of Science and Technology, Taiwan
Prof. Tzong-Chen Wu, National Taiwan University of Science and Technology, Taiwan
Dr. Ke-Jin WANG, Iowa State University, China
Dr. Arash Kamran-Pirzaman, University of Science and Technology of Mazandaran, Behshahr. Iranian
Dr. Fares A. Almomani, College Of Engineering, Qatar University, Jordanian
Dr. Hazem M. Elzarka, College of Engineering and Applied Science, Qatar
Dr. A M Mabrur Ahmad Rashedi, Nanyang Technological University, Singapore
Dr. Lin-Na DU, Wenzhou Academy of Agricultural Science, China
Dr. Sheng-Fu YANG, National Taiwan University, Taiwan
Dr.Xu-Jie LU, Hainan tropical Ocean University, China
Dr. Stig Morling, Institution Royal Institute of Technology, Stockholm, Swedish
Dr. Bader S. M. AlAnzi, Loughborough University, UK, Kuwait
Prof. Hua ZHONG, School of Water Resources and Hydropower Engineering science, Wuhan Univer sity, China
Dr. Alessandra Criscuoli, Institute on Membrane Technology (ITM-CNR), Italy
Dr. Peng-Yuan SHEN, University of Pennsylvania, China
Dr. Yun DUAN, Taiyuan University of Technology, China
Prof. Ebrahim Babaei, University of Tabriz, Iran
Dr. Constantinos S. Psomopoulos, Piraeus University of Applied Sciences, Greece
Dr. Gheorghe Grigoras, Technical University of Iasi, Romania
Dr. Jian-Wu ZENG, Minnesota State University, Mankato
Dr. Fabio Mottola, University of Naples Federico II, Italy
Prof. João António Esteves Ramos, Polytechnic of Leiria, Portugal
Prof. Hassan Abdullah Kubba, University of Malaya, Iraq
Dr. Meriem Bensmira, Mascara University, Algeria
Prof. Hosam El-Din Mostafa Saleh, Cairo University, Egyptian
Prof. Wen-Tsai Sung, National Chin-Yi University of Technology, China
Prof. Dan Dobrotă, Constantin Brâncusi University of Târgu Jiu, Romanian
Dr. Javad A. Esfahani, Ferdowsi University of Mashhad, Iran
Prof. Chi-Wen Lin, National Yunlin University of Science and Technology, Taiwan
Dr. Dina Nath Tewari, University of Allahabad, India
Dr. Hedayat Omidvar, Affairs with Science & Research Centers, Iranian
Dr. Abdallah BARAKAT, University of Poitiers-France, French
Dr. Xiao Enrong, Institute of Hydrobiology, Chinese Academy of Sciences, China
Prof. Carmelo Maria Torre, Engineering at the Polytechnic of Bari, Italy
Dr. Marcela Elisabeta BARBINTA- PATRASCU, University of Bucharest, Romanian
Dr. Zhi-En ZHANG, Chongqing University of Technology, China
Dr. Hajime Hirao,City University of Hong Kong, China

Prof. I-Shyan Hwang

Keynote Speech I
Decentralized SIEPON-Based ONU-Initiated Tx/TRx Energy-Efficiency Mechanism in EPON

Abstract: Ethernet passive optical network (EPON) has the least energy consumption among access network technologies and is the best candidate to improve the energy consumption by time utilization. Current studies have proposed OLT-initiated schemes to turn off transmitter/receiver of the optical network unit (ONU) for substantial time to achieve the energy-saving. In this paper, a new SIEPON based ONU-initiated energy-saving mechanism is proposed that ONU calculates the transmitter (Tx) sleep duration based on the current queue state and maximum boundary delay requirements, and forwards it to the OLT; then, the OLT calculates the ONU's receiver (Rx) sleep duration and decides the ONU's sleeping mode in Tx or TRx sleep mode. Furthermore, the Sleep manager and Green DBA and TRx controller components are proposed in the OLT and ONU architectures to manage the energy-saving mechanism more precisely. Simulation results show that the proposed energy-saving mechanism significantly improves the energy-saving up to 47 % and 42 % in upstream and downstream direction, respectively, and guarantee the QoS requirement in terms of mean packet delay, packet loss, throughput and jitter. Moreover, the proposed energy-saving mechanism has better delay performance compared to the OLT-initiated energy-saving mechanism such as upstream centric scheduling (UCS). of end-to-end delay, jitter, system throughput, fairness, and packet dropping rate.

Biography: Prof. I-Shyan Hwang received B.S. and M.S. degrees in Electrical Engineering and Electronic Engineering from Chung-Yuan Christian University, Chung-Li, Taiwan, in 1982 and 1984, respectively, and M.S. and Ph.D. degrees in Electrical and Computer Engineering from the State University of New York at Buffalo, NY, in 1991 and 1994, respectively. In Feb. 2007, he was promoted to Full Professor in the Department of Computer Science & Engineering at the Yuan Ze University, Chung-Li, Taiwan. He was Chairman of Department of Computer science & Engineering and Department of Information Communication from 2007 to 2013, respectively. Prof. Hwang has published more than 120 journal and 180 conference peer-reviewed papers, and 10 book chapters. His current research interests are fault-tolerant computing, high-speed networks, fixed mobile convergence, heterogeneous multimedia services over fiber optic networks, next generation networking, green computing and optical-network based infrastructure over cloud computing, software-defined optical networking, optical networks for Internet of Things. He serves as a member of the Editorial Board for the Springer Photonic Network Communications Journal.

DOI 10.1515/9783110540048-204,

Prof. Yong Wang

Keynote Speech II
A Glass Half Full or Half Empty: Water Resouces in Western China – an Earth Observation and Data Analytics Perspective

Abstract: Water cycles through the Earth's atmosphere, lithosphere, and hydrosphere redistributing fresh water globally. As civilization evolves on the planet, the increase of human population and activities demands additional food, infrastructure, housing, manufacturing, and commerce. The ever-increasing stresses on water resources and significant change in the global hydrologic regime would entail serious and negative consequences in many places where water resources are already scarce and are strained. Possession of adequate water and managing it wisely are elevated to the level of the national security and prosperity strategically. Precipitation and melting of snow and glacier in mountain ranges of Tibet Plateau and southwest of China are sources of water infiltrating into ground as ground water, and flowing through vast land surface feeding major river systems (e.g., Changjiang). With the population growth, and rapid social and economic development, the water in western China is more precious than ever. Moreover, the climate change can alter the water cycle in the region. Thus, it is critical for us to have abilities to observe and quantify the amount of available ground water and surface water in liquid and solid forms, to analyze the resulting large data flows and long time-series, and ultimately to forecast the future availability of fresh water. Capabilities of Earth observation and data analytics are well suited to undertake these tasks. The challenge, recent progress, and crucial future need are presented.

Biography: 06/2012-present Professor, School of Resources and Environment, University of Electronic Science and Technology of China. 08/1994–present Assistant, Associate, and Full Professor in Dept. of Geography, Planning, and Environment East Carolina University (ECU). Summers of 95, 96, & 97 Visiting Scientist, Institute for Computational Earth System Science (ICESS), UCSB. 12/1992–08/94 Postdoctoral Researcher, ICESS, UCSB. 06/1991–08/1991 Visiting Scientist, Radar Science and Engineering Group, NASA/JPL NASA - National Aeronautics and Space Administration JPL – Jet Propulsion Laboratory. 09/1987–11/1992 Graduate Student Researcher and Teaching Assistant, Department of Geography, UCSB. 09/1986–06/1987 Visiting Scholar, Department of Electrical and Computer Engineering, UCSB. 07/1982–

Prof. Yong Wang, School of Resources and Environment, University of Electronic Science and Technology of China

DOI 10.1515/9783110540048-205,

08/1986 Electrical Engineer, Radar Science Group, Chengdu Aircraft Corp, China.

Prof. Teik-Thye Lim

Keynote Speech III Heterogeneous Catalysis and Environmental Nanomaterials for Water Decontamination: Promises, Challenges and Research Needs

Abstract: Heterogeneous catalysis for environmental remediation has been a fertile research field in the recent decades. For environmental applications of the heterogeneous catalysis, environmentally-benign, efficient and robust catalysts are desired for the removal of hazardous substances in waste streams. This presentation will cover an overview of applications of the heterogeneous catalysis in environmental remediation with specific focus on water treatment. Several examples of nanomaterials and nanocomposites which have been developed for water remediation will be presented. Among these include photocatalyst/adsorbent composite for synergistic adsorption-photocatalysis, photocatalytic ceramic membrane for simultaneous separation and degradation of organic pollutants, bismuth ferrite for switchable multiplex catalysis under light and dark conditions, metal oxide catalysts for catalytic activation of peroxymonosulfate and generation of sulfate radical, and carbocatalysts. Despite immense researches and significant progress in developing advanced nanomaterials, there remain several challenges for field-scale adoption of heterogeneous catalytic processes in water decontamination. The final part of this presentation will provide a snapshot of these challenges and the future research needs to propel the research toward achieving practical adoption of heterogeneous environmental catalysis in water treatment.

Biography: Associate Professor, Nanyang Technological University, 2008 – present. Assistant Professor, Nanyang Technological University, 2000 – 2008. Design Engineer, BBR (S) Pte Ltd, Singapore, 1997 – 1999. Research Assistant, Nanyang Technological University, 1992 – 1997

Prof. Teik-Thye Lim, Nanyang Technological University

DOI 10.1515/9783110540048-206,

Dr. Hajime Hirao

Keynote Speech IV
Computational Studies of Molecules and Materials Pertaining to Environmental Science and Sustainable Energy

Abstract: The physical principles used in computational chemistry underlie all branches of chemistry; as such, computational chemistry has unlimited potential to contribute to the advancement of fundamental chemistry in every different subdiscipline as well as to finding solutions to critical challenges that humankind faces today, such as healthcare and energy/environmental issues. With this in mind, our computational exploration of chemistry applies quantum chemistry, multiscale QM/MM and QM/QM approaches, and many other advanced computational chemistry techniques to a broad range of complex molecular systems such as metalloenzymes, transition-metal catalysts, drugs/drug targets, metal-organic frameworks, and nanomaterials. In particular, using computational approaches and often with experimental collaborators, we seek to derive information about chemical reaction mechanisms and bonding patterns of these complex molecules.

Biography: 03/1998: BEng, Kyoto University, Faculty of Engineering, School of Industrial Chemistry. 03/2000: MEng, Kyoto University, Department of Molecular Engineering, Graduate School of Engineering. 11/2004: PhD, The University of Tokyo, Graduate School of Pharmaceutical Sciences. 06/2008 – 03/2009: Postdoctoral Research Associate. Cherry L.Emerson Center for Scientific Computation and Department of Chemistry, Emory University (Supervisor: Prof. Keiji Morokuma). 04/2009 – 10/2010: FIFC Fellow, Fukui Institute for Fundamental Chemistry, Kyoto University (Supervisor: Prof. Keiji Morokuma). 11/2010–02/2011: Project Associate Professor, Graduate School of System Informatics, Kobe University. 03/2011–06/2011: Assistant Professor, Division of Chemistry and Biological Chemistry, School of Physical and Mathematical Sciences, Nanyang Technological University. 07/2011–01/2017: Nanyang Assistant Professor, Division of Chemistry and Biological Chemistry, School of Physical and Mathematical Sciences, Nanyang Technological University. 01/2017–Associate Professor, Department of Biology and Chemistry, College of Science and Engineering, City University of Hong Kong.

Dr. Hajime Hirao, Department of Biology and Chemistry, College of Science and Engineering, City University of Hong Kong.

DOI 10.1515/9783110540048-207,

Awards:

2007: JSPS Postdoctoral Fellowship for Research Abroad
2011: Nanyang Assistant Professorship 2010

Table of Content

Session 3 Energy Science and Technology

Session 4 Control Theory and Application

Session 5 Materials

Session 6 Artificial Intelligence and Computer Science

Session 7 Sensors

Session 8 Control System and Automation

Session 10 Emerging Fields and Other Related Fields

Xia-Fei Duan, Xiao-Mu Chen, Jian-Peng Lv, Ou-Jing Li, Jin-Wang Li, Ling Wang and Zhong-You Pei*

Study on Immature Embryos of Sorghum Tissue Culture

Abstract: The callus induction and differentiation of sorghum were studied in three different treatments. The results showed that the callus induction rate of 2,4–D and ZT was different when the concentration of 2,4–D was 1.0 mg/L and the concentration of ZT was 0.1 mg/L and the genotype of Xin Liang 52 had the highest induction rate of 59.3 %. The callus differentiation rate of three genotypes was different. Among them, 07–27 differentiation rates were the highest, 16.2 %. Followed by Xin Liang 52, 13.1 %. The differentiation rate of M81-E was the worst, 10.3 %.

Keywords: Sorghum, ZT, immature embryos, 2, 4–D

1 Introduction

Sorghum is an annual herb of sorghum. It is a very important crop which is the world's sixth largest crop [1]. Sorghum, a drought-tolerant crop that grows on barren land where other crops cannot grow, is widely cultivated globally. Besides, sorghum is a major contributor to some 50 billion people in semi-arid tropics of Africa and Asia, more than 30 countries Diet [2]. Sorghum can be used as human food and animal feed, but also can be used as raw materials for industrial alcohol production and bio-energy [2].

Sorghum tissue culture is late in development of plant tissue, the earliest sorghum explant culture research is *Strogonol* in 1968 using the root and section, callus was induced in the additional 2,4-D 1 mg/L, KT 1 mg/L MS medium, and the callus were compared with callus of *Salicornia europaea L*, *Melilotus suaveolens Ledeb*, and *Brassica oleracea L.*. So, the history of sorghum tissue culture research has been unveiled. Plant tissue culture provides the appropriate recipient cells for gene transfer, and also provides favorable conditions for plant regeneration and trait performance. It is helpful to solve the problems of low genetic variability, low heritability and long time for breed improvement in sorghum breeding research [3].Compared with other crops, sorghum is difficult to obtain plant regeneration through tissue

Xia-Fei Duan, Agriculture Resourde and Environment College, Tianjin Agricultural University, Tianjin, China. Email: 2529138995@qq.com
Xiao-Mu Chen, Agriculture Resourde and Environment College, Tianjin Agricultural University, Tianjin, China. Email: zhongyoupei@tjau.edu.cn

DOI 10.1515/9783110540048-001,

culture. The reason is mainly high dependence on genotypes and phenolic compounds [Shi Taiyuan 2003]. First of all, studies have shown that in the process of tissue culture, callus induction rate and differentiation rates of different genotypes are significantly different. So the genotype of sorghum tissue culture is very important [4]. Secondly, phenolic compounds can inhibit the callus of sorghum induction and differentiation, thus reducing callus induction rate and differentiation rate. Through continuous optimization of sorghum regeneration system, sorghum tissue culture has also made some progress. The regenerated plants were obtained from in vitro culture of immature embryos, anthers, young spikes and young leaves. Although the regenerated plants were obtained from the above explants, the induction and regeneration rates of sorghum tissue culture were very different. And it dependent on the genotype. Therefore, it is necessary to study and identify genotypes, and further improve and optimize sorghum regeneration system. Experiments of Zhao et al. (2009) showed that M81-E is a better genotype in mature embryo culture. In this experiment, we used XL52, 07–27, M81-E as experimental materials to study the induction and differentiation on immature embryo of sorghum.

2 Materials and Methods

2.1 Materials

Immature embryos of sorghum: XL52, 07–27, M81-E were supplied by Crop Genetics Key Laboratory of Tianjin Agricultural, planted in the crop specimen garden of Tianjin Agricultural University. Young seeds are pollinated 12–15 days later. The young seed surface was wiped with 75 % ethanol and then packaged in plastic wrap and placed at 4 °C for 12–24 h.

2.2 Methods

2.2.1 Medium

(1) Induction medium of sorghum callus
(2) Subculture medium of sorghum callus: MS +1.5 mg/L 2, 4-D, pH 5.8.
(3) Differentiation medium of sorghum callus: MS + 0.1 mg/L ABA + 0.25 mg/L NAA + 0.1 mg/L 1.0 mg/L ZT + 8 g/L D-Sor + 30.0 mg/L hygromycin + 1000.0 mg/L Cef + 50.0 mg/L Amp, pH 5.8

Tab. 1: Induction medium of sorghum callus

Number	Medium	2,4-D(mg/L)	ZT(mg/L)
1	MS	1.0	0.0
2	MS	1.0	0.1
3	MS	1.0	0.5
4	MS	2.0	0.0
5	MS	2.0	0.1
6	MS	2.0	0.5
7	MS	4.0	0.0
8	MS	4.0	0.1
9	MS	4.0	0.5

2.2.2 Callus Induction of Immature Embryo

The pre-treated young seeds were carefully removed, washed with distilled water for 1 times, and then sterilized with 75 % alcohol for 1 minute, and then sterilized with $HgCl_2$(0.1 %) for 20 minutes, and finally rinsed with distilled water for 7–10 times. The treated seeds were placed on sterile filter paper and allowed to dry. The young embryos of the seeds were extruded in the clean bench and inoculated on the configured induction medium. Finally, the cultures in a constant temperature incubator at 28 ± 1 °C for 10 to 14 days in the dark, and then to obtain statistical induction rates.

2.2.3 Sorghum Callus Subculture

After 10-14 days of induction, callus was formed on the small scutellum of immature embryo and the callus was inoculated into subculture medium with tweezers. And then the cultures in a constant temperature incubator at 28 ± 1 °C for 2 weeks in the dark.

2.2.4 Sorghum Callus Differentiation, Rooting and Transplanting

The callus of good growth was inoculated on the differentiation medium, the cultures in a constant temperature incubator at 25 ± 1 °C, the daily light 16h, light intensity 2000LUX. After 4 to 5 weeks of differentiation, the callus differentiation rate (3 cm height) was calculated. When the regenerated shoots grow to about 5 cm, they are transfered to the rooting medium. When the root of the regenerated seedlings is strong, the plants are transplanted into pots containing vermiculite and nutritive soils, and until the plants mature in the greenhouse.

2.2.5 Statistical Analysis of the Data

Callus induction rate = (callus number / number of immature embryos) × 100 %;

Callus differentiation rate = (the number of regenerated shoots / the number of callus inoculation) × 100 %.

After the data were transformed, the variance analysis was performed with SPSS statistical software, and the significance level was analyzed with Duncan's new multiple range method.

3 Results and Analysis

3.1 Callus Induction of Immature Embryo

The results showed, the morphological characteristics of callus were different in different genotypes. The callus induced by XL 52 immature embryo was white and the brown matter was secreted. The callus induced by M81-E immature embryo was deep yellow and a little black matter was secreted. The callus induced by 07–27 immature embryo was deep yellow and no secretions (Figure 1).

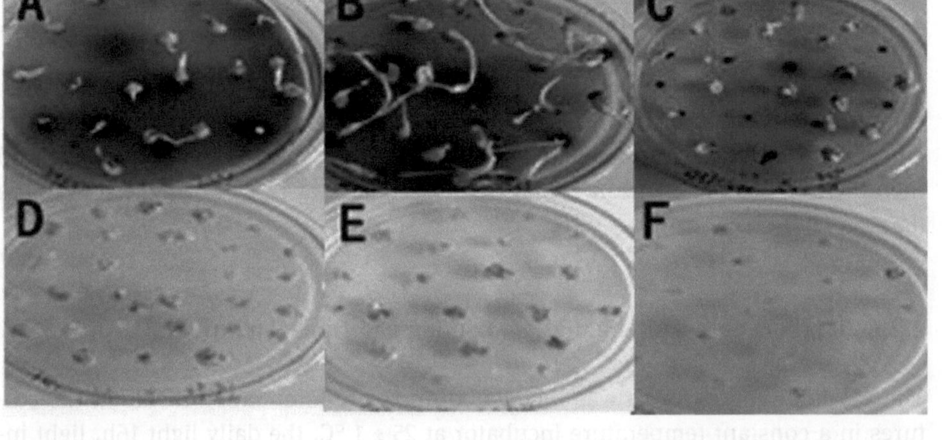

Fig. 1: Sorghum callus

A: XL52 immature embryos were induced for 10 days; B: M81–E immature embryos were induced for 10 days; C: 07–27 immature embryos were induced for 10 days; D: X L52 callus on the subculture medium; E: M81–E callus on subculture medium; F: 07–27 callus on subculture medium.

3.1.1 Effects of 2,4-D, ZT and Genotypes on Callus Induction of Sorghum Immature Embryos

In this experiment, the variance analysis of callus induction rate was carried out. The results showed that the callus induction rate of 2,4–D, ZT, and the genotype is different. The highest callus induction rate for 2,4–D was at a concentration of 1.0 mg/L, induction rate was 54.7 %.The highest callus induction rate for ZT was at a concentration of 0.1 mg/L, induction rate was 52.7 %. The induction rate of xl52 was the highest among the three genotypes, induction rate was 59.3 %. (Tables 2, 3 and 4).

Tab. 2: The effect of 2,4–D on callus induction

2,4-D concentration (mg/L)	Induction rate (%)
1.0	54.7±0.03a
2.0	52.9±0.03a
4.0	26.8±0.03b

Table 3: The effect of ZT on callus induction

ZT concentration (mg/L)	Induction rate (%)
0.0	35.7±0.03b
0.1	52.7±0.03a
0.5	45.9±0.03a

Table 4: The effect of ZT on callus induction

Genotypes	Induction rate (%)
07-27	31.7±0.03c
M81-E	43.3±0.03b
XL52	59.3±0.03a

Table 5: Interaction effect of callus induction

Variation source	Quadratic sum	DOF	F value	The confidence level of 5 %
2,4-D*ZT	0.8	4	12.1	0.0
Genotypes*2,4-D	0.1	4	1.9	0.1
Genotypes *ZT	0.9	4	14.5	0.0
2,4D*ZT*Genotypes	0.4	4	3.0	0.008

3.2 Interaction Effect of Callus Induction

2, 4-D and genotypes were no interaction effect. Interaction effect were obviously among the .ZT * 2, 4–D, genotypes * ZT, and 2,4–D * ZT * genotypes. (Table 5) Therefore, a combination of significant interaction effects should be analyzed in detail

3.2.1 The interaction effect of 2,4-D and ZT on callus induction

Because of the interaction effect of 2,4–D and ZT, it is necessary to analyze the effect of 2,4–D and ZT on callus induction. The results showed that the induction rate of three genotypes was significantly different when the concentration of 2,4–D was 1.0 mg/L, the highest induction rate was 64.9 % when concentration of ZT was 0.5 mg/L. At the concentration of 2,4–D were 2.0 mg/L and 4.0 mg/L, the induction rates of the three genotypes were significantly different, and the highest induction rate were 67.2 % and 34.7 % when concentration of ZT was 0.1 mg/L (Table 6).

Tab. 6: The effect of 2,4-D and ZT on callus induction

2,4-D (mg/L)	ZT (mg/L)	Induction rate (%)
1	0.0	43.0±4.3c
	0.1	56.1±4.3b
	0.5	64.9±4.3a
2	0.0	30.7±4.3c
	0.1	67.2±4.3a
	0.5	60.8±4.3b
4	0.0	33.6±4.3a
	0.1	34.7±4.3a
	0.5	12.1±4.3b

3.2.2 The Interaction Effect of ZT and Genotypes on Callus Induction

Because of the interaction effect of ZT and genotype. So in this experiment, the interaction effect between ZT and genotype on callus induction was analyzed. The results showed, the callus induction rate of the three genotypes was significantly different when concentration of ZT was different. In addition, with the increase of ZT concentration, the callus induction rate of 07–27 and M81–E was increased and then decreased, the highest callus induction rate was 41.0 % and 58.9 % when ZT was 0.1 % mg/L. With the increase of ZT concentration, the callus induction rate of XL52 was decreased, the highest callus induction rate was 67.9 % when ZT was 0.0 % mg/L. (Tables 7 and 8)

Tab. 7: The effect of ZT and genotypes on callus induction

ZT (mg/L)	Genotypes	Induction rate (%)
0.0	07-27	24.7±4.3b
	M81-E	14.7±4.4c
	XL52	67.9±4.5a
0.1	07-27	41.0±4.6b
	M81-E	58.9±4.7a
	XL52	58.1±4.8a
0.5	07-27	29.6±4.9b
	M81-E	56.4±4.1a
	XL52	51.8±4.1a

Tab. 8: The effect of ZT and genotypes on callus induction

Genotypes	ZT(mg/L)	Induction rate (%)
07-27	0.0	24.7±4.3b
	0.1	41.0±4.6a
	0.5	29.6±4.9b
M81-E	0.0	14.7±4.4b
	0.1	58.9±4.7a
	0.5	56.4±4.1a
XL52	0.0	67.9±4.5a
	0.1	58.1±4.8b
	0.5	51.8±4.1b

3.2.3 The Interaction Effect of ZT, 2,4-D and Genotypes on Callus Induction

In this experiment, the effects of ZT, 2,4–D and genotypes on the callus induction were studied. The results showed, when the concentration of 2,4–D was 1.0 mg/L, the callus induction rate of 07–27 and M81-E were increased with the increase of ZT concentration, and the difference was significant. And when the ZT was 0.5 mg/L, the highest induction rate of 07–27 was 48.7 % and the highest induction rate of M81-E was 76.0 %. When the concentration of 2,4–D was 1.0 mg/L, the callus induction rate of Xl52 was decreased and then increased, and the highest induction rate was 70.0 % when ZT concentration was 0.5 mg/L. When the concentration of 2,4–D was 2.0 mg/L, the callus induction rate of 07–27, M81-E andXL52 was increased and then decreased with the increase of ZT concentration, and the difference was significant. The highest induction rate was reached at 0.1 mg/L, which was 58.3 %, 87.7 % and 72.7 %, respectively. When the concentration of 2,4–D was 4.0 mg/L, the callus induction rate of 07–27 and M81-E was increased and then decreased with the increase of ZT concentration, and the difference was significant. When ZT concentration was 0.1mg/L, the callus induction rate was the highest, which was 19.0 % and 42.3 %, respectively. When the concentration of 2,4–D was 4.0 mg/L, the callus induction rate of XL52 was decreased with with the increase of ZT concentration, the difference was significant, the highest induction rate was 86.3 % when ZT concentration was 0.0 mg/L .(Table 9)

Tab. 9: The interaction effect of ZT, 2,4–D and genotypes on callus induction

Genotypes	2, 4-D (mg/L)	ZT (mg/L)	Induction rate (%)
07-27	1.0	0.0	37.3±7.4b
		0.1	45.7±7.4a
		0.5	48.7±7.4a
	2.0	0.0	31.3±7.4b
		0.1	58.3±7.4a
		0.5	32.7±7.4b
	4.0	0.0	5.3±7.4b
		0.1	19.0±7.4a
		0.5	7.3±7.4b
M81-E	1.0	0.0	22.3±7.4c
		0.1	63.7±7.4b
		0.5	76.0±7.4a
	2.0	0.0	12.7±7.4b
		0.1	87.7±5.8a
		0.5	82.3±7.4a
	4.0	0.0	9.0±7.4b
		0.1	42.3±7.4a
		0.5	11.0±7.4b
XL52	1.0	0.0	69.3±7.4a
		0.1	59.0±7.4b
		0.5	70.0±7.4a
	2.0	0.0	48.0±7.4b
		0.1	72.7±7.4a
		0.5	67.3±7.4a
	4.0	0.0	86.3±7.4a
		0.1	42.7±7.4b
		0.5	18.0±7.4c

3.3 Callus Differentiation of Immature Embryo

3.3.1 The Effect of Genotypes on Callus Differentiation

By analyzing the effect of genotype on callus differentiation. The results showed: the callus differentiation rate of three genotypes was different. Among them, 07-27 callus differentiation rates were the highest, 16.2 %. Followed by XL52, 13.1 %. The differentiation rate of M81-E was the worst, 10.3 %. (Table 10)

Tab. 10: The effect of genotypes on callus differentiation

Genotypes	callus differentiation number	Inoculation number of calluses	Differentiation rate
07-27	52	321	16.2 %
M81-E	38	368	10.3 %
XL52	90	689	13.1 %

3.3.2 The Effect of ZT on Callus Differentiation

The results showed that all of the genotypes were differentiated to obtain regenerated shoots, but the callus differentiation rate was very different (Fig. 2). The differentiated results shown, when ZT concentration was 0.5 mg/L, the highest differentiation rate of 07–27 and M81- E was 21.9 % and 14.6 %, respectively. The highest callus differentiation rate was 10.2 % when ZT concentration was 1.0 mg/L. (Table 11)

Tab. 11: The effect of ZT on callus differentiation

Genotypes	ZT (mg/L)	Callus differentiation numbers	Inoculation number of calluses	Differentiation rate
07-27	0.0	19	93	20.4%
	0.5	23	105	21.9%
	1.0	10	123	8.1%
M81-E	0.0	6	122	4.9%
	0.5	18	123	14.6%
	1.0	14	123	11.4%
XL52	0.0	6	132	4.5%
	0.5	12	150	8.0%
	1.0	10	98	10.2%

A.XL52; **B.M81-E;** **C.07-27**

Fig. 2: Callus differentiation of immature embryo

3.4 Rooting and Transplanting of Regeneration Seedlings

When the regenerated shoots were grown to 3-5 cm, the rhizoid was cut and transferred to 1/2 MS rooting medium for rooting (Fig. 3). To grow a lot of roots, the bottle will open, hardening 7 days. After washing the medium of the roots, and cut most of the leaves, transplanted to pots in the dark growth for 1 week. And then transplanted to the greenhouse.

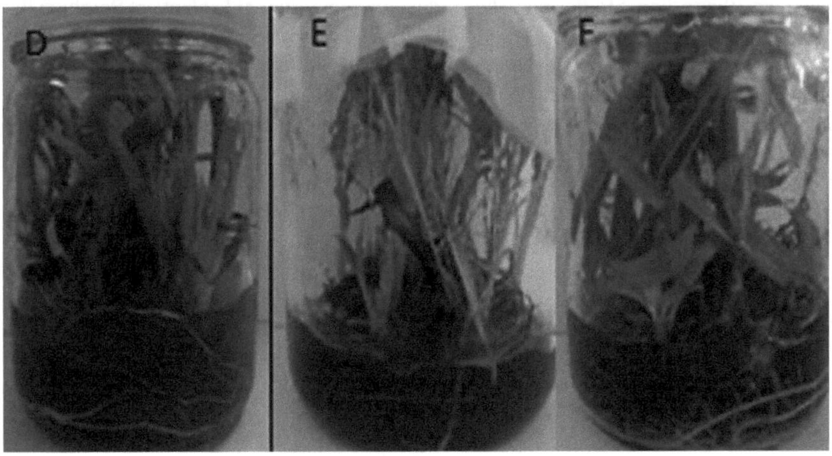

D.XL52; **E.M81-E;** **F.07-27**

Fig. 3: Regeneration seedlings of sorghum

4 Discussion

Immature embryo is an important material in tissue culture of sorghum [6–9]. From the experiment, the callus induction rate of XL52 was the highest among the three genotypes; the callus differentiation rate of 07–27 was the highest. So, XL52 and 07–27 were the better genotypes in sorghum tissue culture. M81-E is a better genotype in mature embryo culture (Zhao Liming 2009), but in this experiment, M81-E is not the highest in callus induction and callus differentiation. In addition, in this study, we found that the ratio of ZT and 2, 4-D is also very important in sorghum tissue culture.

References

[1] Shi Taiyuan, Yang Liguo, Xiao Jun.2003. Genetic bombardment of sorghum callus to obtain transgenic plants. Liaoning Agricultural Sciences, (6): 9 ~ 10
[2] Dahlberg J, Berenji J, Sikora V, Latkovic D. 2011. Assessing sorghum [Sorghum bicolor (L.) Moench] germplasm for new traits: food, fuels and unique uses. Maydica, 56: 85–92
[3] YANG Li-heng, 3 HAO Xiu-ying, WANG Xiao-jun, CAO Yu-jin, KANG Xi-liang, ZHANG Shu-min. Advances in tissue culture of sorghum [J]. Beijing Agriculture,
[4] Shi Taiyuan, Yang Liguo, Zhang Hua, et al. 1995. Sorghum somatic cell culture in different genotypes and explants reaction. Foreign agronomy - grains crops, 6: 26 ~ 28
[5] Zhao Liming, Liu Shujun, Song Songquan. 2008. The establishment of sweet sorghum regeneration system. Botany Bulletin, 25 (4): 465 ~ 468
[6] Han Fuguang, Zhang Ying. Studies on Callus Induction from Different Explants of Sorghum [J]. LiaoNing Agricultural Sciences, 1993 (1): 45–48.
[7] Bai Zhiliang, Wang Liangqun, Zheng Liping, et al. In vitro culture of different explants of sorghum [J]. China North Agricultural Journal, 1995, 10 (1): 60–63
[8] Wei Zhiming, Xu Zhihong. Sorghum protoplast culture regeneration plant [J]. Plant Physiology Communications, 1989 (6): 45–48.
[9] Chen Xiaomu, Li Oujing, Shi Lili, et al. Studies on establishment of young panicle regeneration system in sorghum [J]. Jiangsu Agricultural Sciences, 2016, 44 (11): 40–43.

Xiang-Yan Zhang, Juan Cao, Ke Bi, Hao-Yan Wang, Zi-Qiang Zhou and Guo-Hong Dai

PSO Algorithm Based Shredding Parameter Optimizing for the Body of ELV

Abstract: In order to improve the utilization ratio of scrap iron and steel in the shredding material of the body of End of life vehicle after disassembly, it is necessary to determine the optimized parameter in order to reduce the energy consumption of the crushing equipment. With the software of EDEM, the crushing material accumulation process is simulated. The porosity of the broken material pile is also obtained, and the optimal fracture radius model of waste steel is established. In the end, the Particle swarm algorithm for optimization is used to solve the optimal shredding material size.

Keywords: ELV; EDEM; PSO Algorithm; Shredding parameter

1 Introduction

With the development of the automobile industry, China has become the world's third largest automobile manufacturer. According to the relevant agencies predicted, China's annual number of ELV (end of life vehicle) will reach 6 million until 2020. The rapid increase in the number of scrapped vehicles is accompanied by the waste of land due to the accumulation of scrapped cars and the threat to the surrounding environment, and the scrapped cars themselves contain large amounts of metallic and non-metallic resources. If the resources are discarded without effective recycling, they will cause serious waste of resources, so the dismantling of scrap cars,

Xiang-Yan Zhang, School of Mechatronic Engineering, China University of Mining and Technology, XuZhou, China. Email: zhangxiangyancumt@163.com
Juan Cao, School of Mechanical Engineering, Jiangsu University of Technology, ChangZhou, China. Email: caolaichen@yeah.net
Ke Bi, School of Mechatronic Engineering, China University of Mining and Technology, XuZhou, China. Email: bikeatsuzhou@cumt.edu.cn
Hao-Yan Wang, School of Mechatronic Engineering, China University of Mining and Technology, XuZhou, China. Email: wanghaoyanjy@163.com
Zi-Qiang Zhou, Jiangsu key laboratory of recycling and reuse technology for mechanical and electronic products, Changshu Institute of Technology, ChangShu, China. Email: zzq_hefei@163.com
Guo-Hong Dai, Jiangsu key laboratory of recycling and reuse technology for mechanical and electronic products, Changshu Institute of Technology, ChangShu, China. Email: dgh@cslg.edu.cn

DOI 10.1515/9783110540048-002,

broken and sorting on environmental protection, resource recovery, energy saving is of great significance. In the process of crushing scrap cars, if scrap metal is too large to be fully melted in the furnace. When undissolved iron accumulates in the bottom of the furnace, the long run, the bottom of the furnace need to be regularly cleaned up, which not only reduces the utilization of scrap steel but also consumes a lot of manpower and material resources in the process of cleaning up the bottom of the accumulation; although the scrap iron that is too small can be fully melted, it still consumes too much energy in the crushing process. Thus, the optimal size of the crushed material in reducing energy consumption and improving the utilization of scrap has an important significance. At present, there is no literature on the shredding materials optimization of the study.

2 Porosity of Shredding Materials for The Body of ELV

ELV are packed, broken and sorted to form shredding materials, the shredding material is output and piled up by the belt conveyor. Porosity, which is the ratio of pore volume to bulk volume, is an important particle accumulation parameter. However, in the study of particle accumulation, due to the porosity is irregular and computationally cumbersome, EDEM is used to simulate the shredding material to obtain the porosity.

Simply speaking, particle accumulation is place particles in a finite space, and cannot occur deformation in the process of accumulation. Because of the simplicity of the sphere itself and the application background in the scientific research、engineering practice and daily life, the spherical particle are the most studied objects in the study of particle accumulation. We will simplify the simulation, select the cone limited space to create a sphere of normal distribution random accumulation simulation.

The sphere model established by the simulation is different from the shape of the steel detritus in the engineering practice. So the density of the sphere can't be simply considered equal to the density of steel, there is a transition relationship between the density of the sphere density and the density of steel. Through the continuous measurement and calculation of scrap iron, the average conversion ratio between sphere and scrap iron density is obtained $\rho_{sphere}=\eta\rho_{Fe}$, where $\eta=0.25$.

Since the particles are randomly stacked, the sphere size is set to a normal distribution; the mathematical expectation is the sphere radius r, sphere drops at 3 m/s, until the volume is filled with 1 m^3 cone. In the process of material crushing and sorting, as the diameter of more than 85 mm of shredding material is a large material, which need to be crushed manually, and those less than 20 mm in diameter non-metallic materials need to be buried, so the best crushing size is between 20 mm–85 mm. The spherical particles with radius r=15, 20, 25, ..., 50 mm were selected for particle packing simulation, and the pore volume was obtained under the corresponding sphere radius. Figure 1 shows the random packing of particles with radius of 15 mm and 50 mm, respectively. Table 1 shows the pore volume under different particle radius.

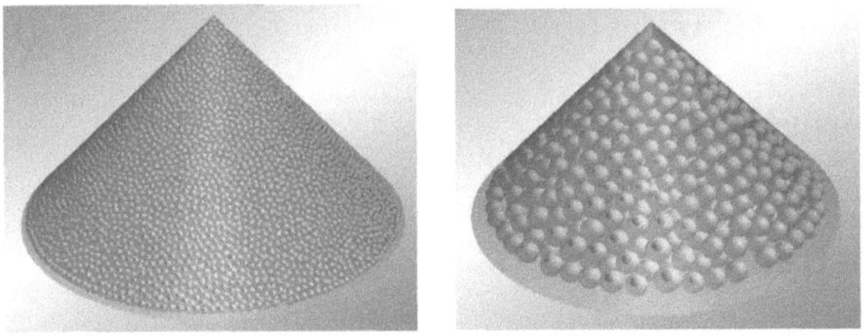

a) r=15mm **b)** r=50mm

Fig. 1: Random accumulation of spheres

Tab. 1: Sphere radius and the total volume of the particles

Radius r/(m)	Volume/(m³)	
	Sphere volume V	Pore volume (1-V)
0.015	0.617108	0.382892
0.020	0.603545	0.396455
0.025	0.601792	0.398208
0.030	0.579737	0.420263
0.035	0.56673	0.43327
0.040	0.551526	0.448476
0.045	0.542735	0.457269
0.050	0.539019	0.460981

We use MATLAB to fit the curve of discrete points in Table 1 to get the mathematical relationship between the porosity and the sphere radius r. The porosity is

$$\phi = 1 - V = e^{-1.08817 + 9.21369r - 55.90788r^2}$$

3 Model for Optimal Shredding Parameter

In order to maximize the degree of melting of scrap iron and steel, we establish the optimal radius calculating model of iron and steel scrap, showing as follows:

3.1 Objective Function

$$A=Max\left(r\right) \tag{1}$$

Where: r is the radius of the crushed material.

3.2 Bulk Density Constraint

According to the national standard GB4223–004, which is stipulated that the vehicle shredding material bulk density should be no less than 1100 kg/m³, which is:

$$\rho_{\text{pile}}=\frac{\sum_{i=1}^{n} m_i}{V}=\frac{\eta\rho_{\text{Fe}}\sum_{i=1}^{n} V_i}{V}\geq 1100 \tag{2}$$

Where:

ρ_{pile} —bulk density;

V—volume of scrap iron and steel shredding material pile, assume V=1m³;

$V_i=\frac{4}{3}\pi r_i^3$;

$r_i\sim\left(r,0.1\right)$.

3.3 Volume Constraint

The sum of the volume of the shredding material (scrap iron and steel) and the volume of the material gap is equal to the total volume of the material, which is:

$$\sum_{i=1}^{n} V_i+\phi\cdot V=V \tag{3}$$

3.4 Decision Variable Constraint

In the process of material crushing and sorting, as the diameter of more than 85mm of shredding material is a large material, which need to be crushed manually. Therefore, the upper limit of not more than 80mm, and those less than 20mm in diameter non-metallic materials need to be buried, so the range of shredding material radius is:

$$10mm \le r \le 40mm$$

4 PSO Algorithm Based Solving Approach for Optimal Shredding Parameter

PSO is a group search algorithm. We assume in such a target search space D, the position of the i-th (i=0,1, ... ,N) particle in the population can be represented as a D dimensional vector $X_i = (x_{i1}, x_{i2}, \dots, x_{iD})^T$, meanwhile, using $V_i = (v_{i1}, v_{i2}, \dots, v_{iD})^T$ (i=0,1, ... ,N) indicates the flight speed of the i-th particle to update the location of the particle. Using $P_i = (p_{i1}, p_{i2}, \dots, p_{iD})^T$ represents the best point of the i-th particle itself, which is optimal location of individual history. Mark $P_g = (p_{g1,} p_{g2,} \dots, p_{gD})^T$ is the best point to search for the current population, which is global optimal location of population.

Updating the velocity and position of every particle according to the following formula:

$$v_{ij}^{k+1} = \lambda v_{ij}^k + c_1 r_1 \left(p_{ij}^k - x_{ij}^k\right) + c_2 r_2 \left(p_{ij}^k - x_{ij}^k\right) \tag{4}$$

$$x_{ij}^{k+1} = x_{ij}^k + v_{ij}^k \tag{5}$$

Where i = 0, 1, ... , N, j represents the j-th dimension of the particle, k represents the number of iterations. λ Is the inertia coefficient, assigning in(0,1). c_1,c_2 are acceleration constant, which take values generally between (0,2). c_1 is used to adjust the flying step of the optimal position of their own particles, c_2 is used to adjust the step of the particle flying to the global best position. r_1, r_2 is the random number between (0,1).

In the solution space, particle will keep track of the individual extremum and global extremum to search until it reaches the specified maximum number of iterations or minimum error standard position. Figure 2 shows the flow chart of PSO algorithm.

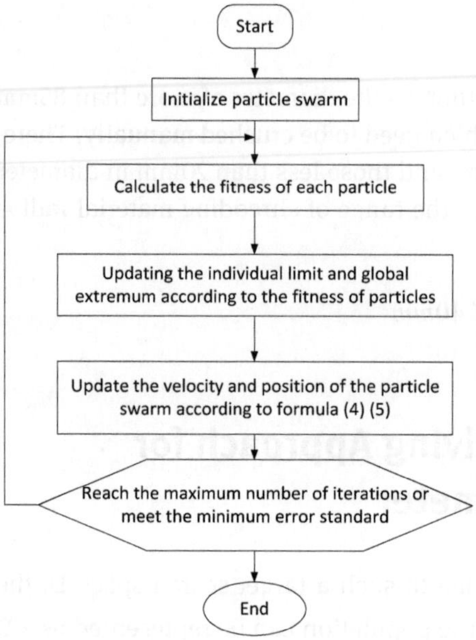

Fig. 2: Flow chart of PSO algorithm

In this paper, the random number is used to represent the radius of the shredding material particle. The basic process of the algorithm is as follows:

Step 1: Produces a random number of particles between (0.015, 0.05) to calculate the current total particle volume;

Step 2: Determine whether to meet the volume constraint, if meet, and then determine whether to meet the bulk density constraints, if meet, then record the particle location;

Step 3: If not meet the volume constraint, go to step 1.

The results of the calculation are shown in Figure 3, the optimal size radius is 0.036m, or 36mm.

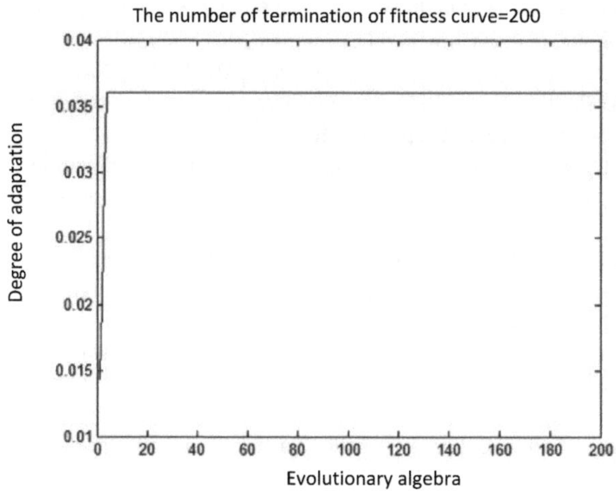

Fig. 3: MATLAB solution results

5 Conclusion

In order to improve the utilization rate of scrap iron and steel, we have researched the optimization method of crushing material parameters based on PSO. EDEM software is used to simulate the accumulation process of shredding material, and calculate the porosity of material pile. This paper establishes a solving model of the optimum crushing size of scrap iron and steel, and use PSO algorithm to solve the problem. The solution of optimal crushing size can greatly improve the utilization rate of scrap iron and steel, which is of great significance for saving resources and environmental protection.

Acknowledgement: This work was financially supported by Science & Technology Pillar Program of Jiangsu Province (No. BE2013060). Key Project of Natural Science Research of Higher Education Institutions of Jiangsu Province (No. 15KJA460001). Postdoctoral Research Program of Jiangsu Province (No.1401071C).

References

[1] Ziqiang Zhou, Guohong Dai, Hanmo Tan. The development and research status of dismantling and recycling of scrapped automobiles. Journal of Changshu Institute of Technology, 2011, pp. 107–111.

[2] Ziqiang Zhou, Hanmo Tan, Guohong Dai. Research of Value Analysis Oriented End of Life Vehicle Dismantling and Recycling Process.2012 International Conference on Energy and Environmental Protection, 2012, Hohhot.

[3] Ziqiang Zhou, Guohong Dai. Research of flexible dismantling cell for ELV recycling.2012 International Conference on Ecology, Waste Recycling, and Environment, 2012, Macao.

[4] Ziqiang ZHOU, Guohong DAI, Juan CAO, Guolin GUO. A NOVEL APPLICATION OF PSO ALGORITHM TO OPTIMIZE THE DISASSEMBLY EQUIPMENT LAYOUT OF ELV. International Journal of Simulation Systems, Science & Technology, 2016, pp. 161–165.

[5] Ming Chen, Fan Zhang, End-of-life vehicle recovery in china: consideration and innovation following the EU ELV Directive. JOM Journal of the Minerals, Metals and Materials Society, 2009, pp. 45–52.

[6] Emilio Brahmst, Copper in End-of-life Vehicle Recycling, Copper Development Association.

[7] Tony Weatherhead, David Hulse, A Study to Determine the Metallic Fraction Recovered from End of Life Vehicles in the UK, Report to the Department of Trade and Industry, 2015.

[8] Hsing-Chih Tsai, Yaw-Yauan Tyan, Yun-Wu Wu, Yong-Huang Lin. Isolated particle swarm optimization with particle migration and global best adoption. Engineering Optimization, 2011, pp. 1405–1424.

[9] Deniz Lnan, Erol Egrioglu, Busenur Sarica, Oykum Esra Askin, Particle Swarm Optimization Based Liu-Type Estimator. Communications in Statistics-Theory and Methods, 2016, pp. 1–21.

[10] Lingling Li, Licheng Jiao, Jiaqi Zhao. Quantum-behaved discrete multi-objective particle swarm optimization for complex network clustering. Pattern Recognition, 2017, pp.1–13.

[11] Moayed Daneshyari, Gary G. Yen. Cultural-based particle swarm for dynamic optimisation problems. International. Journal of Systems Science, 2012, pp. 1284–1304.

Rui Wang, Chun-Lin Li, Xia-Xia Zhou, Hui-Fen Liu and Jian-Chao Hao

The Effects of Organic Manure Produced by TWR Technology on Rape Growth, Heavy Metals Accumulation and Soil Physical and Chemical Properties

Abstract: Using new type of organic fertilizer that was produced by total waste recycling (TWR) technology as the experimental materials, its effects on rape growth and quality were compared with the usual organic manure and chemical fertilizer in the pot experiments. Five treatments were designed: no fertilizer (CK), new type of organic manure (T1), new compound fertilizer (T2), common organic manure (T3) and urea (T4). Each treatment was repeated 3 times. The results showed that the fresh weight, the dry weight, the photosynthetic rate and the chlorophyll of rapeseed were significantly affected by application of different fertilizers. The fresh weight and the dry weight of rape treated with T2 were the highest and had significant differences with T3 and T4. The content of nitrogen and potassium in the treatments of T2 and T1 was significantly higher than CK. The accumulation of heavy metals in roots part of the rapes was higher than the leaves part and the contents of heavy metals in rape were also different under different fertilization treatments. The content of as in rape leaves of T1 and T2 was significantly lower than T3 while there was no marked difference in Cr^{6+}, Hg and Pb contents in rape leaves. Soil electrical conductivity, soil total N, available P, available K increased with different fertilization treatments.

Keywords: TWR technology; New type of organic manure; Rape; Quality; Heavy metal

1 Introduction

In recent years, the garbage encirclement phenomenon occurs frequently in many cities. Urban garbage has become an important cause of urban pollution in China [1, 2]. Since the 1970s, our country has developed many new technologies of domestic waste compost and has been widely promoted and applied [3, 4]. Domestic waste

Rui Wang, College of Agronomy and Resources & Environment, Tianjin Agricultural University; Tianjin Engineering Research Center of Agricultural Ecological & Remediation; Tianjin 300384, China; Email: 1047951695@qq.com

DOI 10.1515/9783110540048-003,

can be composted to the fertilizers which can promote the growth and development of crops. The fertilizers not only can prevent the environmental pollution caused by the random packing of a large amount of domestic garbage, but also can improve the soil physical and chemical properties, crop yield and quality [5,6]. However, the fertilizer produced from waste compost has many impurities and heavy metal exceeds, making it difficult for compost to be used in agricultural production [7–9]. Ma Kun [10] studied the effect of compost on spring wheat, with the highest yield at 150 t/ha in the study area, and excessive compost will reduce the production. Different amounts of fertilizer applied would lead to different nitrogen, phosphorus and potassium content in wheat grain and straw. Zhao Fenglian [11] researches show that, compared with cow manures, when using chicken biogas manures, rape yield, nitrate content and nitrogen use efficiency had increased, while vitamin C content had no significant differences. Wang Jian xiang and Zhou jie liang [12] showed that Phenol, sugar and aldehyde in the organic fertilizer could inhibit the nitrification of ammonium nitrogen and reduce nitrate nitrogen to control the accumulation of nitrate. Urban garbage compost will had an impact on soil physical and chemical properties and heavy metal accumulation [13]. The results of Fan Haihua [14] showed that the application of compound fertilizers and composts would change the soil physical and chemical properties in the turf grassland soil and the organic matter, nitrogen, phosphorus and potassium in the soil layer would increase, at the same time caused the accumulation of Cu in the soil.

The new organic fertilizers and new compound fertilizers that using the whole resource waste treatment technology (TWR) produced by Tianzi Environment Protection Investment Holdings Co.,Ltd. were chosen as the research objects. Through the pot experiments of small rapes, these new organic manures were compared with common urea and common organic fertilizers. And the effects on the soil physical and chemical properties and the growth and quality of rapes would be discussed in order to provide the theoretical basis for the safety evaluation and reasonable application of these new fertilizers.

2 Materials and Methods

2.1 Experimental Materials

The 0~20 cm of soil was collected from Tianjin Agricultural University experimental fields. After natural drying and sieving, the basic physical and chemical properties of the soil were determined. The bulk density was 1.32 g/cm^3, pH was 8.4, the content of organic matter was 18.7 g/kg, total nitrogen was 2.1 g/kg, effective phosphorus was 7.5 mg/kg, available potassium was 107.9 mg/kg. The rape seed “Shanghai qing” was purchased in the market. The new type of organic fertilizers and compound fertilizers produced by TWR technology were provided by Tianzi Environment Protection Investment Holdings Co., Ltd and the ordinary organic fertilizers and urea were provided by Tianjin Green Plant and Animal Nutrition Technology Development co., LTD.

2.2 Experiment Design

Five treatments were designed in the experiment: no fertilizer (CK), new type of organic manure (T1), new compound fertilizer (T2), common organic manure (T3) and urea (T4). Each treatment was repeated 3 times. A small pot with an inner diameter of 27 cm and a depth of 25 cm was chosen and filled with 3.5 kg soil. In addition to the blank control, each tub was applied to different fertilizers at a dose of 0.15 g N/kg soil. And one-time fertilization was operated before sowing and let it balance for two weeks after fully mixed. Twenty seeds were sowed per pot on September 21st, 2015 and thinned to ten after ten days, weighed and watered every two to three days in the growing process to keep the soil content being 70 % field moisture capacity. After eight weeks, the plants were harvested, the soil samples were collected and the basic physical and chemical properties were determined. Besides, the aboveground and underground biomass, the quality of rape and heavy metal content were determined. Three kinds of organic fertilizers nutrient and heavy metal contents are shown in table 1.

Tab. 1: Three kinds of organic fertilizers nutrient and heavy metal contents

Type of manure	N	P_2O_5	K_2O	As	Cd	Cr^{6+}	Hg	Pb
	(%)			(mg/kg)				
T1	2.4	4.8	0.6	0.3	0.0	0.0	0.3	9.2
T2	3.3	6.2	1.5	1.5	0.0	0.0	0.7	12.0
T3	1.1	2.5	1.1	0.7	0.0	0.0	0.4	7.3

2.3 Measurement Methods

After the harvest, the rapes were rinsed with tap water and then distilled water, dried and packed the above and underground parts of plants separately in paper bags, blanched under 105 °C, dried under 60 °C, weighed the dry weight of the aboveground and underground parts. The dried rapes were crushed and mixed. The soil was naturally dried, ground and sieved.

The contents of N, P and K in rape were determined by national standard method (NY/T 2017–20119). The plant samples were digested with sulfuric acid-hydrogen peroxide until the solution was colorless and clear. Nitrogen content was determined by full auto azotometer, phosphorus content was determined by using spectrophotometry and the content of potassium was determined by using flame atomic absorption spectrometry. The rape samples were digested using nitric acid and perchloric acid mixture. The content of As, Hg, Pb, Cd, Cr^{6+} were determined by atomic fluorescence spectrometry (AFS), soil pH was measured by pH meter and the conductivity by conductivity meter. The available potassium was measured by using flame spectrophotometry, the available phosphorus by using Olsen-P and total nitrogen by using Kjeldahl method.

2.4 Data Processing

All data were processed using Microsoft Excel 2013, and each value of mean and S.D. (standard deviation) in the tables represents 3 replications of measurements. One-way ANOVA with Duncan test for multiple sample comparison was used by GLM procedures of SPSS16.0 software.

3 Results and Analysis

3.1 The Effect of Different Fertilizers on Rape Biomass

The influence of different fertilizer treatments on rape biomass was shown in table 2. The effect of new compound fertilizer (T2) on the fresh weight of rape was the highest, followed by the new organic manure (T1). The fresh weight in T2 increased by 34 % compared with T3, the treatment T2 increased by 39.7 %–40.6 % compared with T4, and significant difference was observed between T2 with T3 and T4. T1 increased by 1.0 %–7.5 % than that of T3, there was no significant difference between T1 and other treatments. There were significant differences in leaf dry weight between T1 with CK and T4. T1, T2 and T3 increased significantly than CK. The root dry weight of T2 increased significantly by 11 %–43 % than CK, while T1 significantly increased by 43 % compared with CK, while there was no significant difference between T4 and CK. There was no remarkable difference in root dry weight among T1, T2, T3 and T4.

Tab. 2: The effects of different fertilizer treatments on the rape biomass (g)

Treatment	leaf fresh weight	leaf dry weight	root fresh weight	root dry weight
CK	205.9±24.14b	19.6±1.82c	13.87±2.32a	1.63±0.25b
T1	252.86±42.02ab	24.1±2.06ab	16.3±1.61a	2.03±0.06ab
T2	292.33±12.67a	24.57±1.74a	16.67±2.97a	2.17±0.30a
T3	241.47±32.76b	22.4±2.38b	17.4±2.55a	1.97±0.15ab
T4	208.57±8.32b	20.37±2.30c	15.2±0.26a	2.13±0.40a

Tab. 3: The influence of different fertilizer treatments on the quality of rape

Treatment	N (%)	P2O5 (%)	K2O (%)
CK	1.29±0.06c	0.25±0.45c	0.63±0.01c
T1	2.27±0.06a	0.83±0.22c	0.84±0.10b
T2	1.77±0.25b	0.76±0.09bc	1.37±0.11a
T3	1.57±0.02bc	0.95±0.19b	0.78±0.11bc
T4	1.79±0.47b	1.79±0.47a	0.39±0.05bc

3.2 The Influence of Different Fertilizer Treatments on the Quality of Rape

He effects of different fertilizer treatments on N, P, K contents in rapes were shown in table 3. There was significant difference in nitrogen content in treatment T1 T2 and T4 compared with CK. In the treatment T1, the nitrogen content of rape was the highest, 43 %–46 % higher than that of T3. Treatment T2 increased by 27 %–30.8 % than T3. The phosphorus content of rape leaves in T1, T2 and T4 had significant differences. The phosphorus content in rape followed the order: urea (T4) > common organic manure (T3) > new type of organic manure (T1) > new type of compound fertilizer (T2) > no fertilizer (CK). Treatments T3 and T4 were significantly different compared with CK. Potassium content in treatment T2 was significantly higher than that of T4. There was a significant difference in the content of potassium between treatment T2 and T3 while there was no difference among T1, T3 and T4. The contents of potassium in T1 and T2 were significantly higher than that in CK. There was no significant difference among T3, T4 and CK.

3.3 Effects of Different Fertilization Treatments on Heavy Metal Accumulation of Rape

The contents of heavy metals accumulated in rape leaves under different treatments were shown in table 4. The content of As was the highest in rape with organic fertilizer (T3), followed by T2, and T3 had significant differences with T1, T2, T4 and CK. The as content of rape leaves in CK, T1 and T4 were below the detectable limit. The Cd content of rape leaves was also below the detectable limit. The highest content of Cr^{6+} in rape leaves was the application of new compound fertilizers (T2), there was no significant difference in Cr^{6+} and Hg content among the other treatments. The Hg content of rape leaves among CK, T1, T2, T3 and T4 had no significant difference. The highest content of Pb in rape leaves was T1, while no significant difference was found among the other treatments and Pb content was lower than the detectable limit.

The content of as in the rape root followed the order: T4>T3>T2>T1>CK, but there was no significant difference among these treatments. Cd content in the rape root was below detectable limit. After the application of new type of organic manure (T1), the Cr^{6+} content in rape root was higher than other treatments, the Cr^{6+} content in T1, T2, T4 had significant differences compared with CK. The new organic fertilizers (T1) and the application of organic manures (T3) had significant effects on Hg content, and there was significant difference between T2 and CK. There was no significant difference in Pb content of the rape root among the treatments.

Tab. 4: Effects of different fertilization on heavy metal accumulation of rape (mg/kg)

Treatment		As	Cd	Cr6+	Hg	Pb
CK	leaf	0.000±0.000b	0.000±0.000a	0.004±0.000a	0.003±0.006a	0.000±0.000a
	root	0.095±0.165a	0.000±0.000a	0.000±0.000b	0.067±0.026a	13.605±2.096a
T1	leaf	0.000±0.000b	0.000±0.000a	0.003±0.006a	0.000±0.000a	1.596±2.257a
	root	0.591±0.836a	0.000±0.000a	0.012±0.001a	0.028±0.003ab	7.979±3.494a
T2	leaf	0.066±0.115b	0.000±0.000a	0.009±0.005a	0.017±0.024a	0.000±0.000a
	root	1.420±1.162a	0.000±0.000a	0.007±0.004a	0.013±0.010b	3.631±4.718a
T3	leaf	0.550±0.576a	0.000±0.000a	0.005±0.000a	0.009±0.017a	0.000±0.000a
	root	1.695±0.901a	0.000±0.000a	0.005±0.006ab	0.018±0.003b	10.013±4.036a
T4	leaf	0.000±0.000b	0.000±0.000a	0.005±0.000a	0.000±0.000a	0.000±0.000a
	root	2.875±0.971a	0.000±0.000a	0.007±0.005a	0.043±0.040ab	18.137±0.284a

Tab. 5: The effect of different fertilizers on soil physical and chemical properties

Treatment	pH	Electrical conductivity (ms/cm)	Available K (mg/kg)	Available P (mg/kg)	Total N (%)
CK	8.76a	104.13 b	291.9 c	18.7 b	0.4 b
T1	8.74a	107.96 b	330.9 bc	28.1 b	0.4 b
T2	8.72ab	122.67 ab	365.3 b	61.6 a	0.4 b
T3	8.26 b	146.07 a	454.4 a	79.0 a	0.5 a
T4	8.65ab	113.83 ab	336.7 bc	14.8 b	0.3 c

3.4 The Effect of Different Fertilizers on Soil Physical and Chemical Properties

As shown in table 5, the soil pH decreased after fertilization which was consistent with the results of Ma et al. [10]. There was no significant difference among T1, T2, T4 and CK; however the pH value in T3 significantly decreased compared with T1 and CK. The soil electrical conductivity in T3 was significantly higher than CK. And there was no significant variation among other treatments with CK. The content of available K in T3 was the highest, and T3 and T2 were significantly higher than CK. The content of available P in T2 and T3 was much higher than those in other treatments. Total N content in T3 was 0.5 %, significantly higher than those of CK, T1, T2 and T4.

4 Conclusions

The application of different fertilizers can increase the fresh weight and dry weight of rape leaves and roots. Compared with common organic fertilizer, the application of new organic fertilizer and new compound fertilizer increased the rape fresh weight and dry weight.

N content in rape leaves with new organic fertilizer was significantly higher than that of common organic fertilizer. K content in rape leaves with the new compound fertilizer was much higher than that applying ordinary organic fertilizer and urea.

Compared with other treatments, the application of urea enhanced the accumulation of as and Pb in roots. There was no significant difference in the content of Cr^{6+}, Hg and Pb among different fertilizer treatments.

The application of fertilizer decreased the soil pH, but increased the electrical conductivity. Compared with urea, application of the new type of organic fertilizer, new type of compound fertilizer and common organic fertilizer could increase the content of soil nutrient and improved soil fertility.

References

[1] Wang Linqing, Li Xiaoming, Zhu fahua. Current situation of municipal solid wastes disposal and development proposals in China [J]. Environmental Pollution and Control, 2015, 37(2): 106–109.

[2] Specialized Committee of Urban Domestic Refuse of CAEPI. Development Report on China Treatment Industry of Urban Domestic Refuse in 2013[J]. CHINA ENVIRONMENTAL PROTECTION INDUSTRY, 2014, 12, 18–24.

[3] ZHANG Yu-shuai, WANG De-han, LIANG Wen-tao, et al. Application of city garbage compost on agriculture [J]. Guangdong Agricultural Sciences, 2006 (12): 93–96.

[4] Zhao Miao, Ren Lian-hai, Wang Pan. Applied analysis of disposal technologies of municipal solid waste in China [J]. Journal of Green Science and Technology, 2013, (12): 146–149.

[5] Kong Yue. The effects of biological organic fertilizer on the growth and quality of tomatoes and pakchoi [D]. Wuhan, China: Huazhong Agricultural University, 2007.

[6] Ge Chun-hui, Yang Xin-hua, Sun Jiu-sheng, et al. Influence of Living Garbage Compost Applied on Yield and Quality of Corn, and Physical-Chemical Character in Soil [J]. Chinese Agricultural Science Bulletin, 2013, 29(33): 237–241.

[7] Agaassi M, A Hadas, Y Benyamini, et al. Mulching effects of Composted Msw on Water Percolation and Compost Degradation Rate [J]. Compost Science and Utilization, 1998, 6(3): 34–41.

[8] Yang Xin-hua, Sun Jiu-sheng, Wang Xin-yong, et al. Influence of living garbage compost on heavy matals in soils, yield and quality of maize [J]. Xinjiang Agricultural Sciences, 2012, 49(11): 2096–2101.

[9] H Nacke, A C concalves Jr, D Schwantes, et al. Availability Heavy Metal (Cd, Pb and Cr) in Agriculture from Commercial Fertilizers [J]. Arch Environ Contam Toxicol, 2013, 64(4):537–544.

[10] MA Kun, LI Xin-ping, WANG Zhao-qian, et al. Effect of Municipal Refuse Compost to the Growth of Spring -Wheat and the Soil [J]. Agro-environmental Protection, 2000, 19(5): 312–314.

[11] ZHAO Feng-lian, SUN Qin-ping, LI Ji-jin, et al. Effects of Different Biogas Fertilizers on Yield, Quality and Nitrogen Use Efficiency of the Rape [J]. Journal of Soil and Water Conservation, 2010, 24(3): 127–130.

[12] Wang Jian-xiang, Zhou Jle-liang. Effects of Different Organic Fertilizers on Quality and Yield of pakchoi [J]. Shanghai Vegetables, 2007(1): 63–64.

[13] Zahra A, Alireza M, Jafar N, et al. Effect of fertilizer application on soil heavy metal concentration [J].Environ Monit Assess, 2010, 83–89.

[14] Fan Hai-rong, Hua Lu, Wang Xue-jiang. Effects of Municipal Waste Compost and Its Compound Fertilizers on Grass and Soil [J]. Journal of Agro-environment Science, 2007, 26(1): 188–192.

Jian-Peng Lv, Xian-Hua Wu, Xia-Fei Duan, Ou-Jing Li, Jin-Wang Li, Ling Wang and Zhong-You Pei*

Study on Establishment of Regeneration System of Sorghum Seed

Abstract: The plant regeneration system was established through callus regeneration using mature seeds and mature embryos of sweet sorghum as explants. The results showed that mature seeds could be induced to grow well in MS medium supplemented with 1.38 g L^{-1} proline, 500 mg.L^{-1} hydrolyzed casein and 3.0 mg.L^{-1} 2, 4–D The callus was transferred to MS + 1 mg.L^{-1} IAA + 0.5 mg.L^{-1} 6–BA medium to induce buds, and then transferred to MS + 3 mg.L^{-1} IBA culture After rooting on the base, can develop into a complete plant.

Keywords: Sweet sorghum, callus, mature seeds, 2, 4–D

1 Introduction

1.1 Sweet Sorghum Research

Sweet sorghum (Sorghum bicolor (L.) Moench) is a family of Poaceae (Poaceae) sorghum family (Andropogoneae) Sorghum, also called sugar sorghum which is a variant of common sorghum. Sweet sorghum are C4 plants, with high photosynthetic efficiency, high biological yield, drought resistance, waterlogging tolerance, barren, salt tolerance and wide adaptability characteristics, known as the "camel". Sorghum can be planted in the tropical, subtropical and temperate, mainly cultivat-

Jian-Peng Lv, Agriculture Resourde and Environment College, Tianjin Agricultural University, Tianjin, China. Email: 1363833717@qq.com
Xian-Hua Wu, Agriculture Resourde and Environment College, Tianjin Agricultural University, Tianjin, China. Email: 512189448@qq.com
Xia-Fei Duan, Agriculture Resourde and Environment College, Tianjin Agricultural University, Tianjin, China. Email: 2529138995@qq.com
Ou-Jing Li, Agriculture Resourde and Environment College, Tianjin Agricultural University, Tianjin, China. Email: 594826682@qq.com
Jin-Wang Li, Agriculture Resourde and Environment College, Tianjin Agricultural University, Tianjin, China. Email: 934675163@qq.com
Ling Wang, Agriculture Resourde and Environment College, Tianjin Agricultural University, Tianjin, China. Email: 704661418@qq.com
Zhong-You Pei, Agriculture Resourde and Environment College, Tianjin Agricultural University, Tianjin, China. Email: zhongyoupei@tjau.edu.cn

DOI 10.1515/9783110540048-004,

ed in tropical arid, semi–arid areas, especially in high altitude low temperature mountainous areas and saline areas and soil barren areas [1–2].

Sweet sorghum is a multipurpose grain crops, the grain can be edible and be made as wine; it's stem riches juice and high sugar content which can be used for processing and production of sugar and syrup can also be used as silage or processed into scarification of straw feed, the most important can be used to produce ethanol, instead of fossil fuels such as oil to ease the global energy crisis [3–5]. Especially in recent years, with the national ban on corn as wheat and other food crops as to ethanol production, ethanol production has developed rapidly attributed to sugar cane, sugar beet and other non–food energy crops. In many non–grain energy crops, sweet sorghum has the advantages of abundant germplasm resources, strong resistance to stress, wide adaptability, high biological yield, high sugar content of stem, high ethanol conversion efficiency and low cost of ethanol production. So it's considered the most potential energy crop [6–8].

1.2 Research Progress on Tissue Culture of Sweet Sorghum

Sorghum is considered to be one of the difficult plants tissue culture, after decades of development, although has made great progress, but still behind the rice, wheat, corn and other monocotyledon crops. The research of tissue culture of sweet sorghum is late and few, and the related research reports are few too.

Tissue culture of sorghum began in 1968 with Strogonov (1968) using sorghum roots and tiller nodes with 1 mg/L 2,4–D, 1 mg/L KT, 5 mg/L calcium pantothenate, 1 mg/L VC, 500 mg/L CH on the MS medium, but no regenerated plants were obtained. And then opened a prelude to the study of sorghum tissue culture. Subsequently, Masteller et al. (1970) induced the callus on the medium supplemented with 1–5 mg/L, 4–D with the bud primordium as the explant, and the regenerated plants were obtained. This is the first report on the success of sorghum tissue culture. Bhaskaran et al. (1983) succeeded in obtaining regenerated plants in tissue culture of mature sorghum. After that, Cai et al. (1987), Bai Zhiliang et al. (1995) also obtained regenerated green seedlings in mature sorghum embryos.

2 Materials and Methods

2.1 Test Material

Mature seeds: mature seeds of Lv Neng No.2 and LiaoTian No.3, Lv Neng No.2 mature seeds were harvested from Jinghai Teaching Base of Tianjin Agricultural College, and LiaoTian No.3 mature seeds were bred by LiaoTian Provincial Academy of Agricultural Sciences.

2.2 Culture Medium

Tab. 1: Different treatments of callus induction of mature seeds of sorghum

Treatment NO.	Genotype	Components of basic media	2,4–D (mg/L)	KT(mg/L)
T1	LvNengNo.2	MS abundant + MS trace + MS vitamin + iron salts	3	0.3
T2	LvNengNo.2	MS abundant + MS trace + MS vitamin + iron salts	5	0.5
T3	LvNengNo.2	MS abundant + MS trace + B5 vitamin + iron salts	2	0
T4	LvNengNo.2	N6 abundant +B5 trace + B5 vitamin + iron salts	2.5	0
T5	LiaoTianNo.3	MS abundant + MS trace + MS vitamin + iron salts	2	0
T6	LiaoTianNo.3	MS abundant + MS trace + MS vitamin + iron salts	3.5	0.3
T7	LiaoTianNo.3	MS abundant + MS trace + MS vitamin + iron salts	5	0.5
T8	LiaoTianNo.3	MS abundant + MS trace + B5 vitamin + iron salts	2	0

Tab. 2: Subculture medium of sorghum seeds

Treatment NO.	Genotype	Components of basic media	2,4–D (mg/L)	KT (mg/L)
T1	LvNengNo.2	MS abundant + MS trace + MS vitamin + iron salts	3	0
T2	LvNengNo.2	MS abundant + MS trace + MS vitamin + iron salts	5	0
T3	LvNengNo.2	MS abundant + MS trace + B5 vitamin + iron salts	2	0
T4	LvNengNo.2	N6 abundant +B5 trace + B5 vitamin + iron salts	2.5	0
T5	LiaoTianNo.3	MS abundant + MS trace + MS vitamin + iron salts	2	0
T6	LiaoTianNo.3	MS abundant + MS trace + MS vitamin + iron salts	3.5	0

Treatment NO.	Genotype	Components of basic media	2,4–D (mg/L)	KT (mg/L)
T7	LiaoTianNo.3	MS abundant + MS trace + MS vitamin + iron salts	5	0
T8	LiaoTianNo.3	MS abundant + MS trace + B5 vitamin + iron salts	2	0

Tab. 3: Differentiation media of calli from different explants of sorghum

MediaNO.	Explants	Components of basic media	KT (mg/L)	6–BA (mg/L)	NAA (mg/L)	Asp (g/L)	PVP (g/L)
R1	Mature seeds	MS abundant + MS trace + MS vitamin + iron salts	2	0.5	0.25	0.2	–
R2	Mature seeds	MS abundant + MS trace + B5 vitamin + iron salts	2	0.5	0.25	0.2	–

Rooting medium: 1/2 MS (MS substantial halved, the rest unchanged) basic medium + NAA 0.2 mg/L + Sur30 g/L +Plant gel 3.2 g/L.

2.3 Sterilization of Mature Seeds and Callus Induction

The complete seeds of Lv Neng No.2 and LiaoTian No.3 were selected, soaked with 70 % ethanol for 1min, then with 0.1 % mercuric chloride for 30min, and then rinsed 5–6 times with sterile water and put the seeds on super clean bench and use filter paper to dry the seeds, and next were inoculated to the corresponding induction medium. The inoculated seeds were placed at 25 ± 2 °C for 16 / 8h light / dark daylight conditions. Each glass dish with a diameter of 90 mm was inoculated with 25–30 seeds of mature seeds, one for every 2 dishes, and each treatment was repeated three times.

2.4 Subculture and Differentiation of Callus

After 2–3 weeks of induction, callus was formed on the hypocotyls of the mature seeds and the incision of the young spike. The number of calli formed and the explants inoculated number and calculate the rate. The calli were isolated and transferred to subculture medium, and cultured under the same culture conditions. The subculture medium was changed every 2–3 weeks. After 2–3 weeks of subculture, most of the calli could grow to a diameter of about 5 mm in size, the selection of good growth of light yellow texture close to the granular callus inoculated to the corresponding differentiation medium for differentiation. Differentiation culture conditions were 25 ± 2 °C / day 16/8 h light / dark. After callus differentiation, the numbers of callus were counted and the differentiation rate was calculated. If the medium browning occurs during the differentiation process, the callus of good growth and the formation of buds and the differentiation of regenerated seedlings could be transferred to the fresh differentiation medium to continue the culture.

2.5 Rooting, Transplanting and Regeneration of Regenerated Plants

When the regenerated seedlings grow to 3–5 cm, they are transferred to the rooting medium to take root. When they are rooted, the roots need to be cut off, and the callus of the roots is removed and the culture conditions are differentiated. After about 3 weeks of rooting, regeneration of the main root length up to 3–5 cm or longer, more fibrous roots and more robust root system which can be transplanted. Before transplanting in order to adapt to the external environment needs regeneration seedlings 3–5 of the hardening. When transplanting, remove the regenerated seedlings from the culture flask, wash the root of the medium with warm water, but careful not to hurt the root hair, and then transplanted to sterile vermiculite and nutrient soil 1: 3 mixed culture pots, and ensure the soil in the basin is moist. After the transplanted seedlings were planted in the greenhouse, they were grown or transferred to the field to grow and mature, and the agronomic traits of the obtained regenerated plants were investigated to find the favourable mutant materials.

3 Results and Analysis

3.1 Analysis on the Difference of Callus Different Explants of Sweet Sorghum

After 3 weeks of differentiation, some callus began to turn green, some callus gradually browned, and most of the callus grew long or hairy (Fig. 1E. 1 F). The regenerated seedlings could be seen at about 7 weeks of differentiation (Fig. 1 G). But different calli differentiation rates were different between different genotypes. During the process of differentiation, the callus of green–2 could secrete brown matter, and the culture medium became black. The callus of LiaoTianNo.3 secreted brown substance, which caused the medium to turn yellow (Fig.1F).LiaoTianNo.3 could differentiate A green point and regeneration seedlings, and green on the 2nd cannot be divided into a green point, it cannot differentiation emergence. The results of callus differentiation of mature seeds are shown in Table 4.

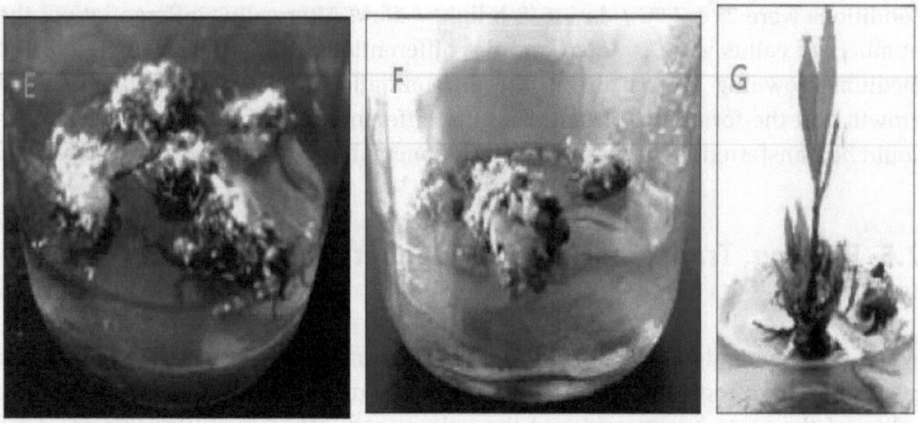

E. calli differentiation of Lvneng2; **F–G. Calli differentiation of Liaotian3**

Fig. 1: Calli differentiation of mature seeds

Tab. 4: Calli differentiation of mature seeds of sorghum

Genotype	Media	CD	GN.	CR	DR
LvNeng No.2	R1	65	0	0	0
LvNeng No.2	R2	42	0	0	0
LiaoTianNo.3	R1	36	2	1	2.78
LiaoTianNo.3	R2	50	8	0	0

3.1.1 The Effect of Different Genotypes on Callus Differentiation of Mature Seeds

From Table 4, the callus differentiation rate of different genotypes of sweet sorghum was different which was significant. The callus of LiaoTian No.3 had 10 green spots, one of them was differentiated and the differentiation rate was 2.78 %. There was no difference between the green point and the green one. The differentiation rate was 0 and the difference was significant. The results showed that genotype had a significant effect on callus differentiation, and the differentiation effect of LiaoTian-No.3was better than that of Lv Neng NO.2.

3.1.2 The Effect of Different Medium on Mature Seed Callus Differentiation

From Table 4, the differentiation rate of the same genotype callus in different differentiation medium is different, and the difference is significant. LiaoTianNo.3 had two green spots on R1 differentiation medium, and the differentiation rate was 2.78 %. Eight green spots were differentiated on R2 differentiation medium, but could not different. The differentiation rate was 0, the difference was significant. The results showed that the medium had a significant effect on the callus differentiation of mature seeds, and the differentiation effect of R1 was better than that of R2.

In conclusion, the callus differentiation of mature seeds was significantly affected by genotype and culture medium, and the differentiation effect of Laotian 3 was better than that of Green No.2, and the differentiation effect of R1 was better than that of R2.

3.2 Rooting, Transplanting and Regeneration of Sweet Sorghum Seedling

In the mature seed culture of sweet sorghum, only one regenerated green plant was obtained from LiaoTian No.3. After 3 weeks of rooting, the main root grew more than 5 cm, more fibrous roots, more robust roots (Fig. 2a), transplanted to vermiculite and nutritive soil in the pot placed in the greenhouse growth, and can flowering and fruiting normally , But plants were short, narrow, and short (Fig. 2b–c). In this experiment, LiaoTian3 regenerated seedlings were transplanted to pots with nutritive soil and vermiculite in mid–March 2013, heading to mid–May, normal flowering in late May, grain maturation at the end of June, Growth Period. Because LiaoTian No.3 is a hybrid of sweet sorghum, dwarf plant in tissue culture may be the result of gene isolation, but it cannot exclude the influence of clonal variation.

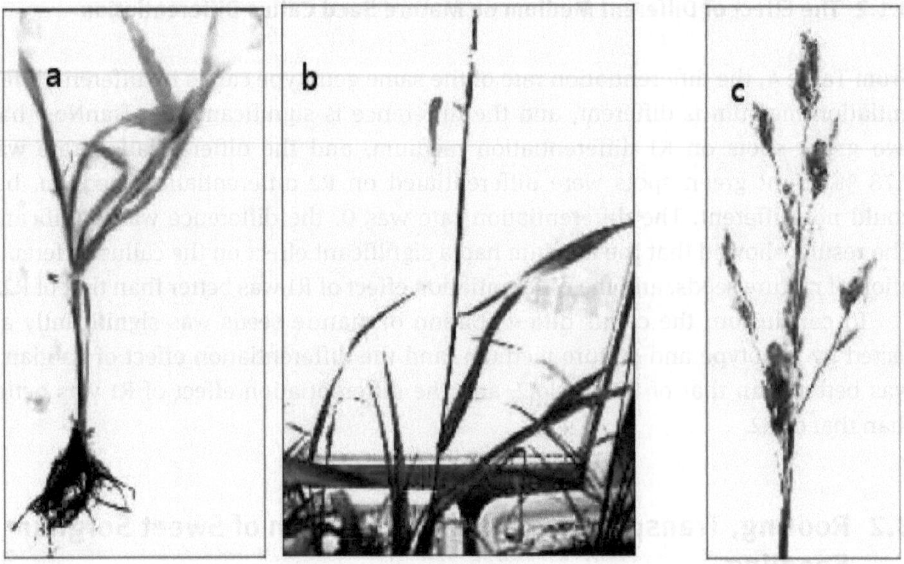

a. Rooting of regenerated plant; b. Heading of regenerated plant; c. Flowering of regenerated plant

Fig. 2: Rooting and growing of regenerated plant of Liaotian3

4 Discussion

In this study, the tissue culture of mature seeds, young spike and immature embryo of sorghum had browning, but the degree of browning was different. In the mature seed culture, the callus of LiaoTianNo.3 in the process of differentiation secrete tan material, the medium yellowing, differentiation rate was very low.

It is well known that genotype is one of the important factors affecting plant regeneration, and has significant effect on callus induction and differentiation. In this study, the regeneration ability of LiaoTian No.3 matured seeds was stronger than that of green No.2 mature seeds.

5 Conclusion

In the callus induction of mature seeds, different basic media and different hormone ratio treatments had a significant effect on callus induction in the process of callus differentiation. Different medium and different genotypes had different effects on callus differentiation significantly affected. The results showed that the induction rate was 56.34 %, the differentiation rate was 2.78 %, the induction rate was 58.79 % and the differentiation rate was 0. The optimal medium for LiaoTian No.3 was MB L, K + 0.3 mg/L, and 2,4–D 3.0 mg/L, respectively. The optimal induction medium for MS medium was 2,4–D3.0 mg/L + KT 0.3 mg/–D and KT with the use of induction than the low concentration of 2,4–D alone is superior to high concentrations of 2,4–D and KT with the use.

References

[1] Li Dajue, Liao Fusun.Positive sorghum and its utilization [M]. Science Press, 1992.

[2] Yang Wenhua. The Status of Sweet Sorghum in China's Green Energy [J]. Chinãs Sugar, 2004, 3: 57–59.

[3] Wang Yanqiu, Zhu Cuiyun, Lu Feng, et al. Application of sweet sorghum and its development prospect [J] .Crops of Crops, 2004, 24 (1): 55–56.

[4] Zhang Zhipeng , Yang Zhen , Zhu Kai, et al. Development and utilization of sweet sorghum [J] .Crops of Crops, 2005, 25 (5): 334–335.

[5] Yan Jinyue, Zhao Lixin, Wang Gehua, et al. Study on Sustainable Development Strategy of Energy Crops in China [M] .Beijing: China Agricultural Press, 2009.

[6] Cao Wenbo.A new approach to develop sweet sorghum production to exploit and utilize energy [J] .Chinese Seed Industry, 2002, 1: 28–29.

[7] Gao Shijie, Liu Xiaohui, Li Yufa, et al. Resources and Utilization of Sweet Sorghum in China [J]. Seed, 2005, 24 (11): 46–47.

[8] Lu Qingshan. Sweet Sorghum [M]. China Agricultural Science and Technology Press, 2008.

Huan Yu and Xian-Lin Meng

Air Pollutants Emitted by Coal Resources City: Health Risks for the Population Living in the Neighborhood

Abstract: The air pollution that can lead to the health risks for human beings is a hot issue in environmental science research. By investigating the changes of air pollution spatiotemporally caused by human activities in a coal resource based city in Heilongjiang province, one systemic research on long - term residents about health risk was studied with the health risk assessment model. The results showed that, BaP were the main factor of regional health risks, rather than noncarcinogenic risk. The region health risk probability changes along with time which is mainly related to the development and utilization of coal resources in the region. Health risks differed in different areas because of the distance to the coking plant.

Keywords: Coking plant; Environmental sensitive; Health risks; Pollution factor; BaP

1 Introduction

Studies on the potential impact on health risk associated with the coal resources development activities have mainly focused on two classes of compounds: (1) criterion pollutants such as carbon monoxide (CO), nitrogen and sulfur oxides (SO_x and NO_x), total suspended particulate matter (TSP), sulfur compounds of hydrogen (H_2S), hydrogen and nitrogen compounds (NH_3) and (2) noncriterion pollutants such as benzo pyrene (BaP), etc.

In recent years, quantitative risk assessment tools have been used to evaluate various site development activities caused by local as well as the pollution of the environment of the area around. Characterization of environmental risk assessment for one or more physical, chemical or biological agents to human health (or possible) effect probability.

Marta Schuhmacher study on the human health risk for long - term living in cement plant, set up the mathematical model for personal health risk and the health

Huan Yu, Harbin Institute of Technology, Harbin, China. Email: 550649804@qq.com
Xian–Lin Meng, State Key Laboratory of Urban Water Resource and Environment, School of Municipal and Environment Engineering, Harbin Institute of Technology, Harbin, China. Email: mengxlll@sina.com

DOI 10.1515/9783110540048-005,

risks of cancer research through by SO_2, NO_x and dioxins from cement production, although the research results shows that the cement plant nearby residents affected by cement plant emissions of pollutants is lower, the paper has made valuable exploration from the methodology [1]. Bingheng Chen use the "four steps" to assess NOx more than ten years of monitoring data in the urban atmosphere residents, the results show that NOx pollution in the atmosphere caused some losses to residents' health, and with the increase of NO_x, the health risk is on the rise [2]. Zhenyuan Zang assess the abandoned chemical crowd also with "four steps", the results show that the venue cancer risk and non–cancer risk over the acceptable values, the chemical plant site for adults and children have risks, so the government need to make urban plan reasonably [3]. Ping Li use mainly adopts domestic health risk assessment of soil health risk model to evaluate heavy metal pollution in atmospheric dust, the results show that the heavy metal pollution of atmospheric dust in October to March of next year is relatively serious, the risk of no cancer in children are bigger than adults, and the maximum has most risk in no carcinogenic risks [4].

Unfortunately at the moment, the monitoring data and the study of population environmental health risk of coking plant is less. Therefore study around the crowd of coking plant health risk is very necessary. This article will divide a develop long–term production release of air pollutants into the carcinogenic risk and no carcinogenic risks, use "four steps" to make assessing system of the health risk of the residents around the coking plant, in order to providing scientific basis for avoiding potential health risk factors and planning the coal resources city development .

2 The Research Object

2.1 The City of Coal Resources General Situation and Coking Industry Development Situation

The coal resource city is a city with coking industry mainly and multi–industry simultaneously. The city has proven recoverable reserves of 400 million tons of coal resources, with the main coking coal, fat coal, gas coal, anthracite coal, especially the main coking coal, accounts for 82 % of total reserves, has a very low sulfur, low phosphorus, high calorific value, high ash melting point, etc., known as "industrial powder", is one of the rare coalfields state in three protective mining. Therefore, the development and production of coke will lead industry of the city's economic development.

Until 1996, the coal resource city set up the first coking plant, the region is given priority to with "small primitive coke". After 2002, coke market demand soared with developing smoothly mutations leading to the city coking plant quantity spurt. The city had 27 coking plant, with production capacity between 1 million tons/year.

Beginning in 2003, the central government began to control small coking plant. By the end of the of June 2004, more than 60 “small primitive coke” have been shut down, banned and destroyed, but large–scale coking plant has built, its coking capacity reach 7 million tons/year. However, the coking severe overcapacity by the steel industry downturn. Beginning in 2011, several coking enterprises had entered coke unsalable, falling prices, operating loss serious trouble, the region is the production of coking enterprises are to extend industrial chain continuously, at present only the four coking plant in the region to avoid the risk of the furnace, capacity only retain the ability of 4 million tons/year. Since May 2016, the prices of coke began to rise, so the coking plant management situation improved markedly. Coking enterprise production operation is worth to pay close attention to the health risks of pollution emissions.

2.2 Pollutant Profile Distributions of Meteorological Conditions and Population Distribution Characteristics

The level of pollutant concentration distribution in environment particularly around coking enterprise is closely related to meteorological conditions in the region. According to the meteorological statistical data of 20 years in the region, the primary wind direction was WSW, with a frequency of 18 %, while the secondary wind direction was W, with a frequency of 14 %. The percentage of calms was more or less constant throughout the year around 7 %.

The city coking enterprises is relatively concentrated, focused on the west of the city. The region main pollutants come from major industrial pollution sources in coking enterprises and its industry chain emissions.

In the present study, the concentration distribution around 1 km in coking enterprises outside the scope, mainly in the east of #3 village, with a population of more than 3600 people. Fig. 1 shows location of sampling points.

Fig. 1: Location of sampling points

2.3 The Source of the Monitoring Data

This study collected the study area from 1999 to 2011. Shown from table 1 to table 4.

Tab. 1: The region's monitoring data in 1999 (mg/m^3)

	BaP	TSP	SO_2	NO_2	CO	H_2S	NH_3
#1	0.0009	0.2540	0.0334	0.0241	1.7400	0.0005	0.0070
#2	0.0005	0.1090	0.0254	0.0173	0.9750	0.0005	0.0070
#3	0.0013	0.1280	0.0263	0.1980	1.1300	0.0005	0.0070

Tab. 2: The region's monitoring data in 2003(mg/m^3)

	BaP	TSP	SO_2	NO_2	CO	H_2S	NH_3
#1	0.0005	0.1574	0.0324	0.0288	1.4000	0.0010	0.0150
#2	0.0010	0.2538	0.0442	0.0318	2.3333	0.0010	0.0150
#3	0.0005	0.2588	0.0488	0.0340	2.3333	0.0017	0.0150

Tab. 3: The region's monitoring data in 2007 (mg/m^3)

	BaP	TSP	SO_2	NO_2	CO	H_2S	NH_3
#1	0.0087	0.1574	0.0488	0.0374	1.0000	0.0200	0.0950
#2	0.0053	0.2492	0.0336	0.0234	1.0000	0.0030	0.0680
#3	0.0036	0.2538	0.0480	0.0362	2.0000	0.0050	0.0900
#4	0.0012	0.2546	0.0316	0.0198	1.0000	0.0010	0.0400
#5	0.0004	0.2636	0.0396	0.0396	1.0000	0.0020	0.0600
#6	0.0005	0.2498	0.0368	0.0204	1.0000	0.0010	0.0200

Tab. 4: The region's monitoring data in 2011 (mg/m^3)

	BaP	TSP	SO_2	NO_2	CO	H_2S	NH_3
#1	0.0005	0.1574	0.0324	0.0288	1.4000	0.0010	0.0150
#2	0.0010	0.2538	0.0442	0.0318	2.3333	0.0010	0.0150
#3	0.0005	0.2588	0.0488	0.0340	2.3333	0.0017	0.0150

3 Method of Study

Health risk assessment have many methods such as four steps from the National Academy of Sciences (NAS) , life cycle analysis, method of MES and beta - Possion model assessment of viral infection, etc. In these methods, four steps has the most common of the national academy of sciences to use includes Hazard Identification, Toxicological Evaluation, Exposure Assessment, and Risk Characterization. The method is widely used in environmental medium such as air, water and soil in the human health risk assessment of pollutants in the toxic chemical [5]. This article will use the "four steps" health risk assessment on the city.

3.1 Hazard Identification

In fact, as the coal resource development activities of coking enterprises, the main harmful substances released into the atmosphere are poison and dust, belongs to the chemical hazard factors, chemicals directly inhaled into the body through the air from the pollution emissions to the atmosphere [6]. In dozens of chemical toxicants of coking enterprise, the largest amount of emissions and the most harmful substances mainly are BaP, benzene soluble matter, sulfur dioxide, carbon monoxide, nitrogen dioxide and ammonia. They all are high toxic substances except the sulfur dioxide. Environmental chemical damage to the human body is mainly poisoning effect. According to the course of disease development speed and characteristics, they can be divided into acute, chronic and chronic poisoning. Chronic special poisoning is embodied in: the body's basic genetic material DNA is induced by poison making its sequence and structure change down to genetic code changes, eventually cause cancer and birth defects. And the follow dust's harm to human body: contacting or inhalation of dust will make cornea, skin mucous membrane on the local stimulation first and cause a series of pathological changes. Long - term inhalation of high concentrations of dust can cause lung diffuse, progressive fibrosis of systemic disease, namely pneumoconiosis. Due to long - term inhalation of productive dust of pneumoconiosis, it is a common occupational disease which has more destructive [7].

As mentioned earlier, the main harmful substances in coking enterprises released into the atmosphere are poison and dust, belongs to the chemical hazard factors. This study, CO, SO_x, NO_x, TSP, NH_3, H_2S and BaP will be to evaluated.

3.2 Toxicological Evaluation

Air pollutants caused people to disease and death in the related research reports earlier. The EPA of China provides some reference dose of pollutants, table 5, table 6 discussed for the reference value of pollutants [8].

Tab. 5: Cancer - causing pollutants reference dose

Pollutants	BaP
$q_i(mg/(kg{\cdot}d)^{-1})$	7.3

Tab. 6: Noncancer - causing pollutants reference dose

pollutants	TSP	SO_2	NO_2	CO	H_2S	NH_3
$D_{IRF}(mg/(kg \cdot d))$	0.04	0.023	0.029	0.046	0.003	0.97

3.3 Exposure Assessment

Coking enterprise have land for homes, schools and so on, the main sensitive crowd is people around, including the adult and the child. Therefore, residents living around the coking enterprise may cause adverse health effects through direct contact with air pollutants in daily life breathing.

3.4 Risk Characterization

Coking enterprises have various kinds of pollutants. However according to the influence on population health risk can be divided into carcinogenic effect (BaP) and carcinogenic effect (SO_2, NO_x, TSP, CO, H_2S and NH_3).This paper carried out in accordance with the carcinogenic effect and noncarcinogenic effect of health risk assessment, and assess health risk on the basis of the comprehensive.

3.4.1 Carcinogenic Risks

Cancer risk expresses the likelihood of occurring cancer due to a definite daily intake of a pollutant [1]. Cancer risk use URF_i to estimate. URF_i is defined as the air pollutant concentration of 1μg. m^{-3} when the carcinogenic risk of cap.

$$R^c = \sum_{i=1}^{k} R_i^c \tag{1}$$

$$R_i^c = [1 - \exp(-D_i q_i) / 70 \tag{2}$$

$$Di = C \times M/70 \tag{3}$$

Where R_i^c is genes toxic substances i through breath way for the average individual in risk of cancer, a^{-1}; D_i is genetic toxic substances i through breath way units daily exposure dose, mg/(kg· d); q_i is genetic toxic substances i through breath way carcinogenic factor, mg/(kg· d) $^{-1}$; M take 20 m^3/d by breathing; 70 a is human life expectancy.

3.4.2 Noncarcinogenic Risks

In the risk of noncancer, all not carcinogenic chemicals are related to noncarcinogenic effect. As a result, noncarcinogenic risk attribute has always been a dose - response analysis, are calculated by calculation, hazard index of the human body to accept the exposure dose and the ratio of reference dose (D_{iRF}) of pollutants, D_{iRF} defined as not because health risks of dose of pollutants in the atmosphere.

$$R^n = \sum_{i=1}^{k} R_i^n \tag{4}$$

$$R_i^n = (D_i / D_{iRF}) \times 10^{-6} / 70 \qquad (5)$$

R_i^n is noncarcinogen i personal average annual risk through breathing, a^{-1}; Di is noncarcinogen i through breath way units daily exposure dose, mg/(kg · d); D_{iRF} is noncarcinogens i through breath way reference dose, mg/(kg · d); 70a is human life expectancy.

3.4.3 Total Risk Characterization

Assuming all the toxic effects of toxic substances harmful to human body health relationship are not antagonism relationship, we characterize the total risk as:

$$R^T = R^c + R^n \qquad (6)$$

International Commission on Radiation protection (ICR) recommends maximum acceptable risk is 5×10^{-5}/a [9].

4 Results and Discussion

4.1 The Analyses of Main Factors which Influence the Health Risk Probability in the Region

Using monitoring data of coking enterprise sensitive area around 1999, 2003, 2007, 2011years, for each sensitive crowd , we assess the overall health risk on the basis of the risk of cancer and noncancer according to the the health risk assessment method. Figure 2 shows the probability of the risk of cancer and noncancer, the blue part of the figure is carcinogenic effect probability, the yellow part is noncarcinogenic effects probability. Table 7 shows the percentage of carcinogenic effect probability and noncarcinogenic effect probability.

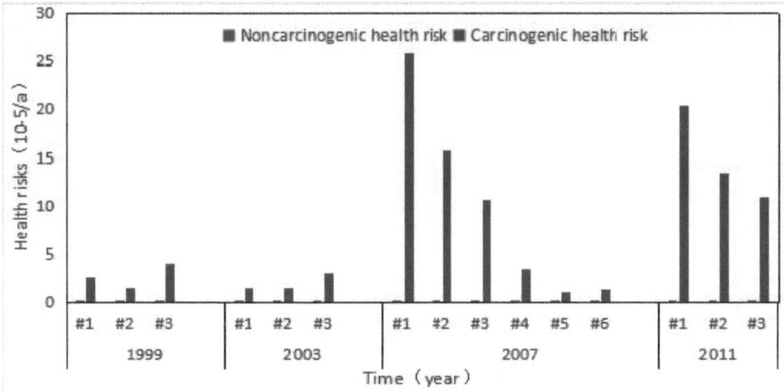

Fig. 2: The probability of the risk of cancer and noncancer

Through the monitoring data of different years about all sensitive evaluation results in Fig.2, The probability of causing cancer by BaP is much higher than that of SO_2, NO_x, TSP, CO and H_2S, NH_3.

Tab. 7: The percentage of carcinogenic effect probability and noncarcinogenic effect probability

Time	#1		#2		#3	
	noncancer	cancer	noncancer	cancer	noncancer	cancer
1999	0.738	99.262	0.666	99.334	0.368	99.632
2003	1.019	98.981	1.666	98.334	0.813	99.187
2007	0.057	99.943	0.081	99.919	0.211	99.789
2011	0.155	99.845	0.136	99.864	0.241	99.759

It can be seen from Table 7 that the percentage of the probability of the risk of cancer are over 98 %, explaining that BaP pollutant is the main factor of regional health risk, rather than noncarcinogenic risk.

4.2 The Region Health Risk Probability Analysis in the Region Through Time

Fig. 3 shows the regional vibration of health risk probability along with time. The blue part of the figure represents #1 location, the red part of the figure represents #2 locations, and the green part of the figure represents #3 locations.

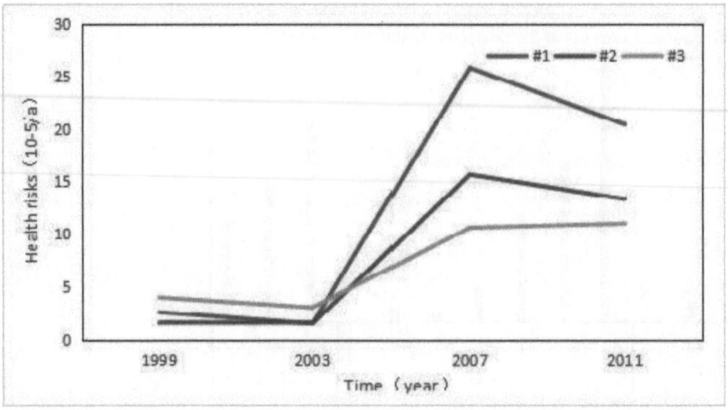

Fig. 3: The regional variation of health risk probability along with time

From Fig. 3, the regional health risk probability goes flatly or slightly down from 1999 to 2003. This is because the region is given priority to be with "small primitive coke", which changes after the birth of first 50000 Ton/Year in 2000. It includes a complete set of the corresponding pollution control engineering measures. From 2003, the government began to control small coking enterprises, "Small primitive coke" was shut down, banned and destroyed, as well large–scale coking enterprises were gradually built. Thus during the period in 2003, the concentrations of pollutants in air distribution are basically flat or slightly down due to construction of a complete set of environmental protection facilities. Sensitive health risk probability was not changed too much. However from 2003 to 2007, its coking capacity reach 7 million tons/year. Pollutant emissions increases in the atmosphere also showed a trend of increase. Its health risk evaluation results in 2007 #1, #2, #3 are even more than the maximum acceptable risk recommended by the ICR. After 2011, several coking enterprises had entered coke unsalable, falling prices, operating loss serious trouble, the region is the production of coking enterprises are to extend industrial chain continuously, to 2016 only the four coking plant in the region to avoid the risk of the furnace, capacity only retain the ability of 4 million tons/year, so health risk probability fell down in 2011.

4.3 The Regional Health Risk Probability Analysis of Characteristics of Different Regions

Since the health risk probability which coking enterprise environment caused changing with the time very consistent on the whole, this section only take an example in coking enterprises around the health risk probability analysis of the research in different areas in 2007.

Fig. 4 shows the characteristic figure of regional health risk probability in 2007. The yellow part of the graph represents the size of human settlement's health risk probability.

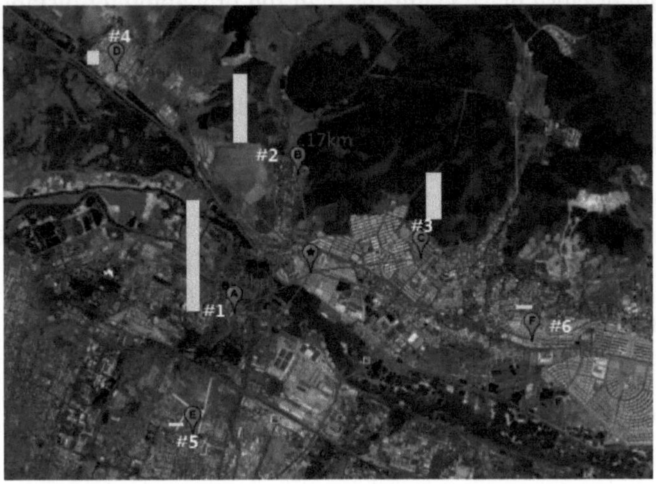

Fig. 4: The regional health risk probability characteristic figure in 2007

From Fig. 4, #1, #2, #3, the health risk probability is much higher than human settlement, they are 0.000258203a^{-1}, 0.000156882 a^{-1}, 0.000106201 a^{-1} respectively, the distance #1, #2, #3 to the coking plant are 1.03 km, 1.17 km, 1.28 km. Since #4, #5, #6 is far away from coking plant, health risk probability is not big because of pollutant concentration are lower than #1, #2 #3, and the farthest #4 is only 0.134 of #1. As the coking enterprise has the low pollution origin, which leads to spread of the pollutant, it always causes the higher concentrations of atmospheric pollutants near by the coking plant. Therefore the probability of close place is greater than the probability of distant location.

5 Conclusion

Results of Environmental sensitive health risk assessment around coking clusters are follows:

The major air pollutants causing the regional health risks is BaP, whose health risk probability is more than 98 % of the total health risk probability;

The coking production has much influence for the atmosphere pollution in the region. The regional health risk probability is well linked to the changes of the coking capacity and the different methods of coking production. During the period around 2007, the health risk probability in the region of environment sensitive which close around the coking enterprises are even more than the maximum acceptable risk recommended by the ICR. So measures must be taken to reduce the emissions of pollutants to reduce health risk probability in that region.

Finally, the probability of close place is greater than the probability of distant place. Thus the coking enterprise has the low pollution origin, which leads to spread of the pollutant making higher concentrations of atmospheric pollutants. Full consideration must be given in the urban planning, and development, because coking enterprises should not be within the scope of people live for a long time.

References

[1] S. Marta, L. D. Jose, and G.Josepa, "Pollutants emitted by a cement plant: health risks for the population living in the neighborhood," J. Environmental Research, 2004 (95:198–206).

[2] C. B. Heng, H. C. Jie, and Z. H. Gang, "Quantitative Assessment of Impact of Ambient Air Nitrogen Oxides on Human Health in Shanghai," J. Shanghai Environmental Sciences, 2002 (21:3).

[3] Z. Z. Yuan, Z. Yi, and W. LI, "Human Health Risk Assessment of an Abandoned Chemical Plant in Beijing," J. Asian Journal of Ecotoxicology, 2008.

[4] L. Ping, X. L. Yin, and W. S. LI, "Pollution Evaluation and Health Risk Assessment of Heavy Metals from Atmospheric Deposition in Lanzhou," J. Environmental Science, 2014.

[5] H. Y. Peng, "A Study of Health Risk Assessment on Contaminated Sites and Method of Remediation Goal Calculation in China," D. China University of Geosciences, 2012.

[6] Q. Z. Ning, "Study on System and Application of Atmosphere Environmental Risk Assessment of Coking Industry," D. Harbin Institute of Technology, 2010.

[7] Y. Jing, "Analysis and Control Strategy for Harmful Materials Atmosphere of Xinxing Coking Industry Park," D. Harbin Institute of Technology, 2013.

[8] L. X. Li, "Population health risk assessmengt of atmospheric SO_2, PM10 and NO_x pollutants in Urumqi," D. Xinjiang Medical University, 2014.

[9] M. X. Lin, S. L. Xin, and L. X. Ying, "Assessment of the human health risk of heavy metal in fly a shand dregs from medical and pharmaceutical waste incineration disposal," J. Journal of Harbin in Institute of Technology, 2008.

Li-Xia Jiang, Gui-Lin Ren, Hui-Ying Zhao*, Jia-Jia Lv, Hai-Xia Zhu and Li-Juan Gong

Evolution of Cold Day Index Based on Minimum Temperature and Its Impact on Crop Yield

Abstract: Meteorological data from 80 meteorological stations, data on per unit area yield of corn and rice from 38 stations in Heilongjiang province were used, current research results and the recognized agrometeorological index were referred to determine calculations on cold day index (CDI) during growth period of crops (May 20 to September 30), mathematical statistics methods were used to study the trends of CDI, so as to analyze its correlation with crop yield. The results showed that: (1) from the year 1971–2014, 1972 and 1976 were typical years with cold days. CDI could accurately indicate the difference of low temperature among different years. (2) From 1971–2014, average CDI of crop growth period within research area declined significantly, with a rate of 1.5d/10a.The high incidence of cold day events lied in 1970s, and declined from 1980s to 1990s; continued decreasing since 21st century, no change or rebound occurred in mountain area. (3) During research period, the spatial distribution of CDI within research area presented a trend of more in north and less in south, more in mountain area and less in plains, with CDI declining from north to south, mountain area to plains. (4) Within growth period, CDI of 89 % of stations within research area has obvious or extremely obvious negative correlation with per unit area yield of corn and rice. When CDI increased, the per unit area yield of corn and rice will decline at different extent, indicating the negative effect of low temperature on crop yield. CDI has significant impact for yield of corn and rice in east of Songnen Plain, Sanjiang Plain and mid-level in Mudanjiang, in which the impact on rice is greater than that on corn. (5) CDI has agronomy and meteorological significance and is able to better reflect the mechanism of impact of low temperature events on crops.

Keywords: cold day index; chilly area; the minimum temperature; corn and rice yield

---

Li-Xia Jiang, Jia-Jia Lv, Hai-Xia Zhu, Li-Juan Gong, About the Author: Li-Xia Jiang, senior engineer, master degree, principally engaged in study of applied Meteorology and climate resources, Heilongjiang Institute of Meteorological Science, Harbin 150030, China; E-mail:hljjlx@163.com
Gui-Lin Ren, Heilongjiang Meteorological Bureau, Harbin 150030, China
Hui-Ying Zhao, Corresponding Author, Heilongjiang Institute of Meteorological Science, Harbin 150030, China; E-mail:zhaohhyy2008@aliyun.com

DOI 10.1515/9783110540048-006,

1 Introduction

As locates in high latitude and cold area, the total grain output of Heilongjiang Province has ranked to No.1 in whole nation in 2011, playing a vital role in guaranteeing national food security. However, due to special geographical environment, limited and variable heat resources, it is one of the areas with most cold days in China, and most often of low temperature events, which even in short term, could affect growth and yield of crops at different extent during crop growth period [1]. Continuous low temperature events, namely low temperature chilling injury has an un–ignorable negative effect on stable and high yield of crops. Thus, a deep study of the rules and evolution characteristics of low temperature events during crop growth period, to determine its relation with crop yield, has precious reference value for agricultural disaster prevention and mitigation and planting structure adjustment, as well as an important practical significance in boosting sustainable development of modern agriculture and guaranteeing national food security.

Since the 1980s, studies related to the relation between climate change and agricultural production have been under a period of upsurge [2], with diversified types and some of which were focus on impact of disasters on agriculture, In this field, various domestic and foreign experts and scholars have conducted a lot of researches on low temperature chilling injury [3–6]. We can see from those studies results that, the chilling injury identification index could meet different analysis demands at different views. Since the point cut of each index is different, they all have their own characteristics, while the overall trends indicated by chilling injury identification results based on different indexes are consistent [1].

In fact, the daily minimum temperature is actual measured value, which has larger change during growth period of crops in high altitude and cold area, and has closer relation with crop yield. However, up to now, there are still rare reports based on updated meteorological data, applied minimum temperature to study cold day index (CDI) and its relation with crop yield in cold area. Thus, based on CDI of minimum temperature, targeted in Heilongjiang region, where chilling injury occurred most, the analysis on temporal and spatial variation characteristics of delayed chilling injury during growth period of crops and study of its relation with crop yield, are expected to deeply identify essential disaster–causing factor in chilling injury, to further provide technical supports to revealing of mechanism of low temperature chilling injury, so as to provide meteorological reference for drawing on advantages and avoiding disadvantages in crop production, for planting structure adjustment and for guaranteeing national food security.

2 Material and Methods

2.1 Data Source and Climate Zones

Daily minimum temperature from May to September from 1971–2014 from 80 meteorological stations in Heilongjiang Province, measured data of corn and rice developmental period from 1971–2014 from 38 meteorological stations, all of which coming from Archives Bureau of Meteorological Bureau of Heilongjiang Province, are used. Data about per unit area yield of corn and rice from 1971–2014 are from Statistical Bureau of Heilongjiang Province.

Based on the geographic of studied stations, the research area was divided into four climate zones (as shown in Figure 1), which includes Daxing'anling and Xiaoxing'anling, Songnen Plain, Sanjiang Plain and Mudanjiang mid-level.

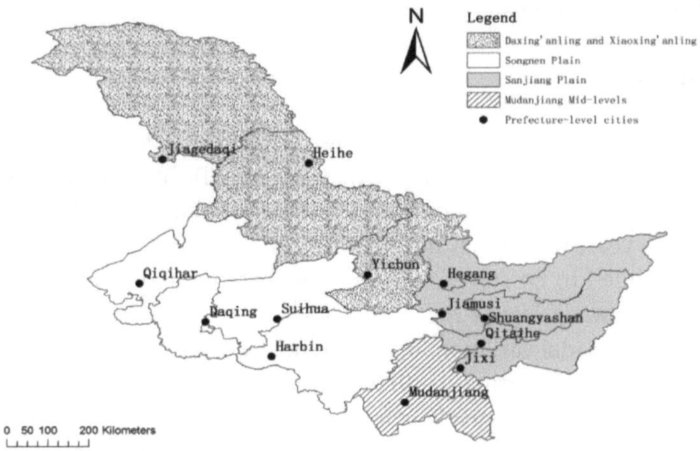

Fig. 1: Spatial Distribution of 4 Climate Zones in Research Area

2.2 Research Methods

CDI could better reflect the integrated impact of low temperature events on crops. Based on research results of Wang Yanhua et al [7], Xu Changshen et al [8], in this Article, CDI is defined as, during crop growth period (May 20 to September 30), the total number of days with daily minimum temperature for consecutive 3 days 3 °C lower than the average of years of daily minimum temperature at that day, and the period applied in the Article is years from 1971–2000. CDI will be identified and calculated as follows:

$$D_0 = \begin{cases} 1, T_i - \bar{T}_i \prec -3 \ \text{Above consecutive 3 days} \\ 0,\ T_i - \bar{T}_i \geq -3 \ \ or \ \ T_i - \bar{T}_i \prec -3 \ \text{Below consecutive 2 days} \end{cases} \tag{1}$$

$$CDI = \sum D_0 \tag{2}$$

In which, T_i is the minimum temperature of the i day during the crop growth period in a certain year from a certain station, T_i is the 5d sliding value of sequence of daily minimum temperature from 1971–2000 at the i day during the crop growth period a certain year from a certain station, $D_0$is the number of days of CDI during the crop growth period in a certain year from a certain station and CDI is the total number of days of CDI during the crop growth period a certain year from a certain station.

3 Results and Analysis

3.1 CDI Time Change

During crop growth period from 1971–2014, average CDI in research area declined obviously, with tendency rate of 1.5d/10a (P<0.01), in 1970s, CDI had violent vibration, and average CDI in 1972 and 1976 continued to be more than 13d, which was typical low temperature year, which that in 1973, 1974 and 1975 did not exceed 5d. And after 1980s, CDI tended to be stable with less rangeability and most of years were kept within 2–4d (shown in Figure 2).

We can see that from decadal variation, overall trends of CDI in each station was declined, with maximum CDI occurring in 1970s, average value of each area were above 7d; CDI in 1980–1990s was relatively stable, declined a lot as compared with 1970s, with rate of more than 3d; since 21st century, as compared with 1980–1990s, CDI in plains continued to decline, which is different in mountain areas, CDI has no change in Daxing'anling and Xiaoxing'anling, and the trends of CDI in Mudanjiang mid-level was rising(shown in Table 1).

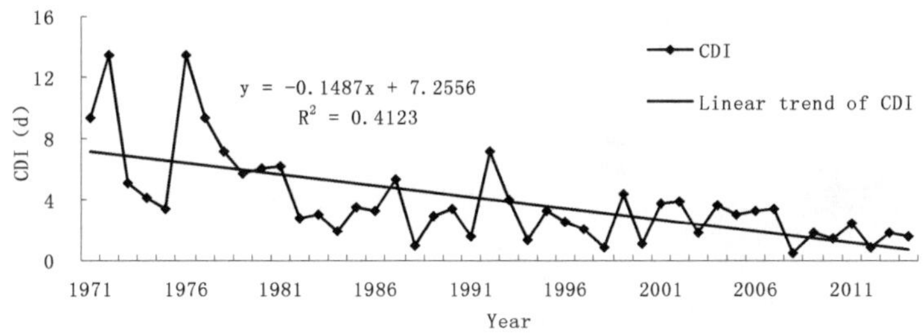

Fig. 2: Time Variation of CDI in Research Area from 1971 To 2014

Tab. 1: Decadal Variation of CDI in Research Area from 1971 to 2014 (d)

Time period	Research area	Daxing'anling and Xiaoxing'anling	Songnen Plain	Sanjiang Plain	Mudanjiang Mid–levels
1971–1980	8	8	8	7	8
1981–1990	3	5	3	3	3
1991–2000	3	4	3	3	2
2001–2014	2	4	2	2	3
1971–2014	4	5	4	3	4
1971–2000	5	6	4	4	4

3.2 CDI Spatial Distribution Characteristics

We can see from Figure 3 that, CDI during crop growth period has an overall trend of more in north, less in south, more in mountain areas and less in plains, in which maximum value 9d, occurred in Mohe, north of Daxing'anling; the minimum value 2.6d, occurred in Harbin at south and Jidong at east, CDI declined from north to south, mountain areas to plains progressively, during the period, a maximum value area occurred at middle of Xiaoxing'anling, at center of it, CDI in Wuying Station, reached to 7.7d, which is close to north of Daxing'anling. We can see from area change that, CDI became from more to less from Daxing'anling, Xiaoxing'anling, Mudanjiang Mid–level, to Songnen Plain and to Sanjiang Plain. The decadal change showed that CDI of each area from different years differed in 1–2d (Table 1).

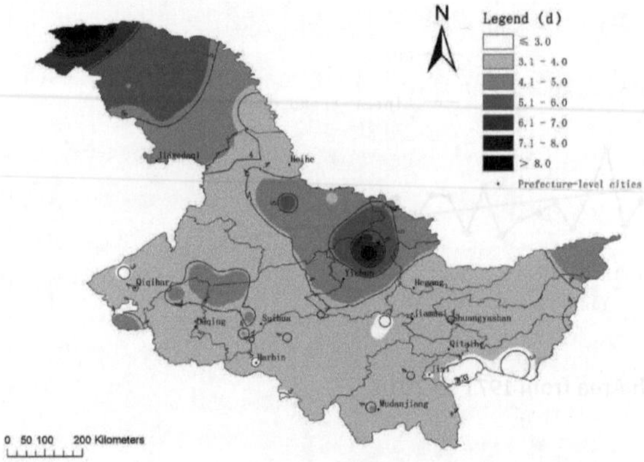

Fig. 3: Spatial Variation of CDI in Research Area from 1971 to 2014

3.3 Impact of Minimum Temperature on Crop Field

In order to further explore the relation between climate conditions and crop yield, in this Article, 38 stations from 4 climate zones are selected for related analysis on official per unit area yield of corn and rice with CDI. We can see from Table 2 that, per unit area yield of corn and rice has obvious or extremely obvious negative correlation with CDI ($P<0.05$ or $P<0.01$). At a certain extent, the lower temperature the area has, the greater the correlation between yield and CDI will be. We can see from Table 2 that, correlation will be different on different crops, in research area, 71 % of stations have R^2 of rice higher than corn, in Mudanjiang Mid–level, Daxing'anling, Xiaoxing'anling, Songnen Plain, the R^2 of rice was higher than corn, and that in Sanjiang Plain was almost the same.

Tab. 2: Correlation between CDI and crop yield in the research area

area	station	Corn R^2	Rice R^2	a_1	a_2	$a_1 - a_2$
Daxing'anling, Xiaoxing'anling	Bei'an	0.152^{**}	0.113^{*}	−229.1	−238.8	9.7
	Xunke	0.091^{*}	0.106^{*}	−172.3	−230.7	58.4
	Jiayin	0.173^{**}	0.276^{**}	−176	−455	279
	Yian	0.249^{**}	0.235^{**}	−175.3	−191.4	16.1
	Logjiang	0.215^{**}	0.386^{**}	−270.1	−331.2	61.1
	Tailai	0.100^{*}	0.160^{**}	−92.9	−133.5	40.6
	Nehe	0.298^{**}	0.458^{**}	−305.3	−337.2	31.9

Tab. 2: Correlation between CDI and crop yield in the research area (Continued)

area	station	Corn R^2	Rice R^2	a_1	a_2	$a_1 - a_2$
Songnen Plain	Hailun	0.196**	0.280**	−368.2	−407	38.8
	Zhaodong	0.317**	0.469**	−383	−405.8	22.8
	Zhaoyuan	0.106*	0.163**	−202.4	−254.2	51.8
	Wuchang	0.263**	0.350**	−640.9	−695.5	54.6
	Fangzheng	0.098*	0.268**	−160.3	−328.9	168.6
	Mulan	0.121*	0.185**	−184.4	−246.3	61.9
	Suihua	0.214**	0.311**	−351	−417.2	66.2
	Shuangcheng	0.245**	0.310**	−362.5	−482.6	120.1
	Shangzhi	0.328**	0.434**	−479.4	−518.7	39.3
	Yilan	0.104*	0.091*	−285.8	−272.1	−13.7
	Mingshui	0.167**	0.165**	−206.3	−225.1	18.8
	Fuyu	0.251**	0.223**	−179.6	−247.6	68
	Qinggang	0.181**	0.239**	−302.1	−326.5	24.4
	Baiquan	0.222**	0.390**	−227.4	−282.4	55
	Wangkui	0.235**	0.434**	−271.5	−424.4	152.9
	Lanxi	0.118*	0.150*	−150.1	−193.6	43.5
	Binxian	0.135*	0.283**	−251.1	−403.6	152.5
Sanjiang Plain	Tangyuan	0.171**	0.188**	−311.5	−292.2	−19.3
	Jiamusi	–	0.113*	–	−265	–
	Jidong	0.379**	0.420**	−536.1	−543.3	7.2
	Huanan	0.097*	0.103*	−203.7	−180	−23.7
	Mishan	0.243**	0.287**	−460.5	−412.8	−47.7
	Raohe	0.286**	0.276**	−340.1	−334.9	−5.2
	Boli	0.159*	0.152*	−258.2	−245	−13.2
	Baoqing	0.097*	0.089*	−219.6	−197.7	−21.9
Mudanjiang Mid-levels	Linkou	0.279**	0.396**	−272.3	−287.2	14.9
	Ning'an	0.197**	0.292**	−353.6	−336.7	−16.9
	Mulin	0.270**	0.374**	−455.7	−402.7	−53

Note: * means significance test of 0.05, ** means significance test of 0.01, "–"means failed in 0.05 test, a_1 and a_2 means the amount of declined yield of corn and rice while CDI increased by 1d (Kg/hm²).

4 Discussion

CDI functions as monitoring such consecutive low temperature events during crop growth period and could fully reflect the occurrence of consecutive low temperature events in a certain year of a certain place, even could reflect the comprehensive impact of extreme low temperature events on crops. But the effect of low temperature monitoring of CDI at an average extent was not distinct.

As for negative correlation between crop yield and CDI, when CDI increased, per unit area yield of corn and rice would decline at different extent, once CDI increased by 1d, the per unit area yield in research area would decline by 289.4 Kg/hm^2 in average, and per unit area yield of rice would decline by 329.9 Kg/hm^2 in average, in which reduction of rice will be 40.5 Kg/hm^2 greater than that of corn, which means, at the overall level of average in research are, the impact of low temperature on rice was greater than that on corn. But in different climate zones, CDI impact on corn and rice has difference. In the two major plains in Heilongjiang Province, the decline yield of corn in most of stations in Songnen Plain was less than that of rice, while it was opposite in Sanjiang Plain, which means that at some extent, CDI impact on rice in Songnen Plain was greater than that of corn; while the CDI impact on rice yield in Sanjiang Plain was smaller than that of corn. CDI definition shows the change states of daily minimum temperature during crop growth period, with concept of "negative accumulated temperature" contained, which means the accumulated value of negative anomaly of minimum temperature. The greater the low margin of daily minimum temperature, the more the CDI accumulated days will be, means the greater the "negative accumulated temperature" will be, shows that the worse the heat states was, the greater the negative effect on crop yield will be. We can see from CDI expressions that, when CDI reached to 1d, the accumulated value of anomaly of minimum temperature during the period will be 9 °C·d greater, thus, the accumulated value of anomaly of minimum temperature has the same indicating functions on low temperature events, so it could be used in evaluation of impact of low temperature on crop yield. During crop growth period, if the accumulated value of anomaly of minimum temperature for consecutive 3d was greater than 9°C·d, it means the low temperature indicated has negative effect on crop yield.

5 Conclusion

(1) CDI based on low temperature has agronomic and meteorological significance, is able to express change of minimum temperature, accurately reflect difference of low temperature in different years and even characterize the mechanism of extremely low temperature events impact on crops.

(2) During 1971–2014, CDI in crop growth period showed a declined trend, in 2010s, CDI in pars of stations showed a trend of rebound, means one the one hand, CDI change has an obvious effect on climate warming, and on another hand, regional CDI has no consecutive declining, which means there is uncertainty of occurrence of low temperature chilling injury, which has further verified the results of Zhao Junfang et al [9] and Jiang Lixia et al [3].
(3) CDI could accurately characterize minimum temperature impact on crop yield, during crop growth period, results in 89 % of stations in research area showed obvious or extremely obvious negative correlation ($P<0.05$ or $P<0.01$) between CDI and per unit area yield of corn and rice, which means low temperature events have negative effect on formation of corn and rice yield. When CDI increased, per unit area yield of corn and rice would decline different extents, the greater the CDI was, the more the declining would be, showing accumulating function of negative effect of CDI. CDI has different impact in different climate zones and different crops. In general, CDI has greater impact on rice and corn in east of Songnen Plain, Sanjiang Plain and Mudanjiang mid–levels, in which CDI impact on rice is greater than that on corn.
(4) CDI expresses a definition of "negative accumulated temperature", during crop growth period, if the accumulated value of anomaly of minimum temperature for consecutive 3d was greater than 9 °C·d, it means the low temperature indicated has negative effect on crop yield.
(5) There is no detailed analysis of CDI impact on crop in essential breeding period, and CDI needs further optimization to promote scientificity, applicability and accuracy of indexes. Above works are still needed to be deeply studied in future.

References

[1] Guo Jianping, Ma Shuqing. Theory and Practice of Monitoring and Forecasting of Low Temperature Chilling Injury of Crops [M] China Meteorological Press, Beijing, 2009: 1–62.
[2] Wei Zhudeng, Fang Xiuqi, Su Yun, et al. A review of climatic impacts on Chinese socio–economic development over the past 2000 years [J]. Advances in Earth Science, 2014, 29(3): 336–343. (in Chinese)
[3] Jiang Lixia, Ji Shengtai, Li Shuai, et al. Relationships between rice empty grain rate and low temperature at booting stage in Heilongjiang Province [J]. Chinese Journal of Applied Ecology, 2010, 21(7): 1725–1730. (in Chinese)
[4] Lyons J M. Chilling Injury in Plants [J]. Annual Review of Plant Physiology, 2003, 24(3): 445–466.
[5] J. Kościelniak,J. Biesaga–Kościelniak. The Effect of Short Warm Breaks during Chilling on Water Status, Intensity of Photosynthesis of Maize Seedlings and Final Grain Yield [J]. Journal of Agronomy & Crop Science, 2000, 184(1): 1–12.
[6] Wang Chunyi, Cai Jingjing, Zhang Jiquan. Risk assessment of drought and chilling injury of maize in Northeast China [J].Transactions of the Chinese Society of Agricultural Engineering (Transactions of the CSAE), 2015, 31(6): 238–245. (in Chinese)

[7] Wang Yanhua, Ren Chuanyou, Huang Ruidong, et al. Spatiotemporal change of consecutive cold index in China in recent 45 years [J]. Journal of Natural Disasters, 2013, 22(2): 116–123. (in Chinese)

[8] Xu Changshen. Agricultural Meteorological Indicators [M] China Meteorological Press, Beijing, 2000: 1–16.

[9] Zhao Junfang, Yang Xiaoguang, Liu Zhijuan. Influence of climate warming on serious low temperature and cold damage and cultivation pattern of spring maize in Northeast China [J]. Acta Ecologica Sinica, 2009, 29(12): 6544–6551. (in Chinese)

Wen Gao and Hong-Yan Shen*

Effect of Oxytetracycline Wastewater on the MDA Contents and SOD Activities in Zebrafish Muscle Tissue

Abstract: In order to obtain eco–toxicity effect of oxytetracycline wastewater on Zebrafish and provide theory basis for oxytetracycline wastewater pollution monitoring indicators in environment, the effect of oxytetracycline wastewater toward Zebrafish were studied with indoor exposure method. In this study, Zebrafish was chosen as test animal and exposed to different volume fraction of 20 %, 40 %, 60 % and 80 % of oxytetracycline wastewater for 15 days. The effect of oxytetracycline wastewater on the MDA contents and SOD activities in Zebrafish muscle tissue were assayed during 15 days. According to the experimental data, the MDA contents and SOD activities in Zebrafish muscle tissue had changed significantly during the period of exposure. The experimental results show that SOD activity and MDA contents showed "Λ" type change during the period of exposure. Different percentage of oxytetracycline wastewater had a great effect on SOD activity and MDA content in Zebrafish muscle tissue. The activity of SOD in Zebrafish muscle was significantly inhibited from 3 to 9 days ($0.01<P<0.05$). On the 12th day, it was highly induced ($P<0.01$). 20 % and 40 % exposure groups reached the maximum of the experimental period (194.05 $U\cdot mgprot^{-1}$, 186.37 $U\cdot mgprot^{-1}$, respectively). On the 15th day, there was no significant difference in SOD activity between each exposure group and control group ($P>0.05$). On the third day, the MDA content in the 20 % exposure groups were inhibited ($0.01<P<0.05$). On the 6th day, there was no significant difference in the MDA contents between the exposed groups and the control groups ($P>0.05$). The content of MDA in each exposed group was mostly induced ($0.01<P<0.05$) from 3 to 9 days. On the 15th day, each exposure group was highly induced ($P<0.01$). 40 % and 80 % groups reached to the maximum (4.42 $nmol\cdot mgprot^{-1}$, 5.41 $nmol\cdot mgprot^{-1}$, respectively). According to the experimental data, the MDA contents and SOD activities in Zebrafish muscle tissue had changed significantly and caused the oxidative damage.

Keywords: Oxytetracycline wastewater, Zebrafish, SOD, MDA

Wen Gao, College of Environmental Science and Engineering, Pharmaceutical Molecular Chemistry Laboratory of Hebei Province, Hebei University of Science and Technology, Shijiazhuang, China.
Hong-Yan Shen, Corresponding author. College of Environmental Science and Engineering, Pharmaceutical Molecular Chemistry Laboratory of Hebei Province, Hebei University of Science and Technology, Shijiazhuang, China. Email: 1178611805@qq.com

DOI 10.1515/9783110540048-007,

1 Introduction

In recent years, antibiotic production continues to rise in our country. In 2013, antibiotics outputs and exports in our country reached 121.2 thousand tons/year and 34.3 thousand tons/year, respectively. Antibiotics is the largest varieties of export crude drugs in our country [1]. It is found that an increasing number of drinking water, surface water, groundwater and soil, sludge, river sediment and faeces can be detected different levels of antibiotics. Its concentration level is $ng\cdot L^{-1}$ or $\mu g\cdot L^{-1}$ in environment [2]. Oxytetracycline (OTC) belongs to the antimicrobial agents of tetracycline class which is mainly used for the treatment of human and veterinary drugs. In the EU countries, its dosage is more than 2500 tons/year, of which is the most widely used terramycin [3]. High concentration of organic matter, composition complex, and poor biodegradability are main characteristics of oxytetracycline wastewater. In recent years, many scholars have studied some toxicological effect of oxytetracycline. Qu Mengmeng et al. studied the acute toxicity of oxytetracycline to Zebrafish and the effects of oxytetracycline on the DNA damage of carp kidney cells. The results show that the oxytetracycline 96–h LC_{50} of Zebrafish is 1,341.764 $mg\cdot L^{-1}$. DNA damage of carp renal cell could result from the oxytetracycline [4–6].

Aquatic organisms are exposed to a wide variety of environmental contaminants to assess the possible biological impacts. Fish have been extensively used as toxicity test organisms. Zebrafish is a small tropical fish, cyprinidae, Danio. Zebrafish have becomed model organism which recommended by the international organization for standardization, because they are sensitive to the change of environment. Besides, it has been widely used in the areas such as gene embryos [6–8]. Oxytetracycline wastewater used in this experiment comes from a terramycin production workshop. This experiment regards Zebrafish as the research object. By discussing effect of oxytetracycline wastewater on the MDA contents and SOD activities in zebrafish muscle tissue, test provide scientific basis for ecological risk assessment of oxytetracycline wastewater.

2 Materials and Methods

2.1 Experimental Material

2.1.1 Experimental Equipment and Instrument

Pipette (Pipet–Lite; TopPette Pipettor); glass homogenizer; hypothermia refrigerator (BBC–226STV); UV–visible light spectrophotometer (UV–2550); High purity water distiller (SYZ–A); analytical balance (EL204) ; water bath kettle (DK–S26) ; microplate reader (SpectraMax190) ; high speed centrifuge (TG16–WS) ; aeration pump;

dissolved oxygen meter (JPBJ–608); whirlpool mixer (XW–80A) ; hydrotimeter (YD300).

The kit of coomassie brilliant blue (CBB), MDA and SOD were purchased from a biological engineering research institute.

2.1.2 Test Organisms

Zebrafish were obtained from aquafarm in Shijiazhuang. In this experiment, mean length of fish used was 25±3 mm. Mean weight of fish used in this experiment was 0.20±0.05 g. Before entering the laboratory, disinfection treatment employ 5 % salt water, and then domesticate a week in 48h of aeration tap water. Feeding was halted 24h prior to testing. Healthy individuals of fish with similar size and weight were employed for test.

2.2 Experimental Method

Subacute experimental set up five experimental groups (0, 20 %, 40 %, 60 %, and 80 %, respectively). Experiments effect of different concentration oxytetracycline wastewater on the MDA contents and SOD activities in Zebrafish muscle tissue at different periods. Each group set up three parallels, each group randomly put in 25 fish. The subacute toxicity experiments lasted 15d. Continuous aeration used aeration device during the experiment. The experiment uses the static displacement method and changes the water every day. Besides, experiment assay on 3, 6, 9, 12, and 15d.

2.3 Assay Method

Zebrafish was dissected. 0.18g muscle tissue was obtained and rinsed in physiological saline. Muscle tissue was blotted with water and weighed. Samples was added to the mass ratio of 1:9 saline and grinded in the ice–water bath for 6–8min. 10 % of the tissue samples were removed for experimentation and centrifugated for 10 minutes at speed of 3000 $r \cdot min^{-1}$ to get supernatant for future determination. The supernatant was collected for the experiment. Activities of Superoxide Dismutase (SOD) and Malondialdehyde (MDA) content in samples were estimated by the method of SOD and MDA kit.

2.4 Statistical Analysis

All statistical analyses were expressed as mean ± standard deviation (SD). One-way Anova of all data were checked using SPSS 17.0. Significant difference analysis uses the method of least significant difference (LSD). It analyzes the data in the groups on the same day. P value of <0.05 was considered statistically different, $0.01<P<0.05$ statistically significantly different, and $p<0.01$ statistically very significantly different.

3 Results and Discussion

3.1 Effect of Oxytetracycline Wastewater on the SOD Activities in Zebrafish Muscle Tissue

Effect of oxytetracycline wastewater on the SOD activities in Zebrafish muscle tissue was shown in table 1.

Tab. 1: Effect of oxytetracycline wastewater on the SOD activities in Zebrafish muscle tissue

Concentration	SOD activities (U/mgprot)				
	3d	6d	9d	12d	15d
0	96.05	91.61	116.38	122.86	66.09
20%	40.95**	86.60	62.56**	194.05**	64.09
40%	37.10**	60.30*	72.95**	186.37**	78.29
60%	83.05	107.59*	97.77*	109.57	72.41*
80%	48.94*	78.83	56.93**	45.08**	65.87

The same day all the data are compared with control group, *p <0.05, ** p<0.01.

As shown in table 1, only 60 % exposed groups had significant induction of SOD activity in the muscle of Zebrafish on the 6th day ($0.01<P<0.05$) and other exposure groups were subjected to different degrees of inhibition at the exposure period (3~9d). On the third day, the activity of SOD in the four exposed groups was inhibited ($0.01<P<0.05$). In 20 % and 40 % exposure groups, SOD activity reached the minimum during the experiment (40.95 U·mgprot^{-1}, 37.10 U·mgprot^{-1}, respectively). The activity of SOD in other exposed groups was also lower than that in the control group. On the 6th day, the inhibitory level was alleviated and the activity of SOD was increased compared with that of 3rd. Compared with the control group, the activity of SOD was not significant. At the 9th day, 60 % exposed groups were significantly inhibited ($0.01<P<0.05$), and the other three exposed groups were very significantly inhibited ($P < 0.01$). On the 12th day, SOD activity of 20 % and 40 % exposure groups suddenly increased to reach the maximum of the entire experiment (194.05 U·mgprot^{-1}, 186.37 U·mgprot^{-1}, respectively), compared with the control groups reached a very significant difference ($P<0.01$); there was no significant difference between the 60 % exposure groups and the control groups; 80 % exposure groups were significantly inhibited ($P<0.01$). At the 15th day, there was no significant difference between the exposed groups and the control groups except the 60 % exposure groups ($0.01<P<0.05$).

On the whole, compared different oxytetracycline wastewater exposure groups with the control groups, SOD activity showed a trend "Λ" type. Ultimately, there was no significant difference in SOD activity between the exposed groups and the control groups. The body is in equilibrium, and the SOD activity of different exposure groups fluctuates up and down during exposure. This phenomenon can be considered as the body's self-defense response to defend foreign substances [9]. In the early stage of the experiment on 3rd to 9rd days, too much superoxide anion radical ($O^{2-}\cdot$) consumed a large amount of SOD in order to maintain the balance of the antioxidant defense system, so that all exposure groups (except for the 6th day) showed different degrees of inhibition of SOD. When fish exposed to 3rd, SOD activity in the four concentrations was lower than that in the control groups. This may be due to exposure to exogenous contaminant stress at the initial stage. The body produces oxygen free radicals to make the body's own enzyme material consumption. It may also be caused by excessive free radicals antioxidase direct damage. In the late (12–15d), Zebrafish adapt oxytetracycline wastewater environment, $O^{2-}\cdot$ not being increased, no longer consume a lot of SOD, that SOD activity increased, reaching a maximum value during the experiment, the final balance of the body is oxidized, SOD activity increased, reaching the maximum value during the experiment, the final body in the oxidation balance, SOD activity at normal levels [10–11].

3.2 Effect of Oxytetracycline Wastewater on the MDA Contents in Zebrafish Muscle Tissue

Effect of oxytetracycline wastewater on the MDA contents in Zebrafish muscle tissue was shown in table 2.

Table 2 show, in the initial exposure on 3rd, the experimental exposure groups compared with the control groups MDA content, has declined, subject to different degrees of inhibition, MDA levels down to the minimum duration of the experiment. There was no significant difference in MDA content between the exposed groups and the control groups at the 6th day ($P>0.05$). The content of MDA in 20 % and 60 % exposed groups were higher than that in control groups. The content of MDA in each exposed group increased from 9 to 15 days compared with the control groups. And there was significant difference ($P<0.01$). At the 9th day, the 20 % exposed groups suddenly reached the maximum value during the experiment (5.27nmol·mgprot^{-1}) and the maximum amount of MDA content that day. Compared with the control group, the content of MDA in other exposed groups was also different. At the 12th day, the content of MDA in 20 %, 40 % and 80 % exposed groups decreased, while the 60 % exposure groups increased compared with the 9th day. The 60 % exposure groups reached the experimental maximum (5.49 nmol·mgprot^{-1}). On the 15th day, each exposure group was significantly induced (P<0.01) compared with the control groups. 40 % and 80 % exposure groups reached the maximum (4.42 nmol·mgprot^{-1}, 5.41 nmol·mgprot^{-1}, respectively).

Tab. 2: Effect of oxytetracycline wastewater on the MDA contents in Zebrafish muscle tissue

Concentration	MDA contents (U/mgprot)				
	3d	6d	9d	12d	15d
0	1.31	1.47	1.53	1.74	2.14
20%	0.69*	1.81*	5.27**	4.03**	3.37**
40%	0.87	1.45	3.21**	2.80**	4.42**
60%	1.03	1.77	3.41**	5.49**	4.30**
80%	0.66**	1.50	2.56*	1.98	5.41**

The same day all the data are compared with control group, *p <0.05, ** p<0.01.

During the exposure period, each exposure group showed "Λ" type change in different time periods, and MDA significantly reflected the severity of the free radical attack. In the middle of the experiment (3–6d), MDA content was inhibited in different degrees. There was no significant difference between the exposure groups and the control groups at 6 days. SOD enzymes also play a role in scavenging oxygen free radicals. The increase of MDA content in fish resulted in lipid peroxidation. But with the extension of time, the phenomenon of peroxide has been lifted, it may be because the SOD enzyme in the body play a great role. On the 9th~15th day, the MDA content in each exposed group was mostly induced. At the 9th day, the maximal population (5.27 nmol·mgprot^{-1}) appeared in the 20 % exposure groups. This shows that oxytetracycline wastewater has entered the Zebrafish muscle tissue, the fish's antioxidant tissue is undergoing a series of changes, which may be due to the body to produce a large number of reactive oxygen free radicals, making the degree of lipid peroxidation significantly rise [13]. On the 12th day, the 60 % exposure groups reached the maximum. Feng Tao et al. studied the effects of benzo (a) pyrene on antioxidant defense system in the liver of large bombarded fish. It was also found that exposure time and concentration had significant effects on MDA concentration [14]. On the 15th day, it was significantly different from the control groups (P<0.01). This indicates that the antioxidant system in the fish has suffered a greater degree of damage, making oxygen free radicals increased, combined with unsaturated fatty acids caused by lipid peroxidation process, MDA content increased but not digestion, and ultimately lead to serious membrane system damage. From the experimental results, the body's antioxidant defense system was a certain degree of damage.

4 Conclusion

Through the subacute toxicity test, this paper study effect of the different volume percent oxytetracycline wastewater on the MDA contents and SOD activities in Zebrafish muscle tissue. The conclusion is that SOD activity and MDA contents showed "Λ" type change during the period of exposure. Different percentage of oxytetracycline wastewater had a great effect on SOD activity and MDA content in Zebrafish muscle tissue. The activity of SOD in Zebrafish muscle was significantly inhibited from 3 to 9 days ($0.01<P<0.05$). On the 12th day, it was highly induced ($P<0.01$). 20 % and 40 % exposure groups reached the maximum of the experimental period (194.05 $U\cdot mgprot^{-1}$, 186.37 $U\cdot mgprot^{-1}$, respectively). On the 15th day, there was no significant difference between each exposure group and control group ($P>0.05$). On the third day, the MDA content in the 20 % exposed groups were inhibited ($0.01<P<0.05$). On the 6th day, there was no significant difference in the MDA contents between the exposed groups and the control groups ($P>0.05$). The content of MDA in each exposed group was mostly induced ($0.01 <P <0.05$) from 3 to 9 days. On the 15th day, each exposure group was highly induced ($P<0.01$). 40 % and 80 % groups reached to the maximum (4.42 $nmol\cdot mgprot^{-1}$, 5.41 $nmol\cdot mgprot^{-1}$, respectively). According to the experimental data, the MDA contents and SOD activities in Zebrafish muscle tissue had changed significantly and caused the oxidative damage.

Acknowledgement: This study was supported by the National Natural Science Foundation of China (41373096), National Environmental Protection Project (201509041–05), Hebei Province Natural Science Fundation (B2014208068), Hebei Province Pharmaceutical Molecular Chemistry Laboratory Open Foundation, Hebei Province Environmental Protection Public interest Project, and Hebei Province Key Disciplines Construction Fund Project.

References

[1] Zhu Pei, Zhang Jianbin, Chen Daijie. The research situation and suggestions of antibiotics residue. Chinese Journal of Antibiotics. Vol.38 (9), pp. 647–651, 2013.

[2] Wang Qia n. Investigation and analysis of aminoglycoside antibiotics in our hospital. Lishizhen Medicine and Materia Medica Research. Vol.15 (9), pp. 637, 2004.

[3] Chen Guixiu, Wu Yinbao. Environment behavior research progress of Veterinary drugs terramycin. Progress in Veterinary Medicine. Vol.32 (5), pp. 102–107, 2011.
[4] R. HIRSCH, T. TERNES, K. HABERER, et al. Occurrence of Antibiotics in the Aquatic Environment. Science of the Total Environment, 1999, 225(1): 109-108.
[5] R. UENO, A. KINOSHITA, J. WAKABAYASHI. Comparative Pharmacokinetics of Oxyt – etracycline Ineel and its Fate in a Aquatic Environment. Aquaculture, 2004, 235: 53–63.
[6] QuMengmeng, Sun Liwei, Chen Yun, etal. Toxicology studies of veterinary drugs arsanilic acid and terramycin. Journal of Agro–Environment Science. Vol.23 (2), pp. 240–242, 2004.
[7] Aragues R, Medina E T, Martinez–Cob A,etal. Effects of deficitirrigation strategies on soil salinization and sodification in a semiarid drip–irrigated peach orchard [J]. Agricultural Water Management, 2014, 142: 1–9.
[8] Fan Zili, Ma Yingjie, Ma Yingjun. Saline soil and its improved utilization in the western region of China [J]. Arid Zone Research. Vol.18 (3), pp. 1–6, 2001.
[9] Wang Lixin. Oxidative stress reaction of streptomycin wastewater induced reaction in zebrafish and ecological risk assessment. [Master thesis, Hebei University of Science and Technology]. Shi Jiazhuang. Hebei University of Science and Technology, 2004.
[10] Zhao Yue, Cao Zhihui, Wang Bing, et al. Effect of streptomycin wastewater effluent after biochemical method treatment on the MDA levels and SOD activities in zebrafish. Jiangsu Agricultural Sciences. Vol.43 (8), pp. 348–351, 2015.
[11] Shen Hongyan. Wang Lixin. Yang Jindi. Gao Jixi. Jiao Xiaohui. LiuJunwei. Effect of streptomycin wastewater on the MDA contents and SOD activities in Zebrafish muscle tissue. Journal of Hebei University of Science and Technology. Vol. (03), pp. 303–308, 2014.
[12] Li Xiuyin. Wang Yuebing, Li Mingchuan. The principles of classification and treatment of antibiotics. Chinese Journal of Clinical Rational Drug Use. Vol.5 (28), pp. 12, 2012.
[13] Zhang Fan. The discovery of penicillin. Biology Teaching. Vol.33 (7), pp. 70–71, 2008.
[14] Bilgili A V. Spatial assessment of soil salinity in the Harran Plain using multiple kriging techniques [J]. Environmental Monitoring and Assessment, 2013, 185 (1): 777–795.

Dong-Chao Xu, Lei Yu*, Shen-Peng Chen and Jia-Nan Hou

Thermal Perceptions Responding to a Diversity of Urban Ecological Environments

Abstract: To describe thermal perception differences in terms of a diversity of urban ecological environments, influences of dry bulb temperature, wet bulb temperature, and black globe temperature on individual thermal perceptions responding to various ecological environments have been explored. After relationships of these temperatures on subjective thermal environment evaluations have been studied, the quality of human settlements could be predicted in advance and the results will be used to guide urban designers and planners theoretically in their practices.

Keywords: thermal perceptions; temperature parameters; urban ecological environment

1 Introduction

It has caused great concern about cities increasingly hot issue because of the expansion of the city and serious ecological destruction. Different urban ecological environment has different thermal comfort effects. However, the correlation between them is not very clear. The factors affecting the human body thermal comfort include temperature, humidity and wind speed. Correspondingly, three measurable indices reflecting the overall thermal comfort of the environment conclude dry bulb temperature, wet bulb temperature and black globe temperature. At present, people mainly use dry bulb temperature to judge the thermal comfort, which is quite different from the actual situation. From the temperature measurement process, dry-bulb temperature is the measured temperature of thermometer in normal air, avoiding radiation and moisture interference, that is, the general weather forecast temperature [1-2]; wet bulb temperature is equal to dry bulb temperature at the same enthalpy value of the air, when the air is saturated with water vapor [3]; black globe tem-

Dong-Chao Xu, Harbin Institute of Technology (Shenzhen), Shenzhen University Town, Nanshan Shenzhen, China. Email: 626401805@qq.com
Lei Yu, Harbin Institute of Technology (Shenzhen), Shenzhen University Town, Nanshan Shenzhen, China. Email: Leilayu@hit.edu.cn.
Shen-Peng Chen, Meteorological Bureau of Shenzhen Municipality, Shenzhen, China. Email:306709482@qq.com
Jia-Nan Hou, Chongqing Longhu real estate, Xingyi Square, Dazhulin Street, Yubei Chongqing. Email 87887729@qq.com

DOI 10.1515/9783110540048-008,

perature refers to the actual sensory temperature under the radiation heat environment exposed to the effect of radiant and convective heat combined, namely effective temperature [4]. Thus, the three temperatures reflect thermal comfort of human body under the conditions of temperature, humidity, and wind speed [5]. Understanding the relationship between the three kinds of temperature and thermal comfort, the thermal comfort in different environments can be inferred, and then objectively express the thermal comfort of different ecological environments.

In order to compare the thermal conditions of different urban ecological environments under the same climatic conditions, the selection of comfort index is very important. In this regard, scholars at home and abroad have carried out lots of research work [6–9]. Based on the monitoring data of dry bulb temperature, wet bulb temperature and black globe temperature of Shenzhen National Climate Observatory in three different ecological environment areas, this article studied the thermal comfort of people, explored the difference of thermal comfort in different urban ecological environmental conditions and revealed the relationship between urban ecological environment and thermal comfort, then provide guidance for urban planning, design and weather forecast.

2 Research Method

2.1 Field Studies

To explore the urban thermal comfort in different ecological environments, this study selects three typical eco-environmental areas, Caiwuwei, Longgang and Xichong, based on the environmental characteristics and climate of Shenzhen. The three regions respectively present ecological environment area of less urban construction, urban development area under construction and development, and the built-up area of urbanization completed. Caiwuwei is located in the core commercial area in Luohu district of Shenzhen. The development intensity and urban density of the area are larger, reflecting the characteristics of urbanization and high density of ecological environment (the floor area ratio is about 8); Longgang is located in the northeastern part of Shenzhen, the secondary development area. And its urbanization degree is not too high, with low density and development intensity of the ecological environment characteristics (the floor area ratio is about 3); Xichong is the southernmost of Dapeng Peninsula of Shenzhen. It is an eco-tourism area, with no city-level development, reflecting the pure nature of the ecological environment characteristics (the floor area ratio is less than 1). The data of dry bulb temperature, wet bulb temperature and black globe temperature of the meteorological substations in the three typical areas are got as the thermal environmental indicators.

To study the relationship between ecological environment and thermal comfort in various urban environments, the surveys on thermal comfort need to conduct. Therefore it is the key issue to determine the survey time. According to the data of different periods of a day from May to October, the dry bulb temperature, wet bulb temperature and black globe temperature were statistically analyzed. It is found that temperature in June, July, August and September is higher than those in May and October. From the various periods of different regions, the temperature is relatively high in July. Therefore, the study selected a day of early July for researching high-temperature thermal comfort, and selected a day in June and August for researching summer general thermal comfort. Specific research time based on the actual weather conditions was determined in the July 12 (high temperature days) and June 28 and August 22 (summer general days) in 2015.

2.2 Social Survey Design

The questionnaire concludes two parts, one part is to collect personal information, including: age, gender, wear, diet, behavioral characteristics etc. This part of the survey content is to grasp the social background of the respondents. Then the subjective hot comfort differences can be excluded, caused by the individual differences in the statistical analysis subsequently. The other part is to explore the thermal comfort of the respondents through six questions. A 7-scale measurement is used to define subjective evaluations of thermal comfort in various environments, in which 7 stands for the worst and 1 stands for the best. The filed survey questionnaire is conducted in the area of 1.5 km on the meteorological observation points in Caiwuwei, Longgang and Xichong on July 12, June 28 and August 22, 2015, including three periods of a day: morning (9:00–12:00), afternoon (14:00–17:00) and evening (18:00–20:00). In addition to field survey, network questionnaire is also applied. Since most of the network questionnaire fillers are indoors, the network questionnaire adds the questions of "indoor or outdoor", "if in the room, whether having the cooling measurement, what kinds of ventilation cooling measurement" and other issues. In addition, for the respondents indoor, in addition to the evaluation of instant experience of thermal comfort, it requires the evaluation of thermal comfort outdoor. Besides, questionnaires are tested in advance, and then the revised questionnaires are used as formal questionnaire. And the thermal comfort and perceptions of ecological environment in different cities is analyzed by combining the field survey and social questionnaire.

3 Result Analysis

3.1 Thermal Environment Analysis

Fig. 1 shows the changes of dry bulb temperature, wet bulb temperature and black globe temperature in the three monitoring sites of Caiwuwei, Longgang and Xichong on June 28, July 12 and August 22. The data of Caiwuwei and Xichong on July 12 are missed due to instrument failure, as well as the date of wet bulb temperature in Longgang on June 28. It can be seen that the variation of black globe temperature is the largest in one day. This is related to solar radiation. As there is no sun at night, black globe temperature is low. In addition, in Caiwuwei, the range of wet bulb temperature is too large on June 28. The daily variation of dry bulb and other wet bulb temperature are not significant. Overall, dry bulb and wet bulb temperature is low in Xichong, black globe temperature is high. In Longgang and Caiwuwei, dry bulb and wet bulb temperature is close to, but higher than the Xichong at dry bulb temperaure. The black ball temperature in Caiwuwei is lower. The dry bulb and wet bulb temperature of Xichong change little in one day. And the variation of black ball temperature in Longgang is complicated.

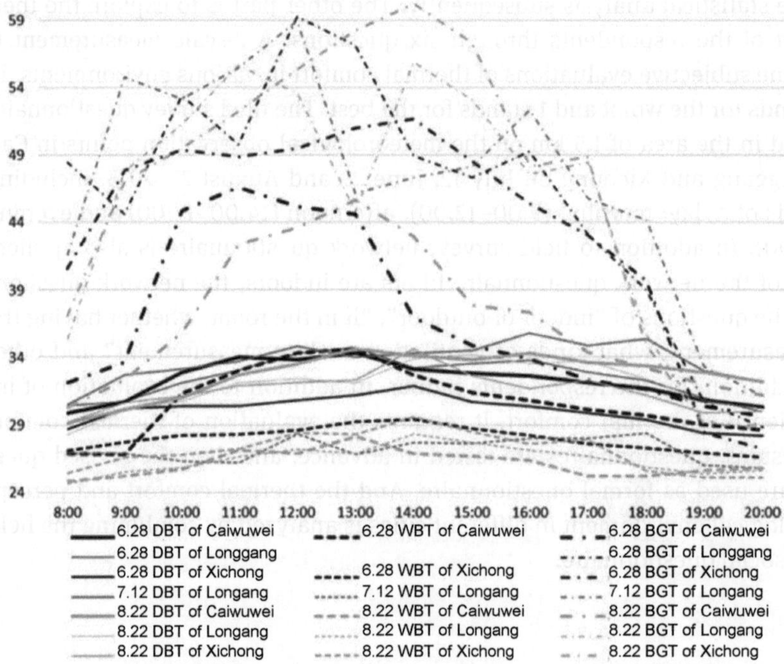

Note: DBT-Dry Bulb Temperature, WBT-Wet Bulb Temperature, BGT-Black Globe Temperature

Fig. 1: Temperature changes of research points in the research day

Table 1 analyzes the correlation of dry bulb temperature, wet bulb temperature and black globe temperature at three research sites in Caiwuwei, Longgang and Xichong. According to the table, the correlation of three kinds of temperature is very high. There is a significant correlation between the dry bulb temperature, wet bulb temperature and black globe temperature, except for dry bulb temperature and black globe temperature, wet bulb temperature and black globe temperature at the Longgang on July 12. And the correlation is positive. It indicates that the dry bulb temperature, wet bulb temperature and black globe temperature have a consistent tendency, namely when one temperature is high, the other temperatures are also high. From the correlation coefficient, the correlation of dry bulb temperature and wet bulb temperature is generally higher than their correlation with the black globe temperature. It indicates that the relationship between dry bulb temperature and wet bulb temperature is more closely. From the three sites, the correlation of dry bulb temperature, wet bulb temperature and black globe temperature in Xichong is obviously higher than other sites, while Longgang is the lowest.

Tab. 1: The correlation between wet bulb temperature, wet bulb temperature and black globe temperature

	Research Time	DBT/WBT	DBT/BGT	WBT/BGT
Caiwuwei	6.28	0.973**	0.879**	0.819**
	8.22	0.600*	0.714**	0.860**
Longgang	7.12	0.793**	0.280	0.259
	8.22	0.884**	0.556**	0.698**
Xichong	6.28	0.990**	0.956**	0.941**
	8.22	0.942**	0.865**	0.934**

Note: DBT-Dry Bulb Temperature, WBT-Wet Bulb Temperature, BGT-Black Globe Temperature

3.2 Thermal Environment Perceptions Evaluations

Thermal comfort has a great relationship with suitable temperature expected for people. Figure 2 analyzed suitable temperature required from field survey and network survey. The figure shows, the suitable temperature suggested by the field survey respondents is lower than the temperature from network respondents. Generally, most people think the suitable temperature is between 25 °C–31 °C. According to the average value of suitable temperature from field and network survey. But it is found that the suitable temperature of field survey is 27.03 °C and the suitable temperature of network survey is 27.31 °C. This result is higher than the results of related studies [10].

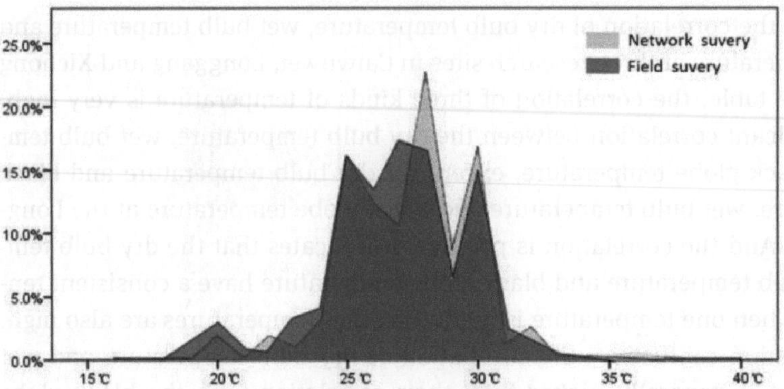

Fig. 2: Analysis of suitable temperature

Fig. 3 shows the sum of thermal comfort evaluation in three research sites for over three survey days. And it also shows that the thermal comfort evaluation of network questionnaire over time. It can be seen from the table that the data of thermal comfort in field survey is basically normal distribution, but the network's is non-normal distribution. And the evaluation results of field and network are different. According to the results of the field research, the thermal comfort of Xichong is better than Caiwuwei and Longgang. From the analysis results of the network evaluation, thermal comfort on June 28 is slightly better than the other days. In addition, the outdoor thermal comfort evaluation of network questionnaire is lower than field survey. This is related to the network respondents who are most indoors.

In addition to thermal comfort perception, the thermal physiological feeling is also an important indicator of thermal environment quality. Table 2 shows the proportion of heatstroke, dehydration and thirst in the three survey sites. The table shows that the proportion of heatstroke and dehydration is less, and most people feel thirsty. In addition, more than 10 % people get heatstroke feelings on June 28 in Longgang and Xichong, and there are also about 10 % people who have heatstroke feelings in Longgang and Xichong on July 12. Overall, people who get heatstroke are more than dehydration. Except for June 28, people with heatstroke and dehydration feelings in Caiwuwei are less than Longgang and Xichong.

1: Very comfortable; 2: Comfortable; 3: More comfortable; 4: Tolerable; 5: More uncomfortable; 6: Uncomfortable; 7: Very uncomfortable

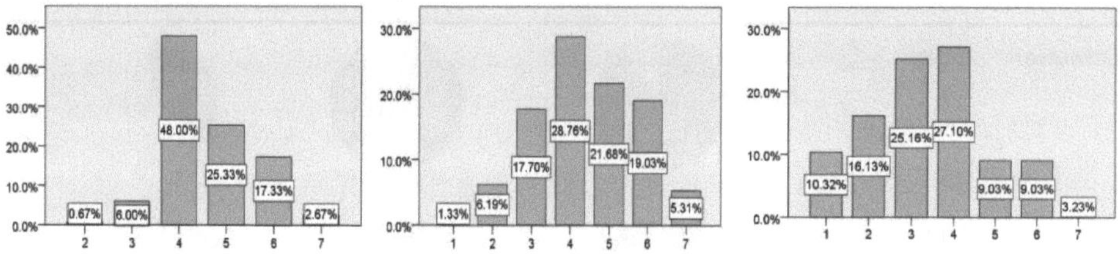

(a) Caiwuwei (b) Longgang (c) Xichong

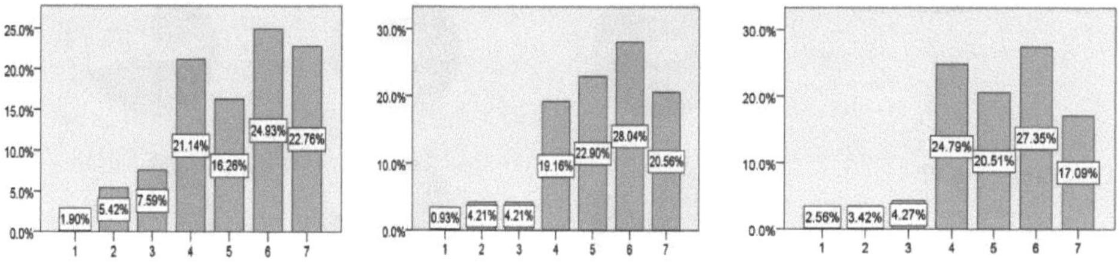

(a) June.28 (b) July.12 (c) August.22

Fig. 3: Survey analysis on thermal comfort evaluation

Because the evaluation of thermal comfort may be related to the subjective background, the study analyzes age, gender, health, clothing, diet, whether in direct sunlight, indoors or outdoors (for network survey), whether to use air conditioning (for the network survey) and other factors. The result shows that the proportion of male and female is close to, both in the field and network survey. The majority is young people in the field research. The proportion of young people increases in network survey respondents. The health of respondents in field and network survey is the same basically. Most of them are in good health and less than 1 % is in poor health. The result reflects the healthy condition of people. In addition, from the clothing, most of the respondents are wearing summer outdoor uniforms, such as shorts or trousers, short-sleeved or long-sleeved. People wearing little or a lot is little. Slightly different is that more people wear less (such as: swimsuit) in Xichong. From the diet, only 1 drinks hot drink, which can be ignored. Most of respondents are without diet. In addition, the respondents with diet are slightly more in Xichong. The result of the analysis on whether or not in the sun shows that most of the respondents are not in direct sunlight. But nearly 20 % is in direct sunlight, most in Xichong. The above analysis doesn't include network survey because network suverys is basically indoors, and clothing, diet, and whether in direct sunlight have no effect on the thermal environment evaluation outdoor.

Tab. 2: Analysis of evaluation on thermal physiological feelings

Site	Research time	Dehydration	Heatstroke	Thirsty
Caiwuwei	6.28		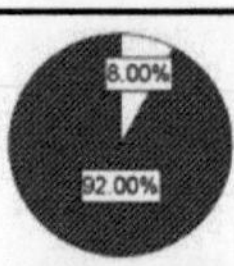	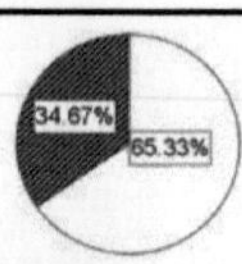
	7.12			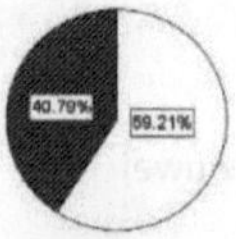
	8.22	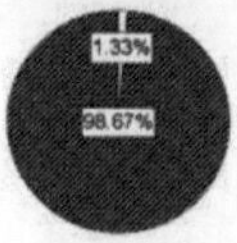	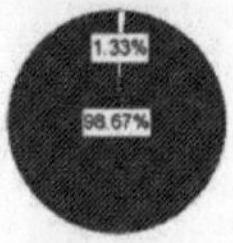	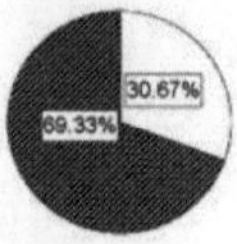
Longgang	6.28		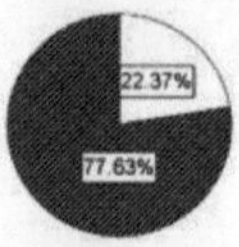	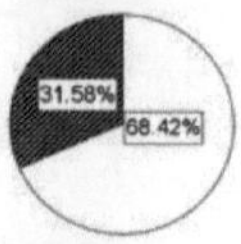
	7.12			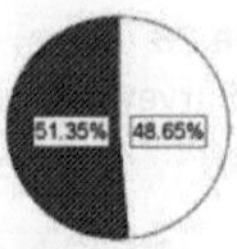
	8.22			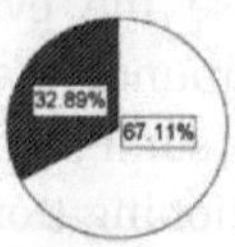
Xichong	6.28			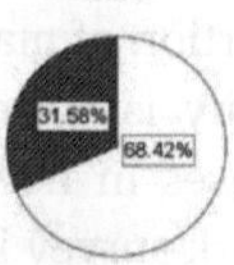
	7.12			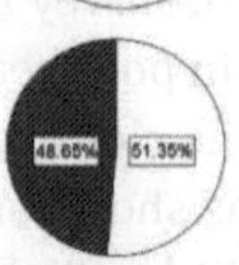
	8.22	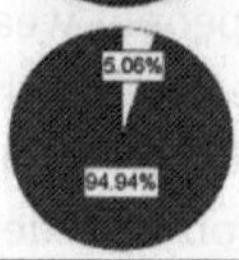		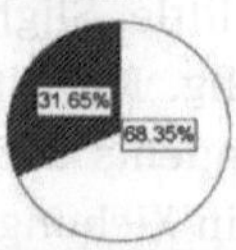

Note: Yes□; No▨

Table 3 analyzes the relationship between respondents' subjective factors and thermal environment evaluation. It can be seen, there are little impact on the thermal environment evaluations for age, gender. The thermal environment evaluation of respondents is only related to the air conditioning rather than subjective factors in network Questionnaire. Air conditioning factor and thermal environment evaluation is negatively correlated, indicating that in an air-conditioned environment, people's heat sensitivity is higher, and the evaluation of outdoor thermal environment is worse. Therefore the colder in the indoor environment, the hotter people feel in the outdoor. The thermal environment evaluation of the respondents in the field survey is associated with health, clothing, diet and exposure to direct sunlight, indicating the healthier, the less clothing, and people eating cold drink have good thermal perceptions.

Tab. 3: Influence of subjective factors on thermal environment evaluation

Respondent Situation	**Field Survey**		**Network Survey**	
	Thermal comfort	Thermal perceptions	Thermal comfort	Thermal perceptions
Age	−0.03	0.03	0.05	0.04
Gender(difference)	−0.16	−0.12	−0.21	−0.11
Healthy condition	0.11**	0.13**	0.07	0.01
Wear condition	0.22**	0.23**	–	–
Diet condition	0.14**	0.15**	–	–
Direct sunlight(difference)	0.15**	0.14**	0.49	0.42
Indoor and outdoor(difference)	–	–	-0.23	-0.01
Air conditioning and ventilation	–	–	−0.12**	−0.19**

Note: In addition to the relationship between variables indicated and evaluations of thermal comfort and thermal perceptions using the difference analysis, the rest are Spearman correlation analysis in the table

3.3 Analysis of Influence Factors on Thermal Environment

From the analysis of section 3 we can find that, for different urban environment, dry bulb temperature, wet bulb temperature and black globe temperature are not the same at the same time. And the thermal environment evaluation results also show that the thermal environment evaluation is not the same at the same time for different research sites. In order to understand the influence of temperature parameters (dry bulb temperature, wet bulb temperature and black globe temperature) on thermal environment evaluation, table 4 analyzes the relationship between temperature parameters and thermal environment evaluation. Among them, data of wet bulb temperature on June 28 in Longgang is not analyzed because of measurement problems.

According to table 4, the impact of three kinds of temperature on thermal comfort evaluation is small. But the impact on thermal perceptions is larger. There is significantly correlation between the temperature parameters and thermal comfort only in two research days, and the time and sites are not the same. From the thermal comfort evaluation, the correlation between the dry bulb temperature and the thermal comfort evaluation of Caiwuwei is relatively high, and the wet bulb temperature has some correlations with thermal comfort evaluation. But the time is different. There are also some correlations between the black globe temperature and thermal comfort evaluation of Longgang and Xichong, and the time is not the same, too. It can be seen that the change of temperature parameter has no significant effect on thermal comfort during the high temperature in the summer. Since only Longgang have the data of July 12, and there is significant correlation between the black globe temperature and thermal comfort, indicating in high temperature days (July 12 is the hottest day at three research day, see Figure 1), the effect of black globe temperature on thermal comfort evaluation is significant.

Tab. 4: Analysis of influence factors on thermal environment evaluation

Site	Date	Thermal comfort evaluation			Thermal perceptions evaluation		
		DBT	WBT	BGT	DBT	WBT	BGT
Caiwuwei	6.28	0.17	0.29*	0.17	0.17	0.28*	0.17
	8.22	0.48**	−0.09	0.06	0.26**	−0.10	0.03
Longgang	6.28	0.27*		0.06	0.13		0.12
	7.12	0.12	−0.02	0.53**	0.09	−0.08	0.50**
	8.22	0.14	0.07	-0.13	0.36**	0.25*	−0.07
Xichong	6.28	0.15	0.11	0.17	0.26*	0.23*	0.31**
	8.22	0.17	0.20	0.26*	0.40**	0.47**	0.50**

Note: DBT–Dry Bulb Temperature, WBT-Wet Bulb Temperature, BGT-Black Globe Temperature

And different from thermal comfort evaluation in table 4, the effect of temperature parameters on thermal perceptions evaluation is significant. The effect of the dry bulb temperature on the thermal perceptions evaluation is significant at the three research sites on August 22, and the wet bulb temperature at Longgang and Xichong has significant effect on the thermal perceptions evaluation. In addition, the dry bulb temperature of Xichong on June 28 has a significant effect on the thermal perceptions evaluation. The wet bulb temperature of Caiwuwei and Xichong has significant effect on the thermal perceptions evaluation. And the black globe temperature in Xichong has a significant effect on the thermal perceptions. Besides, black globe temperature has a significant effect on the thermal perceptions in Longgang on July 12, which is consistent with the thermal comfort evaluation. Because the correlation between thermal comfort and thermal perceptions evaluation is found to be very high, correlation coefficient reaching 0.714, it can be concluded when thermal perceptions is poor in environment, thermal comfort is also poor.

Table 5 analyzes the effects of dry bulb temperature, wet bulb temperature, and black ball temperature on dehydration, heatstroke and other physiological feelings. As weather is not only causing for thirsty, there may be other reasons. And the proportion of thirsty is far larger than heatstroke and dehydration, and there has no significant relationship between them. So there is no need to explore the relationship between thirsty and temperature. Table 5 shows the mean value of dry bulb temperature, wet bulb temperature, and black globe temperature when people have heatstroke and dehydration feelings. From the analysis results of the three research time periods in three research sites, it is found that when having the feeling of heatstroke, dry bulb temperature is 30 °C–35 °C; wet bulb temperature is 26 °C–34 °C; and black globe temperature is 37 °C–54 °C. When having the feeling of dehydration, the dry bulb temperature is 29 °C–35 °C; the wet bulb temperature is 26 °C–34 °C; and the black globe temperature is 37 °C–52 °C. Therefore, compared to dehydration, dry bulb temperature, wet bulb temperature, and black globe temperature of heatstroke are higher.

Tab. 5: Effects of temperature parameters on physiological feelings of thermal environment

T(°C)	6.28						7.12		8.22					
	Caiwuwei		Longgang		Xichong		Longgang		Caiwuwei		Longgang		Xichog	
	He	De	He	De	He	De	He	De	He	De	He	De	He	De
DBT	33.0	32.3	32.8	32.7	30.1	29.3	26.7	27.1	34.3	34.3	26.4	26.8	31.9	31.7
WBT	32.0	30.1	34.4	34.0	27.8	27.9	26.7	27.1	26.4	26.4	26.4	26.8	27.7	27.1
BGT	38.6	36.5	46.8	44.6	45.0	42.9	50.5	51.6	37.4	37.4	45.8	51.2	53.7	45.5

Note: DBT-Dry Bulb Temperature, WBT-Wet Bulb Temperature, BGT-Black Globe Temperature, He-Heatstroke, De-Dehydration

4 Conclusion

This study analyzes the thermal environment of summer hot weather in three different urban ecological environments: Caiwuwei, Longgang and Xichong of Shenzhen. First, from the data of dry bulb temperature, wet bulb temperature and black globe temperature, the dry bulb temperature is lower, and black globe temperature is higher in Xichong. In Caiwuwai, wet bulb temperature and black globe temperature is lower. From the thermal environmental perceptions evaluation, thermal comfort and thermal perceptions in Xichong is the best, and Caiwuwei and Longgang have not obvious difference. Overall, the urban ecological environment has a certain impact on the thermal environment. The thermal environment, where the ecological environment is intact, without urban development such as Xichong, is better than urban construction and development area. However, as the urban construction and development zones for Caiwuwei and Longgang, whose construction density is very different (the floor area ratio respectively is 8 and 3), the thermal comfort of them is not very different. Thus, urban development and construction approach have a great impact on thermal environment. From analysis, dry bulb temperature, wet bulb temperature, and black globe temperature have consistency. But dry bulb temperature and wet bulb temperature have greater relevance, reflecting thermal environment comfort. While the black globe temperature more reflects the impact of thermal environment on people's physiology. It is found that the better the ecological environment is, the closer the dry bulb and the wet bulb temperature are to thermal comfort. However, the black globe temperature is different. It is more controlled by shading elements (such as building density, tree, forest, mountain, etc.). Although the thermal environment is mainly affected by external conditions, the study also shows that environmental background and people's subjective behavior can play a role. For example, people indoor on the thermal comfort temperature requirement is lower than the people outdoor. And clothing, food, direct sunlight and other subjective alternative behaviors also have an impact on the thermal perceptions.

References

[1] Yaglou CP, Minard D. Control of heat casualties at military training centers. Archives of Industrial Health, 16: 302–316, 1957.

[2] Hot environments-Estimation of the heat stress on working man, based on the WBGT-index (wet bulb globe temperature). ISO. International Organization for Standardization 7243.

[3] Yongxiang Gan, Zizhu Dai, QingZhang, et al. Discussion on several problems of WBGT index testing and Implementation on "Classification standard for high temperature work". Chinese Journal of occupational health and occupational diseases, 23(6): 484–485, 2005.

[4] Rong Cheng, Zhengquan Cheng, Jiancong Hang. Study on Forecast of WBGT Index of Guangzhou. Weather, 8(5): 623–628, 2012.

[5] Dahai Xu, Rong Zhu. Study on temperature, humidity and wind speed and the sensory index of cloth. Journal of Applied Meteorology, 11(4): 430–439, 2000.
[6] Thom EC. The discomfort index. Weatherwise, 12(1): 57–60, 1959.
[7] Moran D.S, Pandolf FKB, Shapiro Y, et al. An environment stress index (ESI) as a substitute for the wet bulb globe temperature (WBGT). Journal of thermal biology, (26), 2001.
[8] Nagano K, Horikoshi T. New index indicating the universal and separate effects on human comfort under outdoor and non-uniform thermal conditions [J]. Energy & Buildings, 43(7): 1694–1701, 2011.
[9] Doan Q V, Kusaka H, Ho Q B. Impact of future urbanization on temperature and thermal comfort index in a developing tropical city: Ho Chi Minh City [J]. Urban Climate, 17: 20–31, 2016.
[10] Ng E. 2009a. Designing High-density Cities for Social and Environmental Sustainability. Earthscan: London, Sterling, VA, 2010.

Yi-Min Sun*

A Health Assessment of the Ecosystem in Shuangtaihekou Wetland

Abstract: Shuangtaihekou Wetland is typical estuarine wetland in the north of China. In order to understand the current health status of the ecosystem in Shuangtaihekou Wetland, Pressure-state-response assessment model is adopted and analytic hierarchy process is used to carry out a comprehensive assessment of it. The assessment results show that the ecosystem health index of the wetland is 0.646, indicating a sub-health level. In general, 25.9 % of the entire ecosystem in the region is in excellent health; 21.4 % of it is healthy; 19.6 % of it is in sub-health state; 15.3 % of it is in morbid state and 17.8 % of it is in severe condition. So all in all, Shuangtaihekou Wetland is in a sub-health condition.

Keywords: Shuangtaihekou wetland; Health assessment; Ecosystems; P-S-R assessment model; Analytic hierarchy process.

1 Introduction

Wetland, along with forest and ocean is one of the major ecosystems on the earth. It has great potential for natural resources and is performing vital environmental functions [1]. However, with the rapid development of economy and society, human being is making revolutionary changes about the nature, wetland included [2]. As a consequence, its ecological function is declining and the kinds of species are shrinking. Facing this problem, how to develop and exploit wetland at the minimum cost and realize the harmonious coexistence of human and nature is inevitably a major subject for study in this world [3].

Shuangtaihekou wetland is at the southern tip of Liaohe River delta and is the biggest wetland along the beach in Liao Ning province, including the shallow water regions (the depth of water is below 6 meters in dry seasons), the tidal flats, the salt marshes, the estuarine waters, the permanent river and artificial wetland [4]. This whole region is rich in plantation and animal species and is a relatively well-preserved natural reserve [5] (Figure1). However, the agricultural and industrial production activities in the upstream reaches in recent years have seriously affected the ecological environment here. This study is to make scientific assessment of the ecosystem condition in this region so as to specify the health condition in this eco-

Yi-Min Sun, Shenyang Institute of Engineering, Shenyang 110136 China. Email: 379303565@qq.com

DOI 10.1515/9783110540048-009,

system and take proper measures to restore and rebuild a healthy wetland ecosystem and secure the sustainable development of this region.

Fig. 1: Panoramic view of the Shuangtaihekou wetland

2 Methodology

2.1 P-S-R Assessment Model

P-S-R (pressure-state-response) assessment model was first employed in analyzing the relationship between environment pressure, its reflections and people's response to them. This model was originally put forward by Tony Friend and David Rapport. In view of its rationality, the environmental reports given by OECD started to adopt this model in 1970s. After that, it has been accepted by many governments and organizations [6]. The P-S-R assessment model is able to reflect the cause and effect relations that result in a certain ecological state in a region. It can indicate the pressure on the ecological environment caused by natural disasters or human production activities. Under this pressure, the environment may change its original nature or quality due to the co-effect of physical, chemical and biological influence. In reaction, human society should take correspondent administrative or management strategy to reduce the stress on environment in case the environment should degenerate.

2.2 The Principle of Assessment Index Selection

To evaluate the ecosystem health of the wetland, the essential characteristics of the wetland ecological system must be reflected by the index system that is established. And the system must be systematic, complete, comparable and operable. The fundamental principles that must be followed while constructing this index system are as follows [6,7]:

(1) The principle of systematization and integrity. The index selected must form a complete assessment system and can reflect the essential nature of the wetland ecosystem in an all rounded way. That is, the index must be able to indicate the ecological status, the socio-economical effect and human's influence with clear hierarchical structure.
(2) The principle of scientific rationality. The index selected must be rational and scientific and can truly reflect the health condition of the ecosystem, so as to ensure the veracity and objectivity of the assessment result.
(3) The principle of representativeness. The index selected must be representative. In view of the important position that Liao He Oilfield (the third biggest oilfields in China) holds in Pan Jin, the assessment index must include the effect that oil exploitation has on the wetland so that the selected index can better reflect the health status of the ecosystem in Shuangtaihekou Wetland.
(4) The principle of operability. The index selected must be simple and operable and the statistics can be acquired in the working period. The operability can ensure scientific and effective information for the assessment result. So the assessment index includes the original data obtained by investigation or surveillance, secondary data obtained by processing original data and some state index of qualitative description.

2.3 The Constitution of Assessment Index System

Based on the above principles, and in reference to the assessment system in item 10 of the reference list, the assessment index system framework can be established. The assessment system of the current health status of the ecosystem in Shuangtaihekou Wetland can be divided into 4 layers: 1. the aim layer: the comprehensive assessment of wetland ecosystem health. 2. The criterion layer: pressure index, state index and response index. 3. The assessment element layer. 4. The index layer: the specific index for each assessment factor.

The assessment system of the current health status of the ecosystem in Shuangtaihekou Wetland is based on the above mentioned principles and the previous researcher's experience, after the collection of various information data of geographical, cultural, economic, social and other specific circumstances in Pan Jin [8–11]. 32 indexes are selected from the 3 respects: pressure, state and response.

2.4 Basic Steps of Fuzzy Comprehensive Assessment

The basic assessment steps includes: Wetland health classification, division of index weight, construction of membership fuzzy relationship matrix and lastly the calculation and assessment of the results. According to reference [6] and [7], the health status of the ecosystem in Shuangtaihekou Wetland is rated as in excellent health, in good health, in sub health, defected and diseased. Assume the five rates are named in order from the excellent to the worst.

2.5 AHP Weighting

Whether the division of assessment index weight is rational is of vital importance to the assessment result. After making reference to [5–9] and some experts' grading, the assessment index weight is obtained through AHP, and in case of the Shuanghekou Westland, some relevant analysis and adjustment are made.

2.6 Fuzzy Comprehensive Assessment Method for Health

In the course of fuzzy comprehensive assessment, construction of membership fuzzy relationship matrix is a key step. First, the degree of membership according to different standards of various influencing factors must be ascertained according to the standard of Wetland health level as stated above. And based on that, the fuzzy assessment matrix can be established in order to avoid the unreasonable inshot assessment phenomenon resulted from the small differentiation between different levels. To ensure that membership functions can transit smoothly between rates, the membership functions should be fuzzified. The membership function adopted in our study is in reference to Membership function formula in [7].

3 Results and discussion

The comprehensive fuzzy assessment of the health condition in Shuanghekou Wetland ecosystem needs to process each specific index. Because of the 32 assessment indexes, some are qualitative and some are quantitative, and the standards also vary. But each index is involved in the assessment and has some weight influence on the result of the assessment. So the processing methods should be different, so is the result of the assessment result. In this study, improved weighting summation model is employed.

The comprehensive health index in this ecosystem is 0.6464. According to the health assessment standard of Wetland, currently Shuangtaihekou Wetland is in a sub-health condition. In terms of rating system, in general, 25.92 % of the entire ecosystem in the region is in excellent health; 21.42 % of it is healthy; 19.59 % of it is in sub-health state; 15.31 % of it is in morbid state and 17.76 % of it is in severe condition. In terms of sub-system, the health index of the pressure system is 0.565, close to the state of sub-health, which indicates that the wetland's ecosystem is undergoing some pressure, which poses as a threat to the entire system. Supreme importance should be attached to the increasing rate of population, population density, and the intensity of human activities, the amount of chemical fertilizer, water pollution and sewage discharge and so on. Since the average income in Pan Jin is rather high in LiaoNing province, it's advised to increase the investment in the disposal of sewage water. The assessment value of Shuangtaihekou Wetland's ecosystem is 0.724, which indicates a healthy condition. This reflects that in spite of the all the negative influence, the ecosystem itself has a certain sustainable ability. Although the change in the wetland area and dominant plants coverage have been greatly influenced by human activities, the ecological and functional state of the region is still in rather good condition: its response system's health index is 0.764, a sub-health situation. Due to the high quality of the people and the good awareness of environment protection of the wetland, as well as the increased investment made by the city, the enhanced enforcement of the law, the improved policy, the wetland and its ecosystem is better protected.

4 Conclusion

Now, the ecosystem of Shuanghekou Wetland is in sub-heatlth. The future of it depends on many factors like the pressure it has to undergo, the investment it can receive and the enforcement of all the protective and managing methods. With scientific protection of the region, the resources in the wetland can be used sustainably. Regional industrial adjustment must be made. Protection should be as important as exploitation. More investment should be made to increase the protective and management level, restore the impaired area and improve the structure of the ecosystem in the region. To protect Shuanghekou in the long run, a dynamic monitoring should applied. Before any exploitation is made, assessment should be made about the negative effects it may bring about on the wetland. To restore and rebuild the impaired parts, successful cases can be used for reference. Different ecoengineering activities must be done while taking into consideration in the geographical advantage of Pan Jin.

Acknowledgement: This research was funded by Project of National Natural Science Foundation of China (No. 50873046) and Project of Shenyang Institute of Engineering (RWB5–1403).

References

[1] Niu Mingxiang, Wang Jun, Xu Binduo. Assessment of the ecosystem health of the Yellow River Estuary based on the pressure-state-response model [J]. Journal of ecology, 2017, (03): 943–952.

[2] SunYiMin, WangTieLiang, LuXiaoFeng. Studys on protection and sustainable use of wetland. International Conference on Remote Sensing, Environment and Transportation Engineering, 5439–5441, 2011.

[3] Niu Mingxiang, Wang Jun. Study on evaluation index system of ecosystem health in the Yellow River estuary [J]. water resources protection, 2016, (01): 57–63

[4] Wang Tie-Liang,Chen Xi,Su Fang-Li etc. Application of fuzzy mathematics to water evaluation of Suangtaihekou[J]. Journal of shenyang agricultural university, 42, 1, 79–83, 2011.

[5] SunYi-Min, WangTieLiang, JinShengJi. A Study on Functions and Impact Factorsof of Wetland International Forum on Information Technology and Applications in 2010, 3, 336–339, 2010.

[6] Wang rong-juan. A study on assessment of ecosystem health of shijiu lake wetland. Nanjing: Nanjing Forestry University, 2011.

[7] YANG Li-fang, ZHONG Yan-xia, LI Xiao-yu, et al. Health Assessment of the Mingcui Lake Wetland Ecosystem Based on PSR Modeling [J]. Journal of Hydroecology, 2015, (03): 38–43.

[8] Marcus W. Beck, Bruce Vondracek, Lorin K. Hatch.Environmental clustering of lakes to evaluate performance of a macrophyte index of biotic integrity [J]. Marcus W. Beck, Bruce Vondracek, Lorin K. Hatch. Aquatic Botany. 2013.

[9] Wang Wei. Comprehensive Appraisal of Wetland Ecosystem Health in the Yellow River Delta-a case in Kenli City. Taian, Shandong agricultural university, 2007.

[10] 10Xue Liang. The healthy assessment of wetland ecosystem in east Dongtimg Lake. Changsha: Central south university of forestry and technology, 2010.

[11] WU Jiapeng, L Liu, S Wang. Evaluation of Ecosystem Health Status Based on Index System in Miyun Reservoir [J]. Journal of Hydroecology, 2015, (03): 38–43.

Shou-Jiang Wang, He-Ping Zhang, Ya-Nan Hou and Xu-Dong Cheng*

Flammability of Building Thermal Insulation Materials Using Self-designed Adiabatic Specimen Holder in Cone Calorimeter

Abstract: Flammability of typical building insulation materials, such as polyurethane (PU), were investigated experimentally and theoretically in a cone calorimeter using a self-designed adiabatic specimen holder to create an ideal insulation testing condition. Effect of fire retardant level on the ignition and flammability of PU was also analyzed. It is observed that comparing to standard specimen holder the self-designed one intends to increase the heat release rate and decrease the extinguish time. It is also known that the total heat release of PU is rather high. Afterglow happened at different fire retardant levels of PU, which is much dependent on material itself but not the fire retardancy.

Keywords: Ignition, Heat release rate, Thermal insulation material, Adiabatic specimen holder

1 Introduction

Polyurethane (PU) as effective thermal insulation materials are widely used in external construction façade on the background of world energy crunch for the critical needs of materials with light weight, low thermal conductivity, compression and perfect physical properties. However, high flammable behaviors of these polymers are great threat to human lives, which have attracted much attention from engineers and related researchers [1–3]. Cone calorimeter supports a well-defined flaming condition, forced by external radiation, typical for a developing fire scenario. Ignition properties and combustion characteristics of various materials, such as wood [4], PMMA [5], are measured and discussed using Cone Calorimeter. Xu et al. [6]

Shou-Jiang Wang, State Key Laboratory of Fire Science, University of Science and Technology of China, Hefei, China. Email: wsjxj119@sina.com
He-Ping Zhang, State Key Laboratory of Fire Science, University of Science and Technology of China, Hefei, China. Email: zhanghp@ustc.edu.cn
Ya-Nan Hou, State Key Laboratory of Fire Science, University of Science and Technology of China, Hefei, China. Email: hhouyn@mail.ustc.edu.cn
Xu-Dong Cheng, State Key Laboratory of Fire Science, University of Science and Technology of China, Hefei, China. Email: chengxd@ustc.edu.cn.

DOI 10.1515/9783110540048-010,

evaluated the fire safety of EPS foam employing multi-scale methods including the cone calorimeter test. A more ideal insulation in cone calorimeter tests would benefit the analysis of thermal properties of material themselves by avoiding heat losses. Insulation conditions of the standard tests are not good as claimed. For example, aluminum foil was usually used acting as an insulation material. Kim et al. [7] wrapped the tested wood samples using very thin aluminum foil, with the shiny side of the foil facing the sample. Shi and Chew [8] covered the edges and rear surface of the specimens with aluminum foil for insulation for the fire behavior tests of wood samples under autoignition conditions. Patel et al. [9] followed the standard cone calorimeter set up and used a 1.5 cm thickness layer of glass wool for the insulation of bottom surface of polymer samples to minimize heat loss effect. An ideal insulation of testing condition in cone calorimeter enables the test to focus on material itself by reducing the outside influences, such as the heat losses at edges or bottom. It will also provide reference data to the relevant numerical modeling.

In the aspect of fire retardant, previous studies have much focused on formulating the most effective combination of the material and flame retardants from the fundamental. For example, small-scale tests such as thermogravimetric (TG) and differential thermogravimetric (DTG) were usually utilized to examine the outcomes of the combinations [10, 11]. Bench-scale experiments, aiming to show a more general idea about their fire characteristics, were attracted much less attention. Bench-scale experiment is critical important to address the fire-resistance capacity [12, 13] in buildings or related areas, which is a part that may be more effective to address their behaviors in practical applications. It is then significant to have the comparison and mechanism investigation of XPS and PU with the development of fire retardant as well.

To address the fire characteristics of PU under ideal insulation conditions, a self-designed insulation holder was employed in this study, which could minimize heat losses as well as prevent entrance of fresh air (oxygen) from any face other than the heated top face of the tested samples. Characteristics parameters such as ignition time, heat release rate, together with the deduced quantities like fire growth rate index (FIGRA), total heat evolved (THE), were observed to assess fire performance in terms of fire prevention and energy conservation purposes. This study provides guidance to the fire behaviors testing of practical applications of building thermal insulation materials.

2 Experimental Apparatus

2.1 Apparatus

The Cone Calorimeter Test (ISO 5660–1) [14] was employed to investigate the ignition and flammability of the tested materials. A calibrated cone heater above the sample provides a uniform heat flux over surface of sample. The sample was held in a self-designed kaowool holder (an adiabatic specimen holder) for insulation to minimize heat losses as well as to prevent the entrance of fresh air (oxygen) from any face other than the heated top face, seen in Fig. 1. It is critical to examine if the ideal insulation conditions (self-designed specimen holder) could benefit the fire characteristics tests when comparing to the standard holder.

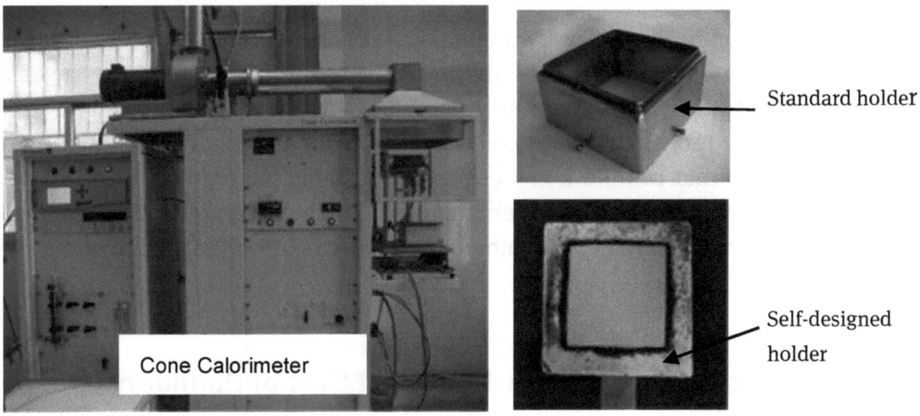

Fig. 1: Cone calorimeter and the standard and self-designed (kaowool) adiabatic specimen holders

Before each test, the sample was placed horizontally on the adiabatic specimen holder, with a fixed distance of 2.5 cm from cone heater to the sample surface. In addition, the edges and rear surface of samples were wrapped with aluminium foil to avoid heat loss and melt dripping. Irradiance levels of 25, 35, 45 and 55 kW/m^2 were chosen, representing various radiation levels in buildings. As soon as the heater and spark plug switch on, the program in the computer was trigged to record the experiment. Once the visible flame appeared, the sample was then considered to be ignited. Each test was repeated at least two times to minimize experimental error and verify the repeatability.

2.2 Materials

The current study involved 25 mm thick PU supplied by a local manufacturer in Hefei, China. PU-B1, PU-B2 and PU-B3 were classified according to the Chinese standard GB 8624-2012 [15]. The bulk densities of PU were 28, 34 and 35 kg•m^{-3} under fire retardant level of B1, B2 and B3, respectively, and Tris (2-chloropropyl) phosphonate (TCPP) was added as flame retardant. Moisture was ignored since the closed-cell structure of these insualtion materials. Thermal conductivity of all test samples remains in the range of 0.030 and 0.033 W•m^{-1}•K^{-1} at the room temperature. The sample size for cone calorimeter tests was 100 mm×100 mm.

3 Results and Analysis

The decomposition of PU in fire is companied with char production and release of combustible. Ignition of PU happens always at the beginning of charring. The char layer cracks and plumps up under sustained heat input. Under a higher external heat flux, burning process becomes much stronger with a higher consumption rate. For certain foams that contain a higher percent of flame retardant, the decompose rate becomes slower, showing a longer ignition time. The decomposition of samples affect both burning behavior and flammability properties, such as ignition time and heat release rate.

3.1 Comparision with Sample Holders (Effect of Boundary Condition)

Heat transfer process of sample can be described by a simplified theoretical model under a specified heat flux. The effective heat flux, $\dot{q}_{net}''$, is a result of external heat flux $\dot{q}_{ex}''$ and the sample's response, including the thermal feedback of flame heat flux $\dot{q}_{flame}''$ and heat loss $\dot{q}_{loss}''$. And the heat loss contains thermal conduction $\dot{q}_{conduc}''$ and re-radiation $\dot{q}_{re-rad}''$:

$$\dot{q}_{net}''=\dot{q}_{ex}''+\dot{q}_{flame}''-\dot{q}_{loss}'' \tag{1}$$

$$\dot{q}_{loss}''=\dot{q}_{conduc}''+\dot{q}_{re-radiation}'' \tag{2}$$

An idealized heat release rate (HRR) during a “steading-state” burning [16] (approximately equals to average steady HRR) can be described as:

$$\mathrm{HRR} = \chi(1-\mu)\frac{h_c^0}{h_g}\dot{q}_{net}'' = \chi(1-\mu)\frac{h_c^0}{h_g}\left(\dot{q}_{ex}'' + \dot{q}_{flame}'' - \dot{q}_{loss}''\right)$$

$$= \chi(1-\mu)\frac{h_c^0}{h_g}\left(\dot{q}_{flame}'' - \dot{q}_{loss}''\right) + \chi(1-\mu)\frac{h_c^0}{h_g}\dot{q}_{ex}'' = \mathrm{HRR}_0 + \mathrm{HRP}\dot{q}_{ex}'' \quad (3)$$

Hence, there is a linear dependency of the HRR from the expected external heat flux. The slope (referred to as the heat release parameter, HRP) is interpreted as a fire response parameter, whereas the intercept (HRR_0) is considered as a flammability parameter.

The self-designed kaowool holder provides an ideal insulation situation through reducing $\dot{q}_{loss}''$ comparing with the standard steel-made holder. The HRR using kaowool holder is theoretically higher than that from standard steel-made holder, indicating that the self-designed holder shows an advantage in describing the intrinsic behaviours of a material. Samples of PU-B3 average-weighting 8.5 g were tested under an external heat fluxof 35 kW/m^2 using these two specimen holders, respectively. The ignition time is almost the same of 3.4 s, while the extinguish time is 112 s for the standard holder and 74 s for the case with self-designed holder. The HRR increases after ignition, reaches the first peak similarly and then reduces because of the blockage of char and consumption of the sample, seen in Fig. 2. Divergence appears as the HRR returns directly back to a transient quasi-steady state with standard holder, and continues declining under the heat conduction at the sample edges. The second peak of HRR occurs in the case of slef-designed holder other than standard one, which is mainly caused by decomposition under the feedback heat from the bottom and the edge as well. It is shown that the adiabatic boundary condition has limited effect on ignition time and the first peak HRR, but shortens the extinguish time and strengthens the second peak HRR. This phenomenon is just followed the expectation that the heat available for unit mass of sample increase when the heat transfer at the sample edges is largely reduced after utilizing the self-designed specimen holder. It is also indicated that the standard way of using aluminium foil could be improved in the aspect of avoiding heat transfer at the sample edges.

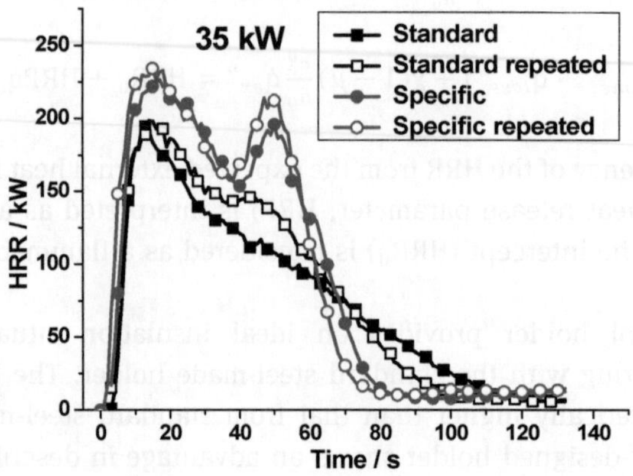

Fig. 2: The HRR curves of PU-B3 along the time under standard and the self-designed specimen holders

3.2 Heat Release Rate (HRR)

Heat release rate history data are much obvious to represent fire behaviors of materials determined by theirspecific properties, influences from the specimen deformation and physical and chemical mechanisms, specifically cracking char layer, endothermic reactions, release of pyrolysis products, residue and so on.

The HRR histories of PU under different external heat fluxes are plotted in Figs. 3. As a thermally thick charring material, PU tends to show the first HRR peak at the beginning of burning, which is because of the increasing thickness of char layer tends to block the heat from outside. The second HRR peak appears later on, which is probably caused by the crack in char or increasing pyrolysis reaction rates of both virgin material and char, seen in Fig. 9. HRR curves of PU-B1, PU-B2 and PU-B3 show two peak values under various heat fluxes, with much clearer peaks under high external heat flux. As for PU-B3, the first peak is higher than the second, while the PU-B2' first peak is lower than the second, but the two peaks of PU-B1 are well-matched except the situation under 25 kW/m^2. It is observed that XPS require limited heat and time for melting process, and a sharp peak is shown in HRR which is mainly the decomposition of the remaining material companied with pyrolysis gases releasing. Fire hazards of PU is mainly reflected during the second peak that showing a sustaining long risk time period.

The peak of heat release rate (PHRR) in the cone calorimeter is strongly dependent on the fire scenario as well as the intrinsic fire properties of the samples, such as thermal conductivity and specific heat capacity. Thermally thin and intermediate thick samples show a larger PHRR than thermally thick samples. The sequencing of

PHRR for different FR level of PU is obtained as: PU-B3> PU-B2> PU-B1. It is indicated that the influence of flame retardant (FR) of PU on its HRR curves cannot be ignored.

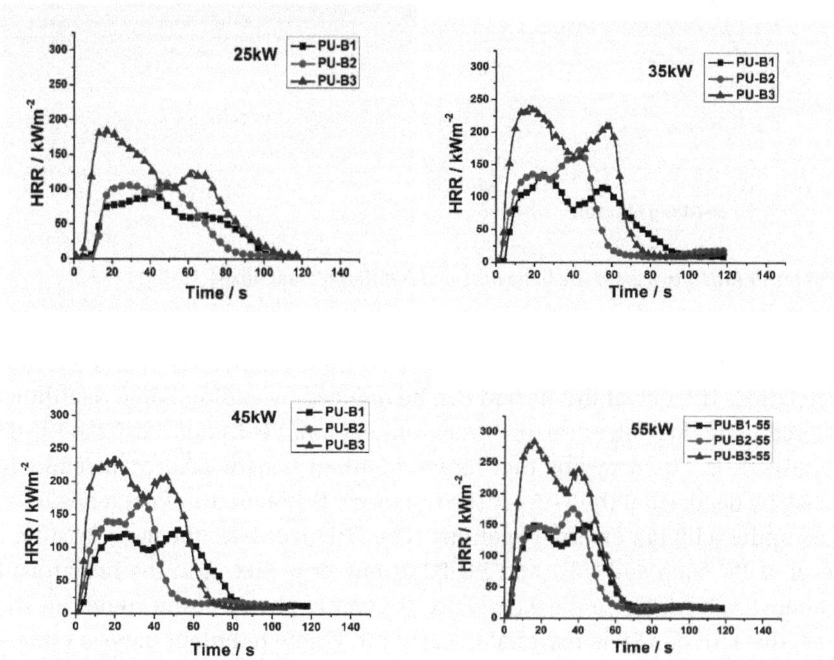

Fig. 3: HRR curves for PU with levels of fire-retardant: PU-B1, PU-B2 and PU-B3 varying with time.

The fire load (THE) and the flame spread or HRR are found having the critical influence on fire hazard. Since flame spread cannot be measured through cone calorimeter tests, Petrella then proposed plotting THE (as fire load) against PHRR/t_{ig} (as fire growth index) to give a comprehensive assessment of the fire hazard. Even though the PHRR/t_{ig} index probably makes a better attempt in addressing flame spread than FIGRA, its application is still hampered because of its disadvantages of oversimplifying empirical indices. Fig. 4 shows the situation of using Petrella's approach to assess the fire hazards of PU series under different heat fluxes. Hence, this is provided to be a useful way to reflect the flame retardancy effect on PU over a range of external heat fluxes.

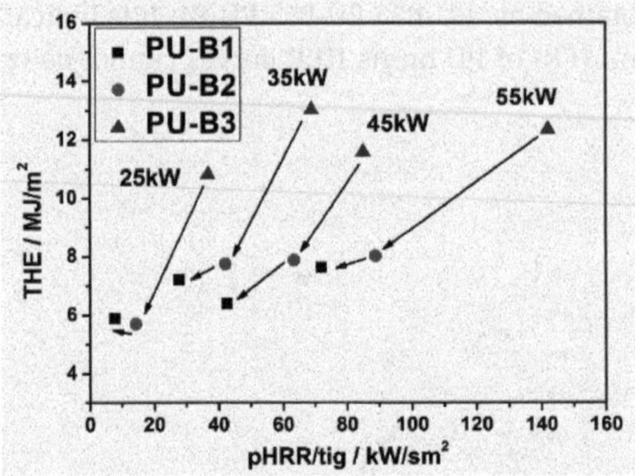

Fig. 4: THE verus PHRR/t_{ig} to assess fire behaviors under different heat fluxes

Fire behavior of each material added with flame retardant is illustrated under different fire scenarios. The risk of fire hazard can be reduced by adding flame retardants such as by reducing the THE or the fire growth rate. From PU-B1 and PU-B2 to PU-B3, the slope growth in Fig. 4 means that flame retardant benefit greatly in reducing their fire risk by decreasing the THE (or the fire load). However, its effect tends to be diminished under a higher external heat flux. The THE decreases the most for different FR level of PU such as PU-B1 and PU-B2 at low heat flux. And the fire growth index is almost halved. Since the XPS series is completely burnt out under all the heat fluxes, the THE of XPS is reasonably constant. Flame retardant plays a critical role in reducing the fire risk of materials.

3.3 Residue Analysis

The residue of PU after flame out is shown in Fig. 5. The fire characteristics of PU are reflected by its char formation, char cracking, melting and bubbling with small range of bubble explosions. Char residue of PU series remains a lot after flame out. The char layer at the top surface of PU intends to block the external heat and mass transfer, which could result in a drop of HRR during the burning. The residue cracks gradually happened around the second peak in HRR. At the low level of radiant heat flux, the surface keeps nearly integrity except edges tear. At a radiance level of 35 and 45 kW/m^2, minute cracks are observed at the charred layer. As for PU-B1 and PU-B2, the charred layer appears to be rougher than PU-B3. The collapse occurs when residual stress is beyond the structure stability.

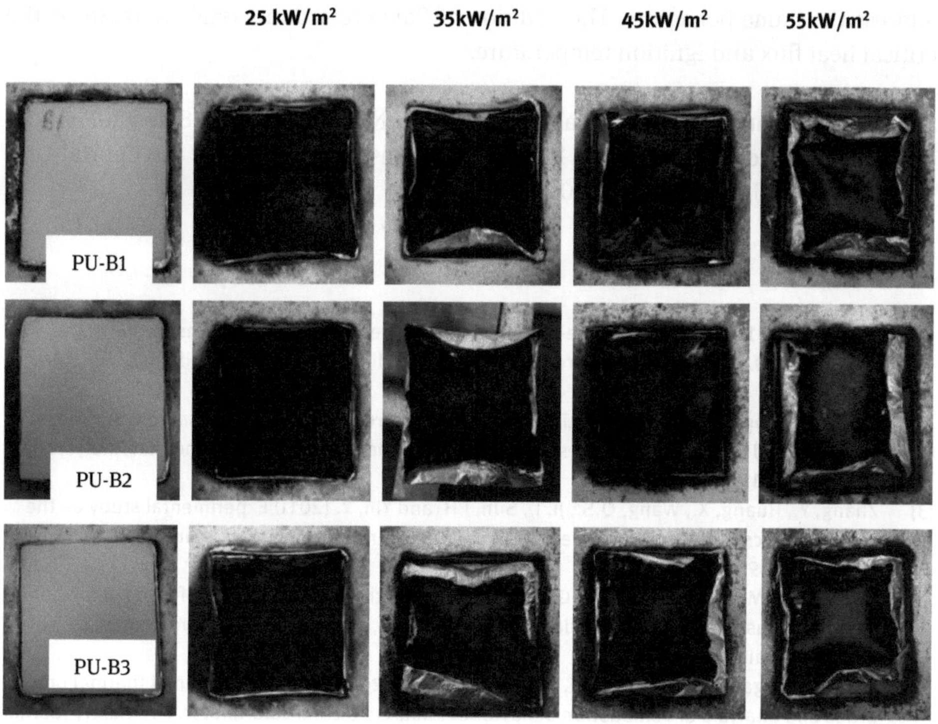

Fig. 5: Comparison of combustion residues with PU series in flame retardant.

4 Conclusion

The flammability investigation of PU with the development of fire retardant in Cone Calorimeter was carried out by using a self-designed adiabatic specimen holder. Both the self-designed holder (kaowool holder) and the standard holder were used to verify the insulation condition of cone calorimeter tests aiming to minimize heat losses as well as prevent entrance of fresh air (oxygen) from any face other than the heated top face. Thermal properties of different kind of thermal insulation materials were obtained through cone calorimeter tests. Thermal conditions, HRR, as well as the peak of heat release rate PHRR and total heat release THR represent the fire behavior controlled by the material specific properties were investigated under a range of heat fluxes and level of fire retardancy. THE (as fire load) against PHRR/t_{ig}(as fire growth index) denotes a comprehensive assessment of fire hazard under the effect of fire retardant. And these solutions can benefit the numerical modeling of current or downstream flame spread. The residue were also analyzed to address the flame behaviors. The addition of flame retardants tends to increase the critical heat flux and ignition temperature.

Acknowledgement: This work was supported by National Natural Science Foundation of China (No. 51323010) and Fundamental Research Funds for the Central Universities under Grant (No. WK2320000035).

References

[1] Jelle, B.P. (2011) Traditional, state-of-the-art and future thermal building insulation materials and solutions – Properties, requirements and possibilities. Energy and Buildings, 43 (10): 2549–2563.

[2] An, W.G., Jiang, L., Sun, J.H. and Liew, K.M. (2015) Correlation analysis of sample thickness, heat flux, and cone calorimetry test data of polystyrene foam. Journal of Thermal Analysis and Calorimetry, 119(1): p. 229–238.

[3] Zhang, Y., Huang, X., Wang, Q.S., Ji, J., Sun, J.H. and Yin, Y. (2011) Experimental study on the characteristics of horizontal flame spread over XPS surface on plateau. Journal of Hazardous Materials, 189(1-2): 34–39.

[4] Spearpoint, M.J. and Quintiere, J.G. (2001) Predicting the piloted ignition of wood in the cone calorimeter using an integral model - effect of species, grain orientation and heat flux. Fire Safety Journal, 36: 391–415.

[5] Luche, J., Rogaume, T., Richard, F. and Guillaume, E. (2011) Characterization of thermal properties and analysis of combustion behavior of PMMA in a cone calorimeter. Fire Safety Journal, 46: 451–461.

[6] Xu, Q., Jin, C., Griffin, G. and Jiang, Y. (2014) Fire safety evaluation of expanded polystyrene foam by multi-scale methods. Journal of Thermal Analysis and Calorimetry, 115 (2): 1651–1660.

[7] Kim, J.Y., Lee, J.H., and Kim S.M. (2012) Estimating the fire behavior of wood flooring using a cone calorimeter. Journal of Thermal Analysis & Calorimetry, 110: 677–683.

[8] Shi, L. and Chew, M.Y.L. (2013) Experimental study of woods under external heat flux by autoignition: Ignition time and mass loss rate. Journal of Thermal Analysis & Calorimetry, 111: 1399–1407.

[9] Pater, P., Hull, T.R., Stec, A.A. and Lyon, R.E. (2011) Influence of physical properties on polymer flammability in the cone calorimeter. Polymers for Advanced Technologies, 22 (7): 1100–1107.

[10] Ullah, S., Ahmad, F., Shariff, A.M. and Bustam, M.A. (2014) Synergistic effects of kaolin clay on intumescent fire retardant coating composition for fire protection of structural steel substrate. Polymer Degradation & Stability, 110: 91–103.

[11] Je, J.Y., Cai, G.P. and Wilkie, C.A. (2014) The effects of several sulfonates on thermal and fire retardant properties of poly (methyl methacrylate) and polystyrene. Polymers Advanced Technologies, 25: 160–167.

[12] Schartel, B. and Hull, T.R. (2007) Development of fire-retarded materials - Interpretation of cone calorimeter data. Fire and Materials, 31(5): p. 327–354.

[13] Zhang, M., Liu, X., Chen, Zhen., Wang, J. and Song, W.H. (2014) Experimental study of the heat flux effect on combustion characteristics of commonly exterior thermal insulation materials. Procedia Engineering, 84: 578–585.

[14] ISO 5660, I., Fire Test - Reaction to Fire - Part 1: Rate of Heat Release from Building Products (Cone Calorimeter Method). International Standards Organisation, 1993. Gevena.

[15] GB 8624–2012: Classification for burning behavior of building materials and products.
[16] Grand, A.F. (1997) Heat release calorimetry evaluations of fire retardant polymer systems. Sample Journal, 33 (4): 47–52.

Qi-Fan Wang, Ya-Li Shao, Tian-Yu Cao and Jun Shen*

Identification of Key Odor Compounds from Three Kinds of Wood Species

Identification and analysis of key odor compounds from three kinds of wood species by Gas Chromatography-Mass spectroscopy-Olfactometry

Abstract: In order to solve the problem of indoor odor and better understand the chemical composition and sources of odor from solid wood, we collected the volatile substances from *Populus ussuriensis Kom, Pinus sylvestris var, mongolica Litv, and Tilia amurensis Rupr,* using the Micro-Chamber/Thermal Extractor and Gas Chromatography-Mass spectroscopy-Olfactometry. Results showed that: The main odor compounds of *Populus ussuriensis Kom* were aromatics (1 kind), aldehydes (1 kind), the main odor compounds of *Pinus sylvestris var, mongolica Litv,* were aromatic (2 kinds), alkenes (8 kinds), alcohols (4 kinds) and aldehydes (1 kind). The main odor compounds of *Tilia amurensis Rupr,* were alkenes (2 kinds), alcohols (1 kind), aldehydes (1 kind). At the same time, 7 kinds unknown compounds were detected from *Populus ussuriensis Kom.,* which appeared at 9.02, 9.7, 20.6, 25.03, 28.82, 32.88, and 34.58 minutes, with the odor of sweet, plastic&spicy, special, stimulating&burning&plastic&spicy, stinky, plastic&spicy and plastic, respectively. 6 kinds unknown compounds were detected from *and Tilia amurensis Rupr.,* which appeared at 5.70, 9.73, 9.92, 30.27, 30.9, and 32.38 minutes and presented as stinky, plastic&stinky, earthy, stinky, delightful, and aromatic. The emergence times of odor compounds focused around 5 to 12 minutes and 18 to 36 minutes. The volatile substances of *Populus ussuriensis Kom, Pinus sylvestris var, mongolica Litv, and Tilia amurensis Rupr,* were 25, 27 and 40 kinds corresponding, with the key odor compounds of alkenes, aldehydes, and alcohols. The odor intensity of VOCs from the three wood species did not show a positive correlation with the concentrations. The aroma of three kinds of wood species are composed of several aroma-active volatile

Qi-Fan Wang, Northeast Forestry University, 26 Hexing Road, Harbin 150040, China. Email: wangqifan66@163.com
Ya-Li Shao, Northeast Forestry University, 26 Hexing Road, Harbin 150040, China. Email: 826995294@qq.com
Tian-Yu Cao, Northeast Forestry University, 26 Hexing Road, Harbin 150040, China. Email: 18504616802@163.com
Jun Shen, Northeast Forestry University, 26 Hexing Road, Harbin 150040, China. Email: shenjunr@126.com

DOI 10.1515/9783110540048-011,

components which can affect the overall flavor for their low threshold.

Keywords: Chromatography-Mass spectroscopy-Olfactometry; Populus ussuriensis Kom; Pinus sylvestris var, mongolica Litv; Tilia amurensis Rupr; Key odor compounds

1 Introduction

In recent years, people are increasingly using wood materials in their interior decorating because of its natural affinity and visual sense of harmony. The release of volatile organic compounds (VOCs) in indoor air is therefore increasing. Organic compounds can release either pleasant or irritating odors into the surrounding environments. These odors are composed of a variety of single compounds, and some of those compounds may be harmful to human's health. Gas chromatography–mass spectroscopy olfactometry (GC-MS-O) is improved from the GC-MS by detecting VOCs emissions from wood and combining the separation ability of GC-MS with the sensitive olfactory ability of humans. GC-MS-O separates the various VOCs components and uses our noses as detectors [1] to evaluate each compound, separately. Based on the sensitivity of the human sense of smell, which can exceed that of other chemical detectors, GC-MS-O was used to select and evaluate active odor substances from complex mixtures of VOCs [2]. This method is helpful in identifying key odor compounds and determining the intensity of those compounds.

GC-MS-O was first proposed in 1964 [3, 4]. It was performed by smelling the outflow components of GC-MS directly. This method combined not only the GC-MS advantages of high separation efficiency, strong qualitative ability, high sensitivity, and quantitative accuracy but also the sensory evaluation of humans. In 1976, Acree et al. improved the original GC-MS-O technology by adding humid air and by smelling the GC outflow after film chromatography processing. In the mid-1980s, Acree and Ullrich [5, 6] used dilution-analysis methods to analyze the intensity of the various odors at the same time, which made GC-MS-O technology widely applicable in many situations. Currently, there are four major GC-MS-O detection methods [7], the dilution analysis, time–intensity analysis, detection frequency methods, and posterior intensity evaluation. With the development of extraction and separation technology, GC-MS-O has been widely used in the research of fruits [8], vegetables [9], dairy products [10], paint [11], gypsum [12], and other fields such as tobaccos, flavoring agents and wines.

In total, there are many researches on odor emissions, but there is few researches about the odor emission from wood materials, especially solid woods. The odor problem of interior decoration has impacted people's life a lot. Usually, we can only smell the overall odor but cannot know the single key odor components that make up the overall odor. This study isolated and indentified the key odor compounds from *Populus ussuriensis Kom, Pinus sylvestris var. mongolica Litv. and Tilia amurensis Rupr*. And at the same time, the characteristics and intensities of the key odor compounds were showed and chemical composition and source of overall odor were found to provide basic data of interior decoration materials.

2 Materials and Methods

2.1 Experimental Materials

The essential parameters of the three wood species are presented in Table 1. The samples were cut into round pieces (60 mm diameter) for the microchamber/thermal extractor apparatus. The exposed area was 5.65×10^{-3} m^2. After the edges of specimen were sealed with aluminum foil to prevent the release of compounds. The samples were stored in Teflon bags and placed in a refrigerator until needed.

2.2 Experimental Equipments

2.2.1 Sampling Devices

Two sampling devices were used. The Micro-Chamber/Thermal Extractor μ-CTE 250 (Markes International Inc., UK) consists of four cylindrical micropools (each with a microcell diameter of 64 mm by a depth of 36 mm). The sampling temperature can be adjusted up to 250 °C. It can test four samples at the same time. The other sampling device was a Tenax TA tube (Markes International Inc., UK). In this device, the material of the pipe body is stainless steel, and the sampling tube contains 200 mg of 2, 6-dibenzofurans porous polymer, which can efficiently adsorp or desorp the volatile organic gases.

2.2.2 Detection and Analysis Devices

Three detection and analysis devices were used.

(i) The Unity thermal analysis desorption unit (Markes International Inc., UK). Nitrogen was used as the carrier gas. The related parameters include: thermal desorption temperature, 280 °C, cold-trap adsorption temperature, –15 °C, and thermal analysis time, 10 minutes; and injection time, 1 minute.

(ii) DSQ II series GC-MS (Thermo Scientific, Germany). Chromatography was performed with a DB-5 quartz capillary column (3,000 mm by 0.26 mm by 0.25 μm). Helium was used as the carrier gas with a constant velocity of 1.0 mL/min by splitless injection. The chromatographic column was initially kept at 40 °C for 2 minutes; then, the temperature was increased to 50°C (in 2 °C/min increments) and was held for 4 minutes. Finally, the temperature was elevated to 250 °C (in 10 °C/min increments) and held for 8 minutes. The injection port temperature was 250 °C. The following GC-MS conditions were applied: ionization mode, electron ionization (EI); ion energy, 70 eV; transmission line temperature, 270 °C; ion source temperature, 230 °C; and mass scan range, 50 to 650 atomic mass units.

(iii) Sniffer 9000 Olfactory Detector (Brechbühler, Switzerland). Combined with GC or GC-MS, the quantitative and qualitative analyses can be made, and the compound's odor intensity can be reflected directly and recorded. The effluent of the GC capillary is divided into two parts, one part enters the mass spectrometer, and the other part is used for sensory evaluation (ratio 1:1). The transmission line temperature was 150 °C, and nitrogen was used as the carrier gas through a purge valve. Humid air was added to prevent dehydration of the nasal mucosa.

2.3 Experimental Method

2.3.1 Sampling Method

The experiment used the Tenax TA sampling tubes tubes to adsorb VOCs from different wood species under constant experimental conditions (sampling temperature, 40 °C; ratio of air exchange rate to loading factor, 0.5; air flow rates, 47 mL/min). Four samples were collected under these conditions, with a sampling cycle of 8 hours. Each sample's contents were 3 liters. After sampling, the Tenax TA sampling tubes were wrapped in teflon plastic bags until needed.

2.3.2 Analytical Method

Two analytical methods were used. GC-MS and its built-in software were used to analyze the VOCs. The MS detection peak used the 2008 spectral library of the National Institute of Standards and Technology (NIST) (matching degrees up to 800 °C or above). An internal-standard method was used in this experiment, with deuterium substituted for toluene at a concentration of 200 ng/μL, which added 2 μL. The internal-standard quantitative-analysis method used the following equation (1):

$$mi = Ai \times (ms/As) \tag{1}$$

Where Ai and as are the peak area of the products tested and the internal standard substance, respectively; and ms is the amount of the internal standard. After detected by GC-MS, the compounds were identified using the NIST (2008 standard spectrum) and Wiley libraries. The primary odor compounds were identified by mass spectrometry, retention index, and odor characterization. The refractive index value was calculated through the retention time of n-alkane under the same conditions [13].

The second analytical method used was GC-O. The time-intensity method was chosen for analysis of the compounds. As the sample was injected and the chromatogram yielded peaks, the sensory evaluation assessors perceived and described the column outflow from the odor port simultaneously. The time of the odor, the odor type, and the intensity of the odor were recorded.

After specific training, four assessors (with ages between 20 and 30 years, no history of smoking, and no olfactory organ disease) formed a GC-O odor-analysis evaluation group. The experimental environment was well ventilated, and there were no peculiar smells in the room. The temperature was kept between 17 and 25 °C during the entire experimental process. Activities, such as eating, which could possibly have an effect on indoor odors, were forbidden for 5 hours before the experiment. Experimental results were recorded when the same odor characteristics were described by at least two assessors at the same time. The intensity value was based on the average value of the different assessors. The grade division of the substance's odor concentration was mainly based on the human sense of smell. The experiment's discrimination of odor intensity was made according to the Japanese standard [14] (see Table 2).

3 Results and Discussion

3.1 Odorous VOCs from *Populus Ussuriensis Kom.*

GC-MS-O was used to detect the VOCs odors from three wood species qualitatively and quantitatively. Figures 1 shows the total ion current of *Populus ussuriensis Kom.*. The corresponding analytical results are shown in Table 3 and Table 4.

From table 3, we can note that there were more components in the alkanes categories than in the others. Components were also found in these categories: aromatics, alkenes, aldehydes and ethers (small amounts). The totally species of volatile substances from *Populus ussuriensis Kom.* were 25. Table 4 shows the relevant information of odor compounds, including retention time, retention index (RI), mass concentration, odor character and odor intensity. We can see that the key odor compounds from *Populus ussuriensis Kom.* mainly appear in aromatics and aldehydes.The odor compounds listed in order of intensity from the largest to the smallest were: hexanal (grass; odor intensity 3), 2-ethyl-furan (special; odor intensity 1), and seven unknown compounds that weren't detected by GC-MS but were detected by the human olfactory system. The odors from these seven unknown compounds appeared at 9.02, 9.7, 20.6, 25.03, 28.82, 32.88, and 34.58 minutes were sweet (odor intensity 2); plastic, spicy; (odor intensity 2.5); special (odor intensity 2); stimulating, burning-plastic, spicy (odor intensity 4); stinky (odor intensity 3); plastic, spicy (odor intensity 2.5); and plastic (odor intensity 3), respectively. These compounds could not be detected by mass spectrometry because their concentration was below the GC-MS detection limit, but they could be detected by the human olfactory system at low thresholds.

3.2 Odorous VOCs from *Pinus Sylvestris var. Mongolica Litv.*

The total ion current of *Pinus sylvestris var. mongolica Litv.* was showed in Figures 2. The corresponding analytical results are shown in Table 5 and Table 6.

From table 5, we can note that there were more components in the aromatics and alkenes categories than in the others. Components were also found in these categories: alkanes, aldehydes, alcohols and ethers (small amounts). The totally species of volatile substances from *Pinus sylvestris var. mongolica Litv.* were 27. Table 6 shows the key odor compounds from *Pinus sylvestris var. mongolica Litv.*. The compounds, ordered by odor intensity from largest to smallest were as following: α-pinene (turpentine; odor intensity3), 3-carene (pine; odor intensity3), toluene (spicy, plastic, stinky; odor intensity2), 1-methyl-5-1-methylethenyl-cyclohexene (turpentine; odor intensity 1.5), d-limonene (lemon; odor intensity 2), copaene (burning-plastic; odor intensity 2), α,α,4-trimethyl-3-cyclohexene-1-methanol (flowers; odor intensity 2), 1-methyl-2-1-methylethyl-benzene (aromatic; odor intensity 1), camphene (camphor; odor intensity 1), α-myrcene (lipids; odor intensity 1), 1-methyl-4-1-methylethyl-1,4-cyclohexadiene (special; odor intensity 1), benzaldehyde (almond bitterness; odor intensity 1), exo-fenchol (small fennel; odor intensity 1), borneol (camphor; odor intensity 1), and 4-methyl-1-1-methylethyl-3-cyclohexen-1-ol (stinky; odor intensity 1).

3.3 Odorous VOCs from *Tilia Amurensis Rupr.*

The total ion current of *Tilia amurensis Rupr.* was showed in Figures 3. The corresponding analytical results are shown in Table 7 and Table 8.

From table 7, we can note that there were more components in the aromatics, alkanes and alkenes categories than in the others. Components were also found in these categories: aldehydes, alcohols and ethers. Aldehydes have the fewest species, only two species were found in it. The totally species of volatile substances from *Tilia amurensis Rupr.* were 40. Table 8 shows the key odor compounds from *Tilia amurensis Rupr.* The compounds, by odor intensity from largest to smallest were 1-methyl-5-1-methylethenyl-cyclohexene (turpentine; odor intensity 3), 3-

carene (pine; odor intensity 1), α,α,4-trimethyl-(s)-3-cyclohexene-1-methanol (flower; odor intensity 1), hexanal (grass; odor intensity 1), and 6 unknown compounds that were not detected by GC-MS but were detectable by the human olfactory system. They unknown compounds appeared at 5.70, 9.73, 9.92, 30.27, 30.9, and 32.38 minutes and presented as stinky (odor intensity 3); plastic, stinky (odor intensity 2.5); earthy (odor intensity 2); stinky (odor intensity 2); delightful (odor intensity 1); and aromatic (odor intensity 2).

3.4 The Analysis of Percentage, Odor Intensities and Emerging Times of Different Components

Figure 4 shows the percentage of different components in three kinds of wood species. The percentages of aromatics in *Populus ussuriensis Kom, Pinus sylvestris var. mongolica Litv. and Tilia amurensis Rupr.* Were 36, 9, and 28 percent, respectively compared with the percentages of odor aromatics, which were 1, 3, and 0 percent, respectively. The percentages of alkenes were 1, 84, and 41 percent, compared with the percentages of odor alkenes, which were 0, 80, and 30 percent, respectively. The percentages of alkanes were 21, <1, and 11 percent, compared with the percentages of key odor alkanes, which were all 0. The percentages of aldehydes were 17, <1, and 4 percent, respectively, compared with the percentages of odor aldehydes, which were 15, close to 0, and 4 percent, respectively, and there 25, <1, and 10 percent in the "other" category, respectively. Meanwhile, the alcohols were about 6 percent in both *Pinus sylvestris var. mongolica Litv. and Tilia amurensis Rupr.*, compared with the percentages of odor alcohols, which were 6 % and 4 % (Fig. 4). In these three wood species, the key odor compounds were alkenes, aldehydes, and alcohols, and the alkanes always had no odor. The percentages of the various odor compounds showed significant differences from the percentages for the total compounds in these three wood species because most compounds in solid wood had no odor or contributed little to the odor. The contribution of a substance to the overall atmosphere is related to the threshold of the species [15]. Just as in food industry, the compounds which contribute to the overall flavor are <5 percent [16].

Figure 5 shows the times of emergence (the time it takes before the odor becomes apparent) and the intensities of the various odor compounds in the three wood species. Results showed emergence times focusing around 5 to 12 minutes and 18 to 36 minutes. The odor intensity of VOCs from the wood species did not show a positive correlation with the concentrations, because some compounds (such as butylated hydroxytoluene and 2,2,7,7-tetramethyloctane) found in high concentrations had no characteristic odor or odor intensity.

Tables 3 through 5 also show that when concentration decreased, the odor for some key odor compounds disappeared, such as benzene, 1-methyl-21-methylethylbenzene, α-pinene, copaene, and undecanol. When the concentration from the compounds fell below certain limits, they were too diluted to smell them. Different woods release compounds with different mass concentrations, so some key odor compounds detected in one wood species may not have key odor characteristics in another tree species.

4 Conclusion

Three wood species were analyzed with GC-MS-O in this experiment. The results showed that the volatile substances of *Populus ussuriensis Kom, Pinus sylvestris var. mongolica Litv. And Tilia amurensis Rupr.* were 25, 27 and 40 kinds corresponding, with the key odor compounds of alkenes, aldehydes, and alcohols. The key odor compounds from *Populus ussuriensis Kom.* were hexanal, 2-ethyl-furan. The primary odor compounds from *Pinus sylvestris var. mongolica Litv.* were α-pinene, 3-carene, toluene, d-limonene, copaene, α4-trimethyl-3-cyclohexene-1-methanol, 1-methyl-5-1-methylethenyl-cyclohexene, 1-methyl-2-1-methylethyl-benzene, camphene, α-myrcene, 1-methyl-4-1-methylethyl-1,4-cyclohexadiene, benzaldehyde, exo-fenchol, borneol, and 4-methyl-1-1-methylethyl-3-cyclohexen-1-ol. The key odor compounds from *Tilia amurensis Rupr.* were 1-methyl-5-1-methylethenyl-cyclohexene, 3-carene, hexanal, α,α,4-trimethyl-(s)-3-cyclohexene-1-methanol. At the same time, 7 unknown compounds in *Populus ussuriensis Kom.* and 6 unknown compounds in *Pinus sylvestris var. mongolica Litv.*were found, which had clear odor characteristics. The emergence times of odor compounds focused around 5 to 12 minutes and 18 to 36 minutes. The odor intensity of VOCs from the three wood species did not show a positive correlation with the concentrations for most compounds in solid wood had no odor or contributed little to the odor. The aroma of three kinds of wood species are composed of several aroma-active volatile components which can affect the overall flavor for their low threshold.

5 Tables

Tab. 1: Essential parameters of experimental wood species

Number	Wood	Province	Production date	Size /mm	Thickness /mm	Density /g cm^{-3}	Moisture content	DBH[1]/ cm
1	Populus ussuriensis Kom.	Xiao Hinggan Mountains, China	2015	D=60	14	0.38	11%	42
2	Pinus sylvestris var. mongolica Litv.	Russia	2015	D=60	14	0.43	10%	38
3	Tilia amurensis Rupr.	Russia	2015	D=60	14	0.52	10%	40

[1]DBH: diameter at breast height

Tab. 2: Odor intensity criterion (Japan)

Odor intensity (grade)	0	1	2	3	4	5
Representation method	Odorless	Barely detectable	A little sense of smell	Significant smell	Strong smell	Very strong smell

Tab. 3: The main components of VOCs emitted from *Populus ussuriensis Kom*

Category	Main components
Aromatics	Benzene, Toluene, Butylated Hydroxytoluene, 2-ethyl-Furan
Alkenes	Dimethyl-Diazene, 2-Butenal, 1-Ethylcyclopropanol,
Alkanes	Hexamethyl-Cyclotrisiloxane, 2,2,7,7-Tetramethyloctane, Decane, Octamethyl-Cyclotetrasiloxane, 2,2,4,4-Tetramethyloctane, 3,3-dimethyl-Heptane, 4,6-dimethyl-Undecane
Aldehydes	Pentanal, Hexanal
Others	1-Penten-3-one, Dibutyl phthalate

Tab. 4: Odor compounds of *Populus ussuriensis Kom.* determined by GC-MS-O

Compounds	Retention Time (RT)	Retention index(RI)	Mass concentration/μg·m^{-3}	Odor character	Odor intensity	Identification method
Aromatics			6.87			
2-ethyl-Furan	7.51	687	6.87	Special	1	MS, odor, RI
Aldehydes			134.56			
Hexanal	11.31	774	134.56	Grass	3	MS, odor, RI
Others			–			
Unknown	9.02	726	–	Sweet	2	RI, odor
Unknown	9.7	740	–	Plastic, Spicy	2.5	RI, odor
Unknown	20.6	959	–	Special	2	RI, odor
Unknown	25.03	1068	–	Stimulating,Burnning,plastic, Spicy	4	RI, odor
Unknown	28.82	1178	–	Stink	3	RI, odor
Unknown	32.88	1305	–	Plastic, Spicy	2.5	RI, odor
Unknown	34.58	1352	–	Plastic	3	RI, odor

Tab. 5: The main components of VOCs emitted from *Pinus sylvestris var. mongolica Litv*

Category	Main components
Aromatics	Benzene, Toluene, 1-methyl-2-1-methylethyl-Benzene, 1-methyl-4-1-methylethenyl-Benzene, Decahydro-4,8,8-trimethyl-9-methylene--1,4-Methanoazulene, Octahydro-7-methyl-4-methylene-1-1-methylethyl-Naphthalene, Butylated Hydroxytoluene
Alkenes	Dimethyl-Diazene, à-Pinene, Camphene, á-Myrcene, 3-Carene, 1-methyl-5-1-methylethenyl-Cyclohexene, D-Limonene, 1-methyl-4-1-methylethyl-1,4-Cyclohexadiene, 1-methyl-4-1-methylethylidene-Cyclohexene, Copaene
Alkanes	Dimethoxy-Methane, Hexamethyl-Cyclotrisiloxane
Aldehydes	Benzaldehyde, Furfura
Alcohols	Exo-Fenchol, Borneol, 4-methyl-1-1-methylethyl-3-Cyclohexen-1-ol, à,à4-trimethyl-3-Cyclohexene-1-methanol, à,à,4-trimethyl-(S)-3-Cyclohexene-1-methanol
Others	Acetic acid

Tab. 6: Odor compounds of *Pinus sylvestris var. mongolica Litv.* determined by GC-MS-O

Compounds	Retention Time (RT)	Retention index(RI)	Mass concentra tra-tion/μg·m-3	Odor charac-ter	Odor intensi-ty	Identification method
Aromatics			397.19			
Toluene	10.06	747	40.49	Spicy, Plastic, Stink	2	MS, odor, RI
1-methyl-2-1-methylethyl-Benzene	22.96	1014	356.70	Aromatic	1	MS, odor, RI
Alkenes			9930.95			
à-Pinene	19.41	933	924.94	Turpen-tine	3	MS, odor, RI
Camphene	19.97	945	47.17	Camphor	1	MS, odor, RI
á-Myrcene	21.76	985	75.31	Lipids	1	MS, odor, RI
3-Carene	22.78	1009	6119.43	Pine	3	MS, odor, RI
1-methyl-5-1-methylethenyl-Cyclohexene	23.20	1020	8.13	Turpen-tine	1.5	MS, odor, RI
D-Limonene	23.39	1025	2676.50	Lemon	2	MS, odor, RI
1-methyl-4-1-methylethyl-1,4-Cyclohexadiene	24.40	1052	43.60	Special stimulat-ing	1	MS, odor, RI
Copaene	35.68	1383	35.87	Burn-ning, plastic	2	MS, odor, RI
Aldehydes			14.06			
Benzaldehyde	19.25	929	14.06	Almond bitter-ness	1	MS, odor, RI
Alcohols			708.62			
Exo-Fenchol	26.24	1100	121.67	Small fennel	1	MS, odor, RI
Borneol	27.99	1153	80.85	Camphor	1	MS, odor, RI
4-methyl-1-1-methylethyl-3-Cyclohexen-1-ol	28.40	1165	32.68	Stink	1	MS, odor, RI
à,à4-trimethyl-3-Cyclohexene-1-methanol	28.77	1177	473.42	Flower	2	MS, odor, RI

Tab. 7: The main components of VOCs emitted from *Tilia amurensis Rupr*

Category	Main components
Aromatics	Benzene, Toluene, 1,3-dimethyl-Benzene, 1-methyl-2-1-methylethyl-Benzene, octahydro-1,7a-dimethyl-5-1-methylethyl-1,2,4-Metheno-1H-indene, 1,2,3,5,6,7,8,8a-octahydro-1,8a-dimethyl-7-1-methylethenyl-Naphthalene, 1,2,3,4,4a,5,6,8a-octahydro-7-methyl-4-methylene-1-(1-methylethyl)- Naphthalene, Butylated Hydroxytoluene, 1,2,3,5,6,8a-hexahydro-4,7-dimethyl-1-1-methylethyl(1S-cis)-Naphthalene
Alkenes	Dimethyl-Diazene, 3,6,6-trimethyl-Bicyclo[3.1.1]hept-2-ene, á-Pinene, 3-Carene, 1-methyl-5-1-methylethenyl-Cyclohexene, 1-methyl-4-1-methylethylidene-Cyclohexene, Copaene, à-Calacorene
Alkanes	Hexamethyl-Cyclotrisiloxane, 2,2-dimethyl-3-methylene-Bicyclo[2.2.1]heptane, 2,2,4,6,6-pentamethyl-Heptane, Decane, Octamethyl-Cyclotetrasiloxane, 2,2,4,4-Tetramethyloctane, 5-Ethyldecane
Aldehydes	Pentanal, Hexanal
Alcohols	Undecanol, Z-2-Dodecenol, 1,3,3-trimethyl-(1R-endo)-Bicyclo[2.2.1]heptan-2-ol, à,à,4-trimethyl-(S)-3-Cyclohexene-1-methanol
Others	Acetic acid, Decyne, 1,2,3,4,4a,7,8,8a-octahydro-1,6-dimethyl-4-(1-methylethyl)-1-Naphthalenol

Tab. 8: Volatile compounds of *Tilia amurensis Rupr.* determined by GC-MS-O

Compounds	Retention Time (RT)	Retention index(RI)	Mass concentration/μg·m^{-3}	Odor character	Odor intensity	Identification method
Alkenes			319.18			
3-Carene	22.65	1006	271.97	Pine	1	MS, odor, RI
1-methyl-5-1-methylethenyl-Cyclohexene	23.32	1023	47.21	Turpentine	3	MS, odor, RI
Aldehydes			39.79			
Hexanal	11.39	775	39.79	Grass	1	MS, odor, RI
Alcohols			42.97			
à,à,4-trimethyl-(S)-3-Cyclohexene-1-methanol	28.75	1176	42.97	Flower	1	MS, odor, RI
Others			–			
Unknown	5.7	607	–	Stink	3	RI, odor
Unknown	9.73	740	–	Plastic,Stink	2.5	RI, odor
Unknown	9.92	744	–	Earthy	2	RI, odor

Compounds	Retention Time (RT)	Retention index(RI)	Mass concen-tra-tion/μg·m-3	Odor character	Odor inten-sity	Identification method
Unknown	30.27	1223	–	Stink	2	RI, odor
Unknown	30.9	1243	–	Delightful	1	RI, odor
Unknown	32.38	1290	–	Aro-matic	2	RI, odor

6 Figures

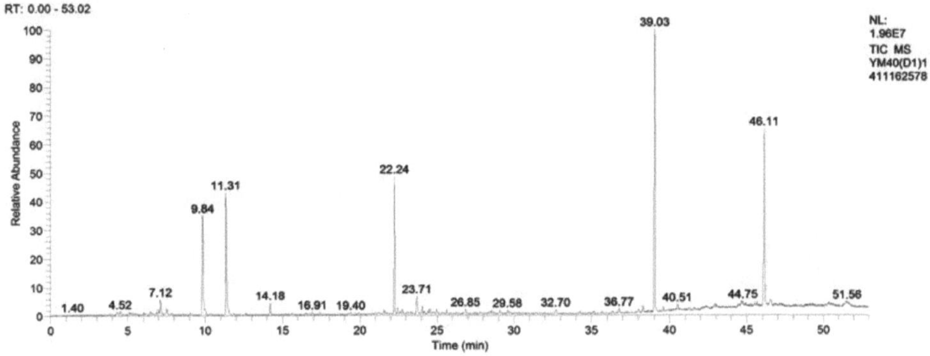

Fig. 1: GC total ion chromatogram on the volatile components of *Populus ussuriensis Kom.*

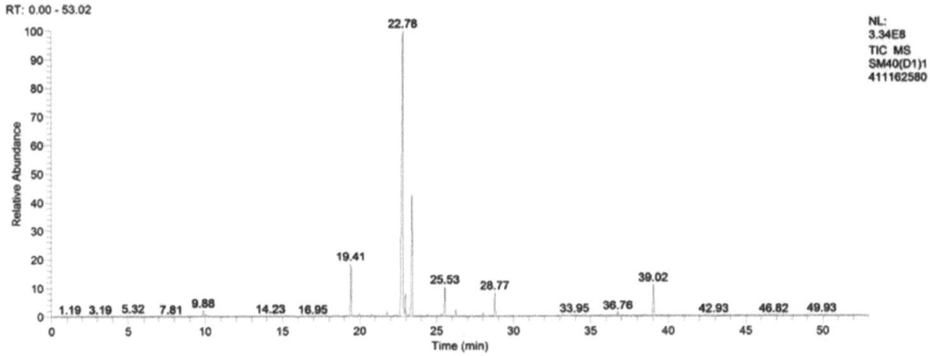

Fig. 2: GC total ion chromatogram on the volatile components of *Pinus sylvestris var. mongolica Litv.*

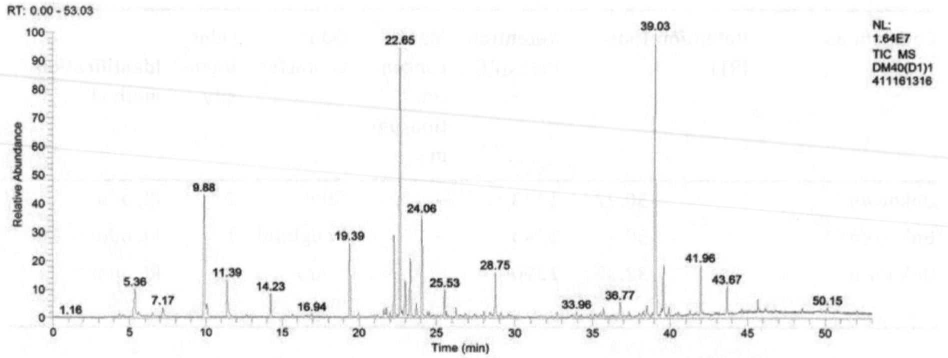

Fig. 3: GC total ion chromatogram on the volatile components of *Tilia amurensis Rupr.*

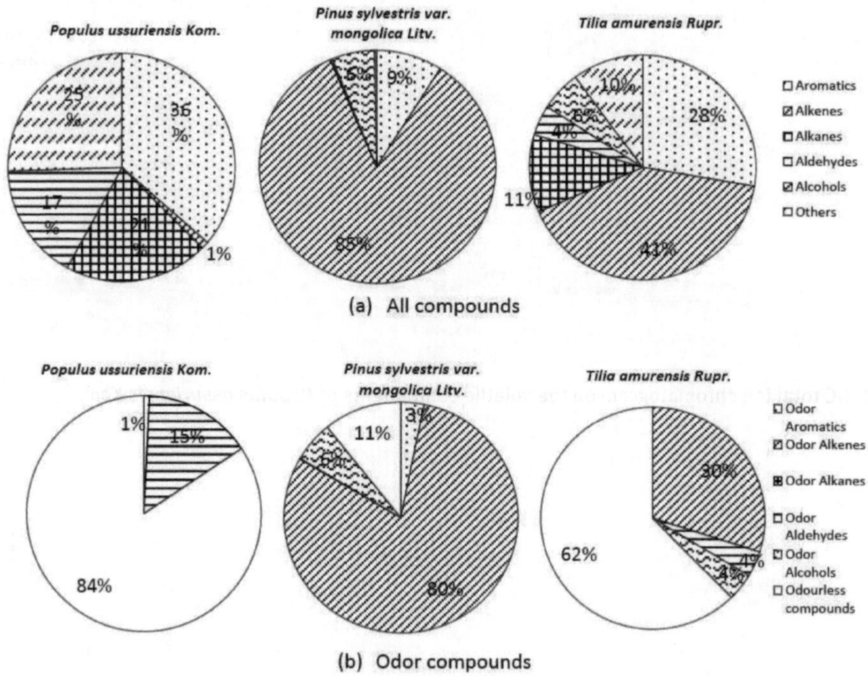

Fig. 4: The percentage of different components in three kinds of wood species

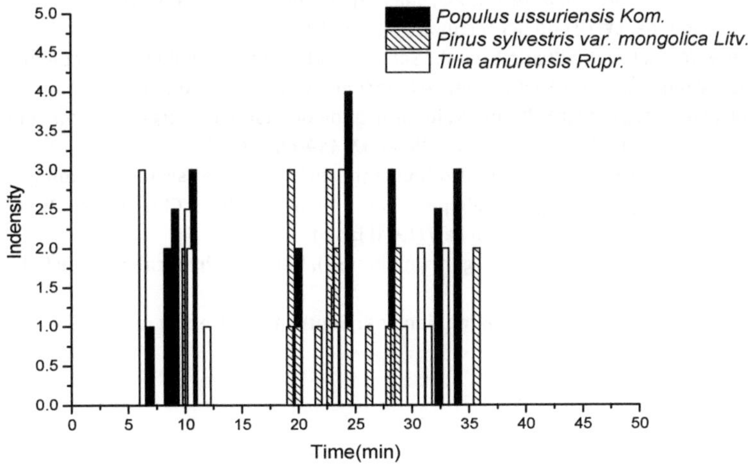

Fig. 5: The odor intensities and emerging times of different odor compounds in three kinds of wood species

Acknowledgement: This study was supported by The National Key Research and Development Program of China (2016YFD0600706-2).

References

[1] Xia LJ, Song HL (2006) Aroma Detecting Technique-Application of the GC- olfactometry (In Chinese). Food Fermn Ind, 32 (1): 83–87

[2] Zhang Q, Wang XC, Liu Y (2009) Applications of Gas Chromatography-Olfactometry (GC–O) in Food Flavor Analysis. Food Sci, 30 (3): 284–287

[3] Fuller GH, Stellencamp R, Tisserand GA (1964) The gas chromato-graph with humen sensor: perfume model. Acad Sci, 116: 711–724

[4] Acree TE, Butts RM,Nelson PR (1976) Snifferto determine the odor of gas chromatographic effluents. Anal Chem, 48 (12): 1821–1822

[5] Acree TE, Barnard J, Cunningham DG (1984) A procedure for the sensory analisis of gas chromatographic effluents. Food Chem, 14 (4): 273–286

[6] Ullrich F, Grosch W (1987) Identification of the most intense volatile flavor compounds formed during autoxidation of linoleic acid. Z Lebensm Unters Forsch, 184 (4): 277–282

[7] Haarse H, Vandeheijd G (1994) Trends in flavour research. Am sterdam: Elsevier Saence Publishers, 211–220

[8] Guillot S, Deylavi L, Bureau S (2006) Odor characterization of various apricot varieties using headspace–solid phase micro extraction combined with gas chromatography-mass spectrometry and gas chromatrography-olfactometry. Food Chem, 96 (1): 147–155

[9] Ruth SV, Boscaini E, Mayr D (2007) Study on the method of gas chromatography and mass spectrometry of dried chili. J. Chin. Capsicum, (1): 41–47

[1] Frank DC, Owen CM, Patterson J (2004) Solid phase micro extraction (SPME) combined with gas-chromatography and ol-factometry-mass spectrometry for characterization of cheese odor compounds. LWT-Food Sci Technol, 37(2): 139-154
[2] Gallagher M, Dalton P, Sitvarin L, et al., Sensory and analytical evaluations of paints with and without Texanol, Environmental Science & Technology 42, 243-248, 2008.
[3] Burdack-Freitag A, Mayer F, Breuer K, Identification of odor-active organic sulfur compounds in gypsum products, Clean — Soil, Air, Water 37, 459-465, 2009.
[4] Dool HVD, Kratz PD (1963) A generalization of the retention index system including linear temperature programmed gas-liquid partition chromatography. J. Chromatogr, 2: 463-470
[5] Law No. 91: Offensive Odor Control Law (1971) Japan
[6] Sun BG (2003) Edible flavoring surgery (In Chinese). Bei Jing: Chemical Industry Publishing House. 24-32
[7] Ma YK, Liu XG (2007) Food Chemistry. Nan Jing: Southeast University press, 305

Zhang-Huang Ye, Yi-Ke Li* and Jia-Huan Guo

A Study on the Global Carbon Emission

Abstract: Since the Industrial Revolution starting from 1860s, the dramatic emission increase of carbon emission has led to the greenhouse effect, and the consumption of fossil fuels contributed 80 % to the carbon emission. Based on the global carbon emission, this article has listed top 13 emitters, as well as the carbon emission per capita. Four major economies including China, European Unions, the United States and Japan, have been singled out to analyse the emission structure. In order to achieve the goal of environmental protection, it is an obligation for every country, especially for China and the United States, to continue their efforts in carbon reduction. It is urgent for China to enhance the energy utilization efficiency in the field of manufacturing and building.

Keywords: carbon emission; structure; climate change

1 Introduction

Since the Industrial Revolution starting from 1860s, the dramatic surge of carbon emission, predominantly, leads to the greenhouse effect, and the consumption of fossil fuels contributed 80 % to the carbon emission [1]. In 2013, the average CO^2 concentration in atmosphere has rose to 393×10^{-6} litres (i.e. 278ppm), going up 41.4 %, compared with that of 278×10^{-6} litres before the Industrial Revolution. The world carbon emission has reached a record of 36 billion tons in 2013, an increase of 2.1 %, compared with that of 2012, of 61 % with 1990.

The greenhouse gas, mainly produced by fossil fuel consumption, leads to a global average temperature rise by nearly 2 °C. The change would cause severe ecological environment destruction. However, environmental protection is the responsibility for all countries.

Zhang-Huang Ye, Jiangxi Science and Technology Normal University. No. 605, Fenglin Road, Nanchang, Jiangxi. Email: chuckverna@sina.com
Yi-Ke Li, Institute of Mineral Resources, CAGS. No. 26, Baiwanzhuang Road, Xicheng district, Beijing. Email: like430@cags.ac.cn. Corresponding author
Jia-Huan Guo, Institute of Mineral Resources, CAGS. No. 26, Baiwanzhuang Road, Xicheng district, Beijing. Email: 345860925@qq.com

DOI 10.1515/9783110540048-012,

2 Global Carbon Emission

Carbon emissions in the thirteen major countries were calculated to reach 25.566 billion tons, accounting for 72.85 % of the world's total emission. China ranked first, followed by United Stated and India. As far as carbon emission per capita is concerned, Saudi Arabia ranked first, followed by United States and Canada, while China was a little bit higher than global average level (Fig. 1). The above-mentioned countries can be classified into the following three types. Firstly, countries with large population: China, India, America, Indonesia, Brazil, Russia and Japan. Their carbon emissions are in highly proportional to their population sizes. Among the thirteen countries, eleven are densely populous (the population ranking top 25 in the world). Secondly, industrial developed countries: the United States, United Kingdom, Japan, Germany, Korea and Canada. The six countries are OECD members. Thirdly, countries with huge energy production: Iran, Saudi Arabia.

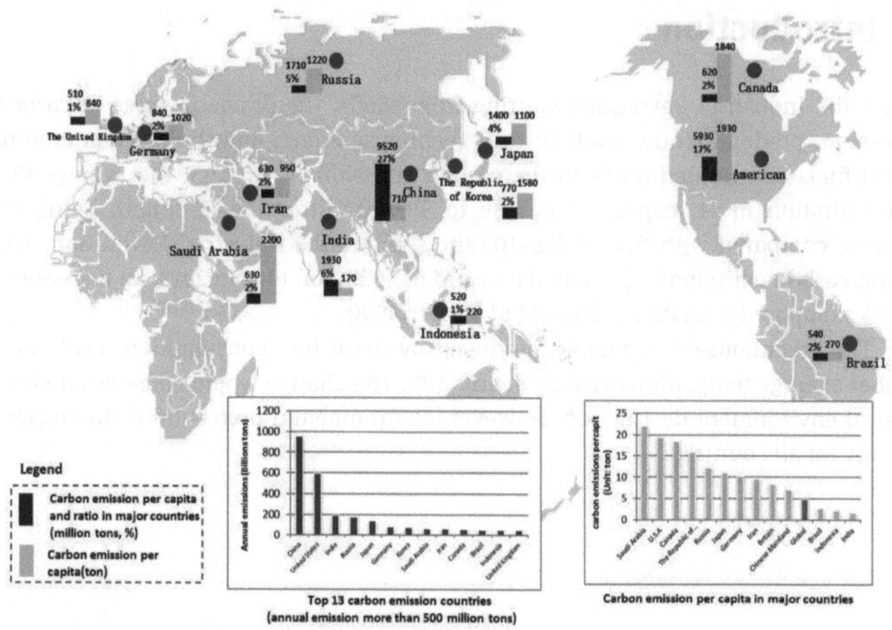

Fig. 1: Major carbon emission economies and carbon emissions per capita in 2013 (Data Sources: BP, 2014 [2])

3 Major Carbon Emission Economies

Among the major economies, China's carbon emission ranks first, exceeding that of European Union since 2004 and of America in 2007. In 2013, China's total carbon emission is about 9.52 billion tons, accounting approximately for 27.1 % in the global emission total. By coincidence, the amount is equivalent to the total emission of all economies ranking the fourteenth and the followings. What's more, the carbon emission increase amount ranks first in the world. According to China's Communique on Sea Level In 2012 released by the State Oceanic Administration, the coastal sea level goes up at the rate of 2.9 millimeters per year in 1980–2012, higher than that of the global average. The direct economic loss caused by the oceanic catastrophe has increased by 150 % in 2012, compared with that of 2011. The ascent of sea level will pose a threat to the Yangtze River Delta, the Pearl River Delta and the Yellow River Delta. If we look at it from the perspective of carbon emission accumulation 1965–2013, the OECD (merely 34 members, representing developed countries) and Non-OECD countries account for 55 % and 45 % respectively among the world's total emission (Fig. 2). In 2013, global carbon emission per capita was 4.97 tons, while OECD countries reached 13.85 tons, much higher than Non-OECD members with its emissions being 8.2 tons.

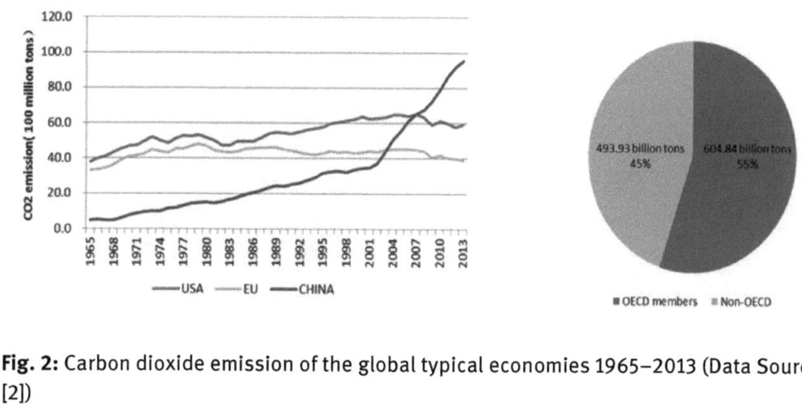

Fig. 2: Carbon dioxide emission of the global typical economies 1965–2013 (Data Sources: BP, 2014 [2])

4 Carbon Emission Structure Analysis

According to UNFCC's latest report, China's carbon emission is produced primarily from energy activities, accounting for 77.27 %, and secondly from agriculture and industry, for 10.97 % and 10.26 % respectively, and the refuse disposal for 1.5 %. Among the energy activities related to carbon emission, the energy industry makes up 44.52 % in the total emission, manufacturing and architecture of 39.13 %, transportation of 7.17 %, residence of 4.87 %, commercial industry of 2.53 %, agriculture, forestry and fishery of 1.24 %. The ratio of the carbon emission from manufacturing industry and architecture in China is much higher than that of other countries (USA: 14.78 %, Japan: 27.36 %, Europe Union: 15.17 %); energy industry slightly higher than other countries; the transportation industry significantly lower than developed countries (USA: 33.65 %, Japan: 17.83 %, Europe Union: 25.52 %), and so is the case with commercials and public facilities (Fig. 3). The quantity and structure of carbon emission in China are mainly decided by its economy development and industrial structure.

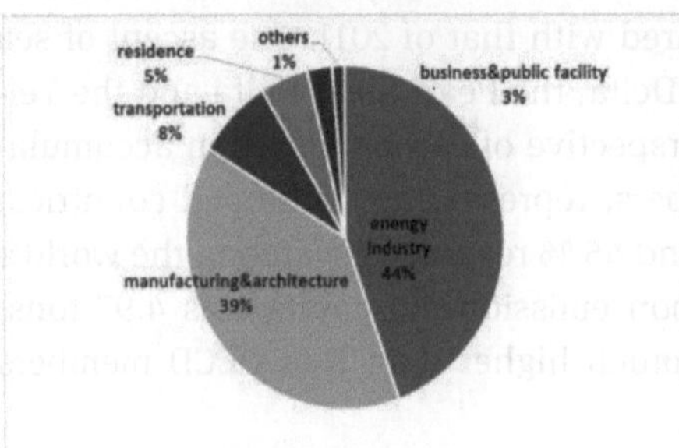

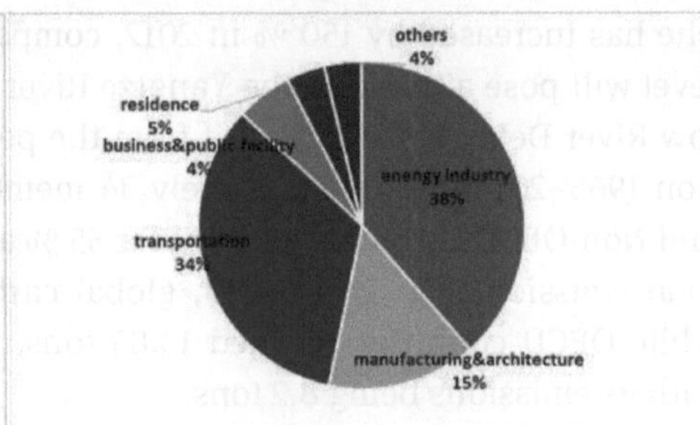

CHINA **AMERICA**

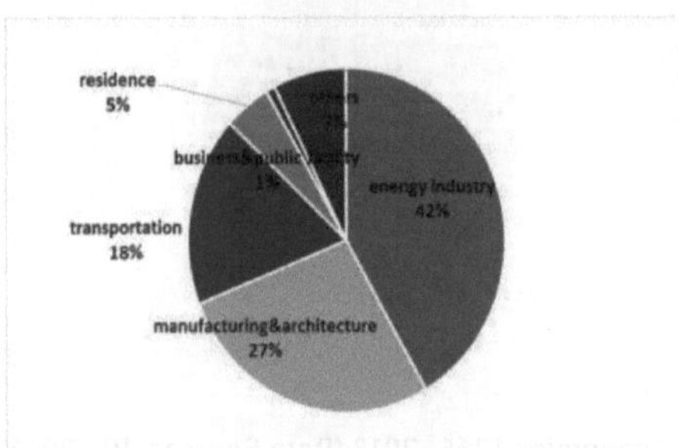

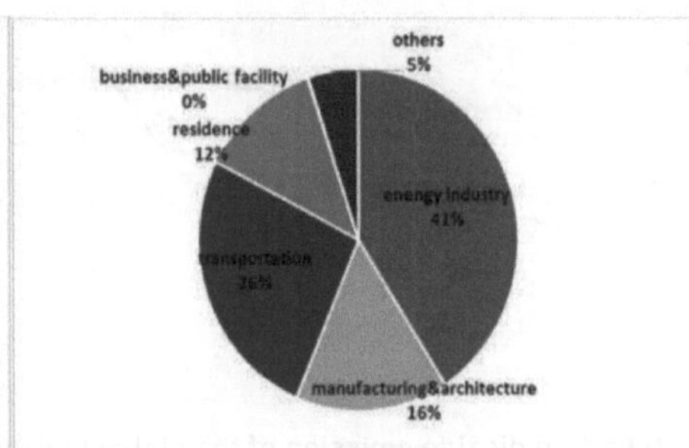

JAPEN **EU**

Fig. 3: The structure of carbon emission in major countries of the world (Data Sources: National Bureau of Statistics [3] and Energy Section of National Bureau of Statistics [4])

5 Suggestions

Faced with the issue on climate change, WMO and UNEP have set up IPCC. Its evaluation, upon which the UNFCC bases its negotiations, has provided a scientific basis for governments of all levels in formulating the policy related to climate change. *Kyoto Protocol* was published in 1997 in order to limit the greenhouse gas emission. It requires that developed countries shoulder their responsibility in reducing the carbon emissions from 2005, so do the developing countries from 2012. Confronted with the international pressure of the greenhouse gas reduction, China has signed the *Kyoto Protocol* in May 1998 and approved it in August 2002. *Paris Protocol* was approved by nearly 200 parties On December 12, 2015, which is a historic step for handling the global climate change.

On November 12th, 2014, China and America issued *the Joint Announcement on Climate Change*, stating that the two countries should jointly deal with climate change. Continuous low-carbon transformation is needed in order to control the global temperature increase within 2 °C. America plans to achieve the aim of carbon emission reductions by 26–28% by the end of 2025, on the basis of 2005, trying to reduce emission by 28 %. Meanwhile, China plans to reach the peak of carbon emission around 2030 and will try every effort to achieve it as soon as possible. Moreover, the proportion of non-fossil energy in primary energy resources consumption will rise about to 20 % [5]. Both sides intend to sustain their efforts. Therefore, as a populous developing country where coal is the main primary energy, China will face an arduous task of carbon emission. It is urgent for China to enhance the energy utilization efficiency in the field of manufacturing and building.

Acknowledgement: This study was financially supported by the programs (K1327) and (DD 20160124).

References

[1] Lu Chuanyi.2004. Resources and environmental economics. Beijing: Tsinghua University Press

[2] British Petroleum Company. 2014. BP statistical review of world energy 2013. London: British Petroleum Company

[3] National Bureau of Statistics. 2015. [DB/OL]http://data.stats.gov.cn/workspace/index?m=hgnd.

[4] Energy Section of National Bureau of Statistics. 2013. Chinese Energy Statistical Yearbook. Beijing: China statistical Publishing House

[5] Xinhua News Agency. 2016. [DB/OL]http://news.xinhuanet.com/world/2016-04/01/c_128854045.htm

Xiao Wang*, Xin-Ming Pu and Ping Liu

Seasonal Variability of *Calanus Sinicus* Brodsky in the Western South Yellow Sea

Abstract: Inter-disciplinary investigations were conducted in the WSYS in 2006–2007 to study the spatial and temporal variations of *Calanus sinicus* and its influencing factors. The distribution of *C. sinicus* in summer varied from that in spring in accordance with the change of the temperature. The species aggregated in the YSCWM to elude the high abundance in the surface and coastal waters. In autumn, the bottom-up effect played a crucial role in the population size. The abundance at the stations in and adjacent the oceanic front was also relatively high. In winter, the highest abundance presented in Haizhou Bay which was related to the entrainment of the YSWC. The abundance was low south of 35 °N thanks to the transportation of the YSWC and the YSCC and relatively high abundance north of 35 °N was related to the jacking effect of the YSWC.

Keywords: Distribution pattern; Seasonal variability; *Calanus sinicus*; South Yellow Sea

1 Introduction

Calanus sinicus Brodsky is comprehensively studied in the northwest Pacific Ocean thanks to its extensive distribution[1]and common dominance in zooplankton community [2–3]. Meanwhile, it is the main part of diets of predators with different sizes. Since it was chosen as the target species of China GLOBEC, its distribution in summer [4], life history strategy [5], feeding [6] and reproduction [7] were studied in details in the western South Yellow Sea (WSYS).

The Yellow Sea (YS) is a main population centre of this copepod. The complicated hydrographical conditions affected the ecosystem in different seasons in the WSYS, i. e., the coastal currents, including Yellow Sea Coastal Current (YSCC), the Lubei Coastal Current (LCC) and Subei Coastal Current (SCC), Yellow Sea Cold Water Mass (YSCWM), Changjiang River Diluted Water (CDW) and Yellow Sea Warm Cur-

Xiao Wang, Xianxialing Road, Laoshan District, Qingdao, China, 266061,E-mail: laoshanxiaoqu@163.com
Xin-Ming Pu, Xianxialing Road, Laoshan District, Qingdao, China, 266061,E-mail: pxm501501@163.com
Ping Liu, Xianxialing Road, Laoshan District, Qingdao, China, 266061,E-mail: liuping112411@163.com

DOI 10.1515/9783110540048-013,

rent (YSWC) [8]. Researches have shown that the presence of the YSCWM plays an important role in maintaining the population in the WSYS [4,9], while little has known about the effects of hydrographical conditions in other seasons. In this study, we investigate the distribution and the influence of environmental factors, especially the hydrographical conditions, on the spatial and temporal variations of *C. sinicus* based on the yearly inter-disciplinary surveys in 2006–2007 which has been the most detailed in the WSYS until now to our knowledge.

2 Materials and Methods

We conducted an annual multidisciplinary comprehensive investigation from July, 2006 to November, 2007 in the western SYS (Fig. 1). Sampling methods of zooplankton and environmental factors refered to Wang et al. 2012 [10].

To better understand the regional distribution and the effects of different environmental conditions on *C. sinicus*, we further divide the sampling area into four subregions, that is the coastal area of Shandong Peninsula (region I), Jiangsu coastal area (region III), mixing area in the south of the surveyed area (region IV) and offshore area (region II)(Fig. 1). The offshore area was determined by the sea bottom temperature (SBT) in summer 2006 where the SBT was < 12 °C.

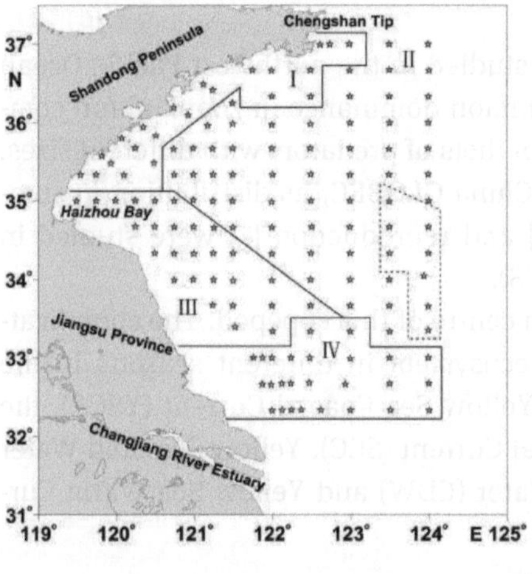

Fig. 1: Surveyed area (subarea) and sampling stations in the western SYS in 2006–2007

3 Results

3.1 The Abundance of *C. sinicus* in Different Seasons

The average of was 99.59±13.15 ind.m^{-3}, accounting for 38.95 % of total zooplankton abundance. In summer, the mean abundance was 89.70±8.26 ind. m^{-3} lower than that in spring, but the proportion of the total abundance was 54.17 %, higher than that in spring. Its average abundance and the proportion in total abundance were even lower in autumn, 44.30±5.89 ind. m^{-3} and 28.33 %, respectively. The abundance was lowest in the wintertime, with the an average merely 21.16±2.83 ind. m^{-3}, but due to the low total abundance, the proportion in total abundance was higher, up to 37.17 % (Table 1).

Tab. 1: The abundance of *C. sinicus* in different seasons

Seasons	Abundance (mean±SE) (ind. m^{-3})
Spring	99.59±13.15
Summer	89.70±8.26
Autumn	44.30±5.89
Winter	21.16±2.88

3.2 Distribution and Seasonal Variations of *C. Sinicus*

The distribution pattern of the species showed apparent distinction in different seasons and sea area. Higher abundance occurred in the coastal area of Shandong Peninsula, the south and southeast part and the abundance was lower than 50, and even 10 ind. m^{-3} in other areas in spring (Fig. 2a). In summer, the abundance was generally >100 ind. m^{-3} where the YSCWM existed except that in the northeastern part. The isoline of 50 ind. m^{-3} was just about in accordance with the isotherm 12 °C (Fig. 2b). The distribution pattern in autumn was more or less the same as that in summer. In the coast of Shandong and Jiangsu Province, the abundance was generally lower than 10 ind. m^{-3}. Higher abundance occurred in the central, southern and southeastern part of the study area (Fig. 2c). During the wintertime, highest abundance was found in the Haizhou Bay. In the south of 35 °N, the abundance was < 30 ind. m^{-3} and even lower than 10 ind. m^{-3} in most sea area. Conversely, it was generally higher north of 35 °N except the sea area off Shidao (Fig. 2d).

Abundance in different investigated area varied as the season. That in region III was minimum in spring, summer and autumn, but was second only to region II in winter. It was always the second of the region IV in spring, summer and autumn;

however, it was the lowest in winter. The abundance in region I was maximum in spring, but invariably lower than the other two region in summer, autumn and winter. The abundance in region II was maximum in summer, autumn and winter, however, it was evidently lower than region I and region IV in spring (Fig. 3).

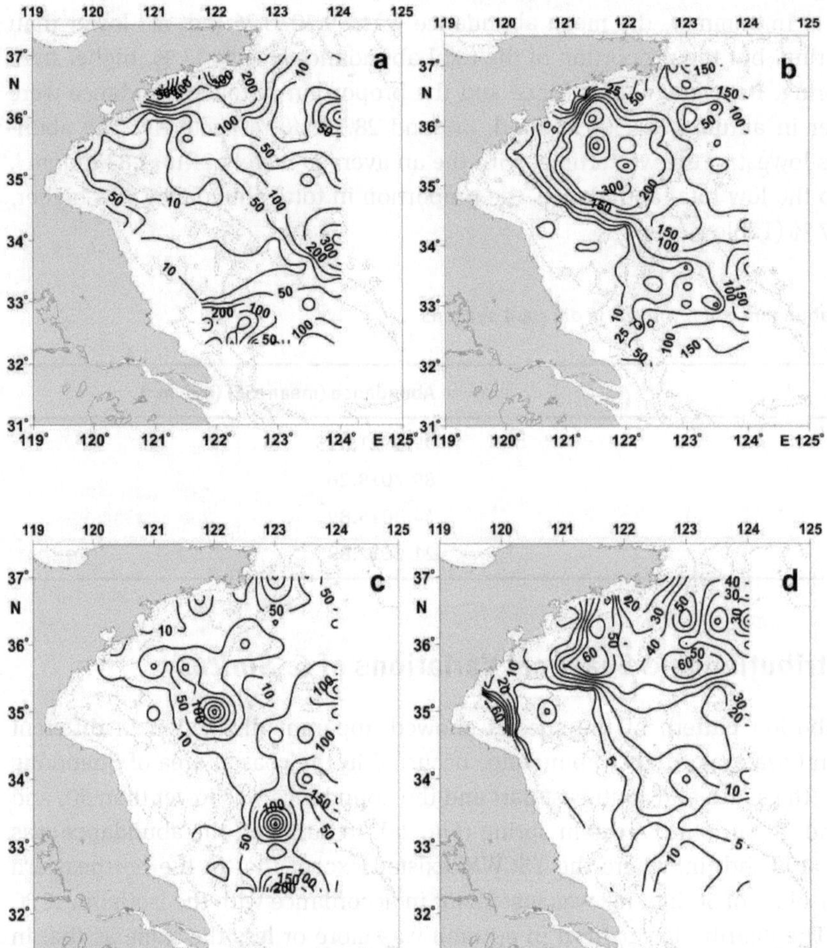

Fig. 2: Horizontal distribution of C. sinicus in the western SYS in 2006–2007 (a-spring; b-summer; c-autumn; d-winter)

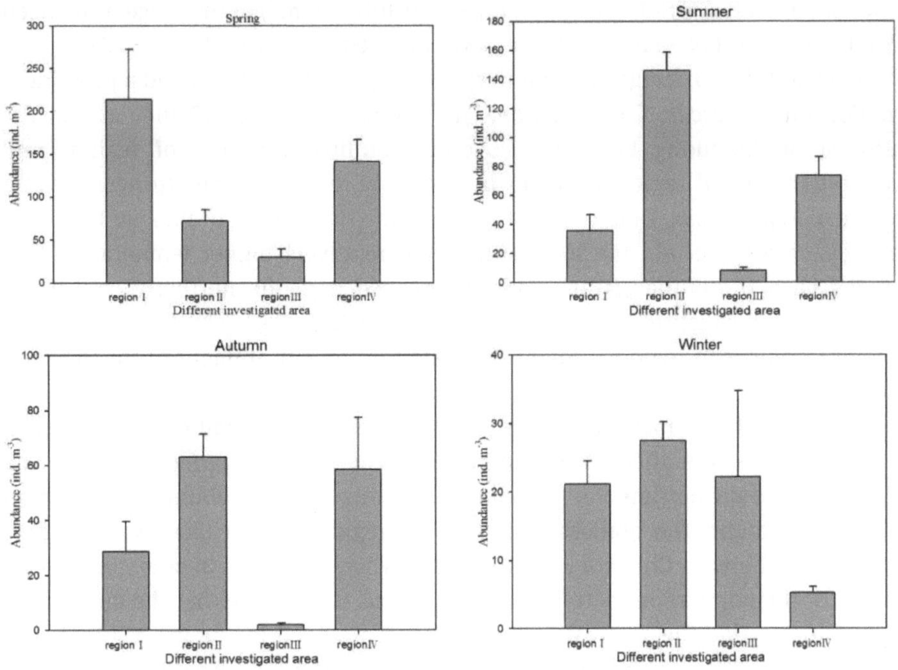

Fig. 3: Abundance of *C. sinicus* in different investigation area

4 Discussion

The water temperature was 10–18 °C when the species was in its flourishing period of propagation [11]. The average temperature in surveyed area in spring was about 10 °C (Fig. 4a) which was suitable for the proliferation of the species except region I where the water temperature was < 10°C. Together with the comparatively higher Chl a concentration produced by the phytoplankton spring bloom (Fig. 4a), the environmental factors in most sea area were appropriate for the reproduction of *C. Sinicus*. The distribution of *C. sinicus* was quite uneven in spring: The average abundance was highest in region I, especially in nearshore area and the northeastern part with the abundance > 100 ind. m^{-3} (Fig. 2a). Oceanic fronts usually have boundary effects and other influences on the the pelagic biogeography [12]. The existence of the oceanic front during the springtime might well play a boundary effect on the expansion of *C. sinicus* population from northeast of region I to region II (Fig. 2a). Besides the local recruitment, the entrainment by the LCC from northern seas contributed to the high abundance in the northeast of region I.

The abundance in region IV was also high where was affected by the northerly SCC, YSCC and the southerly CDW, TWC. The abundance was high (> 100 ind. m^{-3}) in the transitional area of SCC and CDW, lower (50-100 ind. m^{-3}) in the place influenced by YSCC and lowest (< 50 ind. m^{-3}) where was affected by SCC or TWC (Fig. 2a).

Zhang and Jin [13] suggested that in spring, *Engraulis japonicus* had a prominent predation on *C. sinicus*. Consequently, the low abundance (< 10 ind. m^{-3}) in the southwest of Shandong Peninsula covering southwestern part of region I and northwestern part of region II, might be probably related to the stronger feeding pressure of *E. japonicus*. Due to the influence of the river runoff of Jiangsu Province, like Sheyang River et al., the low salinity and relatively higher temperature SCC extended east- or southeastward and confluenced with the northerly cold water (Fig. 2a-b) which constituted the main body of the YSCC. Therefore, the lower abundance of *C. sinicus* in regionIII might be related to the entrainment of the SCC and YSCC.

In summer, the temperature exceeded the upper thermal limit of *C. sinicus* (Fig. 4b) except where was affected by the YSCWM. The species aggregated inside the YSCWM to avoid the detrimental temperature [4], therefore, the abundance in region II was undoubtably the highest (Fig. 3b). The region II was a place with highest abundance and lowest Chla concentration. The whole sea areas in summer was not suitable for the population recruitment of *C. Sinicus* thanks to either the deleterious temperature outside the YSCWM or the low food provided inside the YSCWM. The existence of YSCWM is merely provide an over-summering place for the population maintenance, not for recruitment, which could also be speculated by the decreased abundance in summer than that in spring (Fig. 2b).

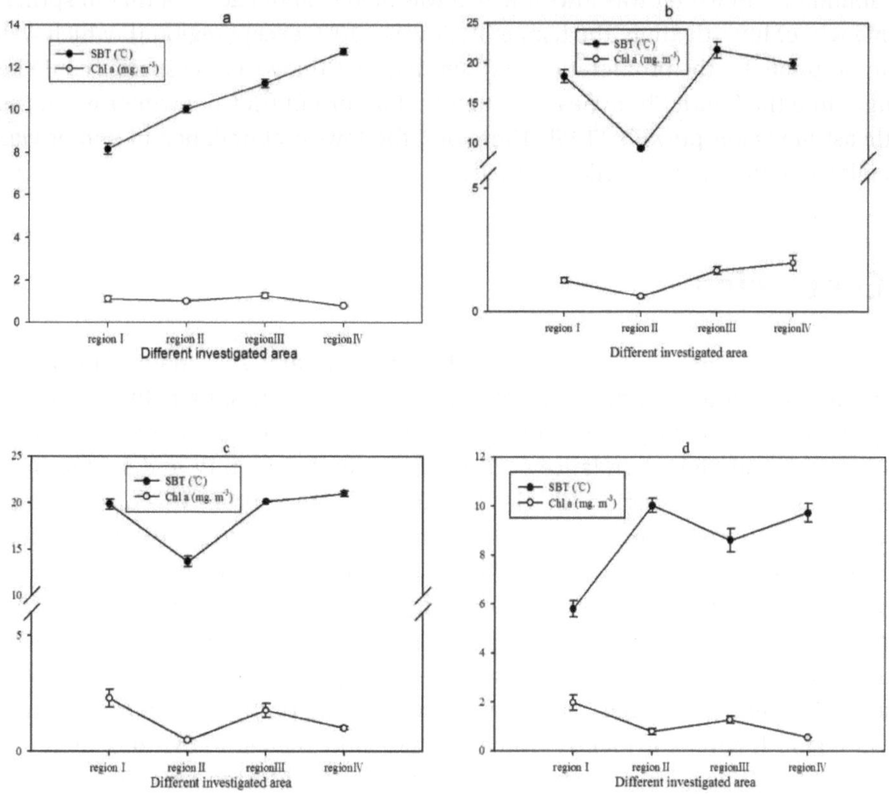

Fig. 4: Changes of SBT and sea bottom Chl a concentration in different investigation area (a-spring; b-summer; c-autumn; d-winter)

Autumn is a transitional period between warm and cold season when the YSCWM shrinks, the stable vertical temperature structure begins to vanish and the embryonic form of the YSWC shapes [14]. The abundance in autumn decreased definitely comparing that in spring and summer (Fig. 2a-c). The abundance in region I and region IV varied little, and the distinction was primarily from that in region II affected by the YSCWM. The Chl a concentration remained low (< 0.5 mg. m^{-3}) in region II in autumn (Fig. 4c), which was adverse to *C. sinicus* that exposed to the environment with low food availability since summer. Otherwise, external food supply was related to the gonad mature and reproduction of *C. sinicus* [7], therefore, the population could not be recruited immediately with the rising temperature in autumn under the circumstance of low food availability, but might have been threatened by low food concentration since summer which was the leading cause for the decline of the abundance.

The abundance in winter was lowest which was merely about 20 % of that in spring. During winter investigation, the temperature was < 10°C except region II, which was not appropriate for the proliferation of *C. sinicus* [11]. Otherwise, the species could be brought into the South China Sea by China Costal Current (CCC) in winter when the northeast monsoon prevails [3,15]. Therefore, the lowest abundance in winter was also related to the hydrographical conditions.

5 Conclusion

The distribution pattern of *C. sinicus* varied spatially and seasonally which was determined by both biological and physical conditions. In spring, thanks to the adequate food availability and favorable water temperature, the proliferation of *C. sinicus* began and high abundance occurred in region I, region IV and southeast of region II and lowest in region III. The low abundance in spring was attributed to top-down effect (predation by *E. japonicus*) and physical conditions (barrier by oceanic front; entrainment by SCC and YSCC). The abundance decreased in region I thanks to the relatively high temperature in the surface and coastal area. The presence of the YSCWM in region III provided an ideal refuge for *C. sinicus* to avoid the high temperature in surface and coastal area. The southerly CDW and TWC was closely correlated to the high abundance area in region IV. In autumn, the population in region II shrank evidently which was probably related to the long-term low level of Chl a. Otherwise; the abundance at the stations in the oceanic front was usually high in autumn. As the prevail of the north wind, the YSWC was prosperous which was relevant to the higher abundance in region III, especially in Haizhou Bay. The entrainment of the YSWC and YSCC was related to the low abundance south of 35 °N and the relatively high abundance north of 35 °N was probably the jacking effect of the YSWC.

Acknowledgement: This work is supported by the Fundamental Scientific Research Fund (GY0214G10)

References

[1] Hulsemann, K., 1994. *Calanus sinicus* Brodsky and *C. jashnivi*, nom.nov. (Copepoda: Calanoida) of the North-west Pacific Ocean: acomparison, with notes on the integumental pore pattern in *Calanus* s. str. Invertebrate Taxonomy 8, 1461–1482.

[2] Xu, Z. L.,Chen, B. J.,2007. Seasonal distribution of *Calanus sinicus* (Copepoda, Crustacea) in the East China Sea.Acta Oceanologica sinica, 26(3), 150–159.

[3] Yin, J. Q., Huang, L. M., Li, K. Z., Lian, S. M., Li, C. L., Lin, Q., 2011. Abundance distribution and seasonal variations of *Calanus sinicus* (Copepoda: Calanoida) in the northwest continental shelf of South China Sea. Continental Shelf Research, 31, 1447–1456.

[4] Wang, R., Zuo, T. and Wang, K. 2003. The Yellow Sea Cold Bottom Water- an oversummering site for Calanus sinicus (Copepoda, Crustacean). Journal of Plankton Research, 25 (2), 169–183.

[5] Pu, X. M., Sun, S., Yang, B., Zhang, G. T., Zhang, F. 2004. Life history strategy of *Calanus sinicus* in the southern Yellow Sea in summer. Journal of Plankton Research, 26 (9), 1059–1068.

[6] Huo, Y. Z., Wang, S. W., Sun, S., Li, C. L., Liu, M. T. 2008. Feeding and egg production of the planktonic copepod Calanus sinicus in spring and autumn in the Yellow Sea, China. Journal of Plankton Research, 30 (6), 723–734.

[7] Wang, S. W., Li, C. L., Sun, S., Ning, X. R., Zhang, W. C. 2009. Spring and autumn reproduction of *Calanus sinicus* in the Yellow Sea. Marine Ecology Progress Series, 379, 123–133.

[8] Sun, X. P., 2006. Regional seas of coastal waters of China. Ocean Press.

[9] Sun, S. 2005. Over-summering strategy of *Calanus sinicus*. GLOBEC international newsletter, 11 (1), 34.

[10] Wang, X, Wang, Z. L, Pu, X. M., et al. 2012. Analysis of the distribution of zooplankton in the Southern Yellow Sea in summer. Haiyangxuebao, 35(5), 147–155.

[11] Chen, Q. C., 1964. A study of the breeding periods, variation in sex ratio and in size of *Calanus sinicus* Brodsky. Oceanography Et Limnology Sinica 6 (3), 272–288.

[12] Sournia, A. 1994. Pelagic biogeography and fronts. Progress in oceanography, 34, 109–120.

[13] Zhang, B. and Jin, X. S. 2010. Seasonal variations of the functional groups of fish community and their consumption of zooplankton in the Yellow Sea. Journal of Fisheries of China, 34(4), 548–558.

[14] Wang, H. W. 2008. Study on seasonal and inter-annual variations of the Yellow Sea Warm Current. Master Thesis, the First Institute of Oceanography, SOA.

[15] Hwang, J. S. and Wong, C. K., 2005. The China Coastal Current as a driving force for transporting *Calanus sinicus* (Copepoda: Calanoida) from its population centers to water off Taiwan and Hong Kong during winter northeast monsoon period. Journal of Plankton Research 27, 205–210.

Xin Shen and Xiao-Feng Xu

Application of Indigenous Microbial Activation System in Ecological Remediation of Nanfeihe River Water Body

Abstract: Healthy water body environment in cities is considered as an important safeguard for the sustainable urban development. As the integrated treatment of urban water environment is being widely launched, the water quality in watercourses has been improved; however, such work demands sustainable efforts. For urban water environment treatment, it is of great significance to keep long-term, good water quality at a low cost and to prevent the undesired relapse. The paper discusses the IMA technology used in the ecological remediation of Nanfeihe River Water Body, as well as the regular testing and the results analyzed. The technology, with characteristics demonstrated in the application, such as efficient construction, quick effect, short cycle, low cost, easy utilization and high safety, possesses a broad application prospect.

Keywords: Water ecological remediation, IMA technology, black and odorous water

1 Introduction

With the fast development of the society, a large number of pollutants are disorderly discharged, causing more severe eutrophication of water bodies in lakes and rivers. To protect the water environment and tackle the disadvantages in conventional water body remediation technologies, like a long-term treatment cycle, high investment and management difficulty, microbial water purification technology can be applied to flowing water bodies based on the experience gained in static water body remediation, so as to restore the biologic chain of the water bodies and keep ecological balance. The purification of water body for ecological remediation adopts indigenous microbial activation technology (IMA technology for short).

Xin Shen, No.180, Xiangzhang Ave., Hefei, China, Email: 1010331469@qq.com
Xiao-Feng Xu, No.180, Xiangzhang Ave., Hefei, China, Email: 1187052872@qq.com

DOI 10.1515/9783110540048-014,

2 Basic Principle of IMA Technology

Through the water microcirculation between IMA equipment and the water body, indigenous microbes which reproduce fast and in a large amount in the IMA equipment are released to the water body; as the polluted water body is rich in the nutriments needed by the microbes for growth, the microbes in the water body can increase quickly, resulting in perfected and optimized population. The microbes can form a large amount of zoogloea sticking to the surfaces of sediment, aquatic plants, and the shells of revetment and demersal molluscs. The microbes have natural characteristics typical of pollutants, such as, short generation time, fast reproduction, strong sticking capability, tropism of movement & expansion (tendency to the areas with high concentration of pollutants for reproduction), strong adaptive capacity; they can decompose multiple kinds of pollutants including oil, protein, cellulose, starch and polysaccharide, thus able to quickly convert the pollutants in the water body (including those in the sediment) into biomass; the microbes, as initial consumers & decomposers, can consume the pollutants (C, N & P) in the water body during mass reproduction, and the biomass in the process can activate superior consumers in the food chain level by level, to gradually improve the growth environment of the aquatic animal and plant system. The pollutants in the water body can be quickly transferred to the aquatic organisms, boosting the recovery of the self-purification capacity of the ecological system in the water body, and the stable ecological balance attained can contribute to water quality purification, clearance of water body and ecological remediation; in addition, the sediment can be degraded gradually, and the biological desilting is thus achieved.

Compared with the traditional bypass water treatment technology, IM technology is characterized with lower investment, less occupancy, lower energy consumption, more environmental friendliness and freedom from desilting and secondary pollution; it can not only purify water but also effect the long-term maintenance at a low cost, thus possessing good social and environmental benefits.

3 IMA Technology System Composition

IMA technology system comprises: (1) IMA equipment and the activation and culture equipment for indigenous microbes; (2) microbe activation agent for IMA; (3) water circulation pipeline system, in which the pipelines and water distribution outlets shall be arranged on the bottom of the water body, and effluents in multiple points; (4) micro-nano reoxygenation system, the water microcirculation between IMA equipment and the water body can not only release a large number of indigenous microbes cultivated and activated, but also convey plenty of blisters of about 30μm to the water distribution outlets.

4 Application Scope of IMA Technology

1) Water body types: Natural rivers & lakes; artificial landscape water bodies
2) Pollution control: IMA technology applies to the ecological remediation and water quality improvement of water bodies whose pollution has been effectively controlled;

Limits in water body to receiving pollutants:

The daily discharge amount of domestic sewage shall not exceed 0.5 ‰–1.0 ‰ of the water amount in the water body; if the self-purification capacity of the water body is good, the assimilative limit can approach the upper limit of 1.0‰, or otherwise, the lower limit of 0.5 ‰. No industrial sewage toxic and harmful to aquatic animals, plants & microbes shall be discharged into the water body.

It is not recommended to use the recycled water chlorinated for ecological water compensation, or the amount of single water compensation is limited to 0.5‰ of the total water amount; no recycled water chlorinated shall be used for the change of water body.

5 Remediation Effect Analysis of Nanfeihe River Water Body

IMA technology was applied to the Nanfeihe River Black and Odorous Water Body Rehabilitation and Purification Project in Luyang District, Hefei. Water quality in the three river sections (Qianshan North Road section, Mengcheng Road Bridge section and Gujing Bridge section) was monitored for five consecutive months from January to May 2017. The results of water quality monitored are shown in Tables 1 to 3.

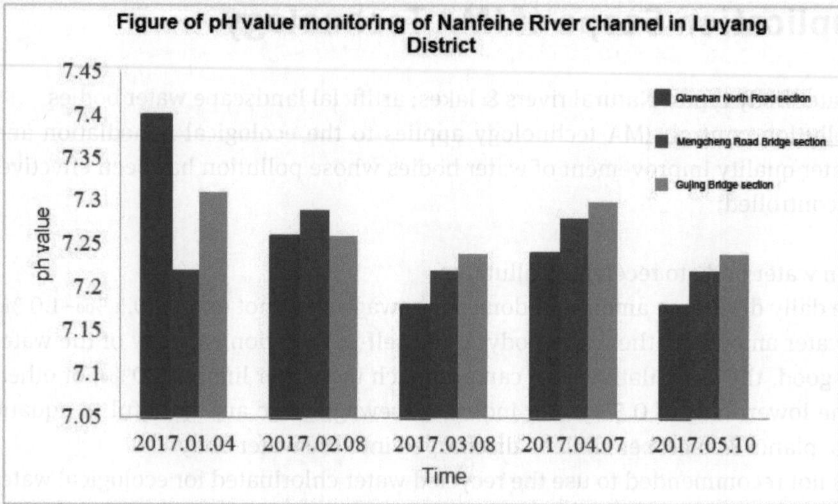

Fig. 1: PH value monitoring at Nanfeihe River channel in Luyang District

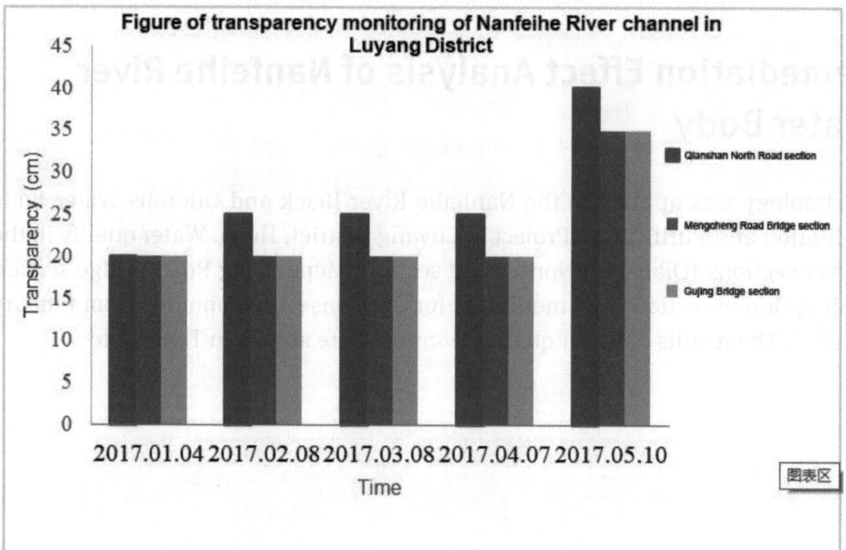

Fig. 2: Transparency monitoring at Nanfeihe River channel in Luyang District

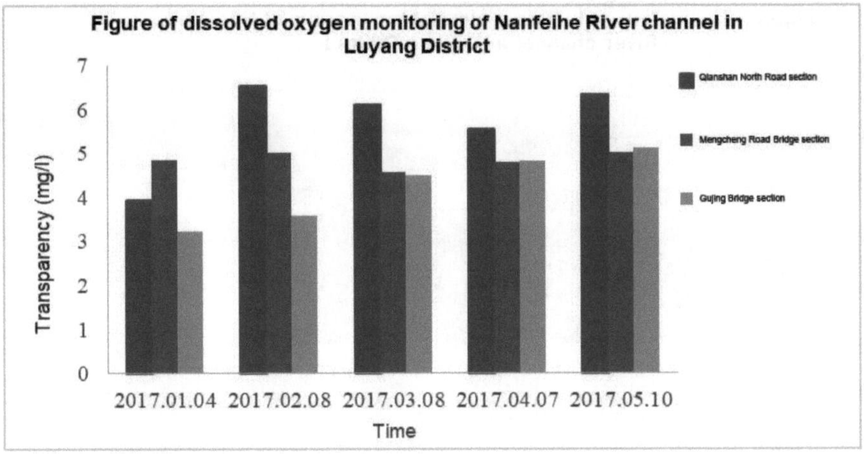

Fig. 3: Dissolved oxygen monitoring at Nanfeihe River channel in Luyang District

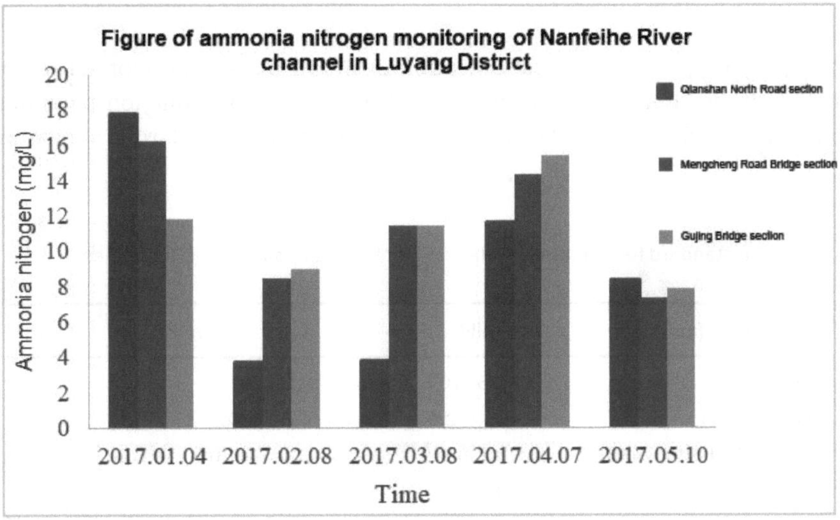

Fig. 4: Ammonia nitrogen monitoring at Nanfeihe River channel in Luyang District

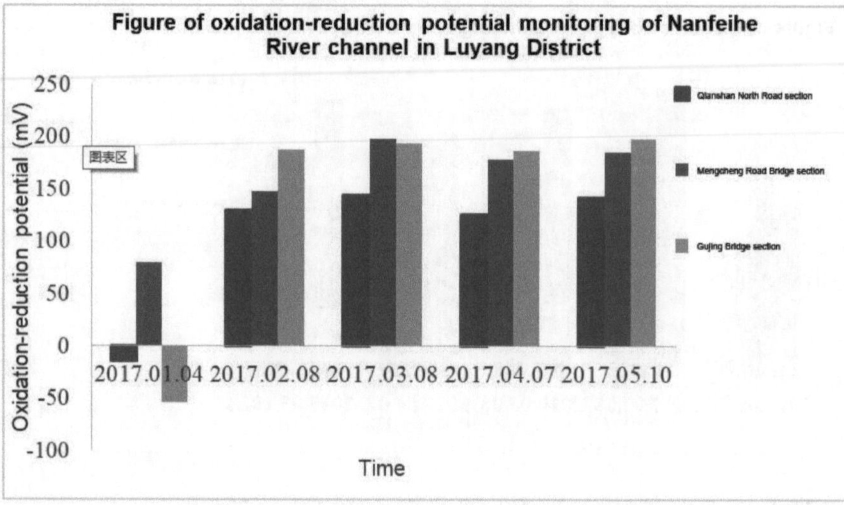

Fig. 5: Oxidation-reduction potential monitoring at Nanfeihe River channel in Luyang District

The evaluation indicators for the classification of urban black and odorous water bodies include transparency, dissolved oxygen (DO), oxidation-reduction potential (ORP) and ammonia nitrogen (NH3–N), and the classification standard is shown in Table 1.

Tab. 1: Classification standard for pollution levels of urban black and odorous water bodies

Characteristic indicator (unit)	Mild	Severe
Transparency (cm)	25~10	<10
Dissolved oxygen (mg/L)	0.2~2.0	<0.2
Oxidation-reduction potential (mV)	-200~50	<-200
Ammonia nitrogen (mg/L)	8.0~15	>15

Among 4 physical and chemical indicators for a detection point, if more than 60 % of the data under one indicator or more than 30 % under not less than 2 indicators reach the "severe" level, the detection point should be identified as "severe", otherwise as "mild".

If more than 3 detection points are continuously identified as "severe", the area among them should be identified as "severe"; if more than 60 % of the detection points in the water body are identified as "severe", the entire water body should be identified as "severe".

Analysis on the above monitoring data:

5.1 Qianshan North Road section

5.1.1 Transparency

As can be seen from the monitoring data, transparency in May was higher, while that in other months remained relatively stable and on the verge of being mild black and odorous and the attention should be brought to it.

5.1.2 Dissolved Oxygen

As can be observed from the monitoring data, the results of dissolved oxygen monitoring in January were low and close to the mild black and odorous level, while the level of dissolved oxygen gradually increased from February and fell slightly in April. In other words, the water quality, on the whole, showed a trend of improvement.

5.1.3 Ammonia Nitrogen

As can be seen from the monitoring data, the ammonia nitrogen monitoring results in January were higher and reached the severe level, then the data significantly decreased, but reached the mild level after April and no optimistic trend was found in the water quality.

5.1.4 Oxidation-reduction Potential

As can be observed from the monitoring data, the oxidation-reduction potential in January was lower and reached the mild level, and then the data increased significantly, the little change occurred in monitoring results from February to May and free from being black and odorous.

5.2 Mengcheng Road Bridge section

5.2.1 Transparency

As can be noticed from the monitoring data, transparency from January to April was on the verge of the mild level and higher transparency was shown in May.

5.2.2 Dissolved Oxygen

From the monitoring data, it can be observed that the dissolved oxygen level from January to May was higher, with the water quality in a good condition, on the whole.

5.2.3 Ammonia Nitrogen

As can be seen from the monitoring data, the ammonia nitrogen level in January became higher and reached the mild level, then the data significantly decreased; on the whole, a trend of improvement emerged.

5.2.4 Oxidation-reduction Potential

From the monitoring data, it can be found that the oxidation-reduction potential in January remained lower, approaching the mild level, and then the data increased significantly; little change occurred in the monitoring results from February to May, and free from being black and odorous.

5.3 Gujing Road Bridge Section

5.3.1 Transparency

As can be seen from the monitoring data, transparency in May was higher, while that in other months remained relatively stable and on the verge of being mild black and odorous and the attention should be brought to it.

5.3.2 Dissolved Oxygen

As can be seen from the monitoring data, the dissolved oxygen level in January remained low, and gradually increased from February. In other words, the water quality, on the whole, showed a trend of improvement.

5.3.3 Ammonia Nitrogen

As can be seen from the monitoring data, the ammonia nitrogen monitoring results in January were higher and reached the severe level, then the data slightly decreased, but reached the severe level after April and were improved in May, without black and odorous situation being detected.

5.3.4 Oxidation-reduction Potential

As can be observed from the monitoring data, the oxidation-reduction potential in January was lower and reached the mild level, and then the data increased significantly, the little change occurred in monitoring results from February to May and free from being black and odorous.

According to the above principle for the detection of black and odorous water body and the water quality situation in various months, the black and odorous rating for Nanfeihe River sections from January to May is shown as follows:

Tab. 2: Water quality monitoring results of Nanfeihe River in Luyang District from Jan. to May, 2017

Section **Black and odorous** **B&O level** **Time**	**Jan. 2016**	**Feb. 2017**	**Mar. 2017**	**Apr. 2017**	**May 2017**
Qianshan North Road section	Severe	Severe	None	Mild	Mild
Mengcheng Road Bridge section	Severe	Mild	Mild	Mild	None
Gujing Road Bridge section	Mild	Mild	Mild	Severe	None

Based on the analysis on monitoring results of the water quality in three sections of Nanfeihe River from January to May, it is shown that, after IMA technology was applied to the rehabilitation and purification of black and odorous water body, the quality, transparency, re-oxygenation capacity of the water body have been improved and the survival environment of aquatic animals and plants has been improved.

Through the control and induction of the IMA system, a large number of native microbes breed, which inhibit the mass multiplication of anaerobic microorganisms and effectively control stinking and floating of the sediment in black and odorous water; with the absorption of the nutrients such as CPN, turbidity is effectively eliminated, water transparency improved, and photosynthesis increased, which promote the restoration of the underwater ecosystem; in addition, both oxygen dissolution and re-oxygenation capacity of the water body are enhanced. As a result, the water ecosystem is completely improved and water purification capacity boosted.

6 Prospect

Transparency, re-oxygenation and self-purification capabilities of the water body can be improved through the application of IMA technology. In addition, when the water body restores high quality, the service of IMA system can be stopped, which can be restarted once the water body shows an omen of deterioration, thus able to keep the water quality in a good state. To sum it up, for the urban watercourses with pollution controlled, the application of IMA technology, which has been proved to be conducive to maintaining the quality of water body at a low cost and for a long term, and to a healthy and stable ecological environment, has gain great significance and will offer a broad prospect.

References

[1] Xue Wang, Tao Tian and Jianying Yang, Summary one cological treatment of urban river [J]. Science of Soil and Water Conservation, 6, 5, 105–111, 2008.

[2] Yue Zhao, Ruihua Yao and Min Xu, Study on the Practice and Route of Combating Urban lack-and-malodorous Water Body[J].Environmental Protection,43, 13, 27–29, 2015.

[3] Yongfeng Gu, Mei Li and Shufen Chen, Research Progress on Urban River Ecological Restoration Technology [J]. Environmental Science and Management, 38, 4, 25–29, 2013.

Xing-Fang Zhang, Hong Luo*, Bao-Liu Zhang and Jian Wang

Climate Change and Externality: Empirical Evidences from the EKC of CO_2 and PM_{10}

Abstract: It is widely acknowledged that climate change is a global problem which needs to be treated seriously. Externality is an important part in the field of theoretical research of climate change. However, how much does the externality really matter in climate change? There are almost no intuitive feelings from empirical evidences, and this paper is aimed at revealing the externality through empirical research tentatively. The empirical studies have been made based on the Environmental Kuznets Curve (EKC) of CO_2 and PM_{10} in this paper. The results illustrate that: 1) Compared with the regional air pollutants PM_{10}, the global atmospheric pollutants CO_2 has more significant externality. That is the main reason why people from different countries have more appeal but less action on the issue of carbon-emission reduction. 2) With the growth of the per capita income, PM_{10} seems to be better controlled, but CO_2 shows no signs of descending. It is unlikely that carbon emission and global warming will be under better control with the development of the economy, unless externality problem can be addressed properly through international cooperation.

Keywords: climate change; externality; carbon emission; EKC; international cooperation

1 Introduction

In recent decades, climate change issues are increasingly concerned by the international community with the rapid growth of greenhouse gas emissions. Global CO_2 emission has been increasing from 22.2 billion tons in 1990 to 34.65 billion tons in 2011 with the annual growth rate of 2.1 %. Kyoto Protocol (as the supplementary provisions of "United Nations Framework Convention on Climate Change") came

Xing-Fang Zhang, Department of Environment & Economics, Chinese Research Academy of Environmental Sciences, Beijing, China. Email: zhangxingfang2008@126.com
Hong Luo, Department of Environment & Economics, Chinese Research Academy of Environmental Sciences, Beijing, China. Email: luohong@craes.org.cn
Bao-Liu Zhang, Department of Environment & Economics, Chinese Research Academy of Environmental Sciences, Beijing, China. Email: zblsmile@126.com
Jian Wang, Department of Environment & Economics, Chinese Research Academy of Environmental Sciences, Beijing, China. Email: wangjian01@craes.org.cn

DOI 10.1515/9783110540048-015,

into effect in 2005, and countries promised to control CO_2 emissions (as the representative of greenhouse gas emissions). The Fifth Assessment Report (AR5) issued by the Intergovernmental Panel on Climate Change (IPCC) in May 2014 reaffirmed that human-induced greenhouse gas emissions are the main cause of global warming, and suggested that early implementation of the global emission reductions path is a must to realize the warm-up temperature control target of 2 °C. China and the United States issued a "U.S.-China Joint Statement on Climate Change" in November 2014, and Paris Climate Conference reached a historic agreement on emission reductions in December 2015.

The studies on the theories of externality and public goods are almost the sources of climate change economics.[1] Climate stability is public goods, and the externality of climate change is global. The characteristics of public goods mainly include three aspects: the indivisibility of utility, the non-competition of consumption and the non-competition of benefit.[2] It is obvious that the utility and consumption of the climate change is global, and the public goods attributes can lead to free-rider phenomena and overuse problems. Externality theory can be traced back to the "lighthouse" economic phenomenon. As the "lighthouse" problem reveals, the public goods provided by individuals do not pay directly. Individuals can hardly profit directly from the provision of public goods because of externality.[3] Similarly, each single country can hardly profit directly from the reduction of greenhouse gas emission. Everyone on earth can breathe air for free without willingness to pay for clean air, but polluted air will threaten the life of each person eventually. The atmosphere and climate are the largest commons, and the "tragedy of the commons"[4] is difficult to avoid if externality fails to be treated wisely. That is the reason why externality in climate change issues should be highly concerned.

As externality theory is so universal and abstract, we have few intuitive feelings about the externality in climate change issues. Both CO_2 and PM_{10} are common atmospheric pollutants, but a big difference between them is that the externalities and the impact scope on the atmospheric environment are different: CO_2 is global, while PM_{10} is regional. The empirical studies and comparative analysis on the EKC of CO_2 and PM_{10} will be made in the following chapters, and it is desirable that empirical evidences may reveal the severity of external problems in climate change issues intuitively.

2 Empirical Studies on the EKC of CO_2 and PM_{10}

2.1 Description of Data and Method

For empirical research, data source is the Little Green Data Book from the World Bank (WB), and the annual data from more than 200 countries and regions from

2000 to 2008 has been selected (Because of limited availability of data, and the carbon dioxide emission data during 2000–2008 performs good in data integrity and consistency). Three main variables have been chosen and been denoted as *GNI*, CO_2 and PM_{10}. The specific data indicators are as follows:

a) "*GNI*" (whose SI is "$") indicates the gross national income per capita (current US$, Atlas method). It is caculated in national currency and converted to U.S. dollars using the World Bank Atlas method (at official exchange rates for comparisons across economies), then divided by the midyear population. Gross national income is the sum of value added by all resident producers plus any product taxes (less subsidies) not included in the valuation of output plus net receipts of primary income (compensation of employees and property income) from abroad.
b) "CO_2" (whose SI is "t") indicates the carbon dioxide emissions per capita (metric tons). Carbon dioxide emissions are those stemming from the burning of fossil fuels and the manufacture of cement. They include carbon dioxide producing during consumption of solid, liquid, and gas fuels and gas flaring.
c) "PM_{10}" (whose SI is "$\mu g/m^3$") indicates the concentration of the inhalable particles (micrograms per cubic meter, country level), whose particle diameter is less than 10 microns. This kind of floating particles are always suspended in the air and have a great impact on the visibility of atmosphere.

The Kuznets curve is the doctrine proposed by the economist Kuznets in 1955 when analysing the relationship between per capita income levels and the degree of the income distribution fairness. [5] It presents the phenomenon that income inequality increases first and then decreases with economic growth, showing an inverted U-shaped curve. Grossman and Krueger studied the environmental impact of NAFTA (North American Free Trade Agreement) in Mexico in 1991, and the relationship between environmental quality and per capita income was studied empirically for the first time. [6] The results indicated that the environmental pollution increases first and then decreases with the increase of per capita income, which is inverted U type. In 1996, Panayotou first called this inverted U-shaped relationship between environmental quality and per capita income the Environmental Kuznets Curve (EKC). [7]

The Environmental Kuznets Curve (EKC) has been getting applications widely in the study of the relationship between environmental quality and economic development. EKC hypothesis argues that environmental quality will first deteriorate and then improve with economic growth, which conforms with the classic inverted-U type Kuznets curve. As to the empirical research on EKC currently, the conclusions are different: Some of them support inverted U-shape, while others show N-type, monotonically increasing type and so on.[8] However, one thing is undeniable that the empirical research on EKC is becoming more and more abundant in the field of environmental economics.

For this part, based on EKC, we will study how CO_2 emission and concentration of PM_{10} will change with the economic growth. The performances of carbon emission reduction and economic development may be revealed from historical data, and some enlightenment about greenhouse gas emission reduction and sustainable economic growth can be drawn from the empirical evidences.

2.2 EKC of CO_2

Typically, EKC presents the relationship between environmental quality and economic development visually. In general, the indicators of economic development (such as per capita income or per capita GDP) are taken as the abscissa (*x*-axis), and the indicators of environmental quality (such as the amount of pollution or pollutant's concentration), are used as ordinate (*y*-axis).

For this part, the regression analysis has been made by referring the idea of EKC, and double logarithmic model has been selected finally because of its good properties. The logarithm of the gross national income per capita ($\ln GNI$) and the logarithm of the carbon dioxide emissions per capita ($\ln CO_2$) were taken as independent variables (explanatory variables, *x*) and dependent variables (interpreted variables, *y*) respectively. It is suggested that economic growth is the most important factor of CO_2 emission, and economic growth has one-way impact on long-term CO_2 emission.[9] Economic quantity as an explanatory variable is suitable for *x*-axis.

Using SPSS software for regression analysis, we can get the regression equation as Eq.(1).

$$\ln(CO_2)=-6.024+0.862*\ln(GNI) \tag{1}$$

The regression coefficients can be seen in detail as Table I. In the significance test of regression coefficients, the coefficients passed t-test at significance level of 5 %, which can be seen from the significance value of t-statistics is less than 5 %. The "Sig.=.000" showed in the Table I means the P-value equals "0.000".

Tab. 1: Coefficients for Model I

Model I $\ln(CO_2)=\alpha+\beta*\ln(GNI)+\varepsilon$	Unstandardized Coefficients		Standardized Coefficients	t Statistics	
	B	Std. Error	Beta	T value	Sig.
(Constant)	-6.204	0.1		61.807	.000
ln(*GNI*)	0.862	0.012	0.855	69.147	.000

According to the regression equation, the diagrammatic sketch of EKC of CO_2 can be showed as Figure1. This is just a simple sketch to show the Eq. (1) visually, so the calibration range of the *x*-axis and the *y*-axis does not represent the accurate range of actual range values of GNI and CO_2 respectively.

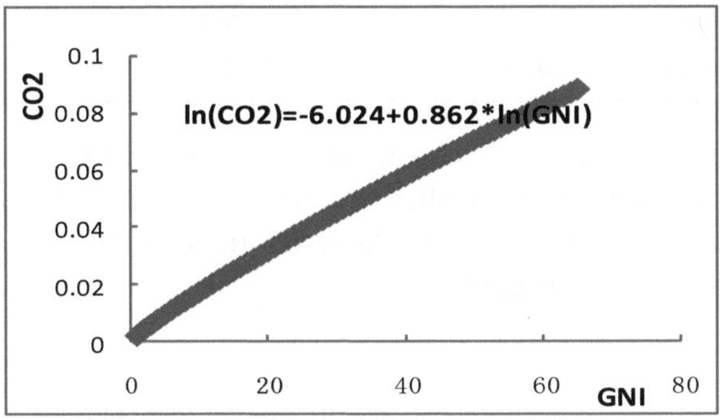

Fig. 1: The diagrammatic sketch of EKC of CO_2

2.3 EKC of PM_{10}

Similarly, the logarithm of the gross national income per capita (ln*GNI*) were taken as independent variables (explanatory variables, *x*), and the logarithm of the concentration of the inhalable particles ($\ln PM_{10}$) were taken as dependent variables (interpreted variables, *y*).

According to the results from the econometric regression analysis based on EKC, the regression equation can be seen as Eq. (2).

$$\ln(PM_{10})=5.256-0.203*\ln(GNI) \qquad (2)$$

Correspondingly, the coefficients can be concluded as Table II. As to the significance test of regression coefficients, the coefficients passed t-test at significance level of 5 %, because the P-value of t-statistics is equal to "0.000", which is showed as "Sig.=.000" in Table II.

Tab. 2: Coefficients for Model II

Model I ln(PM_{10})=α+ β*ln(GNI)+ ε	Unstandardized Coefficients		Standardized Coefficients	t Statistics	
	B	Std. Error	Beta	T value	Sig.
(Constant)	5.256	0.071		73.808	.000
ln(GNI)	-0.203	0.009	-0.491	-22.918	.000

According to the regression equation, the diagrammatic sketch of EKC of PM_{10} can be showed as Figure2. It is just a simple sketch to show the Eq. (2) visually, and the calibration range of the *x*-axis and the *y*-axis does not represent the accurate range of actual range values of GNI and PM_{10} respectively.

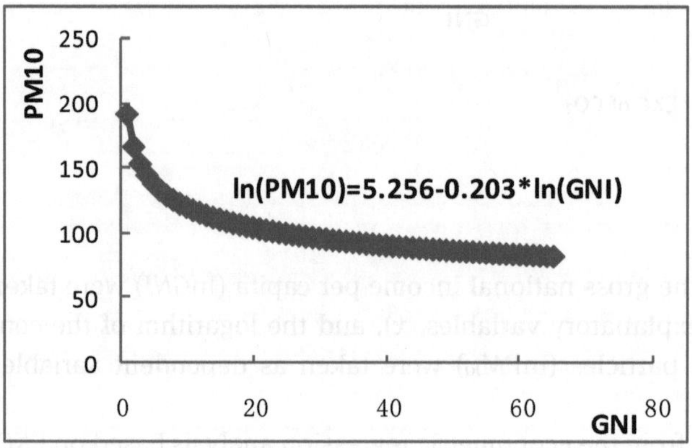

Fig. 2: The diagrammatic sketch of EKC of PM_{10}.

3 Comparison Analysis on CO_2 and PM_{10}

Although both the CO_2 and PM_{10} are commom air pollutants, their variation tendency with economic growth can be quite different. As we can see from the Eq.(1) and Eq.(2), CO_2 and GNI have positive correlation, while PM_{10} and GNI have negative correlation. From Fiure1 and Figure 2, it can be seen obviously that: As GNI increases with the development of economy, the concentration of PM_{10} decreases at a decreasing rate, while the CO_2 emission tends to increase.

The reasons can be mainly summarized as follows:

From the aspect of the difficulty of pollution control, it is more difficult to control CO_2 than PM_{10}. Generally, PM_{10} is much easier to observe and control, and its harm to human health (such as respiratory disease) is more obvious. However, the issue of CO_2 is hard to figure out and control. It is still controversial that whether CO_2 should be listed as real pollutant (because appropriate amount of CO_2 is good, and it can be potentially threatening only when excessive CO_2 causes serious problems such as global warming). Besides, it is still not clear enough that how CO_2 will damage human health because of complex and obscure mechanism. Moreover, CO_2 emission is difficult to control technically because of its universality, and it is concerned that too strict control may cause too much harm to economic development. Only some island countries have a sense of urgency about carbon emissions reduction and climate change mitigation, other developing countries pursue economic growth more urgently.

From the viewpoint of the willingness of pollution control, people always have more willingness to control PM_{10} rather than CO_2. PM_{10} belongs to regional pollutants, and its scope of the spread is smaller. The victims affected by PM_{10} pollution are mainly local people. With the development of the local economy, the local government and residents are increasingly willing to control PM_{10} pollution as long as the condition is available. However, things are quite different since CO_2 is global not regional. When a country emits a large amount of CO_2, even if there is negative externalities definitely, the scope of externalities will spread to the world. The rest of the world have to help the emission country to share the externalities, while the emission country may benefit from their own CO_2 emissions and the domestic economic growth. Thus, all the country have no strong sense of urgency and desire to control CO_2 emission.

In a word, technological and financial constraints can be the reasons, but the most essential reason lies in the externalities.

4 Conclusions and Suggestions

Through the trend of the regression curve, it can be predicted that:

According to the existing conditions, if there is no significant improvements in technological progress (i.e. energy efficiency) and energy consumption structure (i.e. new energy alternative), then economic growth must be supported by energy consumption. Extensive economic growth patterns are common in developing countries, which will inevitably lead to continued energy consumption growth, and this will be accompanied by sustained growth of CO_2 emissions. In other words, for most countries, CO_2 emissions is hard to decrease with economic growth without great reform currently.

The amount of PM_{10} seems to be effectively controlled with economic development, but we had better maintain moderate vigilance from the perspective of environmental protection. Because the harmfulness of PM_{10} is not only related to its concentration, but also connected with its adsorption of harmful substances (e.g. a small amount of $PM_{2.5}$ carrying toxic substances may cause great harm to human health).

It can be concluded that CO_2 emissions will tend to increase globally without special efforts made by the international community.

Considering that carbon emission reduction is very difficult to implement, some recommendations can be made as follows:

Against this global pollution of CO_2, international co-operation should be taken to solve the problem of externality.[10] Through international negotiations and consultations, the probability to solve the problem of externality can be prosperous. The behavior of every country can be restricted, and global carbon emission reduction can be promoted. Through international exchanges and cooperation, the technological advantages of developed countries can spill over to the less developed countries. The technological progress of all countries can be accelerated, and the sustainable development of the global green economy can be motivated.

For the sustainable low-carbon development, the mode of economic development and the structure of energy consumption should be changed. We should try to reform the traditional mode of economic development as well as decrease the dependence on high energy consumption. The enterprises should improve energy efficiency, and the government should encourage the development of new clean renewable energy. Technological progress on low-carbon and energy-saving should be highly encouraged. Meanwhile, industrial structure and energy consumption structure should be adjusted wisely. The economic growth should depend less on fossil energy. Every country should make corresponding preparation for sustainable development and energy economy security.

In order to control CO_2 emissions and promote low-carbon economy, combined efforts from all parties should be highly appreciated. Developed countries should support the reduction of carbon emissions in developing countries in terms of funding and policies. Developing countries should seek a path of sustainable and low-carbon development. Government should strengthen policy guidance, and enterprises should focus on innovative research and development. Low-carbon lifestyles should be advocated widely, and citizens should embrace energy-saving and eco-friendly lifestyle.

To sum up, faced with increasing carbon emssions and rising global warming, it is a must for the international community to cooperate with each other and restrict nagative externality together.

References

[1] XIE Huai-zhu, YU Li-na. The Economics of climate change: a literature review [J]. Journal of Shandong University (Philosophy and Social Sciences), 2010(2): 72–80.
[2] ZHU Chao. Public goods, externality, and climate change - a case of cooperation and game between China and America [D]. East China Normal University, 2011.
[3] JIANG Yue. Global problems of externalities and the way of internalization - example of greenhouse gases [D]. Tianjin University of Finance and Economics, 2013.
[4] Hardin G. The Tragedy of the Commons [J]. Science, 1969, 162(5364): 1243–8.
[5] Kuznets S. Economic growth and income inequality [J].American Economic Review, 1955, 45(1): 1–28.
[6] Grossman G, Krueger A. Environmental Impacts of a North American Free Trade Agreement[C].National Bureau Economic Research Working Paper 3914 ,NBER, Cambridge MA .CEPR Discussion Paper, No 644,1991.
[7] Panayotou T. Empirical Tests and Policy Analysis of Environmental Degradation at Different Stages of Economic Development [C]. Working Paper WP238.
[8] GUI Xiao-dan, LI Hui-ming. Empirical Study of Environmental Kuznets Curve [J]. China Population, Resources and Environment, 2010, S1: 5–8.
[9] E Hong-bing. Factor analysis of influencing China's carbon emissions [D]. Wuhan: Huazhong University of Science and Technology, 2012.
[10] ZHU Yi-bin, LIU Jia-shun. Carbon Emissions and Market Failure--Carbon Negative External Effect and Governance Path [J]. Journal of Low Carbon Economy, 2014, 03: 7–14.

Jian-Feng Luo, Zhen-Yu Yang, Hai-Qing Yi and Zhen-Dong Yang

Study on the Reflection Characteristics of Ceramic Tile Building Materials to Solar Radiation

Abstract: The development of economy and science and technology has accelerated the process of urbanization. In the hot summer season, the absorption of the solar energy by the outer surfaces of the city high buildings has an important influence on the greenhouse gas emission, and it should be as far as possible to be reduced. In this paper, the solar energy absorptive characteristics of ceramic tiles, a common building outer surface decorative material, are studied. With 5780 K's black body to simulate the solar radiation, spectral reflectivity of the tiles in the wavelength range of 190 nm ~ 2600 nm was measured by spectrophotometer. Based on the spectral band model theory, the average reflectivity of ceramic tile is finally obtained. The study shows that the reflectivity of ceramic tiles to the solar energy is generally not high. The maximum is about 60 %, and the lowest reaches to 17.3 %. Thus, it is a kind of strong solar energy absorbing material. Thus, the extensive use of the tiles as outer decorative material should be avoided to save too much solar energy in order to cool the building in a hot summer.

Keywords: greenhouse effect; spectral reflectivity; ceramic tile material; urban building

1 Introduction

With the development of social economy, and science and technology, the process of urbanization has been accelerated. A large number of monster high-rise buildings spring up like mushrooms all over the cities. The absorption of solar energy in the outer surface of a building in summer will result in the increase in indoor temperature, and then it needs air conditioner to cool the indoor temperature by taking

Jian-Feng Luo, College of Mechanical Engineering, Hunan International Economics University, Changsha 410205, China. Email: Luo-jianfeng@263.net
Zhen-Yu Yang, College of Mechanical Engineering, Hunan International Economics University, Changsha 410205, China. Email: Yangzhenyu@163.com
Hai-Qing Yi, College of Mechanical Engineering, Hunan International Economics University, Changsha 410205, China. Email: Yihaiqing@163.com
Zhen-Dong Yang, College of Mechanical Engineering, Hunan International Economics University, Changsha 410205, China. Email: Yangzhendong@163.com

DOI 10.1515/9783110540048-016,

away the heat from the indoor to the outdoor, which is equal to the solar energy absorbed by the outer surface of the building and transferred to the indoor of the building. And as a result, the atmosphere eventually is heated up. On the other hand, the electricity consumption for keeping the air conditioner running is generated by the boiler through burning coal, As we know, this causes carbon dioxide and other greenhouse gas emissions, and directly heating up the atmosphere through coal burning. As a result, the greenhouse effect of the earth is further intensified. Therefore, the absorption of solar energy in the outer surface of the buildings has multiple influences on the greenhouse effect. We should raise people's awareness of the solar energy absorption properties of building materials. Thus, in the construction process of urban buildings, in addition to considering the architectural beauty and practicality, it should be taken into account to reduce the absorption of solar energy in the building outer surface simultaneously.

Energy saving and emission reduction, making full use of wind energy and solar energy to generate electricity and to heat for buildings in winter, etc., are the common measures to control the earth's greenhouse effect by greatly reducing the consumption of fossil fuel. In 2014, Sergio and Wendell [1] investigated that the water solar heating had been used as an alternative for the consumption of natural gas in chemical reactors from cosmetic industries in order to decrease in greenhouse effect gases emission. In 2015, Zuberi and Ali [2] investigated the methane emissions by applying First Order Decay model to landfills in Pakistan over the latest data available by Pakistan Environmental Protection Agency. In 2014, Stanisław et al. [3] summarized important factors that were argued for and against the dangers connected with the emission of carbon dioxide. In 2008, Lam and Chu [4] aimed at solar energy applications in industry for the goal of replacing fossil fuels and studied the absorbing properties of the annual global solar radiation on various inclined planes facing different orientations based on measured sunshine hours. In 2014, Wang et al. [5] reviewed the status of the application of the greenhouse gas reduction policies in China. In 2015, Philippe and Nicks [6] reviewed the studies of the greenhouse gas emissions from pig houses. Energy consumption in residential buildings is one of the major sources of greenhouse gas emissions, and in 2014, Estiri [7], Onat et al. [8] and Heinonen and Junnila [9] studied this problem. In 2015, Wang's studies [10] showed that the steel slag as a soil amendment could enhance crop yield and reduce greenhouse gas emissions in subtropical paddy fields in China.

Ceramic tiles with its colorful decorative effect, smooth surface, convenience of producing and processing and other characteristics, have won the favor of the majority of building designers, and they become very common building exterior decorative materials. In this paper, the radiation of the sun is simulated by a 5780 K's black body's radiation. The thermal process of the exterior wall of building is analyzed. Based on the thermal process analysis, the relationship between the solar energy absorption and greenhouse gas emission is analyzed. Spectral reflectivity of ceramic tile is measured by spectrophotometer. The average reflectivity of the ceramic tile material to the solar radiation is obtained by dividing the spectral band intervals. Thus, the absorption characteristics of the ceramic title to the solar energy can be understood.

2 Analysis of Heat Transfer Process of Building

2.1 Black Body Radiation

Planck's law reveals the relationship among the radiation of black body, the wavelength, and the temperature. It can be expressed as:

$$E_{b\lambda} = \frac{c_1 \lambda^{-5}}{e^{c_2/(\lambda T)} - 1} \left(W / \left(m^2 m \right) \right) \tag{1}$$

Where, λ is the wavelength, m . T Is the black-body temperature, K . C_1=3.742×10^{-16} Wm^2 Is the first radiation constant; $C_2 = 1.4388 \times 10^{-2}$ WK is the second radiation constant.

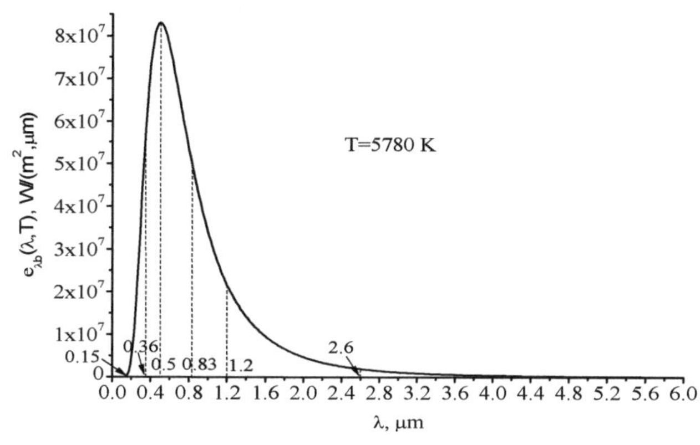

Fig. 1: Variation of the monochromatic radiation of the 5780 K's black body with the wavelength according to the Plank's law.

The Stefan-Boltzmann law can be deduced from Planck's Law:

$$E_b = \int_0^{\infty} E_{b\lambda} d\lambda = \sigma T^4 \quad \left(W / m^2 \right) \tag{2}$$

Where, $\sigma = 5.67\times10^{-8}\,W/(m^2K^4)$ is the Stefan-Boltzmann constant.

If the sun is modeled as a black body with temperature of 5780 K, and then according to the Planck's law, the solar radiation energy can be drawn with the change of wavelength, as shown in Fig.1. As shown in the figure, the maximum value of solar radiation energy is located in the visible light wavelength band (0.38 μm ~ 0.76 μm), which is about at the wavelength of 0.5 μm.

2.2 Radiation Characteristics of Real Objects

2.2.1 A Very Complex Function with the Wavelength for the Radiation Characteristic Parameters of the Real Object

The black body as well as the gray body is ideal object. The radiation characteristic parameters are independent of the wavelength. However, the radiation characteristic parameters of the actual object, such as reflectivity ρ, show a very complex change with the change of wavelength.

The radiation energy, emitted by a black body of temperature T and near the unit wavelength interval at the wavelength λ, is $E_{b\lambda}$, i.e., the Plank'slaw Eq.2. When this energy is incident on the opaque surface of the object, a part q'_λ is reflected, and the other part $q''_\lambda(=E_{b\lambda}-q'_\lambda)$ is absorbed. Thus, the average spectral reflectivity ρ_λ and the average spectral absorptivity α_λ of the opaque surface of the real object can be respectively defined as

$$\rho_\lambda = q'_\lambda / E_{b\lambda} \tag{3}$$

$$\alpha_\lambda = (E_{b\lambda} - q'_\lambda)/E_{b\lambda} = 1 - \rho_\lambda \tag{4}$$

According to the Kirchhoff's law, the spectral emissivity of an object is equal to its spectral absorptivity, that is:

$$\varepsilon_\lambda = \alpha_\lambda = 1 - \rho_\lambda \tag{5}$$

The spectral reflectivity of the real object, ρ_λ, is not only related to the incident direction, but also related to the reflection direction. Thus, Eq. 3 is usually used to define the average spectral reflectivity, which is equivalent to assume that the surface of the actual object is a diffuse reflection surface.

The total energy reflected by the actual object to that emitted by a black body within full wavelength region ($\lambda \in (0,\infty)$) is:

$$q_r = \int_0^\infty q'_\lambda d\lambda = \int_0^\infty \rho_\lambda E_{b\lambda} d\lambda \tag{6}$$

The energy emitted by the black body within full wavelength region is the Eq. 2, and the average reflectivity of the real object to the radiative energy emitted by the black body at full wavelength region can be expressed as follows:

$$\rho = q_r / E_b = \int_0^{\infty} \rho_\lambda E_{b\lambda} d\lambda \Big/ \sigma T^4 \tag{7}$$

The average absorptivity of the real object to the radiative energy emitted by the black body within full wavelength region can be derived:

$$\alpha = 1 - \rho = 1 - \int_0^{\infty} \rho_\lambda E_{b\lambda} d\lambda \Big/ \sigma T^4 \tag{8}$$

According to the Kirchhoff's law, the emissivity of the object is equal to its absorptivity:

$$\varepsilon = \alpha = 1 - \rho \tag{9}$$

2.2.2 Using Spectral Band Model to Describe the Radiation Characteristics of Real Object

The radiation characteristic parameter of the actual object is a very complex function of wavelength λ. It is very difficult to deduce a mathematical expression to describe this relationship. Thus, the Eq. 7 in fact cannot be analytically solved. In order to solve this problem, the concept of spectral band model is usually used, as shown in the following.

Assume in the whole wavelength region ($\lambda \in (0, \infty)$) the radiation characteristic parameters of the object can be divided into N_1 bands. Within the *k*th band, the spectral reflectivity ρ_λ can be considered as constant ρ_k , where the subscript k represents the *k*th region of the spectral band model. Then, in the spectral band *k*, the radiative energy emitted by the black body at this spectral band and reflected by the object is:

$$q_{r,k} = \int_{\lambda_1}^{\lambda_2} \rho_\lambda E_{b\lambda} d\lambda = \rho_k \int_{\lambda_1}^{\lambda_2} E_{b\lambda} d\lambda \quad \left(W/m^2 \right) \tag{10}$$

In Eq. 10, for the reason that the spectral reflectivity ρ_λ can be considered as constant within the wavelength interval $[\lambda_1, \lambda_2]$ of the *k*th spectral band. It can be extracted from the inner of the integral sign to the front of the integral sign. Then inside the integral sign is the Planck's formula, which can be integrated. The spectral band model greatly simplifies the complexity of the problem.

Define A_k to be the quotient of the radiative energy emitted by the black body at the spectral band k to that emitted at the whole wavelength band, i.e.

$$A_k = \int_{\lambda_1}^{\lambda_2} E_{\lambda b} d\lambda \Big/ \left(\sigma T^4\right) \tag{11}$$

In above equation, as long as the upper limit of integral λ_1 and lower limit of integral λ_2 are known, the value of A_k can be solved, which is between 0 and 1. With A_k, Eq. 10 can be rewritten as follows:

$$q_{r,k} = \rho_k A_k \sigma T^4 \quad \left(W/m^2\right) \tag{12}$$

The radiative energy emitted by the black body within full wavelength band and reflected by the real object can be expressed as:

$$\int_0^{\infty} \rho_\lambda E_{b\lambda} d\lambda \approx \sum_{k=1}^{N_1} q_{r,k} = \sum_{k=1}^{N_1} \rho_k A_k \sigma T^4 \tag{13}$$

Thus, substitution of the above results into the Eq. 7 gets the average reflectivity of the object to the radiative energy emitted by the black body within the full wavelength region:

$$\rho = \sum_{k=1}^{N_1} \rho_k A_k \tag{14}$$

2.3 Analysis of Building Heat Transfer Process

In hot summer, and in the southern part of China, the building interior usually runs air conditioner to cool, as shown in Fig. 2. Assume the radiative energy emitted by the sun at wavelength λ to the external surface of the outer wall of the building is $Q_\lambda \left(W/\left(m^2 m\right)\right)$. However, assume part $q_{\lambda,a}$ of this energy to be absorbed by the outer surface, and the other part $q_{\lambda,r}$ to be reflected. So we have:

$$q_{\lambda,r} = \rho_\lambda Q_\lambda \quad \left(W/\left(m^2 m\right)\right) \tag{15}$$

$$q_{\lambda,a} = \alpha_\lambda Q_\lambda = \left(1-\rho_\lambda\right) Q_\lambda \quad \left(W/\left(m^2 m\right)\right) \tag{16}$$

where, ρ_λ and α_λ are the spectral reflectivity and absorbtivity of the outer surface of the building wall respectively. Integrating the two equations above with wavelength λ, we can get the radiative energy reflected and absorbed within the full wavelength region:

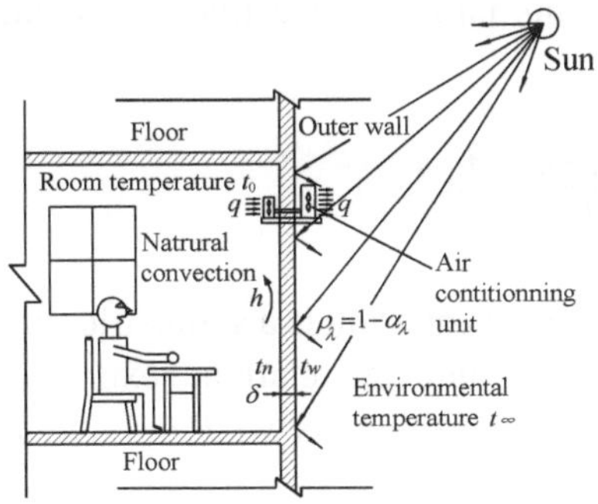

Fig.2: Diagrammatic sketch of heat transfer process of building in summer

$$q_r = \int_0^\infty q_{\lambda,r} d\lambda = \int_0^\infty \rho_\lambda Q_\lambda d\lambda \quad \left(W/m^2\right) \tag{17}$$

$$q_a = \int_0^\infty \alpha_\lambda Q_\lambda d\lambda = \int_0^\infty (1-\rho_\lambda) Q_\lambda d\lambda \quad \left(W/m^2\right) \tag{18}$$

Although the sun can be seen as a black body of 5780K, the radiative energy emitted by the sun will be absorbed and scattered by ozone, carbon dioxide, carbon monoxide, water vapor, aerosol particle etc. of the atmosphere of the earth, when it transfers through the atmosphere. Thus, when it reaches the ground of the earth, it cannot satisfy the Planck's law, Eq. 1, and the specific expressions need to be further analyzed and deduced.

The more the solar energy absorbed by the outer surface of the outer wall, the more the heat transferred to the interior of the room. For simplicity and without loss of generality, we assume that there is no heat exchange between the outer surface of the outer wall and the surrounding environment, that is, the solar energy q_a absorbed by the outer surface of the outer wall is totally transferred to the room, and then we get:

$$q_a = k(t_w - t_n)/\delta \quad \left(W/m^2\right) \tag{19}$$

Where, k is the conductivity of the wall, $W/(mK)$. δ is the thickness of the wall, *m*. t_w and t_n are the outer and inner surface temperatures of the wall, respectively.

When heat is transferred to the inner surface of the wall, the temperature of the inner surface is increased, and the heat will further be transferred to the indoor air of the room from the inner surface by natural convection, i.e.:

$$q_a = \frac{k(t_w - t_n)}{\delta} = h(t_n - t_0) \quad \left(W/m^2\right) \tag{20}$$

where, t_0 is the indoor temperature. h is the natural convective coefficient. Eq. 20 can be rewritten into the following form:

$$q_a = \frac{t_w - t_n}{\delta/k} = \frac{t_n - t_0}{1/h} \quad \left(W/m^2\right) \tag{21}$$

The two equations on the right end of the above equation is a geometric series. Thus, summing the numerator and the denominator yields the relational expression of the solar energy q_a absorbed by the outer surface of the outer wall and transmitted to the interior of the room:

$$q_a = \frac{t_w - t_0}{\delta/k + 1/h} \quad \left(W/m^2\right) \tag{22}$$

As can be seen from the above formula, and in the case of other parameters unchanged, the more solar energy q_a the outer surface of the outer wall absorbs, the higher the temperature t_w of the outside surface of the outer wall.

In order to maintain the comfortable indoor temperature t_0, one runs the air conditioning unit that releases heat to the outdoor environment, as shown in Fig. 2. How much heat into the room, how much heat is transferred out of the room. So, in Fig. 2, the heat q discharged by the air conditioner to the outside environment must be equal to the heat q_a transferred into the room, i.e.:

$$q = q_a \quad \left(W/m^2\right) \tag{23}$$

Air conditioning unit operation needs to consume electric energy, which is generated by the boiler burning coal to produce high temperature and high pressure steam to drive turbine generator. The boiler burning coal discharges the heat to the atmosphere environment, and directly heat the atmosphere, and at the same time, carbon dioxide, sulfur dioxide, water vapor and other greenhouse gases are also to be created, which prevent the earth to loss heat into space environment through radiation. The more solar energy transferred into the room, the more the coal will be burnt to get rid of the heat, and the more the combustion heat will be produced, and the more the greenhouse gas will be emitted. Therefore, the solar energy q_a transferred into the interior of the room will eventually affect the earth's greenhouse effect in the following three aspects:

(1) Make the heat q_a transferred into the indoor to be entirely released to the atmosphere environment.
(2) Directly heat the atmosphere by coal combustion, which is needed for the electric energy consumption of the air conditioner.
(3) Produce greenhouse gases, which prevent the earth releases radiative energy to the outer space.

The influence of the above three aspects leads to the increase of atmospheric temperature, and the q_a of the article (1) plays a decisive role and determines the effect of the article (2) and (3). So it should be as much as possible to reduce the absorbed solar energy q_a transferred into the interior of the room. And from Eq. 18 transferred into the room. While, there is $\alpha_\lambda = 1 - \rho_\lambda$, therefore, we should try to increase the spectral reflectivity of the outer surface of the outer wall ρ_λ as far as possible.

In general, the spectral absorption and reflection characteristics of the building outer surface material to the solar energy are not considered in the construction of the building, which only the beauty and practicability of the building are considered. The purpose of this paper is to make the sample of common building outer surface material, to measure the change of spectral reflectivity with wavelength and so as to improve people's understanding of the solar energy absorption characteristics of the outer surface decoration materials of the building. So then realize which material absorbs more solar energy, and which material absorbs less solar energy.

3 Average Reflectivity of External Wall of Ceramic Tile to Solar Energy

Ceramic tile is one of the most common building exterior decoration materials, with its colorful decorative effect to make the building to look very beautiful. In fact, ceramic tile is made from refractory metal oxide and semi metal oxides by grinding, mixing, pressing, glazing and sintering, with resistance to acid and alkali.

Appendix Fig. 1 shows 33 kinds of ceramic tiles with different colors, commonly used building materials in china and numbered from t1 to t33. The last 3 kinds of tiles are different from other ceramic tiles, in which the surface is not smooth and there are concave and convex ups and downs, is used for the study of the spectral absorption characteristics of the uneven surface tiles to solar energy.

The instrument used for measuring the spectral reflectivity is Spectrophotometer U-4100 (Hitachi). The measurement wavelength range of the instrument is 190 nm~2600 nm. Calculations show that the radiation energy emitted by the black body of 5780K (simulated solar radiation) in this wavelength range is 96.8 % of the total radiation energy emitted in the full wavelength range ($0 \sim \infty$). Thus, the wavelength range which we have studied, has fully covered the radiation wavelength range of the sun.

3.1 Average Reflectivity of the Light White Ceramic Tiles

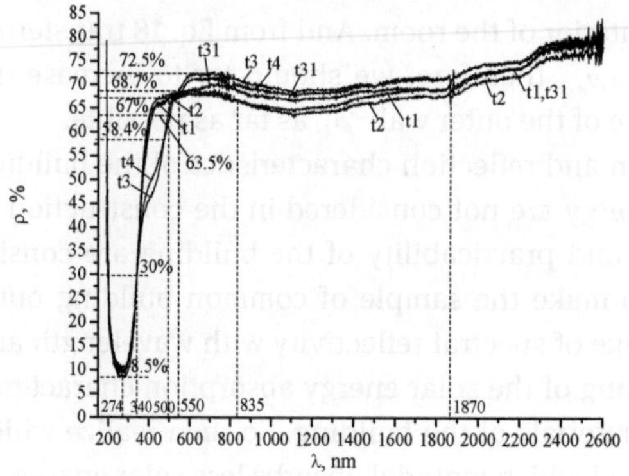

(a) t1~t4, t31

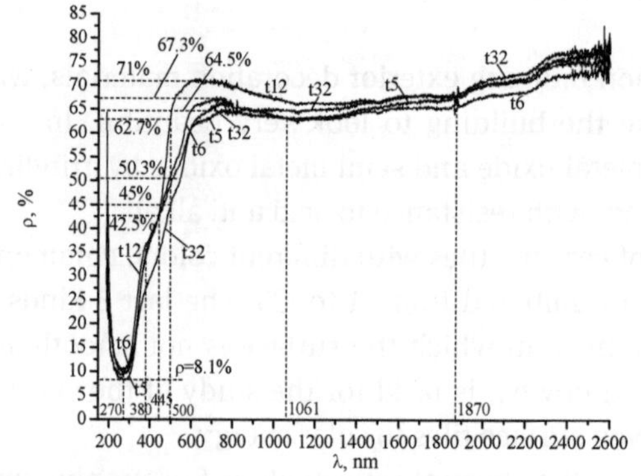

(b) t5, t6, t12, t32

Fig. 3: Spectral reflectivity of light white ceramic tile (t1~t6, t12, t31, t32)

As shown in Fig. 1 of Appendix A, the color of the nine ceramic tiles, t1 to t6 as well as t12, t31 and t32, is light white, and the changes of the spectral reflectivity with wavelength of these nine kinds of ceramic tile are shown in Fig. 3, which are similar to each other. Here we take tile t1 as an example to illustrate the solving process of average reflectivity of these nine ceramic tiles. The spectral band interval division of ceramic tile t1 is shown in Fig. 4 and Table 1. A_k can be calculated by Eq.11(the same as this in the following), as shown in Tab. 3. The average reflectivity of ceramic tile t1 to the solar energy can be obtained from Eq. 14:

$$\rho_{t1} = \sum_{k=1}^{12} \rho_k A_k = 60.5\% \tag{24}$$

The solving process of average reflectivity of the other eight ceramic tiles is similar to that of tile t1, and for simplicity we don't give the detailed solving process here, and their results are shown in Tab. B-1 of Appendix B. As can be seen from the above analysis that, the average reflectivity of various ceramic tiles of Fig. 3 (b) to the solar energy is less than the average reflectivity of the ceramic tiles of Fig. 3 (a) to the solar energy.

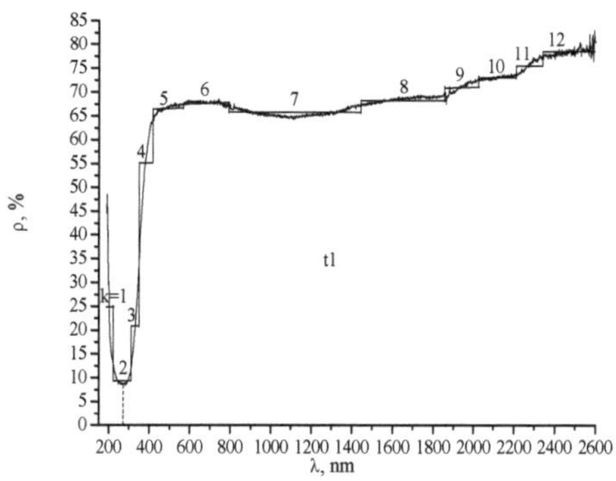

Fig. 4: Spectral band interval division of spectral reflectivity of ceramic til t1

Tab. 1: Spectral band interval division of spectral reflectivity of ceramic tile t1

Spectral band *k*	**Wavelength range, λ_1~λ_2, nm**	**Reflectivity ρ_k, %**	**A_k, %**
1	190~230	25.0	0.4
2	230~310	9.5	3.3
3	310~350	20.8	3.2
4	350~417	55.0	7.2
5	417~567	66.4	19.3
6	567~790	67.7	23.9
7	790~1443	65.8	29.5
8	1443~1855	68.4	5.8

Spectral band k	Wavelength range, $\lambda_1 \sim \lambda_2$, nm	Reflectivity ρ_k, %	A_k, %
9	1855~2025	71.0	1.4
10	2025~2211	73.0	1.2
11	2211~2340	75.5	0.6
12	2340~2600	78.4	1.0

3.2 Average Reflectivity of the Light Gray Ceramic Tiles

As shown in Fig. A-1 of Appendix A, the color of the three ceramic tiles, t7, t8 and t9, is light gray. The changes of the spectral reflectivity with wavelength of these ceramic tiles are shown in Fig. 5, which are similar to each other. The calculation of the average reflectivity of these three ceramic tiles is similar to the above and the results are shown in Appendix b.

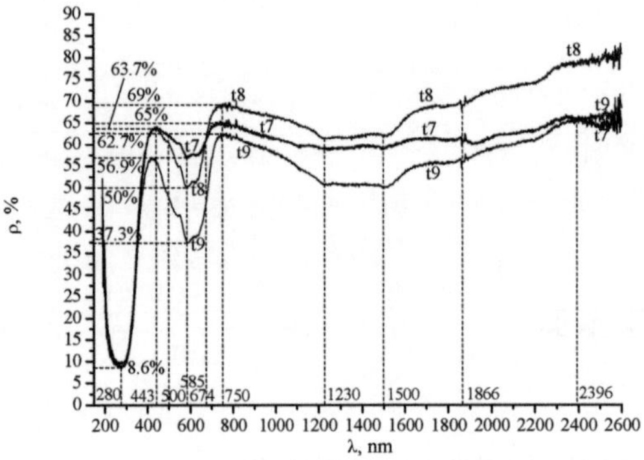

Fig. 5: Spectral reflectivity of light gray ceramic tiles (t7, t8, t9)

3.3 Average Reflectivity of the Dark Gray Ceramic Tiles

As shown in Fig. A-1 of Appendix A, the color of the three ceramic tiles, t10, t11 and t28, is dark gray, and the changes of the spectral reflectivity with wavelength of this three ceramic tiles are shown in Fig. 6. The calculation of the average reflectivity of these three ceramic tiles is similar to the above and the results are shown in Appendix b.

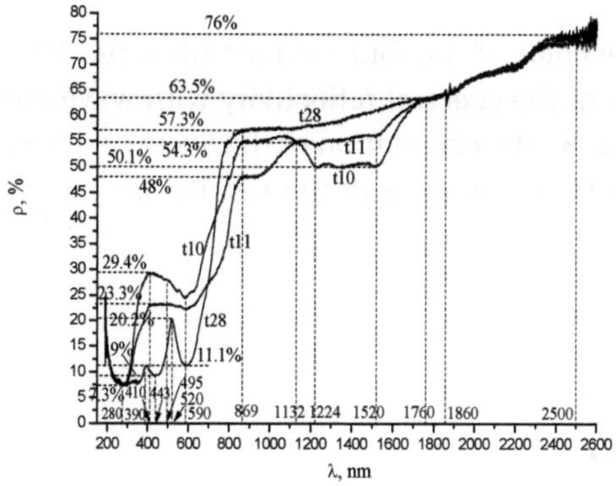

Fig. 6: Spectral reflectivity of dark gray ceramic tiles (t10, t11, t28)

3.4 Average Reflectivity of the Dark Light Yellow Ceramic Tiles

As shown in Fig. A-1 of Appendix A, the color of the three ceramic tiles, t13, t19 and t20, is light yellow, and the changes of the spectral reflectivity with wavelength of this three ceramic tiles are shown in Fig. 7. The calculation of the average reflectivity of these three ceramic tiles is similar to the above and the results are shown in Appendix b.

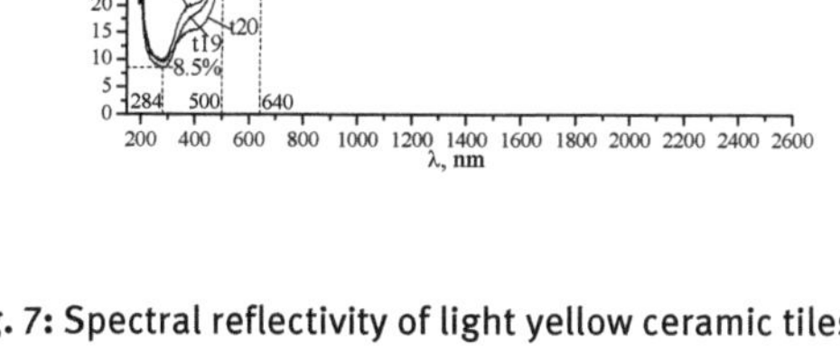

Fig. 7: Spectral reflectivity of light yellow ceramic tiles (t13, t19, t20)

3.5 Average Reflectivity of the Light Brown Ceramic Tiles

As shown in Fig. A-1 of Appendix A, the color of the four ceramic tiles, t14, t15, t21 and t22, is light brown, and the changes of the spectral reflectivity with wavelength of this four ceramic tiles are shown in Fig. 8. The calculation of the average reflectivity of these four ceramic tiles is similar to the above and the results are shown in Appendix b.

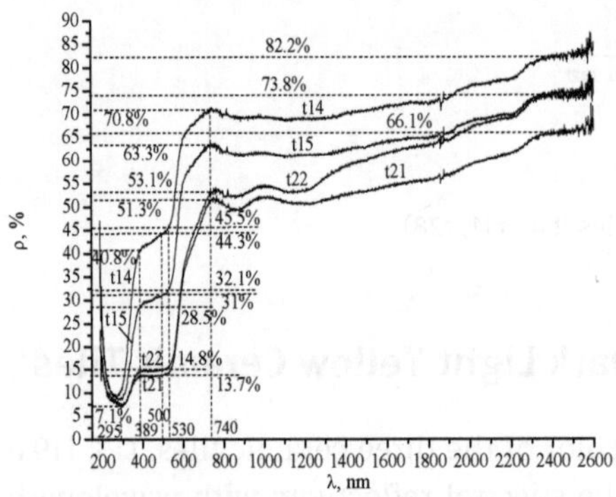

Fig. 8: Spectral reflectivity of light brown ceramic tiles (t14, t15, t21, t22)

3.6 Average Reflectivity of the Light Brown Ceramic Tiles

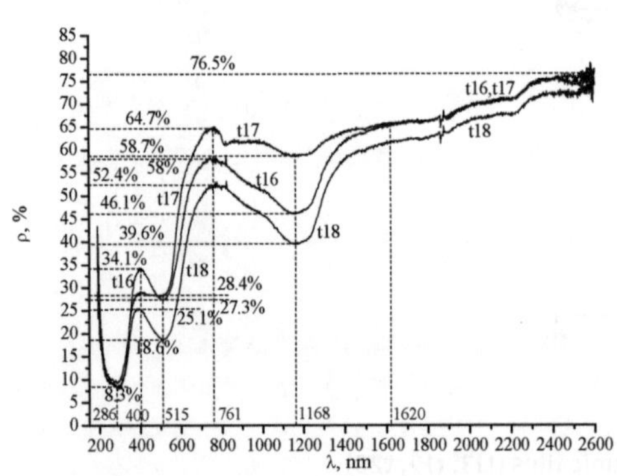

Fig. 9: Spectral reflectivity of light brown ceramic tiles (t16, t17, t18)

As shown in Fig. A-1 of Appendix A, the color of the four ceramic tiles, t16, t17, and t18, is light brown, and the changes of the spectral reflectivity with wavelength of these three ceramic tiles are shown in Fig. 9. The calculation of the average reflectivity of these three ceramic tiles is similar to the above and the results are shown in Appendix b.

3.7 Average Reflectivity of the Dark Brown Ceramic Tiles

As shown in Fig. A-1 of Appendix A, the color of the three tiles, t25, t26, and t33, is dark brown, and the changes of the spectral reflectivity with wavelength of this three ceramic tiles are shown in Fig. 10. The calculation of the average reflectivity of these three ceramic tiles is similar to the above and the results are shown in Appendix b.

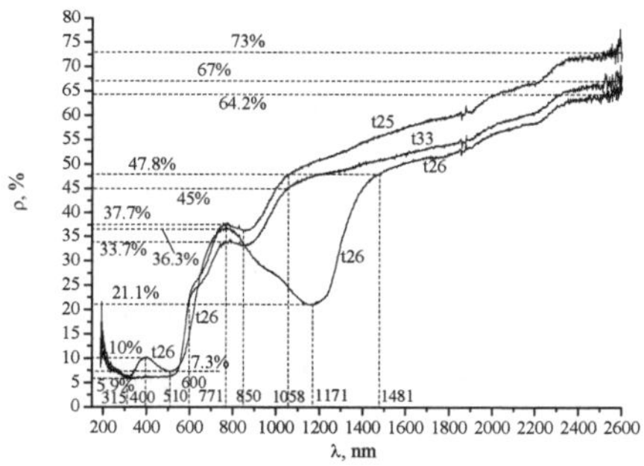

Fig. 6: Spectral reflectivity of dark brown ceramic tiles (t25, t26, t33)

For simplicity, the study of average reflectivity of the other left ceramic tiles to solar energy is not shown here, and the results are shown in Appendix b.

3.8 Summary of This Section

The average reflectivity of 33 kinds of common ceramic tiles sold in the market to solar energy is studied in this section, and the summary of the study is shown in Tab. B-1 of Appendix B. As shown in the table, although ceramic tile is a kind of material which is frequently used to decorate the outer surface of a building, its reflectivity to solar energy is not high. The highest is about 60 %. The lowest reaches 17.3 % respectively for ceramic tile t29 and t30, and the color of the two ceramic tiles is also the darkest, where t29 is black ceramic tile. From the point of view of color, the color of ceramic tile is shallower, its reflectivity is higher; the color is darker, the reflectivity is lower. The black tile is of the lowest reflectivity, which is in line with people's common perception.

4 Conclusions

In this paper in the process of urbanization, absorption characteristics of common external surface materials of urban buildings to solar energy has been studied, in order to improve people's understanding of the relationship between the absorption of solar energy of external surface of buildings and greenhouse gas emissions, to provide a theoretical basis to adopt more measures to reduce greenhouse gas emissions. In the summer, it should be possible to reduce the solar energy absorption of the external surface of urban buildings. For the sake of simplicity, this study does not consider the interaction between the solar energy and the atmosphere. It can be seen from the research of this paper: the reflectivity of ceramic tiles to the solar energy is not high, the maximum is about 60 % and the lowest reaches 17.3 %; from the point of view of color, the color of ceramic tile is more light, its reflectivity is higher; the color is darker, the reflectivity is lower.

Acknowledgement: This research is supported by the National Natural Science Foundation of PR China (Project 51276193), and the authors thank the Center of Material Science of the National University of Defense Technology for providing the measuring and test instruments.

References

[1] S.A. de Souza, and W. de Queiroz Lamas, "Thermoeconomic and ecological analysis applied to heating industrial process in chemical reactors," Renewable and Sustainable Energy Reviews, vol. 29, pp. 96–107, 2014.
[2] M.J.S. Zuberi, and S.F. Ali, "Greenhouse effect reduction by recovering energy from waste landfills in Pakistan," Renewable and Sustainable Energy Reviews vol. 44, pp. 117–131, 2015.
[3] S. Gumuła, K. Pytel, and M. Piaskowska-Silarska, "Polemical Remarks to the Claim that Carbon Dioxide Strengthens the Greenhouse Effect in the Atmosphere," Pol. J. Environ. Stud., vol. 23, no. 6, pp. 2321–2325, 2014.
[4] D.H.W. Li, T.N.T. Lam, and V.W.C. Chu, "Relationship between the total solar radiation on tilted surfaces and the sunshine hours in Hong Kong," Solar Energy, vol. 82, 1220–1228, 2008.
[5] T. Wang, G. Foliente, X. Song, J. Xue, and D. Fang, "Implications and future direction of greenhouse gas emission mitigation policies in the building sector of China," Renewable and Sustainable Energy Reviews, vol. 31, pp. 520–530, 2014.
[6] F.X. Philippe, and B. Nicks, "Review on greenhouse gas emissions from pig houses: Production of carbon dioxide, methane and nitrous oxide by animals and manure," Agriculture, Ecosystems and Environment, vol. 199, pp. 10–25, 2015.

[7] H. Estiri, "Building and household X-factors and energy consumption at the residential sector A structural equation analysis of the effects of household and building characteristics on the annual energy consumption of US residential buildings," Energy Economics, vol. 43, pp. 178–184, 2014.

[8] N.C. Onat, G. Egilmez, and O. Tatari, "Towards greening the U.S. residential building stock: A system dynamics approach," Building and Environment, vol 78, pp. 68–80, 2014.

[9] J. Heinonen, and S. Junnila, "Residential energy consumption patterns and the overall housingenergy requirements of urban and rural households in Finland," Energy and Buildings, vol. 76, pp. 295–303, 2014.

[10] W. Wang, J. Sardans, D.Y.F. Lai, C. Wang, C. Zeng, C. Tong, Y. Liang, and J. Peñuelas, "Effects of steel slag application on greenhouse gas emissions and crop yield over multiple growing seasons in a subtropical paddy field in China," Field Crops Research, vol. 171, pp. 146–156, 2015.

Appendix a Ceramic Tile Building Material

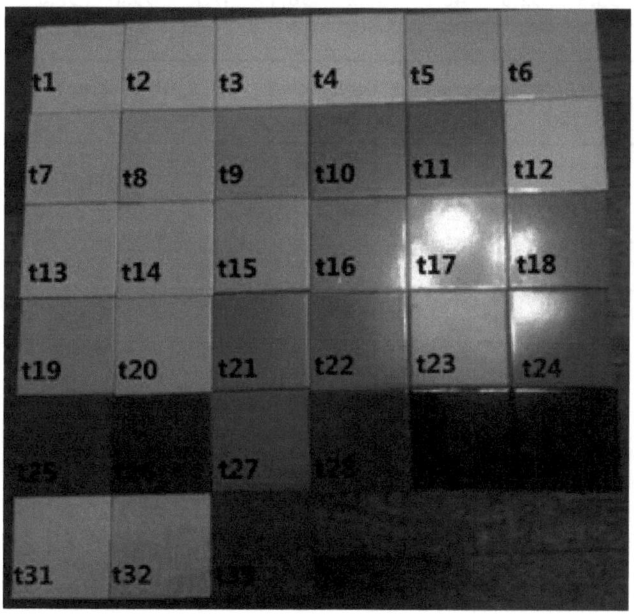

Fig. A-1: 33 Ceramic tiles with different colors bought from the building materials market

Appendix b the Average Reflectivity ρ (%) of Ceramic Tiles to Solar Energy

	1, Light white ceramic tile									2, Light gray ceramic tile		
Ceramic tile, tn	t1	t2	t3	t4	t31	t5	t6	t12	t32	t7	t8	t9
Reflectivity, %	60.5	60.4	60.9	60.1	61.5	53.4	54.3	58.5	52.5	55.6	57.1	48.1

	3, Dark gray ceramic tile			4, Light yellow ceramic tile			5, Light brown ceramic tile			
Ceramic tile, tn	t10	t11	t28	t13	t19	t20	t14	t15	t21	t22
Reflectivity, %	38.0	34.2	33.5	52.5	48.1	52.2	56.4	47.7	35.4	37.6

	6, Light brown ceramic tile			7, Dark blue ceramic tile			8, Dark brown ceramic tile			9, Other ceramic tile		
Ceramic tile, tn	t16	t17	t18	t23	t24	t30	t25	t33	t26	t27	t28	t29
Reflectivity, %	43.1	48.3	37.2	36.6	35.1	17.3	28.4	26.4	23.4	38.4	33.4	17.3

Xiao-Qian Bai, Feng Luo, Peng-Pai Yu, Zhong-You Pei and Shou-Jun Sun*

Research Status of Sorghum Straw Utilization

Abstract: With the development of agriculture, although the level of China's grain production and production continues to increase, but the crop straw utilization is still in the lower case. This paper describes the present stage of utilization of crop stalks and straw at home and abroad for the treatment, as well as for domestic utilization of sorghum.

Keywords: Sorghum, stalks, Utilization status, Renewable energy

1 Introduction

Sorghum is one of the earliest cultivated cereal crops in China. It originated in Africa and has the characteristics of drought resistance, water logging tolerance, barrenness, salinity tolerance, disease resistance and insect pest. It is an annual herbaceous plant with a long history of cultivation in China. Can be planted in arid climate and barren soil. Because of its high photosynthetic efficiency, high biological yield and wide adaptability, it is widely used as feed and energy crops in China. For a long time, sorghum is one of the main food crops of the people in the north [1–2]. Sorghum can be classified into sorghum, sweet sorghum, and sudangrass according to its characteristics and uses. In China, grain sorghum is mainly used for food, wine and feed; sweet sorghum biological yield, sugar rich stem, not only can provide high-quality feed, but also as raw material production of industrial ethanol, "high-energy crops" Said it is a high quality feed crop, sugar crop and renewable energy crops [3–6]. Sudan grass is a kind of high-quality pasture, with high protein content, nutrient-rich, good palatability characteristics of livestock [7–9].

2 Overview

Straw is a general term for the stem and leaf part of mature crop. It usually refers to the remaining stem and leaf part of wheat, rice, maize, potato, rape, cotton, sugar-

Xiao-Qian Bai, Agriculture Resourde and Environment College, Tianjin Agricultural University, Tianjin, China. Email: 1044235123@qq.com
Shou-Jun Sun, Agriculture Resourde and Environment College, Tianjin Agricultural University, Tianjin, China. Email: sunshoujun@tjau.edu.cn

DOI 10.1515/9783110540048-017,

cane and other crops. The straw quality traits include crude protein (CP), crude fat (EE), soluble sugar (WSC), acid detergent fiber (ADF), acid lignin (ADL), neutral detergent fiber [10]. Improve the feeding quality of stems, straw can be used as ruminant feed, can effectively alleviate the problem of feed shortages.

3 The Purpose and Significance of the Study

Crop stalks are the fourth largest energy source in the world after coal, oil and gas [11–13]. Due to the waste and shortage of energy resources in recent years, the use of waste materials such as crop stalks to produce energy has been extensively studied Concern, not only increased the production of new energy sources, reducing production costs, but also reduce the process of dealing with straw pollution generated by environmental problems.

As a big country of agricultural production, China has abundant straw resources, the annual output of straw is about 900 million tons, rational use of straw is conducive to the rational development of agriculture, industry and rural economy. If the straw is directly burnt, not only release a lot of harmful gas pollution Environment, kill the soil microorganisms to change the soil structure, but also a direct result of waste of resources. At present, China's stalks for stalks stagnated in the way, livestock feed and building materials, for the innovative use of straw to be further developed.

4 Research Status at Home and Abroad

4.1 Overview of Utilization of Straw in Foreign Countries

According to the main crop harvest index and its annual agricultural production statistics, estimated by the Food and Agriculture Organization (FAO), the total output of straw in 2012 is 5.081 billion tons, of which the total output of straw is 940 million tons, The United States, India, Brazil and other countries 15 more than 500 million tons of straw in the country, the total output of 2.875 billion tons of straw, accounting for 56.58 % of the world's total. Straw production in the world's first straw production, accounting for 18.50 % ; Straw yield of less than 50 million t of other countries, the total output of 1.266 billion tons of straw, accounting for 24.92 % of the world [14].

Abroad, the main way for the use of straw is straw recycling, developed countries are focusing on fertilization structure, basically formed a direct return of straw and other systems for the application of straw more fully, basically avoid the straw waste and open burning The problem.

In the developed countries, through scientific and technological progress and innovative technologies, a variety of ways have been found for the comprehensive utilization of crop straws. In addition to the traditional methods of crushing straw into organic fertilizers, there are concentrated in emerging energy sources, such as straw feed, straw Power generation, and straw ethanol.

4.2 Overview of Domestic Straw Utilization

In China, crop straw is very rich, about 900 million t each year, of which straw, wheat straw, corn stalks accounted for more than 70 % of these three stalks. In the domestic straw, livestock is a very traditional straw, the use of high efficiency, straw through silage, ammoniation, micro-storage treatment and other methods to feed livestock to promote the development of animal husbandry while the straw Return to field, increased soil fertility, is conducive to the next season crop growth and development [15–17].

In southern China, people use rice straw and other crops for storage, can be used for weaving brooms, cushions, mattresses and other household goods, you can also feed livestock, poultry compost to the field, and even as a simple housing roof, However, due to the popularity of coal, electricity, natural gas and increasingly abundant daily necessities in China in recent decades, there has been a decrease in the demand for straw, and a large amount of straw has been reduced in quantity, and the production of straw has been reduced. Straw residue, and even some farmers in the field directly to the straw burning, on the one hand a large number of harmful gases released on the environment causing serious pollution, easily lead to fire, on the other hand a serious waste of resources.

5 Development and Discussion

In recent years, the main grain-producing areas in China there have been more serious phenomenon of open burning straw, this approach is not only a waste of natural resources, causing environmental pollution, but also seriously endanger human health and safety; the state has enacted relevant laws, the relevant departments Much work has been done to order the prohibition of open burning, although some success has been achieved, but such problems remain serious.

Crop straw is a valuable renewable energy, in addition to straw, power generation, building materials to do, effectively improve the content of nutrients in straw and feed quality, improve palatability, can effectively promote the development of animal husbandry, but also improve straw utilization, reduce waste.

The main quality of various sorghum for different purposes is mainly concentrated in sugar hammer degree, stalk energy and so on. The content of reducing sugar can reflect the product quality and economic value of sorghum to some extent, which is one of the important indexes to evaluate the quality of sorghum. The crude protein of sorghum mainly accumulates in grain and stalk, the current research on stem is relative the research on crude protein of stalk is in the primary stage. At present, most of the research on the crude protein of sorghum is focused on sorghum grain. With the transformation of sorghum use, people pay more and more attention to the production and utilization of sorghum. More and more attention has been paid to the nutrient composition of stem. Therefore, a large number of sorghum varieties with different genotypes were selected to determine the crude protein content in stems, and the comparative analysis showed that the heritability of crude protein content of stems was higher, which could provide better and more choices for breeding new sorghum .

China sorghum stalks yield more each year, but the utilization rate is low, a serious waste of resources, especially sweet sorghum after the production of industrial ethanol residue of sweet sorghum residue, which contains large amounts of nutrients, sweet sorghum is also a very important feed Crops are not only of high nutritional value but also of high biological yield, and their yields of silage are much higher than those of other crops such as corn [18]. Is the preferred crop for the production of silage in water-scarce areas [19–20].

Sorghum stalks are rich in nutrients, including crude protein, reducing sugar and other nutrients, can be used as ruminant roughage to be used to alleviate the shortage of ruminant feed problems. Green feed is rich in nutrients, ruminants like to feed, but because of its high water content is not easy to store. In order to be able to save the green for a long time the nutritional content of feed, you can take two methods to deal with green feed to achieve the purpose of storage. One is the green feed dehydrated hay made; the other is the addition of silage to green feed into the silage. This can adjust the supply of green feed seasonal problems. Silage after the green feed, not only to maintain a good green feed nutrients and water, and in silage fermentation, will produce large amounts of aromatic compounds, it has an aromatic smell and soft and juicy, improved palatability Sex, feeding effect than the green feed is more superior, is a long-term preservation of green feed a good way.

References

[1] Tang Chaochen, Luo Feng, Gao Jianming, Pei Zhongyou, Li Xinyu, Sun Shoujun. Effects of fertilization on yield, sugar brix and related traits of sweet sorghum [J]. Heilongjiang Agricultural Sciences, 2014, (7): 46–50.

[2] LI Xin-yu, LUO Feng, LIU Hui-fen, PEI Zhong-you, GAO Jian-ming, JIANG Yi-wei, SUN Shou-jun. S tudy on the relationship between different types of sorghum and environment, Jiangsu Agricultural Sciences.

[3] Li Guiying, Tu Zhendong, Zou Jianqiu. Research and utilization of sweet sorghum in China [M]. Beijing: China Agricultural Science and Technology Press, 2008: 40–42.
[4] Lu Qingshan. Sweet sorghum [M]. Beijing: China Agricultural Science and Technology Press, 2008: 103–104.
[5] Zhang Fuyao, Zhao Weijun, Ping Junai. High energy crops - sweet sorghum [J]. Chinese Journal of Agricultural Science and Technology, 2006, 8 (1): 14–17.
[6] ZHANG Xiao-ying, ZHAO Wei-jun.Studies on sugar content of sweet sorghum stalks [J]. Shaanxi Agricultural Sciences, 2011, 39 (6): 616–618.
[7] Zhang Huihui, Zhang Xiuli, Hu Yanbo, et al. Growth characteristics and photosynthetic function of sorghum-sudangrass hybrids [J]. Journal of Grassland Science, 2012, 9, Volume 20, Number 5: 881–887.
[8] Zhang Xiaoyan, Liu Feng, Wang Fengyun, et al. Effects of nitrogen fertilization on nitrate accumulation and yield of hybrid sudangrass [J]. Journal of Grassland Science, 2009, 17 (3): 327–332.
[9] Yu Zhuo, Liu Yongwei, Zhao Xiaojie, et al. Agronomic characteristics and cytogenetics of sorghum 11A and three sudangrass hybrids F1 [J]. Chinese Journal of Grassland Science, 2006, 15 (1): 90–96.
[10] LUO Feng, SUN Shou-Jun, PEI Zhong-You, et al. Correlation analysis of forage quality traits and morphological traits in sweet sorghum [J] North China Journal of Agricultural Science, 2009, 24 (6): 202–204.
[11] Ningxia Agroforestry Science and Technology [J], 2012, 53 (3): 113–114.
[12] HU Dai-ze. Current Situation and Prospects of Utilization of Crop Straw Resources in China [J] .Resources Development & Market Research, 2000, 16 (1): 19–20.
[13] HANG Wei-qi, CHEN Jian-jiang. Effects of Field Combustion Straw on Environmental Quality and Its Prevention and Control [J]. Acta Environmental Monitoring and Management, 2000, 12 (2): 36–37.
[14] Wang Hongyan, Wang Fei, Sun Renhua, et al. Review of Policies and Regulations of Crop Straw Utilization in Foreign Countries and Its Experience and Revelation [J]. Transactions of the Chinese Society of Agricultural Engineering, 2016.
[15] Yang Lijuan. The way and technical requirement of crop straw returning to field [J]. Modern Agriculture, 2011 (5): 165.
[16] Liu Changyue. Straw return to the benefits and disadvantages [J]. Agricultural environment and development, 2004 (5): 37–38.
[17] Sha Wenfeng, Li Shijiang, Zhu Juan. Study on the Development and Utilization of Crop Stalks [J]. Journal of Jinling College of Science and Technology, 2005, 21 (4): 73–76.
[18] GONG Jiang-li, JIANG Chuang-dao, SHI Lei, et al. Advances in sweet sorghum [J] .Anhui Agricultural Sciences, 2006, 34 (22): 5815–5816.
[19] Analysis of silage quality of maize and sweet sorghum with different sugars in southern Xinjiang [J]. Chinese Journal of Grassland Science, 2014, 23 (3): 232–240. [Article in Chinese] Zhang SJ, Ai JJ Jiang Wusman, XUE Xingzhong.
[20] Oliver A L, Grant R J, Pedersen J F, et al, Comparison of brown midrib-6 and -18 forage sorghum with conventional sorghum and corn silage in diets of lactating dairy cows [J]. *J Dairy Science,* 2004, 87 (3): 637–644.

Yan-Hui Bi, Guo-Dong Yang and Yan-Hong Zhang*

The Technical Method and Key Issues of the Delimitation of Ecological Protection Red Line at County Scale

A Case Study of Zhenlai County, Baicheng City

Abstract: The ecological protection red line is the base line to maintain the national and regional ecological security with the aim of the integrity and connectivity of the ecosystem in China. Firstly, the ecosystem service function was evaluated by the formulas of habitat quality and habitat scarcity, and the key issues of the delineation of ecological protection red line were discussed. Then, taking Zhenlai County as a demonstration area, the technical solutions of the delimitation of ecological protection red line at county scale under the "combination with up and down" principle were explored. Finally, the area of ecological protection red line was designated as about 1420 km^2, accounting for 30 % of Zhenlai County land area combined with Zhenlai County actual situations. The scope and location of the proposed red line is in line with local economic, social development and ecological protection needs after review and investigation. Ecological red line delineation at county scale plays a role in connecting in our country and it's the minimum basic unit delineated by the red line of ecological protection. The discussion on the delineation of ecological red lines at the county scale will lay the foundation for the final mapping of provincial and national ecological protection red lines.

Keywords: ecological protection red line, ecological security, environment protection, red line defined GIS spatial technical method

1 Introduction

With the rapid development of China's economy and the acceleration of the process of urbanization, the pattern of "land for development" has made the urban expan-

Yan-Hui Bi, School of Earth Exploration Science and Technology, Jilin University, Changchun, China. Email: yhbii0301@163.com
Guo-Dong Yang, School of Earth Exploration Science and Technology, Jilin University, Changchun, China. Email: ygd@jlu.edu.cn
Yan-Hong Zhang, School of Earth Exploration Science and Technology, Jilin University, Changchun, China. Email: zhyh98@163.com.

DOI 10.1515/9783110540048-018,

sion too fast and the natural ecological environment has become overwhelmed which led to the increasingly serious ecological problems in China [1]. In order to fundamentally guarantee the harmony and security of the natural ecosystem of the country and the region in this context, China first proposed the concept of "ecological protection red line" [2]. This line is the bottom line of our country to maintain the national and regional ecological security aiming at the integrity and connectivity of the ecosystem. The scientific and rational delineation will provide the important support for the ecological protection and construction, the orderly development of natural resources and the rational distribution of industry of our country.

2 Connotation and Understanding of Ecological Protection Red Line

The ecological protection red line is based on the protection of the integrity, connectivity and stability of the ecosystem and fully considers the national and local planning, so as to determine the red line area which has an important role and basic support in ecological protection construction, ecological security and industrial rational distribution of the study area [3]. The ecological red line was first proposed in Guangdong Province in 2005 [4]. In 2011, the "State Council on strengthening the focus of environmental protection advice" clearly put forward the task of national and local delineation of ecological protection red line which means that the ecological protection of the red line strategy has been raised to the national level, and in 2014,the Ministry of Environmental Protection issued the "National Ecological Protection Red Line - Ecological Function Red Line Demarcation Technical Guide (Trial)" marked the ecological protection of red line delineation work is in full swing in our country [5]. "Ecological protection red line" is the concept of China's ecological environment in the protection of innovation [6] with a clear political background and management content [7]. Although there is no similar concept abroad, but it has a similar approach and practical application in the natural resource protection planning.

3 Research Status of Ecological Protection Red Line

The western region is the most vulnerable areas of the ecological environment of Jilin Province. The ecological protection red line of Zhenlai County in Baicheng County combined with the ecological characteristics of the west of Jilin Province, and then used the ecological red line delineation method of "combining of the up and down, multi-participant". "Up" should be based on the national and Jilin Province as a whole to determine the protection of the skeleton, "down" should be combined with the actual situation of the region, remote sensing and GIS technology to do quantitative evaluation [8]. Jiangsu Province, Shenzhen City, Pearl River Delta and other regions have explored the division of the red line based on local ecological and environmental conditions, ecosystem integrity, ecological space continuity and other indicators [9]. These International similar empirical methods provide important reference and valuable experience for the delimitation of ecological protection in China [7], such as the construction of the EU Natura 2000 Nature Reserve [10] or establishing a system of protected areas for the conservation of biological diversity, natural resources and cultural resources associated with them [11].

4 The Key Technical Methods of Red Line Delineation of Ecological Protection

4.1 Extension Determination of Ecological Protection Red Line

Article 29 of Chapter 3 of the Environmental Protection Law of the People's Republic of China stipulates: "The state shall impose strict protection on the red line of ecological protection in areas such as key ecological functional areas, sensitive areas of ecological environment and fragile areas. Article 29 of Chapter 3 of the Environmental Protection Law of the People's Republic of China stipulates that the state shall designate the ecological protection red line in the key ecological functional areas, the sensitive areas of ecological environment and the fragile areas and implement strict protection. It has determined that the area of ecological protection red line is mainly divided into four key types: key ecological functional area, ecologically sensitive area / fragile area, prohibited area and protected area based on the actual situation of Baicheng County. The details are as follows.

4.1.1 Key Ecological Functional Area

According to the "National Main Functional Area Planning" in 2011 stipulates that the restricted development area is divided into the main producing areas of agricultural products and key ecological functional areas. The national key ecological functional area is divided into four types: water conservation type, windbreak and sand fixation type, soil and water conservation type, biodiversity maintenance type [12].

4.1.2 Ecologically Sensitive Area / Fragile Area

The sustainable development of ecologically sensitive areas / fragile areas needs to be based on ecological and environmental protection [13] and its environmental threshold is small, and the environmental carrying capacity is weaker than other regions. It is highly affected by human activities, and the ecological environment is extremely difficult to restore the original state after the destruction. It is important to protect the ecologically sensitive area/ fragile area.

4.1.3 Prohibited Area

Establishing prohibited area fundamentally prevents the deterioration of the ecological environment and it is the basic guarantee for the region ecological security[14]. The prohibited area of Baicheng County is established in accordance with the relevant laws and regulations, mainly including national and provincial nature reserves, state-level scenic spots, the national forestpark and the national wetland park and other types.

4.1.4 Protected Area

In addition to the above three range of areas, other areas are not included but have the ecological elements with ecological value to the region or their ecological environment is more sensitive and fragile should be divided into protected areas. The precipitation in the western part of Jilin Province is extremely uneven during the year and there is continuous occurrence. Jilin Province Water Resources Department carried out River and Lake Connectivity Project to ensure adequate water which uses the construction of large water conservancy projects based on gravity. It means that lead the surplus water in flood period and the drainage of irrigation area of natural river to the dry or semi dry lake in the west of Jilin Province [15]. River and Lake Connectivity Project has improved the wetness of the lake, wetland degradation, soil desertification and salinization. Therefore, River and Lake Connectivity Project as the region's special projects is classified as protected area.

4.2 Evaluation of The Importance and Sensitivity of Ecosystem Service

The evaluation of ecosystem service function is the qualitative or quantitative analysis of the services provided by the ecosystem to human beings in order to reveal the supporting role of the ecosystem to the ecological security and the economic and

social development, as well as the evolution of the interaction between human and ecosystem and the evolution of the interaction between human and ecosystem. Habitat quality and habitat scarcity are the comprehensive evaluation indicators of ecosystem services, which can reflect the importance and sensitivity of ecosystem of Zhenlai county. The specific evaluation methods are as follows.
Habitat quality assessment model:

$$Q_{xj} = H_j \left(\frac{D_{xj}^Z}{D_{xj}^Z + k^z}\right) \quad (1)$$

Where $Q_{xj}$is the habitat quality of grid x in land use and land cover j, H_j is habitat suitability of land use and land cover j, k is a unsaturated constant and K value is equal to D when $1 - \left(\frac{D_{xj}^Z}{D_{xj}^Z + k^z}\right) = 0.5$
D_{xj} is the habitat stress level of land use and land cover or habitat type j and grid x.

$$D_{xj} = \sum_{r=1}^{R} \sum_{y=1}^{Y_r} \left(\frac{W_r}{\sum_{r=1}^{R} W_r}\right) r_y i_{rxy} \beta_x S_{jr} \quad (2)$$

Where W_r is the weight of the stress factor, which indicates that the relative destructive of a certain stress factor to all of the habitat, β_x is the accessibility level of grid x, 1 of which indicates that it is easy to achieve, and S_{jr} is land use and land cover (or habitat types) j sensitivity to the stress factor r, the value is more sensitive when it is closer to 1.
The effect of stress factor r (r_y) in the grid y on the habitat in the grid x isi_{rxy} .

$$i_{rxy} = 1 - \left(\frac{d_{xy}}{d_{rmax}}\right) \text{ (linear).} \quad (3)$$

$$i_{rxy} = \exp\left(-\left(\frac{2.99}{d_{rmax}}\right) d_{xy}\right) \text{ (index).} \quad (4)$$

Where d_{xy} is the linear distance between the grid x and the grid y, d_{rmax} is the maximum influence distance of the stress facto r.
Habitat scarcity evaluation model:

$$R_x = \sum_{x=1}^{X} \sigma_{xj} R_j \quad (5)$$

Where R_x is grid x in land use and land cover j, $\sigma_{xj} = 1$.

$$R_j = 1 - \frac{N_J}{N_{j,baseline}} \quad (6)$$

Where N_J is the grid number of current land use and land cover j and $N_{j,baseline}$ is the grid number of land use and land cover j in the baseline landscape pattern.

Land use and land cover are more likely to be protected when the value of land use and land cover $R_j$is closer to 1 and $R_j = 0$, if the land use and land cover j disappeared under the baseline landscape pattern.

4.3 Overlay Analysis and Comprehensive Mapping Technology

The division of ecological protection red line is based on the current situation of the natural ecological environment in the region, and the needs of the regional ecosystem integrity and connectivity [16]. After the evaluation of ecosystem service function, the area with higher evaluation level is divided into ecological protection red line area [6]. In the region of the two kinds or more than two kinds of overlapping types, Determine the overlapping area of dominant ecological function and highlight the most valuable ecological protection elements when two types or more than two types overlap. Use GIS software to gather the aggregation of adjacent or adjacent spots and weed out the independent pots which area is under $1km^2$ to ensure the integrity and continuity of ecological protection red line. After these treatment, manually draw the ecological protection red line area if the degree of fragmentation of the region is still high. Technical team investigated for several times to Zhenlai County to develop ecological protection red line area in order to identify the controversial ecological protection red line patches for further.

5 A Practical Case of Zhenlai County in Baicheng

5.1 Regional General Situation

Zhenlai County is located in the northeast of Baicheng City and located in the middle latitude inland with an area of 4719 km^2. Zhenlai County is located in the east of Da Hinggan Mountains, the territory of the “Three Rivers” through Nenjiang, across the county east, Tao’er River, river, Tao River was called long passing through the rich wetland resources. Nenjiang acrosses the east of the county and Tao’er River, Huerda River,Erlongtao River pass through it. Zhenlai County has abundant wetland resources and its main type of ecosystem is wetland grassland ecosystem. But serious destruction of wetland ecosystem appeared due to the local climate change, land salinization and other factors. Therefore, Zhenlai County can be the ecological protection red line delineated representative demonstration area of the west of Jilin Province at scale County.

5.2 Delineated Content

According to the present situation of natural ecological environment and the demand of ecological protection in Zhenlai County of Baicheng city and consistented with the national special planning and local urban and rural planning, the classification of ecological protection red line area of Zhenlai county is shown in Table 1 [17] referring to the specific requirements of the Jilin Province Ecological Protection Red Line Delineating Plan (Baicheng). Merge prohibited area and protected area of Zhenlai County in Table 1 and use GIS software to dospatial overlay processing to the overlapping regions to determine the dominant ecological function of overlapping regions. Oil development zones of Yingtai Oil Production are located in the experimental area of Momoge Nature Reserve and need to the development and utilization of oil resources should be reserved according to the Jilin Province Ecological Protection Red Line Delineating Plan (Baicheng). At the same time, the economic development zone and the scope of small towns development planning of Zhenlai County should be deducted according to the current land use types and the special planning of all departments. After the polymerization processing, remove independent figure spots and then form the ecological protection red line area of Zhenlai County.

Tab. 1: Area Type of Ecological Protection Red Line in Zhenlai County

Serial Number	Control Type	Red Line Type	Dominant Ecological Function
1	Prohibited Area	Momoge National Nature Reserve	Biodiversity Conservation
2	Prohibited Area	Groundwater Drinking Water Source Protection Area	Water Quality Protection
3	Prohibited Area	Zhenlai National Wetland Park	Biodiversity Conservation
4	Protected Area	Nenjiang National Fish Germplasm Resources Conservation Area of Zhenlai County	Germplasm Conservation
5	Protected Area	River and Lake Connectivity Project	Water Quality Protection
6	Protected Area	National Key Public Welfare Forest	Water and Soil Conservation

5.3 Results and Analysis

The ecological protection red line area of Zhenlai County of Baicheng city is according to the local ecological environment, the social and economic development plan and the special planning of all departments. The results of visual interpretation of land use status of Zhenlai County by using GIS software are shown in Fig. 1 with the high resolution image (4.9 m spatial resolution) map in 2014. The results are divided into 11 categories; the main types of land use are forest land, paddy field, dry land and residential land.

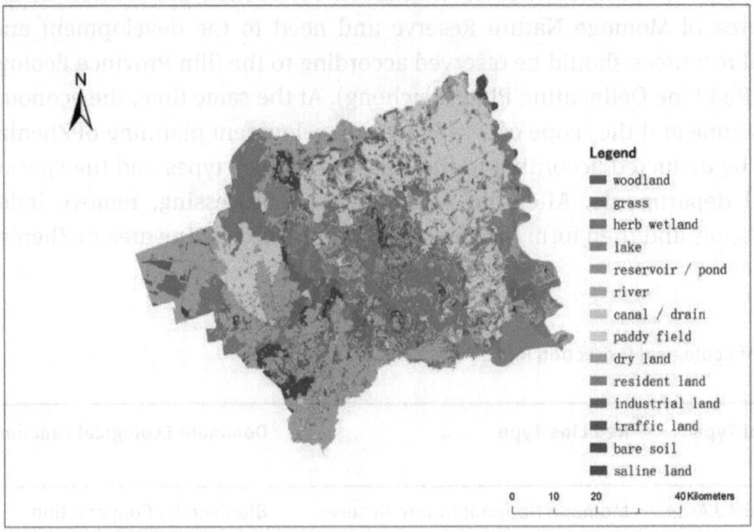

Fig. 1: The current Land Use Type map for Zhenlai County

Zhenlai ecological protection red line area is about 1420 km^2, accounting for about 30 % of the land area of Zhenlai County, as shown in Fig. 2. The red line area is relatively concentrated and closely linked with the relevant planning of Zhenlai County.It is mainly distributed in the southeast of Momoge National Nature Reserve in Zhenlai County, as far as possible to maintain the original reserve boundary. The red line is according with the principle of ecological protection red line delineation of Zhenlai County and provides basic ecological security for the local ecological security, and sustainable development of human settlements and economic society.

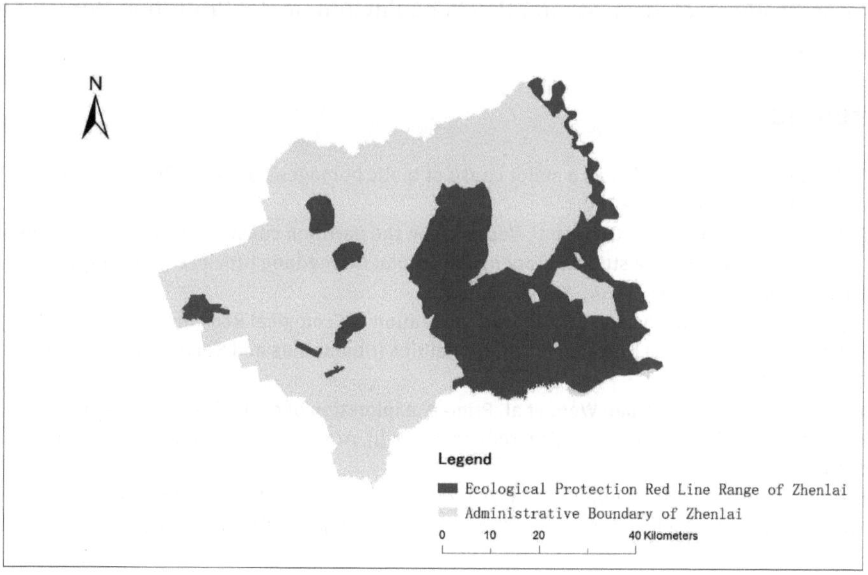

Fig. 2: Regional Planning Map of Ecological Protection Red Line of Zhenlai County

6 Conclusions

Ecological protection red line is a major innovation system in the construction of ecological civilization in China. This paper takes Zhenlai County in the west of Jilin Province as an example and discusses the technical methods and key problems of the red line delineation of ecological protection at county level. Subsequent red line controls should be further perfected in order to ensure the effective implementation of ecological protection red line. At present, it is necessary to make use of the existing control system and regulations to hierarchically control the red area of ecological protection and protect strictly. The county level is the smallest basic unit in the ecological protection red line delimitation work. It lays the foundation for the provincial ecological protection red line into a map, but also provides a reference for other areas of ecological protection red line delineation work.

Acknowledgement: This work was financially supported by the Ecological Protection Red Line Delineation Project, "Production and Research" Cooperation Project of Jilin Provincial Bureau of Surveying, Mapping and Geoformation, Wetland Ecological Security Project of Jilin Momoge National Nature Reserve and Application Project of National Conditions Census of ilin Provincial Bureau of Surveying, Mapping and Geoformation. Especially thanks Northeast Institute of Geography and Agroecology of Chinese Academy of Sciences and Baicheng Environmental Protection Bureau for supporting.

References

[1] Mckinney M L. Urbanization as a major cause of biotic homogenization [J]. Biological Conservation, 2006, 127 (3): 247–260.

[2] Bende Yu, Youhua Ye, Wei Guo, et al. Research on the partition construction mode for ecological essential line: A case study in Dapeng Peninsula, Guangdong province [J]. Ecology and Environmental Sciences, 2014, 23 (6): 962–971.

[3] Yejia Shi, Xianjing Discussion on the Standardization of Ecological Red Line Concept [J]. Wu. Journal of South-Central University for Nationalities (Humanities and Social Sciences), 2016, 36 (3)149–153.

[4] Yong Lin, Jingfeng Fan, Quan Wen, et al. Primary exploration of ecological theories and technologies for delineation of ecological redline zones [J], Acta Ecologica Sinica, 2016, 36 (5): 1244–1252.

[5] Ji Yu, Tonghui He. Understanding and Obsenration Of Ecological Red Line–Problem and thinking of Dawning Ecological Red Line in Ningxia [J]. Environmental Science and Managment, 2015, 40 (1): 173–176.

[6] Dalin Jiang, Xiaofeng Cao, Honghai Kuang,et al. Ecological red line planning and related key issues analysis for China [J]. Resources Science, 2015, 37 (9): 1755–1764.

[7] Dong Liu, Naifeng Liu, Changxin Zou, et al. Development of foreign ecological protected areas and linkages to ecological protection redline delineation and management in China [J]. Biodiversity Science, 2015, 23 (6): 708–715.

[8] [8] Shifa Ma, Mei Ma, Yumei Cai, et al. Delimitating Red Line of Ecological Protection for Territorial Spatial Planning: A Case Study of Hunan Province [J]. Tropical Geography, 2015, 35 (1): 43–50.

[9] Liyuan Fan. Research on Shandong Pronvince ecological redline classification and ecological space control [D]. Shandong Normal University, 2015.

[10] Fengchun Zhang, Liucai Zhu, Ning Peng. The EU Natura 2000: A Model of Nature Reserve [J]. Environmental Protection, 2011 (6): 73–74.

[11] Dearden P, Bennett M, Johnston J. Trends in Global Protected Area Governance, 1992–2002 [J]. Environmental Management, 2005, 36 (1): 89–100.

[12] Xin Liu, Yongqing Wang. Analysis of Current Research on Key Ecological Function Zones in China [J]. World Forestry Research, 2014, 27 (1): 93–96.

[13] Zhuo Ren. The Study on the Economy Sustainable Development in Ecological Sensitive Area–Shiyan city, the water source of middle routes of the South-to-North Water Diversion Project as an example [D]. Wuhan University, 2014.

[14] Zhi Wang, Mingkang Jiang, Weihua Wing, et al. Consideration in Planning and Management of Areas Prohibited to Exploitation [J]. Journal of Ecology and Rural Environment. 2009. 25 (4): 110–113.

[15] Jianwei Dong, Junyu Yang, Weiping Yang, et al. Eco-environmental Issues Related to River-lake Connection in Western Part of Jilin Province [J]. China Water Resources, 2014 (22): 26–30.

[16] Shouguang Yan, Naifeng Lin, Weitao Shen, et al. Delineation and Protection of Ecological Red Lines in Jiangsu Province [J]. Journal of Ecology and Rural Environment, 2014, 30 (3): 294–299.

[17] Jia Yao, Min Wang, Yuchi Huang, et al. Three-dimensional Frame Exploration of the Ecological Protection Red Line in China: A Case Study of the Ningde Region [J]. Acta Ecologica Sinica, 2015, 35 (20): 6848–6856.

Cai-Fen Zou and Qian Yu

Impact of Environmental Regulation on Technological Innovation: Differences among Industries Categories

Abstract: This paper examines the relationship between environmental regulation and technological innovation by the categories of industry, using 34 industries from 2010 to 2013 as the study sample. The results show the environmental regulation has a certain role in technological innovation, if not distinguish industries. Based on the principle of energy-saving and emission reducing, we classify the industries. The main conclusions are: environmental regulation has positive effect on innovation in medium pollution and energy consumption industries, while the positive correlation is not significant for the industries with low pollution and energy consumption industries and environmental regulation has negative effect on innovation in high pollution and energy consumption industries, but it is also not significant. The paper carries out a specific analysis and put forward the corresponding policy recommendations.

Keywords: Environmental regulation, Technological innovation, Energy-saving and emission reducing, Industrial classification.

1 Introduction

With the economic development, the industrial enterprises grows rapidly, environmental pollution has become the focus again. Chinese pollution presents a serious tendency, the overall environmental regulation intensity in the world is at a weak level. Chinese economy has entered a new normal, the economic structure will be further optimized, industrial transformation and upgrading, quality and efficiency will be the main task of the next stage. How to coordinate the ecological construction and economic growth is an important problem. For enterprises, environmental regulation is an important factor of the enterprise's long-term stable growth and the basic guarantee of sustainable development. Technological innovation is the key to improve business performance. In the background of such environmental protection,

Cai-Fen Zou, Accounting School, Wuhan Textile University, Wuhan Hubei, China. Email: ahicarol@163.com
Qian Yu, Accounting School, Wuhan Textile University, Wuhan Hubei, China. Email: b710432257@qq.com

DOI 10.1515/9783110540048-019,

whether uses environmental regulation to promote technological innovation becomes a worth exploring topic.

There are many literatures on environmental regulation and technological innovation. Porter (1991) argued that a rational environmental regulation policy can stimulate technological innovation, generate compensation for innovation, and compensate for the cost of environmental regulation. Making the environmental performance and corporate economic performance become the “win-win” situation. Many scholars have supported the Porter hypothesis by examining such aspects as industry, regional, performance and system. Greenstone (2002) argued that environmental regulation increases the firm’s production costs, crowding out other resources, reducing the competitiveness of firms and denying the Porter hypothesis. The existing literature does not deeply discuss the differences between different subsectors of industry. Is there a common need for environmental regulation across the industry? Is the same impact of environmental regulations on technological innovation at different industries? The study of these problems will help to discuss these questions from the perspective of the industry, and give some reference for the enterprises to control the environmental regulation.

2 Theoretical Analysis and Research Hypothesis

Resource industry ecology theory considers the environmental problem is increasingly becoming a key factor in restricting the industrial sustainable development. On the control of environmental regulation, business uses two ways normally: First, by increasing pollution control costs to achieve environmental governance; the second is to increase investment in technological innovation, improve the level of output and corporate profits, reduce or even offset the negative impact of environmental regulation.

From the static point of view, “environmental regulation” and “technological innovation” is likely to fall into a dilemma, when the one realizes, we must sacrifice the other to make up. Under the certain conditions of technology, manpower and demand, enterprises will make the best judgments, and environmental regulation will increase the cost constantly, which may greatly weaken the competitiveness of enterprises and adversely affect the innovation of enterprises. Technological innovation requires capital investment, and environmental regulations require enterprises to invest in pollution control, under the condition of limited funds, these investments will be squeezed out of the original enterprise prepared for investment in technological innovation, making a certain hindrance on technological innovation. Barbera and Mcconnell (1990) tested the US data found that environmental regulation will lead to increase lager cost, this is harmful to corporate R&D investment [1]. From the micro-mechanism point of view, Wang wenpu (2013) selected the

30 provinces of medium and large enterprises data for empirical analysis, and found the environmental regulation has a negative impact on technological innovation in general [2].

From the dynamic point of view, whether environmental regulation and enterprise technology innovation form a common role? Porter's theory (Porter, 1991) threw the "win-win" view that environmental regulation promotes technological innovation [3]. Environmental regulation will increase the cost of production, the increasing cost will promote enterprises to make technological innovation, and further optimize the allocation of resources to form a remedial effect. Enterprises adopt technological innovation to reduce production costs, ensure the profits to keep same or even increase. Under the pressure, enterprises are forced to innovate and compensate for environmental costs by forcing them to improve product R&D. Therefore, environmental regulation can not only reduce pollution control costs, save environmental resource use costs, but also may improve production efficiency, and ultimately lead to increased profits, stimulate enterprises to carry out technological innovation. From the perspective of the industry as a whole, Jaffe (1997) used panel data from the manufacturing industry, found a significant positive correlation between environmental regulation and R&D spending in the control industry [4]. From a particular industry perspective, Ambec (2007) found that environmental regulation would encourage firms to invest in research and development through polluting enterprises [5]. Considering the relationship between environmental regulation and technological innovation from the regional perspective, the data of eastern China and inter-provincial panel showed that there was significant correlation between environmental regulation intensity and technological innovation Positive correlation and causality, environmental regulation has a significant positive impact on the core innovation indicators [6].

Inspired by the environmental Kuznets Curve, there may also be U-shaped features between environmental regulation and technological innovation. With the increase of environmental regulation intensity, the effect of "offset effect" develops to "compensatory effect" [7].

There will be a policy consensus, companies need to carry out environmental regulation to promote technological innovation. Are all industries using the "one size fits all" approach to deal with? The existing literature has not been explained in detail. Based on the principle of energy-saving and emission reducing, we carry out the study.

According to Chinese actual situation, each enterprise's environmental requirements are different, which directly determines the degree of attention to environmental regulation, investment is inconsistent, so the impact of environmental regulation on technological innovation produced a selective effect. Based on the above analysis, this paper proposes the following assumptions:

2.1 Hypothesis 1: Environmental Regulation Has a Positive Correlation with Innovation Input.

1a: Environmental regulation has a negative correlation with innovation input in high energy consumption and pollution industries.
1b: Environmental regulation has a positive correlation with innovation input in medium energy consumption and pollution industries.
1c: Environmental regulation has a positive correlation with innovation input in low energy consumption and pollution industries.

2.2 Hypothesis 2: Environmental Regulation Has a Positive Correlation with Innovation Output.

2a: Environmental regulation has a negative correlation with innovation output in high energy consumption and pollution industries.
2b: Environmental regulation has a positive correlation with innovation output in medium energy consumption and pollution industries.
2c: Environmental regulation has a positive correlation with innovation output in low energy consumption and pollution industries.

3 Research Design

3.1 Sample Selection and Data Collection

The Samples from 34 industries listed on China Industrial Statistical Yearbook. The data comes from China Science and Technology Statistical Yearbook, China Energy Statistical Yearbook in 2011–2013. In this paper, we select industries above designated size and deal with a small number of industries with incomplete data, and finally get 136 observations.

3.2 Variable Measurement

Dependent variable: Innovation input; Innovation output
Independent variable: Environmental regulation intensity
Control variable: Size, Lev

Tab. 1: Variable definitions

symbol	variable name	definition	Data Sources
R&D	Innovation input	Industry R&D/Operating income	China Statistical Yearbook of Science and Technology
Patent	Patent output	Number of patents granted , taking the natural logarithm	China Statistical Yearbook of Science and Technology
EI	Environmental regulation intensity	Industry GDP (gross output value) / industry Energy(energy consumption)	China Energy Statistical Yearbook
SIZE	Industry scale	LN (industry total assets), taking the natural logarithm	China Industrial Statistical Yearbook
LEV	Industry debt ratio	Liabilities / assets	China Industrial Statistical Yearbook

4 Measurement model

$$R\&D=\alpha+\beta_1EI+\beta_2SIZE+\beta_3LEV+\varepsilon \tag{1}$$

$$Patent=\alpha+\beta_1EI+\beta_2SIZE+\beta_3LEV+\varepsilon \tag{2}$$

Model (1) is used to test the relationship between environmental regulation and innovation input. Model (2) is used to test the relationship between environmental regulation and innovation output.

5 Regression Analysis

5.1 Descriptive Statistics

Tab. 2: Descriptive analysis

variable	Minimum	maximum	Mean	Standard
R&D	6.76	16.18	10.61	4.445
Patent	2.99	11.49	7.39	1.76
EI	0.06	192.24	10.47	17.08
SIZE	2.54	11.54	9.14	1.51
LEV	0.054	0.98	0.53	0.11

From the analysis of Table 2, we found that for environmental regulation, the difference between the minimum and maximum values is very large, the average is very small. This shows that the regulatory intensity of the various industries vary greatly, the average level is not high. The number of patent applications varies from industry to industry, and the average level is high. R & D input is also at a relatively high level.

6 Regression Analysis

6.1 The Whole Industry Regression

Tab. 3: Whole industry regression

	Innovation investment	Innovation output
C	−104.03*	-3.963***
	(−2.496)	(−4.276)
EI	0.012*	0.024***
	(1.469)	(4.216)
SIZE	16.917***	1.272***
	(3.890)	(12.964)
LEV	−17.210	−1.596
	(−0.391)	(−1.622)
Adjusted R^2	0.082	0.558
F statistic	5.306	61.208

Note: ***, **, * represent at 1 %, 5 % and 10 % levels of significant value in parentheses represent t statistic p-value corresponding coefficients.

The regression analysis shows that there is a positive correlation between environmental regulation and innovation input, and the significance is not particularly high, and there is a positive correlation with innovation output ($\beta = 0.024$, $p < 0.01$), which proves Hypothesis 2. At the same time, it is found that the scale of enterprises has a great influence on the input and output of technological innovation, showing a significant positive correlation. ($\beta = 16.917, 1.272$; $p < 0.01$). However, we believe that there is a strong difference between different industries and cannot distinguish the relationship among the whole industry regressions. The study will further examine the impact of environmental regulation on innovation input and innovation output.

6.2 Different Industries Regression

6.2.1 Industry Classification

Based on the classification of Sun Gennian (2009) from the resource-saving and environment-friendly society, this paper uses the survey data of 40 industrial sectors in China in 2013. The data comes from China Environmental Statistical Yearbook 2014[8], see in the attached table.

6.2.2 Industry Regression Analysis

Tab. 4: Regression on industry division

	Low pollution and energy consumption industries		Medium pollution and energy consumption industries		high pollution and energy consumption industries	
	Innovation input	Innovation output	Innovation input	Innovation output	Innovation input	Innovation output
C	−0.913 (−0.466)	−6.454*** (−4.972)	−0.517 (−0.174)	−7.308*** (−4.243)	6.915 (1.94)	4.143 (2.601)
EI	0.009 (0.999)	0.015* (2.524)	0.176 (1.406)	0.257*** (3.526)	−0.062 (−0.579)	−0.074 (−1.525)
SIZE	1.529 (7.601)	1.570*** (11.785)	1.333*** (4.347)	1.434*** (8.053)	0.826* (2.340)	0.395* (2.504)
LEV	−1.674 (−0.91)	−1.129 (−0.927)	−0.816 (−0.185)	−0.789 (−0.308)	−3.481 (−1.210)	0.101 (0.078)
Adjusted R^2	0.453	0.678	0.313	0.651	0.149	0.352
F statistic	19.462	47.98	8.152	30.211	2.108	4.443

The regression analysis shows that there is a positive correlation between the environmental regulation and the innovation output in medium pollution and energy consumption industries (β=0.257, $p<0.01$), environmental regulation has a strong promotion effect on technological innovation. The innovation input and output coefficient for the industries in low pollution and energy consumption are positive, but not significant for innovation output ($p<0.05$).The correlation between environmental regulation and innovation input of high pollution and high energy consumption are negative and not significant, and the coefficient of innovation output is negative. Indicating that the correlation between environmental regulation and innovation is the most obvious.

7 Conclusion and Policy Suggestion

The empirical results show that environmental regulation has a positive impact on the technological innovation output of enterprises in whole industry. However, the environmental regulation of low pollution and energy consumption industries has no significant positive correlation with innovation input and output, it is significantly positively correlated with innovation output in medium pollution and energy consumption industries, environmental regulation and innovation output is not significantly negatively correlated in high pollution and energy consumption industries, further illustrates the differences between the industries.

The policy implications of this paper are two aspects. First, distinguishing the industry to carry out the environmental regulations, to medium pollution and energy consumption industries should strengthen the environmental protection technology, development of energy-saving emission reduction system, make full use of industry advantages, so as to promote the output of technological innovation. Medium pollution and energy consumption industries can strengthen the industry through increasing scientific and technological personnel input. For high energy consumption and pollution enterprises should control budget and allocate the environmental regulation and R&D costs reasonably. Second, formulating the environmental policies to promote technological innovation. It should be based on the nature of the incentive policy of the environment. By improving the production of some of the conditions, reducing the adverse effects of enterprises, and promote enterprises to improve competitiveness and the quality of the environment.

References

[1] Barbera A J, McConnel V D. The impact of environmental on industry productivity: direct and indirect effects [J]. Journal of Environmental Economics and Management, 1990, 18 (1): 50–65.

[2] Wang Wenpu. Environmental regulation, Spatial spillover and Regional industrial Competitiveness [J]. China Population, Resources and Environment, 2013, (8): 123–130.

[3] Porter Michael. America's Green Strategy [J]. Scientific American, 1991, 264 (4).

[4] Jaffe, A.B, Palmer, K. Environmental Regulation and Innovation: A Panel Data study. Review of Economics and Statistics, 1997, 79 (4): 610–619.

[5] Ambec, S.Barla, P.Can Environmental Regulations be good for business? .An Assessment of the Porter Hypothesis, 2007, (1): 99–103.

[6] Jiang Ke, Lu Xianxiang. Environmental Regulation and Technological Innovation—Analysis of Provincial Panel Data from 1997 to 2007[J]. Scientific Research Management, 2011, (7): 60–66.

[7] Jiang Fuxing, Wang Zhujun, Bai Junhong. The Dual Effects of Environmental Regulations on Technological Innovation——An Empirical Study Based on Dynamic Panel Data of Jiangsu Manufacturing Industry [J]. China Industrial Economy, 2013, (7): 44–55.
[8] Sun Gennian, Wang Meihong, Kang Guodong. Evaluation of Chinese industrial environmental friendliness based on energy saving (water) - emission reduction [J]. Journal of Shaanxi Normal University (Natural Sciences Edition), 2009, (1): 87–92.

Tab. 5: Industry Classification

Low energy consumption and pollution industries	**Medium energy consumption and pollution industries**	**High energy consumption and high pollution industries**
Subsidiary activities in the mining of agricultural and sideline food processing industry Food industry tobacco industry Textiles & Garments, Apparel Industry Fabricated Metal Products Leather, fur, feathers and their products and footwear industry Wood processing and wood, bamboo, rattan, brown, grass products industry Furniture manufacturing industry General equipment manufacturing industry Printing and Recording Media Reproduction Special equipment manufacturing Cultural, educational, industrial, sports and entertainment products manufacturing industry Rubber and Plastic Products Industry Comprehensive utilization of waste resources Transportation Equipment Manufacturing Electrical machinery and equipment Computers, communications and other electronic equipment manufacturing equipment Instrumentation Manufacturing Metal products, machinery and equipment repair industry Gas Production and Supply	Coal mining and washing industry Petroleum and natural gas industries Mining and Dressing of Ferrous Metals Nonferrous Metals Mining and Dressing Nonmetalliferous Mining and Dressing Wine, beverages and refined tea manufacturing Textile industry Petroleum Processing, Coking and Nuclear Fuel Processing Pharmaceutical Manufacturing Chemical fiber manufacturing Water production and supply Chemical Raw Materials and Chemical Products	Other mining industry Paper and Paper Products Non - metallic mineral products industry Smelting and Pressing of Ferrous Metals Smelting and Pressing of Nonferrous Metals Other Manufacturing Industries Electricity, heat production and supply

Hong-Bing You*, Meng-Yun Xu and Fang Sun

On the Relationship between Industrial Growth and Environmental Pollution in Jiangsu Province and Policy Recommendations

Abstract: Jiangsu Province has achieved spectacular economic growth in the past decades, but also has been faced with increasing environmental pressure. The paper draws on the decoupling framework proposed by Tapio to examine the relationship between Industrial Growth and environmental pollution for the period between 2006 and 2015 in Jiangsu Province. The data used include gross industrial output, industrial wastewater, industrial sulfur dioxide and solid waste. The results show that Jiangsu's Industrial Growth is not consistently decoupled from environmental pressure. In particular, the decoupling relationship between Industrial Growth and solid waste is markedly inconsistent. The study finds that Industrial Growth has been strongly decoupled from both industrial wastewater and industrial sulfur dioxide for the period. To achieve sustained growth in the industrial sector for Jiangsu Province, the paper concludes with the following recommendations: Spreading the application of cleaner production technologies; Optimizing industrial structure and distribution, promoting the use of third-party environmental services; Implementing and updating environmental protection policies.

Keywords: Industrial Growth, environmental protection, decoupling theory

1 Introduction

The international situation is undergoing profound and complex changes. At the same time, China has a major task for achieving a sustainable economic growth. Against this backdrop, China has been endeavoring to look for new ways to manage macro-economic issues. The year 2016 marked a good start for the 13th Five-Year Plan. Jiangsu Province has always been a leading economic player in China, and the provincial government has been relentlessly optimizing and upgrading the local

Hong-Bing You, School of Economics and Management, Decision-making and Consultation Research Base for Jiangsu Industrial Cluster, Nanjing University of Science & Technology. Email: nustyou@126.com
Meng-Yun Xu, School of Economics and Management, Nanjing University of Science & Technology. Email: sunfang28@hotmail.com
Fang Sun, School of Economics and Management, Nanjing University of Science & Technology. Email: 602685517@qq.com

DOI 10.1515/9783110540048-020,

economy. The economic growth rate has been down to single-digit levels from a decade of double-digits. In terms of ecological improvement and environmental protection, Jiangsu Province has outperformed many of its Chinese counterparts and achieved substantial results. However, problems persist: unreasonable industrial structure, excessive development, high waste emissions. Meanwhile, Jiangsu Province is confronted with a range of tough challenges: several lower-than-average environmental indicators, slower environmental progress, degrading ecological systems, increasing environmental risks and concerns.

Decoupling refers to the ending or removal of coupling between two interacting physical objects in the field of physics. Organization of Economic Cooperation and Development (OECD) applied the notion of decoupling to the study of agricultural policies in late 1990s. Then, when the notion of decoupling appeared in the study of environmental issues, the notion has referred to breaking the link between economic growth and resource use or that between economic growth and environmental pollution. Based on the definitions of decoupling provided in the OECD and other research, decoupling is defined in the paper as follows: Decoupling will occur when the economy grows faster than pollution in a given time period and that economic growth is ideally decoupled from environmental pollution when the economy grows and environmental pollution decreases during the same period.

OECD (2002) differentiates relative decoupling from absolute decoupling. Vehmas (2003) has separated six forms of de-linking: weak de-linking, strong de-linking, expansive re-linking, strong re-linking, weak re-linking and recessive de-linking. Tapio (2005) uses the concept of decoupling elasticity and distinguishes eight possibilities of decoupling: strong decoupling, weak decoupling, recessive decoupling, expansive coupling, recessive coupling, expansive negative decoupling, strong negative decoupling, and weak negative decoupling.

The research into decoupling started late in China and achieved substantial progress till this century. Research attention in China has focused on decoupling economic growth from carbon emissions, land use, water use, energy consumption and environmental pollution. Chinese researchers often use the decoupling framework constructed by Tapio and elasticity value.

2 Industrial Growth and Environmental Pollution in Jiangsu Province: Current Situation

2.1 Industrial Growth in Jiangsu Province

Since the start of China's economic reforms in late 1980s, Jiangsu Province has always centered the economic growth policies around ensuring the quality of economic growth and economic efficiency. Jiangsu Province is among economically the most advanced and dynamic province in China. In 2015, Jiangsu's GDP was 7.01164 trillion RMB, an increase of 8.5 % against the previous year. The primary sector grew by 3.2 %, the secondary sector by 8.4 %, and the tertiary sector by 9.3 %. The size and output of the industrial sector have been ranked first in China for years. In 2015, value-added industrial output for enterprises above a designated size rose by 8.3 % against the previous year. Advanced manufacturing industries grew robustly. In particular, output in auto manufacturing, computer telecommunications and electrical equipment manufacturing grew over 9 % for the year (see Fig. 1).

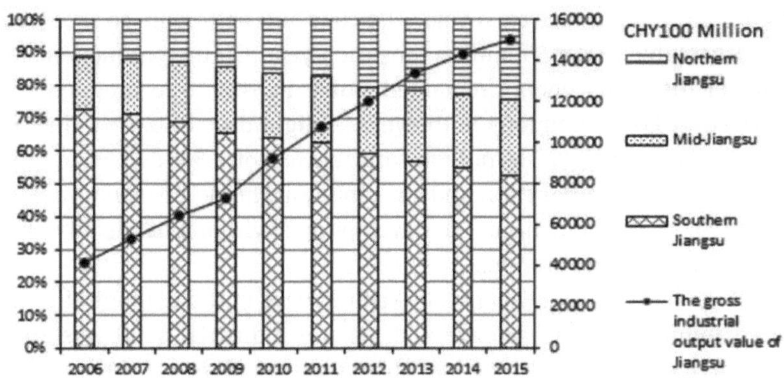

Fig. 1: The Total GDP and GDP for Each Area in Jiangsu Province

The six energy-consuming industries in Jiangsu Province include chemical material and chemical product manufacturing, the nonmetallic mineral product manufacturing, ferrous metal smelting and rolling processing, non-ferrous metal smelting and rolling processing, petroleum processing, coal-coking and nuclear fuel processing. According to Jiangsu Statistical Yearbook, the gross industrial output value for enterprises above a designated size has grown on an annual basis and the growth rate has slowed down. However, the output value for enterprises above a designated size in ferrous metal smelting and rolling processing industry kept declining in the past several years and declined by as much as 9 % in 2015. The output value for enterprises above a designated size in non-ferrous metal smelting and rolling processing industry dropped to 9 % of the gross output value in 2015 from 36 % in 2007.

During the same period, the output value of production and supply of electric power and thermal power always ranked first. But the rate of growth in the industry slowed down to 1.4 % in 2015 from 16 % in 2007.

The industrial sector in Jiangsu Province has sustained a sound growth momentum. With socio-economic development and accelerating urbanization, environmental pollution and resource scarcity are generating substantial impacts on the balanced growth of the local economy. Jiangsu Province is a leading region in heading towards a new sustainable normalcy. Jiangsu Province is advised to stick to. the principles of innovation, collaboration, environmental-friendliness, openness and sharing in growing the local economy.

2.2 Environmental Pollution in Jiangsu Province

According to Jiangsu Statistical Yearbook, the crude coal consumed by industrial enterprises above a designated size in the province has dropped since 2009 and fell somewhere under 250 million tons each year. The coke and petroleum consumption has remained fairly stable under 50 million tons each year. On the production side, the total output of industrial energy remained stable in 2015. The production of standard coal was 158.337 million tons in the year, a slight drop of 0.6 %. The total output of primary energy was 19.558 million tons, a drop of 4.5 %. As energy is a main source of pollution, its production and consumption should be managed effectively. There is also a pressing need for developing environment-friendly, low-carbon and high quality energy products.

Between 2013 and 2015, there was no marked increase in the amount of industrial wastewater, industrial sulfur dioxide and general industrial solid waste produced by the six high energy-consuming industries. And the aforesaid three types of waste was even under control in some industries. For example, the industrial solid waste produced by nonmetallic mineral product manufacturers fell to 81,920 tons from an average of 3 million tons between 2013 and 2014. There remains much room for controlling pollutant emissions in high energy-consuming industries led by chemical material and chemical product manufacturing, ferrous metal smelting and rolling processing, electric power and thermal power generation and supply.

3 Research Methods and Data Collection

3.1 Research Methods

OECD is the first to define decoupling as breaking the link between economic growth and environmental pressure. Decoupling occurs when environmental pressure grew slower than economy. Decoupling includes absolute decoupling and relative decoupling. Absolute decoupling indicates a stable or even declining environmental pressure over a growing economy. Relative decoupling refers to a state when the growth rate of the environmentally relevant variables is less than that of economy over a given time period.

Decoupling elasticity is primarily used to evaluate the extent of decoupling through a comparison and analysis of economic growth variables, environmental pressure variables and the elasticity value between economic growth variables and environmental pressure variables. The paper draws on Tapio's decoupling framework and applies the elasticity value analysis method to examine the emissions of the industrial "three wastes" (i.e. wastewater, industrial sulfur dioxide, and solid waste) and Industrial Growth in Jiangsu Province. The research results are expected to provide future directions for Jiangsu Province to grow a sustainable industrial economy.

The formula is constructed as follows:

$$w = \frac{\Delta d}{\Delta g} = \frac{D_t - D_{t-1}}{D_{t-1}} / \frac{G_t - G_{t-1}}{G_{t-1}} \tag{1}$$

In the formula, w means decoupling elasticity coefficient. Δ d means the rate of change in pollutant emissions. Δ g is the rate of change in economic growth. $D_t$and D_{t-1} represent the pollutant emissions at the end of the given year and at the beginning of the given year respectively. G_t and G_{t-1} represent economic growth at the end of the given year and at the beginning of the given year respectively. Tapio divides decoupling into three subcategories: decoupling, coupling and negative decoupling. To interpret subtle changes in elasticity, Tapio still regards a ± 20 % variation of the elasticity values around 1.0 as coupling and therefore defines coupling as elasticity values of 0.8...1.2. Tapio further divides the state of decoupling into eight subcategories: strong decoupling, weak decoupling, recessive decoupling, expansive coupling, recessive coupling, expansive negative decoupling, strong negative decoupling, and weak negative decoupling as illustrated in Table1. Strong decoupling is the most desired situation which occurs when economy grows and pollution decreases. Strong negative decoupling is the most undesirable situation which occurs when economy decreases and pollution increases.

Tab. 1: 8 subcategories of decoupling between economic growth and pollutants in Tapio's model

Category	Subcategory	Δd	Δg	W	Degree of decoupling
Decoupling	Strong decoupling	<0	>0	<0	A
	Weak decoupling	>0	>0	[0, 0.8)	B
	recessive decoupling	<0	<0	>1.2	C
Coupling	expansive coupling	>0	>0	[0.8, 1.2]	D
	recessive coupling	<0	<0	[0.8, 1.2]	E
Negative coupling	expansive negative decoupling	>0	>0	>1.2	F
	weak negative decoupling	<0	<0	(0, 0.8)	G
	strong negative decoupling	>0	<0	<0	H

A~H indicates declining orders of different degrees of decoupling.

3.2 Data Collection

Based on data availability and data significance, the data on major pollutant emissions in Jiangsu Province between 2006 and 2015 are used. For individual indicators, the value-added of the industrial output represents Industrial Growth. Three industrial pollution indicators including industrial wastewater discharge, industrial sulfur dioxide and industrial solid waste measure environmental pollution. All the data on economic growth and environmental pollution in the study are collected from Jiangsu Statistical Yearbook.

4 Decoupling Analysis of Industrial "Three Wastes" in Jiangsu Province

4.1 Measuring Decoupling

The actual annual growth rate of industrial output and that of pollutant emissions are calculated on the basis of the data on the value-added of industrial output and industrial pollutant emissions between 2006 and 2015 in Jiangsu Province. Then decoupling elasticity value and decoupling extent can be determined from Formula (1) as illustrated in Table 2.

Tab. 2: Decoupling elasticity value and decoupling extent between industrial pollutants and Industrial Growth between 2007 and 2015

Year	industrial wastewater	decoupling extent	sulfur dioxide	decoupling extent	solid waste	decoupling extent
2007	0.23	B weak decoupling	−0.63	A strong decoupling	0.01	B weak decoupling
2008	−0.20	A strong decoupling	−0.60	A strong decoupling	0.33	B weak decoupling
2009	0.19	B weak decoupling	0.69	B weak decoupling	0.80	D expansive coupling
2010	−0.17	A strong decoupling	−0.12	A strong decoupling	1.65	expansive negative decoupling
2011	−0.19	A strong decoupling	0.06	B weak decoupling	0.44	B weak decoupling
2012	−0.57	A strong decoupling	−0.88	A strong decoupling	−0.33	A strong decoupling
2013	−0.98	A strong decoupling	−0.78	A strong decoupling	0.93	D expansive coupling
2014	−1.24	A strong decoupling	−0.76	A strong decoupling	0.11	B weak decoupling
2015	0.19	B weak decoupling	−2.26	A strong decoupling	−0.53	A strong decoupling

4.2 Analyzing Results

Through the analysis of the varying degrees of decoupling between Industrial Growth and pollutant emissions in Jiangsu Province, the findings are summarized as follows:

First, the relationships between Industrial Growth and “three wastes” present the four subcategories of decoupling: strong decoupling, weak decoupling, expansive coupling and expansive negative decoupling. As shown by the four decoupling subcategories, industrial economy in Jiangsu Province maintained growth momentum between 2007 and 2015. The industrial economy slowed down to certain controllable percentage points. Recession did not occur during the period. The Industrial Growth presented a sound momentum and seemed to continue to be absolutely decoupled from environmental pollution in the years ahead.

Second, the results indicate that both industrial wastewater discharge and industrial sulfur dioxide emissions are effectively decoupled from industrial economy. In most years for the period from 2007 to 2015, there existed a strong decoupling. Affected by the 2008 Financial Crisis, the local economy grew slower in 2009. Both industrial wastewater discharge and industrial sulfur dioxide emissions turned out to be weakly decoupled from Industrial Growth rather than strongly decoupled. Industrial solid waste turned out to be expansively coupled with industrial economy. Before 2010, both industrial wastewater discharge and industrial sulfur dioxide emissions were decoupled again from Industrial Growth. But the relationship between industrial solid waste and Industrial Growth before 2010 was expansive negative decoupling. In 2013, the relationship between the two was expansive coupling. For the other years, the relationship between the two was strong or weak decoupling.

Third, the degrees of decoupling Industrial Growth from environmental pressure have not been consistent over the period. In particular, the decoupling of Industrial Growth from industrial solid waste was markedly inconsistent. Strong decoupling of industrial wastewater from Industrial Growth was observed possibly thanks to a lower proportion of industrial wastewater out of the total wastewater discharge in Jiangsu Province and also to the improved wastewater treatment.

5 Policy Recommendations

5.1 Spreading Application of Cleaner Production Technologies

Jiangsu Province has scarce energy resources and low environmental capacity available per person. However, rapid industrial development has led to worsening industrial pollution caused by energy consumption. With ongoing economic transformation and upgrading, it is imperative to spread green, low-carbon, and cost-effective cleaner production technologies. Technological progress is a core element in achieving cleaner production. Main practices include reduction of energy consumption, substitution of poisonous materials, all-process control and reduction of pollutant emissions. Therefore, Jiangsu Province should promote the efficient use of energy and promote the wide use of clean energy. Meanwhile, Jiangsu Province should design policies and incentives on the use of state-of-the art clean production technologies and facilities. The industrial enterprises in key industries should speed up technological transformation for the implementation of cleaner production. For non- electric power industries like steel and cement, the facilities for desulfurization, denitration and dedusting should be upgraded. For metallurgical, chemical and other heavy metal-related industries, cleaner production requirements should be strictly implemented. Throughout the industrial sector, consistent efforts should be made in extraction of clean energy, efficient use of energy, optimal allocation of energy resources, optimization of energy consumption structure and maximum reduction of pollutant emissions.

5.2 Optimizing Industrial Structure and Distribution

For industries where Jiangsu enjoys clear advantages, environmental-friendly and resource-friendly development approaches should be taken. Integration between industrialization and informatization should be promoted. The proportion between light industries and heavy industries is advised to be moderately adjusted. Outdated production facilities should be phased out. Industrial structure in Jiangsu Province needs to be further optimized. Ecological industries and modern service industries should get the incentives to expand and therefore contribute to industrialization. For highly polluting industries and industrial parks, agglomeration of industries and collaboration in pollution-control efforts should be exerted. Therefore, regional resources, energy and wastes can be recycled for use. Meanwhile, non-core industries can be relocated to Middle Jiangsu, North Jiangsu and other economically backward areas on account of complementary factor endowments.

5.3 Promoting the Use of Third-party Environmental Services

A third-party pollution control approach was proposed in China at the end of 2014. This approach operates on the principle that "Polluting enterprises pay; Specialized environmental service contractors clean". The third-party pollution control approach has been in use for years in developed countries and the third-party service accounts for 45 % of all pollution control approaches. In comparison, the third-party pollution control service is still in its fledgling stage in China. Polluting enterprises can raise efficiency in pollution reduction by using specialized technologies and expertise provided by environmental service contractors. The third-party service is now used in some areas in Jiangsu Province and the results have turned out to be satisfactory. Towards better pollution control results, Jiangsu Province should further extend the use of the third-party environmental services and increase investment in the area and grow it into an industry. This particular industry can provide specialized pollution control services to public facilities, industrial parks and other related facilities.

5.4 Implementing and Updating Environmental Protection Tax Policies

Environmental Protection Tax Law of the People's Republic of China was adopted In December 2016 and will be effective as of 1 January 2018. The Law is poised to replace the existing pollutant discharge fee system with a pollution tax as the main economic tool regulating environmental pollution by businesses in China. The law aims to initiate a tax collection and management method that will work like this: Businesses file their tax returns; Tax authorities collect taxes; Environmental protection authorities monitor pollutants; Environmental protection authorities and taxation authorities establish a tax related information sharing platform. In compliance with the Law, Jiangsu industrial enterprises should follow environmental protection tax policies and duly pay up taxable amounts. To prevent possible disputes, the provincial government authorities should specify each party's rights and obligations when they administer taxation. Environmental protection tax evasion or any other form of tax fraud shall carry a hefty punishment. In addition, the environmental protection tax law should also be updated to address changing environmental needs. For example, taxes on CO2 emissions and lighting pollution can be included in taxable pollutants. In this way, the environmental protection tax will literally serve the purpose of green taxation.

6 Conclusion

More and more people have begun to pursuit the green development. It is time to carry out the basic state policy of resource conservation and environmental protection and insist on sustainable development and to form the harmonious development of the human being and the nature.

Acknowledgement: This work was financially supported by National Social Science Fund Major Project: Study on the Environment Effects of the Industrial Clusters and the Regulation & Control Policies in New Normal (15ZDA053).

References

[1] OECD. Indicators to Measure Decoupling of Environment Pressure from Economic Growth[R]. Paris: OECD, 2002.
[2] Tapio P. Towards a theory of decoupling: Degrees of decoupling in the EU and the case of road traffic in Finland between 1970 and 2001[J]. Transport Policy, 2005 (2): 137–151.
[3] Vehmas J, Luukkanen J, Jarikaivo-oja. Material flows and economic growth [Z]. Turku: Turku School of Economics and Business Administration, 2003: 9–11.
[4] Wietze L. Decomposition of CO2 emissions over 1980-2003 in Turkey [J]. Energy Policy, 2006, 34(14): 1841–1852.
[5] Bai Caiquan, Huang Yabao, Song Weixuan, Feng Yaqi. Decoupling effect of industrial Economic Development and Environmental Pollution in Yangtze River Delta [J]. Environmental Science & Technology, 2015, 38(7): 157–163.
[6] Niu Tong, Yuan Changju, Niu Chonghuai, Cheng Lijuan. A study of decoupling between economic growth and energy consumption in China based on elastic decoupling [J]. Science and Technology Management Research, 2015, 18: 236–240.
[7] Xia Yong, Zhong Maochu. Relationship between EKC hypothesis and the decoupling of environmental pollution from economic development: based on China Prefecture-level cities' decoupling partition. China Population, Resources and Environment, 2016, 26(10): 8–16.
[8] Xiong Xi. Decoupling relationship between regional industrial development and carbon emission: a case study of Human (2006–2013) [J]. The Theory and Practice of Finance and Economics, 2015, 36(198): 120–125.
[9] Zhou Changbo, Li Zi, Liu Jingjun, Yu Huajun. Situation, Problems and Countermeasures of the Cleaner Production Development in China [J]. Environmental Protection, 2016, 44(10): 27–32.

Tian-Ping Bi, Qiang Zhang and Pei-Wen Wang

The Research on the Credit Evaluation System of Construction Cost Consultation Enterprise

Abstract: In order to solve the problems such as the lagging of credit management level and credit management system construction of construction cost consultation enterprise. According to the construction characteristics and development needs of the industry, establishing credit evaluation index system and rating calculation model of construction cost consultation enterprise, and using B/S architecture and C#.net language for the credit evaluation system development based on the SQL server database. In the end, the efficient and convenient project cost consulting enterprise credit evaluation system is established, and also have much more functions such as the sorting, querying and publicity for evaluation results. Fundamentally changing the evaluation methods, improving the credit evaluation management efficiency of project cost consulting enterprise by using the "Internet plus credit evaluation system", and providing protection for the healthy development of the industry.

Keywords: Cost consulting, Credit evaluation, System research

1 Preface

The construction cost consulting enterprise is one of the Construction market subject, which undertakes some works such as the investment of construction projects, the determination and control of the project cost, and so on [1]. Up to now, there is no systematic study on the credit evaluation of construction cost consulting enterprise abroad, and the research in domestic is in the stage of establishing evaluation index system and evaluation method. There are some researches such as: Wei Liu built the construction cost consulting enterprise's integrity evaluation index system from the staff quality, enterprise quality and so on 5 aspects, and used the quantification value weighted function method to evaluate the construction cost consultation enterprise [2]; Lei Shen established the suitable for Zhejiang province construction cost consulting enterprise credit evaluation system based on the analysis of the reason of construction cost consulting enterprise credit lack [3]; Xiao-Feng Liu calculated the index weight and index data by using analytic hierarchy process (AHP)

Tian-Ping Bi, Hunnan District, Shenyang, Liaoning, China. 57036475@qq.com
Qiang Zhang, Hunnan District, Shenyang, Liaoning, China. 965892317@qq.com
Pei-Wen Wang, Hunnan District, Shenyang, Liaoning, China. 919800456@qq.com

DOI 10.1515/9783110540048-021,

and fuzzy mathematics subordinate assignment method, and established the construction enterprise credit information announcement platform in the end [4]; Although the above research can realize the evaluation of enterprise credit rating, there are some problems such as low efficiency, waste of resources and difficult filing. On the basis of establishing the evaluation index system and determining the evaluation method, author researched and developed a credit evaluation system for the construction cost consulting enterprise, and introduced the concrete process and method of realizing the credit evaluation on the system.

2 Evaluation Standard

The design of evaluation index system is based on the "Construction cost consulting enterprise credit evaluation standards of Liaoning province", enterprises with good or bad credit behavior record and other government documents. At the same time, considering the implementation of the evaluation index system and striving to make the evaluation index system can entirely reflect the real situation of the participating enterprise. Based on the above, we come to the evaluation criteria, and make it listed in the following table.

Tab. 1: Credit evaluation standard for the construction cost consulting enterprise

first index	second index	scoring rules	basic points	add points limit
basic condition (0.2341)	computer assisted management (0.0228)	The enterprises using mature computational software, valuation software, project management software, enterprise management software plus 0.5 points each.	2	2
	cost engineer (0.0594)	10 basic people in class a enterprises and 6 people in class b enterprises. 0.2 points change with 1 people increase or decrease.	4	3
	intermediate title or above(0.0533)	16 basic people in class a enterprises and 8 people in class b enterprises. 0.2 points change with 1 people increase or decrease.	3	2
	stock structure (0.0549)	The number and the amount of investment of the cost engineer not less than 60%. If the capital wholly invested by cost engineers, the enterprises plus 1 points.	4	1
	revenue growth rate(0.0437)	The average annual revenue growth rate of not less than 5%. Adding 1% plus 0.1	3	2

first index	second index	scoring rules	basic points	add points limit
		points.		
achievements quality (0.5543)	consulting contract management (0.0977)	Without consulting contracts, each item is reduced by 1 point; The consulting contract has no definite agreement on the term of the consultation, the quality of the results, the charging standards and the liabilities for breach of contract. Each of these items is reduced by 0.1 points, and each item is limited by 0.5 points.	7	
	consulting service charges (0.0583)	The actual charge exceeds 20% of the floating rate of the prescribed price, and each item is limited reduced by 0.5 points.	4	
	quality control (0.1215)	The enterprise reduces 2 points if it has no level 3 quality control process list; and the audit data is filled in detailed, each increments 0.1points.	12	4
	trial system of major issues(0.0694)	If the enterprise has written meeting minutes about consultation on major issues, it can plus 1point with each.	5	4
	consultation results (0.1199)	0.5 points will be reduced with each if price behavior does not comply with national, provincial relevant pricing regulations, documents. 1 point will be reduced with each if bidding control price total error rate more than 5%.	11	
	file management(0.0537)	Lacking each archive content reduces 0.5 points; No special record room reduces 0.5 points; No file keeper reduces 0.5 points.	5	
	statistical reports(0.0338)	1 point will be reduced each time if not submitting report of the project cost consultation, and o.5 points will be reduced if submitting later.	5	
operating performance (0.2116)	consulting income (0.0614)	Annual average basic cost consulting income is 3 million Yuan in class a and 800 thousand Yuan in class b. The enterprise plus' 0.5 points every 500 thousand Yuan more than basic income.	4	4
	personal performance(0.0903)	The cost engineer must has two or more work performances during the evaluation period. The enterprise reduces 0.5 points per person without work performance.	5	
	customer satisfaction	Each record of less than 85% of satisfaction	4	

first index	second index	scoring rules	basic points	add points limit
	survey(0.0599)	is reduced 0.2 points.		
	Total		78	22

Note: the weights in the table are derived from the following rating calculation model.

3 The Rating Calculation Model

Analytic hierarchy process (AHP) is adopted to evaluate the credit of construction cost consulting enterprise, and the calculation steps are as follows:

3.1 Determine Index Weight

3.1.1 Construct Judgment Matrix

The analytic hierarchy process requires that the relative importance of each index in each level is judged, and these judgments are expressed by numerical values, which are the judgment matrix [5–6]. Assuming the weight of the first index A_k (K = 1, 2,... r) is a_k, the second index B_i is $b_i^{(k)}$ relative to the first index A_k. In this case, the second index associated with A_k is $B_1^{(k)}$, $B_2^{(k)}$,...$B_n^{(k)}$, the corresponding index weights are $b_1^{(k)}$,$b_2^{(k)}$,...$b_n^{(k)}$.

$$B=\begin{bmatrix} b_{11} & b_{12} & \cdots & b_{1n} \\ b_{21} & b_{22} & \cdots & b_{2n} \\ \vdots & \vdots & \cdots & \vdots \\ b_{n1} & b_{n2} & \cdots & b_{nn} \end{bmatrix} \tag{1}$$

3.1.2 Calculate the Eigenvector and the Maximum Eigenvalue of the Judgment Matrix

3.1.3 Check Consistency

$$CI = \frac{\lambda_{max-n}}{n-1} \tag{2}$$

Considering the possible error in the expert evaluation process, the CI is processed by 1/RI. The results for

$$CR = \frac{CI}{RI} \tag{3}$$

In the formula, RI is the average random consistency index of the same order as CI, the values are taken from the following table:

Tab. 2: Table of values for different levels of RI

levels of RI	1	2	3	4	5	6	7	8	9	10	11	12
values of RI	0.00	0.00	0.58	0.90	1.12	1.24	1.32	1.41	1.45	1.49	1.52	1.54

When CR≤0.1, the judgment matrix is satisfied, otherwise the judgment matrix is adjusted and the analysis is redone.

3.1.4 Calculate the Total Weight

$$b_i = \sum_{k=1}^{r} a_k \times b_i^{(k)} \text{ (i=1,2,...,n)} \tag{4}$$

3.2 Calculate the Credit Scores

$$F = \sum_{i=1}^{n} f_i \times b_i \tag{5}$$

3.3 Rating

The enterprise credit rating standard adopts the hundred percent system, which consists of the basic division and the increment or decrease [6]. The specific classification standard is shown in the following table:

Tab. 3: Credit rating tables

score	level
$90 \le F \le 100$	AAA
$80 \le F < 90$	AA
$70 \le F < 80$	A
$60 \le F < 70$	B
$F < 60$	C

4 Design and Implementation of the System

4.1 Structured Design

The credit evaluation system of this paper is based on SQL server database, adopts B/S framework, develops and designs with C#.net language, and realizes the interaction of browser and server, server and backstage database [7]. The information collection of the government credit management department and the uploading of enterprise information are used as the information input of the system, and the announcement platform is used as the information output of the system.

4.2 Functional Module Design

According to the reality demand analysis, the system is divided into six functional modules: credit declaration information inquiry, query and statistic, credit evaluation, data summary, system management, announcement. Among them, each function module has different sub modules.

4.3 Database Design

The credit archives database of construction cost consulting enterprise is the comprehensive record, which reflects the credit behavior in the course of its operation and the behavior of the project cost consultant in the business dealings [8]. The database needs to cover the basic information of enterprise personnel and condition of enterprise participates in the project, and so on. The content of each part contains many sub items, and the relation is complex and the data is huge. In order to realize the functions of credit rating, query and statistics, the system needs to query the information across the tables and operate across the tables [9-10]. Through the analysis of the functional requirements of the system, we get the part E-R diagram of the database entity and relation, as shown in Fig. 1.

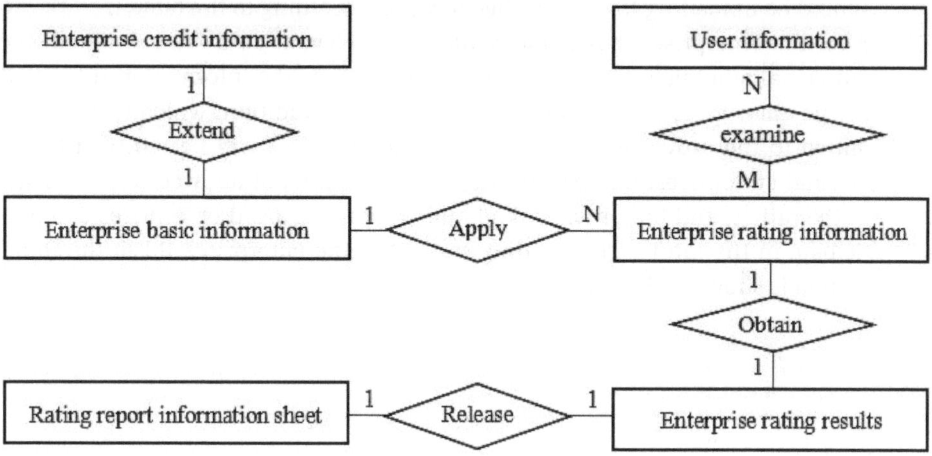

Fig. 1: E-R diagram of the database entity and relation

4.4 System Function Implementation

1) Users log in the system from the main interface (Fig. 2) through the applying user name and password.

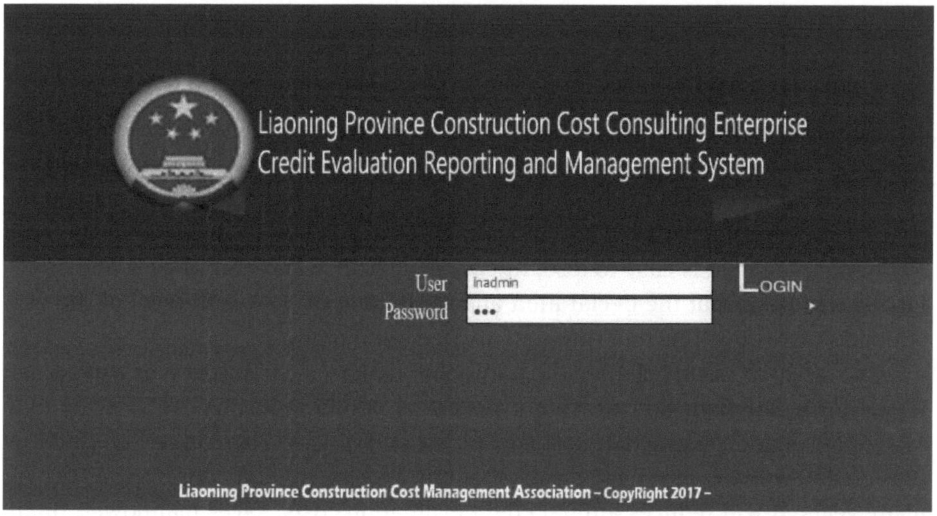

Fig. 2: System main interface

2) According to the credit evaluation requirements, the enterprise organization code, legal representative and other information are entered into the basic information table of the enterprise. At the same time, the required information should be uploading to the specified location according to the table 1.
3) After the enterprises' basic information and relevant data are submitted, the municipal administration organization shall review the information submitted by the enterprises. if the enterprises through the audition, which will be submitted to the province to audit, or else the materials will be returned to the original enterprises, this part of the enterprises are not allowed to participate in the rating, and the provincial administration approved enterprises can participate in the rating. The pending state after the enterprises fill in the information is shown as Fig. 3.

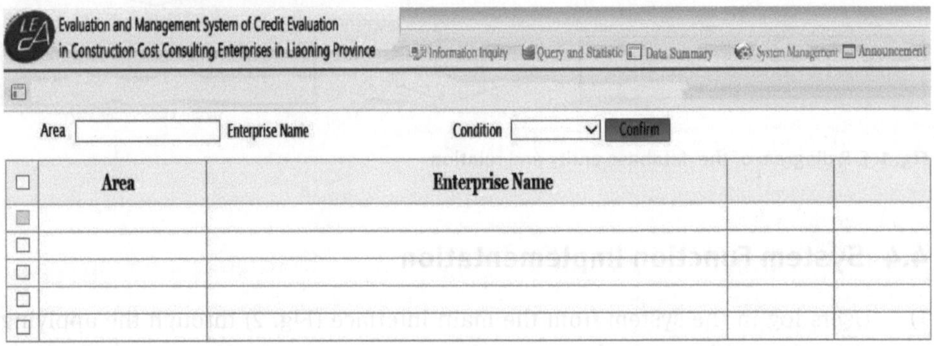

Fig. 3: Declaration of enterprise information to be audited interface

5 Conclusion

Applying modern technology of network information to research and develop the credit evaluation system of construction cost consulting enterprise, and solving the past common problems such as the work efficiency is low in the process of credit evaluation, the data is easy to lose, human disturbance and controversy and so on. Finally, realizing that the declaration and evaluation of the credit level of the construction cost consulting enterprise are carried out online.

The research results change the traditional pattern of submission of written materials for evaluation, and provide a paperless online reporting credit evaluation system for many construction cost consulting enterprises. The method of "Internet plus credit evaluation system" means that one investment and continuously use for many years. It will reduces annual credit evaluation costs, saves manpower, financial and material resources, and further improves the modernization level of the construction cost consulting enterprise management mechanism service for enterprise members. The main functions of this system can be achieved by testing, but the system's responding time is a little longer when querying cross tables and data extraction, and then need to improve the technology to solve the problems.

Acknowledgement: This work was supported by program for discipline training in Shenyang Jianzhu University of China (No.XKHY2–78), National Natural Science Foundation of China (No.51678375), Foundation of Liaoning Educational Committee (No.LJ2016005), and Natural Science Foundation of Liaoning Province (No.20170540759).

References

[1] S.B. Chen, L. Zheng, B. Wang. Research on integrity evaluation of construction cost consulting enterprise, Journal of Construction Management, 4, 149–153, 2015.
[2] W. Liu. Research on integrity evaluation of construction cost consulting enterprise, Xi'an University of Architecture and Technology, 2007.
[3] L. Shen, J.Y. Luo. Research on credit evaluation system of construction consulting industry in Zhejiang Province, Northern Economy, 10, 71–72, 2010.
[4] X.F. Liu, E.S. Qi. Research on credit evaluation system of construction enterprises, Journal of Xi'an Electronic and Science University (Social science edition), 2, 62–66, 2006.
[5] Z.D. Shen. Using AHP to construct performance evaluation system of state owned enterprises, Audit Research, 2, 106–112, 2013.
[6] C.X. Sun, H.Y. Li, R.Q. Li, X. Wang. Application of analytic hierarchy process in comprehensive evaluation of management level, Industrial Technology Economy, 9, 72–78, 2013.
[7] Y.N. Liu. Research on credit evaluation system of small and micro enterprises in China. Harbin University of Science and Technology, 2014.
[8] R.X. Guo. Data processing model of bank credit evaluation system. Journal of Software, 6, 1241–1247, 2011.
[9] Rod Stephens. The database design solution entry classic, Tsinghua University Press, 2010.
[10] C.Q. Jin, W.N. Qian, M.Q. Zhou, A.Y. Zhou. Benchmark for data management systems: from traditional databases to emerging big data. Journal of Computer Science, 1, 18–34, 2015.

Zhang-Huang Ye, Yi-Ke Li* and Bin Zhang

A Review on the Usage of Natural Gas and Its Outlook

Abstract: Natural gas is a gaseous hydrocarbons generated through long-term geologic process. China was the earliest country discovering the natural gas while the UK was the earliest user in Europe, which was 1000 years later than China. Natural gas has a wide use in modern society, mainly in residence, industry, power generation and transportation at present. The essence of the orderly succession (firewood-coal-oil-natural gas) is a process of carbon-decreasing and hydrogen-increasing. Due to the low carbon properties of natural gas, its consumption prospect is very broad, and mankind will enter the era of gas energy as predicted. Therefore, it is very meaningful to make a review on the usage of natural gas and its outlook.

Keywords: natural gas; history; usage; outlook

1 Natural Attributes

Natural gas is a gaseous hydrocarbons generated from ancient living species remains buried deeply in the ground through long-term geological process. It consists of methane hydrocarbons mainly, and a small amount of hydrogen sulfide, carbon dioxide and nitrogen as well as some trace amount of non-hydrocarbon gas like noble gases. Natural gas has the attributes of relatively small density (0.6–0.8 g/cm^3, lighter than air), being prone to fire and explosion, high calorific value, strong solubility, strong corrosivity and being convenient transportation.

Generally natural gas with more than 90 % methane is called dry gas, while that with less than 90 % is called wet gas. Based on the gas source, it can be divided into four types: pure natural gas, oil field associated gas, condensate gas field gas and coal mine gas. According to the existence state, it can be divided into free, dissolved and adsorption state, and only the free state of natural gas can be formed into natural gas reservoir through accumulation, thus it can be mined.

Zhang-Huang Ye, Jiangxi Science and Technology Normal University. No. 605, Fenglin Road, Nanchang, Jiangxi. Email: chuckverna@sina.com
Yi-Ke Li, Institute of Mineral Resources, CAGS. No. 26, Baiwanzhuang Road, Xicheng district, Beijing. Email: like430@cags.ac.cn
Bin Zhang, Institute of Geology and Geophysics, CAS. No. 19, Beitucheng West Road, Chaoyang district, Beijing. Email: zhangbinzhb@mail.iggcas.ac.cn.

DOI 10.1515/9783110540048-022,

2 History of Usage

China was the earliest country to discover natural gas. 3000 years ago, it was recorded that there happened fire in the lake and the fire was above the lake in *the Book of Changes*, which described the scene that the flammable natural gas was burnt above the lake surface. There were records about the discovery of natural gas well ("fire well" was called in ancient time) and the use of it long before in China. According to the description in *Han Geography*, the natural outcrops of fire well were discovered in the West River County (located in Shenmu County, Shaanxi Province at present) in the period of the Emperor Wu. Sichuan was the place where natural gas was exploited and used first in China. It was recorded in *Encyclopaedic Narration* in Xi Jin Dynasty that there was a fire well in Lin Hong with the width of five feet and depth of two to three Zhang, an ancient unit of length, above which the salt was being boiled. It was also described in the *Huayang Narration* of East Jin Dynasty (317–420) that working people in Lin Qing of Qin Dynasty drilled and explored natural gas and boiled salt by it: The flame from the fire wells could be watched more than 10 miles away at night. If you light it with fire and it would be exposed suddenly. When filled into the bamboo tube, the fire wouldn't cease all day. During the Ming and Qing Dynasties, the natural gas has been utilized at a large scale. The gravitationally flowing gas field located in Zigong City, Fushun County and Rongxian County of Sichuan Province were the largest and the earliest natural gas field to be put into use around the world. Based on the *Records of Gravitationally Flowing Gas Well*, the depth of drilling wells could be up to 300 feet and a fire well could burn up to 700 salt pots at the same time. Till the mid-Ming Dynasty (the middle of the 16th century), the gravitationally flowing natural gas fields have been greatly exploited, and gas pipes made from bamboo and wood reached the total length of two or three hundred miles. In Europe, Britain was the first country to use natural gas in the year 1668, over 1,000 years later than China.

3 Modern Usages

Natural gas has been widely used in modern society, mainly in residence, industry, power generation and transportation at present (Fig. 1). The natural gas in residential use mainly refers to public utility. As the improvement of people's awareness of environmental protection and the improvement of living standards, the demand of natural gas for residential purpose has been increased significantly, due to its high security, non-carbon monoxide, being lighter than air. Once leaked, it will spread up. Besides, its economic benefits are greater than industrial fuels. In the industrial sector, natural gas is mainly used in chemical raw materials and industrial kiln fuel. Taking the chemical use as an example, nearly 20 varieties of raw materials, mainly including synthetic ammonia, methanol and carbon black, are extracted from natural gas by once processing, while other 50 important chemical products, mainly including formaldehyde, acetic acid and dimethyl carbonate, are from natural gas by being processed twice or third times. The natural gas used in the industrial sector has the advantages of less pollution, wide use and low investment. Natural gas power generation plant has the characteristics of good economic efficiency, short construction period and less environmental pollution. For example, the emission of carbon dioxide from electricity generation by natural gas is only the half of that from the coal power, and the emission of sulfur is nearly to zero. It is an effective way to alleviate the energy shortage, reduce the coal power percentage and decrease the environmental pollution. In the field of transportation, the compressed natural gas can replace gasoline, with the features of low price, less pollution and reliable security.

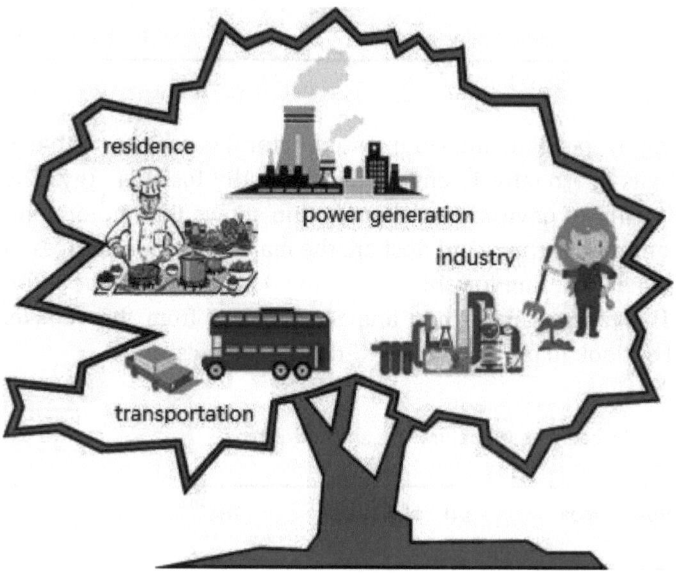

Fig. 1: Illustration of the natural gas modern usages

4 Outlook

It is predicted that natural gas is the energy in 21 Century, and human being will enter into the era of gas energy [3–5]. The reason why natural gas is favored is that natural gas has better environmental, social and economic benefits compared with other fossil fuels such as coal and oil, which is an inevitable trend in the era of low-carbon economy. According to the statistical analysis, the emission of carbon dioxide generated by the same heat unit from natural gas is 67 % from petroleum, and it is 44 % from coal. Compared with the dusts emitted from coal, the natural gas is 1:148, sulfur dioxide, 1:2700, while NOx is 1: 29 (Table 1). It is estimated that the same heat will be generated by 100×10^8 cubic meters of natural gas and 3000×10^4 tons of coal, but it would reduce about 36×10^4 tons of sulfur emission and 30×10^4 tons of dust emission.

Tab. 1: Pollutants from different energy sources

pollutants / fossil energy	SO_2	NOx	CO	CO_2	Dust
coal	2700	29	10	2.27	148
petroleum	400	5	16	1.49	14
natural gas	1	1	1	1	1

Data source: Chen Genliang [1] and Hu Jianyi [2]

The comparison of driving by gasoline and compressed natural gas (CNG) (Table 2) also shows that natural gas is a relatively environment-friendly fossil energy. The latest research by the Ministry of environmental protection shows that motor vehicles, industrial production, coal burning and dust are the major pollution sources in most Chinese urban areas, accounting for about 85 % to 90 %. And the first pollution source in Beijing, Hangzhou, Guangzhou and Shenzhen is from the vehicle. Therefore, it has a broad outlook to promote the CNG-driven vehicles.

Tab. 2: Vehicle emissions driven by compressed natural gas and petroleum

items / energy	octane value	NOx	CO	CO_2	THC	NMHC
petroleum	90	0.14	1.20	303	0.17	0.14
Natural gas	110	0.029	0.16	229	0.068	0.006

Data source: Chen Genliang [1]

From the law of energy consumption, it is inevitable that energy consumption will substitute in an orderly way with the development of society. When looking back the energy history, it has experienced four stages roughly: (1) firewood stage: Before the industrial revolution, people burnt woods to cook or heat.; (2) coal stage: Coal has replaced woods to become the main energy in the first Industrial Revolution during which machines were used to take the place of human; (3) petroleum stage:

Petroleum has replaced coal to become the main energy. The western saying "The oil is black gold, but at the same time it is causing the disaster devil" reflects the complex geopolitics; (4) multi-energy stage: After the second World War, fossil fuels are still dominant, and many kinds of new energy serve as main supplement, forming a coexisting situation of multi-energy consumption structure (Fig. 2).

(1) Firewood stage

(2) Coal stage

(3) Petroleum stage

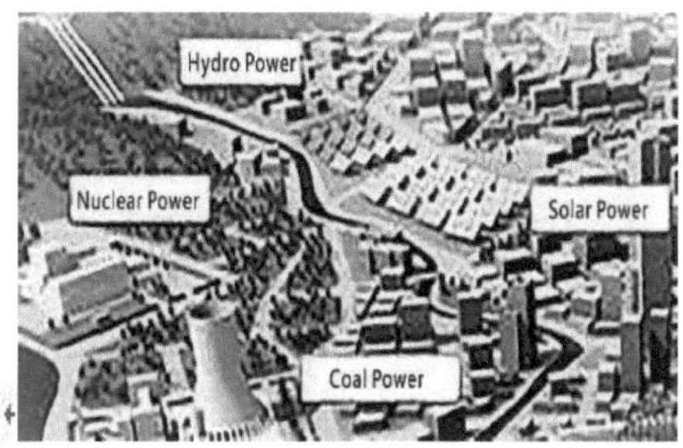

(4) Multi-energy stage

Fig. 2: Illustration of the global energy consumption replacement

5 Conclusion

Natural gas is a gaseous hydrocarbons generated from ancient living species remains buried deeply in the ground through long-term geological process. China was the earliest country discovering the natural gas, while the UK was the earliest user in Europe. The essence of the orderly succession (firewood-coal-oil-natural gas) is a process of carbon-decreasing and hydrogen-increasing. Firewood comprises 10 carbon atoms +1 hydrogen atom; Coal 2 carbon atoms + 1 hydrogen atom; petroleum 1 carbon atom + 2 hydrogen atoms; Natural Gas: 1 carbon atom + 4 hydrogen atoms

[6]. Due to the low carbon properties of natural gas, it is reasonable to predict that natural gas would be the energy of the 21st century.

Acknowledgement: This study was financially supported by the programs (12120114077201) and (GJJ160776).

References

[1] Chen Genliang, Wang Kaiyue, 2004. Comprehensive utilization of natural gas [M]. Beijing: Petroleum industry press
[2] Hu Jianyi, Guo Bincheng. 2011. Natural gas: the important stage and trend of low carbon energy development [J]. Journal of engineering science in China, 13(4):9–14
[3] Li Ning, Wang Binghuai, Zhao Guiying. 2009. Research on China's natural gas consumption prediction [J], Chinese engineering consultation, 102(2): 25–27
[4] San Weiguo. 2011. Global natural gas market development and trends [J]. Journal of oil and gas, 33 (1): 13–16
[5] Scott Montgomery. 2012. Translated by Song Yang and Jiang Wenbo. Global energy trends [M]. Beijing: mechanical industry press
[6] [U.S.] Robert hefner iii. 2013. Translated by Mu Yuanchun, Li Boshu. Great energy transformation [M]. Beijing: China citic publishing house

Zhang-Huang Ye, Rui-Ping Li* and Yi-Ke Li

Regional Comparison of Air Pollution in China

Abstract: Low-carbon economy is a big progress after agricultural civilization and industrial civilization in human society. Environment, especially atmospheric environment, is not a closed system. Thus, the whole region should be taken into consideration on air pollution control. In this paper, quantitative analysis and horizontal comparison are conducted on the air pollution in the perspectives of the integration of Beijing-Tianjin-Hebei. The figures show that there is a close relationship between economic structure and air pollution. To control the air pollution of Beijing-Tianjin-Hebei, it is imperative to adjust the industrial structure of Hebei.

Keywords: pollutant emission; energy consumption; industrial structure; region

1 Introduction

Low-carbon economy is a big progress after agricultural civilization and industrial civilization of human society, which will bring win-win vision of social development and environmental protection for human beings. The air pollution in history serves as a wake-up call to us. The air pollution caused by fossil energy consumption mainly includes the Photochemistry Smoke Event of Los Angeles in 1940s–50s and the London Smog Event in 1950s, both of which have caused heavy casualties. In recent years, the smog of Beijing also has raised widespread concern in China, even triggered international diplomatic arguments, which has a serious impact on people's life and health in body and mind.

Environment, especially the atmospheric environment, is vulnerable to the surroundings. Therefore, the whole region should be taken into consideration on air pollution control. The following is a quantitative analysis and horizontal comparison on the air pollution in the perspectives of the integration of Beijing-Tianjin-Hebei.

Zhang-Huang Ye, Jiangxi Science and Technology Normal University. No. 605, Fenglin Road, Nanchang, Jiangxi. Email: chuckverna@sina.com
Rui-Ping Li, Institute of Mineral Resources, CAGS. No. 26, Baiwanzhuang Road, Xicheng district, Beijing. Email: 49323060@qq.com Corresponding author
Yi-Ke Li, Institute of Mineral Resources, CAGS. No. 26, Baiwanzhuang Road, Xicheng district, Beijing. Email: like430@cags.ac.cn

DOI 10.1515/9783110540048-023,

2 The Energy Consumption and Industrial Structure of Beijing-Tianjin-Hebei

In 2013, the day percentage of good quality air in 13 cities in the region of Beijing-Tianjin-Hebei was 10.4 %–79.2 %, with the average of 37.5 %. Among them, days of good quality air in 10 cities were below 50 %, and days with PM2.5 as the primary pollutant accounted for 66.6 %. According to the 6 types of pollutant, all cities in Beijing-Tianjin-Hebei do not meet the standard. In a report released in 2013, there are 7 cities in Hebei province among the worst air quality 10 cities in China, including Xingtai, Shijiazhuang, Tangshan, Handan, Hengshui, Baoding and Langfang [1]. Therefore, the air in Beijing and Tianjin would be inevitably be affected. In China, the monitoring of PM10 and PM2.5 in atmosphere is relatively late, and scientific research on pollutant sources is still in the early stage. Pollutant source analysis has been carried out by the Ministry of Environmental Protection. According to the outcome of the published research of Beijing, 64 %–72 % of PM 2.5 was produced locally, and the rest 28 %–36 % was from regional transmission.

In 2013, the total GDP of Hebei accounts for 45.54 % of Beijing-Tianjin-Hebei. However, its GDP per capita only accounts for 41.54 % of Beijing, 38.87 % of Tianjin. In addition, the proportion of the secondary industry of Hebei was 52.16 %, Tianjin 50.64 % and Beijing 22.32 % (Fig. 1). Disproportionately, the total energy consumption of Hebei accounts for 66.29 % of the total consumption in Beijing-Tianjin-Hebei, and the coal consumption of Hebei accounts for 88.8 % of its energy consumption, which was significantly higher than 25.4 % in Beijing and 59.6 % in Tianjin. The high proportion of coal consumption of Hebei was an important factor in the formation of smog in Beijing–Tianjin–Hebei. The sulfide emission in Hebei accounted for 76.5 % of Beijing–Tianjin–Hebei, nitrogen oxide 83.23 % and dust 88.9 %.

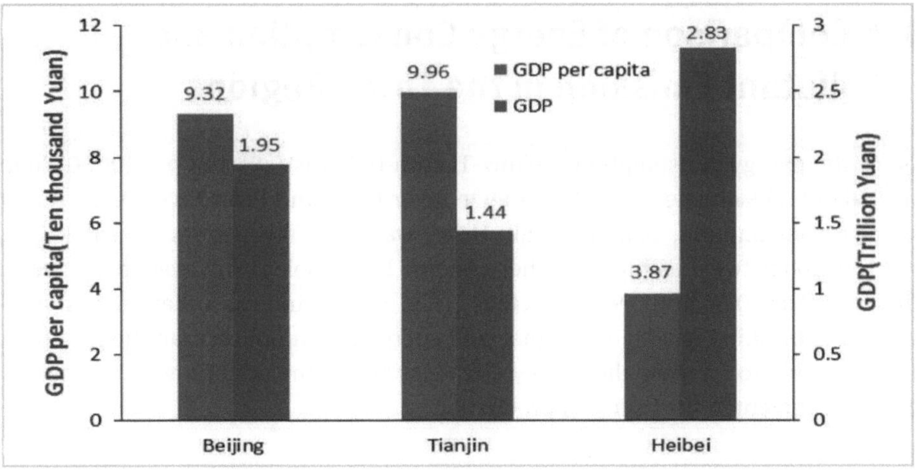

Fig. 1: Economic condition of Beijing-Tianjin-Hebei (Date source: China Statistical Yearbook [2])

There is little difference in the proportion of industrial structure in three regional economies, Beijing-Tianjin-Hebei, Yangtze River Delta and Pearl River Delta. Among them, the secondary industry the in these regions was lower than the average level in China. However, the proportion of secondary industry in Hebei is significantly higher than the rest, accounting for 52 % (Fig. 2). The unbalanced industrial structure would hinder the sustainable development [3].

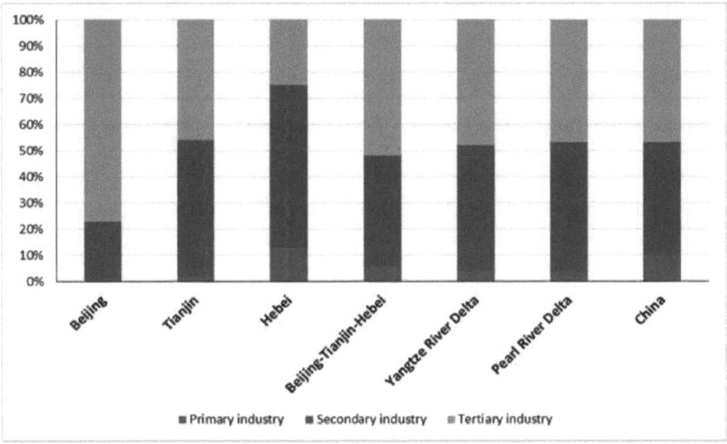

Fig. 2: Industrial structure in different regions of China (Date source: China Statistical Yearbook [2])

3 A Comparison of Energy Consumption and Pollutant Emission in the Three Regions

As for the energy consumption, Beijing-Tianjin-Hebei is 0.51 Toe per 10000 Yuan GDP, about 50 % higher than the Yangtze River Delta and Pearl River Delta. As for the coal consumption, Beijing-Tianjin-Hebei was 0.44 Toe per 10000 Yuan GDP, which is about twice as high as other regions. If we look at the average values in China, they are 0.96 and 0.94 respectively (Fig. 3). This indicates that the vast mid-west areas in China have much higher unit energy consumption than the East part. It is high time to increase the energy-efficiency by means of improving technology and management in the course of production.

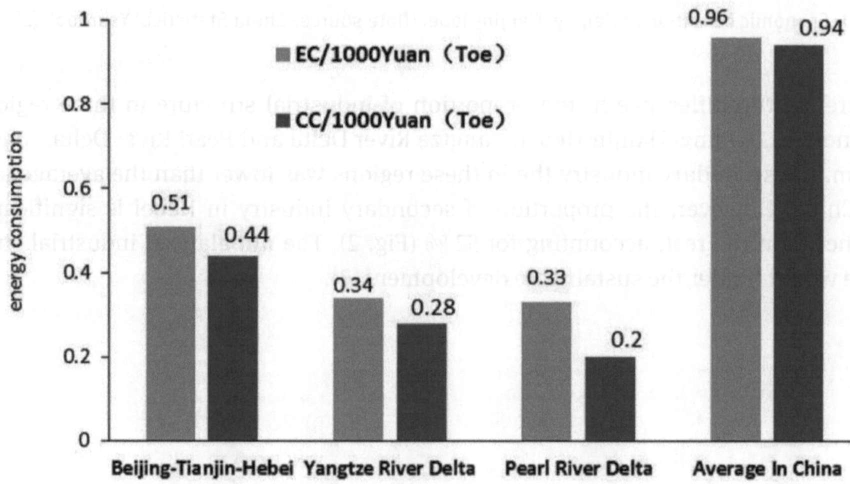

Fig. 3: Energy consumption in different regions of China (EC: energy consumption, CC: coal consumption, Toe: ton oil equivalent) (Date source: Chinese Energy Statistical Yearbook [4])

The pollutant emission is highly concerned as well. The emission of sulfur dioxide, dust and oxynitride were calculated in different regions. The sulfur dioxide emission in Beijing-Tianjin-Hebei reached 13.31 kg, dust 10.64 kg and nitrogen oxide 18.23 kg per 10000 Yuan GDP, which was much higher than Yangtze River Delta and Pearl River Delta. It is remarkable that the dust emission per 10000 Yuan GDP in Beijing-Tianjin-Hebei was 7.2 times of than in Pearl River Delta. Regarding the emission intensity of unit area, the sulfur dioxide and oxynitride emission in Beijing-Tianjin-Hebei were the lowest. However the dust emission intensity in Beijing-Tianjin-Hebei was still the highest among the three regions, which was nearly two times of the other two regions, four times of the average level in China (Fig. 4). It is no doubt that the air pollution in Beijing-Tianjin-Hebei would be the most serious part in China from the above data.

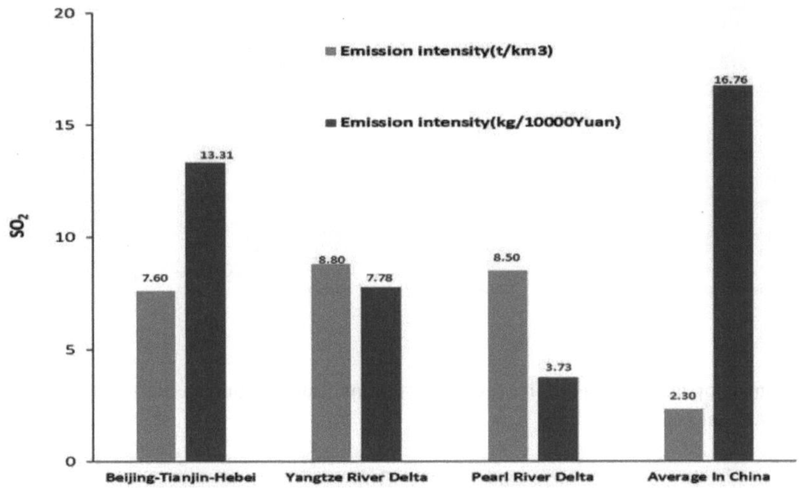

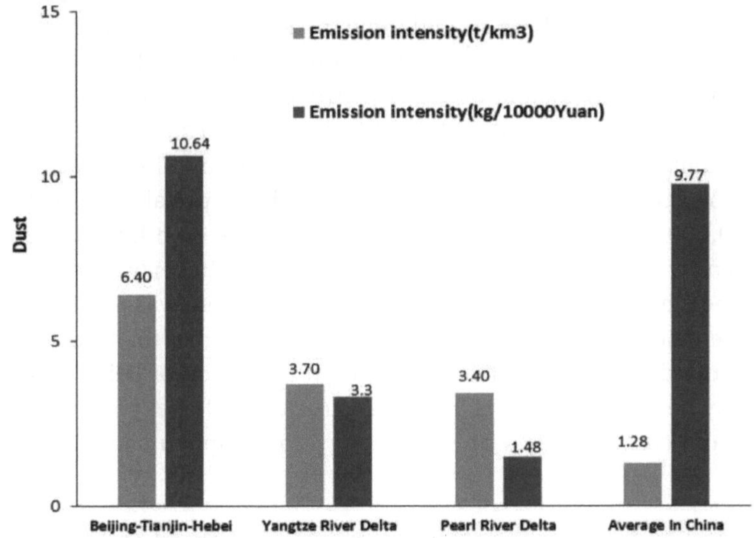

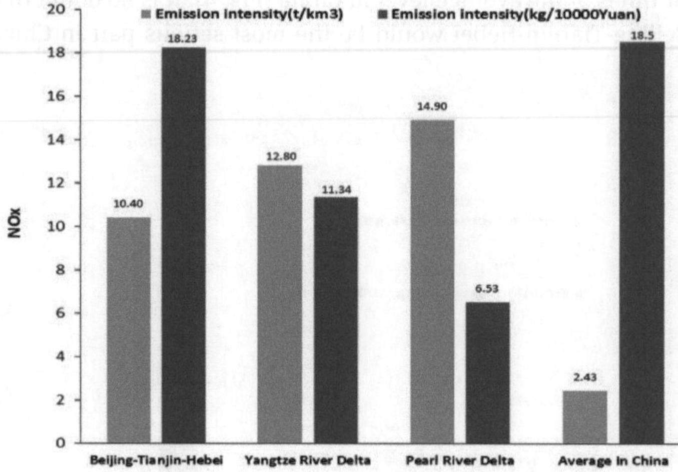

Fig. 4: Pollutant emission intensity in different regions of China (Data Resources: National Environmental Statistics Annual Report [5])

4 Conclusion

From the perspectives of the integration of Beijing-Tianjin-Hebei, the unbalanced economic structure in Hebei, whose second industry accounts for more than 52 %, was the main cause of air pollution in the region of Beijing-Tianjin-Hebei. The three regional economies, Beijing-Tianjin-Hebei, Yangtze River Delta and Pearl River Delta, have little difference in the proportion of their industrial structures. But from the energy consumption per 10,000 yuan GDP, Beijing-Tianjin-Hebei was 50 % higher than the Yangtze River Delta and Pearl River Delta, especially the coal consumption per 10,000 yuan GDP in Beijing-Tianjin-Hebei was even higher than the double. The intensity of pollutant emission in Beijing-Tianjin-Hebei was much higher than that of the other two regions; especially the smoke (dust) was nearly as double as that in the other two regions, which was four times higher than the average level in China. In order to improve the air quality in the region of Beijing-Tianjin-Hebei, it is highly needed to adjust its industrial structure in Hebei province; especially the steel and coal capacity with high energy consumption should be compressed.

Acknowledgement: This study was financially supported by the programs (121201103000161316), (K1327) and (12120114077201).

References

[1] Wang Na. 2016. Research on the causes of air pollution in Hebei and the prevention measures. Modern economic information (7).

[2] National Bureau of Statistics. 2015. [DB/OL]http://data.stats.gov.cn/workspace/index?m=hgnd.

[3] Cui Wenjin, Wang Lijuan. 2015. Hebei industrial structure adjustment and sustainable development. China management information18 (18).

[4] The Energy Statistics Department of National Bureau of Statistics. 2013. Chinese Energy Statistical Yearbook. Beijing: China statistical Publishing House.

[5] Ministry of Environmental Protection of PRC.2016. [DB/OL] http://www.mep.gov.cn/xxgk/xxgk_1/ndbg/

Zhang-Huang Ye, Jia-Qi Liu* and Yi-Ke Li

An Analysis on Chinese Primary Energy Consumption

Abstract: Energy is the material base for human activities, which is closely related to national economic security. Statistics show that there is a close positive correlation between GDP and energy consumption, GDP per capita and energy consumption per capita. In 2015, Chinese primary energy consumption reaches 3.014 billion tons of oil equivalent, which is the largest energy consuming country. The biggest feature of the primary energy consumption structure in China is that the proportion of coal is too high and the proportion of natural gas is too low, which is very irrational. But in recent years, China's energy consumption has shown some positive changes with low increment and slow growth rates. The analysis shows that the Chinese energy consumption per capita is at a level comparable to its economic development.

Keywords: energy; consumption structure; China

1 Introduction

The history of human civilization is a process of learning to use energy, discover new energy resources and consume a great deal of energy. Energy is the material base of human activities, the motive force of economic development and social progress. Energy is closely related to national economic security and the core interest of all countries in the world, which is also the focus of the game of big powers. Statistics show that since 1850, there is a close positive correlation between GDP and energy consumption, GDP per capita and energy consumption per capita [1,2].

2 Total Energy Consumption in China

According to the National Bureau of statistics, the coal accounts for 69.8 %, oil 21.2 %, natural gas 2.5 % (oil and gas 23.7 %) in Chinese energy consumption struc-

Zhang-Huang Ye, Jiangxi Science and Technology Normal University. No. 605, Fenglin Road, Nanchang, Jiangxi. Email: chuckverna@sina.com
Jia-Qi Liu, Institute of Geology and Geophysics, CAS. No. 19, Beitucheng West Road, Chaoyang district, Beijing. Email: liujq@mail.iggcas.ac.cn.
Yi-Ke Li, Institute of Mineral Resources, CAGS. No. 26, Baiwanzhuang Road, Xicheng district, Beijing. Email: like430@cags.ac.cn

DOI 10.1515/9783110540048-024,

ture in 2003. The coal accounts for 66 %, oil 18.4 %, natural gas 5.8 % (oil and gas 24.2 %) in Chinese energy consumption structure in 2013. The proportion of coal has declined slightly, and the proportion of oil and gas has increased slightly (Fig. 1). As China advances the energy supply and consumption revolution, it is expected that the proportion of oil and gas in Chinese energy consumption structure will continue to grow steadily.

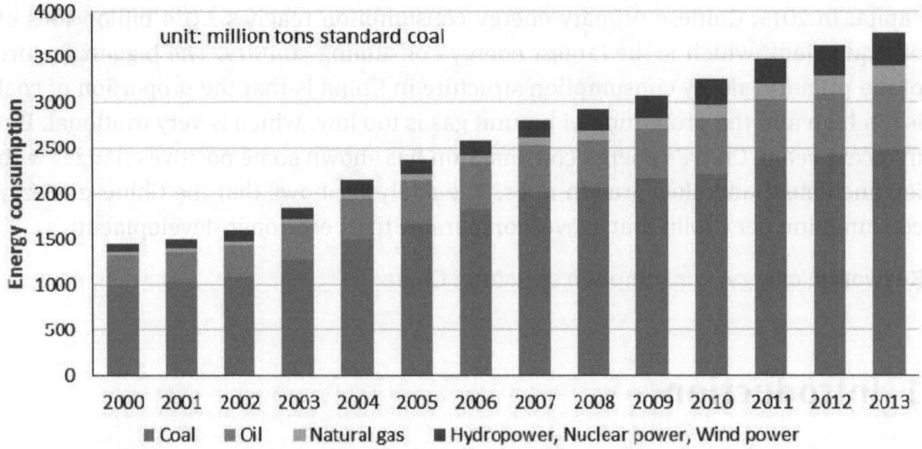

Fig. 1: The Chinese energy consumption structure from the year 2000–2013 (Data Source: National Bureau of Statistics Website [3])

Growing too fast is the most prominent problem of energy consumption in China. Its primary energy consumption surpasses the United States for the first time in 2009, which becomes the largest energy consuming country. The total primary energy consumption is 2322 MTOE, increasing by 1.5 times just in the past 9 years. The energy consumption in 2015 is 3014 MTOE, which increases by 692 MTOE compared to 2009 (Fig. 2). The annual rate of growth of energy consumption as high as 7.6 % from 2000 to 2015. At the same time, the proportion of energy consumption in China has also increased rapidly, which presents an obvious growth from 10.5 % in 2000 to 22.9 % in 2015 (Fig. 3). However, there is a turn point around 2013. The growth of energy consumption in China has slowed down significantly. In 2015, it is only 1.5 % higher than in 2014. The authors believe that this is the performance of new economic norm in the field of energy, and further attention is needed.

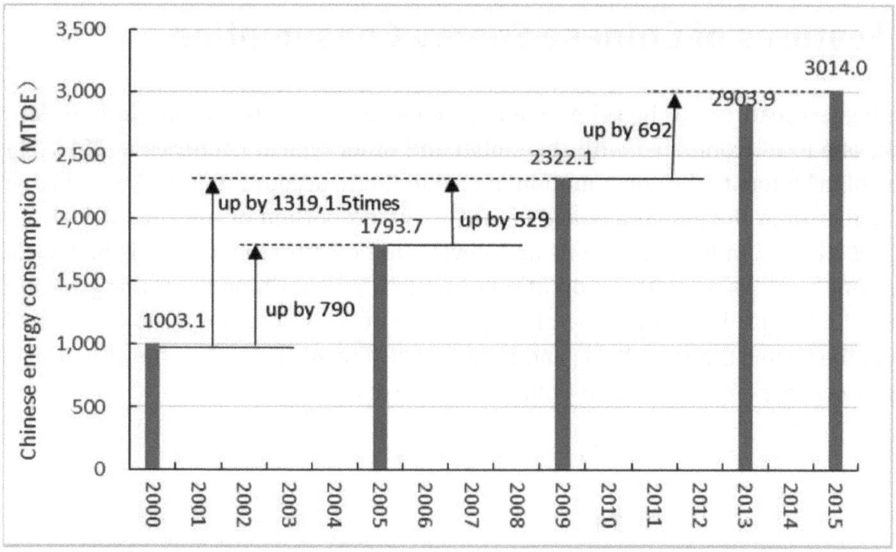

Fig. 2: Chinese energy consumption in China from 2000 to 2015 (Data sources: BP, 2016 [4])

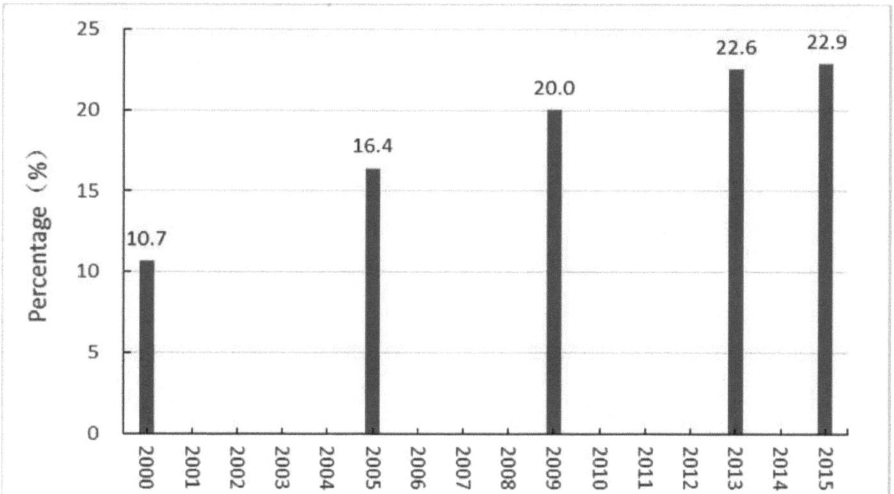

Fig. 3: The proportion of Chinese energy consumption in the world from 2000 to 2015 (Data sources: BP, 2016 [4])

3 Features of Chinese Energy Consumption

The biggest features of the primary energy consumption structure in China is "one high, one low", compared with the global and other typical countries in 2013. The "one high" refers to the consumption of coal in China accounting for 67 %, which is far higher than the global average of 30 %. The proportion of the United States is only 20 %, and India, a populous and developing country like China, is just 55 %. The "one low" refers to the proportion of natural gas in China accounting for just 5 %, which is far lower than the global average of 24 %. The proportion of natural gas in the United States is 30 %, and India is 8 % (Fig. 4). Therefore, the structure of primary energy consumption in China should be optimized urgently.

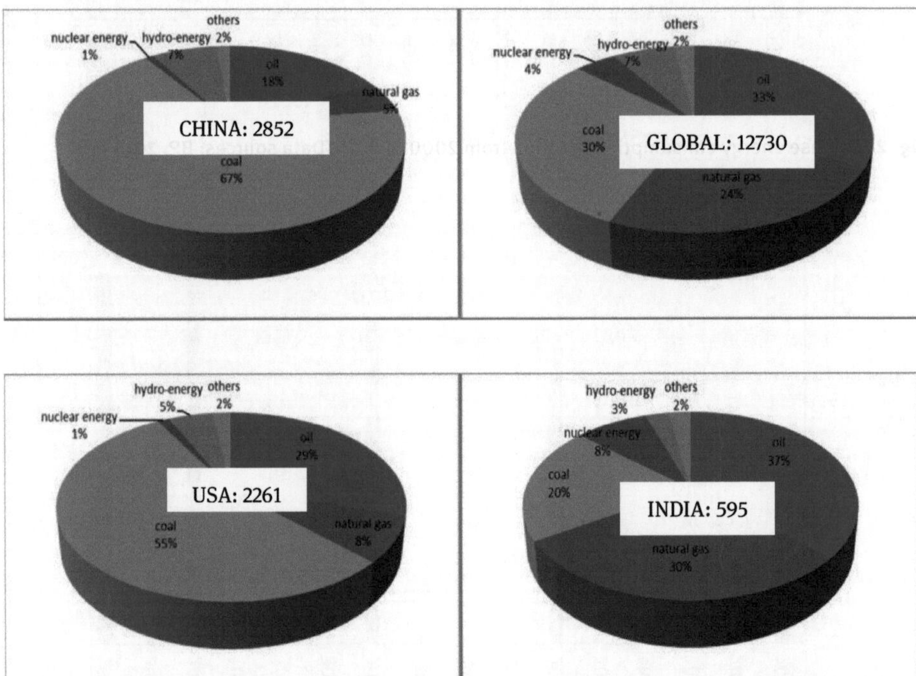

Fig. 4: The primary energy consumption structures in 2013 (unit: MTOE) (Data sources: BP, 2016[4])

But in recent years, Chinese energy consumption has shown some positive changes. One of them is that the annual increment of Chinese total primary energy consumption has declined from 210 million tons of standard coal per year from the year 2003–2011 to 130 million tons from the year 2012–2013 (Fig. 5). The growth of energy consumption has declined from the average of 8.1 % from the year 2003–2011 to 3.7 % from the 2012–2013 (Fig. 6).

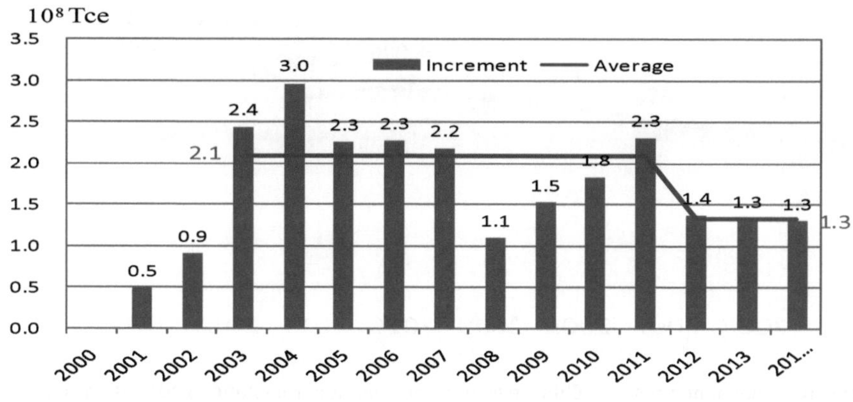

Fig. 5: The average increment of Chinese energy consumption (data sources: China Energy Statistics Yearbook [5])

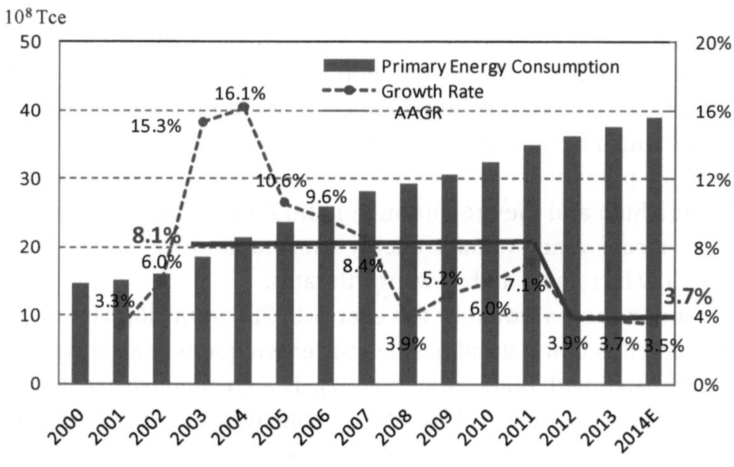

Fig. 6: The average growth rate of Chinese energy consumption

The second positive change is that the proportion of non-fossil fuels in Chinese energy consumption has been on the rise in recent years, which is from 6.7 % in 2006 to 10.1 % in 2013 (Fig. 7). However, in the component of non-fossil energy, the controversial hydropower proportion is high, accounting for about 75 %, and on the contrary, the inexhaustible wind power and solar energy are in low ratio.

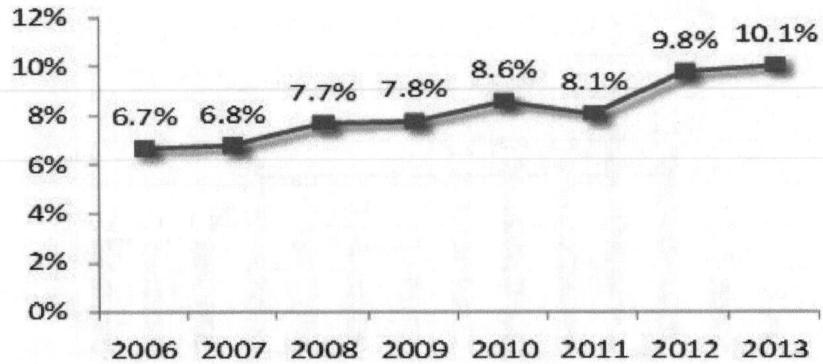

Fig. 7: Percentage of non-fossil in Chinese energy consumption from 2006 to 2013 (data sources: China Energy Statistics Yearbook)

4 Conclusion

In 2015, Chinese primary energy consumption reaches 3.014 billion tons of oil equivalent, which is the largest energy consuming country. However, from the energy consumption law, China's energy consumption is in line with the stage of its economic development. The energy consumption per capita of China is 1.93 tons of oil equivalents in 2013 which is slightly higher than the world average of 1.8 tons. The biggest feature of the primary energy consumption structure in China is that the proportion of coal is too high and the proportion of natural gas is too low, which is very irrational. But in recent years, China's energy consumption has shown some positive changes with low increment and slow growth rates.

It is predicted that natural gas would be the energy in 21 Century [6]. The reason why natural gas is favored is that natural gas has better environmental, social and economic benefits compared with other fossil fuels [7]. The greenhouse gas, mainly produced by fossil fuel consumption, will lead to a global average temperature rise by nearly 2 °C. The change would cause severe ecological environment destruction. In order to limit the greenhouse gas emission, *Kyoto Protocol* was published in 1997, which requires that developed countries shoulder their responsibility in reducing the carbon emissions from 2005, so do the developing countries from 2012. On November 12th, 2014, China and America issued *the Joint Announcement on Climate Change*, stating that the two countries should jointly deal with climate change [8]. *Paris Protocol* was approved by nearly 200 parties On December 12, 2015, which is a historic step for handling the global climate change. It is highly urgent for China to try every effort to enhance the energy utilization efficiency and encourage the more use of natural gas in the daily life.

Acknowledgement: This study was financially supported by the programs (2016ZWH028A–28) and (DD 20160124).

References

[1] Wang Anjian, Wang Gaoshang, Chen Qishen, etc. 2008. Energy and economy development [M]. Beijing: Geological publishing house

[2] Yu Lihong. 2008. Strategy research on energy resources substitution. [M]. Beijing: China modern economy publishing house

[3] National Bureau of Statistics. Data of 2015[DB/OL]. http://data.stats.gov.cn/workspace/index?m=hgnd

[4] British Petroleum Company. 2016. BP statistical review of world energy 2015[R]. London: British Petroleum Company

[5] Energy Section of National Bureau of Statistics. 2013. Chinese Energy Statistical Yearbook. Beijing: China statistical Publishing House

[6] Scott Montgomery. 2012. Translated by Song Yang and Jiang Wenbo. Global energy trends [M]. Beijing: mechanical industry press

[7] Hu Jianyi, Guo Bincheng. 2011. Natural gas: the important stage and trend of low carbon energy development [J]. Journal of engineering science in China, 13(4): 9–14

[8] Xinhua News Agency. 2016. [DB/OL]http://news.xinhuanet.com/world/2016–04/01/c_128854045.htm

Zhang-Huang Ye, Yi-Ke Li* and Qiang Yan

Global Natural Gas Consumption History and Energy Consumption Pattern

Abstract: The increasingly important role which the natural gas plays in the structure of energy consumption is not only reflected in the steady growth of total energy consumption, but also in its yearly increasing proportion of the primary energy consumption. The basic situation of global energy consumption is that fossil energy would predominate for a long time, which would remain a high percentage about 85 %–90 %. However, the consumption percentage of non-fossil energy and new energy would increase gradually. The internal structure of fossil energy would be insistently adjusted, that is, the consumption percentage of coal would decline, and the petroleum consumption would come to the summit. Meanwhile, the consumption of natural gas would increase greatly.

Keywords: natural gas; energy; consumption; structure

1 Introduction

The history of human civilization is to learn to use energy, discover new energy and consume energy greatly. Energy is the material base of human activities, as well as the impetus of economy development and social progress. Energy is closely relevant to the national economy security and is the core benefit of most countries around the world. Therefore, it is of great importance to analyze and prejudge the situation of global energy consumption.

2 Global Consumption of Natural Gas

The increasingly important role which the natural gas plays in the structure of energy consumption is not only reflected in the steady growth of total energy consumption, but also in its yearly increasing proportion of the primary energy consumption.

Zhang-Huang Ye, Jiangxi Science and Technology Normal University. No. 605, Fenglin Road, Nanchang, Jiangxi. Email: chuckverna@sina.com
Yi-Ke Li, Institute of Mineral Resources, CAGS. No. 26, Baiwanzhuang Road, Xicheng district, Beijing. Email: like430@cags.ac.cn
Qiang Yan, Institute of Mineral Resources, CAGS. No. 26, Baiwanzhuang Road, Xicheng district, Beijing. Email: cagsyq@163.com

DOI 10.1515/9783110540048-025,

From the diagram of the overall global consumption of natural gas from the year of 1900 to 2015 (Fig. 1), it can be seen that the global consumption of natural gas have been in an obviously rising trend. It was increased from 0.0067 trillion cubic metres in 1900 to 3.47 trillion cubic metres in 2015. That is to say, it is increased by nearly 520 times and the average annual growth is 5.6 %. The global consumption of natural gas can be divided into two phases: (1) the pre-war slow growth stage (from the year of 1900 to 1945). The global consumption of natural gas was small and growth speed tended to fluctuate significantly. (2) The post-war smooth rapid growth stage (from the year of 1945 till now). The global consumption of natural gas in 1945 was 0.1289 trillion cubic meters. For the next 70 years, the negative growth of natural gas consumption only happened in the year of 1975(−0.25 %), 1997(−0.33 %) and 2009(−2.37 %). Because of the global financial crisis in 2009 and the economic recession, the global natural gas consumption declined than that in year of 2008. In this stage, the average annual growth rate is 4.8 %, which is characterised with a faster and steady growth.

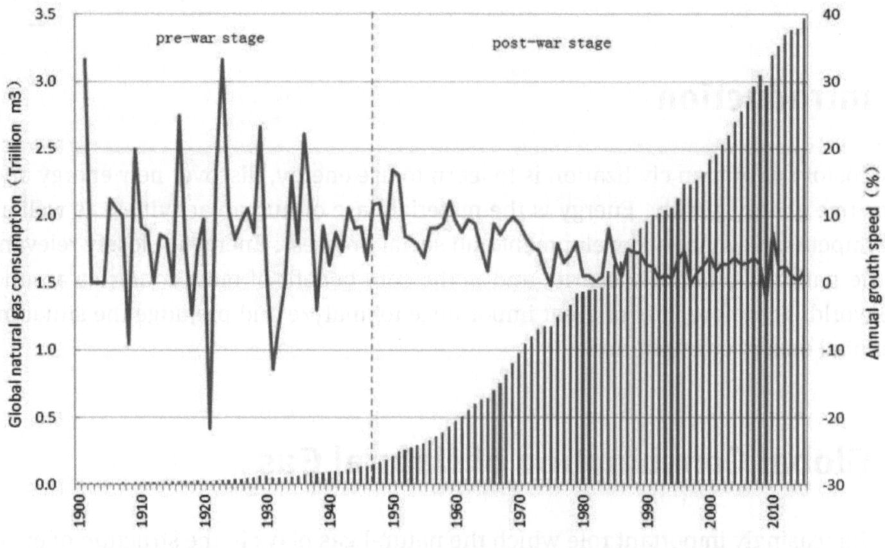

Fig. 1: Global natural gas consumption and the annual growth speed (data source: BP [1])

3 Global Primary Energy Consumption

From the diagram of the global primary energy consumption structure from the year of 1900 to 2015, it can be seen that the proportion of natural gas in the global primary energy consumption structure is increasing annually, from 10 % in 1900 to 23.8 %

in 2015. Among different types of the primary energy, the natural gas is the only one which maintains sustainable growth in the proportion of energy consumption. On the contrary, the proportion of coal was in the declining trend, and the proportion of oil was in the increasing trend before the 1970s, yet it has been in the constant declining trend since 1970s (Fig. 2).

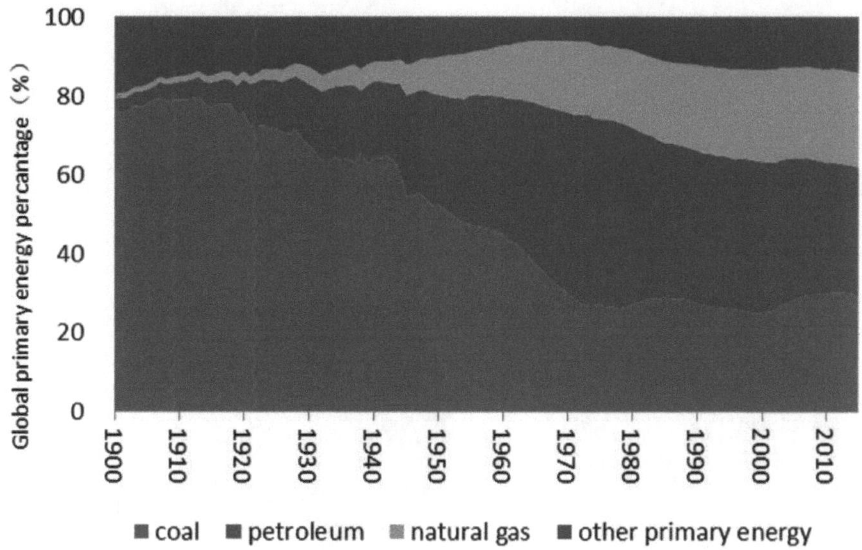

Fig. 2: Global consumption structure of primary energy (data source: BP [1])

The global major primary energy consumption ratio changes are as follows if taken a decade as a cycle. The consumptions of natural gas, oil and coal were 3.93, 10.53 and 13.49 respectively in 1960, taking the 100 mote as the unit (the same below), and the proportions in global primary energy consumption were 13.1 %, 35.0 % and 44.9 % respectively. 8.92, 22.57, 15.5 were consumed respectively in 1970, and the according proportions were 18.1 %, 45.7 %, 31.4 %; 12.94, 30.98, 18.06 were consumed respectively in 1980, and the according proportions were 19.5 %, 46.7 %, 27.3 %; 18.7, 31.63, 22.15 were consumed respectively in 1990, and the according proportions were 23.0 %, 39.0 %, 27.3 %; 21.77, 35.84, 23.43 were consumed respectively in 2000, and the according proportions were 23.1 %, 38.0 %, 24.8 %; 28.68, 40.4, 34.69 were consumed respectively in 2010, and the according proportions were 24.0 %, 33.8 %, 29.0 %. The above data clearly show that the global natural gas consumption was in steady growth trend, and only the natural gas “makes the cake bigger” among various primary fossil energy (Fig. 3).

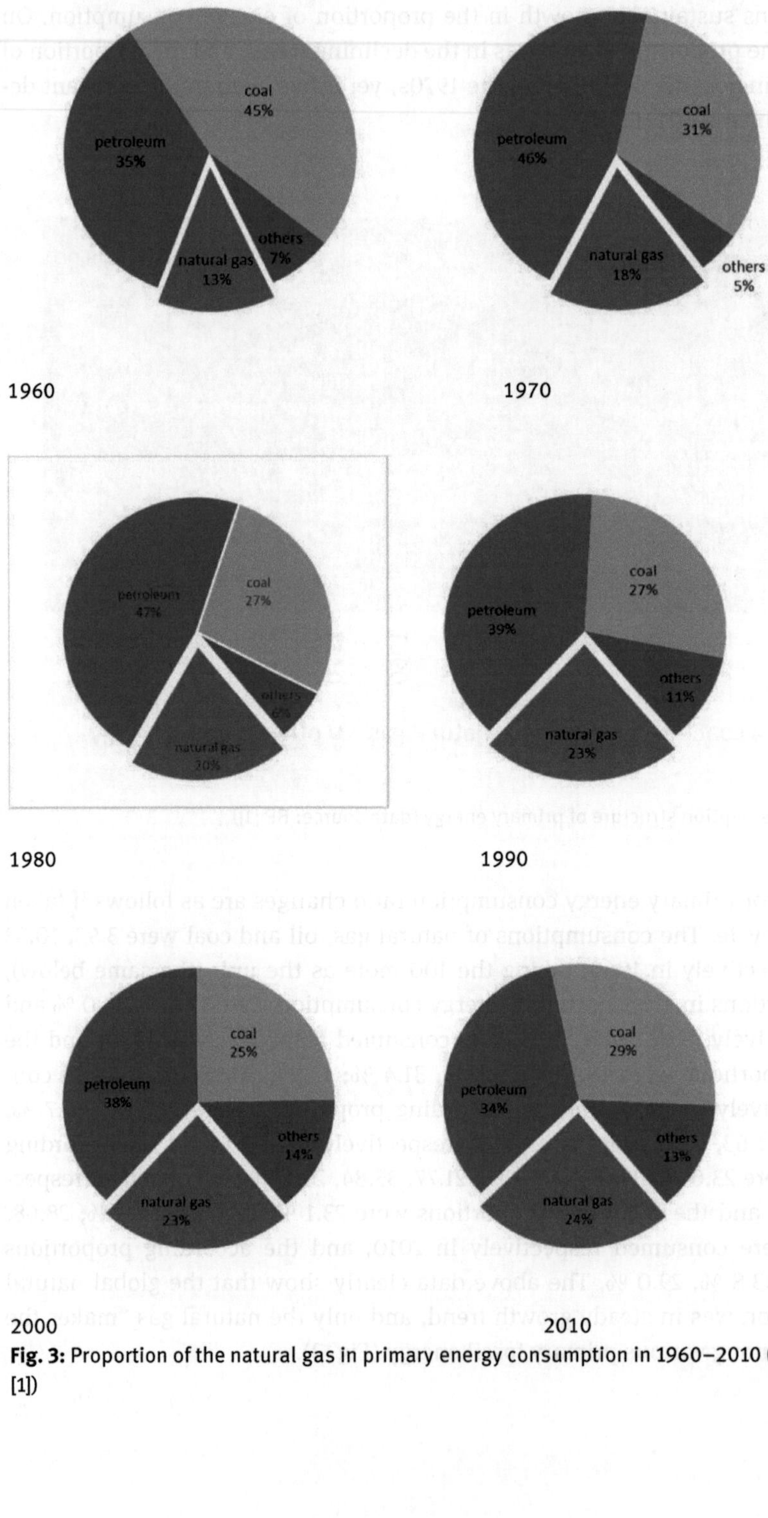

Fig. 3: Proportion of the natural gas in primary energy consumption in 1960–2010 (data source: BP [1])

4 Structure of Global Energy Consumption

According to BP's latest data, the world's total energy consumption in 2015 was 13.147 billion tons of oil equivalents. The fossil energy has accounted for as much as 86 % of total energy consumption. The oil is still the largest energy, accounting for 32.9 %; coal is the second largest energy, accounting for 29.2 %; the natural gas in the third place, accounting for 23.8 %. That is the basic global energy consumption pattern.

It is predicted that natural gas would be the energy in 21 Century. The reason why natural gas is favored is that natural gas has better environmental, social and economic benefits compared with other fossil fuels [2]. The greenhouse gas, mainly produced by fossil fuel consumption, will lead to a global average temperature rise by nearly 2 °C, which would cause severe ecological environment destruction. *Paris Protocol*, approved by nearly 200 parties On December 12, 2015, would exert great effect on the global energy consumption.

5 Conclusion

The basic pattern is that fossil energy will still predominate in the global primary energy consumption for a long time, maintaining 85 % to 90 %. However, the consumption percentage of non-fossil energy and new energy would increase gradually. The internal adjustment in the primary energy consumption will be inevitable, i.e. the consumption percentage of coal would decline, and the petroleum consumption would come to the summit. Meanwhile, the consumption of natural gas would increase greatly. As result, the natural gas will become the "first energy" in the structure of energy consumption [3–6].

Acknowledgement: This study was financially supported by the programs (K1327) and (121201122087).

References

[1] British Petroleum Company. 2014. BP statistical review of world energy 2013 [R]. London: British Petroleum Company

[2] Hu Jianyi, Guo Bincheng. 2011. Natural gas: the important stage and trend of low carbon energy development [J]. Journal of engineering science in China, 13(4):9–14

[3] Tong Xiaoguang. 2010. Significance and possibilities of improving the percentage of natural gas in energy structure in China. Natural gas industrial (10)

[4] Zhang Weizhong. 2011. Global development trend of natural gas. International Petroleum Economy, (6)

[5] International Energy Agency. 2012. Golden Rules for a Golden Age of Gas. Paris, France: International Energy Agency
[6] David Victor, Amy Jaffe and Mark Heyse. 2010. Natural Gas and Geopolitics: From 1970 to 2040[M]. (Translated by Wang Zhen and Wan Hongyan). Beijing: Petroleum Industrial Press

Xin Shen, Shao-Lin Pan, Xiao-Feng Xu and Gui-lin Zhou

Study on Design Idea of ADB-funded, Integrated Treatment Project of Huainan Urban Water System

Abstract: The paper is based mainly on the design and planning concepts summarized from the implementation of ADB-funded, Integrated Treatment Project of Huainan Urban Water System. The integrated treatment of the water system follows the principle of "sewage interception for point sources, control of plane sources, treatment of internal sources, and remediation of ecological system" to improve the sewage collection and transport system, urban water environment, flood management, and urban lake works, and to ultimately enable an effective management of surface water resources. Through such means as sewage interception, water system interconnection, improvements in flood discharge standards and surrounding environments, and ecological remediation for water bodies, a safe and healthy water environment has been established, beautiful and coordinated water landscapes and harmonious waterfront spaces created, and local unique humanistic connotations promoted.

Keywords: Urban water system, Sewage interception, Flood discharge, Lakes, Integrated treatment

1 Introduction

The Integrated Treatment Project of Huainan Urban Water System was established by Huainan Municipal Government to solve urban water environmental issues with the loan provided by the Asian Development Bank (ADB). The project, with a total investment of about RMB 1.9 billion, covers the main itemized projects like sewage interception, urban water system treatment, flood drainage pump station upgrading as per improved standards, ecological remediation for lakes and water quality monitoring; by taking a series of engineering and non-engineering measures, Huainan's standards for urban flood prevention and drainage, together with its water environmental functions, can thus be improved. In combination with the survey find-

Xin Shen, No.180, Xiangzhang Ave., Hefei, China 1010331469@qq.com
Shao-Lin Pan, No.180, Xiangzhang Ave., Hefei, China
Xiao-Feng Xu, No.180, Xiangzhang Ave., Hefei, China 1187052872@qq.com
Gui-lin Zhou, No.180, Xiangzhang Ave., Hefei, China

DOI 10.1515/9783110540048-026,

ings of the water system and the concept of "building a sponge city", the paper provides the idea for the integrated treatment of urban water system.

In recent years, with the promotion of scientific development philosophy and the enhancement of ecological protection consciousness, the competent authorities in different places are endeavoring to achieve a win-win situation in flood prevention (including drainage) and the ecological protection in urban water system improvement [1,2]. Due to different characteristics in topography and water system and different levels in social and economic development, the planning of the integrated treatment of urban water systems is still at a stage of exploration with many aspects worthy to probe into [3].

2 Considerations and Treatment Principles for Urban Water System Treatment

2.1 Considerations for Urban Water System Treatment

2.1.1 Survey and Analysis of Present Situation within the Scope of the Project

The planning and design of the project are normally based on the full understanding of present state, and that is especially true for the urban water system treatment. For example, the project design schemes will be directly influenced by such factors as the quantity and positions of existing sewage drainage outlets, the accuracy in general survey of existing pipeline network, the relationship between a river course and buildings on both sides, the surrounding environment, the incoming water quality and the area of urban lakes, as well as lake outlets and their surrounding environment.

2.1.2 Basic Data Analysis

Basic data analysis and derivation are fundamental to the expansion and divergence of the ideas for the integrated treatment of urban water systems; the urban construction is often viewed as a complex, in which the hydrological conditions of urban water system is greatly influenced by external environment, and the selection of parameters play a significant in planning and designing water system, and demonstrating the scale of a pump station.

2.2 Idea of Treatment

From the overall thinking of "treatment of system, processes and basin", the core eight-hierarchy philosophy featuring water-themed safety, resource, environment, ecology, water landscape, culture, management and economy can be drawn, which goes further to develop the itemized treatment idea characterized with "discharge reduction at internal sources, dredging at external sources, water quality purification and ecological remediation".

In the integrated remediation planning, overall considerations should be given to the safety of flood prevention and discharge, environmental ecology, smooth transportation and other requirements; with the water system treatment as a link, the existing water system should be fully utilized to completely improve urban flood prevention and discharge capabilities; through the separation of rainwater from sewage, the water quality can be improved; and urban water system landscapes can be established. When all these means are taken, a network of urban water systems, composed of "streams–wetlands–lakes–pump stations", can be finally achieved, and thus make come true a beautiful city featuring "interdependent water system and city, interconnected water systems, harmony between human and nature, and clear water and green gardens".

2.2.1 Improvement of the Sewage Collection and Transport System

The sewage design should be directed by the guideline which highlights area and systematic discharge while observing the sequence from sewage interception to pollution control and then to river regulation, and bringing into practice green design and sustainable utilization; through sewage interception at the discharge outlets along the river course in the project areas, it should be secured that no sewage will flow into the river course.

2.2.2 Urban Water Environment and Flood Management

On the precondition that the basic functional requirements of an urban water system, including flood prevention and discharge, are met, the integrated treatment for the urban water system is expected to be planned and designed to reflect the following features:

1) Perspectiveness. The water system treatment is combined with a general urban development program, tourism demands, and urban construction; through overall consideration and integrated coordination, the water system resources are developed and utilized comprehensively, which can lay a consolidated foundation for sustainable urban development.

2) Coordination. The relationship between flood prevention and natural environment creation should be well coordinated; global considerations are required for rivers, city, human and nature; while the requirements for flood prevention and discharge are to be met, the demands for landscape environment should not be ignored, either.
3) Diversity. A diversified water ecological environment is to be built, and the integrated function of different channels is to be fully exploited for the maintenance of ecological balance, the adjustment of area microclimate, recreation and entertainment, so as to achieve the sustainable development of waterfront green landscapes.
4) Water accessibility principle. The accessibility of the urban river water system is to be highlighted and an approachable waterfront space be created, which can realize good visibility, adjacency and touchablity and promote the integration between human and nature, resulting in a good waterfront landscape environment.
5) Integrated principle. In the planning process of a water system, the water resource allocation, hydrogeology, urban planning, landscape design, environmental aesthetics and other expertise shall be well coordinated, so that a reasonable and feasible scheme for the integrated treatment plan can be put forward.

2.2.3 Urban Lake Improvement Project

Lake regulation works: In line with principle of "ecological treatment and introduction of clean water through natural and harmonious means", different but specific ecological schemes (scientific desilting, ecological remediation, ecological slope protection, long-term management and other measures) are designed to cut down inflowing pollutants and hold back pollutant sources through the interception of slope protection, ensuring the introduction of clean water, the improvement in both water quality and regional environment, and the creation of self-regulating and healthy ecological systems in lakes. In addition, a combination of aeration, plant ponds, subsurface wetlands and ecological floating islands is adopted to ensure stable water quality in lakes [4].

3 Issues in Design Process of Urban Water System Treatment and Relevant Measures

3.1 Hydrological Computation

The calculation of flood flows to be controlled is rather complex, as the towns are mixed with crop fields and villages, and the flood from upstream mountainous area need to be taken in during drainage period. Different calculation methods for flood control are to be adopted on the basis of specific conditions in each drainage area, including terrain, landform and ground cover.

3.2 Improvement Works for Sewage Collection and Transport System

The urban sewage collection and transport system feature a broad coverage and long pipelines. The primary and secondary trunk pipes for the sewage collection should be combined closely with the urban planning, and the framework for sewage collection systems in areas is to be built up, so that the collection rate can be improved and the treatment and reuse be facilitated.

3.3 Treatment of Urban Water Systems

The modern, integrated treatment of urban rivers is expected to establish a design philosophy for river course treatment with ecological water conservancy and the conventional ideas regarding engineering water conservancy may have to be renovated. The new philosophy should reflect the ideas as follows: an idea concerning the harmony between human and nature, sustainable development of rivers, and humanistic features of rivers. On the condition that flood prevention and discharge are met, the design should include "safety, nature, ecology, water accessibility and landscape" into its objectives. Aesthetic conception, design and practice are to be utilized to construct urban rivers with integrative, multiple functions, which will further contribute to a clean, beautiful, ecologically healthy living environment with convenience and comfort.

Diversified forms of a river plane. The plane form of a natural river normally has its twists and turns; while the requirements for flood prevention and discharge are met, the multiple forms of the existing river plane are to be kept to the greatest extent by adapting the design to the natural bends.

Diversification and naturalization of a river course section. A river space is to be reasonably utilized; single, compound, or anomalous and asymmetrical sections can be adopted, according to the conditions in a specific river segment, to meet human desire in sharing the river body as well as their demands for leisure, water accessibility and closeness to nature.

Ecologization and landscaping of bank protection along the river course. Simple, hard bank protection is not to be used preferably, and the ecological bank protection materials and structures are to be selected according to the actual local status, while considerations are also given to the demands for safety, ecologization and landscape.

Characteristics of river waterfronts. Different segments of a river have different natural scenes, land usage, history and cultures, so it is required to create featured and pleasant river environment.

The improvements in urban channels, flood prevention and discharge facilities should be combined with a number of functions, including flood prevention, discharge and ecological remediation. Therefore, a safety lifeline for flood prevention and discharge along the river is to be built in a scientific and reasonable manner, engineering and ecological measures are to be taken to control the soil loss along the bank, while the ecological environment is to be actively restored, which should fully demonstrate the organic combination of water-related safety, environment, landscape, culture and economy. Ecological water conservancy should be established in present cities, including water accessibility, healthy water quality in rivers and lakes with the function of flood and disaster control, ecological water system remediation, as well as landscape waterfront space.

Given the fact that most of the designed urban channels in the project are located in the built-up areas of the city, the plane layout and section design of river courses followed the principle that made adjustments according to local conditions and fully considered surrounding premises and migrant status in an aim to cut down workload and demolition. For some river segments with dense residential quarters and a large amount of demolition, upright ecological retaining walls were adopted on the cross sections of the river course, and the planes were appropriately re-routed according to the house removal and actual surrounding conditions.

In combination with the administrative requirements for flood prevention and discharge, an open channel was reserved with a 10 m-wide boundary line for management on both sides, and a 4m-wide colored flood prevention channel, made of pervious concrete, were set up, which was coupled with a green belt. For flood prevention channels, pervious ecological pavement was adopted, which were reasonably connected with municipal roads. On the first place, the flood prevention can be kept unblocked in a flood period; secondly, no ponding or light reflection exists on the pavement, as the rainwater can infiltrate directly; finally, the flood prevention channels (also for flood discharge) can form a green corridor together with the green belt along the river, thus capable of intercepting initial rainwater, and by forming a green urban system for pedestrians and bicycles, providing surrounding residents with a good place for rest and exercise. The main technical difficulties were found in structure design of the pervious pavements and the allocation of signs and green belts.

3.4 Non-engineering Measures

The non-engineering measures for a urban water system mainly include general survey of flood disasters, designation of dangerous areas, determination of early-warning indicators, including critical rainfall and water level, setting-up of monitoring and early-warning system, establishment of responsible organization system, perfection of prevention plans, publicity, training and drillings. These are non-engineering measures mainly targeted at the monitoring and early-warning of a flood caused by the rainfall in the internal water system, and at mass prediction and disaster prevention, so as to reduce the casualties and property loss.

In addition, they also include general survey of flood disasters in an urban area, designation of dangerous areas of flood disasters, construction of automatic collection and manual observation sites to form a monitoring site network for rain status, construction of a monitoring and early-warning center integrating networks, database and geographic information technology; establishment of alarming system consisting of the early-warning center and key prevention areas; preparation of flood forecast schemes reflecting the hydrological characteristics of the area, as well as scientific and operable flood prevention plans, establishment of an early-warning mechanism and an organization system with mass prediction and disaster prevention, in-depth publicity of knowledge about flood prevention and disaster reduction, thus setting up a non-engineering measure defense system incorporating technologies and management.

4 Conclusion

As a water system is seen as an important subsystem in the urban ecological system, a good urban water system becomes one of the foundations for a "resource-saving and environment-friendly", harmonious community; the treatment of urban water environment, like the revitalization of blood capillaries of a city, lays a solid foundation for healthy development of other urban construction works. The improvement of a water system not only serves as a key content in urban construction, but also becomes an important way to improve urban ecological environment and reconstruct urban ecological system. Therefore, it is of great significance to explore and perfect main principles, final plans and realization approach for the integrated treatment of urban water systems which is based on actual conditions in a local and on the experience of completed projects. Such an endeavor will definitely beneficiary to the construction of a "harmonious and healthy" urban water system, to the development of regional advantages and to the realization of a faster and better growth.

References

[1] Zhou Yibing, Li Wanbin et al, Function of Urban Water System Planning Control for Establishment of Ecological City [C], Proceedings of Annual National Planning Conference (2008)

[2] Pan Shaolin, Shen Xin et al, Thoughts about Beautiful and Compatible Urban Inland River Design [J], Urban Construction Theory Research, 2013.12

[3] Lei Yu, Ran Chunwang et al, Initial Study to Ecological Treatment of River Courses [J], Water Conservancy Science and Technology and Economy, 2009.4

[4] Shen Xin, Brief Introduction of Application of Constructed Wetlands in Treatment of Longhu Huainan in Huainan City [J], Basic Level Construction, 2016.33

Hong-Na Liu and Hui Xiao

Research on Popularization and Application of Agricultural Water Saving Technology in China

Abstract: As the ecological environment and global climate change, more and more countries have realized the existence of water crisis and urgency, and then adopted a series of measures to improve the existing water resources utilization. China is an agricultural country, and farmland is water conservation, according to some statistics that water consumption accounts for about 80 % of the total water, therefore, developing water saving agriculture is vital for effective utilization of water resources, it is also an important part of agricultural modernization, is an organic part of the rural sustainable development. In order to improve the spread of water-saving irrigation technology in agriculture, the diffusion process of water-saving irrigation in agriculture was analyzed in this paper. By using questionnaire and analytic hierarchy process (AHP) to evaluate the main influencing factors for spreading China's water-saving agricultural technology, and determine the sorting of factors affecting to make clear the emphasis ,and finally, build a reasonable system of promoting water-saving agricultural technology.

Keywords: Agricultural; Water saving technology; Popularization and application

1 Introduction

With the increasing supply and demand of global water supply, the whole world has made the development of water-saving and efficient agriculture an important measure for the sustainable development of agriculture. Agricultural water use is also the largest in China, in which agricultural irrigation water amount accounted for 90.2 percent in agriculture. At the same time, due to the effect of farming in traditional agriculture, agricultural irrigation water use coefficient has been low, the current national average of only between 0.4~0.5, agricultural irrigation utilization rate is much lower than the developed countries, which results in a significant portion of the water being lost in the process due to leakage or evaporation. Therefore, under the condition of the lack of water resources, how to improve the utilization rate of agricultural irrigation water, the implementation of water saving agriculture to alle-

Hong-Na Liu, Harbin University of Commerce, No. 1 Xuehai Street, Songbei District, Harbin, China. Email:hljliuhongna@163.com
Hui Xiao, Heilongjiang Institute of Technology, No.999Hongqi Street, Daowai District, Harbin, China. Email:xhlhn@sohu.com

DOI 10.1515/9783110540048-027,

viate the water shortage situation is critical.

The adoption of agricultural water-saving irrigation technology, can improve agricultural irrigation water use ratio to in a certain extent, therefore, agricultural water-saving irrigation technology diffusion is a key factor to improve agricultural irrigation water use ratio. Although the state has increased its support for water-saving irrigation since 1995. improved agricultural production conditions and comprehensive production capacity of grain by developing water-saving irrigation techniques, and improved the utilization ratio of litigation water, but there is a big gap when compared with developed countries, both in the area of the water-saving irrigation technology and water utilization,. Therefore, how to further improve the technology of water-saving irrigation in agriculture and promote the diffusion of water-saving irrigation technology is the important factor in the sustainable development of agriculture in China.

2 Process of Agricultural Water-saving Irrigation Technology Diffusion

Agricultural water-saving irrigation technology diffusion is the interactive process that agriculture technology adopters to source learn from agricultural water-saving irrigation technology and apply the technology, which is the process that agricultural technology adopters cognize, study and adopt from water-saving irrigation technology and relevant experience. The ultimate goal of the spread of water-saving irrigation in agriculture is to use water-saving irrigation techniques to reduce the use of irrigation water without reducing crop yields, or increase crop yields by increasing the utilization of irrigation water under constant water supply. Therefore, the effect of the spread of water-saving irrigation technology will be directly affected by the spread of water-saving irrigation technology.

In many countries, water households are actively engaged in irrigation management. The practice is to transfer some or all of the irrigation management functions from state agencies to nongovernmental organizations or groups organized by farmers. Research on the federation of water in India or farmers organization's operation shows that government agencies and non-governmental organizations in the financial, technical and management support for farmers' water organization and organization management together with water irrigation system can improve the publicity of water resource utilization and effectiveness to ensure that the water in a timely manner, so that the irrigation department received operation management fees for facilities maintenance and maintenance(Thankhiwale, 2001). Privatization of small-scale irrigation projects in Bangladesh can fully mobilize farmers' incentives and promote investment. Open markets have led to continued growth in irrigation, higher farm incomes and higher employment opportunities (Aartvannes,

2001). To eliminate the uncertainty for the influence of the farmers' cooperative behavior, at the same time of promoting irrigation district management responsibility diversification, the government should be in solving conflict between area or association play a proper role (Hongyun Han, 2004)

In China, current practice of agricultural water-saving irrigation technology diffusion are influenced by many factors, including natural environment, social condition, the farmers themselves, water-saving technologies, five main factors influencing the government behavior, each kind of specific factors can be further divided into multiple influence factors, according to the final goal, primary subgoal, secondary subgoals few level will influence the spread of water-saving agricultural technology promotion factor analysis and hierarchical, as shown in Table 1 :

Tab. 1: The influencing factors of agricultural water-saving irrigation technology diffusion

The Influencing Factors Of Agricultural Water-saving Irrigation Technology Diffusion A	Nature Environment B1	Time and space distriCution of water resources and water and soil resources matching condition C1
		The sustainaCle development of ecological environment and construction C2
		Cultivated land quality and scale C3
		Ground water depth C4
	social condition B2	Food safety issues caused Cy increasing population C5
		The necessity of agricultural structure adjustment caused Cy the increase of farmers' income C6
		Demands of speeding up the traditional water conservancy to modern water conservancy, traditional agriculture to modern agriculture C7
	Peasant B3	Years of land contract C8
		Degree of education C9
		Proportion of laCor force in the family C10
		Accept the new technical aCility and creativity C11
		Investment aCility C12
		Recognition and acceptance of new technology of water saving C13
		Agrotype C14
		Farmer ideas C15

Government B4	Institutional innovation leading Cy government C16
	Extension agency C17
	Expenditure C18
	Water price C19
	Enthusiasm of extension worker C20
	The property of water-saving facilities C21
	Convenient degree of information C22
	Traffic convenience degree C23
Water-saving Irrigation Technique B5	Productivity effect C24
	Costs of water saving technology C25
	Complexity of operation on water saving technology C26
	Adaptation degree to local conditions C27
	The degree of all kinds of water saving technology research and development C28
	DuraCility and service life C29
	The form of water-saving irrigation equipment C30
	Investment and earning C31
	Market price of crops C32

3 Influence Factor Analysis on Agricultural Water Saving Technology Diffusion in China

Water-saving agricultural technology popularization diffusion factor evaluation is to determine the factors affect the importance of the sort, for the major in water-saving agricultural technology popularization. Methods of influencing factors include subjective scoring method, analytic hierarchy process, decision tree method and network model method. The current adoption of the "expert scoring comparison method", which is susceptible to the subjective consciousness, has certain limitations. There are many factors influencing the spread of water-saving agricultural technologies, some can be measured by quantitative measures, others are vague and difficult to quantify. In order to reduce the arbitrariness of judgment and enhance the credibility of the evaluation results, this paper adopts the AHP method (AHP) to judge the method of the questionnaire survey.

3.1 Construct Judgment Matrix

This questionnaire survey is mainly in colleges and universities across the country irrigation and water conservancy expert, the area of Heilongjiang province water conservancy department workers, the agricultural sector workers, department of science and technology workers, agriculture extending stations at the grass-roots level staff. The questionnaire was determined by reading a large amount of literature and conducting interviews with experienced agricultural water experts. In the questionnaire investigation stage, and by respondents feedback to modify the questionnaire, researchers surveyed generally reflect the influence factors of the listed in the questionnaire is quite comprehensive.

The questionnaire was asked to evaluate the importance of the affected factors identified by the experience. In order to facilitate statistical analysis, the respondents were asked to compare the factors in the questionnaire according to the scale (see table 2). You can construct the judgment matrix as shown in table 3.

Tab. 2: 1~9 Scaling scoring criterion

Scale	Definition
1	Factor i is equal important as j
3	Factor i is slightly important than j
5	Factor i is obviously important than j
7	Factor i is much important than j
9	Factor i is extreme important than j
2, 4, 6, 8	The importance of factor i compares with j is among the above results
Reciprocal	The importance of factor j compares with i is the reciprocal of i compares with j

The questionnaire was asked to evaluate the importance of the affected factors identified by the experience. In order to facilitate statistical analysis, the respondents were asked to compare the factors in the questionnaire according to the scale (see Table 2). You can construct the judgment matrix as shown in Table3.

Tab. 3: Judgment matrix

A	B1	B2	......	Bj	......	Bn
B1	b11	b12	......	B1j	......	b1n
B2	b21	b22	......	b2j	......	b2n
⋮	⋮	⋮		⋮		⋮
Bi	bi1	bi2	......	bij	......	bin
⋮	⋮	⋮	......	⋮	......	⋮
Bn	bn1	bn2	......	bnj	......	bnn

Here bij is Bi compares with Bj, and for A, Bi relative importance

3.2 Investigation Situation

The survey is a total of 320 questionnaires, the questionnaire distributed and recycled concrete are shown in Table 4.

Tab. 4: Questionnaire recovery rate

Respondents type	Number of questionnaire	Recovery number	Recovery rate(%)
Cropland irrigationist	90	25	28
Water department staff	60	15	25
Agricultural sector staff	50	11	22
Staff at Science and technology department	60	18	30
staff at the grass-roots agricultural promotion station	60	13	22
Total	320	82	26

The rate of recovery of 26 per cent of the survey was compared with 15 per cent (R.L.B ingetal, 1996), and the results were acceptable.

Experience the data shows that respondents are quite rich, not only understand the national policy, the situation of water resources in China and the mechanism of water-saving agricultural technology to farmers at the same time also have a fairly clear understanding of technology adoption situation, this ensures the reliability of the survey results.

3.3 Estimation Matrix

According to the results of the questionnaire survey, expert opinion after summary, weighted average, we can get the Estimation Matrix that each factors of the second floor B corresponding to the first floor A, and the three layers C corresponding to the second floor B (Ren Dong flourishing,2003). As the following Table 5~9.

Tab. 5: A–B Estimation Matrix

A	B1	B2	B3	B4	B5
B1	1	4	3	1/2	1/2
B2	1/4	1	1/2	1/6	1/6
B3	1/3	2	1	1/4	1/5
B4	2	6	4	1	1
B5	2	6	5	1	1

Tab. 6: B1–C Estimation Matrix

B1	C1	C2	C3	C4
C1	1	4	4	4
C2	1/4	1	1	1
C3	1/4	1	1	1
C4	1/4	1	4	1

Tab. 7: B2–C Estimation Matrix

B2	C5	C6
C5	1	3
C6	1/3	1
C4	1/4	1

Tab. 8: B3–C Estimation Matrix

B3	C7	C8	C9	C10	C11
C7	1	3	1/3	1/3	7
C8	1/3	1	1/5	1/5	6
C9	3	5	1	1/2	8
C10	3	5	2	1	8
C11	1/7	1/6	1/8	1/8	1

Tab. 9: B4–C Estimation Matrix

B4	C12	C13	C14	C15	C15
C12	1	2	2	2	3
C13	1/2	1	1	1	2
C14	1/2	1	1	1	2
C15	1/2	1	1	1	2
C16	1/3	1/2	1/2	1/2	1

Tab. 10: B3–C Estimation Matrix

B5	C17	C18	C19	C20	C21	C22
C17	1	2	3	3	2	5
C18	1/2	1	2	3	2	6
C19	1/3	1/2	1	1	1	5
C20	1/3	1/2	1	1	1	5
C21	1/2	1/2	1	1	1	5
C22	1/5	1/6	1/5	1/5	1/5	1

3.4 Determination of Weight Coefficient

According to the judgment matrixa. A-B to calculate the relative importance of coefficient of Bi.(I = 1,2,3,4,5) for water-saving agricultural technology promotion factors A , also known as the weight, which is WBi (i=1,2,3,4,5),expressed by column vector as the following:

$$W_{Bi} = \begin{bmatrix} W_{B1} \\ W_{B2} \\ W_{B3} \\ W_{B4} \\ W_{B5} \end{bmatrix} \text{ and } \sum_{i=1}^{5} w_{Bi} = 1$$

Calculate the product of the elements of each row in judgment matrix, that is

$$M_{Bi} = \prod_{j=1}^{n} bij, (i = 1,2 \ldots n)$$

Calculate the n root of MBi, that is $\overline{W}_{Bi} = \sqrt[n]{M_{Bi}}$(i=1,2...n)

Normalize vector $\overline{W}_{Bi}$, and get the order of weight $W_{Bi} = \frac{\overline{W}_{Bi}}{\sum_{i=1}^{n} \overline{W}_{Bi}}$,(i=1,2...n)

According to the above steps, we can determine the weight of each factor in the second layer on A, and then determine the order of importance of each factor of the second B on A. In the same way, we can compute five judgment matrices of B~C. The detailed calculation procedure is shown in the table:

Tab. 11: The calculation of W_{Bi}

A	B1	B2	B3	B4	B5	M_{Bi}	$\overline{W}_{Bi}$	W_{Bi}	single hierarchical arrangement
B1	1	4	3	1/2	1/2	3	1.2457	0.1913	3
B2	1/4	1	1/2	1/6	1/6	1/288	0.322	0.0495	5
B3	1/3	2	1	1/4	1/5	1/30	0.5065	0.0778	4
B4	2	6	4	1	1	48	2.1689	0.3331	2
B5	2	6	5	1	1	60	2.2679	0.3483	1
Σ							6.5112	1.0000	

According to the above calculation results, can be spread of water-saving agricultural technology popularization in our country the size of the factors affecting the weight sorting, rounding out the top five were government guiding, water saving technology and economic benefits, the cost of supply and demand of water resources, water saving technology and promotion agencies.

4 Conclusion

Firstly, we need to increase the support of research funding for the problem of water saving irrigation in agriculture. Let all kinds of water-saving irrigation techniques be suitable for the conditions of climate, soil condition and water quality of different regions should be adapted to the cultivation pattern of the crop, and the design parameters are suitable for the water consumption patterns of the crops. At the same time, the equipment hardware, the research public relations strength, the research has the quality pass the water saving irrigation equipment.

Secondly, the government should increase investment in water-saving irrigation in agriculture. In the early days of the government's promotion of water-saving irrigation, the adoption of incentives and support policies related to technology adoption has had a significant impact on farmers' adoption. In order to improve the farmers in the early of new technology adoption rate, first of all, the government can undertake some at the early stage of the agricultural new technology promotion preferential policy, provide period longer low-interest loans to farmers, use credit means to strengthen the government's support of water-saving technology; Second, to adopt water-saving irrigation technology of farmers to buy equipment subsidies or lower the price of related resources, to improve its enthusiasm, which can help to increase the rate of farmers to adopt new technology; The third , the government should do a good job in promoting the relevant policies, so that the farmers can understand the goal of water saving irrigation and the concrete implementation of water-saving technologies.

Thirdly, we should encourage people in agriculture and water technology. At present, scientific research institutions, agricultural science and technology enterprises and water-saving technology promotion center of agricultural technology personnel source diffusion: a lack of initiative of agricultural water-saving irrigation technology products. The most effective way is to motivate these technicians, but the way to motivate them is worth exploring. Technical personnel technical guidance to farmers, should as far as possible in the form of easy to understand, as far as possible in the form of visual and operational guidance, in order to minimize the technical potential difference. In addition, the agricultural water-saving irrigation technology adopters in the diffusion process of technology should first focus on agricultural income as the main source of income of farmers, large scale of peasant household and the innovation diffusion with high degrees of farmers, to a model household after using this technology has good practical effect to attract more other farmers use.

Finally, strengthen the training and guidance of farmers' technologies. Farmers are the main force of agricultural development, and their adoption of new technology depends greatly on the knowledge level and ability of the farmers themselves. Therefore, we must increase the training of farmers in various ways, at various levels and in many forms. First, the government decision-making departments and dissemination of scientific research in choosing the spread of water-saving irrigation technology and project, want to consider to farmers for the technical risk, capital capacity and grasp of the technical ability, to help farmers to objective analysis of technical risk, capital and information system. Second, the spread of agricultural water-saving irrigation technology way to give priority to with voluntary consultation from bottom to top type, supplemented by other a variety of ways, change the past top-down driven model method to promote the administrative instruction. To emphasize the role of farmers, not didn't see the people, for the science and technology to further analyze the needs of the farmers and the psychological characteristic, attract farmers participate in actively, support and cooperation, stimulate peasants' innovation with strong motivation and interest.

Reference

[1] Liang Guo, Junqi Hou. Factor analysis on the process of agricultural water-saving irrigation technology diffusion Journal of Xidian University (Social Science Edition), Jan.2011: 50–54 (in Chinese).

[2] Arial Dinar, Dan Yaron. .Adoption and abandonment of irrigation technologies. Agricultural Economics,1992 (04): 315–332

[3] Bass F. M. 1983.Differential Equation Models. New York: Spinger-Verlag Beal. G M. J. M. Bohlen, J. N. Raudabangh. Leadership and dynamic group action .Iowa State University Press, Ames, Iowa, 1960: 65–78

[4] Brown L. 1981. Innovation diffusion. London: Methuen Caswell, Margriet F. and Zilberman, David. The Choices of Irrigation Technology in California. American Journal of Agricultural Economics, 1985 (5): 223–234

[5] Daisuke Satoh. A discrete bass model and its parameter estimation. Journal of the Operations Research Society of Japan, 44 (1): 1-18 Dearden J, Ickes B W, Larry Samuelson. 1990. To innovate or not innovate: incentives and innovation in hierarchies. The American Economic Review, 2001 (3): 1105–1123

[6] Eric C. Schuck, W. Marshall Frasier, Robert S. Webb, Lindsey J. Ellingson, Wendy J. Umberger. Adoption of More Technically Efficient Irrigation System as a Drought Response. Water Resource Developmemt, 2005 (12): 651–662

Nan Wang, Shang Wang, Chen Dou, Guang-Fei Peng, Wei-Hua Tao, Jian Chen, Jie Fang and Zheng-Hao Fei*

Adsorption Properties of Phenol in Aqueous Solution with Different Acidic Adsorption Resins

Abstract: The hypercrosslinked adsorption resins, modified with trimellitic anhydride (named FJ–1), concentrated sulfuric acid (named FJ–2) and p-hydroxy mandelic acid (named FJ–3) respectively, were successfully prepared by the crosslinking and chemical modification reaction and characterized by IR and BET. The adsorption properties and adsorption mechanism of phenol onto the FJ–1, FJ–2 and FJ–3 resins were studied by isotherm adsorption and adsorption kinetics experiments and compared with NDA99 resins. The results shown that the adsorption effect of phenol on the FJ–1 resins was best and the adsorption capacity of phenol on it decreased with increasing temperature. The adsorption process was mainly reversible exothermic physical adsorption. For the FJ–1 and FJ–2 resins, the Langmuir isotherm model can fit their adsorption process well. The Freundlich isotherm model can fit the adsorption process of the FJ–3 resins better than Langmuir isotherm model. The results of ΔH and ΔG shown that the adsorption of phenol on four kinds of resins was a feasible spontaneous endothermic process. ΔS showed the increasing randomness of the solid-solution interface during the adsorption of phenol on resins. Pseudo-first-order model and pseudo-second-order model were both described the adsorption of phenol onto the FJ–1 and NDA99 resins. The particle diffusion was the main control step of entire adsorption process.

Nan Wang, College of Chemistry and Chemical Engineering, Nanjing Tech University, Nanjing, 210009, China. Email: 15805101245@163.com
Shang Wang, College of Chemistry and Chemical Engineering, Yancheng Teachers University, Yancheng, 224002, China. Email:1605993674@qq.com
Chen Dou, College of Chemistry and Chemical Engineering, Yancheng Teachers University, Yancheng, 224002, China. Email:1343067460@qq.com
Guang-Fei Peng, College of Chemistry and Chemical Engineering, Yancheng Teachers University, Yancheng, 224002, China. Email:2732746389@qq.com
Wei-Hua Tao, College of Chemistry and Chemical Engineering, Yancheng Teachers University, Yancheng, 224002, China. Email: taoweihua3@163.com
Jian Chen, College of Chemistry and Chemical Engineering, Yancheng Teachers University, Yancheng, 224002, China. Email:xchenjian@163.com
Jie Fang, College of Chemistry and Chemical Engineering, Yancheng Teachers University, Yancheng, 224002, China. Email: fangjielala@163.com
Zheng-Hao Fei, College of Chemistry and Chemical Engineering, Yancheng Teachers University, Yancheng, 224002, China. Email: feizhenghao@163.com

DOI 10.1515/9783110540048-028,

Keywords: Resin, Phenol, Adsorption, Adsorption isotherms, Kinetics.

1 Introduction

Phenol (C_6H_5OH) also known as carbolic acid, hydroxybenzene, found in coal tar by the German chemist Runge (Runge F) in 1834, is the most simple phenols. It is slightly acidic, appreciably soluble in water, but easily soluble in organic solvents at room temperature. It is an important raw material, which is widely used in the production of certain resins, petrochemical, oil refining, fungicides, preservatives, pharmaceutical and pesticide and the like [1–4]. However, phenol is a toxic organic pollutant in the industrial wastewater, which are harmful to human being and animals even at low concentrations [5–9]. Phenol is corrosive, so it can make partial protein denaturation through skin contact and it make an effect on the water system seriously [10–12]. Therefore, phenol should be removed or decomposed effectively from wastewaters in order to protect the environments and people's lives.

With the development of science and technology, there are many methods to remove the phenol from water, such as adsorption [13–14], catalytic oxidation [15–16], membrane separation [17], chemical coagulation [18], solvent extraction [19–20], and so on, the adsorption is most commonly used and effective in industry among them currently. Various of materials have been used as adsorbents, including polymeric resins [21], activated carbon [22], silica gel, molecular sieves, bentonite [23] and soil [24]. Polymeric resins as adsorbents for adsorption of phenols are very popular due to their mechanical strength, good adsorption capacity and selectivity, easy desorption regeneration and so on [8]. However, they are not fully meeting the needs of people because of some shortcomings. Existing adsorption resins have been made great modification in order to improve their properties, including excellent physicochemical stability, large adsorption capacity, good selectivity, structural diversity [8].

In this paper, the hyper-cross-linked adsorption resins, modified with trimellitic anhydride (named FJ-1), concentrated sulfuric acid (named FJ–2) and p-hydroxy mandelic acid (named FJ-3) respectively, were successfully prepared by the cross-linking and chemical modification reaction and characterized by IR and BET. The adsorption properties and adsorption mechanism of phenol onto the FJ–1, FJ–2 and FJ-3 resins were studied by isotherm adsorption and adsorption kinetics experiments and compared with NDA99 resins in order to provide a theoretical basis for the treatment of wastewater containing high concentrations of phenol.

2 Materials and Methodology

2.1 Reagents and Instruments

Chloromethylcated styrene-divinylbenzene copolymer (CLPs) and NDA99 were purchased from Nanjing Maike Fei Co. LTD. Trimellitic anhydride were purchased from Element Mall. P-hydroxyl mandelic acid and urea were purchased from J & K Technology Co,Ltd. Ethanol, nitrobenzene, anhydrous aluminum chloride and hydrochloric acid were analytical grade, which were purchased from Shanghai Chemical Reagent.

DF–101S collector-type thermostat heating magnetic stirrer(Zhengzhou Great Wall Branch Co, Henan, China.); Pore surface area analyzer (Micromeritics, ASAP2010,USA); High–performance liquid chromatography (ULTIMATE 3000, Thermo-Fisher, USA); Infrared spectrometer (Bruker company Vertex 80 Switzerland).

2.2 The Synthesis of Adsorption Resins

A certain amount of CLPs were dried at 333K for 12h and then fully swollen in nitrobenzene at room temperature for 8h. After that, the mixture of the CLPs and nitrobenzene were stirred while heated. A certain amount of trimellitic anhydride and anhydrous aluminum chloride were added to the reaction system when 353K and then the reaction was kept at this temperature for 8 h (see Fig. 1). After the desired time, the resins modified with trimellitic anhydride were filtered-off, thoroughly washed with 1% w/w HCl in ethanol, 10 % w/w sodium hydroxide solution, 5 % hydrochloric acid and water successively until neutral pH, and dried at 333 K for 12h. Trimellitic anhydride is replaced by concentrated sulfuric acid, p-hydroxy mandelic acid respectively under the same conditions in order to obtain a series of hypercrosslinked adsorption resins containing different functional group or specific surface area (named FJ–1, FJ–2 and FJ–3 respectively).

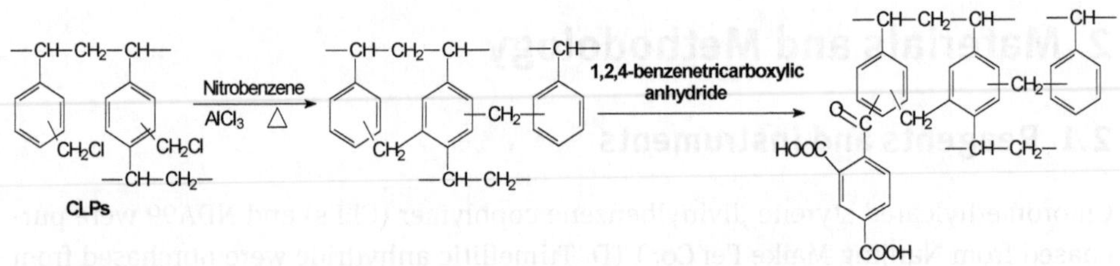

Fig. 1: Postcrosslinked and modified reaction equation of Friedel-Crafts for CLPS

2.3 Isotherm Experiments

Weigh 0.100g FJ–1, FJ–2, FJ–3 and NDA 99 resins respectively which were used as adsorbents in 150mL Erlenmeyer flask and then 100 mL of phenol aqueous solution were also added. The initial concentration of phenol, C_0 (mg/L), was set to be 100, 200, 300, 400 and 500 mg/L in the batch adsorption experiments. At last, a set of conical flasks having a resin sample in different phenol solutions were shaken in a thermostatic oscillator at speed of 110 rpm and 288, 303, 313 and 288K*. After the adsorption equilibrium, the equilibrium concentration, C_e, was analyzed by HPLC. The equilibrium adsorption capacity, Q_e, was calculated according to reference [10, 25]:

$$Q_e = (C_0 - C_e)\, V/W \tag{1}$$

where Q_e is the equilibrium adsorption capacity (mg/g), C_e is the equilibrium concentration (mg/L), C_0 is the initial concentration (mg/L), V is the solution volume (L), W is resins weight (g).

2.4 Adsorption Kinetics

Weigh 0.100g FJ–1 and NDA 99 resins respectively in 150mL Erlenmeyer flask and then 100 mL of phenol aqueous solution were also added. The initial feed concentration of phenol aqueous solution C_0 (mg/L), was set to be 300 mg/L. At last, a set of conical flasks having a resin sample in same phenol solutions were shaken in a thermostatic oscillator at speed of 110 rpm and 288, and 303 K. At a preset time interval, 1mL of the sample solution was taken and the residual concentration of phenol was determined by HPLC until adsorption equilibrium was reached.

2.5 Particle Diffusion Experiments

Particle diffusion equation [10]:

$$Q_t = K \cdot t^{0.5} + C \qquad (2)$$

where K is the intra-particle diffusion rate (mg g^{-1} $min^{1/2}$), and C is a constant. The K and C were obtained by plotting Q_t versus $t^{0.5}$. Q_t is the adsorption amount of time (mg/g), t is the adsorption time (min).

3 Results and Discussion

3.1 Characterization of Adsorption Resins

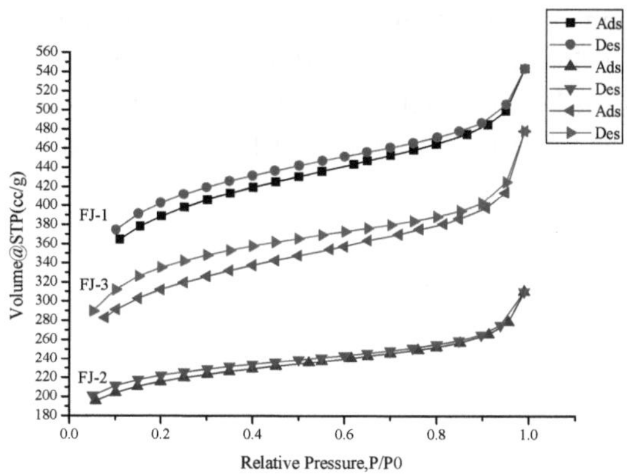

Fig. 2: Adsorption-desorption isotherms of N_2 on the resins

N_2 adsorption–desorption isotherms on the resin sample at 333 K were measured with a Micromeritics ASAP 2010 pore surface area analyzer (see Fig. 2)[5]. Fig. 2 displays the N_2 adsorption–desorption isotherms on the FJ-1, FJ-2 and FJ-3 resins. The N_2 adsorption amounts on the FJ-1 resins are much larger than those on the FJ-2 and FJ-3resins at the same relative pressure (p/p0), indicating the BET surface area and pore volume of the FJ-1 resins being significantly higher than those of the FJ-2 and FJ-3resins [5].

The specific surface area and pore size distribution of adsorption resins were measured by BET (Bruaaures. S-Emmett H-Teller) method (see Table 1).

Tab. 1: Surface property of resins

Resin	Specific surface area/ ($m^2.g^{-1}$)	Microporous area/ ($m^2.g^{-1}$)	Average pore radius/(nm)	Microporous volume/($cm^3.g^{-1}$)
FJ-1	1081.3	870.8	2.05	0.430
FJ-2	670.9	336.5	0.99	0.086
FJ-3	976.5	617.7	1.70	0.140
NDA99	726.5	393.8	1.30	0.180

The Fourier transform infrared spectroscopy (FT–IR) of adsorption resins were collected with a pellet of powdered potassium bromide and resin on a Vertex 80 infrared spectrometer. Characterization results as shown in Fig. 3: there were a C–Cl stretching vibration of CLPs at 676.5cm^{-1}, a C-H outer surface deformation vibrations or C = C in-plane deformation of the benzene derivative at 2383 cm^{-1} of FJ-1 resins, a characteristic absorption peaks of sulfonic acid groups of FJ–2 resins, an O-H outside surface deformation vibration characteristic peaks at 901.03 cm^{-1}.

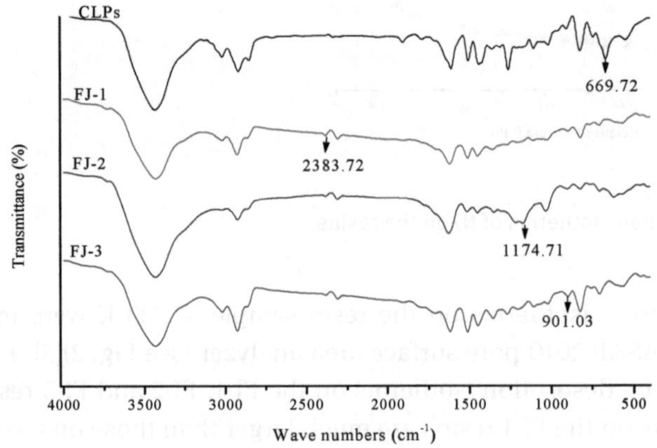

Fig. 3: IR spectra of resins and CLPS

3.2 Static Equilibrium Adsorption Isotherm

The concentration of phenol in aqueous solution was analyzed by UV analysis performed on an ULTIMATE 3000 high-performance liquid chromatography with the wavelength at 270 nm.

The adsorption isotherms of phenol on FJ–1, FJ–2, FJ–3 and NDA99 resin at different temperatures are shown in Fig. 4. As shown in Fig. 3, FJ–1, FJ–2 FJ–3 and NDA99, the overall adsorption capacity of phenol on the FJ-1 resins were better than FJ–2 FJ–3 and NDA99 resins at the same temperature and concentration, this was mainly because that the specific surface area of the FJ–1 resins was larger than the other resins. In addition, this was related to the carboxyl functional groups connected on the FJ–1 resins. The adsorption capacity of phenol on the FJ–1, FJ–2 and NDA99 resins decreased with increasing temperature according to Fig. 3 FJ–1 FJ–2 and NDA99, which indicated that the adsorption process was mainly physical adsorption and the process is an exothermic process. When the temperature dropped from 318K to 288K*, the adsorption capacity were consistent with at 288K, which proved that the adsorption of phenol onto the FJ–1 and NDA99 resins was mainly reversible physical adsorption. For the FJ–2 resins, when the temperature dropped from 318K to 288K*, the adsorption capacity were not far from at 288K, which proved that the adsorption of phenol FJ–2 resins was also mainly reversible physical adsorption. As shown in Fig. 3 FJ–3, the adsorption capacity of phenol on the FJ–3 resins firstly decreased and then increases with increasing temperature, which may because that the physical adsorption is weakening but the chemical adsorption became strong with increasing temperature.

Classic Langmuir equation and Freundlich equation [10] were taken to fit the experiment data, the results shown in Table 2 and Table 3.

The Langmuir equations for the equilibrium adsorption can be written as:

$$C_e/Q_e = 1/(K_L \cdot Q_m) + C_e/Q_m \tag{3}$$

Where Q_e is the equilibrium adsorption capacity (mg/g) with the equilibrium concentration C_e, Q_m is the maximum monolayer adsorption capacity (mg/g) and K_L is a Langmuir constant related to adsorption energy (L/mg).

The Freundlich equations for the equilibrium adsorption can be written as:

$$\ln Q_e = (1/n)\ln C_e + \ln K_F \tag{4}$$

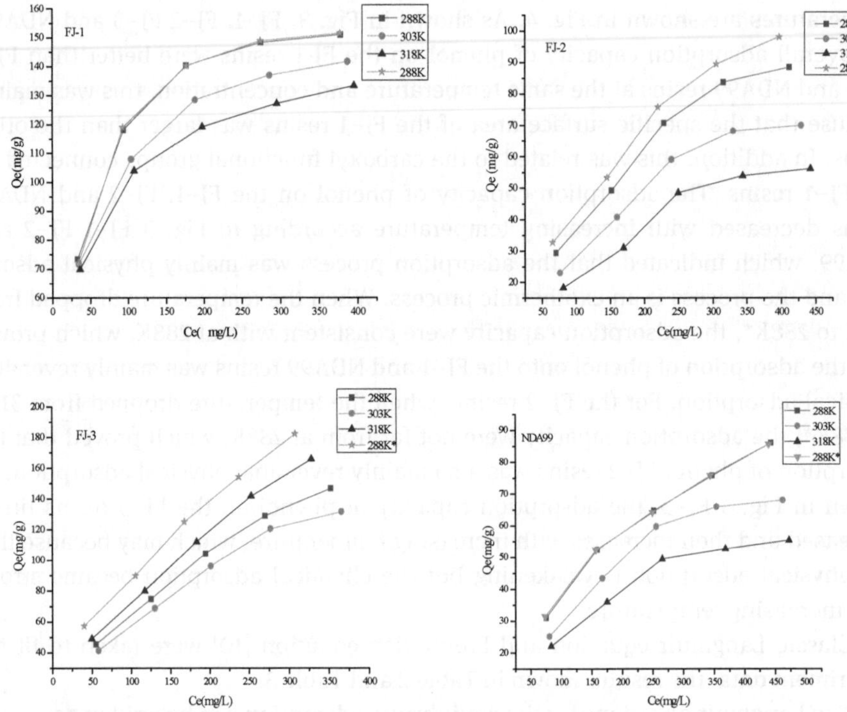

Fig. 4: Equilibrium adsorption isotherms of phenol on resins at 288K, 303K, 318K and 288 K*

Where K_F [(mg/g)(L/mg)$^{1/n}$] and n are the characteristic constants and they are temperature dependent. The parameter n is usually greater than unity. As this value becomes larger, the adsorption isotherm becomes more non-linear as its behavior deviates further away.

As shown in Table 2 and Table 3, the Langmuir isotherm model and Freundlich isotherm model can fit the experimental data well. For the FJ–1and FJ–2 resins, the value of the correlation coefficient (R^2) in the Langmuir isotherm model was a little closer to 1.0 than that in the Freundlich isotherm model, which suggested that the former was more suitable to fit the adsorption equilibrium data [10]. On the contrary, for the FJ–3 resins, the R^2 in the Freundlich isotherm model were bigger than that in the Langmuir isotherm model, which were all greater than 0.99. It is obvious that the former could describe the adsorption equilibrium data more accurately. So, for the FJ–3 resins, the adsorption process fitted the characteristics of heterogeneous surface adsorption better [10]. However, as to the NDA99 resins, the Langmuir isotherm model and Freundlich isotherm model all can fit its adsorption process well. In addition, for the FJ–1, FJ–2, FJ–3 and NDA99 resins, the parameters n in Freundlich isotherm model at different temperatures were all greater than 1, which indicated it was a favorable adsorption [10].

Tab. 2: Fitting result of the Langmuir isotherm equation

Resins	Temperature(K)	$C_e/Q_e= C_e/ Q_m +1/ (K_L \bullet Q_m)$	R^2	Q_m	K_L*10^2
FJ–1	288	C_e/Q_e=0.0053 C_e+0.664	0.9916	188.68	0.79
	303	C_e/Q_e=0.0058 C_e+0.756	0.9922	172.41	0.77
	318	C_e/Q_e=0.0061 C_e+1.005	0.9926	163.93	0.61
	288*	C_e/Q_e=0.0054 C_e+0.502	0.9974	185.19	1.08
FJ–2	288	C_e/Q_e= 0.0065C_e+2.010	0.9956	153.85	0.32
	303	C_e/Q_e= 0.008C_e+2.671	0.9994	125.00	0.30
	318	C_e/Q_e=0.0094C_e +3.768	0.9976	106.38	0.25
	288*	C_e/Q_e=0.0062C_e +1.572	0.9986	161.29	0.39
FJ–3	288	C_e/Q_e =0.0045C_e+1.033	0.9711	222.22	0.44
	303	C_e/Q_e =0.0041C_e+1.235	0.9732	243.90	0.33
	318	C_e/Q_e =0.0036C_e+0.973	0.9814	277.78	0.37
	288*	C_e/Q_e =0.0035C_e+0.751	0.9623	285.71	0.47
NDA99	288	C_e/Q_e=0.0076C_e +1.880	0.9947	131.58	0.40
	303	C_e/Q_e=0.0062C_e +1.373	0.9976	161.29	0.45
	318	C_e/Q_e=0.0104C_e +2.126	0.9954	96.15	0.49
	288*	C_e/Q_e= 0.0037C_e +1.146	0.9944	270.27	0.32

Tab. 3: Fitting result of the Freundlich isotherm equation

Resins	Temperature (K)	$\ln Q_e=(1/n)\ln C_e+\ln K_F$	R^2	n	K_F
FJ-1	288	$\ln Q_e=0.5002\ln C_e+2.059$	0.9389	2.00	7.84
	303	$\ln Q_e=0.4639\ln C_e+2.159$	0.9546	2.16	8.67
	318	$\ln Q_e=0.5215\ln C_e+1.681$	0.9708	1.92	5.37
	288*	$\ln Q_e=0.3952\ln C_e+2.667$	0.9611	2.53	14.40
FJ-2	288	$\ln Q_e=0.6554\ln C_e+0.612$	0.9818	1.53	4.15
	303	$\ln Q_e=0.6778\ln C_e+0.247$	0.9655	1.48	1.77
	318	$\ln Q_e=0.7043\ln C_e-0.153$	0.9600	1.42	1.42
	288*	$\ln Q_e=0.6305\ln C_e+0.863$	0.9902	1.59	7.29
FJ-3	288	$\ln Q_e=0.6550\ln C_e+1.388$	0.9955	1.53	4.01
	303	$\ln Q_e=0.6490\ln C_e+1.131$	0.9972	1.54	3.10
	318	$\ln Q_e=0.6270\ln C_e+1.392$	0.9941	1.60	4.02
	288*	$\ln Q_e=0.5330\ln C_e+2.143$	0.9987	1.88	8.53
NDA99	288	$\ln Q_e=0.5665\ln C_e+1.026$	0.9906	1.76	2.79
	303	$\ln Q_e=0.4518\ln C_e+1.937$	0.9997	2.21	6.94
	318	$\ln Q_e=0.5085\ln C_e+1.118$	0.9907	1.97	3.06
	288*	$\ln Q_e=0.4809\ln C_e+2.140$	0.9995	2.08	8.50

3.3 Adsorption Thermodynamics

To better illuminate the adsorption mechanism of phenol onto the FJ–1, FJ–2 FJ–3 and NDA99 resin, the adsorption thermodynamics parameters ΔH, ΔG and ΔS were evaluated. ΔH were calculated according to Van't Hoff equation [10]:

$$\ln C_e=\Delta H/RT-\ln K \tag{5}$$

where C_e is the equilibrium concentration of phenol (mol/L), T is the absolute temperature (K), ΔH is the isosteric enthalpy change of adsorption (kJ/mol), R is the ideal gas constant (8.314 $J\cdot mol^{-1}\cdot K^{-1}$) and K is a constant. Q_e were respectively set to 60 mg/g, 80 mg/g and 100 mg/g in order to calculate the C_e at different temperatures. ΔH was determined by plotting $\ln C_e$ versus 1/T and could be calculated from the slope of the fitting line.

Adsorption free energy can be calculated as [10]:

$$\Delta G = -nRT \tag{6}$$

Where ΔG is the adsorption free energy (kJ/mol), n represents the Freundlich exponent and R is the ideal gas constant (8.314 $J{\cdot}mol^{-1}{\cdot}K^{-1}$).

The adsorptive entropy change, ΔS ($J{\cdot}mol^{-1}{\cdot}K^{-1}$) were calculated using the Gibbs–Helmholtz equation [10]:

$$\Delta S = (\Delta H - \Delta G)/T \tag{7}$$

Thermodynamic parameters of phenol adsorption onto the four resins were shown in the Table 4. The value of ΔH were all positive indicating an endothermic process. And the positive ΔH value were less than 42 kJ/mol indicated a physical process for phenol adsorption of an endothermic nature. The negative values of ΔG also indicated the feasibility of the process and suggested the adsorption of phenol on the FJ–1, FJ–2, FJ–3 and NDA99 resin were spontaneous. What's more, positive values of ΔS showed the increasing randomness of the solid-solution interface during the sorption of phenol on four kinds of resins. These positive values of ΔS might be due to some structural changes in the FJ–1, FJ–2, FJ–3 and NDA99 resin during the adsorption process [10].

Tab. 4: Thermodynamic parameters of phenol adsorption based on resins at different temperatures

Resins	ΔH(KJ/mol)	ΔG(KJ/mol)			ΔS($J{\bullet}mol^{-1}{\bullet}K^{-1}$)		
		288K	303K	318K	288K	303K	318K
FJ-1	9.44	-4.79	-5.44	-5.08	49.41	49.11	45.66
FJ-2	28.98	-9.94	-4.46	-3.75	135.14	110.36	102.92
FJ-3	4.12	-3.66	-3.88	-4.23	27.01	26.40	26.26
NDA99	17.15	-4.21	-5.57	-5.04	74.17	74.98	69.78

3.4 Adsorption Kinetics

Fig. 5 shows the kinetic adsorption curves of phenol on the FJ–1 and NDA99 resins at the initial concentrations of 300 mg/L and temperatures of 288 K and 303 K. The as seen in Fig. 5, the adsorption of phenol on two resins were very fast in the first 1 h and then slower and reached the equilibrium within 180 min [5]. In addition, the adsorption capacity of phenol on the FJ–1 resins were better than NDA99 resins at the same temperature, which was consistent with the results of static adsorption. This was related to the specific surface area of the FJ–1 resins and the carboxyl functional groups.

Pseudo-first-order and pseudo-second-order kinetic models were used respectively to fit the adsorption kinetics. The pseudo-first-order model [10] can be written as follows:

$$\ln(Q_e - Q_t) = \ln Q_e - K_1 t \tag{8}$$

where Q_t is the amount of phenol adsorbed at time t (mg/g) and Q_e is the amount of phenol adsorbed at equilibrium (mg/g). K_1 is the rate constant of pseudo-first-order model (min^{-1}). The Q_e and K_1 values were calculated by plotting Q_t versus t.

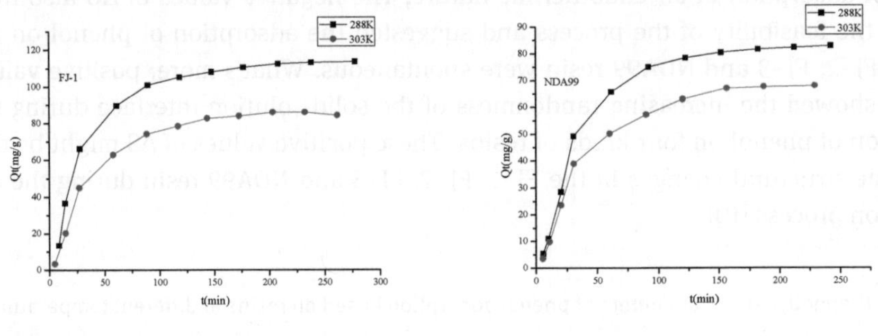

Fig. 5: Kinetic adsorption curves of phenol on the FJ-1 and NDA99 resins

The pseudo-second-order model [10] can be written as follows:

$$t/Q_t = 1/h + t/Q_e \text{ where } h = K_2 Q_e^{\,2} \tag{9}$$

where K_2 is the rate constant of pseudo-second-order model ($g \cdot mg^{-1} \cdot min^{-1}$). The Q_e and K_2 values were obtained by plotting Q_t versus t.

The fitting results of phenol adsorption onto the FJ–1 and NDA99 resins were shown in the Table 5 and Table 6. For two resins, the correlation coefficient R^2 (all greater than 0.98) in pseudo-first-order model were closer to 1.0 than that in pseudo-second-order model, which implied that the adsorption rate of phenol onto the FJ–1 and NDA99 resins were mainly controlled by the liquid film diffusion or particle diffusion or both. The particle distribution was the main control step of entire absorption process because of the ln (Q_e-Q_t) and t showing a good linear relationship.

Tab. 5: Kinetic parameters of phenol onto the FJ-1 resins

Temperature	$\ln(Q_e\text{-}Q_t)=\ln Q_e\text{-}K_1t$		$1/(Q_e\text{-}Q_t)=1/Q_e+K_2t$		
	$K_1(1/min)\times10^2$	R_1^2	$Q_e(mg/g)$	$K_2(g\bullet mg^{-1}\bullet min^{-1})\times10^2$	R_2^2
288K	1.65	0.985	115	0.25	0.865
303K	1.36	0.997	87	0.09	0.879

Tab. 6: Kinetic parameters of phenol onto the NDA99 resins

Temperature	$\ln(Q_e\text{-}Q_t)=\ln Q_e\text{-}K_1t$		$1/(Q_e\text{-}Q_t)=1/Q_e+K_2t$		
	$K_1(1/min)\times10^2$	R_1^2	$Q_e(mg/g)$	$K_2(g\bullet mg^{-1}\bullet min^{-1})\times10^2$	R_2^2
288K	1.60	0.993	85	0.17	0.965
303K	1.51	0.997	70	0.15	0.939

3.5 Particle Diffusion

Fig. 6 shows the particle distribution curves of phenol on the FJ–1 and NDA99 resins. As shown in Fig. 6, the correlation coefficient R^2 of the particle diffusion equation of phenol adsorption onto two resins both were more than 0.98, which indicated that the adsorption rate of phenol onto the FJ–1 and NDA99 resins were mainly controlled by the liquid film diffusion. However, four lines were not through the origin, showing that external liquid film diffusion may make an effect on the adsorption.

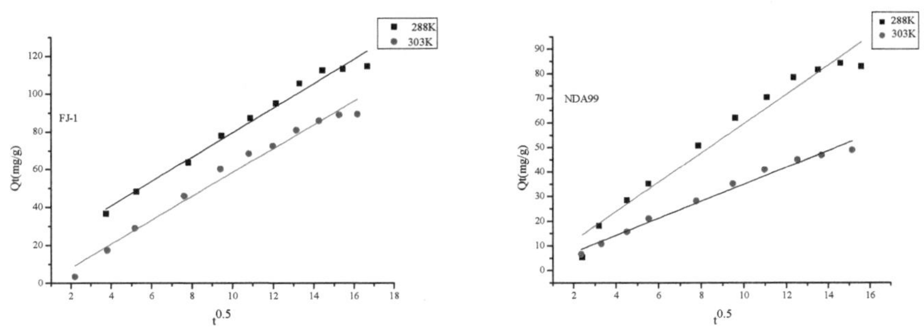

Fig. 6: Particle distribution curves of phenol on the FJ–1 and NDA99 resins

4 Conclusions

IR spectra of FJ–1, FJ–2 and FJ–3 resins show that three kinds of hyper-cross-linked adsorption resins were successfully prepared by the cross-linking and chemical modification reaction. At the same temperature and concentration, the adsorption capacity of phenol on the FJ–1 and FJ–3 resins were better NDA99 resins. But the adsorption capacity of phenol on the FJ–2 resins was smallest among them, maybe about sulfonic acid group and its specific surface area. The adsorption capacity of phenol on the FJ–1 resins decreases with increasing temperature and the adsorption process was mainly reversible exothermic physical adsorption. The adsorption of phenol FJ–2 resins was also mainly reversible physical adsorption. For the FJ–1 and FJ–2 resins, the Langmuir isotherm model can fit their adsorption process well. On the contrary, for the FJ–3 resins, the Freundlich isotherm model can fit its adsorption process better than Langmuir isotherm model. However, the Langmuir isotherm model and Freundlich isotherm model all can fit the adsorption process the NDA99 resins well. The results of ΔH and ΔG shown that the adsorption of phenol on four kinds of resins was a feasible spontaneous endothermic process. ΔS showed the increasing randomness of the solid-solution interface during the adsorption of phenol on resins. Pseudo-first-order model and pseudo-second-order model were both described the adsorption of phenol onto the FJ–1 and NDA99 resins. The particle distribution was the main control step of entire absorption process.

5 Conflict of Interest

This article content has no conflict of interest and this work was funded by the National Science Foundation (No. 21307103), Jiangsu Province Six Talents Peak Project (2011–JY–011), Jiangsu Province 333 High-Level Personnel Training Project, Jiangsu Province Science and Technology Department Foundation (BY2015058–03) and Jiangsu Province Department of Education (Grant Code 13KJB610016).

Acknowledgement: This work was funded by the National Science Foundation (No. 21307103), Jiangsu province 333 high-level personnel training project and Jiangsu provincial department of technology (BY2015058–03).

References

[1] Md. Ahmaruzzaman, "Adsorption of phenolic compounds on low-cost adsorbents: A review", Advances in Colloid and Interface Science, vol. 143, pp. 48–67, November 2008.

[2] C.R. Ispas, M.T. Ravalli, A. Steere, S. Andreescu, "Multifunctional biomagnetic capsules for easy removal of phenol and bisphenol A", Water Research, vol. 44, pp. 1961–1969, March 2010.

[3] Md. Ahmaruzzaman, S.L. Gayatri, "Activated neem leaf: a novel adsorbent for the removal of phenol, 4-nitrophenol, and 4-chlorophenol from aqueous solutions", Journal of Chemical and Engineering Data, vol. 56, pp. 3004–3016, May 2011.
[4] J.H. Huang, X.Y. Jin, SH.G. Deng, "Phenol adsorption on an N-methylacetamide-modified hypercrosslinked resin from aqueous solutions", Chemical Engineering Journal, vol. 192, pp. 192–200, June 2012.
[5] M. Khalid, G. Joly, A. Renaud, P. Magnoux, "Removal of Phenol from Water by Adsorption Using Zeolites", Industrial & Engineering Chemistry Research, vol. 43, pp. 5275–5280, July 2004.
[6] H.B. Senturk, D. Ozdes, A. Gundogdu, C. Duran, M. Soylak, "Removal of phenol from aqueous solutions by adsorption onto organomodified Tirebolu bentonite: Equilibrium, kinetic and thermodynamic study", Journal of Hazardous Materials, vol. 172, pp. 353–362, December 2009.
[7] G. Busca, S. Berardinelli, C. Resini, L. Arrighi, "Technologies for the removal of phenol from fluid streams: A short review of recent developments", Journal of Hazardous Materials, vol. 160, pp. 265–288, December 2008.
[8] U.F. Alkaram, A.A. Mukhlis, A.H. l-Dujaili, "The removal of phenol from aqueous solution by adsorption using surfactant-modified bentonite and kaolinite", Journal of Hazardous Materials, vol. 169, pp. 324–332, September 2009.
[9] X. Qiu, N.J Li, X.SH. Ma, SH. Yang, Q.F. Xu, H. Li , J.M. Lu, "Facile preparation of acrylic ester-based crosslinked resin and its adsorption of phenol at high concentration", Journal of Environmental Chemical Engineering, vol. 2, pp. 745–751, March 2014.
[10] Bahdod, S. El Asri, A. Saoiabi, T. Coradin, A. Laghzizil, "Adsorption of phenol from an aqueous solution by selected apatite adsorbents: kinetic process and impact of the surface properties", Water Research, vol. 43, pp. 313–318, February 2009.
[11] Z.R. Guan, L. Liu, L.L. He, S. Yang, "Amphiphilic hollow carbonaceous microspheres for the sorption of phenol from water", Journal of Hazardous Materials, Vol. 196, pp. 270–277, November 2011.
[12] L.L. Zhu, Y.F. Deng, J.P. Zhang, J. Chen, "Adsorption of phenol from water by N-butylimidazolium functionalized strongly basic anion exchange resin", Journal of Colloid and Interface Science, vol. 364, pp. 462–468, December 2011.
[13] J. Yin, R. Chen, Y. Ji, C. Zhao, G. Zhao, H. Zhang, "Adsorption of phenols by magnetic polysulfone microcapsules containing tributyl phosphate", Chemical Engineering Journal, vol. 157, pp. 466–474, March 2010.
[14] A. Liu, "Adsorption isotherm, kinetic and mechanism studies of some substituted phenols on activated carbon fibers", Chemical Engineering Journal, vol. 157, pp. 348–356, March 2010.
[15] O. Chedeville, M. Debacq, C. Porte, "Removal of phenolic compounds present in olive mill wastewaters by ozonation", Desalination, vol. 249, pp. 865–869, December 2009.
[16] N.A.S. Amin, J. Akhtar, H.K. Rai, "Screening of combined zeolite-ozone system for phenol and COD removal", Chemical Engineering Journal, vol. 158, pp. 520–527, April 2010.
[17] S.F. Shen, K.H. Smith, S. Cook, S.E. Kentish, J.M. Perera, T. Bowser, G.W. Stevens, "Phenol recovery with tributyl phosphate in a hollow fiber membrane contactor: Experimental and model analysis", Separation and Purification Technology, Vol. 69, pp.48-56, September 2009.
[18] V.M. Egorov, S.V. Smirnova, I.V. Pletnev, "Highly efficient extraction of phenols and aromatic amines into novel ionic liquids incorporating quaternary ammonium cation", Separation and Purification Technology, Vol. 63, pp. 710–715, November 2008.
[19] P. Yu, Z. Chang, Y. Ma, S. Wang, H. Cao, C. Hua, H. Liu, "Separation of p-Nitrophenol and o-Nitrophenol with three-liquid-phase extraction system", Separation and Purification Technology, vol. 70, pp. 199–206, December 2009.

[20] B. Burghoff, A. de Haan, "Liquid-Liquid Equilibrium Study of Phenol Extraction with Cyanex 923", Separation and Purification Technology, vol. 44, pp. 1753–1771, September 2009.
[21] Li, Q. Zhang, G. Zhang, J. Chen, Z. Fei, F. Liu, "Adsorption of phenolic compounds from aqueous solutions by a water-compatible hypercrosslinked polymeric adsorbent", Chemosphere, vol. 47, pp. 981–989, June 2002.
[22] Da̧browski, P. Podkos´cielny, Z. Hubicki, M. Barczak, "Adsorption of phenolic compounds by activated carbon—a critical review", Chemosphere vol. 58, pp. 1049–1070, February 2005.
[23] S.K. Nadavala, K. Swayampakula, V.M. Boddu, K. Abburi, "Biosorption of phenol and o-chlorophenol from aqueous solutions on to chitosan–calcium alginate blended beads", Journal of Hazardous Materials, vol. 162, pp. 482–489, February 2009.
[24] Subramanyam, A. Das, "Study of the adsorption of phenol by two soils based on kinetic and isotherm modeling analyses", Desalination, vol. 249, pp. 914–921, December 2009.
[25] Y. Huang, X. Ma, G. Liang, H. Yan, "Adsorption of phenol with modified rectorite from aqueous solution", Chemical Engineering Journal, vol. 141, pp. 1–8, July 2008.

Shang-Wu Yu, Fa-Xin Zhu, Shuai-Jun Wang, Tian Chen, Hai-Lin Zheng and Jin-Chao Yang

Design of the Cabin in the Ultra Low-Temperature Refrigerated Vessel

Abstract: In the paper, some defects or shortcomings existed in the traditional cabin were considered, the new cabin of the globular was designed by means of AutoCAD software, and its three dimension model was built based on CAXA software. It was indicated from the calculation results that heat-transfer areas and heat-transfer capacity of the new cabin were smaller than the traditional one in the cases of the same external condition and volume of the cabin. Therefore, the performance of the heat preservation in the new cabin of the globular is better than the traditional one. At the same time, with the better property of the materials in the cabin wall and the higher technology of the refrigeration plant, the coefficient of the heat convection will be less, the performance of the heat preservation in the new cabin of the globular will be better also.

Keywords: Ultra low-temperature refrigerated vessel, Cabin, Design, Three Dimension model, CAXA, Heat-transfer area, Heat-transfer capacity.

1 Introduction

The ultra low-temperature refrigerated vessel is one of the very high technical standard ship types. The tuna fishery industry has taken up a dominant position in the deep sea fishing because of its high economic value and high nutritive value. The tuna will be frozen rapidly and saved and transported under −50 °C to keep its quality as far as possible. The transfer of the transport link can cause a rise of tem-

Shang-Wu Yu, School of Port and Transportation Engineering, Zhejiang Ocean University, ZhouShan, China. Email:yushangwu2017@163.com
Fa-Xin Zhu, School of Port and Transportation Engineering, Zhejiang Ocean University, ZhouShan, China. Email:zhufax7758@163.com
Shuai-Jun Wang, School of Port and Transportation Engineering, Zhejiang Ocean University, ZhouShan, China. Email:530636620@qq.com
Tian Chen, School of Port and Transportation Engineering, Zhejiang Ocean University, ZhouShan, China. Email:1486596575@qq.com
Hai-Lin Zheng, School of Port and Transportation Engineering, Zhejiang Ocean University, ZhouShan, China. Email:71946318@qq.com
Jin-Chao Yang, School of Port and Transportation Engineering, Zhejiang Ocean University, ZhouShan, China. Email:yangjinchao2016@163.com

DOI 10.1515/9783110540048-029,

perature easily while affect the nutritional quality of tuna, therefore a rather high demand of the tuna' low temperature storage and transportation is proposed. The existing cabin in the ultra low-temperature refrigerated vessel is of generally rectangular shape (or hexahedral type), and is simple to cooperate with other ship's cabin and building layout. But the cabin also has the drawbacks or disadvantages of large heat transfer area and high rate of heat transfer, which need more refrigerating capacities to guarantee the same quality of the fish. Therefore, this study is to design a new type of refrigerated cabin structure and based on the existing refrigerated cabins and from the design requirement of refrigerating module of the ultra low-temperature refrigerated vessel. Its three dimension model was built based on CAXA software and from the heat transfer area and heat transfer in the cabin to verify its rationality.

1.1 Design Requirement Analysis of the Cabin in the Ultra Low-Temperature Refrigerated Vessel

Design requirements of the ultra low-temperature refrigerated vessel are determined by referencing the Rules and Regulations for the Construction and Classification of Sea Going Steel Ships and other ship design materials

1 Quick-freezing cabin and heat preservation cabin are coordinated effectively to reach a high-efficiency heat preservation (or cold insulation).
2 In the cases of the same volume of the cabin, reduce the heat transfer area and heat transfer capacity of the cabin and make good use of the ship cabin arrangement.
3 Quick-freezing cabin and heat preservation cabin are assigned reasonably to insure the ability of heat preservation.
4 Actual needs are met by reasonable design of the cabins in consideration of different volume and quality of goods and capacity of frozen cabinet.
5 In order to improve the cycle efficiency of the refrigeration cycle, the equipment of insulating door, hatch, bulkhead and air blower that are more convenient to transport goods are designed.

2 Design of New Ultra Low-temperature Refrigerated Cabin

2.1 Structure of New Ultra Low-Temperature Refrigerated Cabin

New refrigerated cabin of double-deck structure was designed by use of AutoCAD software as shown in Fig. 1, including the quick-freezing and insulation cabin.

The frozen body was spherical shape, the internal set of a thermal insulation cabin, the insulation compartment for the cylinder structure , And in the frozen body and the insulation compartment is equipped with two insulated doors on the wall to the insulation cabin, while the outer wall of the insulation compartment set up different sizes of quick-freezing cabinet arranged in the cabinet cooling coil, the cabin set more Typhoon fans to strengthen the air flow to ensure uniformity of the refrigeration temperature.

In addition, the quick-freezing cabin and the insulation compartment of the hatch are equipped with open or close the lid of the lifting device. The cabin structure on the use of quick-freezing cabin and the effective cooperation with the insulation can reduce heat loss and strengthen maintenance and management; The frozen body (Or arc-shaped) as a whole, compared with the same volume of hexahedron (or rectangular), spherical body of the frozen body as a whole can reduce the compartment heat transfer area (or outer surface area),.The use of quick-frozen and the insulation box effectively integrated into one can reduce the heat loss and strengthen the maintenance and management. At the same time, the hatch covers are used to start and stop the lifting device for the refrigerated and access to the goods to facilitate [3–4]

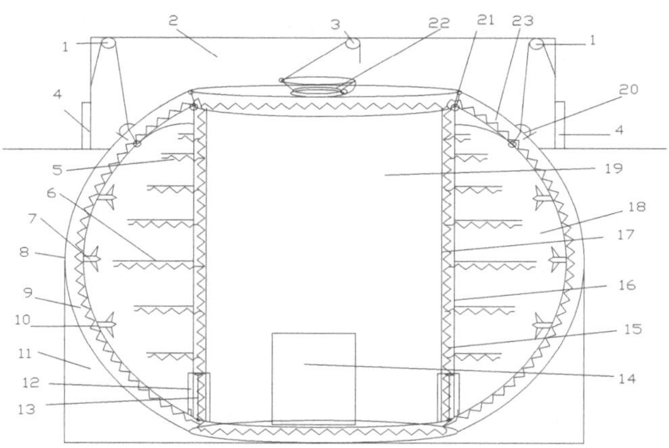

Fig. 1: New cabin main view

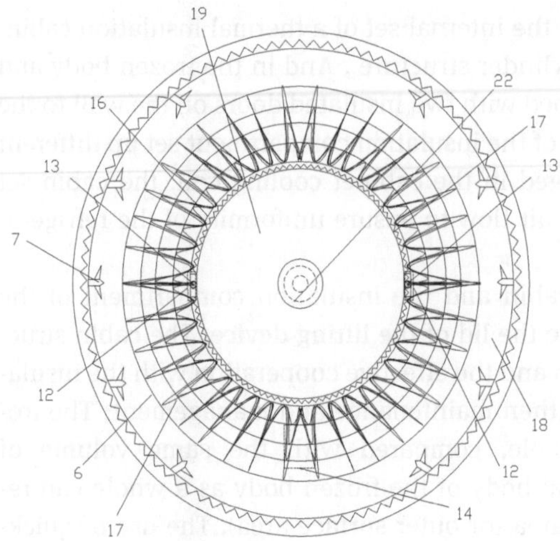

Fig. 2: New cabin top view

1–speed pulley; 2–compartment; 3–speed pulley; 4–compartment compartment; 5–coil; 6–compartment freezer; 7–fan; 8–compartment main body; 9–evaporative coil, 10–frozen chamber wall; 11–small cryogenic refrigerated area; 12–insulated door; 13 –rotating shaft; 14–large bay; 15–evaporative coil; 16–quick freezing compartment; 17–Insulation compartment inner wall; 18 - Frozen tank; 19 - Insulation compartment; 20–Hook; 21–Conveyor equipment; 22–Insulation compartment cover; 23–Frozen cabin cover.

2.2 The Working Principle of the New Ultra-Low Temperature Refrigerated Cabin

As shown in Fig. 1, when the refrigerated tank cooling device is operated, when the frozen fish cargo is to be transported to the ship, the frozen fish stock transported from the deck is fed from the space chamber door 4 into the refrigerated compartment and opened by the pulley device 1 The quick-freezing compartment hatch 23 distributes the articles in a different refrigerated cabinet 6 according to the volume and weight of the fish stock.

After the quick freeze is completed and when the rapid freezing is required, the fan 7 is opened to speed up the freezing rate, and the frozen fish can be transferred directly to the insulation chamber 19 through the heat insulation door 12, where the quick-freezing tank 18 and the inside of the holding tank are common 15 and 17.The quick-freezing chamber fan 7 in the work increases the loss of heat within the quick-freezing compartment. The temperature is lower than the temperature inside the insulation chamber, so that the quick-freezing compartment To the "cold" effect when the thermal insulation performance of the insulation compartment is affected, greatly improving the efficiency of the cabin and the refrigeration cycle of the cycle efficiency; When the frozen fish needs to be taken you will open the insulation compartment cover 22 through the pulley 3 to access to cold fish. When the cabin Internal cargo needs to be transferred quickly, you can open the large bay door 14 and delivery of goods quickly.

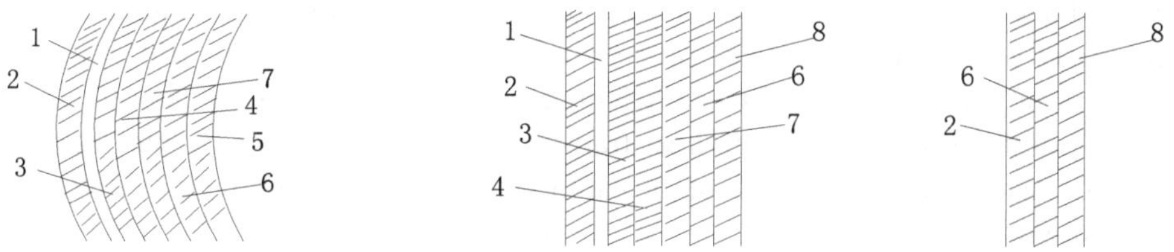

Insulation bulkhead material selection Frozen bulkhead material selection Public bulkhead material selection

Fig. 3: Material distribution of the bulkhead

1–air layer; 2–steel plate; 3–heat resistant plastic; 4–foam; 5–mesh lithium alloy; 6–waterproof plastic; 7–coil;

2.3 Performance of New Ultra-Low Temperature Refrigerated Cabin and Material Selection

By the nature of the work of the cabin, the cabin to meet the cabin, the outer wall material heat transfer coefficient as small as possible, that is, to reduce heat, the wall due to close to the evaporative coil, so it requires the greater the heat transfer coefficient, Therefore, the outer wall of the quick-frozen cabin steel material, the middle layer of foam, the wall selection of good heat transfer performance of lithium alloy, the only difference between the insulation compartment is the use of mesh wood. One of the steel plate fixed support cabin, plastic from the heat, waterproof and so on, the foam has a heat insulation, mesh wood can strengthen the insulation, protection coil, mesh lithium alloy with enhanced refrigeration and other functions [5–8], as shown in Figure 3.

As the cabin side of the wall directly from the heat source radiation and convection, heating up faster, and the bulkhead outside the large temperature difference, a lot of heat will flow through the bulkhead to the surrounding compartment, so its insulation performance should also have special requirement. In principle, Cold storage temperature to meet the design requirements of the minimum temperature, the continued cooling, to maintain the design requirements of the minimum temperature of 8 hours or more to absorb latent heat, while 8 hours pressure rise is not greater than 1333Pa. One of the important characteristics of the wall of refrigerated cabin is its evaporative coil on its surface, and the effect of different types of evaporative heat exchangers on the wall is very different. Ship evaporation heat exchanger is mainly shell and tube or plate heat exchanger two, mostly for the shell and tube heat exchanger. Shell-and-tube heat exchangers have the advantages of strong construction, easy manufacturing, strong adaptability, large heat capacity, small pressure loss and good sealing. For the use of refrigerators (freon, etc.) as the cooling medium of the heat exchanger, its stability is higher. Plate heat exchangers have many uses on board, such as central cooler, piston water cooler, etc. It has the advantages of high heat transfer coefficient, compact structure, easy maintenance, easy to increase or decrease the heat transfer area, but its cost is higher, Easy to leak, so the design requirements of the cabin selection shell and tube heat exchanger is more suitable [9–11].

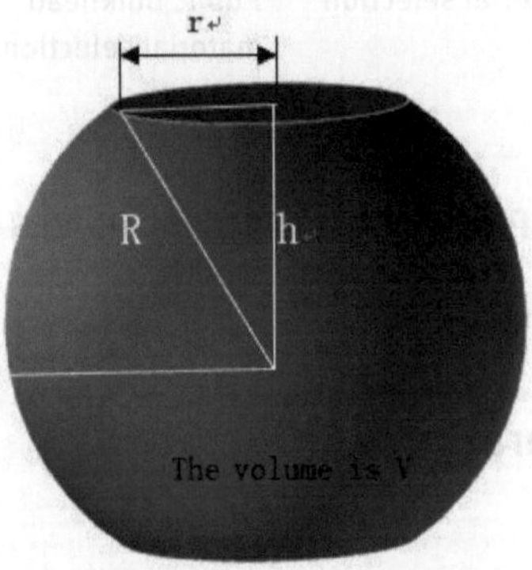

Fig. 4: CAXA model for new refrigerated compartments

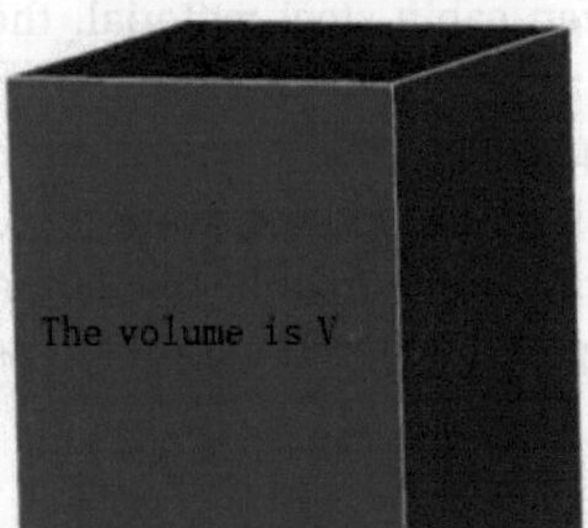

Fig. 5: CAXA model for traditional compartments

3 Modeling and Analysis of CAXA Model for New Refrigerated Cabin

3.1 Modeling of New Refrigerated Cabin CAXA Model

According to the design model of the new type of ultra-low temperature refrigerated cabin, the CAXA model is used as the modeling platform to establish the CAXA model of the new ball-type ultra-low temperature refrigerated cabin, as shown in Fig. At the same time, in the same volume V under the premise of the establishment of traditional hexahedron refrigerated cabin CAXA model, as shown in Figure 5.

3.2 Analysis of Cold Storage Capacity of New Type Refrigerated Cabin

Calculate the heat transfer area and heat transfer of the new compartments and the conventional compartments with the same volume V as the known conditions for the two chamber models to compare the refrigerated and heat preservation capacities of the new compartments and the conventional compartments.

3.2.1 Heat Transfer Area

Surface area of traditional compartments:$S_{tradition} = 6V^{\frac{2}{3}}$ (Assuming that the traditional refrigerated cabin structure is a regular hexahedron)

Volume of new compartments: $V_{newtype} = S_{ball} - 2V_{sector} = \frac{4}{3}\pi \cdot R^3 - 2\int_h^R \pi \cdot r^2 dh$, among them $r^2 = R^2 - h^2$. The surface area of the new compartment is: $V_{newtype} = S_{ball} - 2V_{sector} = 4\pi \cdot R^2 + 2\pi \cdot r^2 - 2\int_h^R 2\pi \cdot r dh = (6\pi - \pi^2)R^2 + 2\pi(h \cdot r - h^2 + \frac{R^2}{2} sin^{-1}\frac{h}{R})$ among them: h for the new compartment height of the half; R for the sphere radius; r for the cylinder radius.

Thus according to the mathematical analysis method and limit the principle of reduction:

(1) At that time $0 \le r \le \frac{\sqrt{2}}{2}R$, hypothesis$r = k \cdot R$;

(2) At that time $\frac{\sqrt{2}}{2}R \le r \le R$, hypothesis $h = j \cdot R$; among them: Respectively, k, j for the constant coefficient.

According to the decentralization method properly and respectively, when$0 \leq k \leq 0.74$, $0.67 \leq j \leq 1$, That is, the cylinder radius and the height of the cylinder are satisfied: $0 < r < 0.74R$ or $0.67R < h < R$, There is $S_{tradition} > S_{newtype}$.

At the same time according to theoretical calculations, at that time *r=0.74R*, which is *h=0.67R*, at this time there are new cabin to meet the conditions in the case, the smallest heat dissipation area, which is *k=0.74*, for the optimal proportion for the new compartment.

3.2.2 Heat Transfer *Q*

The heat transfer of the cabin and the outside of the ultra-low temperature refrigerated transport ship is mainly heat conduction and thermal convection. The heat transferred to the chamber by the cooling capacity of the chamber to the environment conforms to the Newton cooling formula, ie the convective heat transfer is: $Q_{tradition} > Q_{newtype}$. among them: Q for the convective heat transfer, α for the convective heat transfer coefficient, S for the heat transfer area, t_w for the external ambient temperature, t_f for the quick freezing/incubation chamber internal temperature). In the cabin insulation materials and internal and external environmental temperature difference are the same circumstances, according to Newton cooling formula: Q and S in the same external conditions are proportional, that is, to meet the design requirements, $Q_{tradition} > Q_{newtype}$.

4 Calculate the Results

According to the calculation results of heat transfer area S and heat transfer Q in the new cabin and the traditional compartment, in the case of the same external conditions (bulkhead insulation and internal and external temperature difference) and the volume of the cabin, $S_{tradition} > S_{newtype}$, $Q_{tradition} > Q_{newtype}$, That is, the heat transfer area and heat transfer capacity of the new type of refrigerated cabin are smaller than the heat transfer area and heat transfer of the traditional refrigerated cabin. Therefore, the new type of circular cabin is better than the traditional compartment.

5 Conclusion

This new type of spherical structure, to make up for the traditional hexahedron type heat transfer area, heat transfer and other deficiencies, for the ship cabin research and design to provide a strong basis. This kind of cabin has the practical operability, and has wide application range, the remarkable effect of energy saving and so on, and can be widely used in medium, low temperature and ultra-low temperature refrigerated ship. With the further improvement of the inner and outer wall materials and the cooling device technology, the convective heat transfer coefficient is also reduced, and the new type of ultra-low temperature refrigerated cabin is better.

The design of the ultra-low temperature refrigerated cabin is a preliminary study, and the rationality of the new refrigerated cabin is verified from the aspects of cold storage capacity. However, there are some shortcomings such as the selection of the new type of cabin and the reasonable connection with other compartments and ship buildings. The effective subdivision of the cabin and other aspects need further discussion and analysis. Therefore, the later research can be further studied from the research and development of ultra-low temperature refrigeration equipment, the research and development of new chamber experiment model, the experimental comparison of cold storage capacity of new and traditional chamber model, and the optimization design of new type of cabin structure.

Acknowledgement: The present research was partly funded by Zhejiang Province in 2013 major science and technology special, No. 2013C03033,Zhejiang Province higher education classroom teaching reform project, number kg2013194 and Zhejiang Ocean University, Port and Transportation School undergraduate students development of funding projects. Also, the research was supported by Zhejiang Ocean University.

References

[1] G.L. Zhang. Large ship cabin division and layout research [D]. Harbin: Harbin Engineering University, 2010.

[2] D. Zhou. Application of cooling pipe in large-scale freon direct expansion of low temperature cold storage system [D]. Dalian: Dalian University of Technology, 2012.

[3] H.F Han. Marine cabin environment control technology research [D]. Harbin: Harbin Engineering University, 2004.

[4] B. Jiang, H.X Xu, Y.Q Zhang, etc. An ultra-low temperature refrigerated transport ship quick-freezing cabin: China, 201420267906.5 [P]. 2014–12–03.

[5] H.P Fan, B. Chen, D. Li, etc. A kind of transport tank insulation cover device: China, 201320844376.1 [P]. 2014-07–02.

[6] W. XUE, C.Y Zheng, W. Chen. Experimental study on the effect of different condensate air volume on the running characteristics of refrigerated containers [J]. Ship Engineering, 2012, 41 (05): 98–101.

[7] Y.Q. Wu, Z.M. Shang, J.D. Sun, S.K. Wang. Application and advantages of ultra-low temperature refrigerated container transport [J]. Containerization, 2012, 07: 24–25.

[8] Y.F. Wang, X.B. Zhao, Y. Wu. Study on frozen and frozen food packaging of ships based on naval ship escort protection requirements [J]. Packaging Engineering, 2011, 23: 5–8.

[9] S.H Liu. Gantry machine tool feed system thermal-force coupling dynamic modeling analysis and optimization method [D]. Nanjing University of Aeronautics and Astronautics, 2013.

[10] Y.X. Zhong, J.H. Zhou. Optimization design of small ship interior [J]. Ship Engineering, 2012, 41 (4): 75–78.
[11] Y. Li, C.Q. Gong .Application of Improved Genetic Algorithm in Optimized Design of Yacht Cabin Layout [J]. Ship Engineering, 2010, 39 (1): 34–37.

Meng Chen, Tao Jing, Jiang-Liang Han, Zhe Wang and Hui-Cheng Xie*

Removal of an Azo Dye Acid Red 114 by *Iris Pseudacorus* and Its Photosynthetic Response

Abstract: Azo dyes, as the largest group of synthetic dyes released from textile and dyestuff industries, have caused serious environmental problems in China. We investigated the removal ability and physiological response of *Iris pseudacorus* to simulated wastewater containing acid red 114 with concentration similar to real wastewater or above. The percent decolorization of the acid red 114 of different dye concentrations varied greatly within 10 days. The percent decolorization of acid red 114 of different concentrations in average was higher than 66.85 %, indicating it can be effectively removed by *Iris pseudacorus*. Net photosynthesis rate (Pn), transpiration intensity (Tr) and stomatal conductance (Gs) of *Iris pseudacorus* decreased significantly with an increase in concentrations of acid red 114 solutions. But Ci increased significantly with an increase in dye concentration ($p < 0.05$), and Pn behaved in the opposite direction. This result showed that the main reason that Pn declined was the non-stomatal factors. When the concentration of acid red 114 was 120mg /L, it decreased the Pn of *Iris pseudacorus* 44 %. When the concentration of acid red 114 increased to 160 mg/L or more, it decreased the Pn of *Iris pseudacorus* more than 50 %. It is concluded that *Iris pseudacorus* can resist 120 mg/L tested acid red 114, so it can be used to phytoremediation azo dyes.

Keywords: Azo dyes; Phytoremediation; *Iris pseudacorus*; Net photosynthesis rate (Pn)

Meng Chen, Forestry College, Shandong Agricultural University, Taian, 271018, Shandong, China. Email: 1132999869@qq.com
Tao Jing, Forestry College, Shandong Agricultural University, Taian, 271018, Shandong, China. Email: 125498469@qq.com
Jiang-Liang Han, Forestry College, Shandong Agricultural University, Taian, 271018, Shandong, China. Email: 1523021312@qq.com
Zhe Wang, Forestry College, Shandong Agricultural University, Taian, 271018, Shandong, China. Email: 644209770@qq.com
Hui-Cheng Xie, Forestry College, Shandong Agricultural University, Taian, 271018, Shandong, China. Email: xiehc@sdau.edu.cn

DOI 10.1515/9783110540048-030,

1 Introduction

Synthetic dyes have been a serious source of water contamination in many parts of the world. In China the situation is more serious. Total colorant production in China is estimated to be above 420,000 tons per annum, which account for 42 % of the total World colorant production [1]. According to incomplete statistics, daily discharge of printing and dyeing wastewater reached 4 million m^3 in China [1]. In China, a large number of organic dyes have escaped and seeped into the environment due to backward dye-manufacturing technology and dye-using by sporadic and small-scale rural textile and synthetic dye industries. This situation made serious synthetic dyes pollution on both water bodies and soil [2]. The traditional treatment techniques applied in textile waste water such as chemical coagulation/flocculation, membrane separation (ultra filtration, reverse osmosis) or activated carbon adsorption only does a phase transfer of pollutants. Moreover, because activated sludge and other types of bioreactor fail to remove sufficient color, COD, N, surfactants and other micro-pollutants present in the textile effluent, conventional waste water treatment plant relying on it is not adequate for the treatment of textiles. Due to its cost effectiveness, aesthetically pleasing and easy operation, phytoremediation is a potential practical method in removing the synthetic dyes from wastewater. Some researchers have confirmed that constructed wetlands can effectively degrade azo dye wastewater [3, 4]. Aquicultural plant system experiments showed that plants can decolorize wastewater containing dyes [5–8]. Some aquicultural plant system experiments further showed that plants can degrade dyes [7–9].

Iris pseudacorus is an aquatic plant and has been widely used in artificially constructed wetlands for the removal and mineralization of triazophos [10], and the removal of nitrogen, phosphorous and heavy metals [11, 12]. However, research on the effects of water polluted by azo dye on photosynthetic physiology of *I. pseudacorus* and the remediation effects of *I. pseudacorus* on water containing azo dyes have not been reported yet. Therefore, the objective of this study were to study the potential of *Iris pseudacorus* for textile dye removal and to explore the effects of dye on the photosynthesis parameters of *Iris pseudacorus*.

2 Materials and Methods

2.1 Reagent and Chemical Analysis

Acid red 114 (CI, 23635) were purchased from Shanghai Huanlan chemical company. The absorption values of samples and standards were determined by a Hitachi U-1100 Spectrophotometer. The standard curve of each dye was made. The remaining concentration of each dye was calculated through the standard curve regression equation.

2.2 Hydroponic Experiment

The experiment was conducted in the greenhouse at Shandong Agricultural University. Seedlings of *Iris pseudacorus* were purchased from a local market. All plants were washed by tap water. And then the seedlings were transferred to 20 L buckets wrapped in black plastic bag to make the bucket in a dark state .Each bucket was filled with 5 L tapped water, going through two weeks of cultivation and domestication in the artificial greenhouse. After the domestication, the plants were supposed to do the simulation test.

Then, the decolorization of wastewater containing red 114 by *Iris pseudacorus* seedlings was conducted. One seedling was placed in a bucket containing 5L half strength Hoagland's solution only (control treatment) or spiked 20, 40, 80, 120,160, 200, 320 mg/L red 114. Each treatment has three replicates. The total weight of bucket with solution and plant were recorded. During this period the total weight of the bucket with solution and plant was maintained nearly constant by adding fresh half strength Hoagland's solution daily. When the experiment was terminated at 10th day, 10mL contaminate solution samples were taken after the addition of fresh half strength Hoagland's solution to the bucket for dye's concentration analysis.

The parameter percent decolorization was defined as:

$$\text{Percent decolorization} = \frac{\text{Concentration of untreated dye} - \text{Concentration of treated dye}}{\text{Concentration of untreated dye}} \times 100$$

2.3 Leaf Gas Exchange Determination

On the eighth day after dye treatment, the photosynthesis parameters were determined. The youngest fully expanded leaves were used for determination of gas change. Net photosynthesis rate (Pn), transpiration intensity (Tr) and stomatal conductance (Gs) were measured with a CIRAS-2 Portable photosynthesis system (PP system, Hitchin, UK). The measurements were made in ambient [CO_2] (Ca=380 μmol mol^{-1}) on individual leaf enclosed in a leave cuvette under a rate of 0.44 L min-1 air flow, relative humidity within the cuvette at 50–55 %, a leaf temperature of 25 °C and photosynthetic active radiation of 1200 $\mu mol \cdot m^{-2} \cdot s^{-1}$. The difference of between control and azo dyes treatment samples were analyzed by one-way analysis of variance (ANOVA) and Duncan's multiple-range test was used for multiple comparisons.

3 Results and Discussion

3.1 Decolorization of Azo Dyes

The percent decolorization of the acid red 114 of different dye concentrations varied greatly (Fig. 1). The percent decolorization of acid red 114 by Iris pseudacoru in 20 mg/L, 40 mg/L, 80mg/L, 120 mg/L, 160 mg/L, 200 mg/L and 320 mg/L was 83.96 %, 74. 9 %, 70.8 %, 68.12 %, 64.6 %, 57.8 % and 47.9 % respectively, indicating it decreased with increasing dye concentrations. The percent decolorization of acid red 114 in different concentrations in average was higher than 66.85 %, indicating it can be effectively removed by *Iris pseudacorus*. The maximum color removal of Reactive red 141 by Narrow leaved Cattails (*Typha angustifolia* Linn.) was 60 % observed on day 14 [13]. The color removal efficiency of Phragmites australis wetlands was 69 % to 74 % for different influent concentrations [14].

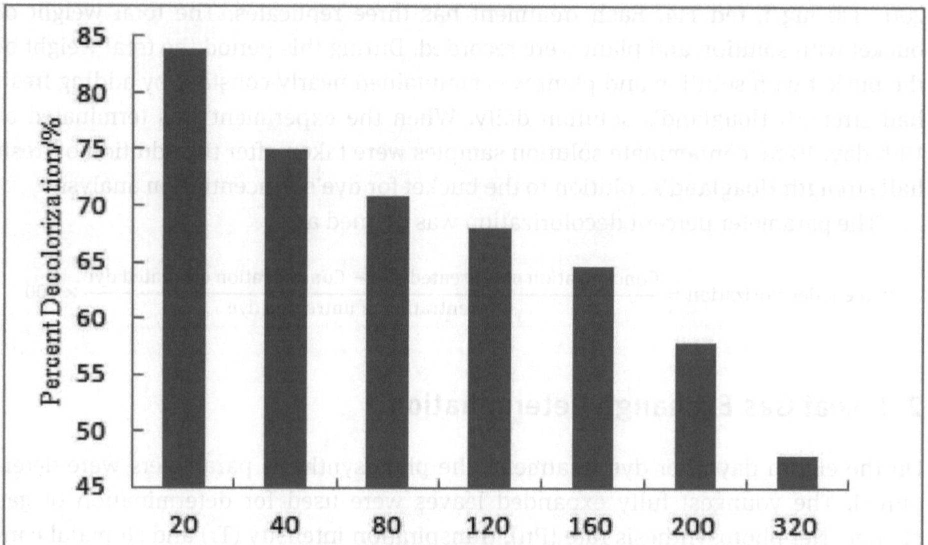

Fig. 1: Percent decolorization of azo dyes wastewater by *Iris pseudacorus*

3.2 Photosynthetic Physiological Response to Azo Dyes

The measured photosynthetic physiological parameters of *Iris pseudacorus* of response to acid red 114 concentrations under the same photosynthetical active radiation (PAR=1,200 μmol m^{-2} s^{-1}) were shown in Fig. 2. It shows that in the fixed PAR (PAR=1,200 μmol m^{-2} s^{-1}) and different concentration of acid red 114, the PN, E, Gs and Ci of the leaves changed significantly but with various trends. Pn, E, and Gs decreased significantly with an increase in concentrations of acid red 114 solutions. E and Gs decreased fast indicating that both were sensitive to acid red 114. But Ci increased significantly with an increase in dye concentration ($p<0.05$), and PN behaved in the opposite direction. According to the criterion of the limitations of photosynthesis of Farquhar and Sharkey (1982) [15], the main reason that Pn declined was the non-stomatal factors. The phytotoxicity must be taken into account if plants are considered for removal of chemicals from soil, surface or groundwater. By visual inspection, Phragmites australis did not show signs of phytoxicity and their growth did not seem to be affected by the presence of Acid orange in solution even for high concentration up to 700 mg/L [3]. The Evans Blue (CI, 23860), Bismark Browny (CI, 21000) and Orange G (CI, 16230) of 100 mg/L decreased the Pn of sunflowers less than 50 % [16]. In this research, 120 mg/L acid red 114 decreased the Pn of *Iris pseudacoru* 44 %. When the concentration of acid red 114 increased to 160 mg/L or more, it decreased the Pn of *Iris pseudacorus more than* 50 %.

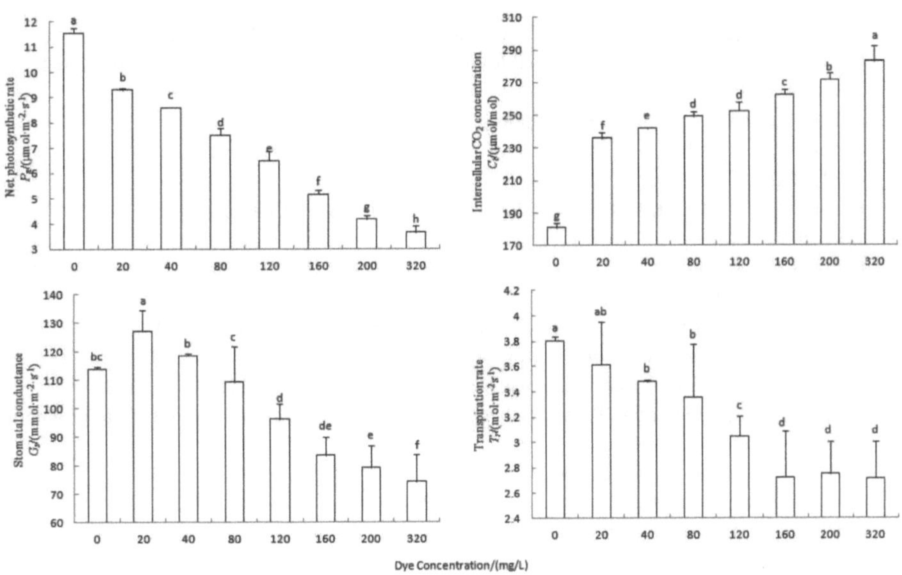

Fig. 2: Photosynthetic physiological response of *Iris pseudacorus* seedlings to acid red 114

4 Conclusion

The percent decolorization of acid red 114 in different concentrations in average was higher than 66.85 %, indicating it can be effectively removed by *Iris pseudacorus*. Because 120 /L acid red 114 decreased the Pn of *Iris pseudacorus* 44 %. When the concentration of acid red 114 increased to 160 mg/L or more, it decreased the Pn of *Iris pseudacorus more than* 50 %. It is concluded that *Iris pseudacorus*r can resist 120 mg/L tested acid red 114, so it can be used to phytoremediation azo dyes.

Acknowledgement: This research was financially supported by Natural Science Foundation of Shandong Province as "The mechanism of purifying dye containing wastewater by phytoremediation" (Approved No., ZR2015CM021).

References

[1] Zi, L., Ding, Z., Water & wastewater Engineering. 34, 1, 236–240, 2008.
[2] Zhou, Q. X., Bull. Environ. Contam. Toxicol. 66, 6, 784–793, 2001.
[3] Davies, L.C., Carias, C.C., Novais, J.M. and Martins-Dias, S., Ecological Engineering. 25, 5, 594–605, 2005.
[4] Tjasa, G.B., Ojstrsek, A., T,Journal of Hazardous Materials. 155, 1/2, 76–82, 2008.
[5] Nilratnisakorn, S., Thiravetyan, P. and Nakbanpote, W., Science of the Total Environment. 384, 1–3, 67–76, 2007.
[6] Aubert, S., Schwitzguebél, J.P., Water Res. 38, 16, 3569–3575, 2004.
[7] Khandare, R.V., Kabra, A.N., Kurade, M.B., Govindwar, S.P., Biores. Technol. 102, 12, 6774–6777, 2011.
[8] Khandare, R. V., Kabra, A. N., Tamboli, D. P., Govindwar, S. P., Chemosphere. 82, 8, 1147–1154, 2011.
[9] Kabra, A. N., Khandare, R. V., Waghmode, T. R., Govindwar, S. P. Journal of Hazardous Materials. 190, 1–3, 424–431, 2011.
[10] Li Z, Xiao H, Cheng S, Zhang L, Xie X, Wu ZA. J Environ Sci. 26, 2, 315–322, 2014.
[11] Yousefi Z., A. Mohseni-Bandpei, Ecological Engineering. 36, 6, 777–782, 2010.
[12] Huang S, Yuan H, Sun Y, Han Y. Chinese Journal of Ecology. 27, 7, 1181- 1186, 2008.
[13] Duc, R., Vanek, T., Sounder, P., and Schwitzguebel, J. P. International Journal of phytoremediation. 1, 3, 255–271, 1999.
[14] Khan, A.A., and Husain, Q., Biores. Technol 98, 5, 1012–1019, .2007.
[15] Farquhar, G.D., Sharkey, T.D.: Ann. Rev. Plant. Physiol.33, 1, 317–345, 1982.
[16] Huicheng, X.; Chongrong, Li; Jihong, Li; Li, Wang. Acta Ecologica Sinica, 32, 5, 240–243, 2012.

Zhen-Bo Bao*, Jian Xiao, Xin-Yuan Liu and Jin-Xin Peng

The Innovative Design of Lift Type Multi-body Straw Biomass Gasifier

Abstract: As clean and renewable energy, straw biomass energy used by gasification combustion can reduce the energy crisis and environmental pollution. At present, small household biomass gasification combustion furnace in use there are some problems: tar and ash in the gasification gas more, the material continuous feeding process complex, and the intermittent feeding inconvenience, which leads to unsatisfactory application. In this paper, the design of a lift type multi-body household straw biomass gasifier is proposed, through three gasification combustion furnace body and removable furnace; gears, racks and other lifting transmission mechanism; air gasifying agent preheating, uniform distribution of supply; the second-class purification system of tar and dust impurities and other structural design, can make biomass material loading conditions to meet the needs of home cooking needs to achieve gasification flammable gas flow without power to ensure gasification reaction process requirements, and the effective removal of the tar and the dust impurities in gasification gas. The lift type multi-body household straw biomass gasifier can effectively utilize the straw biomass energy to solve the problem of straw biomass waste and pollution.

Keywords: Lift type, Multi-body, Straw biomass basifier

1 Introduction

As clean and renewable energy, straw biomass energy used by gasification combustion can reduce the energy crisis and environmental pollution [1–3]. The small household biomass gasifier has many advantages involving simple structure, short pipeline and easy operation. However, there exist some disadvantages, such as more tar and ash in the gasification gas and the complexity of continuous process of

Zhen-Bo Bao, Engineering and Technology College, Tianjin Agricultural University, Tianjin, China, 300384. Email: zhenbobao@sohu.com.
Jian Xiao, Engineering and Technology College, Tianjin Agricultural University, Tianjin, China, 300384. Email: xiaojian0857@163.com.
Xin-Yuan Liu, Engineering and Technology College, Tianjin Agricultural University, Tianjin, China, 300384. Email: liuxinyuan11@163.com.
Jin-Xin Peng, Engineering and Technology College, Tianjin Agricultural University, Tianjin, China, 300384. Email: pengjinxin2005@163.com.

DOI 10.1515/9783110540048-031,

adding the material, which leads to unsatisfactory application [4–6]. To improve the disadvantages of the straw biomass gasification furnace, a lift type multi-body straw biomass gasification gasifier is designed.

2 The Overall Structural Design of Lift Type Multi-body Straw Biomass Gasifier

As shown in Fig. 1 to Fig. 8, the lift type multi-body straw biomass gasifier comprises three furnace bodies, an ignition device, an air gasifier supply system, a gasification gas delivery system, a gasification gas purification system and so on. Each furnace body comprises upper cover plate 4, side wall plate 5, outer furnace body 6, inner furnace body 7, bottom base coaming 11, inner furnace body annular fixing plate 12, bottom plate 13, inner furnace body bottom plate 14, inner furnace body bottom bracket 15, and outer furnace body support block 16 and so on. Ignition device comprises ignition port fixed plate 8, ignition port block 9 and ignition port 10. The air gasification agent supply system comprises bottom and side supply system, the bottom supply system comprises bottom gas inlet damper 17 and bottom gas inlet 18, the side supply system comprises side air inlet damper 34 and side air inlet 35. The gasification gas delivery system consists of two parts: outward output unit and internal circulation unit. The outward output unit comprises sealing pipe nut 1, upper gas port 2 and upper fixing sleeve 3. The internal circulation unit comprises side fixed sleeve 21, side upper air guide port 22, three-way pipe joint 23, short connection pipe 24, tar and dust impurity recovery unit 25, side lower air guide port 26, female-threaded pipe joint 27 and lower fixing sleeve 28. The connection of the two furnaces is effected by means of the side upper gas guide port 22, the three-way pipe joint 23, the short connecting pipe 24 and the side down airport 26.

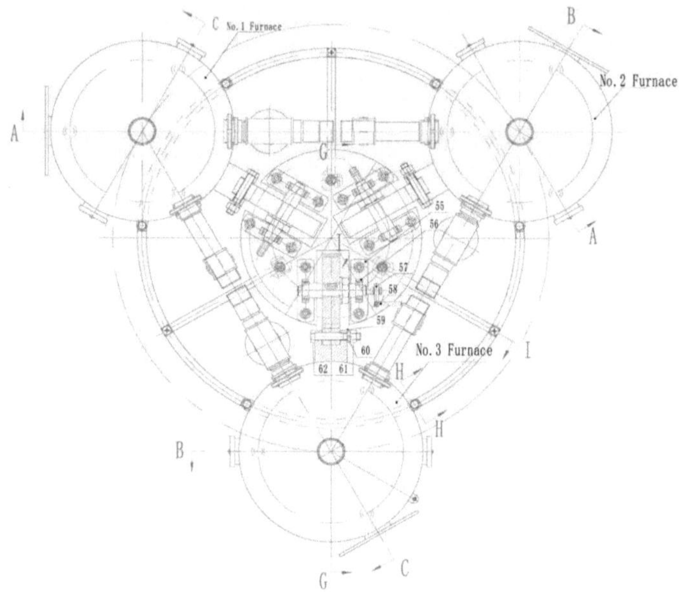

Fig. 1: The overall structure of gasifier

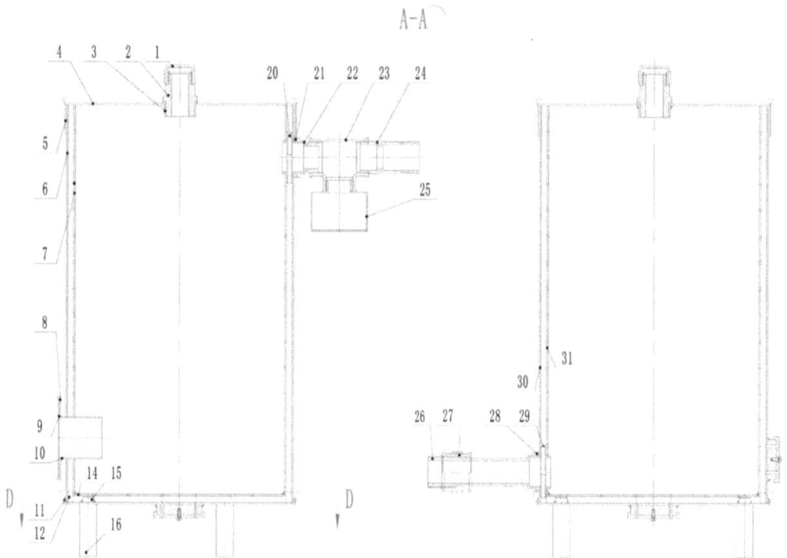

Fig. 2: The A-A cross-sectional view of gasifier

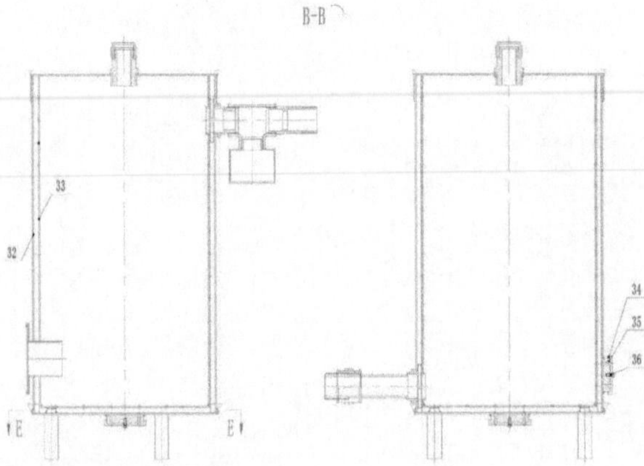

Fig. 3: The B-B cross-sectional view of gasifier

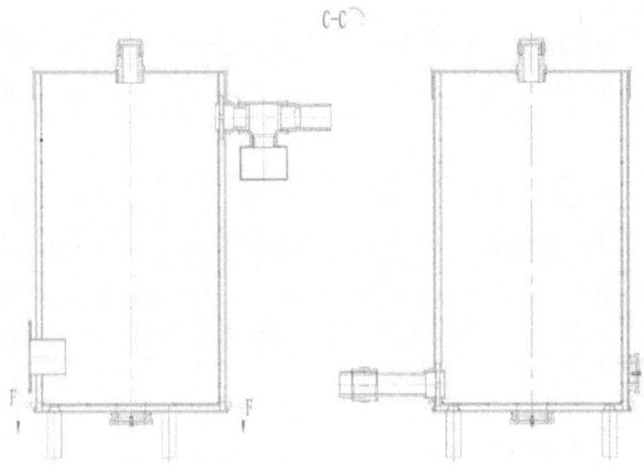

Fig. 4: The C-C cross-sectional view of gasifier

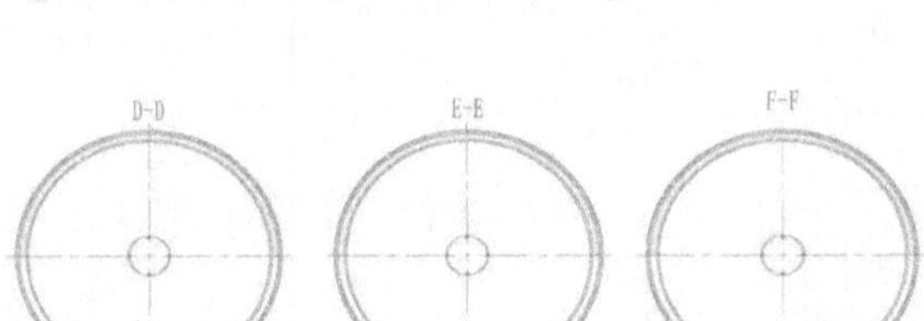

Fig. 5: The D-D, E-E and F-F cross-sectional view of gasifier

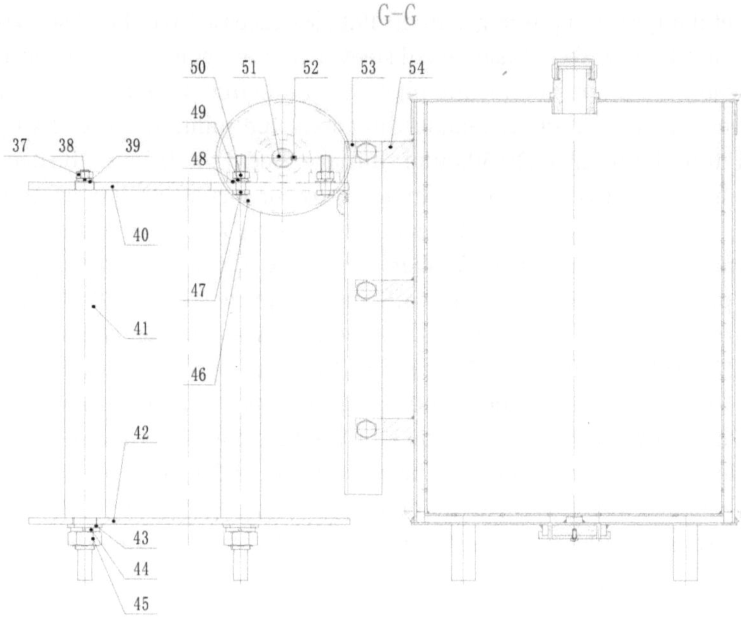

Fig. 6: The G-G cross-sectional view of gasifier

Fig. 7: The H-H cross-sectional view of gasifier

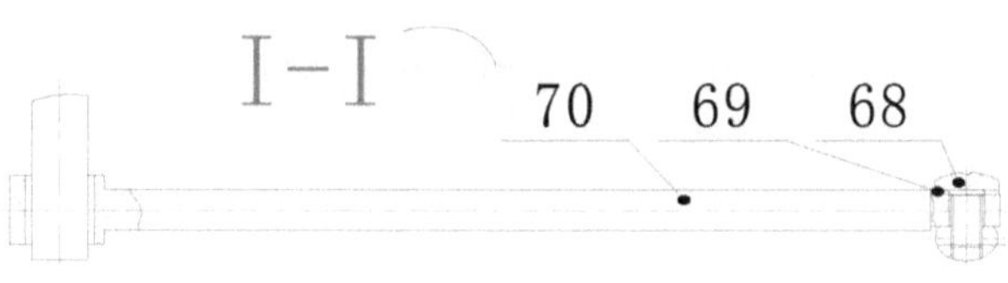

Fig. 8: The I-I cross-sectional view of gasifier

1.sealing pipe nut; 2.upper gas port; 3.upper fixing sleeve; 4.upper cover plate; 5.side wall plate; 6.outer furnace body(1); 7.inner furnace body(1); 8.ignition port fixed plate; 9.ignition port block; 10.ignition port; 11.bottom base coaming; 12.inner furnace body annular fixing plate; 13.bottom plate; 14.inner furnace body bottom plate; 15.inner furnace body bottom bracket; 16.outer furnace body support block;

17.bottom gas inlet damper; 18.bottom gas inlet; 19.Cross recessed pan head screws M5×14; 20.Small round nut M56x2; 21.side fixed sleeve; 22.side upper gas guide port; 23.three-way pipe joint; 24.short connecting pipe; 25.tar and dust impurity recovery unit; 26.side lower air guide port; 27.female threaded pipe joint; 28.lower fixing sleeve; 29.Small round nut M56x2; 30.outer furnace body(2); 31.inner furnace body(2); 32.outer furnace body(3); 33.inner furnace body(3); 34.side air inlet damper; 35.side air inlet; 36.Cross recessed pan head screws M5×12; 37.Hex nuts M12; 38.Standard spring washers 12; 39.Flat Washers 12; 40.fix upper plate; 41.support rod; 42.fix lower plate; 43.Flat Washers 24; 44.Standard spring washers 24; 45.Hex nuts M24; 46.gear; 47.Hexagon head bolts M12x50; 48.Flat Washers 12; 49.Standard spring washers 12; 50.Hex nuts M12; 51.shaft; 52.Ordinary type flat key-A type 6x6x20; 53.rack; 54.rack mounting block; 55.bearing housings; 56.Deep groove ball bearings 6205; 57.rotating lever; 58.handle; 59.Flat Washers 16; 60.Standard spring washers 16; 61.Hex nuts M16; 62.Hexagon head bolts M16x130; 63.support ring; 64.support base; 65.guide post; 66.lower guide ring; 67.upper guide ring; 68.Cross recessed pan head screws M10×25; 69.Flat Washers 10; 70.fixed connecting rod.

3 Innovative Design of the Key Structure of Lift Type Multi - body Straw Biomass Gasifier

3.1 The Design of Gears and Racks Lifting Mechanism

The lever 57 and the shaft 51 are rotated by the rotation of the handle 58 so that the gear 46 is rotated to move the rack 53 up and down, and the rack 53 drives the furnace body up and down by the rack mounting block 54. The connection of the two furnaces is effected by means of the side upper gas guide port 22, the three-way pipe joint 23, the short connecting pipe 24 and the side down air port 26, achieving the goal of the three furnace body in turn recycling, the gasification gas entering into another furnace body, and non-boosting flow of the gasified combustible gas.

3.2 The Design of Gasifier Inner Furnace Body That Can be Demolished

After using the biomass material in the No. 1 furnace, the biomass material in the No. 2 furnace is started, and the upper cover plate 4 is removed to replace the inner furnace body in the No. 1 furnace with another inner furnace body filled with the biomass material, complete the loading of biomass materials.

3.3 The Design of Air Gasification Agent Preheating and Uniform Supplying Structure

Each furnace comprises a bottom gas inlet 18 and two side air inlets 35. The flow of the air gasifying agent is regulated by adjusting the relative position of the bottom gas inlet damper 17 and the bottom gas inlet 18, and the relative position of the side air inlet damper 34 and the side air inlet 35. Combustion heat is passed through the outer wall of the inner furnace body 7 to preheat air gasification agent.

3.4 The Secondary Purification Treatment Structure Design of Tar and Dust

The tar and dust in the gasification gas are first collected and processed by the tar and dust impurity recovery unit 25; next, the gasification gas passes through the side fixing sleeve 21, the side upper gas guide port 22, the three-way pipe joint 23, the short connecting pipe 24, the tar and the dust impurity recovery unit 25, the side lower air guide port 26 and the female threaded pipe joint 27, are introduced into the next furnace, the tar and dust are further intercepted and purified by the biomass material in the furnace to obtain a relatively clean gasification gas, the relatively clean gasification gas are collected or delivered directly to the furnace for combustion.

4 Conclusion

The lift type multi-body straw biomass gasifier can meet the practical needs of farmers with low economic cost, and make effective use of biomass resources in rural areas. It has practical significance to alleviate the mineral energy crisis and improve the rural environment [5, 7–8].

Acknowledgement: The research work was supported by Tianjin College Students' Innovative Entrepreneurial Training Program (No.201610061022 and No.201610061024). The research work was also supported by Tianjin Research Program of Application Foundation and Advanced Technology (No.16JCQNJC08200).

References

[1] X.Zhang, Z.S.Cai, Y.Chen, H.Li, Journal of Agricultural Mechanization Research, 2, 1–5, 2015.
[2] Y.L.Bao, Z.B.Bao, D.C.Jin, Y.Liu, J.X.Peng, Mechanical Engineer, 8, 149–151, 2015.
[3] Z.B.Bao, F.Fang, D.C.Jin, Y.L.Liu, L.Yang, Journal of Agricultural Mechanization Research, 5, 255–258, 2016.

[4] H.Guo, T.Zhu, J.P.Wang, GuangZhou Chemical Industry and Technology, 42, 18, 35–37, 2014.
[5] Z.B.Bao, C.A.Liu, G.C.Xie, J.X.Peng, D.C.Jin, Journal of Agricultural Mechanization Research, 3, 253–257, 2017.
[6] B.Li, H.P.Chen, H.P.Yang, X.H.Wang, S.H.Zhang, Transactions of the Chinese Society of Agricultural Engineering,
[7] 5, 205–209, 2011.
[8] Z.F.Tian, S.Y.Wang, S.P.Li, Journal of Agricultural Mechanization Research, 12, 243–246, 2014.
[9] G.Y.Chen, Z.P.Xia, B.B.Yan, W.H.Zhou, Q.Wang, Renewable Energy, 32, 9, 1395–1399, 2014.

Zhong-Nian Li, Chun-Lei Ding and Lei Zhou

Research & Development on the EEWT Facility

Abstract: Based on the principle of electromagnetic resonance coupling, a novel and advanced electromagnetic resonance type EEWT (Electric Energy Wireless Transmission) transmitter and receiver facility have been researched & developed. In this paper, orchestrated by the "multi-unit, multi direction and concentric circles (i.e. the range of coil diameter equal) spiral coil (spiral coil good direction, suitable for signal directional transmission) as the transmitting coil, thus forming a" non-radiative magnetic field around it, the electric energy is converted into a magnetic field to carry out electromagnetic wave emission. When the natural resonance frequency of the receiving coil is the same as that of the frequency of electromagnetic waves emitted by a transmitting coil, the electromagnetic resonance is caused and the strong electromagnetic coupling is generated, i.e., the conversion from "the magnetic field" to "the electric energy" is realized, thereby realizing the EEWT. The experimental results show that the EEWT has higher transmission efficiency and higher transmission power, and is suitable for medium distance EEWT.

Keywords: EEWT, Magnetic resonance, Transmission efficiency, Transmission power, Transmission distance

1 Introduction

What is known as "EEWT (Electric Energy Wireless Transmission)" means the electric energy transmission of between the transmitter and receiver without the use of electric wires (i.e. wireless), which has so far been without actual engineering cases.

EEWT is an advanced technology for power transmission by radio waves, which is similar to the use of radio waves to transmit mobile phone, radio, television and other communication information, but there are essential differences, namely EEWT and wireless communication in principle, although some similarities, but the technical objectives of the two different requirements, the main focus is on the transmission of the above information frequency signals (not concerned with energy transmission efficiency and transmission power),as long as you can select the desired

Zhong-Nian Li, School of electrical & energy engineering, Nantong Institute of Technology, Nantong China. Email:lzn13914396143@126.com
Chun-Lei Ding, School of electrical & energy engineering, Nantong Institute of Technology, Nantong China. Email: 442889910@qq.com
Lei Zhou, School of electronic information & electrical engineering, Shanghai Jiao Tong University, Shanghai China. Email:1021441955@qq.com

DOI 10.1515/9783110540048-032,

frequency signal at the receiving end; and the former (EEWT) focuses on the pursuit of energy transmission, transmission efficiency is as high as possible / transmission power as large as possible, which meet the requirements of the target technology of high voltage and current of long distance EEWT .

Magnetic resonance type EEWT is one of the most popular ways of EEWT principle, the basic method is to transmit power to high frequency power supply load to the launch coil to generate high frequency magnetic field, the receiving coil coupling high frequency magnetic field produces a current sense, in order to achieve the EEWT.

Now the development of EEWT technology, there are three kinds generally: first, non-contact type charging electromagnetic induction technology used portable terminal, the upcoming two coils placed in the adjacent position on the magnetic flux generated when the current flowing in a coil when the coil EMF leads to another power supply for the electric load. Secondly, the electromagnetic wave energy is used to transmit power to the electric load through the antenna transmitting and receiving technology. This is similar to wireless communication technology. Compared with wireless communication technology over 100 years ago, the efficiency of EEWT technology is greatly improved. Thirdly, the electric energy is supplied by the resonance technique of the electromagnetic field to supply the power load. Note that this EEWT technique is not powered by electromagnetic waves or current, but by means of electric or magnetic fields.

Suitable for the short range "EEWT" facilities, such as electromagnetic induction type micro power (such as a portable terminal Contactless charging for electric power facilities) wireless transmission has been gradually taking and widely used; and the electric coupling type, radio waves, magnetic resonance type suitable for short distance and distance of these facilities is still EEWT stay in the stage of laboratory research, once successfully developed into practical engineering, the electric power facilities (such as: power transmission facilities, new energy electric vehicle charging grid facilities, new energy power generation system) is a technology innovation turn the world upside down, and its social and economic benefits will be very considerable. Obviously, the extraordinary significance.

2 Basic Methods

The research and development of "magnetic resonance" EEWT infrastructure as shown in Figure 1; the facility using LC magnetic coupled resonant circuit for wireless power transmission, because of magnetic resonance technology on the distance of wireless power transmission, with high transmission efficiency, low requirement of obvious advantages. The project group on the basis of mutual inductance theory and coupling theory, this facility conducted in-depth and detailed research (mainly explores how to improve the efficiency of magnetic resonance wireless transmission), through painstaking exploration and careful research, theoretically demonstrates the effect of distance, line diameter, coil winding method of the transmission efficiency. The transmitter of the facility, produced by PWM (Pulse Width Modulation) special microprocessor through the gate signal and PWM signal is converted into two complementary, and then through the optocoupler driving full bridge rectifier, the DC voltage of the inverter AC voltage, LC resonance, and the electrical energy into magnetic energy radiated by the LC resonant magnetic energy receiver; to receive a transmitter.

The “EEWT” facility development scheme and technical route is: first at the transmitter frequency transmission current into high frequency current, and then use the air as the medium for power transmission. The transmission efficiency of the system depends on the coupling coefficient and the quality factor between the transmitter and the receiver. The project is based on the principle of coupled magnetic resonance (MRI), which uses a non-radiated magnetic field to achieve power transmission at specified distances. The transmitter has two coils, one of which is the oscillation ring (oscillation ring and the transmitted electric energy is used to generate a high frequency current), another is to launch the ring, used to emit electromagnetic wave, the formation of “no radiative magnetic field around about electrical energy into magnetic field. When the natural frequency of the receiving end and the received electromagnetic waves of the same frequency, the oscillation current generates the receiving end circuit in the strongest, the completion of the “magnetic field” to “energy conversion”, in order to achieve energy efficient transmission.

On the basis of reference [3], the coupling coefficient $k = \omega_1 - \omega_2$ (ω_1 and ω_2 is the coupling coil two resonant point angular frequency), two resonant frequency is the experimental study as long as the measured coupling coil can be that the coupling coefficient of K. Adjust the frequency of the signal source so that it scans the frequency range of the whole possible resonance, and can find the frequency of two peaks. When the two is in strict accordance with the coil, the coil self-resonant frequency of $\omega_S = \omega_D = \omega_0$, the decay rate of $\Gamma_S = \Gamma_D = \Gamma$, then measure the frequency of the peak point of ω_1 and ω_2[1] [8].

$$\omega_{1,2} = \omega \pm \sqrt{k^2 - \Gamma^2} \tag{1}$$

$$k = \sqrt{(\omega_{1,2} - \omega_0)^2 + \Gamma^2} \tag{2}$$

As long as the corresponding measured two peak frequency and gamma can get the coupling coefficient of K. By analyzing the two cases of the excitation source and the non-excitation source, we can see that the frequency splitting phenomenon exists in the magnetic resonance EEWT technique.

2.1 Systems Tructure of the Magnetic Resonance Type EEWT

The block diagram of the magnetic resonance EEWT system is shown in figure 1. It mainly includes power frequency power supply, frequency converter, electromagnetic launching system, electromagnetic receiving system, inverter and load. The frequency converter turns the frequency power supply (220V 50Hz) into the resonant frequency of the electromagnetic emission system and provides the electromagnetic emission system. The transmitting system transfers the energy to the receiving coil of the electromagnetic receiving system by using a transmitting coil, and then the frequency of the received electric energy is changed into the frequency required by the load by the inverter for the use of the load.

2.2 Teoretical Bsis on Dsign for the Coil

As the transmission medium, the transmitting coil in this paper will be guided wave circuit in high frequency current or feed transmission line effectively into a space electromagnetic wave polarization direction, to the provisions of the emitted electromagnetic wave; a polarization receiving coil will be from a particular direction of space efficient conversion of guided wave in high frequency circuit the current or the transmission line. Its main function is:

(1) Energy conversion for the launch coil, the coil should be in the circuit RF current as much aspossible into the space of the electromagnetic wave energy radiation. For the receiving coil, the coil shall convert the received electromagnetic wave energy to the maximum of the RF current in the circuit to the receiving module. This requires that the transmit coil and the firing module, the receiving coil and the receiving module match the load as best as possible. A pair of highly efficient transmitting and receiving coils is an efficient energy conversion device. [1] [3]

(2) Directional radiation or receiving transmitting coils for its emission of electromagnetic waves should have a direction, such as the transmission of electromagnetic energy transmitting coil this paper requires the coil axial direction, and the radiation of electromagnetic energy in the other direction as little as possible. For the receiving coil, the electromagnetic energy received from the transmitting coil is received only in axial direction, and the receiving efficiency is obviously decreased in other directions. The directional radiation and receiving characteristics of the transmitting coil and the receiving coil are decisive for the transmission efficiency of energy. For the transmitting coil and the receiving coil, the emission and reception of the coil are equivalent. Therefore, the electromagnetic emission system and the electromagnetic receiving system are studied at the same time. The transmitting coil and receiving coil is the core part of magnetic resonance EEWT transceiver module, when the coil design and

production is completed, its only one resonant frequency, that is to say the physical parameters of the coil determines the electrical parameters of coils. The transmission efficiency of the magnetic resonance EEWT system is closely related to the frequency, so the physical parameters of the coil have a great impact on the transmission efficiency of the whole system. Starting from the relationship between the physical parameters and the electrical parameters of the coil, the transmitting coil and the receiving coil are designed according to the relationship between the coils and the electrical parameters. Mainly consider the length of the coil, h, the total length of the wire, l, coil diameter D, coil turns N, wire resistance R0, radiation resistance Rr, inductance L, resonant frequency f0, distributed capacitor C and other parameters.

2.3 Coil Physical Parameter Design

As a medium range energy transmission technology, the magnetic resonance EEWT technology needs to achieve energy aggregation, that is, the transceiver coil has a certain orientation, so the transceiver coil usually adopts spiral coil. The schematic diagram and the equivalent circuit are shown in Figure 1.

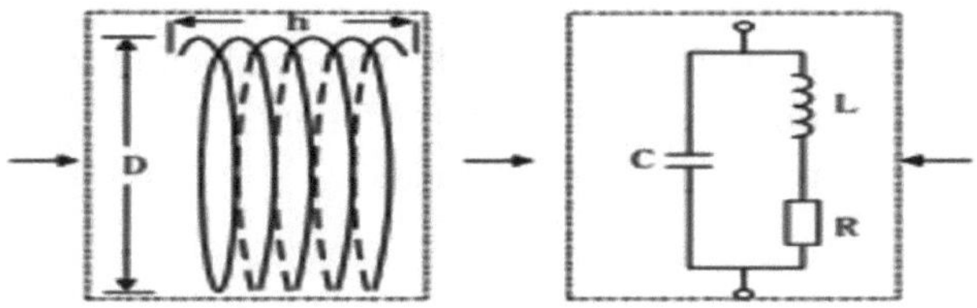

Fig. 1: Iral coil structure and equivalent circuit

In Figure 1, the length of the coil is indicated by Rr. The diameter of the coil is indicated by D, and the C represents the distributed capacitance. The L represents the inductance of the coil, and the R represents the total resistance of the wire. The value is the sum of the wire resistance R_0 and the radiation resistance. The relation between the physical dimensions of the coil [1][3] is shown in the following formula.

$$K = \sqrt[N]{(\pi D)^2 + (\frac{h}{N-1})^2} \quad (3)$$

K is the total length of winding coils. In the upper form, the h/N-1 is the interval between the turns of the coil, D means coil diameter, its unit is “m”.

2.4 Calculation for Electrical Parameters of Coils

2.4.1 Calculation for Inductance

According to the inductance calculation method of spiral coil [2] [4]:

$$L_2 = \tau \mu_0 \mu_S N^2 \frac{S}{h} \tag{4}$$

Among them, L_2represents inductance, its unit is “H”, h represents coil length, its unit is “m”, N represents coil turn, and its unit is “turn”. Among them, L_2 express the inductance , its unit is: H, h express the coil length, and its unit is m, S express the conductor cross-sectional area, the unit is m^2, the relative permeability of the hollow coil S=1, τ is the length coefficient associated with D/h. [2] [6]

2.4.2 Coil Self Resonance Frequency f_0

$$f_0=10^7 \frac{3}{ND}(\frac{h}{D})^{\frac{1}{5}} \tag{5}$$

Here h represents coil length, its unit is: m, D expresses coil diameter, its unit is: m, f_0 expresses resonance frequency, its unit is: Hz. [3] [4] by the formula (5), the resonance frequency of the coil is directly related to the coil structure.

2.4.3 Distributed Capacitor C

Because of the resonant frequency of the LC parallel loop. Therefore, distributed capacitor can be obtained from the next equation.

$$f_0 = \frac{1}{2\pi}\sqrt{\frac{1}{LC} - \frac{L^2}{R^2}} \tag{6}$$

$$c = \frac{1}{4\pi^2 f_0^{\ 2} L + R^2/L} \tag{7}$$

Here the unit of capacitance C is “F”, the unit of resonance frequency f_0 is “Hz”, the unit of inductance L is “H”, When to meet ωL>>Rcondition (here ω is angular frequency),distribution capacitance[4] [5]

$$C = \frac{1}{4\pi^2 f_0^2 L} \tag{8}$$

2.4.4 Calculatethe Equivalent Resistance for the Coil

The wire is DC resistance R_0, the total resistance R, unit is Ω.Here Vacuum permeability$\mu_0=4\pi\times10\text{-}7$H/m, copper wire conductivity$\sigma=5.9\times10^7$s/m,a is wireway radius, unit is "m". [5] [6]

$$R_0 = \sqrt{\frac{\mu_0\omega_0}{2\sigma}}\frac{1}{4\pi\alpha} \tag{9}$$

$$R_0 = \sqrt{\frac{\mu_0\omega_0}{\varepsilon_0}}\left[\frac{\pi}{12}N^2\left(\frac{\omega_0 r}{c}\right)^4 + \frac{2}{3\pi^3}\left(\frac{\omega_0 h}{c}\right)^2\right] \tag{10}$$

Here the radiation resistance Rr, Vacuum dielectric constant $\varepsilon_0=8.85\times10^{-12}$F/m, r is coil radius, unit is m. The relation between the physical parameters of the coil and the electrical parameters is obtained from the above formula. When the physical parameters of the coil are timed, the electrical parameters of the coil are also fixed. On the other hand, the physical parameters of the coil can be deduced by knowing the electrical parameters of the coil. [6] [7]

3 Experiment Facilities

In this paper, the transmitting terminal facilities for experimental research plan and receiving terminal facilities have been carefully planned, respectively, as shown in Figure 2 , the dotted line represents the "non wire connection" all "the two figure of the ellipse is made of high-quality magnetic materials.

Informed by several experimental results, magnetic resonance EEWT facilities for the effective transmission distance is generally 3–6 times the diameter of the transmission coil can get higher transmission efficiency and high transmission power, and EEWT is not affected by non magnetic obstacles. Compared to the EEWT facility, the magnetic resonance type EEWT facility has a longer transmission distance. Magnetic resonance EEWT facilities compared with the radiation type EEWT facilities, the former has little effect on the electromagnetic environment, and the transmission power is larger (because of these advantages, magnetic resonance EEWT received more and more

Attention and research researchers). The two innovative points of the magnetic resonance EEWT facility developed in this paper are as follows.

The transmitting coil and receiving coil are respectively wound on each ceramic stem or epoxy resin columns (equivalent to the hollow coil, because ceramic or epoxy resin are non-magnetic material) and using multilayer concentric circle structure. Obviously, the electrical energy between the transmitting coil and the receiving coil transfers the medium to air. The terminal of the launching facility and the receiving facility can be either a coil or an elliptical conducting plate arranged in an orderly way according to the specified rules. [6] [7]

The experiment in the relatively simple conditions. The experimental results show that the load is the same in the lamp brightness maintain the voltage and current of high frequency power supply and constant transmitter conditions, can coil distance from 20 cm to 50 cm began to increase. [5] [8]

In addition, according to the traditional theory of energy transmission, the transmission of electric energy requires at least two coils, and the EEWT with two resonant coils is called the “two ring framework”. If the increase of two induction coils based on “two ring structure” on (see Figure 1), one coil (called power coil) for isolated power supply and emission ring; another coil (called load coil) to isolate the use of electrical appliances (load) and receive, experiments show that this kind of “two ring architecture better utilization effect.

4 Research Result

With the continuous development and progress of technology, the EEWT technology is getting more and more attention, especially in some special occasions, EEWT has the unique advantage of cable and overhead power transmission than the former, namely power transmission convenience, safety, reliability and economy is obviously better than the latter. The EEWT facility in this paper is to use the “hollow” (i.e. coil wound on the epoxy resin column or ceramic stem) as the transmitting coil and receiving coil (also can replace the magnetic conductive plate hollow coil with rectangular ordered), and the transmission of electrical energy into PWM waveform, alternating current to generate an alternating magnetic field in the space through the hollow coil and related circuit, and then the theory and technology of magnetic resonance, in terminal facilities induced potential and current supplied to the load terminal receiver, which can realize the EEWT.[6] [9]

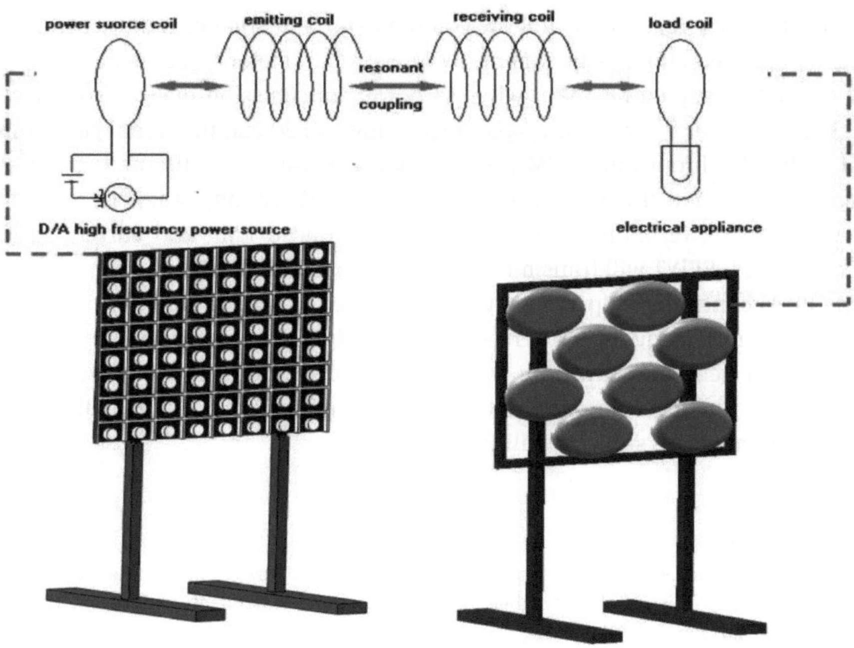

Fig. 2: Magnetic resonance type EEWT facilities

5 Conclusion

In this paper, a novel type of advanced magnetic resonance (also called magnetic resonance type) EEWT facilities are developed, the innovation points and characteristics are as follows: first, the transmitting coil and receiving coils are respectively wound on the respective ceramic stem or epoxy resin columns (equivalent to the "hollow" coil, because ceramic and the epoxy resin were non magnetic materials) and using multiple concentric circles (i.e. each direction equal to the diameter of the coil structure). Obviously, the medium between the transmitting coil and the receiving coil is air. Secondly, the terminal of the launching facility and the receiving facility can be arranged as a coil or an elliptical high quality magnetic guide plate arranged in an orderly way according to the specified rules.

High frequency power supply voltage magnetic resonance type EEWT experimental device planning the transmitter of the effective value is 18 volts (i.e. when the transmitter PWM wave frequency is 500KHz (current) power that is EEWT in the experiment), 100 cm (1 meters) of the distance under the ground the receiver can output AC voltage of about 10 volts alternating current supply of about the load.

Although the research results reach the expected technical requirements, but the EEWT facilities are still many problems to be further improved in space (such as:

resonant frequency regulation; hollow coil diameter adjustment; hollow coil number; each hollow coil range adjustment; PWM wave parameter adjustment......).

A magnetic coupled resonant wireless power transmission device, hollow coil for making diameter for transmitting and receiving coil 20 cm; the signal generating circuit of 15V DC input into PWM pulse signal, generating an alternating current through the coil drive circuit in space to produce an alternating magnetic field, the magnetic coupling resonance principle and the induced EMF and current in the receiver coil end, EEWT will transmit to the receiving coil and the radio.

The PWM controller chip TL494 (which is a fixed frequency pulse width modulation circuit chip), using RC series resonance principle, the internal linear sawtooth wave oscillator generates sawtooth wave, realize pulse width modulation; in addition, TL494 has the drive, monitor and control, the protection function, suitable for the development of this subject facilities.

References

[1] Zhong-Nian LI and Lei ZHOU .Research on Theory Method and Key Technology for Electric Corona Purifying Tail Gas of Automobile. Mechatronics and Industria Informatics. 23, 9, 1637–1640, June, 2013.

[2] Zhong-Nian LI and Lei ZHOU .Research on the Fault Diagnosis System for ICM. Advances in Manufacturing Science and Engineering. 12, 2, 2055–2058, May, 2013.

[3] Zhong-Nian LI and Lei ZHOU .Research on Fault Diagnosis Method for ECS-CNCL Based on Rough Set. Proc. of IEEE ICNC-FSKD'12. 4, 2,254–256, May, 2012.

[4] Zhong-Nian LI, Yun-Feng YUE and Guo-Liang CHEN. Construction and Operation on the Training Base for Application Technology about New Energy Resuorce. Advanced Materials Research. 36, 3, 2895–2898, 2012.

[5] ZHONGNIAN LI, TIANYING LI & YAN QIN. Research on the Control Method with Quick Charging for Lithium-Yttrium Battery. Proc. of ICSEEP.56, 4, 458–463, April, 2015.

[6] Zhong-Nian LI and Lei ZHOU .Research & Development on the MPPT Intelligent Control Device for WPVC Pwoer Generation System. Advances in Engineering Research. 68, 4, 1–5, June, 2015.

[7] Zhong-Nian LI, Tian-ying LI, Yan QIN, Li-juan ZHANG, Qing WANG. Research on the Development for Intelligent Equipment Manufacturing Industry concerning New Energy Resources Power Generation———in Nantong China, Have the Whole World in View. Proc. of IFEESM '2016.2, 2, 307–310, July, 2016.

[8] Zhong-Nian LI and Lei ZHOU .Research on Conditioning and Application for the F^2N^2.Advanced Materials Research. 46, 3. 1895–1898, October, 2012.

[9] Zhong-Nian LI. Electronic control technology and PLC. Beijing: Tsinghua University press, December, 2014.

Hua Zhou, Ming-Jun Zhang, Ren-Guang Wang* and Bin Wang

Further Analysis on Triple Mode Powertrain for HEV

Abstract: With addition of two clutches, a two driving mode powertrain is modified into a triple mode powertrain, which provides three different driving modes including pure electric driving mode, power split driving mode, parallel hybrid driving mode. It can be operated in proper mode according to real vehicle demand. The detail analysis on its operation and mode change are presented for understanding its working principle.

Keywords: Hybrid electric vehicle, Triple mode powertrain, Driving mode change

1 Introduction

In current days, the HEV technology is becoming more and more popular. The THS of Toyota is the most successful HEV system, its total production has been more than 10 million until end of 2016 [1], which uses one or two planetary gear sets and two e-motors to realize full hybrid function. THS system does not use any brake or clutches, which reduces its complicity and malfunction potential, and THS can get excellent fuel consumption by use of advanced ICE technology and optimized system integration approach. In reference [2], a three mode power train for HEV was briefly introduced. In this paper, the authors discuss its operation modes in detail and modify its mode change method from a synchronizer gear shift mechanism to two clutches.

2 Main Structure of Triple Mode Powertrain

As shown in figure one, the main structure are almost same as that in reference [2],

Hua Zhou, CATARC, No.68 East Xianfeng Rd. DongLi Dist.Tianjin 300300,China. Email: zhouhua@catarc.ac.cn
Ming-Jun Zhang, CATARC, No.68 East Xianfeng Rd. DongLi Dist.Tianjin 300300,China. Email: zhangmingjun@catarc.ac.cn
Ren-Guang Wang, CATARC, No.68 East Xianfeng Rd. DongLi Dist.Tianjin 300300,China. Email: wangrenguang@catarc.ac.cn ,Corresponding author.
Bin Wang, CATARC, No.68 East Xianfeng Rd. DongLi Dist.Tianjin 300300,China. Email: wangbin@catarc.ac.cn.

DOI 10.1515 / 9783110540048-033,

they are ICE, first motor M1,second motor M2,first planetary gear set P1, second planetary gear set P2, power battery B and inventors between the B and two motors. The main difference are two clutches C1 and C2.The clutches are added to improve power train function, with these two clutches, the operation modes of the power train are pure electric driving mode(E mode), power split hybrid driving mode (H mode), and parallel hybrid driving mode(P mode).

The C1 is located between the power train case and carrier of P1, which is used as a brake to lock carrier of P1 and the ICE. The C2 is located between the ring gear and carrier of P1, which is used to lock these two parts together. In real design, the synchronizer can be selected to replace the function of clutches.

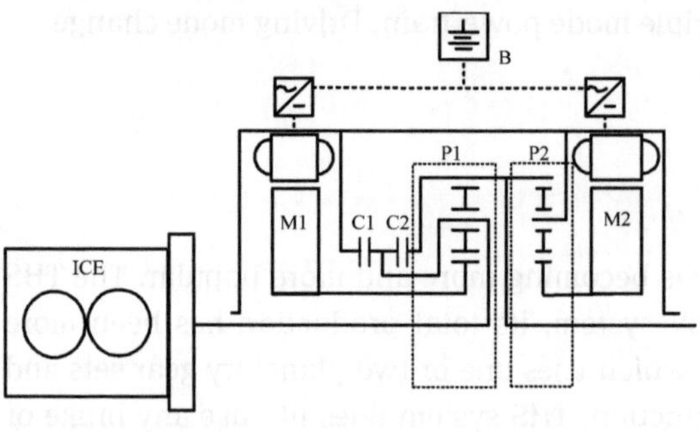

Fig. 1: The sketch of triple mode HEV power train

In the design without mode change mechanism such as clutch, brake, or other synchronizer, its pure electric driving mode is performed by only one motor M2. In the new design, it can combine two motor to drive vehicle simultaneously, which can improve driving torque to get better vehicle dynamical performance such as acceleration and slope climbing capability. Compared with one driving motor, the two motors can both operated as driving motor can reduce the motor size and reduce instalment space.

3 Analysis of Driving Mode Change Process

The triple mode HEV power train can provide three driving modes, two clutches are used to select proper operation mode. The E mode is mainly used in low speed and launching condition with SOC of power train at proper level. In the E mode, the engine is turned off and locked from rotation, and then both motors can be used to drive or brake the vehicle.

The H mode is suit for moderate speed range, which is a power split hybrid mode. In the conventional H mode, the motor M1 is used as a generator and adjust the engine speed. The engine power is distributed between the ring gear and sun gear, which are corresponded to driving through ring gear of P1 and generating electricity through the sun gear of P1 respectively.

The P mode is for higher speed condition, the engine can directly output its power to drive vehicle, and the two motors can boost and subtract the engine torque to drive vehicle as a parallel hybrid configuration.

With these two clutches, the E mode and H mode can be bidirectional changed, H mode and P mode can also be bidirectional changed. But the P mode and E mode can not be switched directly; the H mode should be the intermediate.

During the changing process from the E mode to H mode, the C1 is released and the M1 reduce its speed to 0 rpm and then change its rotation direction to be starter to start ICE. During the reverse process from the H mode to E mode, the engine is turned off and the M1 regulates the speed of carrier of P1 to 0 rpm and then C1 is engaged to lock the engine.

During the changing process from the H mode to P mode, the M1 is also performed as a speed regulator to make the speed difference of carrier and ring gear approaching to zero or nearly zero and then C2 can be engaged smoothly.

In addition to regulate the rotation speed of P1, the motor M1 can start the engine and turn it off very quickly, which can reduce vibration and improve drivability performance. The following parts describe the detail of three driving modes of the powertrain.

4 Detail Analysis of Three Driving Mode Operation

4.1 Analysis on Pure Electric Driving Mode(E mode)

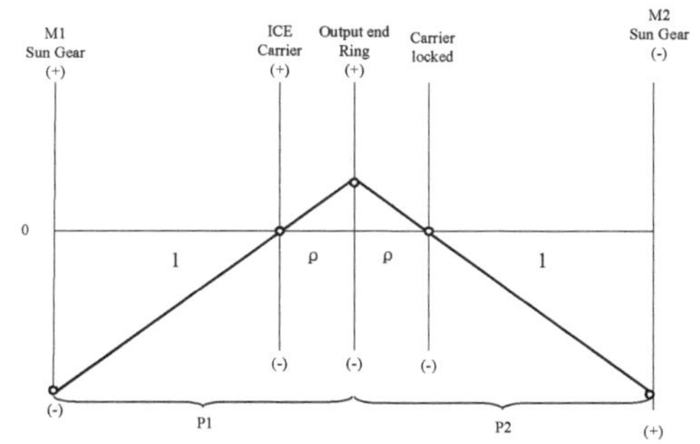

Fig. 2: Speed analogy lever of power train in E mode

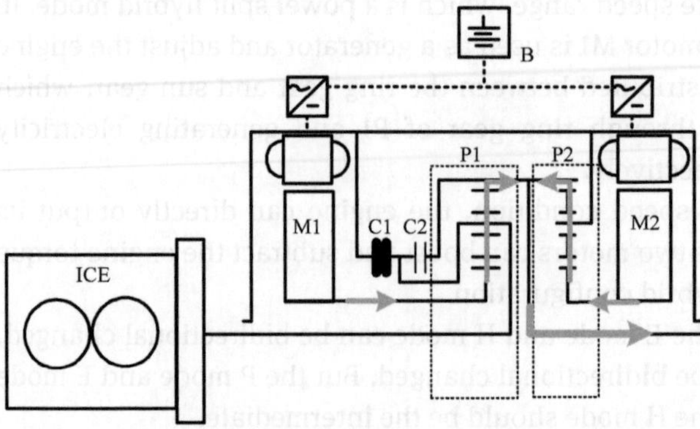

Fig. 3: Sketch of power flow in E mode

In the E mode, the c1 is engaged and the make carrier of first planetary gear set P1 locked with case of power train, which means the engine shaft is mechanically connected to the power train housing and the engine is off. Judged from the figure 2, the E mode is suit for low velocity, because of the speed ratio is mostly between 2 and 3, and higher vehicle speed will make the motors rotate at extreme high speed.

During E mode, the power train supports the launch operation from standstill and at positive and negative driver torque demand. The E mode is generally used at high power battery SOC and lower speed such as under 30km/h. When the power battery is approaching its low SOC limit, or vehicle reaching higher speed or higher driving torque requirements will trigger the operation mode change, usually C1 disengaged and starting the ICE by motor M1, and enter Hybrid operation mode. In this mode, two motor can work together or just one motor to drive in smaller torque requirement. The power flow of E mode is shown in figure 3.

For recuperation in E mode, the M1and M2 can work as generators to charge power battery. But in most conditions, maybe only one motor is used to generate, the other one does not produce and braking torque, the power battery cannot absorb too much charging current because of charging current limitation.

4.2 Analysis on Power Split Hybrid Operation Mode (H mode)

In H mode with engine on, the vehicle is driving in low or medium velocity range, and both Clutch C1 and C2 are disengaged. In normal condition for H mode, the power train operates in power split mode, the mechanical power for driving vehicle is from the ICE and M2, and the M1 is generating electricity from part of the engine power and power the motor M2. Based on the vehicle operation conditions, the power battery is charged or discharged.

In H mode, the system can operated at three different conditions. With the motor M1 as a speed regulator for controlling ICE running at optimum operation line or area, there are three possible working sub-modes. The following section provides a detail discussion on these conditions. For common planetary gear set, the relation of the speed of sun gear ω_1, carrierω_3, and ring gearω_2 are describes as equation 1[3].

$$\omega_1 + k_p\omega_2 - (1+k_p)\omega_3 = 0 \tag{1}$$

Where the k_p is the gear ratio defined by ring gear number divided by sun gear number. When speed of the sun gear is zero, a special transmission ratio k_0 is gotten from equation 1, this is a special mechanical point.

$$k_0 = \omega_2 / \omega_3 = k_p / 1 + k_p \tag{2}$$

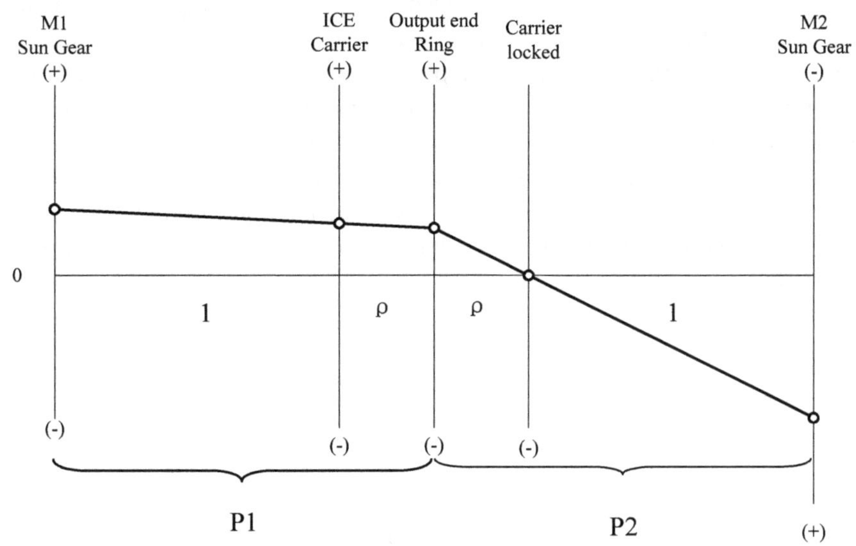

Fig. 4: Speed analogy lever of power train in H mode

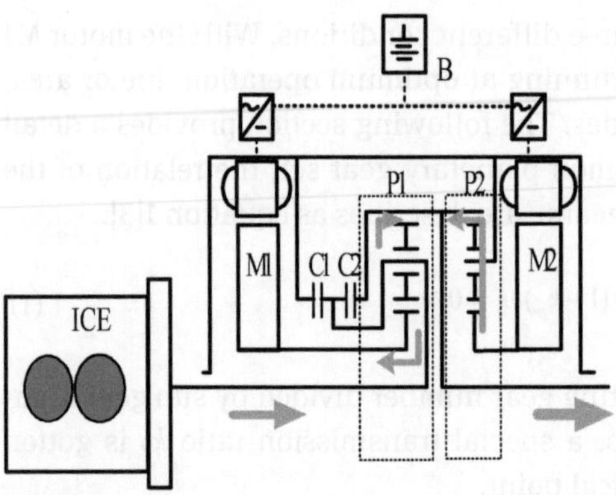

Fig. 5: Sketch of power flow in H mode

If the speed ratio k is defined as ICE speed divided by ring speed(same as the output shaft speed), then when the k is great than k0, the power train is in true power split mode, the speed of different parts are showed in figure 4 (for simplicity, the planetary gear set P1 and P2 consumed the same gear ratio, and ρ is the reciprocal of k). As shown in figure 5 , the engine produce power, most of engine power is transmitted to ring gear through first planetary gear set, a small part is used to drive motor M1 to generate. And the motor M2 in this condition produces driving torque to boost vehicle driving, whose energy is from M1 or/and power battery. With the vehicle speed increasing, the motor M1 reduce its speed to maintain optimal ICE speed. When the M1 speed reduced to zero, a mechanical point is gotten, and at this point, k is equal to k0.

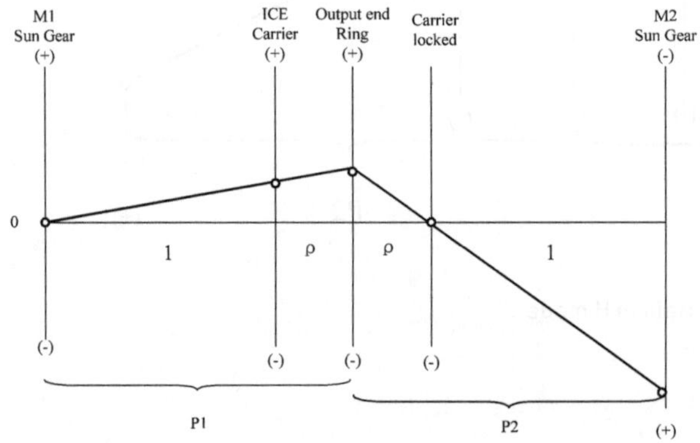

Fig. 6: The analogy lever in H mode with M1 stopping

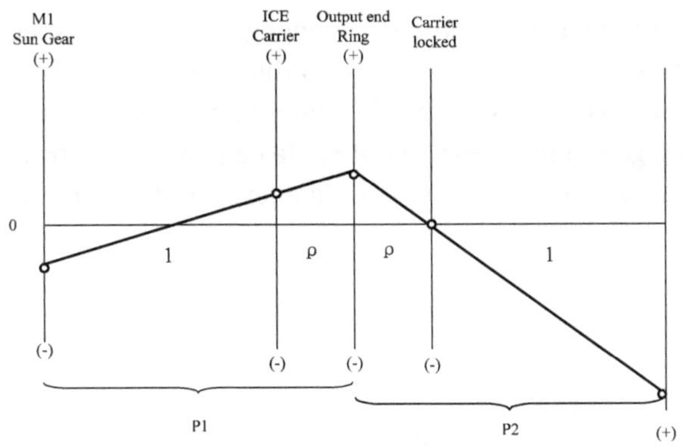

Fig. 7: The analogy lever in H mode with M1 driving

When the k is equal to k0, the speed relation diagram is showed in figure 6, the motor M1 shaft is stopped. Whole torque from ICE is transmitted to drive vehicle without power split. Theoretically, there is no electric power goes between M1 and M2, and the M2 can boost or not. The mechanical point is characterized with top efficiency of the power train. But this point cannot maintain because of vehicle speed changing frequently if the engine speed keep constant, or at small speed range , this point can be kept by regulating engine speed with less efficiency loss. Even in this condition, the motor M2 can still provide additional torque to boost vehicle driving.

In the H mode, with increasing of vehicle speed and the engine speed is kept at constant value for optimal operation, motor M1 will pass the 0 rpm point and change its rotation direction to become a driving motor, as shown in figure 7. And at the same time the M2 operated as a generator to produce electricity to power M1 at very high speed. This sub mode is for only theoretically analysis, and it will not be used in real operation.

During the H mode, when the accelerator pedal is released and the brake pedal is released or pressed, the vehicle enter the reception condition, the engine can be cut off or still on, the M2 will driven by the vehicle kinetic energy to become a generator to charge the power battery. During the reception process in H mode, if the SOC reach its low limit, the engine can be started again to charge power battery.

4.3 Analysis on Parallel Hybrid Driving Mode(P mode)

When the vehicle is driving with speed increment in H mode, it will soon reach a specific speed threshold to mode change, and the power train is needed to be changed into operation of P mode. The motor M1 regulates speed to reduce speed difference between carrier and ring gear of P1, the C2 is engaged at proper speed point, which make the ICE shaft is mechanically coupled to the ring gear. It means the P1 rotates as an integrated one without any speed difference between its three main parts such as sun gear, ring gear and carrier, the P mode can get higher transmission efficiency with energy conversion between electric energy and mechanical energy. As shown as figure 8.

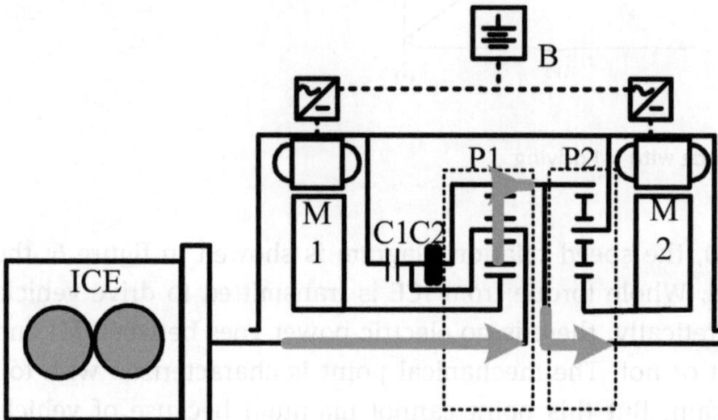

Fig. 8: Sketch of power flow in P mode

In normal P mode, the source of the driving power is from the ICE, and M1 and M2 can work as motor to provide additional torque to boost power train or operated as generator to charge power battery. The real working conditions of M1 or M2 are dependent on the vehicle operation control strategy. The other two conditions in P mode can be called driving with charging and driving with boost respectively.

In the driving with charge condition, the vehicle is driving only by engine propulsion with a constant speed. The source of the mechanical power is only the engine, M1 or M2 is operated in generator mode to supply the HV loads and to charge the power battery. This drive with charge in P mode is used mainly at cruising at medium and higher velocities with medium power demand conditions and low power battery SOC conditions.

In the drive with boost condition, the vehicle is driving mainly by engine propulsion with an increasing velocity. The accelerator pedal input is pressed for more torque. The source of the mechanical power is the engine, M1or M2. The electrical energy is provided by the power battery. Which means this sub mode is used mainly at higher vehicle velocities with higher power demand conditions.

In the P mode, when the brake pedal is released or pressed and the accelerator pedal is released, and the engine is in fuel-cut-off mode, then the vehicle is moving with a decreasing speed. The source of the power is vehicle kinetic energy, which drive the M1 or M2 to generate electricity to charge power battery, during this condition, sufficient energy can be regenerated by the M1 or M2. During the reception process in P mode, if the SOC reach its low limit, the engine can be started again to charge power battery.

4.4 Analysis of Other Nun Driving Mode

There are also some nun driving modes, such as charging mode at standstill. In this mode, both C1 and C2 are disengaged; the vehicle is park mode which means that the ring gear of P1 cannot rotate any more. And the engine can be started by M1 and then it drives the M1 to generate and charge power battery. M2 is at zero torque and zero rpm condition.

For more function extension, the power train system can be a small power plant to provide electricity to power other utility out of vehicle. This system can also be easily extended to become plug-in HEV power train and receive external charging from grid.

5 Summary

The detail operation analysis of a triple mode power train for HEV was presented. With two clutches indifferent states, the pure electric driving mode, power split driving mode, and parallel hybrid driving mode can be selected to meet real vehicle driving condition. And three sub modes for H mode are discussed to benefit understanding its complicated operation. The triple mode power train can improve vehicle fuel and dynamics performance.

Acknowledgement: This research project was financially supported by the China National High Technology Research and Development Program 863 (2011AA11A211), and Product Development Program of China Automotive and Research Center (10120101).

References

[1] Toyota hybrid sales pass 10-million [EB/OL]. http:// www.toyota.com.cn/news,2017-03-01.
[2] Z.Yang, R. Wang, L. Zhang, et al. A New Power Train for Electric Hybrid Vehicle[J]. Advanced Materials Research Vols. 694-697 (2013), p1587-1590
[3] H. Zhao, G. Shi. The Technology of Hybrid Electric Vehicle[M]. Beijing: China Machine Press. 2012.6

Sheng-Yen Hu, Chao-Ping Huang and Yuan Kang*

Static Characteristic of Self-compensated Hydrostatic Bearing

Abstract: The influence of design parameter on the static characteristic of a self-compensated hydrostatic bearing was investigated in this article. This paper is obtaining the relationships of worktable displacement, load capacity, and static stiffness by using flow continuity equations of a self-compensated hydrostatic bearing. The results reveal that the appropriate range of design parameters for self-compensated hydrostatic bearing can obtain the maximum stiffness. We provided and verified a new design parameter which is restriction island height ratio can improve static performance of self-compensated hydrostatic bearing. Further, we also discuss the influence of viscosity ratio for bearing.

Keywords: Self-compensated Hydrostatic bearing, Load capacity, Static stiffness, Restriction island height ratio, Viscosity ratio

Nomenclature

A_i, A_{eff}, $\overline{A}_i$	practical area (m^2), effective area (m^2) and dimensionless effective area $\overline{A}_i = A_{eff}/A_i$, $i=1,2$
h_0	initial film thickness or bearing clearance (mm)
e, ε	worktable displacement (mm) and dimensionless worktable displacement $\varepsilon = e/h_0$
K, $\overline{K}$, $\overline{K}_p$	static stiffness, dimensionless static stiffness and dimensionless static stiffness parameter
P_i	working pressure ($\mathrm{N/m^2}$), $i=1,2$
P_s	supply pressure ($\mathrm{N/m^2}$)
$\overline{P}_i$, $\overline{P}_s$	pressure ratio $\overline{P}_i = P_i/P_s$, $i=1,2$
Q_{ri}, Q_{hi}	flow rates of restriction part and oil pad part ($\mathrm{m^3/s}$), respectively, $i=1,2$

Sheng-Yen Hu, Chung Yuan Christian University, Chung Li 32023, Taiwan, R.O.C. Email: u9211030@yahoo.com.tw
Chao-Ping Huang, Chung Yuan Christian University, Chung Li 32023, Taiwan, R.O.C. Email: b631018@cycu.edu.tw
Yuan Kang, Chung Yuan Christian University, Chung Li 32023, Taiwan, R.O.C. Email: yk@cycu.edu.tw

DOI 10.1515/9783110540048-034,

$\overline{Q}_{ri}$, $\overline{Q}_{hi}$	dimensionless flow rates of restriction part and oil pad part, respectively, $i = 1, 2$
R_1	radius of connecting hole (mm)
R_2	radius of restriction island (mm)
R_3	radius of restriction island surround (mm)
W , $\overline{W}$, $\overline{W}_p$	load capacity (N), dimensionless load capacity and dimensionless load capacity parameter
α	effective area ratio of loading side and backing side $\alpha = \overline{A}_1 / \overline{A}_2$
β	restriction parameter ratio of both pads $\beta = \zeta_1 / \zeta_2$
μ_{ri} , μ_{hi}	viscosity of restriction part and oil pad part ($N \cdot s/m^2$), respectively
$\overline{\mu}_i$	viscosity ratio $\overline{\mu}_i = \mu_{ri} / \mu_{hi}$, $i = 1, 2$
ζ_i	restriction parameter of oil pad $\zeta_i = \gamma_i \ell n(R_2/R_1)/2\pi$, $i = 1, 2$
ξ	restriction island height ratio

Subscript

1, 2	denote for loading side and backing side , respectively of pressures, flow rates, areas, and viscosities

1 Introduction

Hydrostatic bearing is widely used in high precision mechanism and machine for the advantage of low friction, high stiffness, high accuracy, and long life. But the pressure in the hydrostatic bearing is generated by external pumps, and compensation devices are necessary to regulate flow into the recesses. Rippel [1] presented a general manual for designing hydrostatic bearings. In his manual, the single pad and multiple pads with various geometries were provided but only capillary, orifice, and constant-flow devices for the compensation of recess pressures are considered. Rowe [2] has introduced the mechanism design and written elementary theory of control valves for designing hydrostatic bearings. However, who mainly related the relationships between static characteristics of various bearing types and contact restrictions belong to capillary, slot, orifice, flow control valves, and constant flow pump; pressure-sensing valves.

The self-compensated hydrostatic bearing doesn't need additional restrictors, so it can avoid manufacturing errors and reduce costs. Bassani [4] persented that a self-regulated hydrostatic opposed-pad bearing is highly efficient in a constant pressure system. He shown its performance is comparable to that of a conventional opposed-pad hydrostatic bearing compensated with flow control valves, and is superior compensated with orifices or capillaries. Slocum et al. [5] presented a self-compensated, water-hydrostatic bearings for precision machine linear motion axes has been developed and tested. The bearing has many benefits of hydrostatic bearings systems and has less sensitive to manufacturing errors than are traditional hydrostatic bearing systems. Kane et al. [6] proposed a hydrostatic bearing with angled-surface self-compensation. The bearing consists of several round parts which can be easily manufactured and assembled and achieves high stiffness and accuracy. Zuo et al. [7] presented a newly designed self-compensating bearing, which is combining the traditional self-compensation and Kane et al.'s self-compensation [6], and based on the principles of flow balance equations derived its design formula to obtain optimal stiffness and optimal condition. Kang et al. [8] investigate the static stiffness of the hydrostatic bearing compensated with membrane-type restrictor and self-type compensation. We obtained the static stiffness due to self-compensation can be larger than that due to constant restrictors and also larger than that due to double-action membrane-type restrictor with dimensionless deformation coefficient and restriction ratio of bearing to restrictor departing from critical values.

In this paper, we design a new parameter which is the restriction island height ratio can improve static performance and obtain optimal design range. It is the critical technology for mechanical engineers to facilitate the design and application of this type hydrostatic bearing.

2 Theory of Self-compensated Hydrostatic Bearing

2.1 The Design Principles for Self-compensated Hydrostatic Bearing

Self-compensated hydrostatic bearing is belonging to the closed-type bearing, which configured as shown in Fig. 1. This bearing is composed of oil pad part and restriction part. A group bearing has two oil pads and two restriction islands that position are opposite. One is loading side and another is backing side. The design of restriction part must be higher than the oil pad part. The pressures in backing side can reduce load capacity of bearing, but can increase the stiffness to resist the loads. The design principle of restriction part is composed of oil inlet hole, connecting hole, Restriction Island, and Restriction Island surround as shown in Fig. 1(c).

The majority of supply oil flow into connecting hole and only the less flow into surround. If designed improperly such as when the restriction island made higher than the restriction island surround, it can make the most oil overflow surround without flowing into connecting hole. Then, it will make the backing side failure and extremely consume pump energy. So, the restriction island always higher than the restriction island surround, but the surround can easy collide machine when it's too high. Where ξ is the restriction island height ratio, defined as the ratio between the restriction island height difference and initial film thickness or bearing clearance (h_0). Its negative values indicate higher than the surface of oil pad, oppositely positive values indicate lower, and absolute value equals dozens of to hundreds of μm . ξh_0 Is the height difference of Restriction Island as shown the symbol in Fig. 1(b).

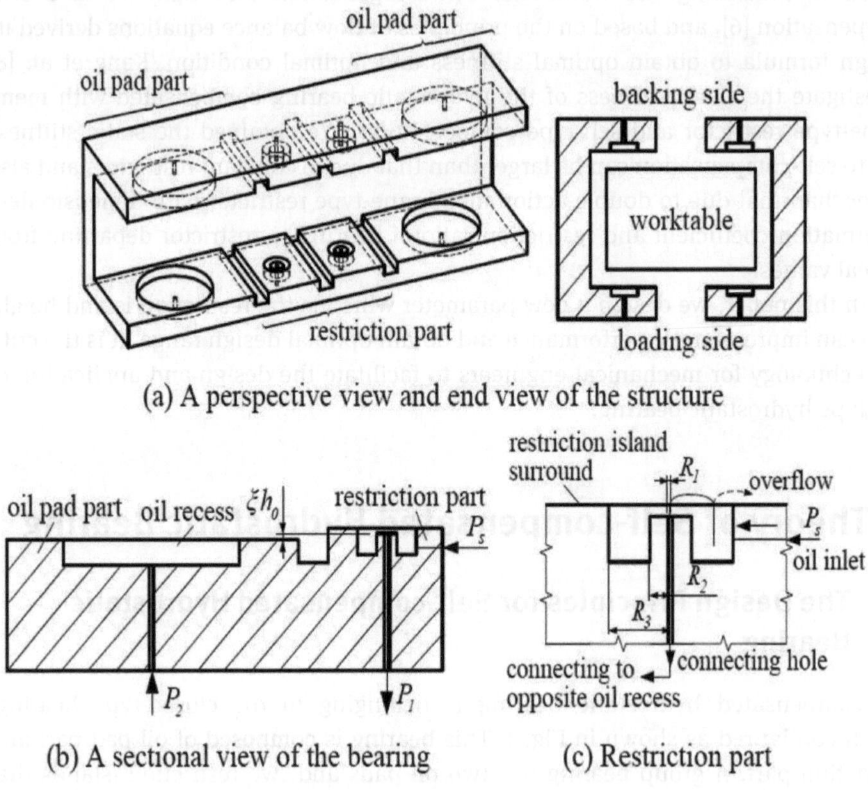

Fig. 1: The self-compensated hydrostatic bearing

2.2 Flow, Static Load Capacity and Stiffness for The Self-compensated Hydrostatic Bearing

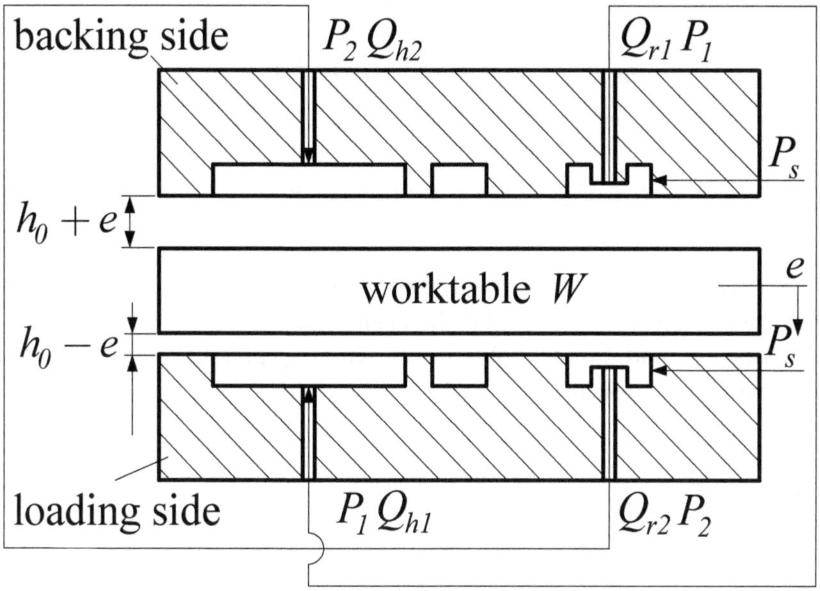

Fig. 2: Self-compensated hydrostatic bearing system

The self-compensated hydrostatic bearing as shown in Fig. 2. The flow through gap of oil pad is equal to through gap of restriction island can be determined

$$Q_{r1}=\frac{\pi(P_s-P_1)(h_0+e+\xi h_0)^3}{6\mu_{r1}\ell n(R_2/R_1)}=Q_{h1}=\frac{P_1(h_0-e)^3}{12\mu_{h1}}\gamma_1 \tag{1}$$

$$Q_{r2}=\frac{\pi(P_s-P_2)(h_0-e+\xi h_0)^3}{6\mu_{r2}\ell n(R_2/R_1)}=Q_{h2}=\frac{P_2(h_0+e)^3}{12\mu_{h2}}\gamma_2 \tag{2}$$

where $Q_{ri}$is flow rate of restriction part, $Q_{hi}$is flow rate of oil pad part, μ_{ri} is viscosity of restriction part, μ_{hi} is viscosity of oil pad part, γ_i is dimensionless geometric constants, in which the subscript 1 denote the loading side of bearing, and the subscript 2 denote the backing side of bearing. R_1 is radius of connecting hole, R_3 is radius of restriction island, R_3 is radius of restriction island surround, and e is worktable displacement.

Eqs. (1) and (2) describe the function of oil film thickness due to the recess pressure and loading caused of flow variation, therefore, they divided respectively by $P_S h_0^{\,3}/12\mu_{r1}$and $P_S h_0^{\,3}/12\mu_{r2}$ to obtain the dimensionless flow equations of the loading side and backing side. They can be expressed as

$$\overline{\mu}_1\zeta_1\overline{P}_1(1-\varepsilon)^3 = (1+\varepsilon+\xi)^3(1-\overline{P}_1) \tag{3}$$

$$\overline{\mu}_2\zeta_2\overline{P}_2(1+\varepsilon)^3 = (1-\varepsilon+\xi)^3(1-\overline{P}_2) \tag{4}$$

Where $\zeta_i = \gamma_i \,\ell n(R_2/R_1)/2\pi$ the restriction parameter of oil pad is, $\overline{\mu}_i = \mu_{ri}/\mu_{hi}$ is the viscosity ratio, and $\varepsilon = e/h_0$ is the dimensionless worktable displacement. Transposing Eqs. (3) and (4) can obtain the working pressure ratios of self-compensated hydrostatic bearing as

$$\overline{P}_1 = (1+\varepsilon+\xi)^3\left[(1+\varepsilon+\xi)^3 + \overline{\mu}_1\zeta_1(1-\varepsilon)^3\right]^{-1} \tag{5}$$

$$\overline{P}_2 = (1-\varepsilon+\xi)^3\left[(1-\varepsilon+\xi)^3 + \overline{\mu}_2\zeta_2(1+\varepsilon)^3\right]^{-1} \tag{6}$$

The pressure ratio which obtained by the simultaneous equations of dimensionless flow is multiplying the dimensionless effective area to get dimensionless load capacity. Thus, the dimensionless load capacity of self-compensated hydrostatic bearing is

$$\overline{W} = \overline{P}_1\overline{A}_1 - \overline{P}_2\overline{A}_2 = \overline{A}_2\left[\alpha\overline{P}_1(\varepsilon) - \overline{P}_2(\varepsilon)\right] = \overline{A}_2\overline{W}_p \tag{7}$$

Thus, the dimensionless load capacity parameter is

$$\overline{W}_p = \frac{\alpha(1+\varepsilon+\xi)^3}{(1+\varepsilon+\xi)^3 + \overline{\mu}_1\beta\zeta_2(1-\varepsilon)^3} - \frac{(1-\varepsilon+\xi)^3}{(1-\varepsilon+\xi)^3 + \overline{\mu}_2\zeta_2(1+\varepsilon)^3} \tag{8}$$

where $\alpha = \overline{A}_1/\overline{A}_2$ is the effective area ratio of loading side and backing side and $\beta = \zeta_1/\zeta_2$ is the restriction parameter ratio of both pads. Differentiating Eqs. (5) and (6) with respect to dimensionless worktable displacement (ε) give

$$\frac{d\overline{P}_1}{d\varepsilon}=\frac{3\overline{\mu}_1\zeta_1(1-\varepsilon)^2(1+\varepsilon+\xi)^2(2+\xi)}{\left[\overline{\mu}_1\zeta_1(1-\varepsilon)^3+(1+\varepsilon+\xi)^3\right]^2},\quad \frac{d\overline{P}_2}{d\varepsilon}=\frac{-3\overline{\mu}_2\zeta_2(1+\varepsilon)^2(1-\varepsilon+\xi)^2(2+\xi)}{\left[\overline{\mu}_2\zeta_2(1+\varepsilon)^3+(1-\varepsilon+\xi)^3\right]^2}$$

The dimensionless static stiffness of hydrostatic bearing is supported the load variation with respect to the eccentricity variation per unit which is defined as follows

$$\overline{K}=\frac{d\overline{W}}{d\varepsilon}=\overline{A}_2\frac{d\overline{P}_1}{d\varepsilon}-\overline{A}_2\frac{d\overline{P}_2}{d\varepsilon}=\overline{A}_2\left(\alpha\frac{d\overline{P}_1}{d\varepsilon}-\frac{d\overline{P}_2}{d\varepsilon}\right)=\overline{A}_2\overline{K}_p \tag{9}$$

Thus, the dimensionless static stiffness parameter is

$$\overline{K}_p=3\zeta_2(2+\xi)\left\{\frac{\alpha\beta\overline{\mu}_1(1-\varepsilon)^2(1+\varepsilon+\xi)^2}{\left[\overline{\mu}_1\beta\zeta_2(1-\varepsilon)^3+(1+\varepsilon+\xi)^3\right]^2}+\frac{\overline{\mu}_2(1+\varepsilon)^2(1-\varepsilon+\xi)^2}{\left[\overline{\mu}_2\zeta_2(1+\varepsilon)^3+(1-\varepsilon+\xi)^3\right]^2}\right\} \tag{10}$$

Using Eqs. (8) and (12) can obtain the relationships of dimensionless worktable displacement (ε), dimensionless load capacity parameter ($\overline{W}_p$), and dimensionless static stiffness parameter ($\overline{K}_p$).

3 Analysis Results and Discussions

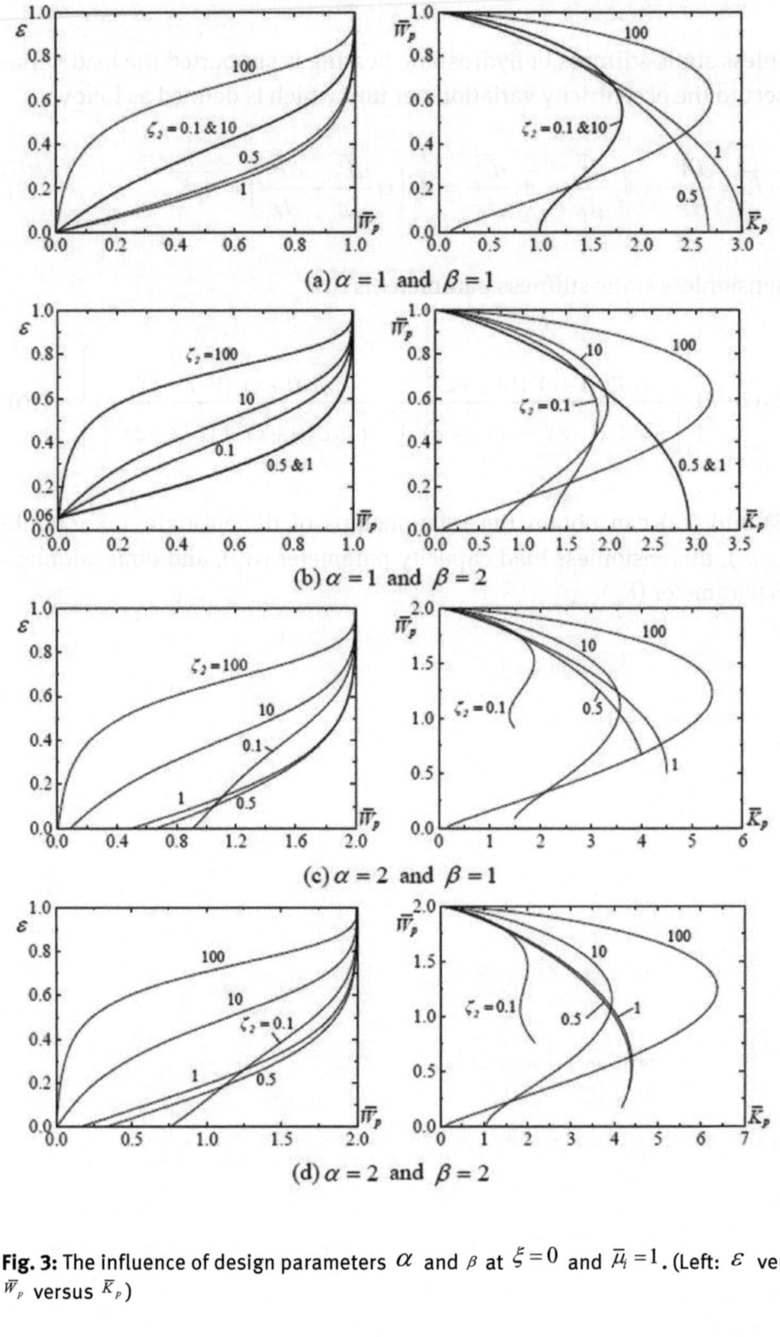

Fig. 3: The influence of design parameters α and β at $\xi = 0$ and $\bar{\mu}_i = 1$. (Left: ε versus $\bar{W}_p$; Right: $\bar{W}_p$ versus $\bar{K}_p$)

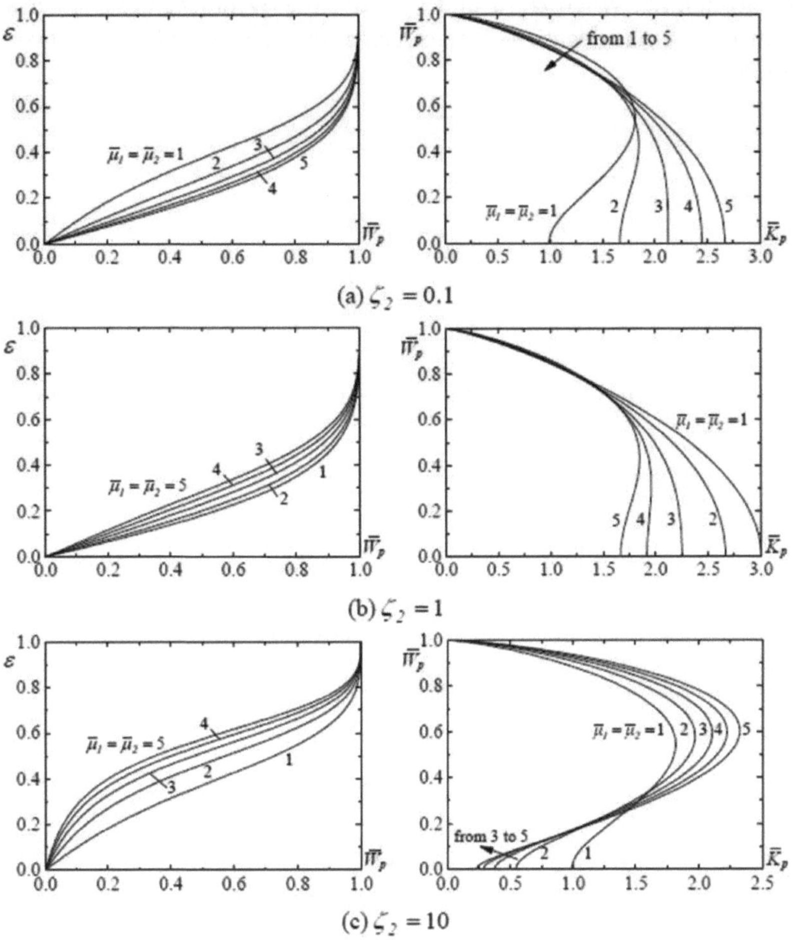

Fig. 4: The influence of viscosity ratios ($\bar{\mu}_i$) using four restriction parameters (ζ_2) at $\xi = 0$, $\alpha = 1$, and $\beta = 1$.(Left : ε versus $\overline{W}_p$; Right : $\overline{W}_p$ versus $\overline{K}_p$)

We analyzed the static characteristics of a self-compensated hydrostatic bearing by four dimensionless parameters which are effective area ratio (α), restriction parameter ratio (β), viscosity ratio ($\bar{\mu}_i$) and restriction island height ratio (ξ). In Figs 3 to 5, the left is dimensionless worktable displacement (ε) versus dimensionless load capacity parameter ($\overline{W}_p$), and the right is dimensionless load capacity parameter ($\overline{W}_p$) versus dimensionless static stiffness parameter ($\overline{K}_p$). The results as shown in Fig. 3 that the load capacity and stiffness of bearing both are proportional to effective area ratio. For raising load capacity and stiffness of a self-compensated hydrostatic bearing, the design on effective area is better than the restriction parameter. In Fig. 3, the ζ_2 decreases as the $\overline{K}_p$increases at low $\overline{W}_p$, but the ζ_2 decreases as the

$\overline{K}_p$decreases at high $\overline{W}_p$. When $\alpha > 1$ and $\zeta_2 < 10$, the $\overline{W}_p$ranges are smaller and smaller because there has negative displacements. But α can obvious improve the static performance.

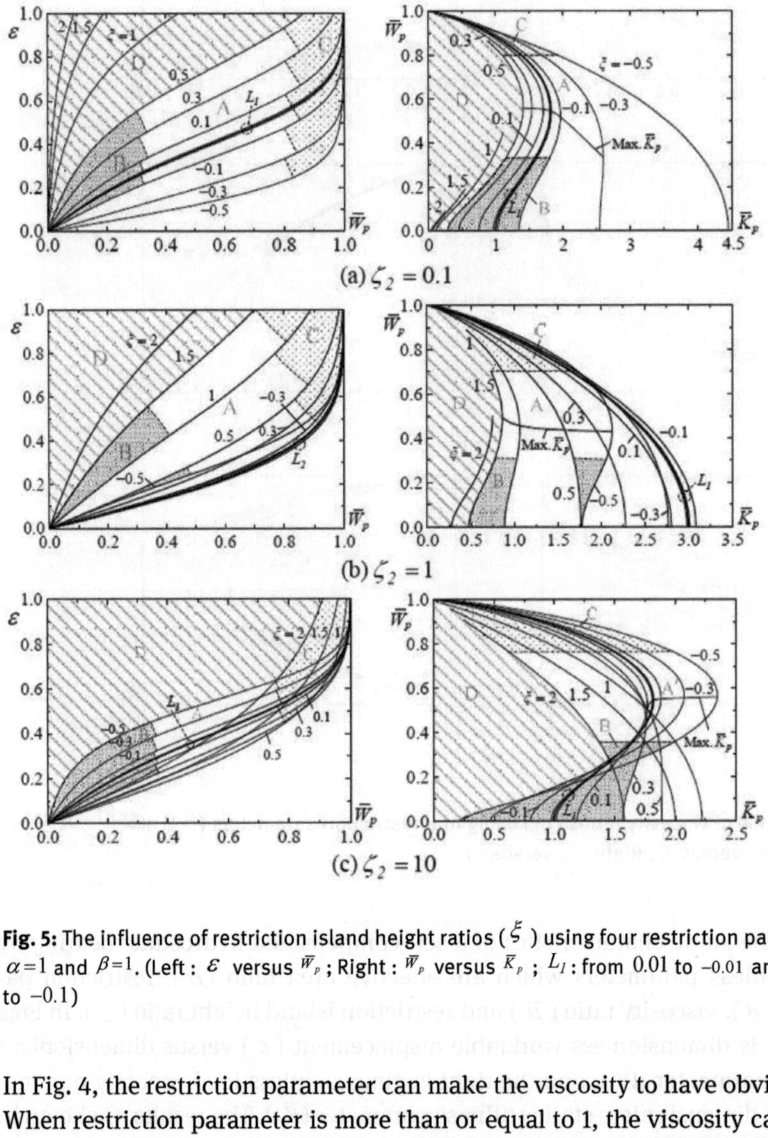

Fig. 5: The influence of restriction island height ratios (ξ) using four restriction parameters (ζ_2) at $\alpha=1$ and $\beta=1$. (Left: ε versus $\overline{W}_p$; Right: $\overline{W}_p$ versus $\overline{K}_p$; L_1: from 0.01 to −0.01 and L_2: from 0.1 to −0.1)

In Fig. 4, the restriction parameter can make the viscosity to have obvious influence. When restriction parameter is more than or equal to 1, the viscosity can make worktable displacement to increase and load capacity to decrease. The stiffness of bearing is not affected by viscosity when high load, but the viscosity can make stiffness to better in specific range of load. Fig. 5 divided into the four zones A, B, C, and D which can know the influences of restriction parameter. For zone of A, it is the optimal operation range because the variation of displacement is lesser when load increased. When ζ_2 is increasing, the zone of A can have more ξ to design and the changes of displacement reduced. When the restriction island height ratio is less than 0 for any restriction parameters, we can get the best stiffness of bearing. The $\overline{K}_p$ has maximum at $\overline{W}_p = 0.4\sim0.6$. In Fig. 5, we can know the optimal $\overline{K}_p$ is in the zone of A and close to maximum $\overline{K}_p$ is the best design.

4 Conclusion

The influence of design parameters on static characteristics of self-compensated hydrostatic bearing is discussed. The effective area ratio and the restriction parameter ratio both can make the stiffness and load capacity to raise, especially effective area ratio can be multiplied. The viscosity is affected by restriction parameter. When the restriction parameter increase, the load capacity and stiffness of bearing and the viscosity ratio are inversely proportional. Only when the restriction parameter is too big or too small, the viscosity can make load capacity and stiffness to better. When the restriction parameter increase, the restriction island height ratio can make the load capacity and stiffness of hydrostatic bearing to better. Only when the restriction parameter is too big or too small, the viscosity can make load capacity and stiffness to better. But the load capacity available range is decreased with restriction island height ratio increasing. The analysis result can get the restriction parameter can affect available values of restriction island height and the influence of effective area on static performance is better than the restriction parameter. For the best stiffness of bearing, the viscosity and restriction parameter have a collocation relationship and the restriction island height must be taller than oil pad.

Acknowledgement: The authors greatly appreciate the support of this work by Grants NSC 104-2221-E-033-061-MY3 from the National Science Council Taiwan R. O. C.

References

[1] H.C. Rippel, Cast Bronze Hydrostatic Bearing Design Manual, Cast Bronze Bearing Institute. Cleveland: Ohio, 1969.
[2] W.B. Rowe, Hydrostatic, Aerostatic and Hybrid Bearing Design. London: Butterworth-Heinemann, 2012.
[3] R. Bassani, Meccanica, 10, 107–113, 1975.
[4] A.H. Slocum, P.A. Scagnetti, N.R. Kane, C. Brunner, Precis Eng, 17, 3, 173–185, 1995.
[5] N.R. Kane, J. Sihler, A.H. Slocum, Precis Eng, 27, 2, 125–139, 2003.
[6] X. Zuo, S. Li, Z. Yin, J. Wang, Int J Rotating Mach, 2013, 638193, 1–10, 2013.
[7] Y. Kang, D.X. Peng, Y.H. Hung, S.Y. Hu, C.S. Lin, Ind Lubr Tribol, 66, 2, 322–334, 2014.

Hai-Feng Wang, Yuan-Hong Qi and Ding-Liu Yan

Process and Equipment Design for the Dry Slag Granulation Technology

Abstract: The granulation of blast furnace (BF) slag with water quenching has many disadvantages, such as high water consumption and difficulty in heat recycling. The dry granulation technology to treat the BF slag is very important. The application of BF slag depends on the content of amorphous solids in the slag after treatment. The research on the activation energy of crystallization of amorphous solids in BF slag is the foundation to find out the reasonable cooling speed during dry granulation of molten slag. So it is of theoretical and practical significance to be analyzed. The activation energy of crystallization of amorphous solids in BF slag (E) was 108.90 kJ/mol. Based on the results; the required cooling rate for BF slag from 1723 K to 973 K was more than 100 K/s. The process and equipment for the dry slag granulation (DSG) technology was desired. The semi-industrial experimental results was that the volume of recycling hot air was 2020 m^3 with 867 K and the content of amorphous solids in granulated slag was about 96 %, which had the similar activity with the slag by water quenching. The problems of the experiment were that the treated slag had a crusting phenomenon, the grain size was not uniform and the slag cotton was produced during BF slag treatment, which needed to be further researched.

Keywords: Dry slag granulation technology, Activation energy of crystallization, Cooling speed, Process and equipment

1 Introduction

Usually a large amount of blast furnace (BF) slag is produced every year and most of them is treated with water in China. Since the composition of BF slag is similar to Portland cement, the major compounds in the molten BF slag become amorphous with chemical energy in the rapid cooling process and the content of amorphous solids after granulation must be over 95 %, it can be further used as the raw material to produce Portland cement [1–3]. China is short of water seriously. The granulation of BF slag with water quenching has many disadvantages, such as high water consumption and difficulty in heat recycling. So developing a new dry granulation technology to treat the BF slag is very important. The applicability of BF slag directly depends on the content of amorphous solids in the slag after treatment [4–6]. The

Hai-Feng Wang, State Key Laboratory of Advanced Steel Processing and Products, Central Iron & Steel Research Institute, Beijing, China 100081

DOI 10.1515/9783110540048-035,

research on the activation energy of crystallization of amorphous solids in BF slag is the foundation to find out the reasonable cooling speed during dry granulation of molten slag. So it is of theoretical and practical significance to be analyzed.

The BF slag of steel plant A was dealt with open cycle process (OCP). The granulated BF slag were almost amorphous solids as 100 % amorphous solids sample. The component of the slag were *w* (CaO) 38.2 %, *w* (SiO_2) 35.7 %, *w* (Al_2O_3) 16.3 %, *w* (MgO) 8.3 %, *w* (FeO) 0.4 %, *w* (S) 1.0 % and *R*=1.070. The activation energy of crystallization of amorphous solids in BF slag (E) was 108.90 kJ/mol.

2 Crystallization Rate of BF Slag at Different Temperatures

According to the Arrhenius equation [7]:

$$\frac{d\ln k}{dT}=\frac{E}{RT^2} \tag{1}$$

where: k is the reaction rate constant of crystallization of amorphous solids in BF slag(1/s), T is the temperature(K), E is the activation energy of crystallization of amorphous solids in BF slag(J/mol), R is the Avogadro's constant, 8.314J/mol·K.
Generally, E would not be changed with temperature. Eq.(1) was integrated as Eq.(2):

$$\ln \mathrm{k} = -\frac{\mathrm{E}}{\mathrm{RT}} + \mathrm{A} \tag{2}$$

where: A is the constant.
The Eq. (2) was rewritten as Eq.(3).

$$k = k_0 \exp(-\frac{E}{RT}) \tag{3}$$

The logarithmic calculation was taken on Eq. (3).

$$-\frac{d\ln(1-\varphi)}{dt} = k \tag{4}$$

So,

$$-d\ln(1-\varphi)=k_0\exp(-\frac{E}{RT})dt \tag{5}$$

For,

$$T=T_0-vt \tag{6}$$

Where: T is the temperature of the BF slag after time t(K), TO is the origin temperature of the BF slag(K), v is the cooling speed (K/s), and t is the time(s).

$$t=\frac{T_0-T}{v} \tag{7}$$

Eq.(7) was taken into Eq.(5),

$$-d\ln(1-\varphi)=k_0\exp(-\frac{E}{RT})d\frac{T_0-T}{v} \tag{8}$$

$$d\ln(1-\varphi)=\frac{k_0}{v}\exp(-\frac{E}{RT})dT \tag{9}$$

The time t was from the origin time to the cooling temperature, φ was from 0 to 0.05, and T was from 1723 K to 973 K,

$$\int_0^{0.05}d\ln(1-\varphi)=\int_{T_0}^{T}\frac{k_0}{v}\exp(-\frac{E}{RT})dT \tag{10}$$

So,

$$v=-7014.4\int_{T_0}^{T}\exp(-\frac{E}{RT})dT \tag{11}$$

The crystallization rate of amorphous solids in BF slag at temperature 1573K to 1723K was calculated with Eq.(3). The results were shown in table 1.

Tab. 1: Crystallization rate of amorphous solids in BF slag at temperature 1573K to 1723K

Temperature(K)	1573	1673	1723
Crystallization rate(1/s)	0.085	0.140	0.175

The crystallization rate of amorphous solids in BF slag at temperature 1573K, 1673K and 1723K were 0.085(1/s), 0.140(1/s) and 0.175(1/s), respectively.

3 Required Cooling Rate of BF Slag at Different Temperature Range

Eq. (11) was used to calculate the required cooling rate for BF slag from 1723K to 973K. As the time t from the origin time to the cooling temperature, φ from 0 to 0.05 and T from 1723K to 973K, the result was shown that the required cooling rate for BF slag from 1723K to 973K was more than 100 K/s (Fig. 1).

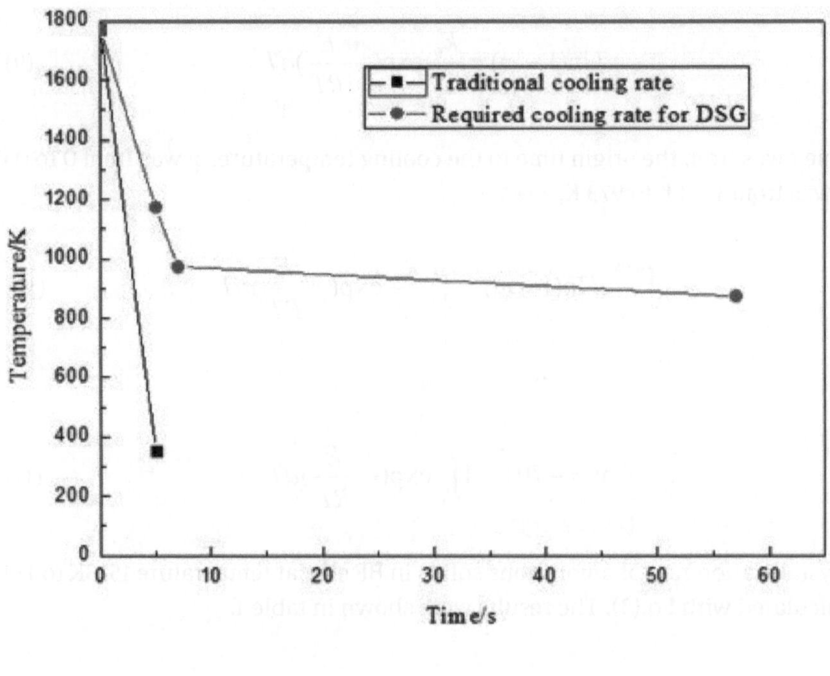

Fig. 1: Required cooling rate for dry slag granulation (DSG)

4 Process and Equipment for Dry Slag Granulation Technology

Based on the results above, the process and equipment of DSG technology was designed by the author. The process was as follows. The BF slag was transported to the slag treatment plant by the slag tank. The slag was poured into the slag ladle by the crane. The metering nozzle was used to control slag flow to the intermediate slag ladle, which always kept the height of the slag. The molten slag was poured into the device of double row roll granulator quantitatively. The granulator was consists of two rows of rollers. The upper row roller was mainly used to granulate slag and the under row roller was used to discharge the slag. The two rows of rollers were cooled by water and the surface temperature was below 573K to ensure the slag at the surface be cooled rapidly. The temperature of cooling water was raised less than 10K. The primary air re-crusts granulated slag with liquid core. The air duct and high-speed nozzle were set on sidewall granulation device to cool the slag with its temperature lower than 1023K and prevent its re-agglomeration. After cooling, the primary air temperature was about 773K. It was collected by the upper collection tube and mixed with the secondary cooling air into the hot cyclone. The fine particle slag was removed. The granulated slag ran into the shaft furnace through the star type feeder to exchange heat with the secondary cooling air. The air temperature was raised and the slag temperature was decreased. Finally, the slag was discharged at the bottom with its temperature lower than 423K. The temperature of secondary cooling air was raised to about 973K. It was emission from exhaust pipe at the top of the shaft furnace and mixed with the primary cooling air. The dust was moved by the bag-type dust remover. After dust removing, the air was got into the waste heat boiler and generated power by the steam turbine. After heat exchanging, the exhaust temperature of the boiler was lower than 423K. It was introduced by draft fan into dust collector chute. After dust removing, part of the air was emission into the stack, and the other part of the air was recycling into the shaft furnace to recover the waste heat as the secondary cooling air, which could reduce the heat dissipation. The slag was discharged from the bottom of the shaft furnace and transported by car to the yard (Fig. 2).

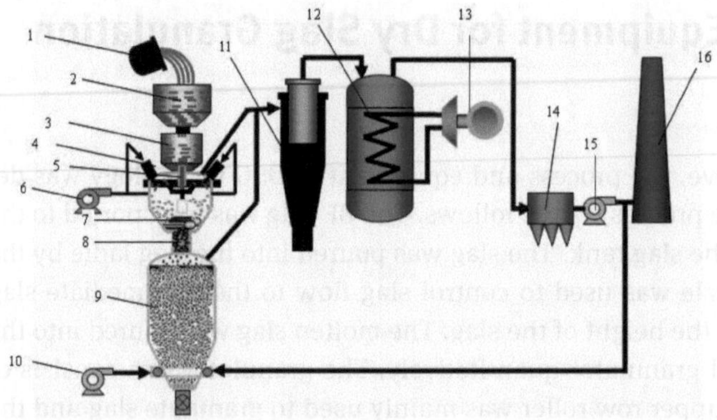

Fig. 2: Flow chart of dry granulation and waste heat recovery process for slag developed by Central Iron & Steel Research Institute (CISRI)

(1)slag tank; (2)slag ladle; (3)intermediate slag ladle; (4)high speed nozzle; (5)metering nozzle; (6)primary cooling fan; (7)double row roll granulator; (8)star type feeder; (9)shaft furnace; (10)secondary cooling fan; (11)hot cyclone; (12)boiler; (13)generating set; (14)bag-type dust remover; (15)circulation fan; (16)stack;

5 The Semi-industrial Experiment and Results

The semi-industrial experiment was carried out. The treatment volume of BF slag was 1.2 ton. The tap speed of slag was 0.2 meter per second and the time was 1 hour. The different volumes of the primary and secondary cooling air were injected, respectively. The volume and temperature of the air and the content of amorphous solids were shown in Fig. 3.

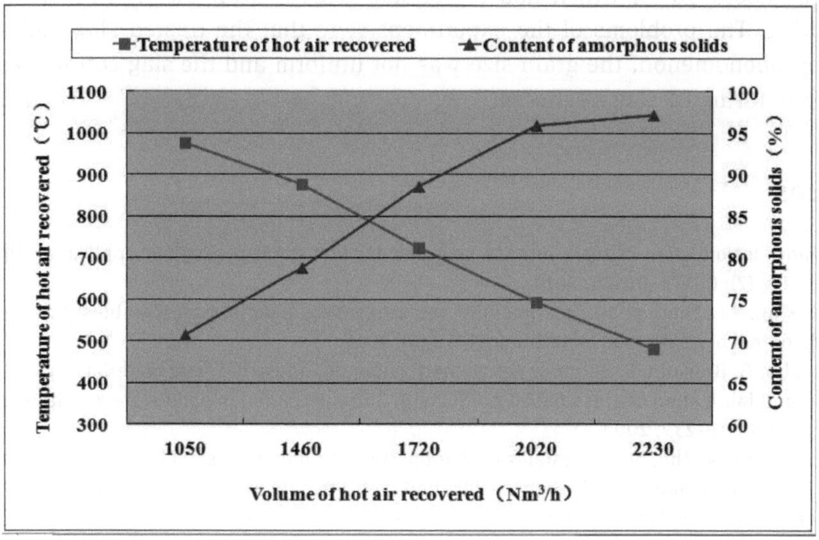

Fig. 3: Semi-industrial experimental results

The theoretical calorific value was 2.1 GJ. The volume of recycling hot air was 2020 m^3 with 867 K. The sensible heat value was 1.6 GJ. The self-power consumption was 10.7 kwh (38.5 MJ), which accounted for 2.4 % recovered heating. Regardless of power generation efficiency, the heat exchange efficiency cold air and slag was about 77 %. The content of amorphous solids in granulated slag was about 96 %, which had the similar activity with the slag by water quenching.

The problems of the experiment were the treated slag had a crusting phenomenon, the grain size was not uniform and the slag cotton was produced during BF slag treatment, which needed to be further researched.

6 Conclusion

The results showed that:

(1) The activation energy of crystallization of amorphous solids in BF slag (E) was 108.90 kJ/mol. The required cooling rate for BF slag from 1723 K to 973 K was more than 100 K/s.

(2) The process and equipment for the dry slag granulation (DSG) technology was desired.

(3) The semi-industrial experimental results was that the volume of recycling hot air was 2020 m^3 with 867 K and the content of amorphous solids in granulated

slag was about 96 %, which had the similar activity with the slag by water quenching. The problems of the experiment were that the treated slag had a crusting phenomenon, the grain size was not uniform and the slag cotton was produced during BF slag treatment.

References

[1] Q. T. Fang. Introduction of blast furnace slag powder in cement production application [J]. Industrial b, (7): 00173–00174, 2016.

[2] H. Li, J. Xing, Y.L. Zhao, et al. Experiment of new cementing material based on blast furnace slag [J]. Journal of nonferrous metals (mine), 68 (6), 2016.

[3] P. Colombo, G. Brusatin, E. Bernardo, et al. Inertization and reuse of waste materials by vitrification and fabrication of glass-based products[J]. Current opinion in solid state & materials science, 7 (3):225–239, 2003.

[4] B. Lin, H. Wang, X. Zhu, et al. Crystallization properties of Chongqing steel blast furnace slag under different cooling rate [C]. Higher education institute engineering thermal physics professional committee of the national academic conference proceedings [A]. 2015.

[5] C. Fredericci, E.D. Zanotto, E.C. Ziemath. Crystallization mechanism and properties of a blast furnace slag glass [J]. Journal of Non-Crystalline Solids, 273 (1–3):64–75, 2000.

[6] Y.C. Wang, W.W. Yu, J.L. Zhang, et al. Crystallization kinetics and microstructure of glass-ceramics prepared from Baotou steel blast furnace slag [J]. Cailiao Rechuli Xuebao / transactions of Materials & Heat Treatment, 35 (1): 88–93, 2014.

[7] Y.J. Liang. Physical chemistry. Beijing: metallurgical industry press. 238, 1983.

Jian-Gang Zhao, Peng Sun, Mei-Nan Li*, Hui-Xiang Li, Rui Yang, Xiao-Yi Zhang and Xue-Zhu Wang

Research and Application of HTHP Dynamic Linear Swell Meter

Abstract: A new type HTHP dynamic linear swell meter is designed and tested in this paper which can be applied to evaluate the effect of expansion inhibitor in HTHP dynamic condition. This instrument is comprised by temperature control system, pressure control system, drilling fluid circulatory system, swell parameter test circuit and the central processing software system.

Keywords: Swell, HTHP, Dynamic, Borehole stability, Shale

1 Introduction

The borehole stability is a key problem in drilling engineering. From 1980s, along with the drilling layer becoming more and more complex, the borehole stability make more effect on drilling speed, cost and quality. Related test methods and instruments are urgent needed.

Shale sample swell test is the main methods to evaluate borehole stability by measuring the increase or decrease in length over time of a reconstituted or intact shale core. Common instruments cannot simulate the HTHP condition under well while the HTHP dynamic linear swell meter can do.

Jian-Gang Zhao, Beijing Institute of Exploration Engineering, Beijing, China. Email: zhaojg@cugb.edu.cn
Peng Sun, Beijing Institute of Exploration Engineering, Beijing, China. Email: 13366155681@189.cn
Mei-Nan Li, Beijing Institute of Exploration Engineering, Beijing, China. Email: limeinanbjiee@foxmail.com
Xiao-Yi Zhang, Beijing Institute of Exploration Engineering, Beijing, China. Email:zhangxiaoyi1001@126.com

DOI 10.1515/9783110540048-036,

2 The Design of a New Type HTHP Dynamic Linear Swell Meter

2.1 Design Idea

Measuring the increase or decrease in length over time of a reconstituted of intact shale core in HTHP condition can evaluate the swell of layer. So the HTHP dynamic linear swell meter should include temperature control system, pressure control system, drilling fluid circulatory system, swell parameter test circuit and the central processing software system.

2.2 Hardware

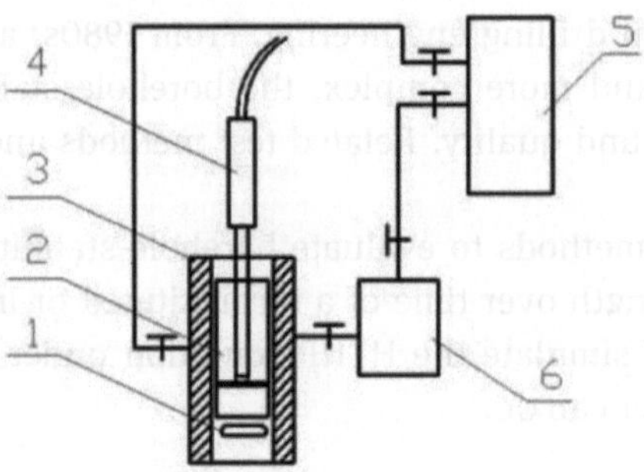

Note: 1. Non-contact dynamic mixing device, 2. Test cell 3.HTHP jacket, 4. Motional pickup transducer, 5. Pressure source, 6. Drilling fluid preheating harbour.

Fig. 1: Structure diagram of the HTHP dynamic linear swell meter

2.3 Software

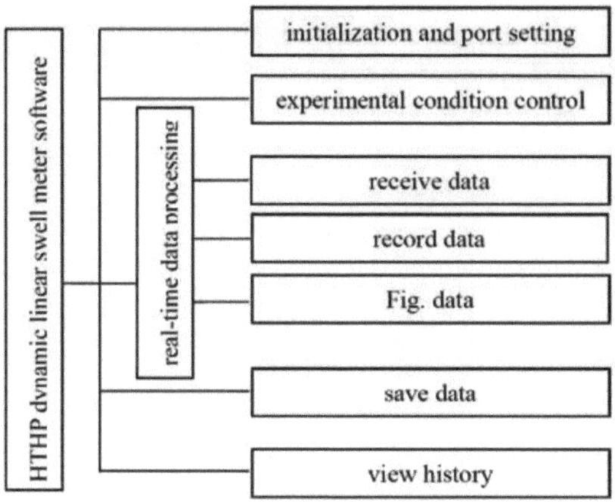

Fig. 2: Software building-block of logic of the HTHP dynamic linear swell meter

2.4 Functions and Features

1) The instrument can test mud shale hydration expansion and expansion rate.
2) The instrument can simulate HTHP condition under well, the max temperature is 260 °C, the max pressure is 7 Mpa.
3) Magnetic drive system speed 1000rpm (=16.667 r/s), Linear velocity is 0.04 m/s.
4) The software can control temperature and pressure and data processing system can collect and save data by GUI.

3 Application

3.1 Comparison of Dynamic and Static Test

Relevant parameters of HTHP borehole:

Caliper: Φ216 mm, Drill pipe: Φ127 mm, Annulus area: 240 cm^2, Mud pump: 25–30 L/s

Annulus velocity: 1.04–1.25 m/s, Sample cup diameter: Φ25 mm

Contact diameter between liquid and rock sample: Φ20 mm

Magnetic rod speed: 1000 rpm(=16.667 r/s), Linear velocity :1.04 m/s

Fig. 3 shows the test results of same sample (Sodium base soil compaction) under same temperature and pressure. Fig. 3 shows the sample swells more quickly under dynamic condition than static condition.

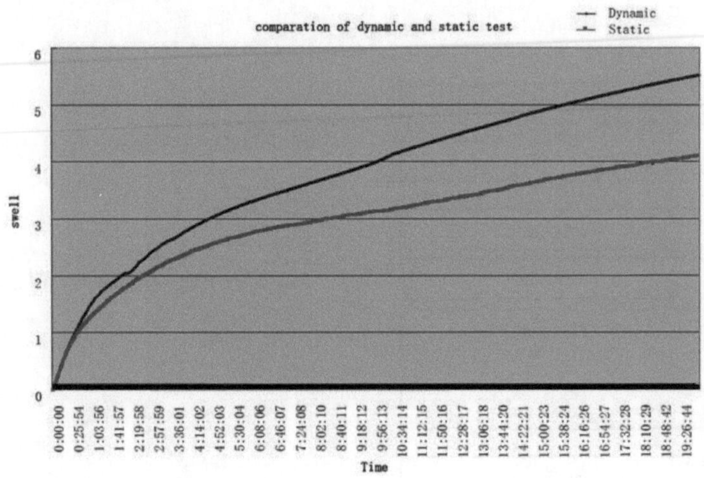

Fig. 3: Comparation of dynamic and static test

3.2 Experimental Results and Analysis

3.2.1 Drilling Fluid Formulation for Experiment

1#, 5 % AnQiu Calcium bentonite + 3 % KCl + 4 % SMP + 0.5 % HPAN;
2#, 5 % AnQiu Calcium bentonite + 3 % sodium silicate + 4 % SMP + 0.5 % HPAN;
3#, 5 % AnQiu Calcium bentonite + 1 % KCl + 3 % sodium silicate + 4 % SMP + 0.5 % HPAN;

3.2.2 Experimental Process

(1) Test condition: temperature≥180 °C, pressure≥7 MPa
(2) Four test should be conducted for every sample: NPT static, NPT dynamic, HTHP static, HTHP dynamic.
(3) Every test time should be controlled in 4–6 h, record data every half an hour.

3.2.3 Test Data Analysis

Fig. 4-Fig. 8 shows the test results:

(1) All sample swell more quickly in static condition than dynamic condition under normal temperature. While all sample swell more slowly in static condition than

dynamic condition under normal temperature. That verifies the necessity of dynamic high temperature test.

(2) The multi-effect of inhibitor KCl and sodium silicate is better than use alone.

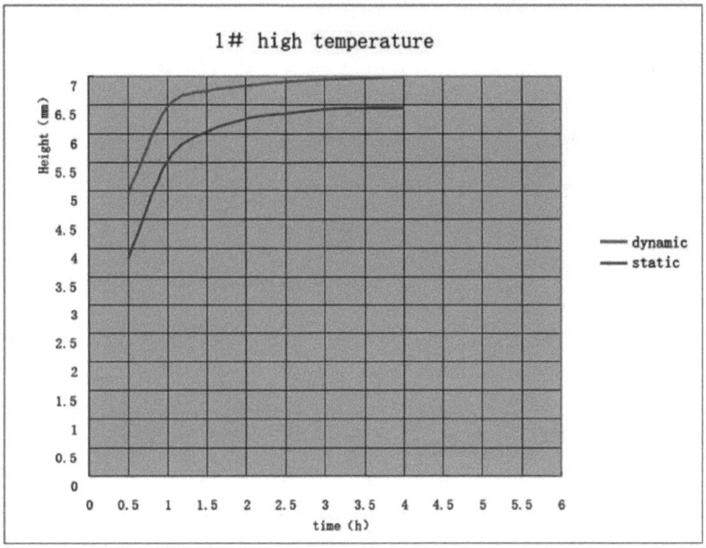

Fig. 4: # sample high temperature test

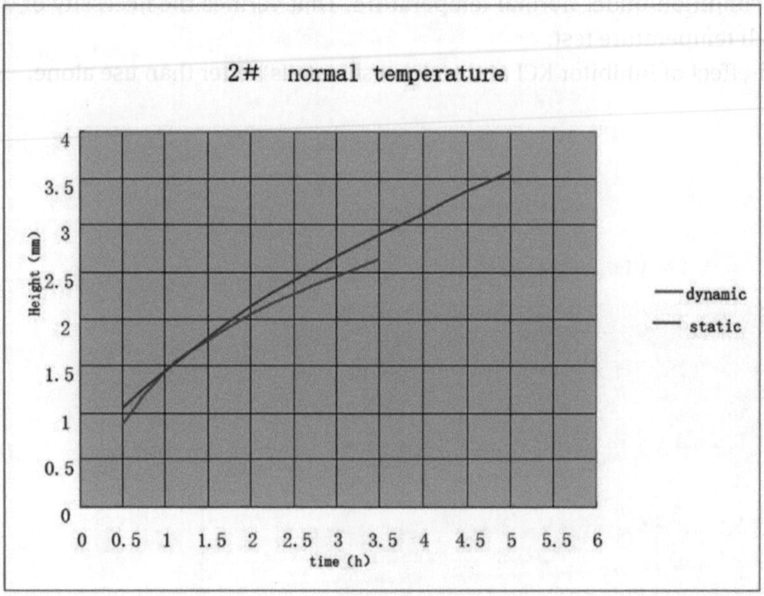

Fig. 5: 2# sample normal temperature test

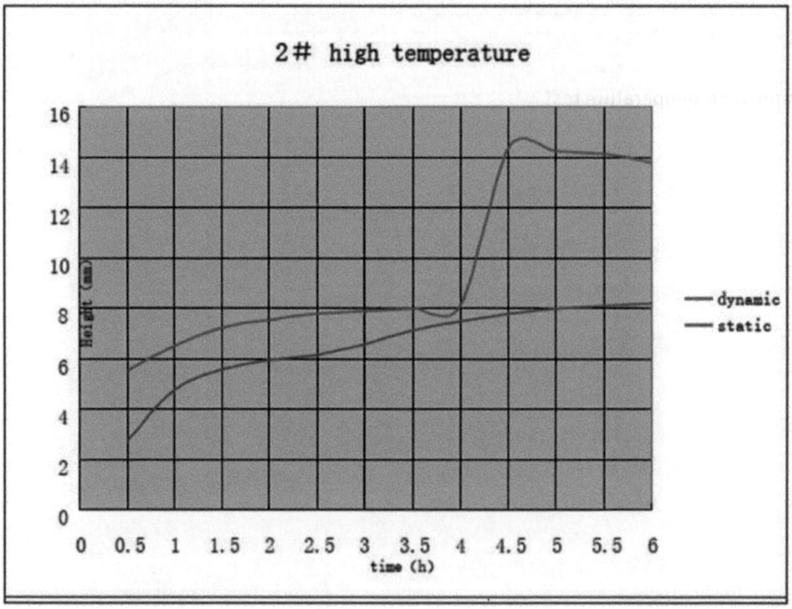

Fig. 6: 2# sample high temperature test

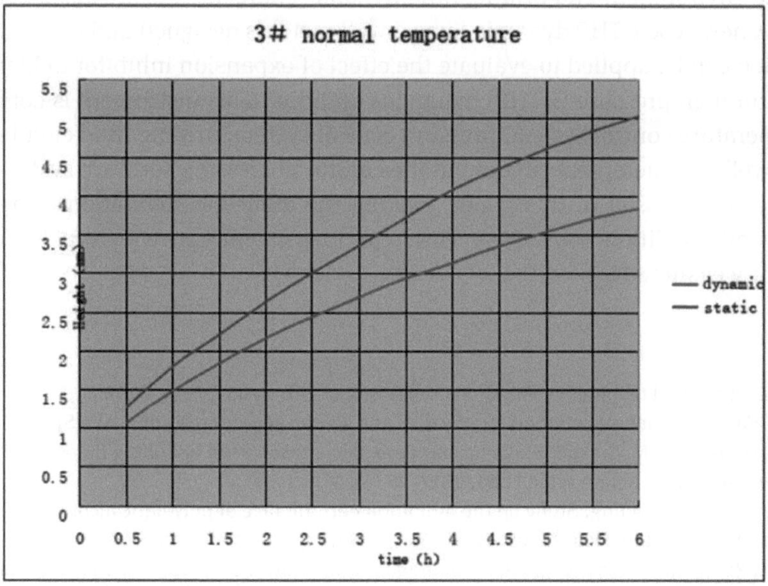

Fig. 7: 3# sample normal temperature test

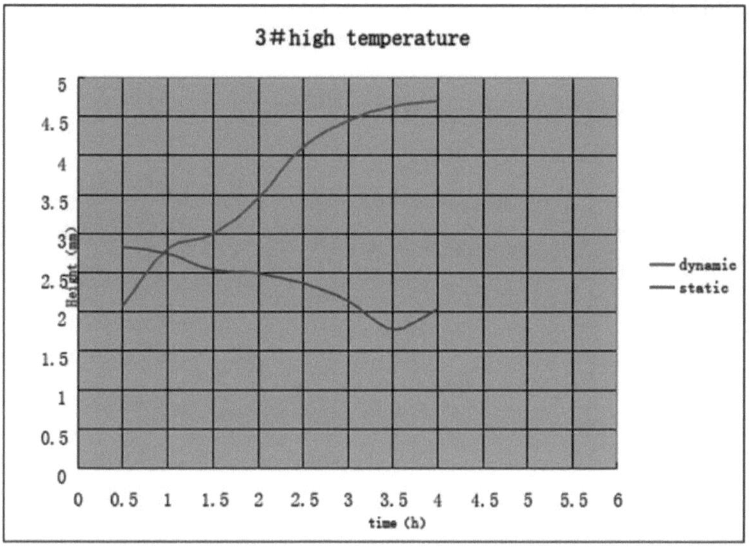

Fig. 8: 3# sample high temperature test

HTHP borehole environment and drilling fluid dynamic characteristic are discussed in this paper. A new type HTHP dynamic linear swell meter is designed and tested in this paper which can be applied to evaluate the effect of expansion inhibitor in high temperature and high pressure (HTHP) dynamic condition. This instrument is comprised by temperature control system, pressure control system, drilling fluid circulatory system, swell parameter test circuit and the central processing software system. Test procedure is discussed in this paper, making the domestic simulation of dynamic HTHP well condition come true. This type instrument can widely used in wellbore stability evaluation.

References

[1] Y. Jienian. Drilling fluid technology [M]. China University of Petroleum press, 2006.

[2] SY/T5613–2000 Experimental method for physical and chemical properties of shale [S].

[3] Y. Junzhen, H. Daquan, M. Cuixue. Study on standard calcium soil used in drilling fluid experiment [J]. Drilling Fluid & Completion Fluid, 2012, 29 (6), 13–16.

[4] G. Baoli, Y. Menglei, W. Ailing. Study on the inhibition performance of polymeric alcohol [J] . Drilling Fluid & Completion Fluid, 2005, 22 (4), 35–36, 39.

[5] L. Huaike, T, Rongjian, L. Jiansheng. A new method for quantitative evaluation of amine inhibitors used in drilling fluids [J]. Drilling Fluid & Completion Fluid, 2012 (6), 13–16.

[6] Z. Jiangang, S. Peng, W. Qi. Study on test method of shale swelling inhibitor [J] . Drilling Fluid & Completion Fluid, 2013, 30 (4), 41–42, 45.

[7] S. Peng, C. Jingge, L. Wei. Research on the software system of the multi joint expansion measuring instrument based on multi serial communication technology. [J]. Geological equipment, 2012, 13(66), P65–68.

Xuan Zheng, Si-Qi Ji, Liu-Tao Xue and Dong-Wei Gao

Research on Thin-Walled Components in High Speed Milling for Machining Distortion Control

Abstract: In order to reduce the machining distortion in the milling process of aeronautical thin-walled component, firstly the theory model of milling force calculation was analyzed and the three dimensional milling force was simulated. The law of machining distortion for thin-walled component was researched deeply based on finite element method. Also the important milling parameter and geometric parameter were specially analyzed.

Keywords: Thin-walled workpiece, Milling, Machining distortion, Finite element method

1 Introduction

The thin-walled parts are widely used in aerospace industry, because of its light weight and overall superior performance characteristics. However, due to its poor rigidity, deformation will easily be caused during the machining process and it is often difficult to ensure the machining accuracy. In the process of machining thin-walled parts, flutter problem seriously affect the quality of the workpiece. In order to solve the problem of machining deformation of thin-walled parts, scholars have done extensive research on Mechanical model of milling [1, 2], error compensation [3] and so on. The above researches are mainly focused on the plate structure, and have a certain distance from the practical application.

Jer-Shyong Tsai, Chung-Li Liao [4, 5, 6] and Zhang Lili made a numerical simulation study for some factors affecting machining deformation, such as residual stress. At present, high speed machining technology has been widely used in the aviation field. How to select the reasonable machining parameters to control the machining deformation and ensure the machining efficiency for Aeronautical Thin-walled workpiece becomes the focus of people's attention. In practice, people often through the qualitative analysis and the experience of the operator to choose, there

Xuan Zheng, Si-Qi Ji, Liu-Tao Xue, School of Mechanical Engineering, Beijing Institute of Petrochemical Technology, Beijing 102617, China. Email: jisiqi@bipt.edu.cn
Xuan Zheng, Beijing Semboo Science and Technology Co., Ltd, Beijing 100086, China Email: as_ben@163.com
Dong-Wei Gao, Jinan Urban Construction Group Co., Ltd., Jinan, China Email: 64512009@qq.com

DOI 10.1515/9783110540048-037,

is certain blindness requires a lot of testing and time investment, parts processing. The finally, quality cannot be guaranteed.

First of all, we discuss the milling machining mechanics model, and the 3D milling force is simulated. On this basis, the finite element method is used to study the typical thin-walled connection characteristics of aeronautical aluminium alloy thin-walled structure and studied the machining deformation law of the process. The single factor analysis of the machining parameters and the geometric parameters of thin-walled are analysed. Finally, the corresponding deformation control scheme is proposed.

2 Milling Force Model and Simulation

2.1 Computational Model of Milling Force

In the machining of thin-walled parts, the cutting force has a significant effect on the deformation due to the low stiffness of the workpiece. Therefore, study of the milling force model is very important. The cutting area is divided into many micro elements, spatial distribution of milling force can be obtained by calculating the force condition of each unit in the cutting area, and then total cutting force [7] is obtained by adding all the forces in the cutting area.

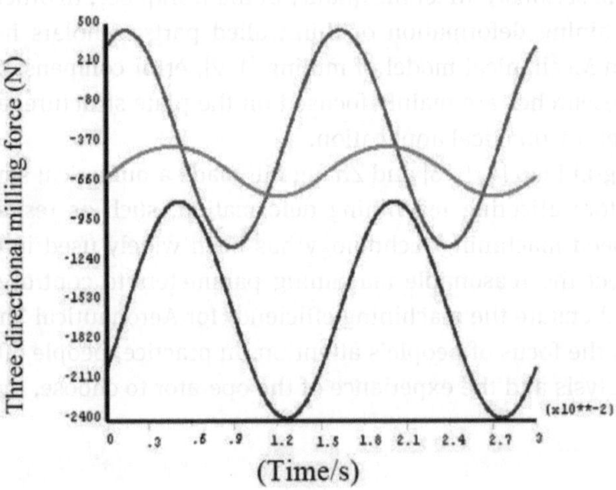

Fig. 1: Three simulation of transient milling force

The tangential, radial and axial cutting forces on the element are similar, the formula is as follows:

$$dF_{t,j}(\phi, z) = K_{tc}h_j(\phi, z)dz$$

$$dF_{r,j}(\phi, z) = K_{rc}h_j(\phi, z)dz \qquad dF_{a,j}(\phi, z) = K_{ac}h_j(\phi, z)dz \tag{1}$$

Where $dF_{t,j}$, $dF_{r,j}$, $dF_{a,j}$ represent respectively tangential force, radial cutting force and axial cutting force; dz is the unit axial width; $h_j(\phi,z)=c\sin\phi_i(z)$ is cutting thickness.K_{tc}, K_{rc}, K_{ac} can be determined by experiment, Neglecting the influence of the cutting force coefficient.

Infinitesimal force is decomposed into feed direction (X), normal (Y) and axial (Z):

$$dF_{x,j}(\phi_j(z)) = -dF_{t,j}\cos\phi_j(z) - dF_{r,j}\sin\phi_j(z) \tag{2}$$

$$dF_{y,j}(\phi_j(z)) = dF_{t,j}\sin\phi_j(z) - dF_{r,j}\cos\phi_j(z) \tag{3}$$

$$dF_{z,j}(\phi_j(z)) = dF_{a,j} \tag{4}$$

The micro force in processing section along the spiral groove is integrate, obtained total cutting force:

$$F_q(\phi_j(z)) = \int_{z_{j,1}}^{z_{j,2}} dF_q(\phi_j(z))dz, \quad q = x, y, z \tag{5}$$

Add the all cutting force of spiral groove at contact angle, and then the instantaneous cutting force acting on the cutter can be obtained.

2.2 Milling Force Identification of Aluminium Alloy

The cutting force coefficient of aluminum alloy 7075 was obtained according to the average cutting force experiment, type of machine tool: Mikron ucp710, cutter is carbide end mill.

Tab. 1: Tool parameters

Parameters	value
Tool diameter[mm]	12
Helix angle[°]	30
Tooth number[z]	3

The results are shown in Table 2.

Tab.2: AL7075/T6 Milling force coefficient

K_{tc}	**796.1N/m2**	K_{te}	**20.70N/m**
K_{rc}	168.8N/m2	K_{re}	30.8N/m
K_{ac}	222.1N/m2	K_{ae}	1.4N/m

In order to facilitate the– subsequent finite element calculation, the milling force simulation program is programmed by APDL technique in the finite element software ANSYS, three direction transient force is calculated. Fig. 1 is a simulation schematic diagram about the three direction transient milling force under the condition of cutting conditions.

It is assumed in this paper that the influence of tool and workpiece deformation on the chip thickness and stress is not considered. In other words, milling force modeling is rigid force. In order to consider the feedback effect of the workpiece or tool deformation on the milling force, it is necessary to consider the influence of cutting thickness on the cutting force and deformation. Until the two numerical value of the milling force is convergent to a certain extent [8, 9]. In this paper, we focus on the macroscopic deformation of Thin-Walled Components by the rigid force model.

3 Analysis of Thin-Walled Components Milling Deformation

3.1 Load Distribution of Thin-Walled Components Milling

When the thin wall frame of thin-walled components is processed by Helical End Mills, because the tool speed is far greater than the feed speed, so we can assume that in a feeding position, the tool axis is fixed, and the cutter tooth by moving up until the cut out of the workpiece, ultimately the formation of the surface [10] axis has been processed, as shown in Fig. 2.

As can be seen from Fig. 2 (a), the dynamic load along the helix direction is applied to the frame wall along the spiral direction at a certain instant. The variation of chip thickness with different position of cutter teeth during milling. In the actual milling process, the chip thickness varies with the position of the cutter tooth. Because of tool cutting edge inclination angle, intersection of the cutting edge and the machined surface of the workpiece is in a certain angle, as a result, machining process simulation is very difficult. For the sake of this simulation, milling process is simplified. As we can see in Fig. 2 (b), the continuous contact between the tool and the machined surface is discretized and the size of the cutting edge angle is controlled by the size of the grid unit [11].

When milling thin-wall parts, cutting depth is very short, so we can assume that the load is linear. Distributed load approximation to grid nodes during Loading on the finite element model. Because the cutting force perpendicular to the thin wall has the greatest influence on the deformation of the thin wall, we only consider the cutting force perpendicular to the thin-wall. Tools and tooling can be regarded as rigid body, no deformation. The deformation of workpiece is pure elastic deformation.

3.2 Establishment of Finite Element Model

The shape of the workpiece in the process is changing, complex and irregular. Generally, analytical algorithm cannot be used to calculate the force deformation. The cutting force of thin-wall is very concentrated in the cutting area. According to Saint Venant principle, we can conclude that the deformation of thin-walled workpiece is mainly related to the local shape of cutting position.

Because of the diversity of the structure and shape of the parts, the local shape of different cutting parts has different characteristics; therefore, it is not suitable to replace all local shapes with regular shapes. Appropriated method is to approximate the different local shapes using different regular shape. The thin-walled structure of these rules is called the basic thin wall element [12]. The connection between the thin-wall element and the other parts of the workpiece plays an important role in the stress and deformation of the thin-walled part, which is also an important factor affected the deformation. When the local thin wall element is used instead of the whole workpiece to analyze the deformation, the connection between thin-walled elements and other parts of the workpiece is reduced to the corresponding deformation boundary condition. For example, the edges connected with other parts are regarded as fixed constraints, and the unconnected edges are called free edges. The vertices of the two adjacent free edges intersect as the free corner points.

Tab. 3: Milling condition

Parameters	Value
Radial cutting depth[mm]	2
Axial cutting depth[mm]	20
Speed[r/min]	14000
Feed rate[mm/z]	0.2
wall thickness[mm]	4

The common types of connections on thin walls are shown in Fig. 3. Selecting one of the side walls of aerospace structure to research. The pedestal is equivalent to clamping and locating in the machine table. The finite element model is shown in Fig. 4. Based on the milling force calculation program, the average milling force can be applied as the load condition.

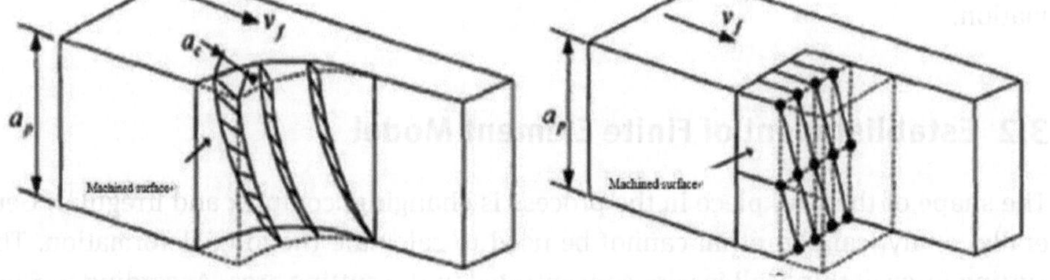

(a)Actual material removal (b) Simplified material removal

Fig. 2: Sketch of instantaneous milling force distribution

(a) The top edge is free and hanging bottom, both ends fixed. (b)Top and end are free, bottom and side fixed. (c)Top edge is free, other three edges are fixed.

Fig. 3: Typical connection type of thin-walled parts

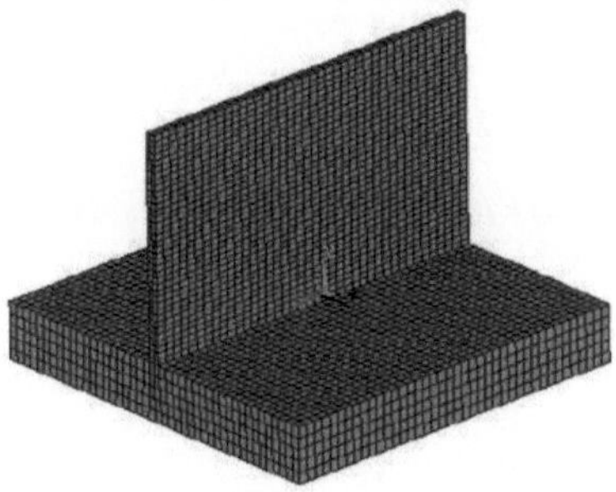

Fig. 4: Finite element model of typical thin-walled workpiece

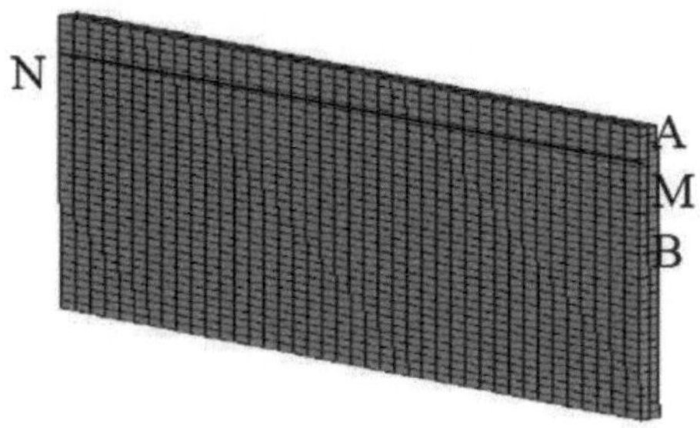

Fig. 5: Sketch map of milling

3.3 Calculation Results Analysis

When we the cutting force is calculated, the continuous cutting process is divided into several cutting positions, each cutting position is a working condition of the workpiece. In order to simplify the analysis process, the milling load of each cutting position can be simplified to a concentrated load located on tool and workpiece contact line along the axial direction [9]. Accordingly, the dynamic milling force of the contact region will be applied to the contact line MN as a moving load. For the first cutting position, Load applied at the midpoint of line AB, see Fig. 5, and so on, to point N. ANSYS technology based on APDL technology to achieve dynamic load cycle. The effect of inertial force and damping force is very small, so we can ignore their influence, doing quasistatic analysis.

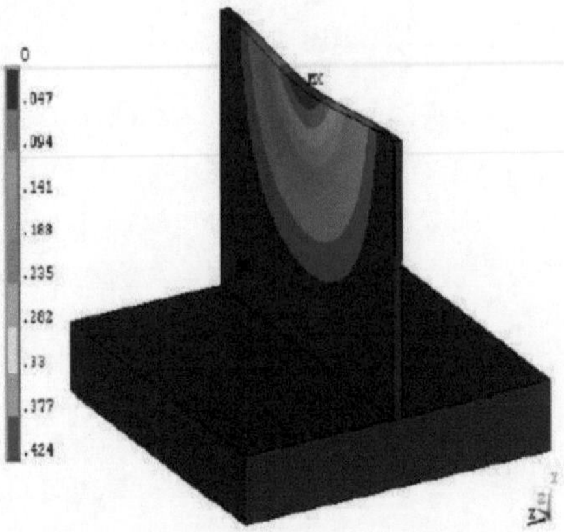

Fig. 6: Deformation nephogram

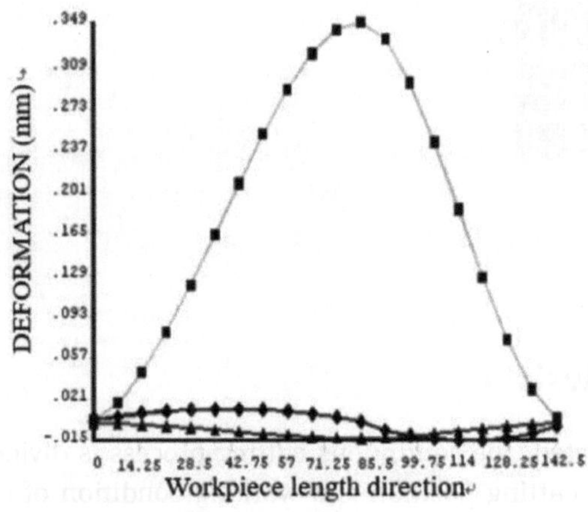

Fig. 7: Deformation curve along length

The milling process simulation analysis is carried out for different constraint modes in Fig. 3. In the three cases, the deformation of the upper part of the thin- wall is inclined; deformation nephogram is shown in Fig. 6. Compare with X normal deformation, Y direction and Z direction deformation is very small as shown in Fig. 7, thus, Y direction and Z direction deformation can be neglect in the course of the study. For workpiece structure, thickness direction has great deformation because it is much smaller than the other two dimensions. For the same cutting load, the constraint method (a) is better, so the deformation is small.

Fig. 8: Deformation curve along the axis

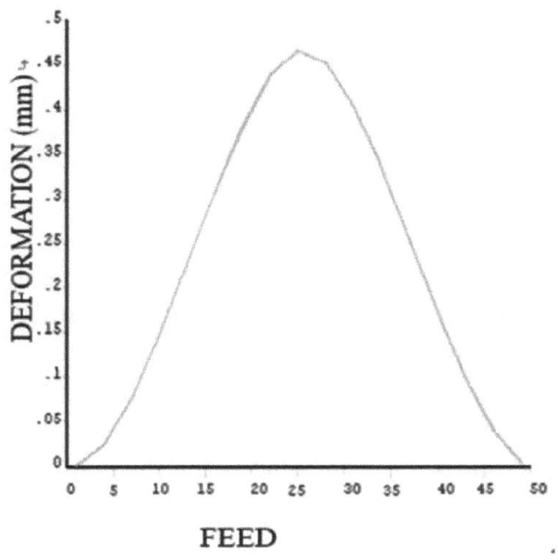

Fig. 9: Deformation-distance curve

According to the deformation trend, middle deformation is large and deformation at both ends is small along the feeding direction at each feed position of the cutter, this is mainly due to the weakest intermediate rigidity two ends of the deformation is small constrained or other parts of the check. Along the direction of the axis, rigidity of the top is weak and deformation is also weak, in a certain range, it can be similar to a straight line (see Fig. 8).

Fig. 9 is the progress curve of a node in the middle part of plate with chip-load movement. It can be seen that the deformation is parabolic. It increases gradually, and then decreases gradually after reaching the maximum value. Analysis of the reasons we believe that because of the cutting force in the normal direction of the distribution is very concentrated, so that the effective width of the plate on the Limited. The node is away from the cutting load when the tool starts cutting, so deformation is small. In the middle of cutting, close to the node, the deformation increases. And then the cutting tool gradually cuts out the workpiece, deformation of the joint is reduced and there is the opposite trend of deformation. Finally, leading to deformation decreases.

Theoretically speaking, terminal deformation is larger than the front the reason is that the whole thin-wall has been completed cutting when cutting to the end, stiffness than the cutting head to be small. But in the calculation process, the influence of wall thickness variation on the stiffness is ignored, so the difference between the two ends cannot be simulated. In general, the last part of the cutting tool is located at the lower end of the plate, and the axial depth is smaller than the height of the plate. This effect is weakened, and the simulation results are acceptable.

3.4 Single Factor Analysis

3.4.1 Effect of Axial Cutting Depth

For the thin-walled parts, the radial cutting depth is smaller at the finishing stage. In order to improve the machining efficiency, large axial cutting depth can be selected. However, increasing axial cutting depth not only the cutting force increases, but also the role of the distribution of high position, which is more likely to lead to deformation of thin-walled workpiece. The single factor analysis of axial cutting depth, main research work is upper workpiece, we only considering the deformation of the middle region processing; cutting load as uniformly distributed is applied along the axial depth of cut. Selecting milling parameters, radial cutting depth is 2 mm, speed is 14000 r/min, feedrate is 0.1 mm/z, wall thickness is 4mm.Axial depth of cut were selected by 4 mm, 8 mm, 12 mm, 16 mm. The results are shown in the Table 4:

Tab. 4: Axial cutting depth and machining deformation

Axial cutting depth (mm)	**4**	**8**	**12**	**16**	**20**
Maximum deformation value (mm)	0.019572	0.079917	0.043529	0.048503	0.062962

As can be seen from Table 4, maximum deformation value does not increase monotonically, but it has a maximum value with the increase of cutting depth. When machining thin-walled parts, larger axial depth of cut should be selected within a certain range, which will improve cutting efficiency and machining deformation will not increase significantly. This is consistent with experimental results of reference [13].

3.4.2 Influence of Wall Thickness

According to theory of plates and shells, bending stiffness of thin-walled plate can be obtained by following formula:

$$D=\frac{Et}{-\mu} \tag{6}$$

In this formula, E is the elastic modulus, t is the thickness, μ is Poisson's ratio. Bending stiffness is proportional to the cube of thickness. For thin-walled parts machining deformation, deformation is greatly affected by its thickness. Select thickness values t=2 mm, 3 mm, 4 mm, 5 mm for calculation, the fitting curve is as follows: $\omega=-0.0265+0.03385 *t-1.4552 *t^2+2.1474 *t^3$.

3.4.3 Influence of Thin Wall Height

In addition, geometric height of thin-walled parts is also an important factor affecting the machining deformation. Fix one side, three edges free. Actually, load only in a limited area, thin-walled parts are reduced to cantilever beams. Under the action of uniform load, the deflection formula is given by:

$$\omega = {}^{qh}\!/_{EI}\left(- {}^{x}\!/_{h} + \frac{x}{h} \right) \tag{7}$$

Thus, we can conclude that height has a great influence on the deformation, it is very easy to produce deformation when machining the deep cavity structure.

4 Mathematical Model between Milling Parameters and Machining Deformation

In order to establish the relationship between milling parameters and machining deformation by finite element simulation, the regression orthogonal method is used to design the combined sequence of milling parameters. Based on the dynamic simulation software developed by the research group, the milling force is calculated, as the load boundary condition of the finite element calculation, the maximum deformation value of the thin-walled parts corresponding to each parameter is obtained. Finally, the mathematical model is established by regression orthogonal method.

Combination of certain cutting parameters, it can be assumed that the machining deformation and n, a_p, a_e, f_z, thickness of the thin-wall conform to the polynomial model. Through the optimization of the regression equation, the polynomial model is determined as follows:

$$U=\beta_0+\beta_1 n+\beta_2 a_p+\beta_3 t+\beta_4 nt+\beta_5 a_p a_e+\beta_6 a_p f_z+\beta_7 a_p t+\beta_8 a_e t+\beta_9 a_e^2+\beta_{10} a_p^2+\beta_{11} f_z^2+\beta_{12} t^2$$

Where u is the maximum value to machining deformation. The selection of milling parameters and the corresponding values are shown in Table 5. Each milling parameter is set to four levels, orthogonal 16 groups are used to simulate the calculation, combination sequence and the corresponding machining deformation values are shown in Table 6.

According to the data in Table 6, based on the orthogonal regression method, we can get the value of the corresponding polynomial:

{−50.0361, 0.0014, 4.6792, −3.3274, −0.0003, 0.123, 0.0836, −0.0148, 0.4152, −4.2245, −0.1124, −0.3191, 0.8108}.

Tab. 5: Milling parameters and the values of thin wall thickness

Parameter \ Level Number	1	2	3	4
Milling speed(n r/min)	10000	12000	14000	16000
Milling depth(a_p mm)	18	20	22	24
Milling width(a_e mm)	0.1	0.4	0.7	1.0
Feedrate (f_z mm/z)	0.2	0.4	0.6	0.8
Thin-wall thickness (t mm)	2	3	4	5

Tab. 6: $L_{16}(4^5)$ Orthogonal tables and corresponding deformation values

Parameter Test number	Milling speed (n /min)	Milling depth (a_p mm)	Milling width (a_e mm)	Feed rate (f_z mm/z)	Thin-wall thickness (t mm)	Deformation value (u mm)
1	1	1	1	1	1	2.257
2	1	2	2	2	2	1.289
3	1	3	3	3	3	0.833
4	1	4	4	4	4	0.546
5	2	1	2	3	4	0.408
6	2	2	1	4	3	1.181
7	2	3	4	1	2	0.532
8	2	4	3	2	1	4.439
9	3	1	3	4	2	2.237
10	3	2	4	3	1	5.254
11	3	3	1	2	4	0.508
12	3	4	2	1	3	0.317
13	4	1	4	2	3	0.400
14	4	2	3	1	4	0.128
15	4	3	2	4	1	9.546
16	4	4	1	3	2	2.377

5 Conclusion

In this paper, we can draw a conclusion that the deformation distribution law of aluminium alloy thin-walled parts is a parabola shape with large deformation and small deformation at both ends. With the increase of axial cutting depth, the deformation value has a maximum value. Therefore, when machining thin-walled parts, the larger axial depth is selected in a certain range, which will improve the cutting efficiency and processing deformation has not significantly increased. In addition, the deformation of thin-walled workpiece is greatly affected by its thickness. Due to the high value of deformation also has a great impact. So there is easy to produce deformation in the machining processing of thin-walled deep cavity structure.

References

[1] Kline W. A, Devor R E, Lindberg R. The prediction of cutting forces in end milling with application to concering cuts [J]. International Journal of Machine Tool Design and Research, 1982, 22 (1): 7–22

[2] Kuang-Hua, Fuh, Ren-Ming Huang. A predicted milling force for high-speed end milling operation [J]. International Journal of Machine Tools and Manufacture, 37 (7): 969–979

[3] S.Ratchev, S. Liu, W. huang, A. A Becker. An advanced FEA based force induced error compensation strategy in milling [J]. International Journal of Machine tools and Manufacture, 2006, 46: 542–551

[4] Jer-Shyong Tsai, Chung-Li Liao. Finite-element modeling of static surface errors in the peripheral of thin-walled workpieces. Journal of Materials Processing Technology, 1999, 94: 235–246

[5] Zhao Lili. Research on Residual Stresses & Machining Deformation of Aeronautic Aluminum Alloy Pre-stretching Board [D]. Beijing: Beihang university, 2006, 6

[6] Wang Zhigang, He Ning, WU Kai, Jiang Chengyu, Zhang Ping, Gong Huimin, Chen Xuemei. Analysis and Control Approach for Machining Deflection of Thin–Walled Workpiece [J]. China Mechanical Engineering, 2002, 13 (2): 114–117

[7] Yusuf Altintas. Manufacturing automation [M]. Beijing: Chemical Industry Press. 2002

[8] Erhan Budak, Yusuf Altintas. Modeling and avoidance of static form errors in peripheral milling of plates [J]. Int. J. Mech. Manufact. 1995, 35 (3): 459–476

[9] S.Ratchev, W. Huang, S. Liu, A. A. Becker. Modelling and simulation environment for machining of low-rigidity comoponents [J]. Journal of Materials Processing Technology, 2004, 153–154: 67–73

[10] Svetan Ratchev, Evan Govender, Stan Nikov, Kevin Phuah, Gregory Tsiklos. Force and deflection modeling in milling of low-rigidity complex part. Journal of Materials Processing Technology, 2003, 143–144: 796–801

[11] Huang Zhigang, Ke Yinglin. Study on Key Technologies of Milling Process Simulation for Aerospace Monolithic Components [J]. Mechanical Science and Technology, 2004, 15 (11): 991–995

[12] Liu Yuanxiang. Research on Techniques of Modeling and Analysis Accelerating in Virtual Machining Process [D]. Nanjing University of Science and Technolog, 2005. 11

[13] Wu Kai, He Ning, Liao Whenhe, Jiang Chengyu. Selection of Cutting Parameters in High Speed Milling of Thin-Walled Structure Components Based on the Machining Errors Control. Mechanical Science and Technology 2005, 24 (7): 778–791

Yun-Yan Xia*, Yong-Ming Xu and Yong-Sen Han

Effect of DC Magnetic Biasing on the Temperature Rise of High-voltage Winding of Power Transformer

Abstract: The DC magnetic biasing can aggravate the core saturation and increase the temperature in the large power transformer. In this paper, a 240000 kVA/330 kA oil immersible power transformer is taken as a research object to analyze the variation of the temperature rise in the high-voltage windings under DC magnetic biasing. The effect of DC magnetic biasing current on the winding loss is analyzed at first. Using computational fluid dynamics and fluid network of power transformer, the fluid field and the temperature distribution in the power transformer are then obtainned. The results show that the temperature rise is greatly affected by the DC magnetic biasing phenomenon. The calculation method proposed in this paper provides a theoretical basis for the structure optimization and the local over-heating suppression.

Keywords: DC magnetic biasing, Winding loss, Temperature rise

1 Introduction

HVDC power transmission over long distance may cause the DC magnetic biasing in power transformer, accelerating the transformer core saturation and forming the local overheating [1]. In addition, the geomagnetic ally induce current generated by the effect of the solar magnetic storm and the high-altitude nuclear electromagnetic pulse may also induce the presence of DC magnetic biasing in power transformer [2].

Many studies have been carried out to research the DC magnetic biasing phenomenon. The magnetization characteristic of transformer laminated core is analyzed under the DC magnetic biasing [3]. According to the field-circuit coupling method, the exciting current and the internal magnetic field distribution are analyzed under different levels of DC magnetic biasing [4–6]. In addition, the effect and its suppression method of DC magnetic biasing on the relay protection equipment in power system and transmission system operation are also considered [7]. However, very few studies have been carried out on the temperature rise of transformer wind-

Yun-Yan Xia, Yong-Ming Xu, Yong-Sen Han, Harbin University of Science and Technology, Harbin 150080. Email: yunyan_x@163.com
Yun-Yan Xia, Jiangsu Huapeng Group Co. Ltd, Liyang 213300. Email: yunyan_x@163.com

DOI 10.1515/9783110540048-038,

ing caused by the DC magnetic biasing.

In this paper, the high-voltage winding loss of a 240000 kVA/330 kA oil immersible power transformer is firstly calculated under different DC magnetic biasing current. A calculation model of temperature field in power transformer is built and is used to obtain to the temperature rise of high-voltage winding. This study provides theoretical basis for the optimization design of winding insulation structure of high-voltage and high-capacity power transformer.

2 Calculation of High-voltage Winding Loss

Taking a 240000 kVA/330 kA oil immersible power transformer as the research object, the internal magnetic density distribution of power transformer is calculated and is used to obtain the loss density of high-voltage winding. According to the loss density, the eddy-current loss in each winding is calculated from the integration along the axial direction of the winding and is 35674 W. In addition, the resistance loss of high-voltage winding is 313602 W based on $P_R=I^2R$. Thus, the high-voltage winding loss of power transformer without DC magnetic biasing is P=349276 W.

The effective value of the rated current of this used power transformer is I_n=514.34 A, so the maximum allowable DC magnetic biasing current is about 2.56 A for calculating the high-voltage winding loss under DC magnetic biasing according to the DL/T5224–2005 technical specification. By setting different magnitude of DC magnetic biasing current, the high-voltage winding loss is obtained, as shown in Table 1.

Tab. 1: High-voltage winding loss (P) under different magnitude of DC magnetic biasing current (Id)

I_d (A)	0	0.256	0.512	0.768	1.024	1.28	1.536	1.792	2.048	2.304	2.56
P (W)	349276	365965	389874	412697	438652	465624	498636	542463	586571	629852	675348

It can be learnt from Table 1 that the high-voltage winding loss is increased with the increasing magnitude of DC magnetic biasing current. This is attributed to the fact that the increasing DC magnetic biasing current accelerating the core saturation and aggravating the distortion of the winding exciting current.

3 Numerical Calculation of Fluid Field

The fluid field in the cooling circuit of power transformer plays an important role in the calculation of the whole temperature rise. According to the flow path of internal fluid and the physical structure of power transformer, the fluid field in power transformer can be divided into the oil circuit in the windings, the cooler and the oil circuit in the connecting structure, etc. By analyzing the fluid field in each physical structure, the corresponding internal flow resistance is obtained and then is lined up via fluid network to calculate the actual flow of oil circuit.

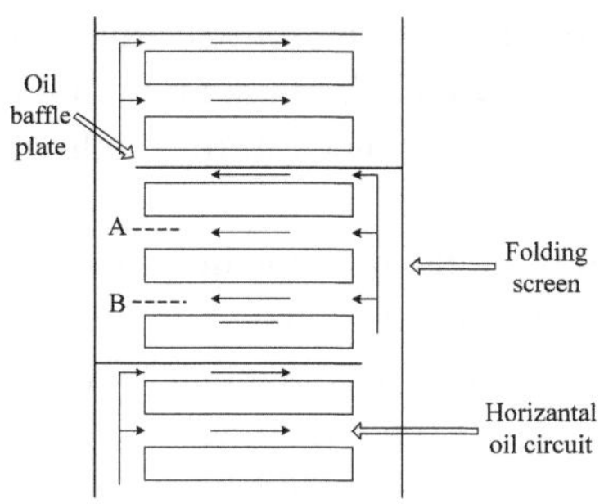

Fig. 1: Schematic of the oil circuit in the windings

The schematic of the oil circuit in the windings is shown in Fig. 1. The oil circuit consists of the vertical oil circuit between the winding and the folding screen, and the horizontal oil circuit between the winding and the insulating block. According to the thermodynamic principle [8], the heat quantity absorbed ($d\varphi$) by the oil in the AB cross section is defined as below,

$$d\varphi = q_m c_{oil} dt \tag{1}$$

Where q_m is the flow velocity of the oil, c_{oil} is the specific heat capacity of the oil and dt is the mean temperature difference of the oil in the cross section. The temperature rise of the winding oil circuit can be considered to be caused only by the winding loss. As the winding loss is not uniformly distributed in the windings and is larger in both ends of the windings, the actual temperature of the oil circuit in the windings is not linearly distributed in the vertical direction.

The heat quantity (ΔQ) escaped from the windings to the surrounding oil is described as below,

$$\Delta Q = k \Delta t \Delta A \tag{2}$$

Where Δt is the mean temperature difference of the winding against the surrounding oil, ΔA is the heat radiating area of the windings and k is the mean convective heat transfer coefficient on the windings.

As the structure of the three-phase transformer windings is exactly the same, the fluid and thermal characteristics of the whole winding oil circuit can be obtained by analyzing the oil circuit in one phase windings. In the calculation of fluid field in the windings, the overall model of the winding oil circuit is simplified according to the repeatability of winding structure. The overall model of the windings is divided into eight sections by the internal oil baffle plate and the external one. The first and final four sections are the same whereas the middle four sections are the same. The whole oil circuit model is converted into two oil circuit models, and it is calculated by taking the circumference of the surrounding block in the circle direction 1/16.

The winding integral model is divided into eight sections, the first and final four sections are the same, whereas the middle four sections are the same. The integral oil circuit model is converted into two oil circuit models, and it is calculated by taking the circumference of the surrounding block in the circle direction 1/16.

The fluid field in the high-voltage windings is then calculated by using the computational fluid dynamics. The fluid field distribution in one section high-voltage winding is shown in Fig. 2 when the entrance velocity is 0.35 m/s.

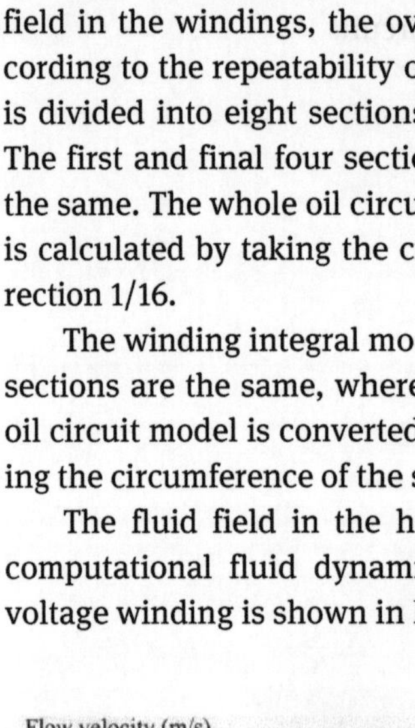

Fig. 2: Fluid field in 1/16 high-voltage winding

4 Temperature Rise of the High-voltage Windings

According to the above winding loss and fluid field in the power transformer, the temperature rise of transformer winding can then be calculated. Due to the axial symmetric of the oil circuit in the windings, only one section winding is modeled.

This model consists of the internal oil baffle plate, the winding and the external oil baffle plate. The oil circuit in the windings is divided into two segments. Moreover, as the repetitive winding along the circumference of a circle can be divided into eight sections, where each section contains 15 winding discs with a mean height of 13.75 mm high, the 1/16 winding model as shown in Fig. 3 is chosen as the analysis model of temperature distribution. On the analysis of the temperature distribution in the winding, this established winding model replaces the whole winding. The temperature distribution of the remaining 7 sections windings can be calculated according to the oil flow and the flow velocity as well as the corresponding parameters obtained from each section winding model.

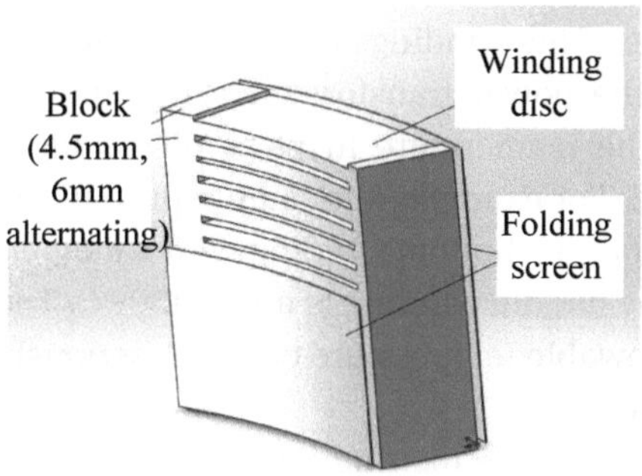

Fig. 3: Winding model

The solving conditions for the temperature distribution are set as follows. The entrance oil temperature is 298 K and is only used for calculating the convective heat transfer coefficient which is modified during the temperature distribution calculation. The winding oil velocity is 0.35 m/s, and the oil flow is 14.26 kg/s. The two sides of the model are symmetry boundary conditions, namely the geometric and physical properties on both sides of the boundary is symmetrical along the cross section. The oil baffle plate is set as an adiabatic boundary (the second boundary condition). The contact surface of the winding and the oil is set as a coupling boundary condition (the third boundary condition).

By solving the established model of temperature distribution with the above solving conditions, the 3D finite element results of single section high-voltage winding are obtained, as shown in Fig. 4, under the effect of different DC magnetic biasing current of 0, 1.28 and 2.56 A.

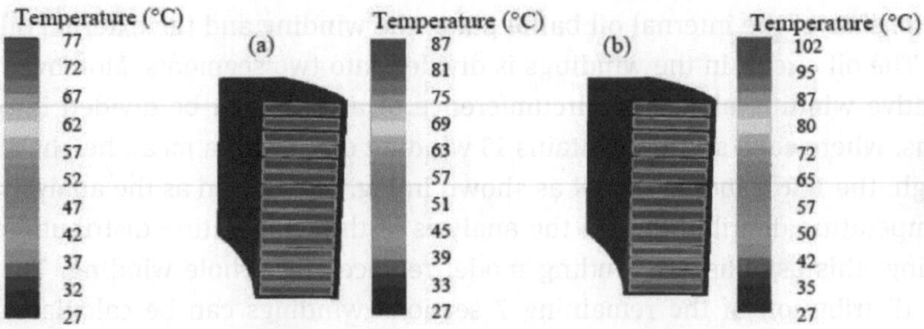

Fig. 4: Temperature distribution in one section high-voltage winding with and without DC magnetic biasing. (a) I_d=0 A; (b) I_d=1.28 A; (c) I_d=2.56 A

It can be seen from Fig. 4 that the temperature of the windings is increased with the increasing DC magnetic biasing current in the power transformer. When the DC magnetic biasing current is set as 2.56 A, the temperature in parts of the high-voltage winding approaches or exceeds 99 °C. For example, the temperature in the top winding is 102 °C which is larger than that of the bottom winding with a temperature of 99 °C. Whereas the temperature in the middle winding is about 100 °C. The above temperature exceeds the maximum allowable temperature of 98 °C, seriously threatening the safe operation of power transformer.

According to the above calculation of temperature distribution under DC magnetic biasing current, the internal heat exchange characteristics of the transformer can be known. The temperature rise of the whole high-voltage winding, as shown in Fig. 5, can be obtained based on the temperature distribution of single winding and the fluid network of power transformer.

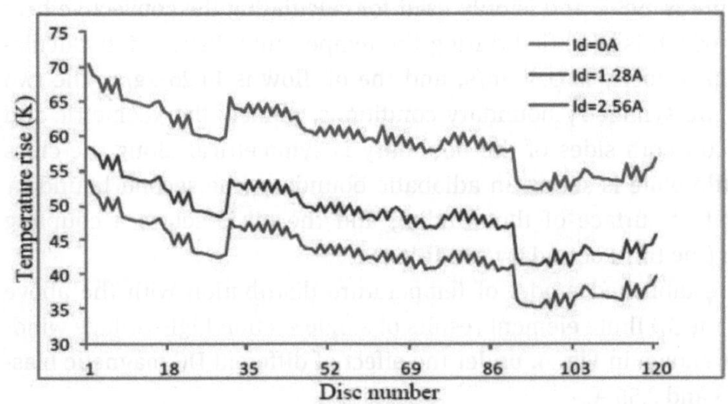

Fig. 5: Temperature rise distribution in the whole high-voltage winding

As the temperature of cooling oil can be consider as constant, the temperature rise of the windings is increased with the increasing DC magnetic biasing current, as shown in Fig. 5. When the applied DC magnetic biasing current is 2.56 A, the temperature rise in parts of the windings has exceeded 71 K, which is larger than the allowable temperature limitation of 65 K proposed in the national standard. Therefore, the local overheating phenomenon will occur in the power transformer under DC magnetic biasing and it is necessary to optimize the insulating structure design of the windings or to take some measures to suppress the DC magnetic biasing phenomenon.

5 Conclusion

The winding loss and temperature distribution of a 240000 kVA/330 kA oil immersible power transformer are calculated by using the finite element method and the computational fluid dynamics. The main conclusions are summarized as below.

1. With increasing the applied DC magnetic biasing current, the winding loss of high-voltage winding is increased. When the applied DC magnetic biasing current is set as the maximum allowable value, the transformer core appears the saturation or over-saturation.
2. The DC magnetic biasing phenomenon will accelerate the core saturation and aggravate the leakage magnetic field, which results in the increasing temperature rise in high-voltage winding. When the applied current is set as the maximum allowable of 2.56 A, the temperature rise in parts of the high-voltage windings have exceeds the allowable temperature limitation proposed in the national standard. This leads to the local over-heating in the windings.
3. The calculation method proposed in this paper can provide a theoretical basis for judging the temperature rise in the power transformer under DC magnetic biasing.

References

[1] X. Yang, F. Wang, R. Duan, and H. Zheng, Power system protection and control, 42, 14, 60–66, 2014.
[2] P. R. Price, IEEE Trans Power Delivery, 7, 4, 1002–1008, 2002.
[3] D. Andreas, M. John and Z. Thomas, IEEE Trans Power Delivery, 24, 3, 1284–1194, 2009.
[4] Z. Zhao, F. Liu and Z. Cheng, IEEE Trans Appl Superconductivity, 20, 3, 1131–1134, 2010.
[5] X. Li and X. Wen, Proceedings of the CSEE, 30, 1, 127–132, 2010.
[6] H. Li, X. Cui, T. Liu, Z. Cheng, D. Liu, and C. Jiao, Proceedings of the CSEE, 29, 25, 119–125, 2009.
[7] X. Dong, Z. Liu, D. Ning, X. Gan, Y. Gao, R. Chen, and X. Mou, Power system protection and control, 40, 23, 12–16, 2012.

[8] F. P. Incropera and D. P. Dewitt, Fundamentals of Heat and Mass Transfer, 1990.

Yong-Ming Ge, Long Shen*, Feng Gao and Liang Liu

Kinematic Analysis of 6-DOF Stewart Platform

Abstract: In this paper, we get the orientation-singularity at a given position, after primary researching singularity for 6-UPS Stewart mechanism. And the effect of the rod length constraint, hinge constraint, link interference and the scale of the radius of circumscribed circle on singularity-free orientation workspace for the mechanism. The minimum inscribed sphere method is also used to express the singularity-free orientation workspace by using the minimum inscribed sphere radius algorithm, which provides the theoretical basis for Stewart mechanism motion analysis and trajectory planning.

Keywords: Stewart platform, Singularity, Sigularity-free workspace

1 Introduction

Compared with the characteristics of serial mechanism of low rigidity, low bearing capacity and difficult to reverse solution, the parallel mechanism has advantages such as high rigidity, high load capacity, and simple anti-solution and high positioning accuracy. However, the singular trajectory of the parallel mechanism is very complex, and its workspace is small, which makes the study of the singularity and workspace of the parallel mechanism very important.

The basic idea of trajectory planning is to plan a path which will not pass the singularity from the beginning to the end. So the study of singularity is very important, many domestic and foreign scholars carry out in-depth and detailed study on this aspect [1–4].

Such as the radius ratio of the move and base platform of the parallel mechanism, the limit length of the parallel link, the limit diameter of the parallel link, the limit swing angle of the moving platform Hooke hinge and the limit swing angle of the fixed platform ball hinge, etc., have a great influence on the workspace of the parallel mechanism. However, in previous studies, only the impact factors were proposed or considered separately, without taking the synthesis of the application process into account. In this paper, 6-UPS Stewart mechanism is taken as an example. The spatial influence factors of the parallel mechanism are calibrated in turn,

Yong-Ming Ge, Long Shen, Liang Liu, School of Mechatronics Engineering and Automation, Shanghai University, P. R. China, E-mail: 673385822@qq.com
Long Shen, Feng Gao, State Key Lab. of Mechanical System & Vibrartion, Shanghai Jiao Tong University, P. R. China, E-mail: shenlong_sjtu@163.com

DOI 10.1515/9783110540048-039,

which provides an important theoretical basis for the selection and application of mechanism parameters in engineering application.

In order to facilitate the study of trajectory planning, the study of singularity-free orientation workspace is great importance, and there are a few people to research on the singularity-free workspace in this purpose. In this paper, we use the minimum inscribed sphere method to determine the rotation range of the model in the fixed position, which facilitates the study of trajectory planning.

The research on the characteristics of the parallel mechanism has laid a solid foundation for the parallel structure of the antenna pedestal structure. In particular, the large range of orientation space for the parallel mechanism is special. And the antenna can be a flexible radiation to 360 ° hemisphere space, or even super-hemisphere space. At the same time, the parallel antenna pedestal can make up the characteristics for the classic pitch-azimuth antenna which cannot continuously track the blind cone area at the top of the antenna position.

2 The Establishment of Mathematical Model

The 6-DOF parallel mechanism consists of move platform, base platform and 6 parallel poles. The parallel poles and the move platform are connected by a spherical hinge and the base platform connects with parallel poles through Hooke hinge. The move platform and the base platform are regular hexagon. β and β_0 are the structural angles corresponding to the long side of the base platform and the short side of the move platform respectively. R and r are the circumcircle radii of the base and move platform respectively.

The move platform model and the base platform model are respectively established in the coordinate systems P-XYZ and O-XYZ, and the platforms are parallel to each other. The hinge point of the move platform is denoted as b_i in the move coordinate system and the hinge point of the base platform is denoted as B_i in the coordinate system, then

$$b_i' = [X_{bi}', Y_{bi}', 0] \quad (i=1, 2, ..., 6) \tag{1}$$

$$B_i = [X_{Bi}, Y_{Bi}, 0] \quad (i=1, 2, ..., 6) \tag{2}$$

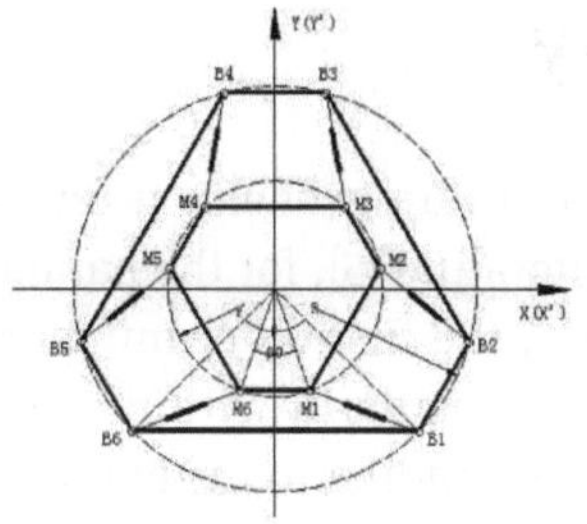

Fig. 1: The structure of the move platform and the base platform

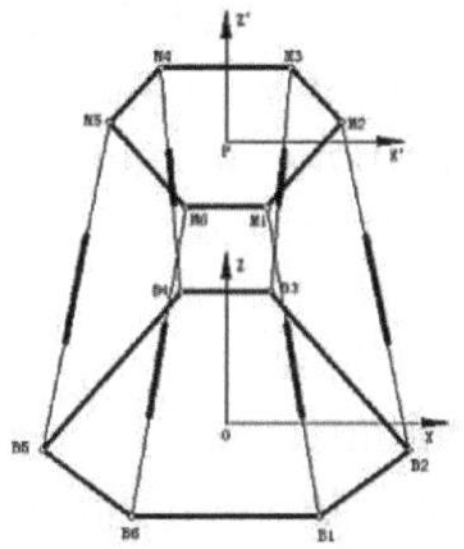

Fig. 2: The structure of Stewart

The coordinate of the geometric center P of the move platform in the base coordinate system O-XYZ is:

$$P=[X, Y, Z]. \tag{3}$$

The parallel pole vector $\vec{l_i}$ and pole length $l_i(i=1, 2, ..., 6)$ in the coordinate system can be expressed as:

$$\vec{l_i}=b_i-B_i=\begin{bmatrix} l_{iX} \\ l_{iY} \\ l_{iZ} \end{bmatrix} \tag{4}$$

$$l_i=\sqrt{l_{iX}^2+l_{iY}^2+l_{iZ}^2}(i=1, 2, ..., 6) \tag{5}$$

3 Research on Singularity Trajectory

Singularity is the inherent nature of the mechanism; it has a variety of effects on the performance of the mechanism. We can know from the literature [5], for the parallel mechanism, when the determinant of Jacobian matrix (ie, the first-order influence coefficient matrix) is zero; the parallel mechanism is in a singularity. At this time, the degree of freedom of the parallel mechanism increases, leading to instability of the mechanism, so that the parallel mechanism out of control. Therefore, the research on singularity trajectory of parallel mechanism is particularly important.

Based on the Jacobian of the parallel mechanism, Gosselin et al. classify singularities into three types [6–7]: inverse kinematics singularity, forward kinematic singularity, and structural singularity. For the first kind of singularity, i.e. inverse kinematics singularity, the Jacobian matrix J of the parallel mechanism is infinite, i.e. det(J)=∞. For the third kind of structure singularity, since the Stewart platforms are two similar hexagon, i.e. $\beta+\beta_0=120°$, and the corresponding point is connected, the structure is a singular structure. The mechanism is singular regardless of the configuration.

It is not difficult to obtain the coordinates of the hinge point in the base coordinate system, which is denoted as $b_i(i=1, 2, ..., 6)$, when the coordinates of the move platform hinge point are known in the move coordinate system. The coordinates of the hinge points of the move and base platforms in the base coordinate system O-XYZ are brought into the Jacobian matrix$[J^T]$, can be:

$$J^T=\begin{bmatrix} \frac{b_1-B_1}{|b_1-B_1|} & \frac{b_2-B_2}{|b_2-B_2|} & \cdots & \frac{b_6-B_6}{|b_6-B_6|} \\ \frac{B_1\times b_1}{|b_1-B_1|} & \frac{B_2\times b_2}{|b_2-B_2|} & \cdots & \frac{B_6\times b_6}{|b_6-B_6|} \end{bmatrix}_{6\times 6}. \tag{6}$$

When the Jacobian matrix is expanded and simplified, the singularity equation can be written as

$$f(a, b)=0. \tag{7}$$

Where, a=[X, Y, Z]means the move platform position, $b=[\varphi, \gamma, \theta]$ means the orientation of the move platform.

Eq. 7 is a function polynomial for the Euler angles φ, γ and θ at fixed tions(X, Y, and Z) in the process of studying the orientation singularity. Because the singularity polynomial is important for the study of trajectory planning, the polynomial is of great significance in the research of singularity of parallel mechanism. However, due to the periodicity of the trigonometric function, Eq. 7 will be expressed as a polynomial with the multiple power of the trigonometric function, so it is difficult to solve the numerical solution. The triangular universal formula can be easily transformed into a trigonometric function contains only a monotone variable function, so here, we may wish to introduce the triangle universal formula

$$\sin\varphi=\frac{2q_1}{1+q_1^2}. \tag{8}$$

$$\cos\varphi=\frac{1-q_1^2}{1+q_1^2}. \tag{9}$$

$$\sin\gamma=\frac{2q_2}{1+q_2^2}. \tag{10}$$

$$\cos\gamma=\frac{1-q_2^2}{1+q_2^2}. \tag{11}$$

$$\sin\theta=\frac{2q_3}{1+q_3^2}. \tag{12}$$

$$\cos\theta=\frac{1-q_3^2}{1+q_3^2}. \tag{13}$$

Where, $q_1=\tan\frac{\varphi}{2}, q_2=\tan\frac{\gamma}{2}, q_3=\tan\frac{\theta}{2}$.

By this method, the singularity of the 6-UPS Stewart parallel mechanism can be easily obtained. As shown in Fig. 3, Fig. 4 are the singularity trajectories for φ, γ, θ and q_1, q_2, q_3 at the fixed position (X, Y, Z) = (0, 0, 800cm) respectively. At this time, the structure parameters of the structure is R=300cm, r=240cm, $\beta=\frac{\pi}{2}$, $\beta_0=\frac{\pi}{6}$.

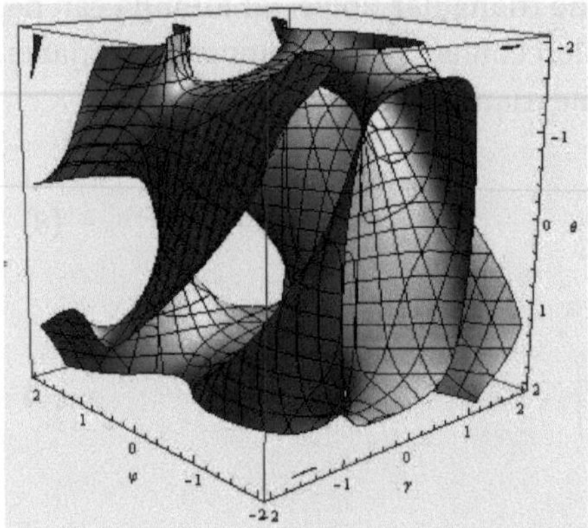

Fig. 3: The singularity trajectory of the mechanism in (0, 0, 800cm) about "φ, γ, θ" n

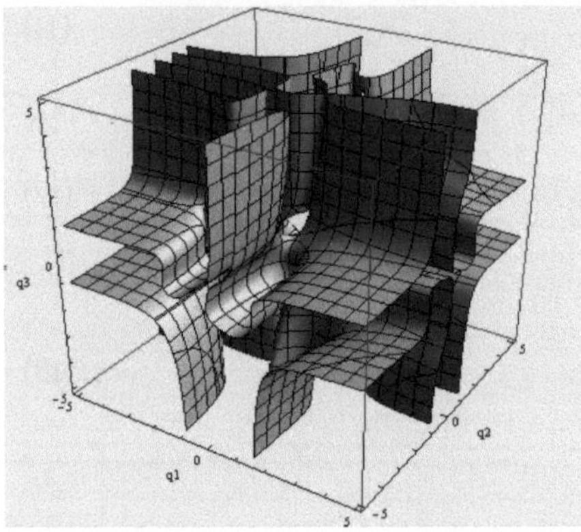

Fig. 4: The singularity trajectory of the mechanism in (0, 0, 800cm) about q_1, q_2, q_3

For any point (X, Y, Z) within the workspace of Stewart mechanism, a singularity-free orientation workspace can be found in the three-dimensional space about the orientation (φ, γ, θ) near the origin. Therefore, the obtaintion of the orientation singularity trajectory is important for the solution of the singularity-free orientation workspace.

4 Research on Singularity-Free Orientation Workspace

4.1 Effect of Structural Parameters on Workspace

4.1.1 Pole Length Constraint

Pose control is achieved by changing the length of the pole, so the pole length constraint is an important constraint for the workspace. Its expression is

$$l_{min} \le |l_i| \le l_{max},\ i=1, 2, \ldots, 6. \tag{14}$$

Where, l_{min} is the shortest pole length, l_{max} is the longest pole length, l_i is the i-th link vector.

4.1.2 Hinge Constraint

In this study, the parallel poles were connected to the upper platform by means of spherical hinges and connected to the lower platform via Hooke hinges, but the swing angle range of the spherical hinge or the Hooke hinge was limited. Therefore, it is necessary to study the swing angle range of the hinge restraint during the movement of the move platform. Suppose the unit vector perpendicular to the move platform is $U_{bi}=[0 \quad 0 \quad 1]^T$, and the unit vector perpendicular to the fixed platform is $U_{Bi}=[0 \quad 0 \quad 1]^T$. So the swing angle of the spherical hinge on the move platform is θ_{bi}

$$\theta_{bi}=\arccos\frac{l_i \bullet U_{bi}}{|l_i|}. \tag{15}$$

The swing angle of the spherical hinge of the base platform is

$$\theta_{Bi}=\arccos\frac{l_i \bullet U_{Bi}}{|l_i|}. \tag{16}$$

While the spherical hinge condition is expressed as

$$0 \le \theta_{bi} \le \theta_{bmax},\ i=1, 2, \ldots, 6. \tag{17}$$

$$0 \le \theta_{Bi} \le \theta_{Bmax},\ i=1, 2, \ldots, 6. \tag{18}$$

Where, θ_{bmax} denotes the maximum swing angle of the spherical hinges on the move platform, and θ_{Bmax} denotes the maximum swing angle of the Hooke hinges on the base platform.

4.1.3 Interference of Connecting Poles

When the parallel pole is a cylinder whose cross-section diameter is D, the constraint condition for avoiding the interference of the parallel pole is

$$D_i \geq D, \; i=1, 2, \ldots, 6 \tag{19}$$

Where, D_i is the shortest distance between the center lines of two adjacent poles.

4.1.4 The Ratio of Circumscribed Circle Radius of Move Platform and Base Platform

The results show that the radius of circumscribed circle of move platform and base platform has a significant influence on the size of working space, and the radius ratio is expressed as

$$u=\frac{r}{R}. \tag{20}$$

Where, r is the circumscribed circle radius of the move platform and R is the circumscribed circle radius of the base platform.

4.2 Representing Singularity-free Orientation Workspace by Minimum Inscribed Sphere Method

The orientation workspace of 6-UPS Stewart parallel mechanism represents the orientation that the mechanism can reach during the course of motion, but because of the search of the orientation workspace, the effect of the mechanism singularity is not taken into account. Therefore, the mechanism may be singular in the orientation workspace. It is shown that the robot should avoid the singularity and should work in the region away from the singularity.

Considering that the parallel mechanism has a singularity-free orientation workspace near the origin (0°, 0°, 0°), therefore, a graph which is well represented by the mathematical coordinates has a significant effect on the representation of the singularity-free orientation workspace.

And the minimum inscribed sphere method, the minimum inscribed sphere radius in the singularity orientation workspace is obtained by spherical coordinate search. The inscribed sphere is shown in Fig. 5. The cross-section at φ = 0 is shown in Fig 6.

In the space represented by the minimum inscribed sphere, the singularity-free orientation workspace of the parallel mechanism can ensure that the mechanism does not occur any motion interference and singularity phenomenon in the whole workspace.

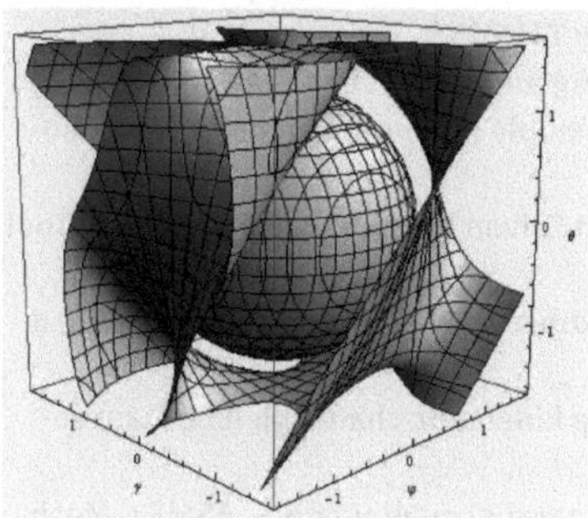

Fig. 5: The minimum inscribed sphere n

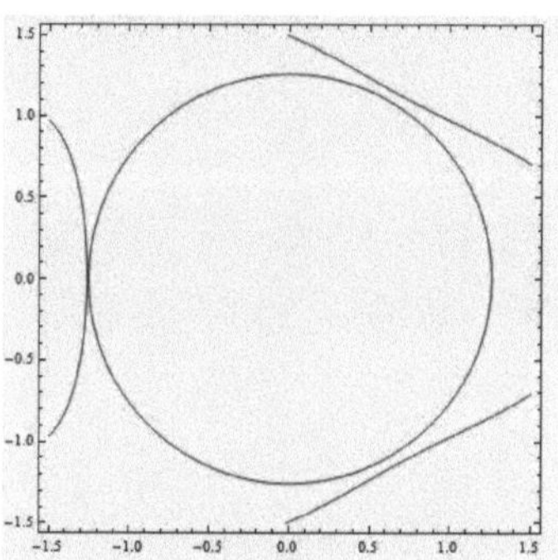

Fig. 6: The cross-section at φ = 0

5 Conclusion

In this paper, the orientation singularity trajectory of 6-UPS Stewart mechanism is analyzed, and the effects of structural parameters such as pole length constraint, hinge constraint, pole interference and circumscribed circle radius ratio on the singularity-free orientation workspace are studied. And the singularity orientation workspace at a given position is expressed by the minimum inscribed sphere method, which provides a theoretical basis for mechanism trajectory planning and engineering application for the parallel mechanism on antenna pedestal.

References

[1] Nguyen C C, Antrazi S S, Zhou Z L, et al. Experimental study of motion control and trajectory planning for a Stewart Platform robot manipulator[C]// IEEE International Conference on Robotics and Automation, 1991. Proceedings. IEEE, 1991: 1873 - 1878.

[2] Dasgupta B, Mruthyunjaya T S. Singularity-free path planning for the Stewart platform manipulato r [J]. Mechanism & Machine Theory, 1998, 33(6): 711-725.

[3] Nguyen C C, Antrazi S S, Park J Y, et al. Trajectory planning and control of a Stewart platform-based end-effector with passive compliance for part assembly [J]. Journal of Intelligent & Robotic Systems, 1992, 6(2): 263-281.

[4] Zhao H, Zhang S. Trajectory Planning of Stewart Platform Driven by Hydraulic [J]. Machine Tool & Hydraulics, 2007.

[5] Hunt K H. Structural Kinematics of In-Parallel-Actuated Robot-Arms [J]. Journal of Mechanical Design, 1983, 105(4): 705-712.

[6] Gosselin C, Angeles J. Singularity analysis of closed-loop kinematic chains [J]. IEEE Transactions on Robotics & Automation, 1990, 6(3): 281-290.

[7] Gosselin C M. Determination of the workspace of 6-DOF parallel manipulators, ASME J. Mech. Design, Vol. 112, pp. 331–336 [J]. Journal of Mechanical Design, 1990, 112(3): 331–336.

Wei Cao, Jian Xu, Juan-Juan San and Ruo-Nan Sun*

Application of Agricultural Sensors Overview

Abstract: The application of sensor technology in agriculture has promoted the rapid development of agricultural modernization, and the new principles, new materials and new technologies of sensors have been emerging. This paper introduces the classification of sensors, the application of sensors in agricultural production, and analyzes the development trend of sensor technology in the future. Wireless sensor, bionic sensor, optical fiber sensor and other new sensors will become the mainstream of the future development of agricultural sensor technology.

Keywords: Sensors, Agriculture, Application

1 Introduction

China is a large agricultural country, China's agricultural development has always affected the national economic development and national security and stability of the pulse. Agricultural modernization is the foundation and support of national modernization. As the two key technologies of precision agriculture, information acquisition and processing technology and intelligent agricultural equipment technology cannot be separated from the development and application of sensor technology [1] Sensor is the source of all information systems, in the measurement, testing, testing, monitoring, and health management system, the sensor is the key component of the system performance. With the rapid development of sensor technology, new sensors and detection methods are increasingly used in agricultural production, and gradually become the new trend of future sensor application in agricultural field, also played a positive role to promote the process of agricultural modernization.

Wei Cao, Beijing City, Changping District Huilongguan Town Road No.7. Email: 979008969@qq.com
Jian Xu, Beijing City, Changping District Huilongguan Town Road No.7. Email: bjnxy@126.com
Juan-Juan San, Beijing City, Changping District Huilongguan Town Road No.7. Email: loveyfb081115@qq.com
Ruo-Nan Sun, Beijing City, Changping District Huilongguan Town Road No.7. Email: 597369409@qq.com.

DOI 10.1515/9783110540048-040,

2 The Types of Sensors Used in Modern Agriculture

2.1 Temperature Sensor

Many kinds of temperature sensor, thermocouple sensor, using the electro thermal effect using the heat resistance effect of the metal thermal resistance sensor and semiconductor thermistor sensor, temperature sensor, integrated semiconductor PN junction temperature sensor and developed in recent years, optical fiber temperature sensor, the application of these sensors, temperature measurement can be implemented timely.

2.2 Humidity Sensor

The humidity sensor can sense atmospheric humidity and convert it into an appropriate electrical signal. According to the principle of the sensor, the commonly used humidity sensor is divided into sintered semiconductor ceramic humidity sensor, porous oxide humidity sensor, junction and MOS humidity sensor, lithium chloride humidity sensor etc.

2.3 Gas Sensor

The gas sensor can be detected specific components in the gas, and transformed into the corresponding electrical signal output, gas sensors can be divided into semiconductor gas sensors, solid electrolyte gas sensors and other types, which is the most commonly used semiconductor gas sensor.

2.4 PH Sensor

PH sensor is based on the electrochemical measurement of the point of work, can be used to measure the pH of the solution. It consists of a measuring electrode and a reference electrode. The measuring electrode generally adopts glass electrode, reference electrode and a calomel electrode of silver chloride electrode two.

2.5 Pressure Sensor

The sensor can measure the pressure of the lot, according to the resistance strain sensor, resistance strain effect made according to the piezoresistive pressure sensor, piezoresistive effect is made based on the piezoelectric effect made of piezoelectric sensor and voltage type capacitive sensor can measure the pressure medium.

3 The Application of New Sensor Technology in Agriculture

3.1 Wireless Sensor Network Technology

Wireless sensor network is composed of sensor technology, MEMS technology, wireless communication technology, embedded computing technology and distributed information technology, has attracted much attention in the world today, interdisciplinary research hotspot because of its wide application prospects. Wireless sensor networks will be the logical information world and the physical world together, will change the interaction between people and nature, with limited resources, dynamic network, and data centric and high density of large number of nodes.

The structure of wireless sensor network is shown in Figure 1, the network is composed of sensor node, sink node, field data collection and processing decision part and decentralized user receiving device. The sensor nodes are randomly deployed in the sensing area, each node can collect data, and using self-organizing multi hop routing in wireless mode to transmit data to the sink node, the sink node can send information to each node, the sink node directly with the Internet PC or wireless mode is to achieve communication between the sensor nodes and task management through the Internet or wireless mode.

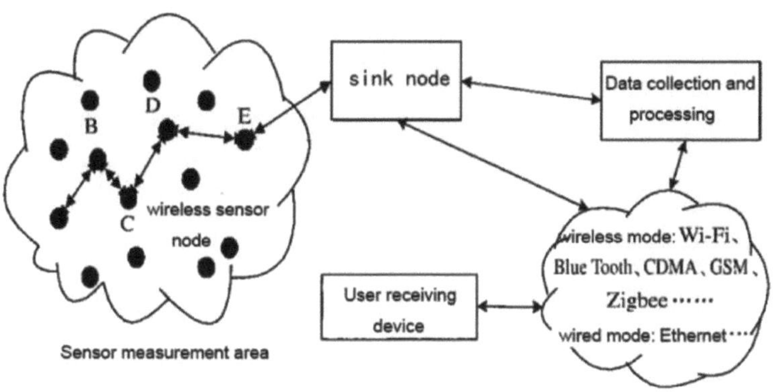

Fig. 1: Structure of wireless sensor network

3.2 Bionic Sensor Technology

Biomimetic sensor technology is a new type of sensor was used to detect the new principle; it adopts the immobilized cells, enzymes or other bioactive substances and is matched with the transducer sensor or sensor array. Bionic sensor is a new kind of information technology, which is developed by the mutual penetration of biomedicine, electronics and engineering.

As a new method of simulating biological senses, the application of bionic sensor technology in agriculture has been reported. In the process of rapid detection and identification of fruits, drinks, wine and meat products, the bionic electronic nose can be used to classify and recognize the odor of different substances effectively. The bionic electronic nose usually consists of several gas sensors. The sensor array, combined detection and identification of volatile gas gas, each sensor in the array of certain types of sensitive components, to identify different components in the gas and give the overall evaluation of the sensor array.

3.3 Optical Fiber Sensor Technology

Optical fiber sensors are mainly sensor type and light transmission type two kinds, two kinds of sensors can be divided into the sensing principle of light intensity modulation, phase modulation, polarization modulation and wavelength modulation in different forms, which constitute different sensors. Optical fiber sensor has high sensitivity, light weight and small volume, multi-use, impact on small, medium resistance to electromagnetic interference and corrosion and intrinsically safe, easy networking and other features, has been rapidly developed in agriculture, aerospace, medical and other fields. Table 1 the principle and main methods of measuring the parameters of optical fiber sensor.

Tab. 1: The principle and main methods of measuring parameters of optical fiber sensor

Measured physical quantity	Measuring method and optical principle		Measured physical quantity	Measuring method and optical principle	
Electric current	Phase	Magnetostriction	Displacement Speed	Phase	Interference phenonmenon
Voltage Electrical		Electrostrictive	Geometric parameters of		Sagnac effect
Parameters	Polariztion	Faraday effect	motion	Light intensity	Reflex, Transmission, Refaction
		Pocket effect		Frequency	Doppler effect
Temperature	Phase	Interference phenonmenon	Vibration Pressure	Phase	Interference phenonmenon
	Polariztion	Birefringence change	Mechanical parameters	Frequency	Doppler effect
	Light intensity	Infrared radiation		Light intensity	Micro bend and scattering loss
		Fluorescent radiation	Radiation	Light intensity	Absorption loss

The typical application of fiber optic sensors in agriculture in the irrigation and water conservancy facilities in the cracks in the optical fiber grating sensor to detect the channel, but because of the cost problem, the more is used in dam safety inspection. In the greenhouse planting and breeding, agricultural products storage, using optical fiber temperature sensor to obtain real-time information with ambient temperature, concentration of optical fiber gas sensor measurement of gas, in order to ensure the environmental conditions required to achieve the best condition.

4 The Typical Application of Sensors in Agriculture

4.1 Agricultural Picking Robot

Agricultural picking robot is the result of the rapid development of robot technology, and is an important symbol of the development of agriculture to automation and intelligence [2]. At present, the developed countries in the field of agricultural harvesting robot in the leading position, has developed tomatoes, cucumbers, grapes, citrus fruits and vegetables harvest robot. Agricultural picking robots working in unstructured, unknown and uncertain environment, the job object is randomly distributed, determines the induction ability of agricultural harvesting robot must have the intelligence to adapt to the complex working environment. The application of the sensor on the agricultural picking robot has a direct impact on the perception ability of the agricultural picking robot to the environment. It is necessary to select a simple, stable and easy to implement sensor to improve the operation ability of different agricultural picking robot. Agricultural picking robot mainly uses the visual sensor, position sensor, force sensor, obstacle avoidance sensor to complete the task of picking robot.

4.2 Water Saving Irrigation Application

Wireless sensor network node has the advantages of simple control, battery powered by power handling DC electromagnetic valve can control the different small and medium power (electric hydraulic solenoid valve, relief valve, pressure regulating valve, safety valve and flow control valve), dormancy state and the nodes of the network to work time if more than one year, powered by solar panels, energy does not need to think too much [3]. Due to the characteristics of sensor network routing, interoperability information delivery, ad hoc network and network communication, time synchronization, so that irrigation area, the number of nodes is not restricted, can flexibly turn group, with nodes with soil and plant and meteorological measurement acquisition device, communication gateway function and Internet RS and GPS technology according to the dynamic management of irrigation district information acquisition and analysis technology, the crop water requirement information acquisition and precision irrigation control technology and expert system technology to build efficient, low energy consumption, low cost, multi-function agricultural water-saving irrigation platform. In the greenhouse, garden green space, the central highway isolation belt, farmland well with irrigation region, realize agricultural and ecological water-saving technology quantification, standardization, modeling, integration, and promote rapid and healthy development of water saving agriculture.

4.3 Application of Agricultural Mechanization

Agricultural machinery is widely used in modern agriculture. Agricultural machinery cannot be separated from the sensor technology in the process of design and test. If in plow design, should ensure that the operation process by the resistance to a minimum [4], in order to reduce the work of agricultural machinery plough wear and power loss, this can strain sensor measurement of plow by working in different soil environment of the resistance, designers can design according to the measured data of sensor the scientific and reasonable to carry out the plow surface; in modern agriculture, to ensure agricultural irrigation, and to avoid the waste of water resources, the temperature and humidity sensor of temperature and humidity on farmland for automatic measurement, automatic irrigation system can be adjusted automatically according to the measuring signal sensor of irrigation water use; in a large planter or the traction power harvester, a temperature sensor to monitor the engine and fuel temperature, oil pressure, pressure sensor composed of piezoelectric sensor acceleration sensor The utility model is used for measuring the stability of the machine operation; in order to ensure the crop emergence rate, the photoelectric sensor installed on a plurality of rows of precision seeding is used to detect the sowing quantity and the fault, and the replay and the seeding are reduced [5]. In order to reduce the loss of harvest, the detection device of grain entrainment loss is installed on the combine harvester. In order to realize the precision of agriculture, the grain flow sensor is installed on the combine harvester, which can effectively measure the grain yield and improve the effective utilization rate of agricultural input.

5 Conclusion

The sensor technology is one of the 3 basic techniques of modern information technology, technology application covering agricultural production from crop planting, growth, to the application of the whole process cannot do without sensor analysis harvesting, processing and testing, covering almost all the scope of agricultural engineering, and effectively support the intelligent agriculture, precision agriculture and green agriculture the technology system. The application of sensors in agricultural machinery not only greatly improves the work efficiency, but also strengthens the automation and intelligence of the work.

1 In the precision seeder installed on the sensor can be at any time on the card species and the phenomenon of lack of detection and early warning, save manpower, and does not delay the operation time.
2 A new type of intelligent omni-directional vision sensor is used in the automatic navigation and positioning system of agricultural machinery.
3 It is helpful to change the current situation of the structural imbalance of agricultural labor force and promote the development of agricultural intensification. Combined with GPS, GIS and RS technology, sensor technology to achieve high yield, stability, energy saving and environmental protection of agricultural sustainable development, from extensive traditional agriculture to modern agriculture fine transition, building a resource-saving and ring.

References

[1] Wei Xinhua, Zhang Jinmin, Dan Zhimin, et al. Signal processing method of impact-based grain flow sensor for predicted yield [J]. Transactions of the Chinese Society of Agricultural Engineering, 2014, 30 (15): 222–228

[2] Wang Ling, Ji Changying. Technical analysis and expectation for cotton harvesting based on agricultural robot [J]. Cotton Science, 2006, 18 (2): 124–128.

[3] Xu Sai, Zhou Zhiyan, Luo Xiwen.Classification and recognition of hybrid and inbred rough rice based in bionic electronic nose [J]. Transactions of the Chinese Society of Agricultural Engineering, 2014, 30 (9): 133–139.

[4] Sun Shenghe. Development trend of modern sensor [J]. Journal of Electronic Measurement and Instrument, 2009, 23 (1): 1–10.

[5] Xiao Kehui, Xiao Deqin, Luo Xiwen. Smart water-saving irrigation system in precision agriculture based on wireless sensor network [J]. Transactions of the Chinese Society of Agricultural Engineering, 2010, 26 (11): 170–175.

Zhang-Huang Ye, Yi-Ke Li* and Xuan Yang

An Analysis on Chinese Graphite Resource and Its Development Suggestions

Abstract: Graphite has been listed as a strategic mineral resource owing to its superior physical and chemical properties and wide application. China is rich in natural graphite resource, which has the proved reserve of 55 million tons, accounting for 42% of the global. The output of graphite product made in China rank the first in the world, as well as the largest exporter. However, the structure of domestic consumption is unreasonable, which is oriented by low-end consumption. China's graphite mining industry is backward in terms of graphite processing technology. Since 1990s, the Chinese graphite industry has experienced the change from prosperity to recession due to the excessive product stock and extreme price disorder. The main reason for the violent change is the rapid expansion of production capacity. It is urgent for us to protect graphite as strategic resources and make scientific development planning.

Keywords: graphite; strategic mineral; resource; suggestions

1 Introduction

Graphite owns many unique physical and chemical properties, including high thermostability, high strength, and good stability. With the development of science and technology, graphite is widely used in all walks of life of modern society as the basic material of high conductivity and high lubrication. Many countries have recognized the strategic value of graphite. For example, the United States and the European Union have listed it as an important strategic mineral resource. China has also realized the importance of graphite and defined it as a strategic mineral product [1].

Zhang-Huang Ye, Jiangxi Science and Technology Normal University. No. 605, Fenglin Road, Nanchang, Jiangxi. Email: chuckverna@sina.com
Yi-Ke Li, Institute of Mineral Resources, CAGS. No. 26, Baiwanzhuang Road, Xicheng district, Beijing. Email: like430@cags.ac.cn
Xuan Yang, Institute of Mineral Resources, CAGS. No. 26, Baiwanzhuang Road, Xicheng district, Beijing. Email: 1959634940@qq.com

DOI 10.1515/9783110540048-041,

2 Abundant Graphite Reserve in China

By 2012, China has ascertained 118 crystalline graphite mining areas and 31 aphanitic graphite mining areas. According to the statistics from the U.S. Geological Survey, the world's natural graphite reserves are 130 million tons. Among them, the reserve in China have reached 55 million tons accounting for 42 % in the world. Together with Brazil, the two countries account for 87 % of the world (Fig. 1). Global distribution of graphite is highly uneven.

The graphite resource in China has wide distribution, and 20 provinces in China have graphite deposits. Heilongjiang province has the biggest reserve, 64 % of the total, followed by Shandong and Sichuan Province. Abundant graphite reserve in China has established a solid foundation for the development of national economy and defence industry. Nevertheless, the reserve of natural graphite in China shows a decreasing trend in recent years, both crystalline graphite and aphanitic graphite, according to statistics of China Non-Metallic Minerals Industry Association.

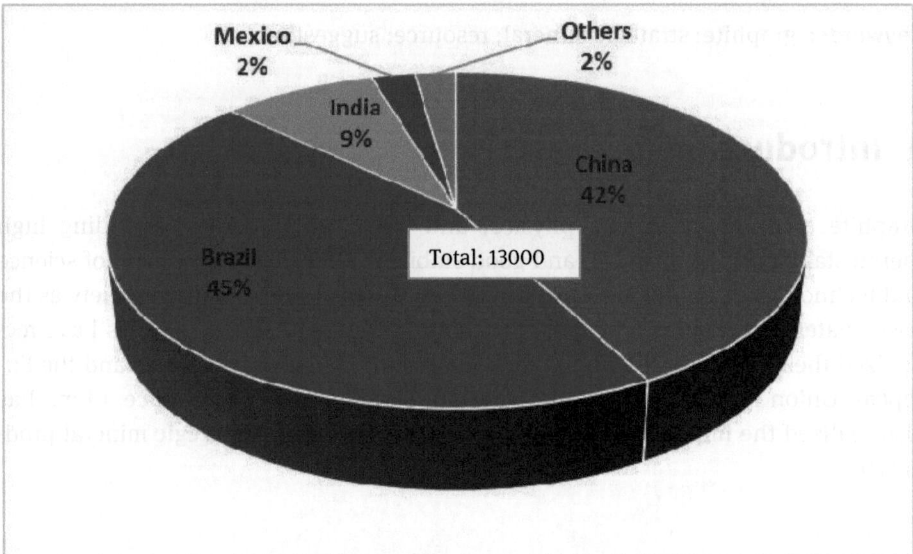

Fig. 1: Global graphite reserve distribution in the year of 2013 (unit: 10,000 tons) Data source: USGS[2]

3 Graphite Production in China

China's crystalline graphite production was basically showing a steady upward trend, from 220,000 tons in the year of 2000 to 700,000 tons in 2011. However, there was a drop in the year of 2009 and 2012. The production decreased to 480,000 tons in 2009 because of the global financial crisis, compared with the 650,000 tons in 2008. The graphite export decreased in 2012 due to the political and economic environment. Besides, the domestic demand in steel industry declined too. Therefore the crystalline graphite production fell to 470,000 tons in 2012 from 700,000 tons in 2011. Since 2006, the production of aphanitic graphite has always been declining (Fig. 2).

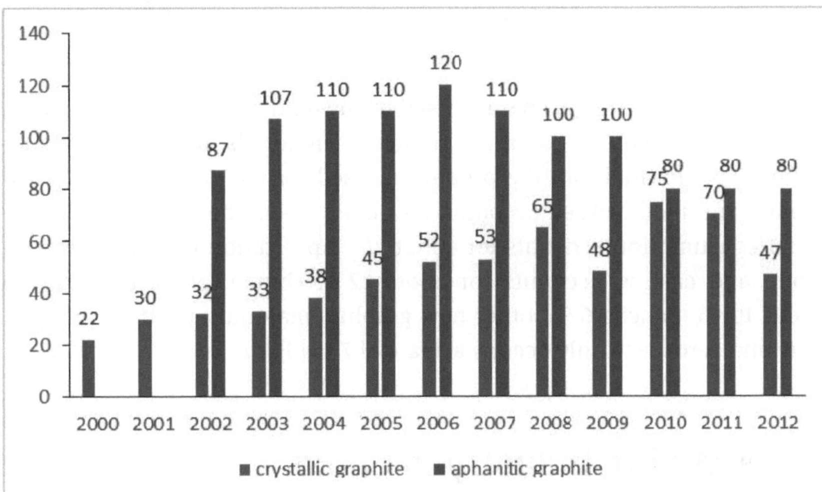

Fig. 2: The graphite production in China in the year of 2000-2012(Unit: 10,000 tons) Data sources: China Non-Metallic Minerals Industry Association

4 Graphite Consumption in China

The consumption of natural graphite in China fluctuates slightly. Its consumption amount was from 602,700 tons in 2000 to 433,900 tons in 2002, from 433,900 tons in 2002 to 703,400 tons in 2006 then declined to 669,400 tons (Fig. 3).

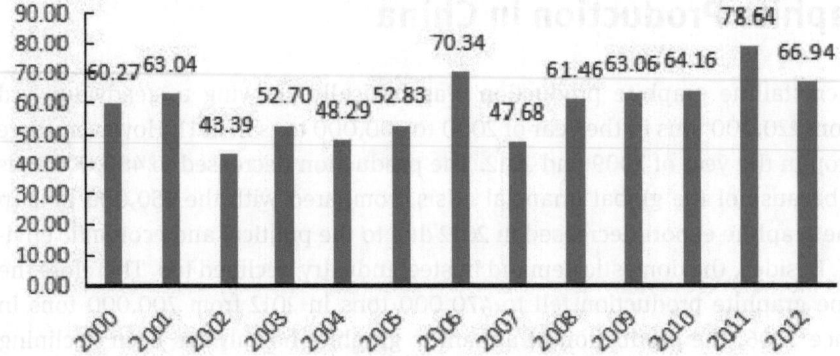

Fig. 3: The graphite consumption in China in the year of 2000-2012(Unit: 10,000 tons) Data sources: China Non-Metallic Minerals Industry Association

The domestic graphite structure is characterised by low-end consumption. The main graphite consumption fields in China are refractory materials and steel industry, accounting for 42 %. In addition, the proportion used for batteries, fuel cells, expanded graphite and motor brush accounts for about 23 %. The above-listed common industrial consumption accounts for about 2/3. Application in lubricants, colloidal graphite and casting accounts for about 12 %, brake facing and friction products about 10 %, pencils 6 %, other new graphite materials, military projects, atomic energy and aerospace only occupy a share of 7 % [3].

5 Graphite Mining Industry in China

Although China ranks the top in both the graphite reserve and the output around the world, the graphite mining industry is relatively backward compared with the countries who own the advanced graphite processing technology, such as the United States, Japan and Germany. At present, raw materials and primary processing products predominate in China, and the long-timed problems have been still remaining solved: disorderly mining, low-price export, the environmental destruction.

The domestic demand for aphanitic graphite is less than 500,000 tons, but the actual output is more than 1 million tons, which indicates that less than half of the aphanitic graphite is used for industrial consumption. To clean the stock, manufacturers were trying to dump to the international market at low prices, where the pricing mechanism was seriously distorted. To make matters worse, the aphanitic graphite was even burned, which was a serious waste of graphite resources.

China is the largest exporter of natural graphite in the world. The export of crystalline graphite has expanded to more than 40 countries and regions, accounting for about half of the total in international market [4]. Since 1990s, the crystalline graphite industry in China has experienced from prosperity to recession. The large new emerging enterprises have led to the rapid excessive expansion of production scale. In the end, the production capacity has brought about the situation of "structural surplus", which resulted in overstocking and extreme disorder of market price, especially low-medium grade crystalline graphite product. Besides, the unbalanced exploitation has caused the serious resources waste of crystalline graphite. The unauthorized mining of private enterprises had no field investigation, nor mining planning.

6 Suggestions

The governments in the United States, Russia, Britain and Germany limit the graphite mining as a strategic resource because of its importance. Instead, they import low-priced graphite from China and other countries and reserve as strategic materials. Japan, who is lack of graphite resources, imports large quantities of raw materials from China and reserves it at sea. Therefore, it is urgent for China to find out the real situation of graphite resources, to evaluate resource potential, to protect the graphite as a strategic resource, and make a specific development planning.

Acknowledgement: This study was financially supported by the programs (2016ZWH028A–28) and (DD 20160124).

References

[1] Yin Liwen, Global graphite development and utilization status, Land and resource information, 2011(6)
[2] U.S. Geological Survey, Mineral Commodity Summaries, January 2015
[3] Shen Wanci, The situation and development of graphite industry, China non-metallic mineral industry, 2013(2)
[3] Cui Yuezhao, China's non-metallic mining industry. Beijing: Geological publishing house, August 2008

Zhang-Huang Ye, Chang-Hui Ke*, Peng Wang and Yi-Ke Li

Isotope Geochemistry of the Shimensi Deposit: Implication for Ore Genesis

Abstract: The Shimensi tungsten deposit in northern Jiangxi is located in the Middle-Lower Yangtze River metallogenic belt. There have been many debates about the origin of metallogenic materials and genetic type. Three mineralization styles are present in the Shimensi deposit. The sulphur isotopic compositions of sulphides in the deposit indicate that the sulphur source was homogeneous and the origin was closely related to the magmatic sulphur. The lead isotopic compositions of sulfides suggest that the lead has multiple sources, and were closely related to magmatism. The carbon isotopic compositions of calcite indicate that hydrothermal fluids were derived from the granitic magma. The Hydrogen and Oxygen isotopic compositions of the quartz veins indicate that the ore-forming fluids had a magmatic-hydrothermal origin, but mixed with meteoric water. Analysis indicates that the mineralization was related to Early Cretaceous magmatism in eastern China. The subduction of the paleo-Pacific plate and subsequent lithospheric delamination resulted in the transition from a compressional to extensional tectonic regime. The upwelling of hot asthenosphere triggered the partial melting of the Neoproterozoic Shuangqiaoshan Group, with high W background concentrations. The ascent of the magma resulted in the exsolution of the ore-bearing magmatic hydrothermal fluid.

Keywords: stable isotopes; ore genesis; Cretaceous magmatism; the Shimensi tungsten deposit

1 Introduction

Several granite-related W and W-Cu-Mo deposits have recently been discovered in the northern Jiangxi, to the south of the Yangtze River, which formed a tungsten-oriented ore concentration area in Wuning County, northern Jiangxi. The Shimensi

Zhang-Huang Ye, Jiangxi Science and Technology Normal University. No. 605, Fenglin Road, Nanchang, Jiangxi. Email: chuckverna@sina.com
Chang-Hui Ke, Institute of Mineral Resources, CAGS. No. 26, Baiwanzhuang Road, Xicheng district, Beijing. Email: kechanghuicags@126.com
Peng Wang, The School of Earth Science and Resources, Chang'an University. No. 126, Yantan Road, Xi'an, Shaanxi. Email: 14020114@qq.com
Yi-Ke Li, Institute of Mineral Resources, CAGS. No. 26, Baiwanzhuang Road, Xicheng district, Beijing. Email: like430@cags.ac.cn

DOI 10.1515/9783110540048-042,

deposit alone is now recognized as a world-class deposit, with an estimated reserve of 742,500 tons WO3@0.195%W, accompanied by 403,600 tons Cu and 28,000 tons Mo [1].

2 Regional Geological Setting

The deposit are located in the northern part of Mt. Jiuling, the center of the Jiangnan Orogen. It is bounded by the Jiurui area of the Middle-Lower Yangtze Valley Mineralization Belt to the north and the Xiangshan district of the Qin-Hang Belt in the south. Its tectonic location is in the middle of the southeast margin of the Yangtze Block (Fig. 1–a).

The Shimensi tungsten deposits lies in southwestern part of the E-W-trending Jiuling uplift area. The Strata of the Shuangqiaoshan Group are partly exposed in the district. They comprise a suite of low-grade metapelite, intercalated with lesser metamorphosed conglomerate. Magmatism was widespread in this region. Abundant felsic magmas, including the Jiuling biotite granodiorite batholith, intruded the metamorphic rocks of the Shuangqiaoshan Group. This batholith, with an area greater than 2500 km^2, is the largest intrusion in southeastern China. Mesozoic granitic rocks in the region occur mainly in two stages. The oldest stage is dominated by porphyritic two-mica granite sills and dikes. It include N-S dikes ranging from 100 m to several hundred meters in length and from several meters to tens of meters in thickness, intruding both the Neoproterozoic granodiorite batholith and rocks of the Shuangqiaoshan Group. The second magmatic stage includes medium- to fine-grained muscovite-bearing granite, porphyritic two-mica granite, fine-grained two-mica granite, and granite porphyry. The relative geological relationships between the phases of this stage are not clear, although the second stage granites cut the first stage granite and the Neoproterozoic granodiorite batholith. A number of SHRIMP U-Pb zircon ages for the late stage Mesozoic granites vary from 144.2±1.3 Ma to 134.6±1.3 Ma [2–3], suggesting that the emplacement in Early Cretaceous (Fig.1–b).

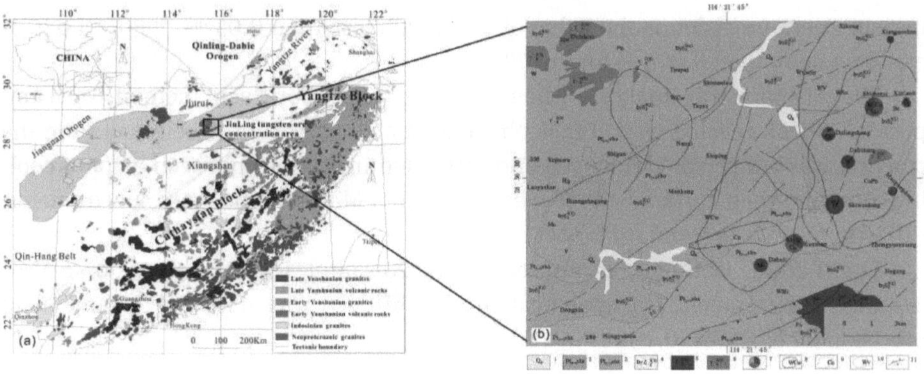

Fig. 1: (a) Tectonic location of the study area (modified after Zhou et al., 2006 [4]). The South China Block is comprised of two blocks: Yangtze Block and Cathaysia Block. The approximate position of the Jiangnan Orogenic Belt is at the edge of Cathaysia Block and belongs to Yangtze Block. The Jiuling ore concentration area is located in the Jiangnan Orogen. (b) Geological and mineral map of the Shimensi deposit, north Jiangxi

3 Sampling and Analytical Methods

Samples for the study were collected from drill cores and underground tunnels in the Shimensi deposit. At least three types of veins have been recognized after careful field study: (1) early barren quartz veins; (2) sulfide-quartz veins, disseminated or concentrated with pyrite, pyrrhotite and chalcopyrite; (3) late calcite veins. Among them, only the sulfide-quartz veins, which account for the majority of W (or Cu, Mo) mineralization, and calcite veins have been used for isotope study. The samples used for S, Pb, H, and O isotope test were collected from sulfide-quartz veins, while samples used for C isotopic analysis from the calcite veins. The samples were crushed and the minerals were handpicked under a binocular microscope. The test has been carried out by the Isotope Laboratory of Wuhan Geological Survey Center.

The S isotope test has been conducted as follows: monomineral (from metal sulphides) and cuprous oxide were exposed to an oxidized reaction after full grinding and mixing according to a certain proportion; generated SO_2 was collect by a freezing method; and then the sulphur isotopic composition was analyzed with a MAT251 gas isotope mass spectrometer. In terms of the standard sample, V-CDT was adopted for measured results, with an analysis precision better than ± 2 ‰.

The Pb isotope samples were firstly decomposed by mixed acid, and then Pb was separated by a resin exchange method. After evaporation, the Pb isotope was measured using the hot surface method by a MAT–261 type mass spectrometry. The measurement precision is the value of $^{204}Pb/^{206}Pb$ less than 0.05 %, $^{208}Pb/^{206}Pb$ generally no more than 0.005 % for 1 μg Pb.

The C isotope was analyzed using the 100 % phosphoric acid method. The samples have been reacted with 100 % phosphoric acid at 25 °C; CO_2 released from the reaction was collected, condensed and separated in a liquid nitrogen-alcohol cooling trap, and then carbon isotope was tested on a MAT251 type mass spectrometer. PDB was taken as a standard for $\delta^{13}C$.

The H and O isotopic compositions of the 20 samples were analyzed on a MAT 252 mass spectrometer. The analytical error for $\delta D_{V\text{-}SMOW}$ was ±1‰ and the analytical precision for $\delta^{18}O$ was±0.2 ‰.

4 Results and discussion

S isotopic composition: The S isotope from molybdenite, pyrite and chalcopyrite are tested. Among them, $\delta^{34}S$ values of 6 molybdenite samples vary from −2.51‰ to −1.12%, with an average value of −1.91 ‰, while $\delta^{34}S$ values of 6 chalcopyrite samples vary from −1.98 ‰ to−0.91 ‰, with an average value of −1.42 ‰; $\delta^{34}S$ values of 5 pyrite samples vary from −1.67 ‰ to −1.27 ‰, with an average value of −1.48 ‰. These data indicate that the variation range of $\delta^{34}S$ sulfide value is small, from −2.51 ‰ to -0.91‰, with an average value of −1.61 ‰.

Most of The $\delta^{34}S$ isotope composition is within the range of −2 ‰ to −1.5 ‰, showing an obvious “peak” distribution, which means the sulphur source was homogeneous. The $\delta^{34}S$ values of sulfides vary from −2.51 ‰ to −0.91 ‰, with an average value of −1.61 ‰, indicating sulphur origin is closely related to the magmatic sulphur [5–6].

Pb isotopic composition: As can be seen from the lead isotope data, the variation ranges from 17 samples are small, indicating that the lead isotope composition in the deposit is fairly homogeneous. Taking the Pb isotope of pyrite as an example, $^{206}Pb/^{204}Pb$: 18.155–18.259, $^{207}Pb/^{204}Pb$: 15.586-15.709, $^{208}Pb/^{204}Pb$: 38.227–38.717. The $^{206}Pb/^{204}Pb$ ratios of sulfides vary from 18.155 to 18.277, with an average of 18.189; $^{207}Pb/^{204}Pb$ ratios vary from 15.586 to 15.709, with an average of 15.640; $^{208}Pb/^{204}Pb$ ratios vary from 38.227 to 38.717, with an average of 38.420.

The parameters of lead isotope, especially μ value and ω value can provide information on the geological process that the geologic body has gone through. Among the 17 samples of metal sulfide lead isotope composition from the deposit, 12 pieces have μ values below 9.58, and 5 pieces have μ values that are slightly higher than 9.58. The calculated ω values are between 37.145–39.621, all of which are higher than the average value of the mantle (36.84). Therefore, the characteristics of low μ, high ω of lead isotope composition in the deposit indicate that the lead has the characteristics of the lower crustal sources, but some may come from the upper crust [7].

C isotopic composition: Calcite veins in this deposit are very common, which can be seen from the outcrops to the deep drillings, with big and pure crystals. The veins do not contain quartz and pyrite, which clearly shows that the calcite was formed after the mineralization stage. The $\delta^{13}C_{V\text{-}PDB}$ values of calcite range from -10.26 ‰ to −3.26 ‰, which mainly focus from −8 ‰ to −7 ‰, and the average value is −7.30 ‰.

The C isotope geochemistry of calcite has been used previously to trace the origin of hydrothermal fluids. As calculated, the $\delta^{13}C_{V\text{-}PDB}$ values of calcite range from −10.26 ‰ to −3.26 ‰, with an average value of −7.3 ‰. All data fall within the magmatic range (−10.3 ‰ to 2.9 ‰), indicating that hydrothermal fluids from which calcites precipitated were derived from the granitic magma. They are mainly distributed from −8 ‰ to −7 ‰, which indicates the ore-forming fluids may come from the lower crust or upper mantle (−5 ‰ to −8 ‰)[8]. But there are two samples whose values are at the edge of −10.3 ‰, suggesting that strong isotope fractionation occurred when deep source fluids migrated in the form of methane. Methane in metallogenic field was oxidized to carbon dioxide, further precipitated to form calcite.

H and O isotopic composition: Hydrogen and oxygen isotopes are important indicators for the evolution of ore fluids. $\delta^{18}O_{SMOW(H2O)}$ values were calculated by using the formulas: $1000\ln\alpha=\delta^{18}O_{SMOW}-\delta^{18}O_{SMOW(H2O)}=3.38\times10^{6}T^{-2}-3.4$[9]. The $\delta^{18}O_{SMOW(H2O)}$ and δD_{SMOW} values of quartz fluid inclusions range from 2.3‰ to 4.6‰ and from −107‰ to −85‰, respectively.

The $\delta^{18}O_{SMOW(H2O)}$ values of the quartz from the deposit vary from 2.3 ‰ to 4.6 ‰, with a peak from 3.0 ‰ to 4.0 ‰, which fall in the range for magmatic water[10] and suggest a magmatic-hydrothermal origin. The δD values of the fluids vary from −107 ‰ to −85 ‰, which are lower than those for magmatic water (−85‰ to −50 ‰). If an H^{2-} or CH^{4-} bearing hydrothermal fluid encounters an oxidizing environment, the H_2O will become depleted in deuterium. However, this effect is small in natural hydrothermal systems because the volume of H_2O involved is enormous. Thus, a magmatic-hydrothermal fluid, but with mixing between the magmatic water and heated meteoric groundwater, can account for the stable isotope data derived from the deposit.

5 Conclusion

The mineralization age of the Shimensi deposit occurred at the same time as other Early Cretaceous deposits of the Middle-Lower Yangtze Valley mineralization belt, Lower Yangtze metallogenic province, ranging from 150–121 Ma. The deposits are spatially close and temporally coeval with the Middle-Lower Yangtze Valley mineralization belt. Mineralization in the Lower Yangtze metallogenic province occurred at the subduction zone of two paleo-blocks and triggered by the tectonic transition.

The subduction of the Paleo-Pacific Plate beneath the Eurasian Plate resulted in lithospheric thickening and subsequent delamination, and consequently the principal N-S stress field in eastern China changed progressively to an E-W orientation from 170 Ma to 135 Ma [11]. Maximum lithospheric thinning occurred in the late Mesozoic, which resulted in the interaction between the asthenospheric mantle and crust. The resulting crustal heating caused intense magmatism, mineralization, and widespread development of extensional structures in eastern China. The eastern part of the Eurasian continent was an active continental margin at *ca.* 180 Ma, as the Izanagi Plate subducted orthogonally beneath the continent. At *ca.* 160 Ma this plate rotated clockwise towards the north, with the subduction orientation becoming oblique with respect to the continental margin. This movement was associated with metallic mineralization controlled by regional NE-SW strike-slip faults. Asthenospheric upwelling along the subducting slab window between 150 and 126 Ma, triggering partial melting of the overlying continental crust with a high W background content in the back-arc setting, resulting in a highly-fractionated, volatile-rich, peraluminous magma.

The highly fractionated S-type granites in the deposit are strongly peraluminous and highly alkalic, with high Ga/Al ratios. These rocks are enriched in LILEs, such as Rb, and are depleted in HFSEs, such as Zr, Nb, Ti. These granites formed at relatively low temperatures (679–760 °C) according to the zircon saturation temperatures. These εNd(t) values of the granites varied from −7.78 to −7.47, with the two-stage modal ages of 1568 Ma to 1543 Ma [2]. The tungsten concentrations of metapelitic rocks of the Shuangqiaoshan Group are as high as 12 ppm [12], which is much higher than that of typical crustal rocks at 1–2 ppm [13]. Tungsten is a strongly incompatible element, and partial melting and extreme fractional crystallization can lead to additional enrichment of the element. These observations indicate that the Shuangqiaoshan Group metapelite, with high W background concentrations, is the most likely source for the ore components in the Shimensi deposit.

The ascent of the volatiles-rich magma resulted in exsolution of the ore-bearing magmatic hydrothermal fluid, large-scale alkaline replacement in the wallrocks, and greisenization, silification and K-feldspar alteration of wall rocks in the deposit. The existence of volatiles, such as F and B, can increase the solubility and enrichment of tungsten in magma. Ore-forming fluid filled joints and fractures, forming the veinlet-disseminated style of orebodies. The Jiuling batholith, which is rich in calcium, contributed the Ca^{2+} that combined with the WO_4^{2-} in the ore-forming fluid to form scheelite along the contact between the Cretaceous biotite granite and the Neoproterozoic biotite granodiorites. When fractures in the region penetrated the top of the porphyritic biotite granites, the exsolving high-temperature liquid released a vapor and a crypto-explosion occurred, resulting in the formation of orebodies of the hydrothermal crypto-explosive breccia style of mineralization. The quartz vein style orebodies formed last, because they cut both igneous rock units and the two other styles of orebodies (Fig. 2).

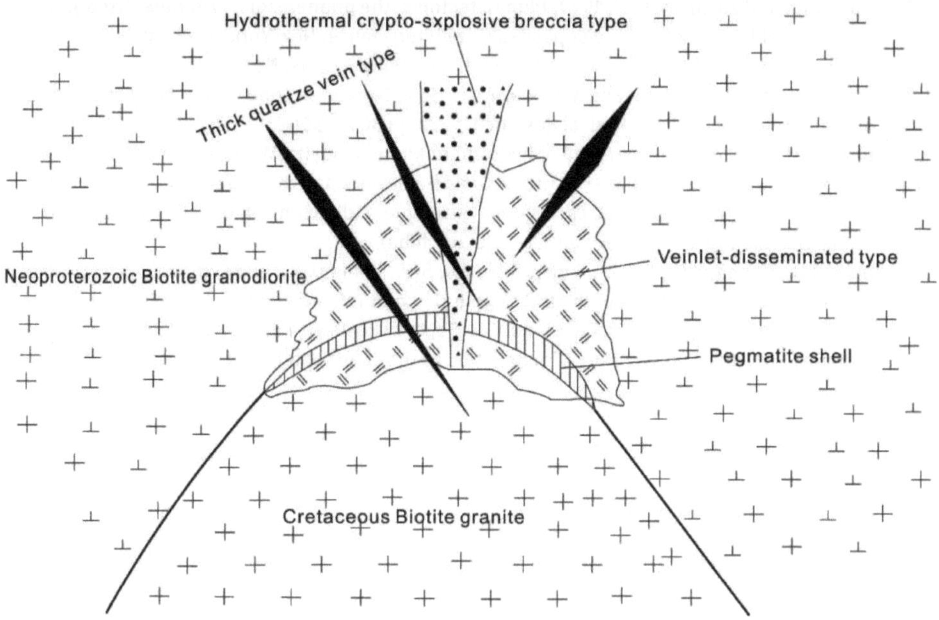

Fig. 2: Metallogenic model of the Shimensi deposit

Acknowledgement: This study was financially supported by the programs (GJJ160776), (k1406) and (1212011220870).

References

[1] Xiang, X.K., Wang, P., Sun, D.M., and Zhong, B., 2013, Re-Os isotopic age of molybdeinte from the Shimensi tungsten polymetallic deposit in northern Jiangxi province and its geological implications: Geological Bulletin of China, v. 32, p. 1824–1831.

[2] Huang, L.C., and Jiang, S.Y., 2012, Zircon U-Pb chronology and geochemistry of porphyritic-like muscovite granite and its genesis from Dahutang tungsten deposit in Jiangxi province: Acta Petrologica Sinica v. 028, p. 887–3900.

[3] Huang, L.C., and Jiang, S.Y., 2013, Geochronology, geochemistry and petrogenesis of the tungsten-bearing porphyritic granite in the Dahuang tungsten deposit, Jiangxi province. Acta Petrologica Sinica 029: 4323–4335.

[4] Zhou, X.M., Sun, T., Shen, W., and Niu, L.S.Y., 2006, Petrogenesis of Mesozoic granitoids and volcanic rocks in South China: A response to tectonic evolution: Episodes, v. 29, p. 26–33.

[5] Harris, A.C., Golding, S.D., and White, N.C., 2005, Bajo de la Alumbrera copper-gold deposit: stable isotope evidence for a porphyry-related hydrothermal system dominated by magmatic aqueous fluids: Economic Geology, v. 100, p. 863–886.
[6] Ohmoto, H., and Rye, R.O., 1979, Isotope of sulphur and carbon. In: Barnes HL. (ed.). Geochemistry of Hydrothermal Ore Deposits (2nd ed.). John Willy & Sons, New York, 509–567 p.
[7] Doe, B.R., and Zartman, R.E., 1979, Plumbotectonic, the phanerozoic. //Barnes HL(ed.). Geochemistry of hydrothermal ore deposits. Wiley Interscience, New York, 22–70 p.
[8] Hoefs, J., 1997, Stable isotope geochemistry. John Willy & Sons, New York, 343 p.
[9] Clayton, R. N., O'Neil, J. R., and Mayeda, T. K., 1972, Oxygen isotope exchange between quartz and water: Journal of Geophysical Research, v. 77, p. 3057–3067.
[10] Barnes, H. L., 1979, Geochemistry of hydrothermal ore deposits. Wiley, New York, 781 p.
[11] Ren, J.S., Niu, B.G., He, Z.J., 1998, The tectonic framework and dynamics evolution in East China// Ren JS, Yang WR (ed.). Lithospheric structure and tectonic magmatic evolution in East China. Seismological Press, Beijing, 1–12 p (in China).
[12] Mao, Z.H., Cheng, Y.B., Liu, J.J., Yuan, S.D., Wu, S.H., Xiang, X.K., and Luo, X.H., 2013, Geology and molybdenite Re-Os age of the Dahungtang granite-related veinlet-disseminated tungsten ore field in the Jiangxin Province, China: Ore Geology Review, v. 53, p. 422–433.
[13] Rudnick, R.L., and Gao, S., 2004, Composition of the continental crust: Treatise on Geochemistry, v. 3, p. 1–64.

Ming-Yang Zhang, Wei Zhang, Bin Lu and Yong Liu

Simulation of β-stabilised γ-TiAl Alloy Rolling and Its Defects Analyses

Abstract: To overcome the poor ductility of this two-phase (γ+γ/α_2) material, novel β-solidifying multi-phase γ-TiAl based alloys were developed to improve the hot workability. In order to study the plastic deformation characteristics of β-stabilised γ-TiAl alloy and find out the feasibility of using computer simulation to analyse the hot deformation process, the rolling process of TiAl billet was simulated by means of DEFORM^{3D}. It is concluded that (1) the changes of the temperatures, stresses and strains at the different regions of the billet were simulated during the entire rolling process, (2) based on the stress and strain distributions of the rolled billet, the forming defects analyse was proceeded and it is concluded that some rolling defects or flaw occur near the head surface or around the side edge during the plastic deformation process. During the rolling process, rolling force decreased with the elevation of rolling temperature and the reduction of rolling rate and pass reduction. Since the cooling rate of the sheet edges is faster with lower temperature, it is easy to generate stress concentrated in edge region during the deformation process. The greater the reduction is and the greater the damage is. In summary, in order to ensure the success of TiAl based alloy sheet rolling, it should be used at higher temperatures, lower rolling reduction rate and slower rolling rate.

Keywords: TiAl alloy, Simulation, Hot rolling, DEFORM^{3D}

1 Introduction

In recent years, high-performance, high-temperature lightweight alloys have been widely required in aerospace field. Traditional Ni-based superalloy has high density, low strength and antioxidant properties at high temperature, thus dissatisfying application. With low density, high melting point, strength and antioxidant property at high temperature, γ-TiAl has been a new light high-temperature alloy with competitiveness [1-4]. It shows great advantages in preparation of aviation structures including turbine engine, low pressure turbine blade and transition catheter beam [5-8]. Excellent properties of γ-TiAl alloy contribute to practical application. As important components of engine in future aerospace, TiAl alloy sheets with excellent organization and performance are more and more emphasized. However, fragility of γ-TiAl alloy results in narrow hot processing temperature window and high deformation resistance at high temperature. Therefore, γ-TiAl alloy is sensitive to process parameters in hot processing, thus increasing the difficulty of sheet preparation [9-10]. To solve the problem, novelβ-solidifying γ-TiAl alloys were developed.

DOI 10.1515/9783110540048-043,

In these alloys, disordered β phase is introduced at elevated temperature by addingβ-stabilizing elements, such as Nb, V, Cr, Mn and Mo [11-12], which acts as a deformation accommodating phase and therefore improve the hot workability [13].

TiAl alloy sheets with excellent organization and performance are prepared to achieve practical process of TiAl alloy. Foreign counterparts have made great progress [14-19] in industrial application of large size TiAl alloy sheet rolling. Plane rolling is one of the most important methods of preparing TiAl alloy sheet. It has more complicated blank stress state, strain rate and temperature distribution in high temperature rolling process, compared with mature hot forging process. At the edge of slab, rolling flow rate is slower than spread degree. Central panel is easy to crack under additional tensile stress. Affected by minimum resistance law, transverse spreading is further smaller than extension of rolling. Multi-pass rolling is used to reduce defects in rolling process because of poor hot processing performance of TiAl alloy. Pass deformation is the key factor of affecting blank deformation uniformity. In rolling process, pass deformation is strictly controlled based on blank plasticity, slab quality, cladding processing and rolling capacity. Rolling temperature is an important processing parameter in TiAl alloy rolling process. Deformation resistance, rolling force, sheet organization and surface state are sensitive to rolling temperature. TiAl alloy has narrow hot processing temperature window and low symmetry degree, thus causing few available slip systems and large deformation resistance. Rheological behavior is sensitive to stress rate, temperature and deformation. Too fast rolling speed results in large pass deformation and cracking of sheet. Too slow rolling speed leads to heat loss and deviation from hot processing temperature window, thus increasing probability of cracking. Therefore, it is difficult to develop hot rolling process of TiAl alloy sheet by traditional method because of narrow hot processing temperature window and poor heat forming performance.

Numerical simulation technology based on Deform^{3D} has been widely used in efficiency improvement of new materials and optimization of process parameters. Rheological instability of TiAl alloy is a multi nonlinear multivariable coupling process in hot deformation. TiAl alloy has narrow temperature window, thus increasing the difficulty of hot processing. Finite element simulation technology in hot processing has an important significance in determining hot processing window, excellent organization and performance of TiAl alloy sheet. The work applied DEFORM^{3D} to simulate package rolling process of TiAl-base alloy, deriving the influence of different rolling rates, reduction ratios and temperatures to deformation of TiAl alloy. After that, the work analyzed distribution law of temperature, stress, strain field and damage coefficient, thus providing theoretical basis for optimal design of rolling process.

2 Experimental Procedures

2.1 Assumption Conditions for Numerical Simulation of Rolling Process

To realize simulation of rolling process by Deform3D, the assumptions are made as follows. (1) The two main drive rollers, with the same diameter and speed, do not produce plastic deformation. (2) The sheet is isotropic plastomer with uniform performance and ignored elastic deformation. (3) The sheet has uniform deformation and metal particle speed. (4) The sheet has incompressible volume. (5) Effect of volume force is neglected.

2.2 Numerical Simulation Model Construction of Rolling Process

Rectangular model 60mm×40mm×10mm is selected as TiAl-base alloy sheet. The sheet has geometric symmetry. Therefore, half of geometric solid is selected as research object to save time and calculation. The model, established by means of mirror constraint, is imported to Deform-3D. The model uses tetrahedral element mesh to achieve automatic meshing function in condition of mesh distortion. Fine grid is applied for accurate simulation. TiAl-base alloy blank consists of 15,000 grids. The sheet is bitten by friction force between roller and strip. The type is set as shear friction; friction factor as 0.3; environment temperature as 25°C; thermal radiation coefficient as 0.02N/sec·mm·°C; thermal conductivity coefficient as 11N/ sec·mm·°C.

Tab. 1: Rolling conditions for β-stabilised γ-TiAl alloy sheets.

Sheet number	Rolling temperature (°C)	Reduction Rate (%)	Rolling speed (mm/sec)
1	1280	25	40
2	1280	15	40
3	1280	10	40
4	1280	5	40
5	1280	10	24
6	1280	10	56
7	1250	10	40
8	1200	10	40
9	1150	10	40

In experiment, engineering condition is used. Industrial titanium sheet with thickness of 1.6mm is taken as canning; cylinder with diameter of 180mm as roller. Rolling parameters are changed to realize rolling simulation in different conditions (See Table 1). To simplify calculation process, rollers are seemed as rigidity, and the simulation experiment is divided into two stages as follows. The first is preparation process that the blank is transferred from furnace to roller; the second is rolling process.

3 Results and Discussion

3.1 Temperature Field Analysis of Rolling Process

Rolling temperature is one of key process parameters affecting rolling deformation. It also affects deformation difficulty, flow resistance, elongation and microstructure of materials. Therefore, temperature distribution and variation trend of sheet should be accurately controlled to understand microstructure evolution, mechanical property and phase transition.

The following figure shows the temperature fields of sheet at different stages by taking rolling simulation based on conditions (T=1,280°C; υ=40mm/sec; Δδ=10%) as an example.

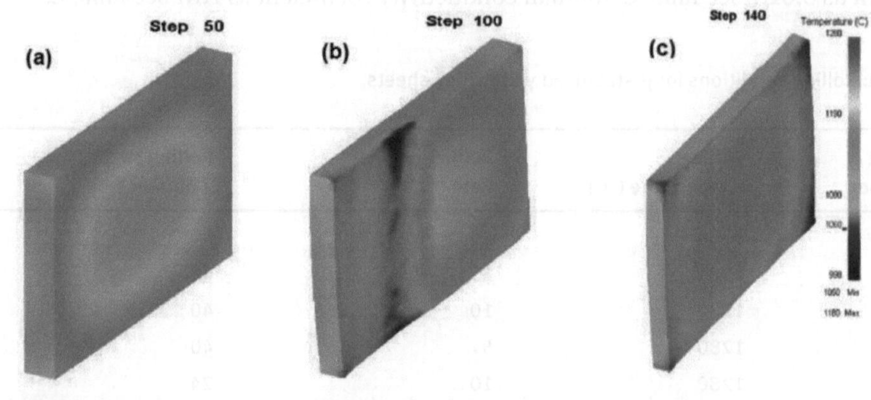

Fig. 1: Temperature distribution of TiAl alloy during rolling process

It takes 10 seconds to transfer TiAl-base alloy from furnace. In this stage, rolling surface radiates by thermal radiation and convection (See Figure 1 (a)). The temperature obviously declines to 1,110°C at the edge of roller. The center temperature of sheet surface is 1,210°C. The temperature inside the sheet is stable. The maximum temperature is 1,230°C. In rolling stage, heat exchange is mainly thermal conduction of rolling sheet and roller, with less heat exchange and radiation of rolling sheet itself (See Figure 1 (b)). The edge of deformation district has the maximum cooling rate, with minimum temperature of 1,030°C. After rolling process, temperature returns because of heat conduction inside rolling sheet. Figure 1 (c) shows temperature field after rolling process. In Figure 1 (c), temperature return occurs inside the sheet. However, the edge of rolling sheet still has the minimum temperature of the sheet.

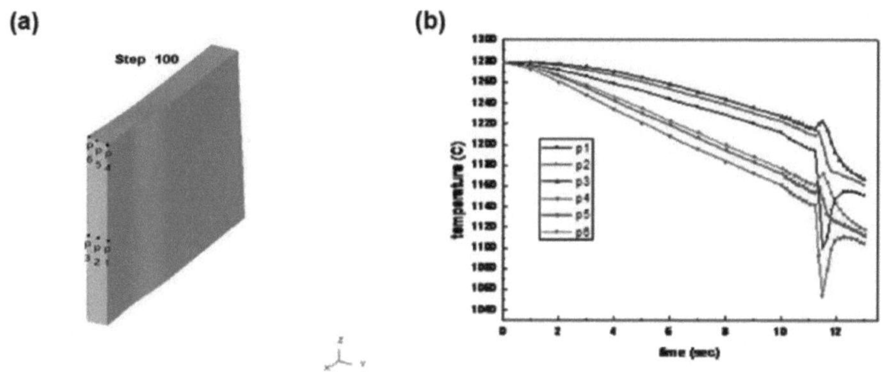

Fig. 2: The variation of temperature of TiAl alloy during rolling process

To observe temperature variation of rolling process, six typical points are selected from cross section of sheet to trace rolling temperature in rolling process (See Figure 2 (a)). Figure 2 (b) indicates that each region has stable heat loss because of thermal transfer and radiation before deformation. Thereinto, P4 temperature at the edge of surface drops fastest; P3 temperature at the center area drops slowest. In rolling process, temperatures at P1 and P2 regions rapidly decline because of heat transfer. Temperatures at the other 4 regions slightly rise because of plastic deformation and friction work. After rolling process, the temperature at surface is lower than that inside the sheet. Therefore, internal heat flows to the surface to increase surface temperature. This is the reason of temperature return at P1 and P4 regions after rolling process.

3.2 Equivalent Strain Field Analysis of Rolling Process

Figure 3 is equivalent strain distribution nephogram of TiAl-base alloy sheet after rolling process. In Figure 3, unstable deformation at biting stage results in unusual equivalent strain at the edges. In stable rolling stage, equivalent strain fluctuates within a certain range (See Figure 3 (b)). In simulation experiment, rollers are not ideal cylinders. They have isogon cross sections. Reduction of rolling sheet changes in a small range, thus causing fluctuation change of equivalent strain at rolling direction.

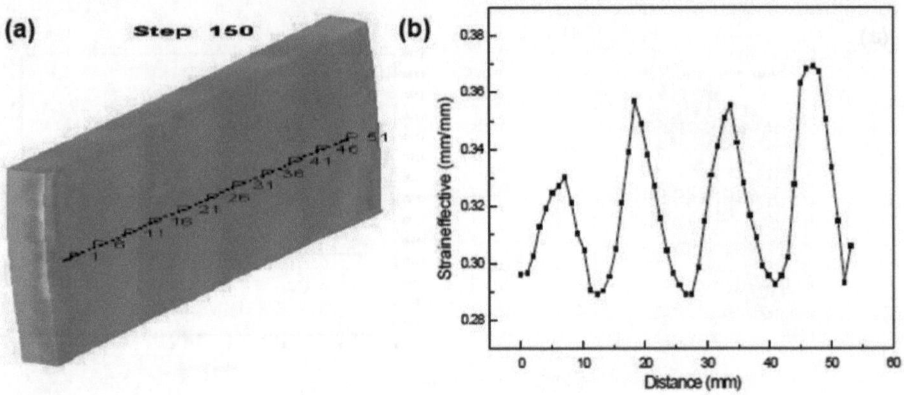

Fig. 3: Effective strain distribution of the as-rolled TiAl alloy sheets

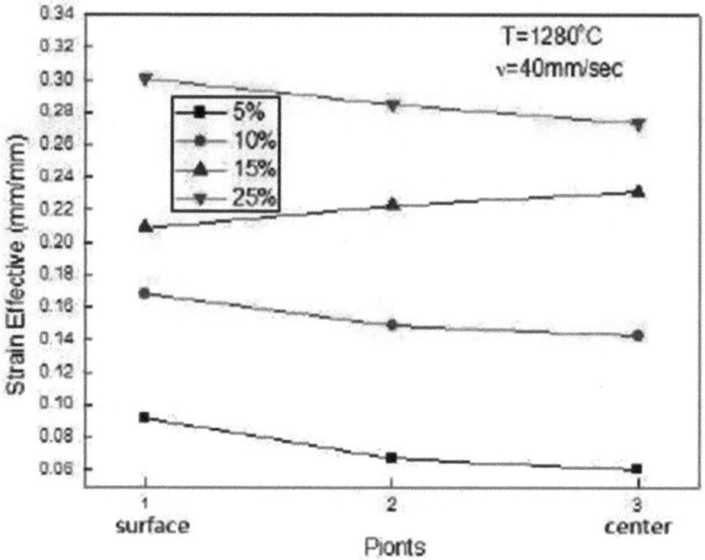

Fig. 4: Effect of reduction ratio on effective strain of cross-section area in the TiAl sheet

Figure 4 shows the influence of reduction percent to equivalent strain of rolling sheet at thickness direction. When reduction percent is small (5% or 10%), the equivalent strain at the surface is higher than that in the center layer. This is because that deformation focuses on surface rather than center area. With the increase of reduction percentage, deformation of center area becomes larger than that of surface. However, the further increase of reduction percentage results in serious interference in deformation region and increase of friction force. Therefore, deformation of surface is larger than that of center area. Figures 5 and 6 indicate the effects of rolling temperature and rate to equivalent strain at thickness direction. For rolling temperature and rate, equivalent strains have the same variation trend. That is, equivalent strain at the surface is larger than that in the center area. Consequently, equivalent strain distribution at thickness direction is mainly affected by reduction percentage rather than rolling temperature and rate. Besides, deformation uniformity of sheet increases with the decline of reduction rate.

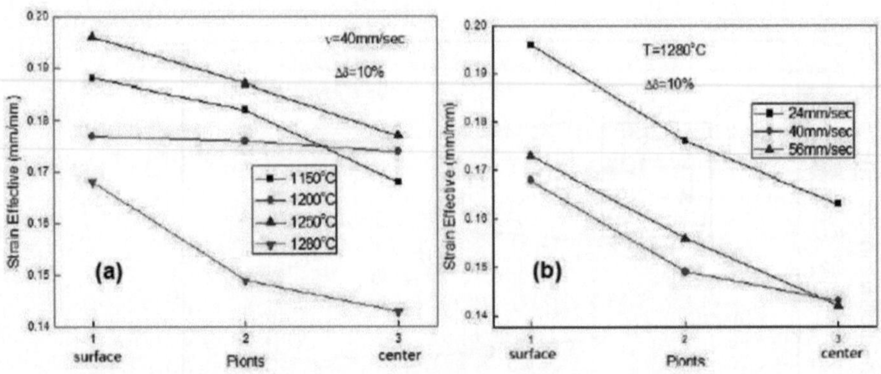

Fig. 5: Effect of temperature on effective strain of cross-section area in the TiAl sheet

3.3 Rolling Force Analysis of Rolling Process

Rolling force is one of the most critical physical quantities in sheet rolling deformation. It affects determination of rolling deformation, commissioning of rolling mill, effective development of equipment potential and automatic control of rolling process. Meanwhile, it is also the key index widely used in strength design and check of mechanical equipment.

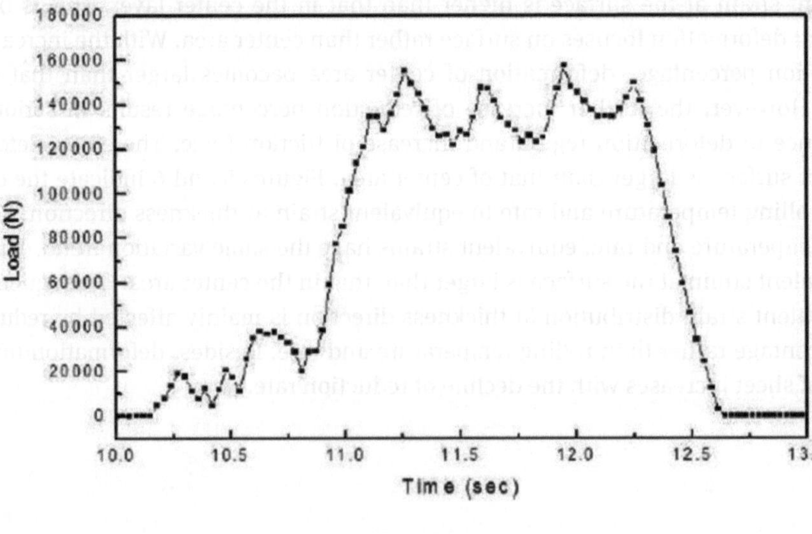

Fig. 6: The changing trend of the rolling force during rolling process

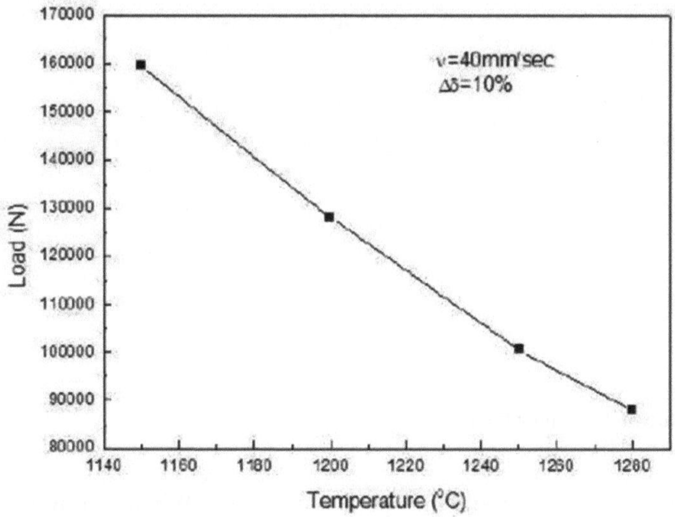

Fig. 7: Effect of reduction rate on effective stress of different area in the TiAl sheet

In rolling process, the trend of rolling force was discussed by taking rolling simulation based on conditions (T=1,280°C; υ=40mm/sec; Δδ=10%) as an example (See Figure 7). Figure 7 show that rolling force begins to rise at biting stage, thus entering unstable rolling stage. When rolling sheet embeds itself into rolling gap, rolling force becomes stable, with slight fluctuation. Therefore, the process comes to stable rolling stage. When rolling process is completed, rolling force sharply drops to zero. At stable rolling stage, rolling model is a prism with polyhedral cross section rather than an ideal cylinder. This model produces slight fluctuation and eccentricity of rolling reduction, thus causing fluctuation of rolling force. Actually, similar results often occur because of roller beat in real rolling deformation.

Figure 7 the changing trend of the rolling force during rolling process. In rolling deformation, rolling force is affected by factors including reduction percentage, rolling rate and temperature. Figures 8, 9 and 10 show the effects of rolling temperature, reduction percentage and rolling rate to rolling force.

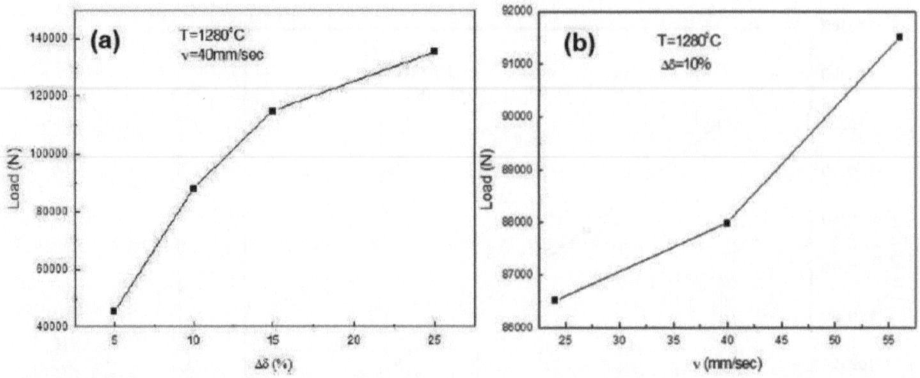

Fig. 8: Effect of reduction rate on the rolling force

In Figure 8, rolling rate=40mm/sec; reduction percentage=10%. When rolling temperature increases from 1150°C to 1280°C, then rolling force will decrease from 160,000N to 90,000N. The reason is that rolling temperature increases the rates of recovery and dynamic recrystallization, thus compensating work hardening of alloy in deformation. The decrease of alloy flow stress results in the decline of rolling force. In Figure 9, rolling temperature=1280°C; rolling rate=40mm/sec. When reduction percentage increases from 5% to 25%, then rolling force will increase from 40,000N to 140,000N. There are two reasons. Firstly, reduction percentage increases to enlarge contact area between sheet and rollers. Therefore, the arc length in deformation area rises to result in severer influence of external friction and the increase of average unit pressure. Secondly, the increase of reduction percentage intensifies work hardening of TiAl-base alloy sheet and the increase of deformation resistance. In Figure 10, rolling temperature=1280°C; reduction percentage=10%. When rolling rate increases from 24mm/min to 56mm/min, then rolling force will increase from 86,500N to 91,500N. The reason is as follows. Rolling rate increases to enlarge deformation rate of TiAl-base alloy sheet, thus driving more dislocation motions. Therefore, deformation resistance of alloy is improved to increase rolling force.

3.4 Damage Coefficient Analysis of Rolling Process

In rolling process of TiAl-base alloy, damage distribution of different parts of sheets is predicted to provide references for formulation of actual rolling process.

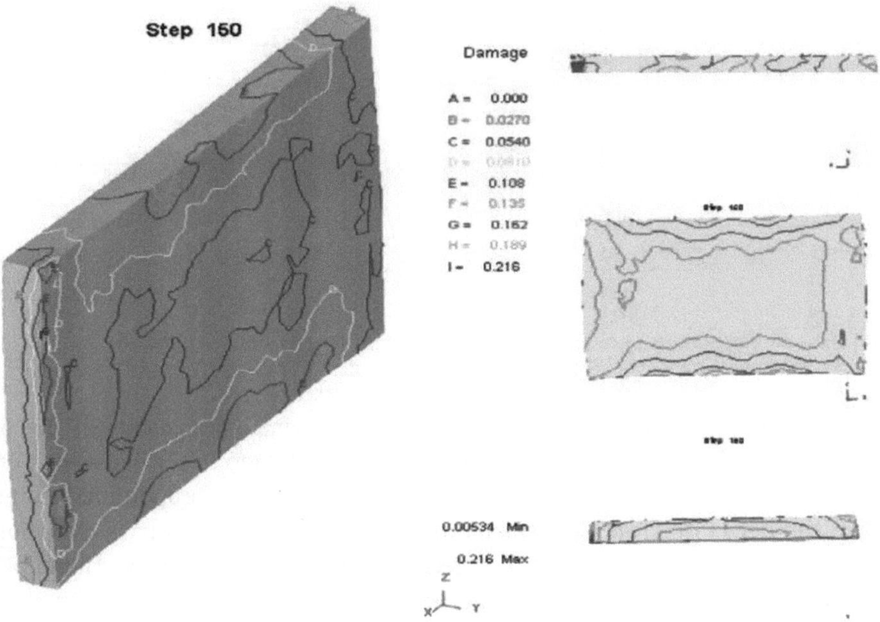

Fig. 9: Damage field distribution of the as-rolled TiAl alloy sheet

Figure 9 is the damage distribution of TiAl-base alloy sheet in rolling process. Figure 11 show that the sheet has smallest damage in central area and large damage in edge and biting areas. The reason is that the edge area subjected to tensile stress has the fastest temperature decline. Besides, central area is subjected to compressive stress. Therefore, edge area is easy to crack, with more non-uniform deformation than central area. The central part inside the sheet has smallest damage. The part closer to the edge has larger damage at width direction. This is the reason why the edge of sheet is easy to produce flaw in sheet deformation process. In addition, damage coefficient gradually decreases from the outside to the inside at thickness direction. Temperature rapidly decreases at contact surface of roller and the work-piece. This edge area is easier to be damaged than central area, with small plastic deformation and large deformation resistance.

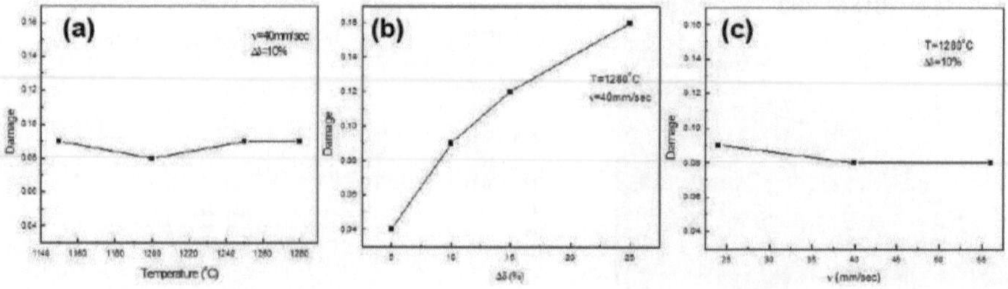

Fig. 10: Effect of temperature (a),reduction rate (b) and rolling rate (c) -+on the damage

Figure 10 shows the influence of reduction percentage to maximum damage value of sheet. In Figure 10, maximum damage value rises from 0.04 to 0.16 when reduction percentage increases from 5% to 25%. Reduction percentage is proportional to the damage and defect of sheet. Conversely, rolling process with small pass reduction percentage reduces the possibility of defects, ensuring sheet integrity. Therefore, rolling process with small pass reduction percentage is suitable for TiAl-base alloy sheet.

3.5 Experimental Verification

Numerical simulation shows that small pass reduction percentage, rolling rate and large rolling temperature contribute to reducing deformation resistance and damage of TiAl-base alloy, improving uniformity of deformation. Research group performs hot rolling experiments of different temperatures and pass reduction percentages to verify this result. Figure 11 indicates flow diagram of TiAl-base alloy sheet when rolling temperature=1280°C; pass reduction percentage=10%. In accordance with simulation result, complete sheet with smooth surface is obtained by experiments.

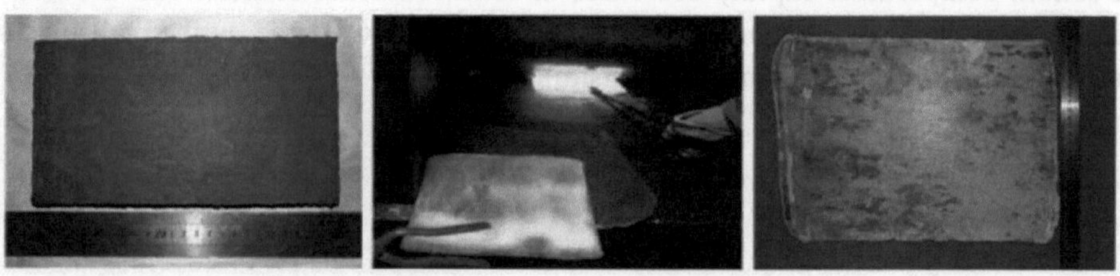

Fig. 11: Hot rolling of large-size PM TiAl alloy sheet (550mm×360mm×1.5mm)

4 Conclusion

(1) Roller model is regular polygonal prism rather than ideal cylinder. After rolling process, equivalent strain of sheet fluctuates along rolling direction. With the increase of reduction percentage, equivalent strain of the rolling sheet increases; deformation uniformity decreases. Small pass reduction percentage is conducive to uniform deformation of sheet.

(2) Equivalent stress and rolling force in each area of sheet have similar variation trend. With the increase of rolling temperature, equivalent stress in each area of sheet decreases. With the increase of pass reduction percentage and rolling rate, equivalent stress in each area rises. At thickness direction, equivalent stress on the surface is larger than that in the central area.

(3) When reduction percentage and rolling rate are certain, then rolling force will be inversely proportional to rolling temperature. The reason is that TiAl-base alloy sheet has small deformation resistance at high temperature. When reduction percentage and rolling temperature are certain, then rolling force will be proportional to rolling rate. Rapid deformation produces more dislocation motions in short time to enlarge rolling force. When rolling temperature and rate are certain, rolling force increases with the increase of pass reduction percentage.

(4) TiAl-base alloy sheet has large cooling speed at the edge. Therefore, stress concentration is produced to cause large damage at the edge of sheet in deformation process. Pass reduction percentage is more sensitive to sheet damage than rolling temperature and rate. The larger reduction percentage is, the larger the damage of the sheet is.

Acknowledgement: The authors acknowledge support from National Natural Science Foundation of China (No. 51201185) and Postdoctoral Foundation Supported Project of CSU (Postdoctoral No.148700).

References

[1] F. Appel, J.D.H. Paul and M. Oering, WILEY-VCH,Weinheim,2011

[2] G. Chen, Y. B. Peng, G. Zheng, Z.X. Qi, M.Z. Wang, H.C. Yu, C. L. Dong and C. T. Liu , Nat. Mater. 15, 8, 876–882, 2016.

[3] N. P. Lavery, D. J. Jarvis and D. Voss, Intermetallics 19, 6, 787–792, 2011.

[4] F. Appel, H. Clemens and F.D. Fischer, Prog. Mater. Sci. 81, 55–124, 2016.

[5] S. L. Semiatin, B. W. Shanahanand F. Meisenkothen, Acta Mater. 58, 13, 4446–4457, 2010.

[6] X. H. Wu, Intermetallics14, 10–11, 1114–1122, 2006.

[7] E. Schwaighofer, H. Clemens, S. Mayer, J. Lindemann, J. Klose, W. Smarsly and V. Güther, Intermetallics 44, 128–140,2014.

[8] H. Clemens, S. Mayer, Adv. Eng. Mater. 15,4,191–215,2013.

[9] S. W. Zeng, A. M. Zhao, H. T. Jiang and Y. S. Ren, Alloys Compd. 698, 786–793,2017.

[10] H. Clemens, W. Wallgram, S. Kremmer, V. Güther, A. Otto and A. Bartels, Adv. Eng. Mater. 10, 8,707–713, 2008.

[11] D. Huber, R. Werner, H. Clemens and M. Stockinger, Mater. Charact. 109, 116–121, 2015.
[12] P. Erdely, P. Staron, E. Maawad, N. Schell, J. Klose, S. Mayer and H. Clemens, Acta Mater. 126, 145–153, 2017.
[13] G. Yang ,H.C. Kou ,J. Yang, J.S Li and H.Z. Fu, Alloys Compd. 663,594–600.2016.
[14] J. G Luo, Acoff L V. Materials Science and Engineering A, 433, 1–2, 334–342, 2006.
[15] J. Oh, S. G. Pyo, S. Lee and N. J. Kim, Journmal of Materials Science, 38, 17, 3647–3651, 2003.
[16] F. Yang, F. T. Kong, Y.Y. Chen and S.L. Xiao, Materials Engineering. 05,98–100,2010.
[17] R. Gerling, A. Bartels, H. Clemens, H and F.P. Schimansky. Intermetallics, 12, 3, 275–280, 2004.
[18] S.H. Clemen, L.F. Apple and S.A. Bartel, WILEY-VCHVerlagGmbH&Co. KGaA. 2123–2136,2003.
[19] R. Gerlin, F. P. Schimansky, A. Star, A. Bartels, H. Kestler, L. Cha, C. Scheu and H. Clemens,
[20] Intermetallics, 16, 5, 689–697, 2008.

Yang Li, Yan-Ling Han, Hong Wang* and Guang-Bin Wu

Polarization Dependence of Defect Mode in One-dimensional Magnetic Photonic Crystal

Abstract: Using 4×4 transfer matrix method, the transmission properties of defect mode in one-dimensional magnetic photonic crystal have been investigated. It is found that the positions of defect mode critically depend on incident light and intensity of external magnetic fields. In the case of system without magnetization, the position of defect mode has no polarization dependence for different incident light, such as linearly polarized light, left-circularly polarized light and right-circularly polarized light, and their peak frequencies synchronously increasing with the incident angle. In the case of system under external magnetic fields, the position of defect mode possesses obviously polarization dependence for different incident light. For the linearly polarized light, the defect mode will be split and move toward to the both edges of photonic band gap respectively with the intensity increasing of external magnetic fields. For the left-circularly polarized light or right-circularly polarized light, their defect modes will move toward to the higher frequency and lower frequency respectively with the intensity increasing of external magnetic fields. Particularly, their peak frequencies for left-circularly polarized light and right-circularly polarized light are completely corresponding to the values of split defect mode on linearly polarized light in this process.

Keywords: Polarization properties, Magnetic photonic crystal, 4×4 transfer matrix method, Polarized light

1 Introduction

Since the original work of Yablonovitch [1] and John [2], Photonic Crystals (PCs) with period modulation of the dielectric constant have attracted considerable attention. It has been confirmed that a Photonic Band Gap (PBG) could be formed as a result of the interference of Bragg scattering in a periodical dielectric structure. In general,

---

Yang Li, Yan-Ling Han, School of Mathematics and Physics, China University of Geosciences, Wuhan 430074, PR China. Email: xiangyangliyang@foxmail.com (Y. Li).
Hong Wang, School of Automation, China University of Geosciences, Wuhan 430074, PR China. Email: wanghong745@sohu.com.
Hong Wang, Hubei key Laboratory of Advanced Control and Intelligent Automation for Complex Systems, China University of Geosciences, Wuhan 430074, PR China.
Guang-Bin Wu, Science and Research Office, Shenzhen Polytechnic, Shenzhen 518055, PR China.

DOI 10.1515/9783110540048-044,

PBG structure strongly depends on the geometry of the lattice and dielectric indices of components. Varieties of materials, such as semiconductors, liquid crystals, and metals have been proposed to acquire unique optical properties.

Magnetic photonic crystals (MPCs) are spatially periodic composite structures with one of the components being a magnetic material, such as a ferromagnet or a ferrite. Due to the anisotropy of magnetic material, the MPCs reveals some new features of electromagnetic properties in periodic mediums and open up new opportunities for practical application, such as enhancement of Faraday rotation [3, 4], nonreciprocity [5, 6] and nonlinear optical effect in microwave and infrared regions [7]. In recent years, magneto-optic effects provide another idea to tune MPCs by changing the refractive index under a external magnetic field, and the preliminary study has shown that the positions of PBG in a photonic crystal can be tunable effectively [8–10]. But the polarization dependence of defect mode for different incident light in MPCs still needs to be discussed in details.

In this paper, we perform a numerical simulation of MPCs system to investigate the polarization dependence of defect mode by 4×4 transfer matrix method. The MPCs is presented as a sandwiched structure that a magnetic gyrotropic material is sandwiched between two Bragg mirrors formed by alternating two isotropic dielectric layers. The transmission spectrum for linearly polarized light TE, Right-Circularly Polarized light (RCP), and Left-Circularly Polarized light (LCP) are calculated. It is found that the position of defect mode critically depends on the properties of incident light and the intensity of external magnetic fields. In the case of system without magnetization, the position of defect mode has no polarization dependence for different incident light, their peak frequencies synchronously increasing with the incident angle. In the case of system under effect of external magnetic field, the positions of defect mode possess obviously polarization dependence for different incident light. For the linearly polarized light TE, the defect mode will be split and move toward to the both edges of PBG with the intensity increasing of external magnetic field. For the incident LCP or RCP light, the corresponding defect mode will move toward to the higher frequency and lower frequency, respectively. Furthermore, we also confirmed that no matter what kinds of polarized light will their defect mode move toward to the higher frequency with the incident angle increasing while the system is under the static magnetic field along the direction of z-axis.

2 Model and Method

A schematic of the magnetic resonance cavity structure is illustrated in Fig.1, which consists of alternating non-magnetic dielectric layers infiltrated with magnetic layer, simply expressed as $(AB)^m C(AB)^m$, m is the periodical number. The indices A and

B correspond to two different dielectric layers with dielectric permittivity ε_1 and ε_2, respectively. The thickness of A, B and C are represent as d_A, d_B and d_C respectively. Component C is the chiral magnetic gyrotropic medium and its dielectric permittivity tensor after applying a static magnetic field along the z-axis can be written as follows[11]

$$\hat{\varepsilon} = \begin{pmatrix} \varepsilon_{xx} & i\varepsilon_{xy} & 0 \\ -i\varepsilon_{xy} & \varepsilon_{xx} & 0 \\ 0 & 0 & \varepsilon_{zz} \end{pmatrix} \quad (1)$$

Where ε_{xx}, ε_{yy} and ε_{zz} are real and $\varepsilon_{xx} = \varepsilon_{yy}$, the ε_{xy} defines the gyrotropy factor responsible for Faraday magneto-optical effect (circular birefringence) and in first approximation linearly with the magnetization vector, corresponding to either uniaxial, biaxial and magnetic material placed in a magnetic field parallel to the z-axis. In the optical wavelength regime, the relative permeability in most practice cases is reduced to a scalar which is the permeability of free space μ_0.

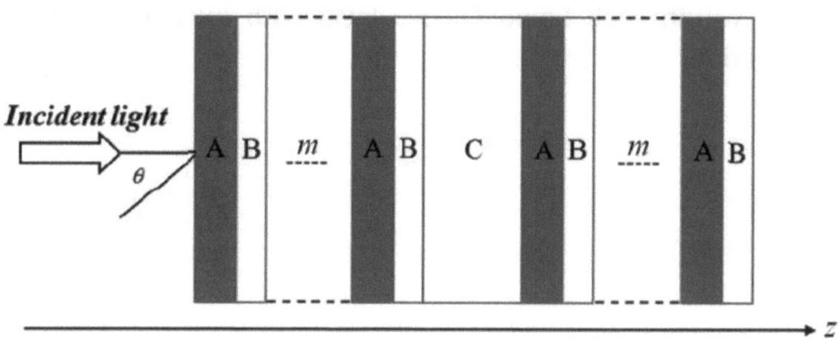

Fig. 1: Schematic diagram of magnetic resonance cavity C sandwiched between a pair of identical periodic non-magnetic stacks (Bragg reflectors)

The 4×4 transfer matrix method is effective in describing the optical properties of multilayered structure, which has been widely applied to uniaxial, biaxial dielectric and magnetic material [12 – 14]. Based on the Maxwell equations for a monochrome light propagating in the medium, we have

$$\frac{d\psi(z)}{dz} = \frac{i\omega}{c}\Delta(z)\psi(z) \quad (2)$$

Where the complex amplitude of electric and magnetic field components are arranged in a column forms: $\psi(z)=\left[E_x(z),E_y(z),H_x(z),H_y(z)\right]^T$, and the subscript T denotes the transpose operator, ω is the angular frequency of the incident light wave and c is the light speed in vacuum. $\Delta(z)$ is Berreman matrix [15] which is the function of permittivity, permeability and incident wave vector, its elements for permittivity tensor given in Eq. (1) is

$$\Delta(z)=\begin{pmatrix} 1 & 1-X^2/\varepsilon_{zz} & 0 & 0 \\ \varepsilon_{xx} & 0 & \varepsilon_{xy} & 0 \\ 0 & 0 & 0 & 1 \\ \varepsilon_{xy} & 0 & \varepsilon_{xx}-X^2 & 0 \end{pmatrix} \tag{3}$$

Where, $X=\sin\theta_{air}$ and θ_{air} is the incidence angle. The total transfer matrix in the present configuration is

$$P=\left(P_A P_B\right)^m P_C \left(P_A P_B\right)^m \tag{4}$$

Where P_A, P_B and P_C are the propagation matrices for A, B and C layers, respectively. The propagation matrix within each homogeneous layer with thickness h can be obtained by direct calculation

$$P(z+h)=SGS^{-1} \tag{5}$$

Where

$$S=\begin{pmatrix} e_{x1} & e_{x2} & e_{x3} & e_{x4} \\ u_{y1} & u_{y2} & u_{y3} & u_{y4} \\ e_{y1} & e_{y2} & e_{y3} & e_{y4} \\ u_{x1} & u_{x2} & u_{x3} & u_{x4} \end{pmatrix}, G=\begin{pmatrix} e^{ik_0\lambda_1 h} & 0 & 0 & 0 \\ 0 & e^{ik_0\lambda_2 h} & 0 & 0 \\ 0 & 0 & e^{ik_0\lambda_3 h} & 0 \\ 0 & 0 & 0 & e^{ik_0\lambda_4 h} \end{pmatrix} \tag{6}$$

Here, the e_σ, u_σ and λ_σ, $\sigma=1,2,3,4$ are elements of the eigenvector and eigenvalue of the matrix in Eq. (3), respectively [16]. With the above results, we can calculate the transmission coefficient T by the following relation [17]

$$T=\frac{\left|E_{xt}\right|^2+\cos^2\theta_{air}\left|E_{yt}\right|^2}{\left|E_{xi}\right|^2+\cos^2\theta_{air}\left|E_{yi}\right|^2} \tag{7}$$

Using the methods proposed above, various cases could be calculated. In this paper, we focus on calculating the transmission spectrum of one-dimensional magnetic photonic crystal to study the polarization dependence of defect mode.

3 Results and Discussion

In order to form the PBG structure, the dielectric permittivity is assumed to be $\varepsilon_1 > \varepsilon_2$, $\varepsilon_1 = 9$ and $\varepsilon_2 = 3$. The corresponding thickness is $d_A = 0.403\mu m$, $d_B = 0.556\mu m$ and the periodical number of dielectric layer is $m = 7$. For magneto-optic material, we assume the $\varepsilon_{xx} = 3$, $\varepsilon_{zz} = 9$ and $d_C = 2d_A$. In the present calculations, we use the values of material parameters $Y_{2.5}Bi_{0.5}Fe_5O_{12}$ films which were given by Inoue et al [18], the off-diagonal dielectric element ε_{xy} varies from 0 to 0.6, and the incident angle varies from 0 to 85degrees.

Before studying the transmission spectrum of defect mode in the presence of external magnetic fields, we start with the characterization of a layered structure without magnetization, i.e. $\varepsilon_{xy} = 0$. We consider the incident lights are linearly polarized TE light, LCP light and RCP light, respectively, and they propagating normally to the above layered structure. Their transmission spectrum is illustrated in Fig. 2, i.e. (a), (c), (e) are the corresponding spectrum of TE, LCP, and RCP light when the incident angle is $\theta = 0$, respectively. We can find that the resonant peaks for different polarized light are overlapping each other, and their peak frequencies are fixed at 0.421, located in the middle of the same photonic band gap. In addition, the transmittance for linearly polarized light TE is 1, but the transmittance for LCP light and RCP light are 0.5. For the Fig.2 (b), (d), (f) are the corresponding spectrum of TE, LCP, and RCP light when the incident angle is $\theta = 55\text{degree}$, respectively. Their peak frequencies are keeping at 0.446. This result indicates that for the system without magnetization, the position of defect mode has no polarization dependence for different incident polarized light, i.e. their peak frequencies are equal, and synchronously move toward to the higher frequency with the incident angle increasing, as shown in Fig. 3.

Next, let us study the nature of the corresponding defect mode in the presence of external magnetic fields. i.e. $\varepsilon_{xy} > 0$. We repeated the calculation of similar structures for different incident light with parameter value $\varepsilon_{xy} = 0.2$ and the incident angle fixed at $\theta = 0$. The corresponding spectrum is shown in Fig.4. (a), (b), (c) are the transmission spectrums of TE, LCP, and RCP light, respectively. Comparing with the corresponding spectrum that system without magnetization listed in Fig.2 (a), (c), (e), we can see clearly that for the incident linearly polarized light TE, its original defect mode will be split when the system is under external magnetic field, and their peak frequency will move toward to the lower frequency and higher frequency, respectively, i.e. for the left-side defect mode TE1, its peak frequency is

$0.4158 < 0.421$, and the peak frequency of right-side defect mode TE2 is $0.4275 > 0.421$.

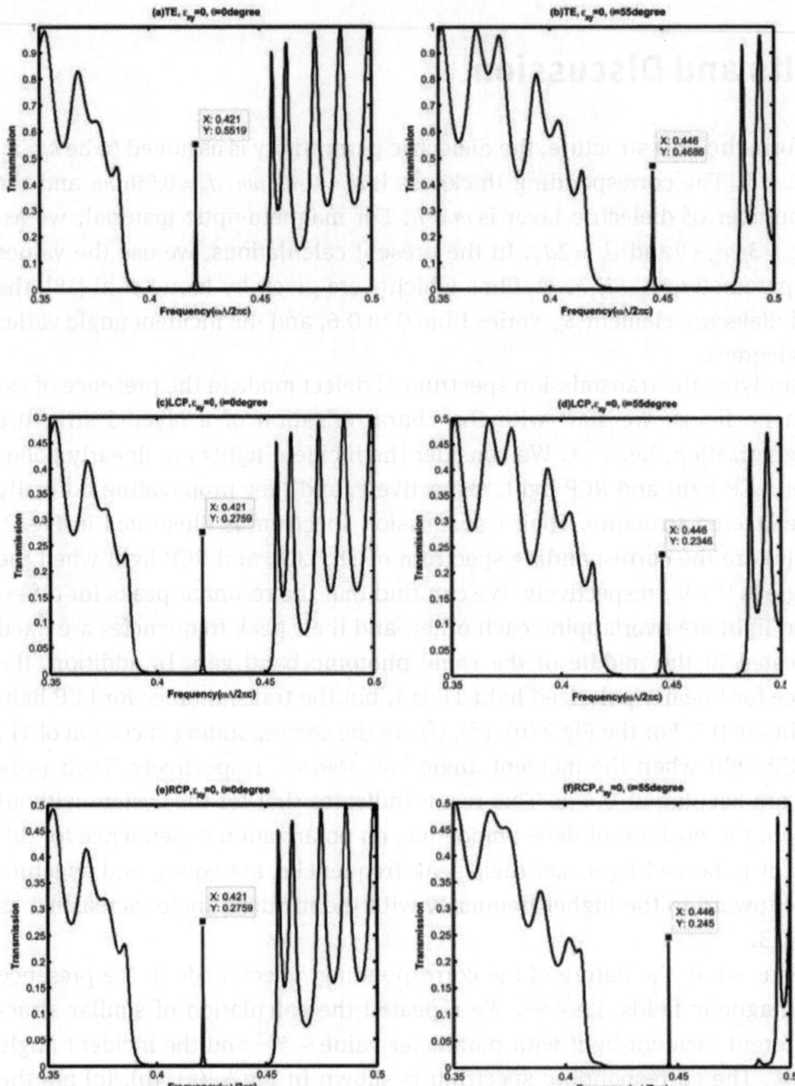

Fig. 2: Transmission spectrum of defect mode for photonic crystal without magnetization depicted in Fig. 1, for different incident polarized light at various incident angle, (a) TE light, (c) LCP light, (e) RCP light at $\theta = 0$, respectively; (b) TE light, (d) LCP light, (f) RCP light at $\theta = 55\text{degree}$, respectively, and other parameters are $\Lambda = d_A + d_B$, $\varepsilon_{xy} = 0$ and $m = 7$

In addition, an interesting feature exhibit in Fig. 4 (b) and (c) is that the corresponding peak frequency for incident light LCP and RCP is 0.4275 and 0.4158, respectively, which are equal to the values of right-side and left-side defect mode illustrated in Fig. 4 (a). In the discussion above, we just consider the particular case, i.e. $\varepsilon_{xy} = 0.2$. For different intensity of external magnetic field, the effect of external magnetic field on defect mode of different incident light is checked in details, which is shown in Fig. 4 (d). It is found that the peak frequency of left-side defect mode TE1 keeps pace with the incident light RCP, and move towards to lower frequency. Meanwhile, the peak frequency of right-side defect mode TE2 keeps pace with the incident light LCP and move towards to higher frequency. It means that in the case of MPCs system under effect of external magnetic fields, the defect modes have polarization dependence for different incident light, and the linearly polarized light can be viewed as the overlapping of LCP and RCP. For the left-side defect mode, it is correspond to the defect mode of RCP. For the right-side defect mode, it is correspond to the defect mode of LCP.

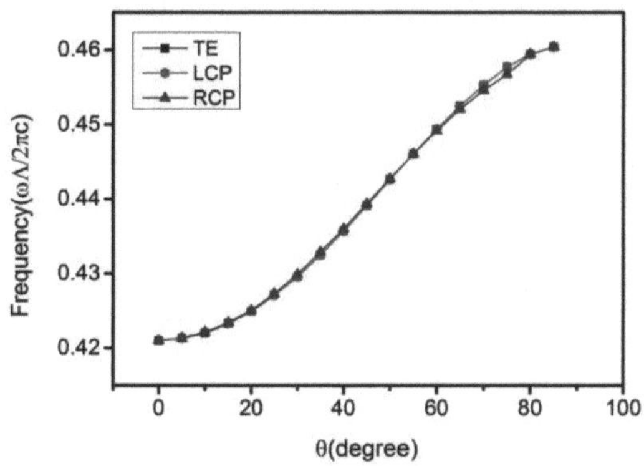

Fig. 3: Effect of defect mode on different incident polarized light with the incident angle increasing, the defect mode for TE, LCP and RCP are overlapping. The corresponding parameters are $\Lambda = d_A + d_B$, $\varepsilon_{xy} = 0$ and $m = 7$

Actually, the polarization properties of defect mode result from the Circular Bragg Phenomenon (CBP) [19]. The electromagnetic eigenmodes of the layered structure with permittivity tensors given in Eq. (1) are all circularly polarized. Generally, for the isotropic medium, the Bragg phenomenon is independent of polarization of a normally incident electromagnetic wave. In this case, the RCP and LCP wave is degenerate, which implies defect modes have no dependence of RCP and LCP wave. However, this conclusion holds only for no magnetization material i.e. $\varepsilon_{xy} = 0$. For magnetized anisotropic material, i.e. $\varepsilon_{xy} > 0$, incident electromagnetic plane waves of

LCP and RCP waves are reflected and transmitted differently in the Bragg wavelength regime. For LCP and RCP wave, in the presence of external magnetic fields, the corresponding refractive indices is $n_{\pm}^2 = \varepsilon_{xx} \pm \varepsilon_{xy}$, where n_+ is for RCP wave and n_- for LCP wave. From this equation, we can see that n_+ increase and n_- decrease in the presence of external magnetic fields. In such a case, according to the fundamental Bragg condition $\lambda = 2\Delta n_{\pm}$, defect mode for RCP and LCP wave move toward to the lower and higher frequency, respectively. When the system has no effect of external magnetic fields, i.e. $n_+ = n_- = \sqrt{\varepsilon_{xx}}$, this is the typical case of isotropic dielectric medium.

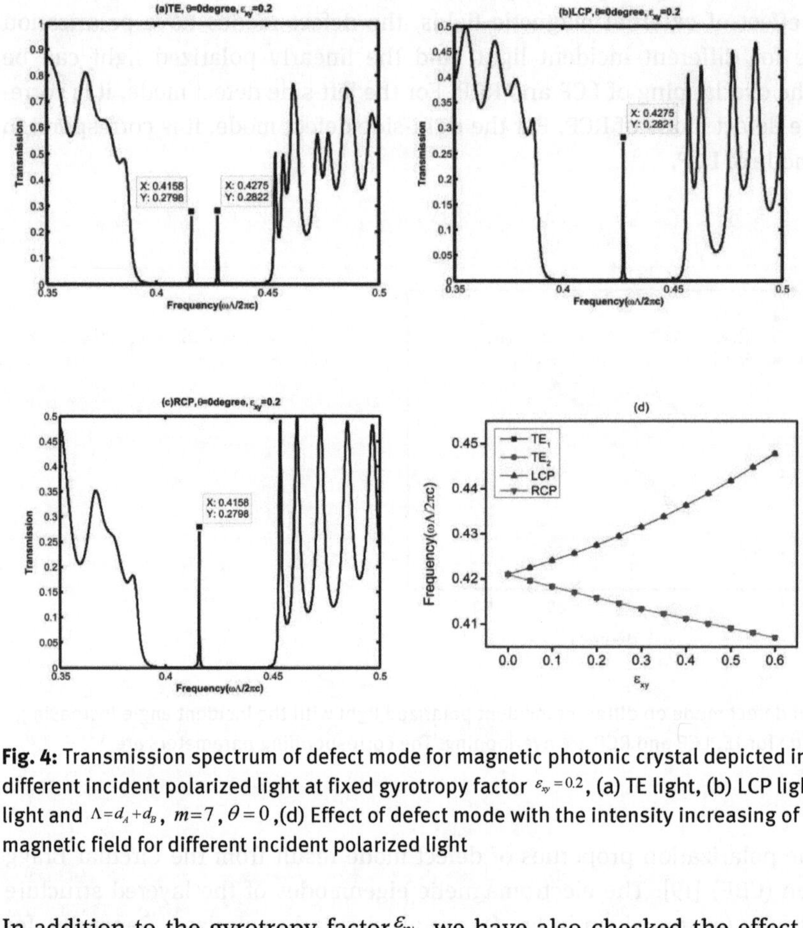

Fig. 4: Transmission spectrum of defect mode for magnetic photonic crystal depicted in Fig. 1, for different incident polarized light at fixed gyrotropy factor $\varepsilon_{xy}=0.2$, (a) TE light, (b) LCP light, (c) RCP light and $\Lambda = d_A + d_B$, $m=7$, $\theta=0$, (d) Effect of defect mode with the intensity increasing of external magnetic field for different incident polarized light

In addition to the gyrotropy factor ε_{xy}, we have also checked the effect of incident angle for different incident light in anisotropic material, and the result is illustrated in Fig. 5. In order to distinguish the difference between them, for the incident light TE, we fixed the factor $\varepsilon_{xy}=0.1$ and the moving of two defect modes TE_1 and TE_2 is marked with black line and red line, respectively. In the case of LCP or RCP, we fixed the factor ε_{xy}=0.01 and the change of their corresponding defect mode under external magnetic fields is also presented. We can see clearly that the defect mode will move toward to the higher frequency with incident angle increasing, variation tendency for LCP wave keep pace with TE_2 and RCP wave keep step with TE_1. From this fig, we can also conclude that for incident LCP light and RCP light, their corresponding defect mode indicates obviously polarization dependence in anisotropic medium.

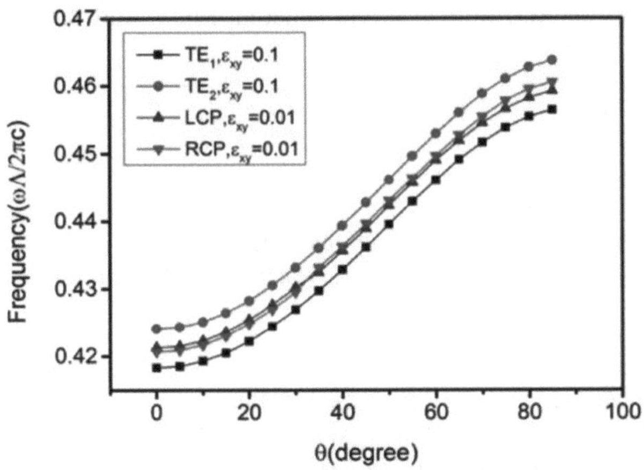

Fig. 5: Effect of defect mode in anisotropic photonic crystal for different incident light with the incident angle continuously increasing. For the incident linearly polarized light TE, the factor $\varepsilon_{xy} = 0.1$, and for the LCP or RCP light, $\varepsilon_{xy} = 0.01$. The corresponding parameters are $\Lambda = d_A + d_B$, and $m = 7$

4 Conclusion

In summary, using 4×4 transfer matrix method, we have investigated the polarization dependence of defect mode in one-dimensional magnetic photonic crystal system. The transmission spectrum for linear polarized light TE, Right-Circularly Polarized light (RCP), and Left-Circularly Polarized light (LCP) are calculated. It is found that the position of defect mode critically depends on the properties of incident light and the intensity of external magnetic field. In the case of system without magnetization, the position of defect mode has no polarization dependence for different incident light, their peak frequency synchronously increasing with the incident angle. In the case of system under external magnetic fields, the positions of defect mode possess obviously polarization dependence for different incident light. Particularly, for the linearly polarized light, its defect mode will be split and move toward to the both edges of photonic band gap respectively with the intensity increasing of external magnetic fields. Meanwhile, for the left-circularly polarized light or right-circularly polarized light, their defect modes will move toward to the higher frequency and lower frequency, respectively, and the peak frequency for left-circularly polarized light or right-circularly polarized light is completely corresponding to the values of split defect mode in this process. Furthermore, we also found that no matter what kinds of polarized light will their defect mode move toward to the higher frequency with the incident angle increasing while the system is under external magnetic field. This study maybe present a significant theory for designing new optical devices, such as spectrometer, optical angle sensors, filters and magnetic field measuring apparatus.

References

[1] E. Yablonovitch, Phys. Rev. Lett. 58, 2059 (1987).
[2] S. John, Phys. Rev. Lett. 58, 2486 (1987).
[3] M. Sadatgol, M. Rahman, E. Forati, J. Appl. Phys. 119, 103105 (2016).
[4] R. Fujikawa, Alexander V Baryshev, Alexander Khanikaev, J. Mat. Sci. Mat. Electron. 209, 493 (2008).
[5] A. Figotin, I. Vitebskiy, Phys. Rev. B. 67, 165210 (2003).
[6] P. A. Belov, S. A. Tretyakov, and A. J. Viitanen, Phys. Rev. E. 66, 016608 (2002).
[7] A. Figotin, Y. A. Godin, and I. Vitebsky, Phys. Rev. B. 57, 2841 (1998).
[8] Z. Wang, Y. D. Chong, J. D. Joannopulos, and S. Martin, Phys. Rev. Lett. 100, 013905 (2008).
[9] H. Wang, G. Wang, Y. Han, Y. Wang, Opt. Commun. 285, 4558 (2012).
[10] H. Wang, Y. Luo, Y. Wang, H. Zhang, Y. Fang, Phys. B, 406, 2977 (2011).
[11] H. Hajian, B. Rzaei, A. Soltani Vala, M. Kalafi, Phys. B. 405, 2996 (2010).
[12] O. Yang, Y. Wang, H. Wang, Y. Han, J. Peng, Optik. 126, 5583 (2015).
[13] G. Wu, Y. Han, Y. Li, H. Wang, Optik. 136, 58 (2017).
[14] H. Wang, G. Wang, Y. Han, F. Chen, Opt. Commun. 310, 199 (2014).
[15] D.W. Berreman, J. Opt. Soc. Am. 62, 502 (1972).
[16] Y. Pochi, J. Opt. Soc. Am. 69, 742 (1979).
[17] C.J. Chen, A. Lien, M.I. Nathan, J. Opt. Soc. Am. 14, 3125 (1997).
[18] M. Inoue, K. Isamoto, T. Yamamoto, Y. Fujii, J. Appl. Phys. 79, 1611 (1996).
[19] A. Lakhtakia, J. A. Reyes, Optik. 119, 253 (2008).

Yan Wang*, Yun-Geng Zhang, Pei-Chen Li* and Cai-Xia Feng

Preparation and Characterization of $TiO_{2-x}N_x$ with High Visible Light Activity

Abstract: Nitrogen doped TiO_2 with high visible light activity were prepared by annealing nanotubular titanic acid ($H_2Ti_2O_4(OH)_2$, denoted as NTA) in NH_3 flow at different temperature for 2h. Diffusion reflectance spectra (DRS) show that all the $TiO_{2-x}N_x$ samples have apparently visible light absorption over the entire region. X-ray diffraction (XRD) patterns indicate that the samples are completely consisted of anatase TiO_2, no rutile phase was detected. The chemical states of nitrogen atoms in $TiO_{2-x}N_x$ samples were investigated by X-ray photoelectron spectroscopy (XPS), and the XPS results show that only one N 1s peak appeares in the spectra, demonstrating that doped-N is present in the interstitial site in the form of Ti–O–N bond. The photoactivities were evaluated by decomposition of propylene under visible light ($\lambda \geq 420$ nm). The results show that the $TiO_{2-x}N_x$ samples annealed at 700 °C possess the best photoactivity.

Keywords: Nanotubular titanic acid, Visible light photocatalysis, Nitrogen doping, Propylene

1 Introduction

Titanium dioxide has been proven to be the most promising photocatalyst for its low cost, non-toxicity, high stability and high efficiency in the field of degradation of difficult-to-remove pollutants. But titanium dioxide is activated only by irradiating with ultraviolet (UV) light because the band gap energies is 3.0 eV for rutile and 3.2 eV for anatase, respectively.To use solar irradiation or interior lighting, several attempts have been made to extend the absorption range of TiO_2 from the ultraviolet to the visible light region using transition metal doping [1–4] or synthesis of reduced TiOx [5,6]. However, the metal-doped materials suffer from a thermal instability [2], an increase of carrier-recombination centers, or the requirement of an expensive ion-implantation facility [3]. Recently, N-doped TiO_2 has attracted widely attention for its photocatalytic activity in the visible light region [7–21]. The recent research in

Yan Wang, Cai-Xia Feng, Institute of Environmental and Analytical Sciences, College of Chemistry and Chemical Engineering, Henan University, Kaifeng 475004, China. E-mail: wangyan8079@vip.henu.edu.cn
Yan Wang, Yun-Geng Zhang, Pei-Chen Li, Department of Scientific Research, Henan University, Kaifeng 475004, China. E-mail: lpc@henu.edu.cn

DOI 10.1515/9783110540048-045,

2001 by Asahi et al [10] has rekindled a great interest in TiO_2 for its high visible light photoactivity, though the visible light response of N-doped TiO_2 was discovered in 1986 by Sato [11]. In conclusion, N-doped TiO_2 has been prepared by a variety of methods such as sputtering, ion implantation, chemical vapor deposition, sol-gel, oxidation of TiN, direct nitridation of TiO_2 and decomposition of N-containing metal organic precursors.

In the present paper, a simple method was found easily to dope nitrogen into TiO_2 and the $TiO_{2-x}N_x$ samples prepared by thermal treatment of NTA at different temperatures in the NH_3 flowing for 2h show apparently visible light photocatalytic activity for decomposition of propylene.

2 Experimental Section

NTA was synthesized from nanotubed sodium titanate ($Na_2Ti_2O_4(OH)_2$) by an ion exchange reaction with 0.1 mol/L HCl solution [22]. Then about 1g NTA was treated in the NH_3 flow for 2h at 400 ~ 700 °C. From the typical procedure, the $TiO_{2-x}N_x$ samples were obtained and denoted as N-NTA(4002 – 7002), respectively. P25-TiO_2 was selected as the reference.

The physicochemical properties of resultant N-doped TiO_2 samples were investigated by means of transmission electron microscopic (TEM), diffusion reflectance spectra (DRS), X-ray diffraction (XRD) and X-ray photoelectron spectroscopy (XPS). TEM images were taken on a JEM-2010 electron microscope. XRD patterns were measured on a Philips X'Pert Pro X-ray diffractometer. DRS were recorded on a Shimadzu U-3010 spectrometer. XPS characterizations were performed using a Escalab210 X-ray photoelectron spectrometer. The energy scale of XPS spectra was corrected using the binding energy of the adventitious carbon as C 1s = 284.8 eV.

The photocatalytic activity of the samples was evaluated by oxidation of propylene. The propylene gas (about 600 ppmV) was passed through the quartz reactor where the glass plate (1 cm × 10 cm) with the photocatalyst was placed at the flow rate of 100mL/h. The photocatalyst was irradiated with visible light from a Xe lamp passed through a cutoff filter ($\lambda \geq 420$ nm) and the propylene concentration were analyzed by the gas chromatograph (Shimadzu GAS CHROMATOGRAPH GC-9A), which was equipped with a GDX-502 colomn and a Fire Ion detector (FID).

Fig. 1: TEM images of NTA and relevant $TiO_{2-x}N_x$ samples prepared by treating NTA at different temperature in NH_3 flowing for 2h, respectively: (a) 400 °C; (b) 500 °C; (c) 600 °C; (d) 700 °C

3 Results and Discussion

Fig. 1 display the morphology of the relevant $TiO_{2-x}N_x$ samples and the inset in Fig. 1 (a) shows the typical TEM image of precursor NTA. As can be seen, NTA is hollow with a diameter of 5 ~ 10 nm and a length of several hundreds nanometers, which is well in agreement with our previous results [22]. However, after thermal treatment of NTA in NH_3 flowing, the resultant $TiO_{2-x}N_x$ samples could not keep the nanotubular morphology but were broken and gradually agglomerated into nanoparticles. With the temperature increasing, the particle size became gradually bigger and bigger.

Shown in Fig. 2 is the diffusion reflectance spectrum of NTA, P25 and the samples produced by treating NTA in NH_3 atmosphere at different temperature for 2h. It is obvious that all the $TiO_{2-x}N_x$ samples showed visible-light absorption over the entire range, while NTA and P25 show no visible-light absorption as a result of the white color. Also it can be obviously seen that the fundamental absorption edge of $TiO_{2-x}N_x$ samples did not shift into visible light region compared with P25, indicating that the N-doping into TiO_2 did not change the band gap of TiO_2. As shown in Fig. 1, it must be pointed out that the visible light absorption of $TiO_{2-x}N_x$ samples has not derect relationship with the thermal treatment temperaure. Samples prepared at 400 °C or 600 °C possess hardly the same light absorption ability.

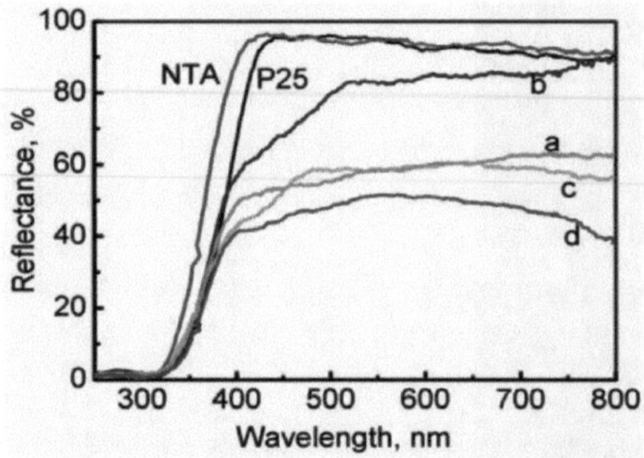

Fig. 2: Reflection spectrum of P25, NTA and relevant $TiO_{2-x}N_x$ samples prepared by treating NTA at different temperature in NH_3 flowing for 2h, respectively: (a) 400 °C; (b) 500 °C; (c) 600 °C; (d) 700 °C

The XRD patterns of N-NTA(4002–7002) samples prepared by NH_3-thermal treatment of NTA were shown in Fig. 3. It shows the effect of calcination temperature on the phase change of the $TiO_{2-x}N_x$ samples. As-prepared NTA materials belong to an orthorhombic system [22] (Figure not given here) As can be seen, the orthorhombic system gradually converted into anatase phase at 400 °C and all the samples were present in the anatase phase accompanying with the treating temperature increasing, no rutile phase corresponding to 2θ around 32.0° appeared even the treatment temperaure enhanced to 700 °C. Furthermore, the intensity of A(101) peaks became more and more sharper and higher accompanying with the treating temperature increasing, meaning that the better crystallization degree of anatase of the $TiO_{2-x}N_x$ samples.

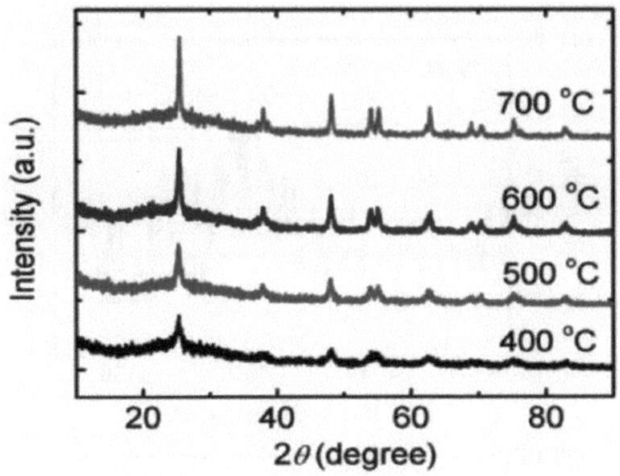

Fig. 3: XRD patterns of $TiO_{2-x}N_x$ samples prepared by treating NTA at different temperature in NH_3 flowing for 2h, respectively

Fig. 4 demonstrates the N 1s XPS spectra of the relevant $TiO_{2-x}N_x$ samples prepared by NH_3-thermal treatment of NTA at different temperatures for 2h. No N 1s XPS signals were detected on the surface of NTA (figures are not given here). After doping nitrogen, one could see that only one N 1s XPS peak with a core level binding energy of 400.0 eV was observed from the spectrum of the every $TiO_{2-x}N_x$ samples. It can be inferred that doped-N was present only in one form. Up to now, the N 1s chemical state in $TiO_{2-x}N_x$ samples was often divided into substitutional-N and interstitial-N depending on the binding energy of 396 ~397 eV and ca. 400.0 eV, respectively [7 – 22]. Therefore, the XPS results show that only one N 1s peak appeares in the spectra, demonstrating that doped-N is present in the interstitial site in the form of Ti–O–N bond.

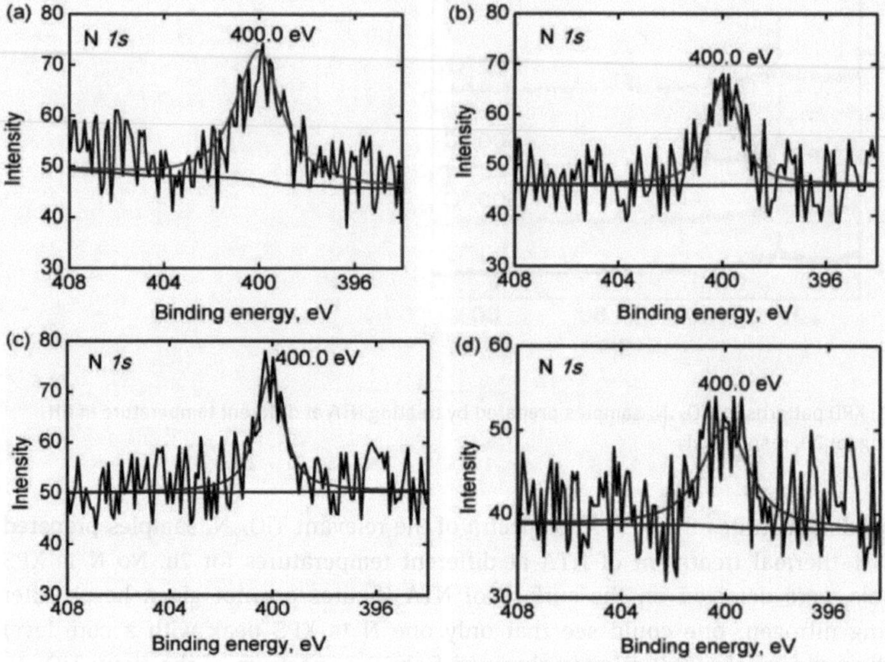

Fig. 4: N 1s XPS spectra of $TiO_{2-x}N_x$ samples prepared by treating NTA at different temperature in NH_3 flowing for 2h, respectively: (a) 400 °C; (b) 500 °C; (c) 600 °C; (d) 700 °C

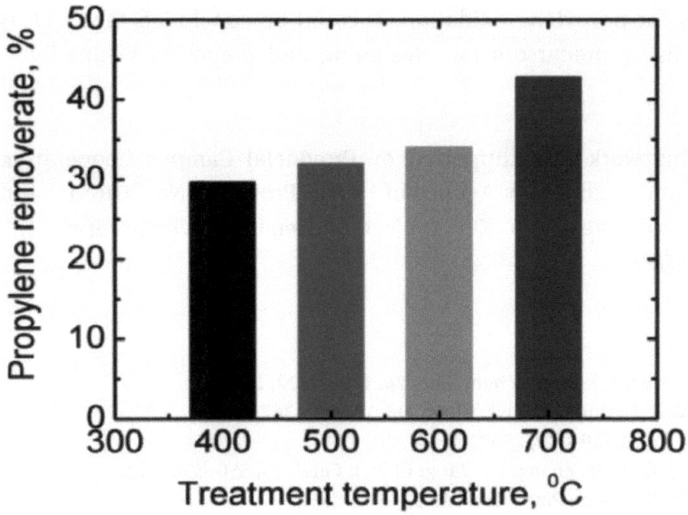

Fig. 5: Photocatalytic degradation of propylene removerate of $TiO_{2-x}N_x$ samples prepared by treating NTA at different temperature in NH_3 flowing for 2h, respectively

Fig.5 shows the photocatalytic ability for the decomposition of propylene under visible light (λ≥420nm). NTA and Degussa-P25 showed no vis-light-photoactivity (not given in Fig. 5). With the calcination temperature increasing from 400 °C to 700 °C, the photoactivity gradually increased from 29.7 % for $TiO_{2-x}N_x$ samples prepared at 400 °C to 42.8 % prepared at 700 °C. It can be seen that the $TiO_{2-x}N_x$ samples prepared at 700 °Cshowed the best photoactivity at this flow system. Furthermore, the removed propylene was converted into CO_2, which can be detected by the gas chromatograph, although we did not show the data here. From the data, it is obviously seen that the $TiO_{2-x}N_x$ samples prepared by anealing NTA in NH_3 flowing showed highly visible light photocatalytic activity.

4 Conclusion

Visible-light-active $TiO_{2-x}N_x$ samples were obtained by treating NTA at different temperature in NH_3 flowing for 2h. The physicochemical properties of resultant $TiO_{2-x}N_x$ samples were investigated by means of TEM, DRS. XRD, and XPS. The results showed that the nanotubular morphology of was destroyed into nanoparticles after thermal treatment, accompanying with its phase converted into anatase $TiO_{2-x}N_x$ samples from orthorhombic system of NTA. After nitrogen doping, doped-N was present in the interstitial site in the form of Ti–O–N bond. It was found that $TiO_{2-x}N_x$ samples obtained by annealing of NTA at 700 °C possessed the highest visible light photocatalytic activity for propylene oxidation. It could be concluded that NTA is one of the most promising precursors for designing and preparing visible-light-active photocatalyst.

Acknowledgement: This work was supported by Provincial Campus Cooperation Projects from Science and Technology Agency of Henan Province (No. 2016S11) and the Science & Technology Research Key Project of Henan province (Grant No. 15A150040, 162102210178).

References

[1] K. Ghosh and H. P. Maruska: J. Electrochem. Soc. 24, 1516–1522, 1977.

[2] W. Choi, A. Termin, M. R. Hoffman: J. Phys. Chem. 98, 13669–13679, 1994

[3] M. Anpo: Catal. Surv. Jpn. 1,169–179, 1997

[4] M. J. Jing, Y. Wang, J. J. Qian, M. Zhang, J. J. Yang: Chin. J. Catal, 33, 550–556, 2012.

[5] R. G. Breckenridge, W. R. Hosler: Phys. Rev. 91,793–802, 1953.

[6] D. C. Cronemeyer: Phys. Rev. 113, 1222–1226, 1959.

[7] A. Hattori, M. Yamamoto, H. Tada, S. Ito: Chem. Lett. 27, 707–708, 1998.

[8] T. Yamaki, T. Sumita, S. Yamamoto: J. Mater. Sci. Lett. 21, 33–35, 2002.

[9] R. Asahi, T. Morikawa, T. Ohwaki, K. Aoki, Y. Taga: Science, 293, 269–271, 2001.

[10] S. Sato: Chem. Phys. Lett. 123, 126–128, 1986.

[11] T. Sano, N. Negishi, K. Koike, K. Takeuchi, S. Matsuzawa: J.Mater. Chem. 14, 380–384, 2004.

[12] H. Tokudome, M. Miyauchi: Chem. Lett. 33, 1108–1109, 2004.

[13] M. Yang, T. Yang, M. Wong: Thin Solid Films, 469, 1–5, 2004.

[14] M. C. Wang, H. J. Lin, C. H. Wang, H. C. Wu: Thin Films. Ceram. Int. 38, 195200, 2012.

[15] O. Diwald, T. L. Thompson, G. Goralski, S. D. Walck, J. T. Jr.Yates: J. Phys. Chem. B 108, 52–57, 2004.

[16] Q. J. Xiang, J. G. Yu, W. G. Wang, M. Jaroniec, Chem. Commun. 47, 6906–6908, 2011.

[17] X. K. Wang, C. Wang, W. L. Guo, J. G., Wang: Mater. Res. Bull. 46, 2041–2044, 2011.

[18] X. Feng, Y. Wang, J. W. Zhang, L. G. Yu, D. L. Li, J. J. Yang: Appl Catal B: Environ, 113–114, 61–71, 2012.

[19] D. Valentin, G. Pacchioni, A. Selloni, S. Livraghi, E. Giamello: J. Phys. Chem. B 109, 11414–11419, 2005.

[20] Y. Wang, C. X. Feng, M. Zhang, J. J. Yang, Z. J. Zhang: Appl Catal B: Environ. 104, 268–274, 2011.

[21] Burda,Y. B. Lou, X. B. Chen, Anna C. S. Samia, J. Stout, and J. L. Gole: Nano Lett. 3, 8–11, 2003.

[22] J. J. Yang, Z. S. Jin, X. D. Wang, W. Li, J. W. Zhang, S. L. Zhang, Z. J. Zhang: Dalton Trans. 20, 3898–3901, 2003.

Ming-Yu Shang, Li-Xi Zhang, Jiao-Jiao Zhang, Cheng Wang and Zhan-Shan Li*

An Optimized MaxRPC Algorithm with a New Search Strategy

Abstract: Max Restricted Path Consistency (maxRPC) is a kind of local consistency for binary constraints in the field. It can achieve noticeably better pruning than arc consistency. Many implementations of maxRPC have been raised in [1]. One of these algorithms (named maxRPC3) has the best time complexity theoretically. However it still has overheads and redundancies for their repeatedly performing constraint checks to some extent. Considering that the process of searching for PC-witnesses can be performed many times and the most constraint checks are performed in this process, the key of improving the performance of maxRPC algorithms is to accelerate this process. In this paper, two algorithms maxRPC3-Alter and maxRPC3^{rm}-Alter with a new search strategy based on bitwise operation to accelerate the computations of this process are proposed. Experimental results demonstrate that both new algorithms and their light versions have good performance and outperform the best ones among existing algorithms by a large margin, which also constitute a more than viable alternative to arc consistency.

Keywords: maxRPC, Constraint Satisfaction Problem, Search Strategy

1 Introduction

Constraint Satisfaction Problem (CSP) is the core technologies in the field of artificial intelligence and is one of the most challenging research directions. At present, it has been widely used in various fields of artificial intelligence, such as qualitative reasoning, model-based diagnosis, scene analysis, task scheduling, system configuration, natural language understanding, and scientific experiment planning and so on. Filtering techniques are always used to remove some local inconsistencies in the search algorithms when used in a preprocessing step or during the search. It com-

Ming-Yu Shang, College of Software, Jilin University, Email: shanggay123@gmail.com
Li-Xi Zhang, College of Software, Jilin University, Email: zhanglx5514@mails.jlu.edu.cn
Jiao-Jiao Zhang, College of Software, Jilin University, Email: zhangjj5514@mails.jlu.edu.cn
Cheng Wang, College of Computer Science and Technology, Jilin University, Email: peter-wang951127 @gmail.com
***Corresponding author: Zhan-Shan Li,** College of Computer Science and Technology, Jilin University, Email: zslizsli@163.com..

DOI 10.1515/9783110540048-046,

presses the variable field of the constraint network without changing the constraint graph by adding new constraints or deleting tuples that do not satisfy the constraints. maxRPC is one of the most important filtering techniques for the two constraint networks [1]. Computational experiments show that maxRPC provides a great compromise between the computational cost and the pruning efficiency [2]. MaxRPC removes both the values that have no PC-supports and AC-supports, which is stricter than AC. MaxRPC has been implemented by many different methods as the most important part of constraint programming, such as: maxRPC1 [1], maxRPC2 [3], maxRPC3 [4], maxRPC^{rm} [4], and maxRPC3^{rm} [5]. According to [5] we know that maxRPC3, lmaxRPC3 and lmaxRPC3^{rm} are the most efficient algorithms when used stand-alone among existing algorithms, and lmaxRPC3^{rm} has the best performance when solving the instances of the CSP.

In this paper, we propose a new method to improve the performance of maxRPC3 algorithm. Traditional algorithms tend to test each values of x_k in $D(x_k)$ one by one when searching for PC-witnesses, as a result of which, some redundant tuples could be repeatedly checked, which will result in a reduction in the efficiency of the algorithm. The new algorithm that we proposed combines maxRPC3/maxRPC3^{rm} algorithms with a more efficient search strategy which exploits special data structures and the bitwise operations during the process of PC-witness searching.

The rest of this paper we talk about these: in the next section, we recall the basic definitions of CSP while fixing the notations used in this paper. Section 3 describes maxRPC3 framework and introduces two implementations of maxRPC3 and their light versions. Then, most importantly, we will expound the new algorithm of maxRPC. In section 4, the algorithm employs a more efficient search strategy. By the end of this paper we will analyze our experimental results and come to conclusion.

2 Preliminary Knowledge

Generally speaking, researchers use a triple (X, D, C) to describe a CSP, in which X, D and C represent a finite set of variables, the set of domains and a finite set of constraints respectively. Each domain in $D(x_i) \in D$ denotes the current domain of x_i, i.e. it denotes the possible values for x_i. For each constraint $c \in C$, there is the corresponding subset of variables of X, which called vars (c) and represent the scope of c. Furthermore, its associated relation is called rel (c) and specifies the values for the variables in vars (c). An assignment of all variables such that each constraint in this problem could find a support could be seen as a solution of a CSP.

2.1 Binary CSP

If each constraint of a CSP involves at most two variables we call it a binary CSP. We assume that all the constraints are binary and binary constraint checks are performed in constant time in our experiment. In a binary CSP, a value $a_i \in D(x_i)$ is arc consistent (AC) *iff* for every constraint c_{ij} there exists a value $a_j \in D(x_j)$ s.t. the pair of values (a_i, a_j) satisfies c_{ij}.

2.2 Related Work

MaxRPC has attracted a lot of research effort. Two categories: coarse-grained algorithms (i.e. constraint-based) and fine-grained algorithms (i.e. value-based) constitute the existing propagators. The following will give a brief description of these maxRPC algorithms which ensure that each value in the domain of each variable is supported by some value in the domain of a variable. The value is constrained.

maxRPC1. maxRPC1 [1] maintains a list, which is used to record values that have been removed from the variable universe but not yet propagated. It is a fine-grained algorithm based on AC6 and has optimal O (end^3) time complexity but requires O (end) space complexity.

maxRPC2. maxRPC2 [3] uses the thought like that of AC2001. MaxRPC2 removes the complicated data structure in maxRPC1 and maintain a list, which is used to record the variables whose fields altered. It is a coarse-grained algorithm having O (end^3) time and O(ed) space complexity.

maxRPC3. maxRPC3 [5] utilizes 2 data structures to record the latest discovered PC-supports and AC-supports for each value on each constraint respectively. MaxRPC3 avoids some redundant constraint checks in maxRPC2.

maxRPC^{rm}. maxRPC^{rm} [4] is based on AC3^{rm} and use a light data structure, which is like that of MaxRPC2. It reduces the maintenance cost of the data structure and can be well used in the search process. It has O (en^2d^4) time and O (end) space complexity.

maxRPC3^{rm}. maxRPC3^{rm} [5] use the same data structures as maxRPC3 to record PC and AC supports. Algorithm maxRPC3 updates the 2 data structures incrementally like maxRPC2 and AC2001 [6] respectively do. However, in maxRPC3^{rm}, these structures are used as residues [7, 8] like maxRPC^{rm} and AC^{rm} do.

Hence it is obvious that there are a lot of improvements have been achieved by: using the lighter data structures to reduce the maintenance cost, using new structures to store more information which is more useful, storing the latest support found to improve the performance of algorithms, etc [9].

3 maxRPC3-Framework and Two Implementations

3.1 maxRPC3:A coarse-grained Framework

MaxRPC3 utilizes data structures LastPC and LastAC which have the following functionalities: For each constraint c_{ij} and each value $a_i \in D(x_i)$, $LastPC_{xi,ai,xj}$ and $LastAC_{xi,ai,xj}$ give (point to) the lastest discovered PC and AC supports of a_i in $D(x_j)$respectively. Initially, all LastPC and LastAC pointers are set to a special value NIL, considered to precede all values in any domain. Algorithm maxRPC3 updates the LastPC and LastAC structures incrementally like maxRPC2 and AC2001/3.1 respectively do. Being coarse-grained, maxRPC3 maintains a propagation list L. Variables that have their domain filtered are inserted in the list L. If the algorithm is used for preprocessing, each value a_i of each variable x_i should be checked whether a_i is maxRPC or not during an initialization phase. If it is not then it is deleted from D (x_i) and x_i is added to L. On the contrary, if the algorithm is used in the search, L will be initialized with the currently assigned variable [5].

3.2 Light maxRPC: The Light version of maxRPC

An approximation of maxRPC is light maxRPC (lmaxRPC). It only propagates the loss of AC-supports but not the loss of PC-witnesses [10]. The obtained algorithm enforces a consistency property that is not stronger than AC is ensured. The description of lmaxRPC is tied to a specific maxRPC algorithm for being a defined local consistency. Experiments with random problems applied algorithms $lmaxRPC^{rm}$ and $maxRPC^{rm}$ showed that the pruning power of lmaxRPC is only slightly weaker than that of maxRPC [11]. Meanwhile, when used during search, it can offer significant gains in run times. A series of experiments on various problem classes lead us to these results.

3.3 Performance Comparison Of Two Algorithms

To begin with, we named the processes of searching and checking PC-witnesses as schPCwit and cckPCwit respectively. LmaxRPC3 and LastPC show the main difference is that LastPC is not updated incrementally, for each seeked PC-support for a value $a_i \in D(x_i)$ in x_j, $D(x_j)$ will be searched from scratch in the worst case. This incurs an extra O (d) cost to the process of searching PC-supports. Hence, the node complexity of $lmaxRPC3^{rm}$ is $O(end^4)$. At the same time, the total cost of searching PC-witnesses in every node cannot be amortized which lead to that the cost of schPCwit within cckPCwit is $O(nd^2)$. Hence, the node complexity of $maxRPC3^{rm}$ is $O(en^2d^4)$. Because the calls to schPCwit are amortized, the branch complexities are the same. The space required for data structures LastPC and LastAC determine the space complexities of the algorithms. This is the node space complexity of (l) maxRPC3 and (l) maxRPC3rm because they both require O (ed) space. With extra space required for the incremental update and restoration of the data structures (l) maxRPC3 has O (end) branch space complexity. (l) MaxRPC3rm avoids part of this, so its branch space complexity is O(ed).

4 MaxRPC-Alter: With a New Search Strategy

In section 3, we have described that among all the performances of maxRPC3, the constraint check needs to be implemented to estimate if $((x_i, a), (x_k, c)) \in rel(c_{ik})$ and $((x_j, b), (x_k, c)) \in rel(c_{jk})$, while testing if a value (x_k, c) is the PC-witness of the value pair $((x_i, a), (x_j, b))$. We have to test the value of x_k separately till the finding of a PC-witness. This method is not quite business-like. For now, for the sake of improving the performance of existing algorithm, we put forward a novel algorithm whose name is (l) maxRPC3-Alter through a more business-like search strategy which is levered. In view of our computers are provided with the 64-bit processor. On the basis of the binary representations in [12], we employ bitwise operation to seek for the PC-witness rather than employ the constraint check, which can test 64 values synchronously every time. Moreover, our novel algorithm still keeps the merits of existing algorithms. Two data structures bs and bd are developed by us to present the constraint and the domain. Bs $[c_{ij}, x_i, a]$ stands for the binary representation of the favors of (x_i, a) which are in the c_{ij}. Every bit, which is in bs $[c_{ij}, x_i, a]$, marks if the corresponding value is a backing of (x_i, a) that exists in the domain of x_j. What's more, every bit, which is in bd [X], records if the corresponding value exists in the domain of X. Provided that we set one bit to 1 in bd [X], then the corresponding value will appear in the domain. The major ways of maxRPC3-Alter are searchPCsup (see Function 1) and checkPCwit (see Function 3). searchPCsup identifies whether (x_i, a) has a PC-support in $D(x_j)$. MaxRPC3-Alter begins to seek from the succeeding value of $LastPCx_i$, because a and x_j are the minimum PC-support of (x_i, a). When the maxRPC3rm-Alter seeks from the beginning value in D (x_j), this is due to LastPC serves as a residual, as well as it does not keep track of the minimum PC-support in maxRPC3^{rm}-Alter for ever. Function IsTrue is used for seeking a value (x_j, b) so that we can get the result of $((x_i, a), (x_j, b)) \in rel\ (c_{ij})$ and return true provided that the corresponding bit of b is 1 which is in bs $[c_{ij}, x_i, a]$. Provided that we find such a value, with regard to every third variable x_k, searchPCwit (see Function 2) is used for establishing if at the fewest one PC-witness of $((x_i, a), (x_j, b))$ consists in $D(x_k)$. Provided that the result is unlike to this, Witness will be set to false; the value will be examined, which is after b in D (x_j). While we find a PC-support b, at that time, maxRPC3-Alter will set $LastPC_{xi,a,xj}$ to b. Just like the maxRPC3rm, it develops the residue's multi-directionality, it also sets $LastPC_{xi,a,xj}$ to b and $LastPC_{xj,b,xi}$ to a. Nevertheless, it does not set $LastAC_{xi,a,xj}$ to b like maxRPC3rm does, just like in line 13, $LastAC_{xi,a,xj}$ is set to the location of b in bs. searchPCwit examines if at the fewest one PC-witness of ((x_i, a), (x_j, b)) consists in D (x_k). During such course, this way which is presented in Section 3 is used. In maxRPC3rm-Alter, we will check the surplus position first. While we detect one PC-witness, the location of it which is in bs will be registered. checkPCwit is used to examine for the reduction of PC-witness. With regard to every variate x_k that is compulsory by not only x_i but also x_j, provided that $LastPC_{xi,a,xk}$ are effective, it will check if there still exists a PC-witness of the value pair ((x_i, a), (x_k, $LastPC_{xi,a,xk}$)) which is in D (x_j). Provided that such a PC-witness exists, $LastPC_{xi,a,xk}$ will still be the PC-support of (x_i, a). Provided that $LastPC_{xi,a,xk}$ is noneffective or it is not the PC-support of (x_i, a) any more, like in line 6 to 8, we have to make sure that if other values in D (x_k) could be sought out for the PC-support of (x_i, a).

By means of removing a number of superfluous constrain inspections, maxRPC3-Alter possesses the best O (end^3) time complexity, and it shows good performance while it is used independently, yet it is costly to use in time of searching. The time complexity of $lmaxRPC3^{rm}$-Alter and $maxRPC3^{rm}$-Alter is alike to the $maxRPC3^{rm}$ and $lmaxRPC3^{rm}$ separately, which are O (end^4) and O (en^2d^4) separately while used independently. Moreover, while $lmaxRPC3^{rm}$-Alter is used in time of searching, it possesses a time complexity of O (end^4) as well. The supererogatory space leads to the overheads of O (ed^2) which is needed by the binary representations.

5 Experimental Results

For the sake of comparing the performance of diverse methods precisely, we have performed all of the proposed algorithms by C++. Moreover, we have also implemented many experiments about series of structured CSP questions which are used as benchmarks and can be seen on the website: http://www.cril.univ-artois.fr/~lecoutre/benchmarks.html, as well as involve not only satisfiable but also unsatisfiable examples. Eliminating examples which were very difficult to all the algorithms, our assessment was accomplished amount to 200 examples from varieties of question classes. Those examples come from diverse backgrounds and tell the structured examples from the arbitrary examples (the entropy of an example will be lower if it is more structured). In our experiments, the performances have been sounded in the light of the CPU time in seconds (t), as well as the amount of constraint checks (cc). Moreover, the consequences of table 1, table 2 and table 3 were implemented on an Intel(R) Core(TM) i5-4200M CPU @ 2.50GHz 3.10GHz which using the Windows 7. Please be advised average consequences for all the examples are ranged into specific question classes.

The performance of the algorithm, which is used independently, is compared in Table 1. We provide average consequences for all the examples which are sorted into particular question classes. Moreover, we contain consequences from the coarse-grained maxRPC algorithm, maxRPC3 and maxRPC3^{rm}, and from the whole coarse-grained algorithm's light versions, and our competitive algorithm (maxRPC-Alter and maxRPC3^{rm}-Alter). The result shows that, in the light of the run time, our algorithm has alike performance and is better than other existing algorithm through an element of three on average. This is because the remove of a number of constraint checks when the cc numbers display. We have compared the performance of search algorithm which applies lmaxRPC3^{rm}-Alter and lmaxRPC3^{rm} as well. This two search algorithms employ the dom/wdeg [11] heuristic.

We can know that, from the results from Tables 2 and 3, lmaxRPC3^{rm}-Alter always transcend lmaxRPC3^{rm}, and in all the 200 examples which we have tried, the situation is the same. When we look at the columns of lmaxRPC3^{rm}-Alter and lmaxRPC3^{rm,} we can clearly see that our approaches are able to decrease the number of constraint check by means of in the ballpark one order of magnitude (e.g. in quasigroup questions qcp and qwh). This is mostly because of the remove of parity check in the function searchPCwit. Cpu time is not reduced by the same amount, yet it can obtain a more than three times' acceleration (e.g. scen2-f25 and scen11-f8). It is important that, the accelerations acquired are able to make search algorithms that effectively apply lmaxRPC3-Alter rival with MAC on a lot of examples. For example, in scen11-f10, we can obtain the same run time like MAC when lmaxRPC3^{rm} is 3 times slower; we start as two times slower than it, and then we become two times faster than it when in the scen11-f7. Moreover, there are some examples in which MAC is transcended (e.g. the graph RLFAPs and most quasigroup questions). Sure, in spite of the ameliorations, there still exist examples where MAC is still fairly fast.

Function 1. searchPCsup(x_i, a, x_j):boolean

```
if ¬ RM and LastPC_{xi,a,xj} ≠ NIL then
  v = LastPC_{xi,a,xj}+1;
else
  v = first value in D(xj);
for each b ∈ D(xj), b ≥ v do
      PCwitness=true;
  if IsTrue(bs[cij, xi, a], b) then
    for each xk ∈ X s.t. cik ∈ C and cjk ∈ C do
      if ¬ checkPCwit(xi, a, xj, b, xk) then
        PCwitness=false; break;
     if PCwitness ≠ false then
        LastPC_{xi,a,xj} = b;
```

Function 1. searchPCsup(x_i, a, x_j):boolean

    if RM **then**
        $LastPC_{xj,\,b,\,xi}$ = a; $LastAC_{xi,\,a,\,xj}$ = b **div** 64;
        return true;
return false;

Function 2. searchPCwit (x_i, a, x_j):boolean

for each $x_k \in X$ s.t. $c_{ik} \in C$ and $c_{jk} \in C$ **do**
    maxRPCsupport = false;
    if $LastPC_{xi,\,a,\,xk} \in D(x_k)$ **then**
        if checkPCwit(x_i, *a*, x_k, $LastPC_{xi,\,a,\,xk}$, x_j) **then**
            maxRPCsupport = true;
    if *maxRPCsupport* and exists $c > LastPC_{xi,\,a,\,xk} \in D(x_k)$ **then**
        if searchPCsup(x_i, *a*, x_k) **then**
            maxRPCsupport = true;
    if ¬ *maxRPCsupport* **then return** true;
return true;

6 Conclusion

In this paper, we presented maxRPC3-Alter and maxRPC3^{rm}-Alter, two new algorithms for maxRPC, and their light versions that approximate maxRPC. These algorithms build on and improve existing maxRPC algorithms, achieving the elimination of some redundant constraint checks. We also investigated heuristics that can be used to order certain operations within maxRPC algorithms. Experimental results from various problem classes demonstrate that our best method, lmaxRPC3^{rm}-Alter, constantly outperforms existing algorithms, often by large margins. Significantly, the speed-ups obtained allow lmaxRPC3^{rm} to compete with and outperform MAC on many problems. In the future, we will do further research to improve the performance of our algorithms by adapting techniques for using residues from [8].

Function 3. checkPCwit(x_i, a, x_j, b, x_k):boolean

```
if RM then
  if LastAC_{xi, a, xk} ≠ NIL then
    i = LastAC_{xi, a, xk};
      if (bs[c_ik, x_i, a][i] AND bs[c_jk, x_j, b][i]
     AND bd[x_k][i]) ≠ 0

      then return true;
  if LastAC_{xj, b, xk} ≠ NIL then
      j= LastAC_{xj, b, xk};
      if (bs[c_ik, x_i, a][j] AND bs[c_jk, x_j, b][j]
AND bd[x_k][j]) ≠ 0
      then return true;
  for each i ∈ {0, ... , bs[c_ik, x_i, a].length-1} do
    if (bs[c_ik, x_i, a][i] AND bs[c_jk, x_j, b][i]
AND bd[x_k][i]) ≠ 0 then
        if RM then
           LastAC_{xi, a, xk} = i; LastAC_{xj, b, xk} = i;
        return true;
  return false;
```

Tab. 1: Average stand-alone performance in all 200 instances grouped by problem class. Cpu times (t) in secs and constraint checks (cc) are given

Problem class		maxRPC3	lmaxRPC3	lmaxRPC3^{rm}	
RLFAP	t	2.356	2.043	2.058	
(Scen,graph)	cc	9M	8M	9M	
Random	t	0.056	0.054	0.053	
(modelB,forced)	cc	0.19M	0.18M	0.18M	
Geometric	t	0.72	0.085	0.086	
	cc	0.36M	0.35M	0.35M	
Quasigroup	t	0.189	0.167	0.161	
(qcp,qwh,bqwh)	cc	0.61M	0.54M	0.54M	
QueensKnights,	t	47.091	45.131	43.736	
Queens,QueenAttack	cc	188M	188M	188M	
Drive,blackHole	t	0.326	0.296	0.303	
Haystacks,job-shop	cc	1.07M	1.02M	1.00M	
Problem class		**maxRPC3-Alter**	**maxRPC3^{rm}-Alter**	**lmaxRPC3-Alter**	**lmaxRPC3^{rm}-Alter**
RLFAP	t	0.821	0.843	0.795	0.803
(Scen,graph)	cc	3M	3M	3M	3M

Problem class		maxRPC3	lmaxRPC3	lmaxRPC3^{rm}	
Random	t	0.054	0.059	0.048	0.05
(modelB,forced)	cc	0.09M	0.10M	0.09M	0.09M
Geometric	t	0.029	0.025	0.02	0.021
	cc	0.13M	0.12M	0.12M	0.11M
Quasigroup	t	0.06	0.064	0.063	0.06
(qcp,qwh,bqwh)	cc	0.19M	0.19M	0.20M	0.19M
QueensKnights,	t	17.13	16.304	15.736	15.121
Queens, QueenAttack	cc	69M	73M	65M	66M
Drive, blackHole	t	0.295	0.338	0.303	0.299
Haystacks, job-shop	cc	0.36M	0.35M	0.34M	0.33M

Tab. 2: Cpu times (t) in secs and constraint checks (cc) from RLFAP instances. The best cpu time among the lmaxRPC methods is highlighted

instance		AC^{rm}	lmaxRPC3^{rm}	lmaxRPC3^{rm} -Alter
scen11-f10	t	11.0	12. 5	4.1
	cc	11M	51M	20.62M
scen2-f25	t	27.1	43.0	9.0
	cc	44M	151M	49.15M
scen11-f8	t	521.1	878. 8	172.7
	cc	638M	3,172M	0.35M
graph8-f10	t	16.4	9.1	5.1
	cc	14M	31M	12.85M
graph14-f28	t	31.4	3.1	3.0
	cc	13M	8M	4.51M
graph9-f9	t	273.5	101.5	30.2
	cc	158M	292M	1.00M

Tab. 3: Cpu times (t) in secs and constraint checks (cc) from various instances

instance		AC^{rm}	lmaxRPC3^{rm}	lmaxRPC3^{rm}-Alter
rand-2-40-8	t	4.0	21.7	5.6
-753-100-75	cc	7M	82M	30.62M
geo50-20	t	102.7	177.5	45.2
d4-75-1	cc	191M	880M	238.15M
qcp150-120-5	t	52.1	50.2	22.7
	cc	27M	53 M	18.35M
qcp150-120-9	t	226.4	238.1	76.1

instance		AC^{rm}	$lmaxRPC3^{rm}$	$lmaxRPC3^{rm}$-Alter
	cc	124M	250M	129.85M
qwh20-166-1	t	52.4	38.9	19.0
	cc	19M	23M	12.51M
qwh20-166-6	t	1673.5	867.1	400.2
	cc	158M	566M	313.00M
qwh20-166-9	t	41.5	25.0	10.5
	cc	15M	15M	9.72M
blackHole	t	1.5	3.8	1.0
4-4-e-8	cc	4M	12M	4.3M
queens-100	t	15.3	106.7	23.6
	cc	6M	377M	163M

Acknowledgement: The authors would like to express sincere thanks to our instructor, Professor Li because of his careful guidance and valuable suggestions in the process of topic selection and paper writing. Besides, our research is supported by the National Undergraduate Training Programs for Innovation and Entrepreneurship as a national project. We are grateful for this opportunity.

References

[1] Debruyne, R., Bessière, C.: From restricted path consistency to max-restricted path consistency. In: Smolka, G. (ed.) CP 1997. LNCS, vol. 1330, pp. 312–326. Springer, Heidelberg (1997)

[2] Debruyne, R., Bessière, C.: Domain Filtering Consistencies. JAIR 14, 205–230 (2001)

[3] Grandoni, F., Italiano, G.: Improved Algorithms for Max-Restricted Path Consistency. In: Rossi, F. (ed.) CP 2003. LNCS, vol. 2833, pp. 858–862. Springer, Heidelberg (2003)

[4] Vion, J., Debruyne, R.: Light Algorithms for Maintaining Max-RPC during Search. In: Proceedings of SARA 2009 (2009)

[5] Balafoutis, T., Paparrizou, A., Stergiou, K., Walsh, T.: Improving the performance of maxRPC. In: Cohen, D. (ed.) CP 2010. LNCS, vol. 6308, pp. 69–83. Springer, Heidelberg (2010)

[6] Y. Zhang and R. Yap, 2001. Making ac-3 an optimal algorithm. IJCAI'01, pages 316–321, Seatle, WA, USA.

[7] Likitvivatanavong, C., Zhang, Y., Shannon, S., Bowen, J., Freuder, E.: Arc Consistency during Search. In: Proceedings of IJCAI 2007, pp. 137–142 (2007)

[8] Lecoutre, C., And Hemery, F.: A Study of Residual Supports in Arc Consistency. In: Proceedings of IJCAI 2007, pp. 125–130 (2007)

[9] Grandoni, F., Italiano, G.: Improved Algorithms for Max-Restricted Path Consistency. In: Rossi, F. (ed.) CP 2003. LNCS, vol. 2833, pp. 858–862. Springer, Heidelberg (2003)

[10] Grandoni, F., Italiano, G.: Improved Algorithms for Max-Restricted Path Consistency. In: Rossi, F. (ed.) CP 2003. LNCS, vol. 2833, pp. 858–862. Springer, Heidelberg (2003)

[11] Vion, J., Debruyne, R.: Light Algorithms for Maintaining Max-RPC during Search. In: Proceedings of SARA 2009 (2009)

[12] Lecoutre, C., Vion, J.: Enforcing Arc Consistency using Bitwise Operations. Constraint Programming Letters 2, 21–35 (2008)

Tian-Jiao Zhao*, Ke Gao, Xing Li and Han Sun

Deep Learning based Design Image Management

Abstract: In the big data era, massive design images widely exist in the Internet. The design images possess the potential abilities of divergent design thinking; stimulate creative idea and patent circumvention. Whereas, there are few professional design image databases and the design images are not well organized to promote the design process. In this paper, we propose an efficient deign image management system (DIMS) based on deep learning techniques. A logo database with 200,000 logo images is collected to provide professional design image retrieval. To validate the superiority of the logo database and the proposed design image management system, we conduct comparisons with the searching results of Baidu and Google. The searching results show that DIMS can find more accurate logo images than Baidu and Google. Even given a sketch image, the related logo images are effectively found from the large-scale logo database by DMIS. This paper effectively bridges the massive design images and design process by deep learning techniques, and can utilize the searching results to inspire more creative design.

Keywords: design, image retrieval, deep learning, creative inspiration

1 Introduction

The idea generation is the first and the most significant step in the design process. A creative idea determines the quality of the final design result. Currently, researchers have developed many idea generation methods that contain brainstorming, metaphors, mind map, storyboards and so on [1]. Among all these creative idea stimulation method, image, text, objects and story are the main content that stimulate people's design idea. In Conclave's' research [2], he has pointed out that image possess the most efficient ability in stimulating creative idea.

In the big data era, data would bring Infinite possibility. In professional design company, intellectual property protection agency and open Internet space, a great amount of images exist. Design images have great potential in design field. Develop-

***Corresponding author: Tian-Jiao Zhao,** 55Building, Beiyang Campus, Tianjin University, Tianjin. Email: zhaotianjiao@tju.edu.cn
Ke Gao,55Building, Beiyang Campus, Tianjin University, Tianjin. Email: gaokeji@vip.qq.com
Xing Li, 55Building, Beiyang Campus, Tianjin University, Tianjin. Email: lxty7399@163.com
Han Sun, 55Building, Beiyang Campus, Tianjin University, Tianjin. Email: 371052213@qq.com

DOI 10.1515/9783110540048-047,

ing a professional design big data management platform is necessary for both generating creative idea and patent circumvention.

Currently, the design images exist in different places, the quality of the image in the open access Internet platform are varies considerably in quality. Mostly, sematic retrieval is used to search for useful images. The design images are divided according to their content, such as natural scenery, character material, biological world and so on. They are also divided according to its format, such as jpg, psd, ai and etc... In some property product images database, the image are arranged according to its mechanical structure, color, and textural. Most of the retrieval is Text-based Image Retrieval (TBIR). However, in design process, as text is a symbol of meaning [3], sematic retrieval has its limitation and shortage. The expression of text is not intuitive as image. Content-based Image Retrieval (CBIR) would bring more oppor-tunities for design. Through Image retrieval to stimulate inspiration would be more efficient and correctly.

In current research, deep learning demonstrates great ability in big scale image retrieval, video analysis and image production. It provides an effective technic for the organization and management of a large number of images. Artificial neural network is a research hotspot in the field of artificial intelligence. However, due to the lack of computing power, training algorithm and data samples, the large-scale deep neural network cannot get a good performance in the 90s of last century. In 2006, the Hinton [4 proposed a fast, greedy algorithm to solve the optimization problem of the multilayer auto encoder and deep belief network. Deep learning achieves the end to end learning by learning the feature from the raw data without the process of feature extraction, i.e. it can automatically learn a complex nonlinear expression with strong discriminant power from the original input data to avoiding the shortcomings of manual design features. At present, there are 4 kinds of network structure: Deep neural network (DNN) [5, Stacked Auto-Encoder (SAE), Deep belief network (DBN) [6, and Convolutional neural networks (CNN) [6].

The identification accuracy of ImageNet Classification is 85% in 2012, which have a milestone promotion compared with 74%, the result in 2011. And in 2013, the deep learning model improves the accuracy to 89%. In 2014, a team form Google win the competition and reduce the mistake rate at 6.66%, just one later, Deep Im-age from Baidu attenuated it at 5.98%, closed to our eyes.

In this study, we firstly analyzed the importance of image retrieval in design process and idea generation stage. Then by adopting deep learning technology, we would promote a new way to manage the design images. Secondly, a professional logo design database is collected as the data sources. A comparative experience is conducted between the professional logo database and Internet database. By ana-lyzing the results, the effect and advantage of DIMS is testified. As the possibility of the images is mining through DIMS, great design possibilities can be found and inspired in the future.

2 Design Image Management And Design Process

The design process contains three steps. The first step is the preliminary work of design. This part contains the design survey and needs analysis, which is used to discover the design opportunities. The medium term of design is a process that concept design is developing. The later period of design is transferring design concept to design product [7]. The three steps of design are the same as the structure of creative in Tim's book "Change by design" [8. He mentioned that creative is a system that composed of overlapping space. These spaces are inspiration—the opportunities that stimulate people to search the solving problems, conception—the process of producing, developing and testing the design idea, and implement—the route that pushing the project to the market. Image as an efficient creative idea stimulates content can be used to stimulate people's creative idea and promote the design process in any steps. In the whole design process, design images can be used to divergent thinking, stimulate inspiration and patent evasion (Figure 1). At the same time, the design result image would be added in the design image database to promote the future design.

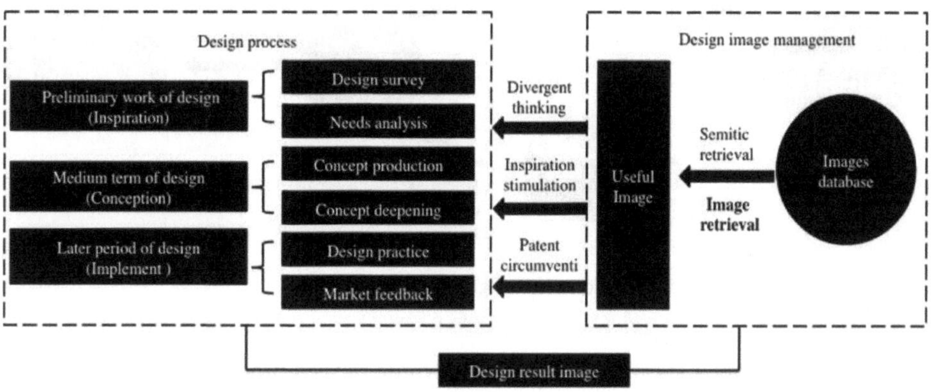

Fig. 1: Design process and design image material management

In these three steps, concept design has close connection with idea generation [9]. In current design research, from design thinking to design concept result, it would firstly experience thinking divergent process. Through the methods of mind map and brainstorm, the design thinking can be transferred to text. Then by adopting current text retrieval technic, related image sources can be obtained through the Internet and all these retrieved images have the potential of stimulating the creative idea. In this process, text as a symbol is used to connect design thinking and design result. This method expands the breadth of thinking and the collected content is emanative and informative. However, the retrieval content is so large and it is not easy to search meaningful images that with strong ability of stimulating creative idea. During the two-transfer process (from design thinking to text, from text to image), some design meanings would be lost and can't be expressed accuracy. To avoid this disadvantage, the connection between design thinking and design result can be constructed through sketching and related images. Sketch is the expression of people's thinking. Designers commonly use sketching to generate, express, and evaluate ideas, producing high-quality creative outcomes in the conceptual stage of design [10]. By adopting image retrieve method to stimulate inspiration would be an efficient and accuracy way for design. Design thinking-text-image model would expand the breadth of divergent thinking while design thinking-sketch-image model would increase the depth of divergent thinking. In design image management, sematic retrieval method and image retrieval method are two ways to search useful images. Only when they work together can design thinking be inspired sufficient.

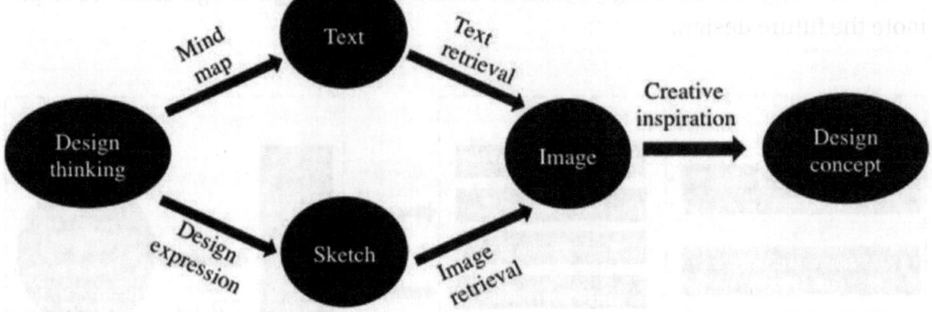

Fig. 2: The process from design thinking to design concept

Design images play a significant role in the whole design process. Besides stimulating the design idea, in both the early stage and the later stage, image retrieval would become a tool for patent circumvention. This work would not only decrease the unnecessary work of designers but also provide intellectual property protection for the designers. In current design image management, image retrieval show great potential in design process.

3 Technical Routes Of Design Image Retrieval

In this research, we use VGG-19 to extract features. VGG-19 is a deep learning network with 19 layers, and the train data should be a 224*224 RGB image, after a series of convolutional layers with the kernel size as 3*3, max pooling layers, and inner product layer, we get a 4096-dimension feature vector. And then, we use the features of all the images to contribute a KD-tree. KD-tree i.e. k-dimensional tree, a data structure segments the k-dimensional space. When we want to search an image in our dataset, we just input the feature we have got from VGG-19, and search the n nearest neighbors in the KD-tree. If we want to search an image from internet or a hand drawing picture, we should extract feature through VGG-19, and then, search the n nearest neighbors in the KD-tree. The KD-tree will give a result of the label of the n images, and then we find and output the image in our dataset according the label.

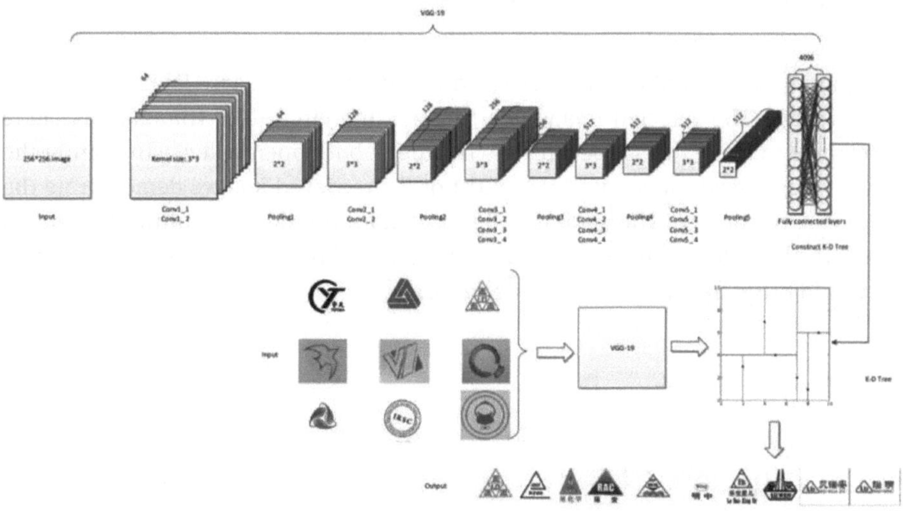

Fig. 3: Technical routes of design image retrieval

4 Experiments Result

In this research, we have conducted the image retrieval on three databases, which are logo image database, google database and Baidu database. In this study, the retrieved image are divided into two parts, which are selected images and sketch images. Figure 4 shows the first 45 images retrieved from the logo database.

Fig. 4: the first 45 images retrieve from the logo database

From figure 4, it is found that the retrieved images have the same features with the selected image. The selected image is a triangle in a circle. The triangle has made some change to make the shape with the feature of both open and closure. From the retrieved image result, it is found that most of the retrieved images demonstrate the feature of permeability. The searched figure is not only demonstrating the related shape with the selected image, but also the style of the logo. These similar style logos would bring inspirations to the designer.

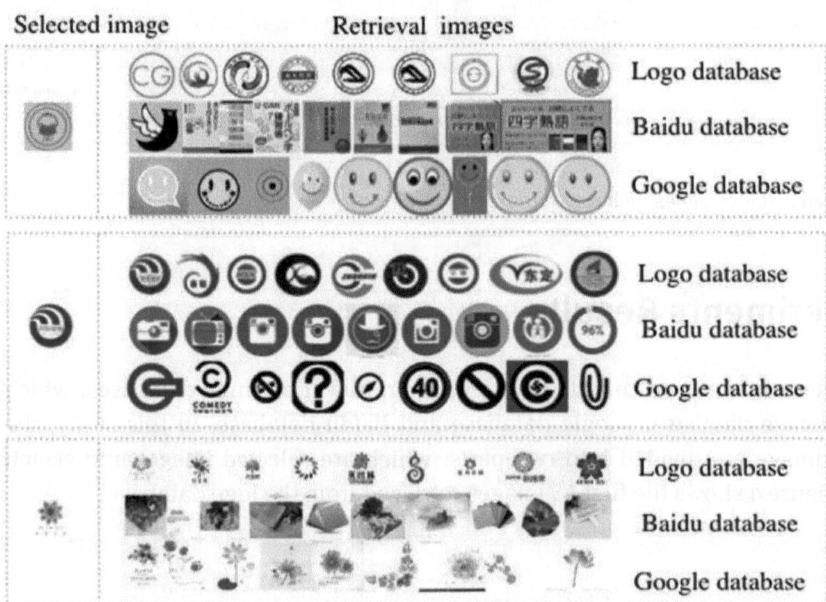

Fig. 5: Selected image retrieval result in different data base

Figure 5 shows the comparison between the retrieved result based on logo database, baidu database and google database. The first 9 retrieved related images are shown together. In figure 5, it is found that the retrieved images are not professional images from the Baidu and Google database. Some of them are posters, icon and even photos. Besides, the unprofessional images have great similarity in shape and color, which is not useful for design process and design thinking stimulating. In construct, the images retrieved from the professional design database demonstrate great advantage in quality and diversity.

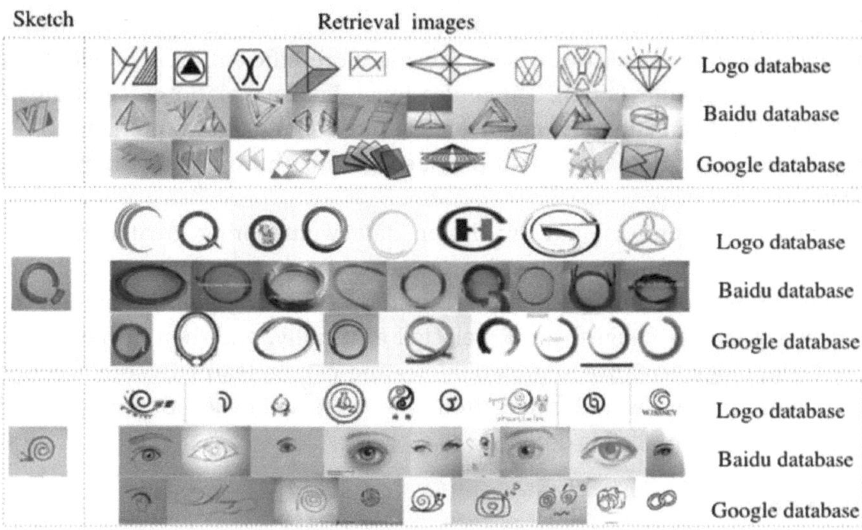

Fig. 6: Sketch image retrieval result in different database

Figure 6 is also a comparison result. All these images are retrieved according to the sketch images from the designers. From the comparison result, it is found that professional logo database not only demonstrate great ability in idea inspiration aspect but also show meaningful value in patent circumvention. For instance, the snail logo design is more or less the same as the one we retrieved from the logo database. In this situation, the sketch should be adjusted and redesign.

5 Discussion and Conclusions

Our purpose of this study is to explore a new design image management system. By connecting the design process with the images' creative potential, the design big data would bring great benefit to the design field in the future. In this research, we have obtained three findings.

Firstly, image retrieval is meaningful in the whole design process. By adopting deep learning technic, this research promotes a design image management system. This DIMS deeply mining the functions of image retrieval. Through this DIMS the design process and design images are connected closely. It would inspire designers' idea from a more intuitive and precise way.

Secondly, this study collected a professional logo database and testified the importance of professional design database. When image retrieval is conducted on professional database, the searched images are with high quality in stimulating creative idea. In design field, professional database should be widely constructed.

Thirdly, the DIMS make the design process more efficient as it provides the function of patent circumvention. The professional database platform is not only a study platform that can stimulate creative idea, but also a platform promotes intellectual property protection. The DIMS provides learning but not copying platform for design.

In conclusion, professional big design data would bring great benefit for the design process. After constructing an efficient image data management system, the future work can be conducted from the following points. Some of the high creative stimulate images can be selected artificially, and then through deep learning tech-nology "creative stimulation" feature can be learned. More accurate and meaningful images would select from the large database. Designer would be inspired easily from the big data.

References

[1] J.J. Shah, S.M. Smith, Vargas-Hernandez, N. Metrics for measuring ideation effectiveness, Design Study, 24:111–134, 2003.

[2] M. Gonçalves, C. Cardoso, P. Badke-Schaub, What inspires designers? Preferences on inspirational approaches during idea generation, Design Studies, 35, 1, 29–53, 2014.

[3] Roland Barthes, Empire of Signs, Hill and Wang, 1983.

[4] G. E. Hinton, R. R. Salakhutdinov. Reducing the dimensionality of data with neural networks, Science, 313, 5786, 504–507, 2006.

[5] [1] H. Larochelle, Y. Bengio, J. Louradour, et al. Exploring strategies for training deep neural networks, Journal of Machine Learning Research, 10, Jan, 1–40,2009.

[6] P. Baldi, Autoencoders, unsupervised learning, and deep architectures. ICML unsupervised and transfer learning, 27, 37–50, 1, 2012.

[7] P. Liu, S. Han, Z. Meng, et al. Facial expression recognition via a boosted deep belief network. Proceedings of the IEEE Conference on Computer Vision and Pattern Recognition, 1805–1812, 2014.

[8] Z. Wang, S. Lyu, G. Schalk, et al. Deep Feature Learning Using Target Priors with Applications in ECoG Signal Decoding for BCI, IJCAI,1785–1791, 2013.

[9] Z. Liu, N. Lu, L.Y. Sun, Method of product innovation knowledge acquisition based on knowledge flow, Computer Integrated Manufacturing Systems, 17,1, 10–17,2011.

[10] Tim Brown, Change by design: How design thinking transforms organization and inspire innovation. Harper Business, 2009.

[11] L. Sun, W. Xiang, C. Chai, C. Wang, Z. Liu, and Impact of text on idea generation: an electroencephalography study, International Journal of Technology and Design Education, 23, 4, 1047–1062, 2013.

[12] L. Sun, W. Xiang, C. Chai, C. Wang, Q. Huang. Creative Segment: A descriptive theory applied to computer-aided sketching, Design Studies, 35, 1, 54–79, 2014.

Xiang-Hui Lv*, Feng-Long Yin, Geng-Bo Zhang, Hao Hu and Ye Sun

The Ultrasonic Detection Terrain Data Processing Based on Adaptive Kalman Filtering

Abstract: In the ultrasonic terrain detection, individual data deviates from the actual data seriously. According to the continuity characteristics of terrain, the scale factor is added to the system process noise for perceiving the statistical characteristics of the noise and signal on the basis of traditional Kalman filtering. The method reached a certain degree matching between estimated value and observed value and achieved the adaptive filtering processing of the data. The algorithms were simulated by adding the noise single factitiously, and the results between traditional Kalman fliting were contrasted, it showed that the adaptive Kalman filtering has better estimation results. In order to test the terrain detection data recovery ability of the algorithms when the noise was added, the virtual data of hard wood terrain and gravelly soil terrain were managed, and the treatment effect of the two algorithms were analysed. The results show that the proposed algorithm can eliminate the adverse effect of noise on the system effectively, and can be used to deal with the real terrain data.

Keywords: Ultrasonic, terrain measurement, Kalman filtering, adaptive, simulation

1 Introduction

When using ultrasonic pulse echo method to detect terrain altitude, ultrasonic echo signal is actually combined signal[1,2] of reverse scattering echo and noise because of influence of the terrain being randomly ups and downs and rough, interference of environment noise, machine noise and noise of the system itself, making detected terrain altitude data abnormal, significantly deviating actual terrain.

Data filtering is a kind of data processing technique to get real data by removing noise; Kalman filtering can estimate the state of dynamic systems from a series of measurement noise in the case of measurement variance known. Because it is

***Corresponding author: Xiang-Hui Lv,** Northwest Institute of Nuclear Technology, Xi'An, 710024, China. Email: yinfenglongwade@163.com

DOI 10.1515/9783110540048-048,

convenient for computer programming, and can real-time update and process data from collection locale, Kalman filtering is the most widely used in filtering method, communication, navigation, guidance and control in many fields has been well used. Kalman filtering estimates the current value of state variable using a previous state estimate value and the latest observation value without requiring reservation of all old data, it can realize effective estimation[3] of past and current state of estimated signal when specific form of the model is unclear. Using recurrence algorithm, realizing smooth processing of data cased on minimum mean square error, Kalman filtering has been successfully used in such areas as automatic control, navigation, remote measurement and remote sensing, geophysics, etc.[4–6]. Yet it is hard to get higher accuracy to use traditional filtering algorithm due to diverse trajectory of target. So, in this paper we propose a kind of adaptive Kalman filtering algorithm that can achieve better data smoothing effect through comparison of simulation and terrain measured data.

2 Traditional Kalman Filtering Algorithm

Description of Kalman filtering consists of several mathematic recurrence formulas. It can realize minimize mean square error of estimated object by getting a kind of method to estimate process state from every recurrence formula. In real terrain measurement, observations are discrete value, so we mainly discuss the Kalman filtering algorithm of discrete system.

In the application of Kalman filter, estimated state value is expressed by $X \in \mathrm{R}^n$, system state model can be expressed as:

$$x[k] = \mathbf{A}x[k-1] + \mathbf{B}u[k-1] + w[k-1] \tag{1}$$

For observed variable use $Z \in \mathrm{R}^m$ to construct measuring equation:

$$z[k] = \mathbf{H}x[k] + v[k] \tag{2}$$

Where, k is time t_k, **A** is state transition matrix, **B** is input matrix, **H** is observation matrix and $u[k]$ is input signal.

$w[k]$ And $v[k]$ represent process noise and observation noise at time k respectively, normally are defined as independent Gauss white noise, witching following natures:

$$P(w) \sim N(0, \mathbf{Q}),\ P(v) \sim N(0, \mathbf{R})$$

Where, **Q** and **R** are square variance matrices of process noise and observation noise.

Kalman filtering adopts recurrence algorithm, first take state variable estimate at the time of k-1 $\hat{x}[k-1]$ as the prior estimate of state at No. K step, the algorithm does not consider the influence of observation noise, formulas (1) and (2) transform into:

$$\hat{x}[k,k-1] = \mathbf{A}\hat{x}[k-1] + \mathbf{B}u[k-1] \tag{3}$$

And

$$\hat{z}[k] = \mathbf{H}\hat{x}[k,k-1] \tag{4}$$

Where, $\hat{x}[k,k-1]$ is the prior state estimate of state variable, $\hat{z}[k]$ is the initial estimate of state variable.

Then use estimation error of output signal to correct estimation of state variable, obtaining estimation result of posterior state $\hat{x}[k]$, see Formula (5).

$$\hat{x}[k] = \hat{x}[k,k-1] + \mathbf{K}(z[k] - \hat{z}[k]) \tag{5}$$

Where, $z[k] - \hat{z}[k]$ is usually called innovation or remains, indicating difference between the estimation in Kalman filtering algorithm and actual value If it is zero, it indicates estimation and actual value are consistent. **K** in the formula is a $n \times m$ matrix, being the gain of innovation, also called Kalman gain matrix. It is typically expressed as:

$$\mathbf{K} = \mathbf{P}[k,k-1]\mathbf{H}^T(\mathbf{V})^{-1} \tag{6}$$

Then

$$\mathbf{V} = \mathbf{HP}[k,k-1]\mathbf{H}^T + \mathbf{R} \tag{7}$$

Where, $\mathbf{P}[k,k-1]$ is the variance matrix of priori estimation error, see Formula (8).

$$\mathbf{P}[k,k-1] = E\left[(x[k] - \hat{x}[k,k-1])(x[k] - \hat{x}[k,k-1])^T\right] \tag{8}$$

The following relationship exists between $\mathbf{P}[k]$ and $\mathbf{P}[k,k-1]$:

$$\mathbf{P}[k] = [\mathbf{I} - \mathbf{KH}]\mathbf{P}[k,k-1] \tag{9}$$

$\mathbf{P}[k]$ Is the error variance matrix at k time, also known as posteriori estimation error

$$\mathbf{P}[k] = E\left[(x[k]-\hat{x}[k])(x[k]-\hat{x}[k])^T\right] \tag{10}$$

Kalman filtering is essentially a kind of feedback control: i.e. while estimating state of a process at one point Kalman filter conducts closed loop feedback control over the estimation at that point using noise contained variable. Hence, Kalman filtering algorithm can be made up of time update and measurement equations. Like standard feedback control system, forward calculation of some process is completed by update equation through operations such as continuously resolving state variable and error estimation; meanwhile, the solution of last process provides reference for prior estimation of next time state. While forward calculation continuously proceeding, measure update equation also continuously instructs feedback mechanism, so posterior estimation is obtained as final calculation result by calculation prior estimate and measurement variable.

Update equation is as follows:

$$\hat{x}[k,k-1] = \mathbf{A}\hat{x}[k-1] + \mathbf{B}u[k-1] \tag{11}$$

$$\mathbf{P}[k,k-1] = \mathbf{AP}[k-1]\mathbf{A}^T + \mathbf{Q} \tag{12}$$

The update equation explains how to calculate state at time *k* through state at time *k-1*. When determining initial value $\hat{x}[0]$ and $\mathbf{P}[0]$, current estimation can be obtained.

In traditional Kalman algorithm, when finishing a process of differential state iteration calculation, it does not focus on the relation between current value and some previous state values. With increase of observed data, errors between state estimation after filtering and actual state exceed its theoretical estimation, resulting in filter divergence that makes filter estimation deviate without achieving filtering effect [7].

3 Adaptive Kalman Filtering Algorithm

Based on continuity of terrain, dynamic ultrasonic terrain detection system can estimate statistic parameters according to observed data, estimating the system using estimated statistic parameters. Thinking of adaptive Kalman filter is: in order to sense statistic characteristics of noise and signal, add scale factor *s* to system process noise, when difference of process noise is updated, the difference in algorithm will change too. Main method is to present a threshold subtracting the variance of previous time from process noise variance and compare the difference with the threshold to determine whether the algorithm needs to reduce influence of probabilistic characteristics of relevant parameters.

Adding scale factor s to system process noise, if input is not considered, then equation (1) will transform to:

$$x[k] = \mathbf{A}x[k-1] + s\mathbf{G}w[k-1] \tag{13}$$

Where, $\mathbf{G}$ is known interference matrix, process noise and observation noise of the system is irrelated, initial state is $x[0]$.

Estimation part of Kalman filtering estimation model changes into:

$$\hat{x}[k] = \mathbf{A}\hat{x}[k,k-1] \tag{14}$$

And

$$\mathbf{P}[k,k-1] = \mathbf{AP}[k-1]\mathbf{A}^T + s^2\mathbf{GQG}^T \tag{15}$$

For calculation of scale factor s, mean estimation of the sum of diagonal of variance matrix can be conducted recurrence calculation with following formula:

$$\hat{b}_1 = \boldsymbol{v}_1{}^T\boldsymbol{v}_1 \tag{16}$$

And

$$\hat{b}_k = (\hat{b}_{k-1} + \alpha\boldsymbol{v}_k{}^T\boldsymbol{v}_k)/(1+\alpha) \tag{17}$$

Where, α is a positive constant. The expansion of formula (17) is:

$$\hat{b}_k = \frac{\boldsymbol{v}_1{}^T\boldsymbol{v}_1}{(1+\alpha)^{k-1}} + \frac{\boldsymbol{v}_2{}^T\boldsymbol{v}_2}{(1+\alpha)^{k-1}} + \frac{\boldsymbol{v}_3{}^T\boldsymbol{v}_3}{(1+\alpha)^{k-2}} + \ldots + \frac{\boldsymbol{v}_k{}^T\boldsymbol{v}_k}{(1+\alpha)} \tag{18}$$

Hence, once estimation matches observed value somewhat, sum of $\hat{b}_k$ must be close to the sum of diagonal elements of covariance matrix $\boldsymbol{P}[k]$. So equation (19) is valid:

$$\hat{b}_k \approx tr(\mathbf{P}_k) \tag{19}$$

Meanwhile, by using equation (7) and (19), the following is concluded:

$$\hat{b}_k \approx tr(\mathbf{R}_k) + tr(\mathbf{HAP}[k-1]\mathbf{A}^T\mathbf{H}^T) + s^2 tr(\mathbf{HGQ}[k-1]\mathbf{G}^T\mathbf{H}^T) \tag{20}$$

Where

$$s^2 \approx \frac{\hat{b}_k - tr(\mathbf{R}_k) - tr(\mathbf{HAP}[k-1]\mathbf{A}^T\mathbf{H}^T)}{tr(\mathbf{HGQ}[k-1]\mathbf{G}^T\mathbf{H}^T)} \tag{21}$$

4 Simulation and Result

With distance of ultrasonic detection unchanged, i.e., in the case of front end sensor having no input data, manually adding Gauss noise, use traditional Kalman filtering algorithm and adaptive Kalman filtering algorithm to estimate the process under the state respectively. Because there is no real sensor data input, traditional Kalman filtering algorithm and adaptive Kalman filtering algorithm can be purely analyzed and compared without signal.

Simulation result is shown in Fig.1, green full line in it indicates actual distance between sensor and measured terrain, namely 800mm, "+" indicates actual distance measured by the detection system between sensor and measured terrain, its value is obtained by adding actual distance to Gauss noise, blue full line indicates estimation of Kalman filtering.

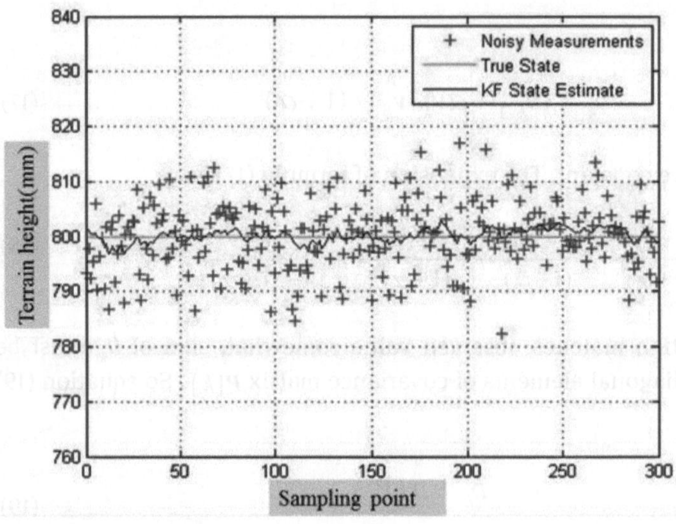

(a) Traditional Kalman filtering

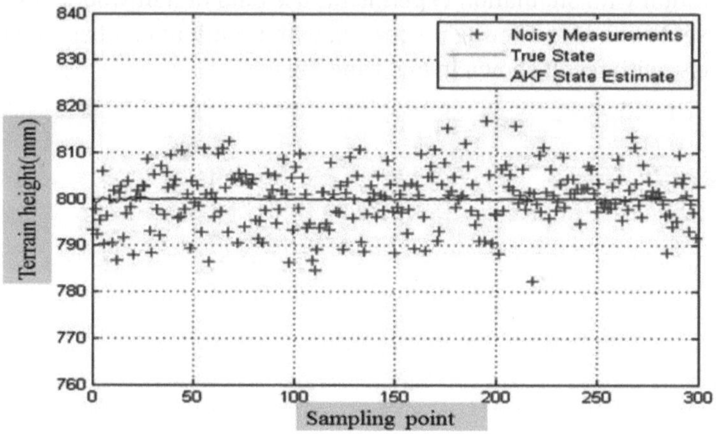

(b) Adaptive Kalman filtering

Fig. 1: The simulation result of Kalman filtering

From Fig.1 it can be found that in this state, actual value is a straight line with amplitude being zero, but measurement value distribute like the form of red "+" in the two figures due to interference of various errors and noise. Through Kalman filtering, measurement value rapidly approached intrinsic distance value between sensor and measured terrain. But comparing Fig. 1(a) and Fig.1(b), effect of traditional Kalman filtering is less good than adaptive Kalman filtering, so in this case adaptive Kalman filtering has better estimation effect.

5 Actual Terrain Data Processing

Measure elevation of hardwood terrain and gravels terrain in laboratory environment with ultrasonic terrain detection system, with sampling interval being 10mm, both are arc terrains with radius of 2,000 mm, as shown in Fig. 2.

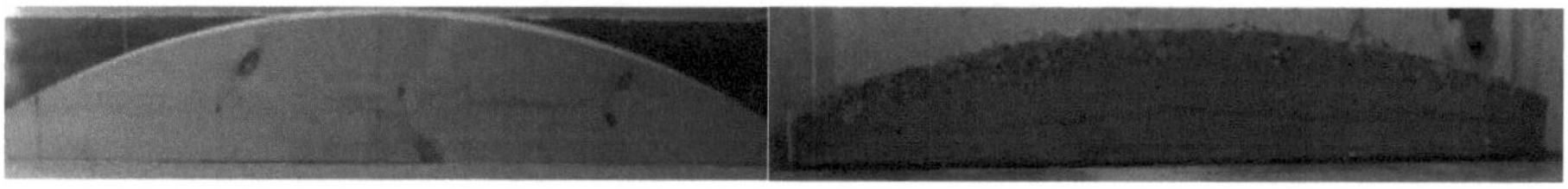

(a) Hardwood terrain (b) Gravels terrain

Fig. 2: Actual terrain in lab

Keeping same condition with simulation experiment, use data of hardwood, gravels terrain as data set to conduct filtering processing with traditional and adaptive methods respectively, with result shown in Fig.3 and Fig.4.

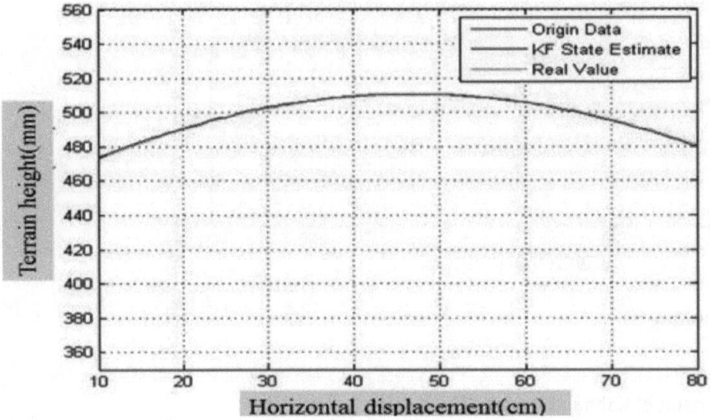

(a) Traditional Kalman filtering

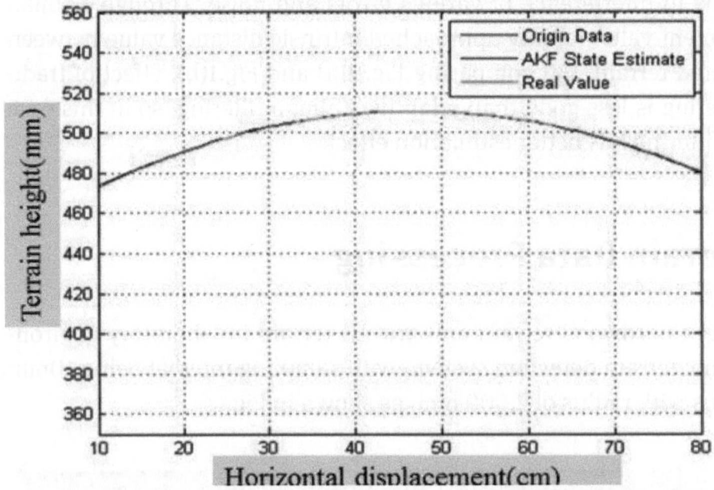

(b) Adaptive Kalman filterin

Fig. 3: Filtering effect of hardwood terrain

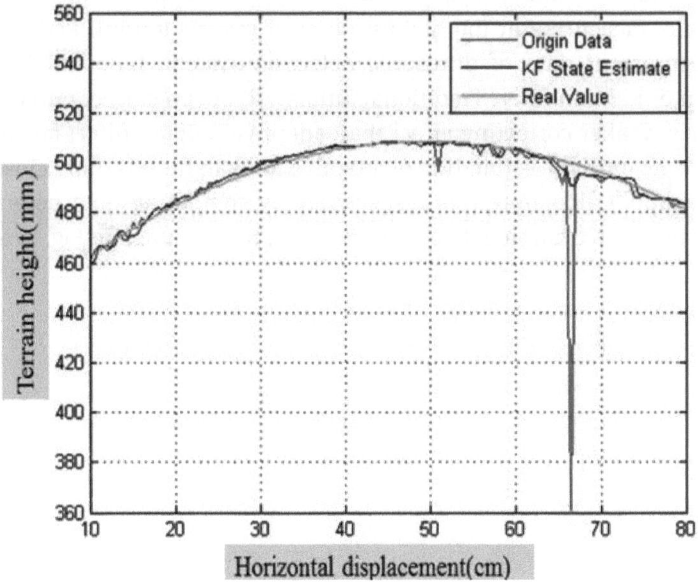

(a) Traditional Kalman filtering

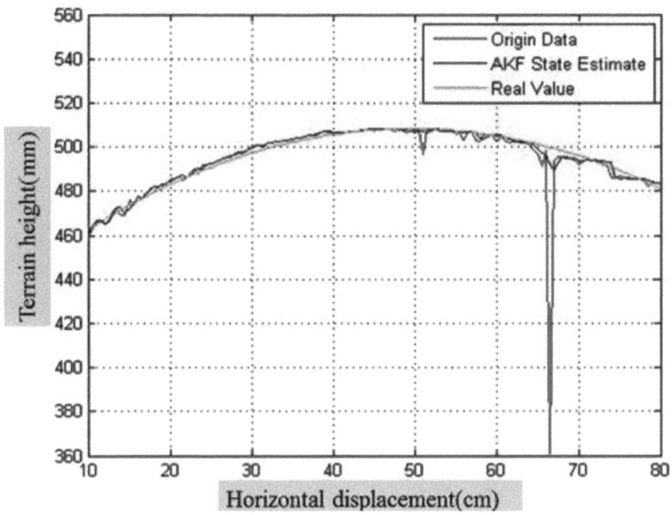

(b) Adaptive Kalman filtering

Fig. 4: Filtering effect of gravels terrain

In Fig. 3, 4, the green full line indicates data of actual terrain elevation which was obtained by manually measuring at the highest point of gravels terrain. Red full line indicates measurement data of ultrasonic detection system; blue full line indicates estimation of Kalman filter. Traditional Kalman filter can correct outliers somewhat, but it has weaker correcting effect than adaptive Kalman filter; meanwhile, where there is no outliers, estimation of adaptive Kalman filter is more close to actual distance value. Calculate errors respectively by comparing data before filtering and actual terrain elevation data of both terrains, result shown in Fig. 5, 6.

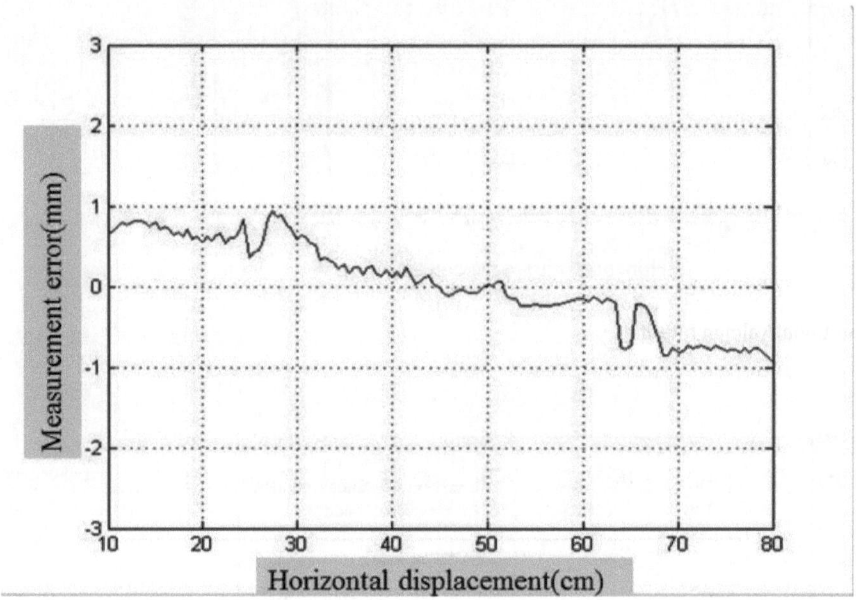

(a) Traditional Kalman filtering

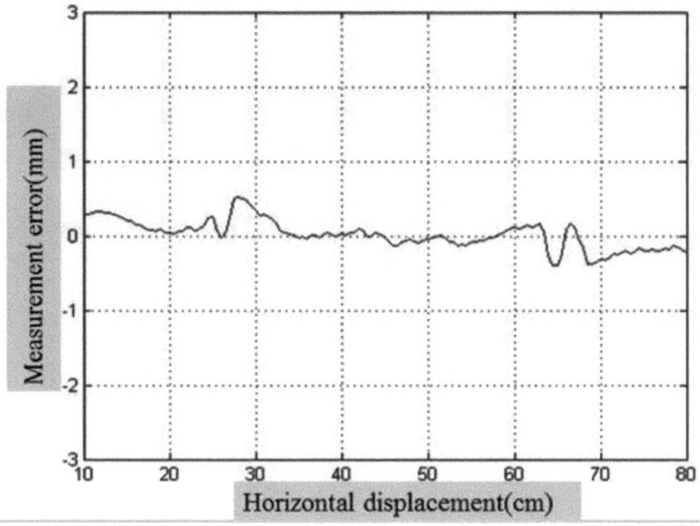

(b) Adaptive Kalman filtering

Fig. 5: Filtering error of hardwood terrain

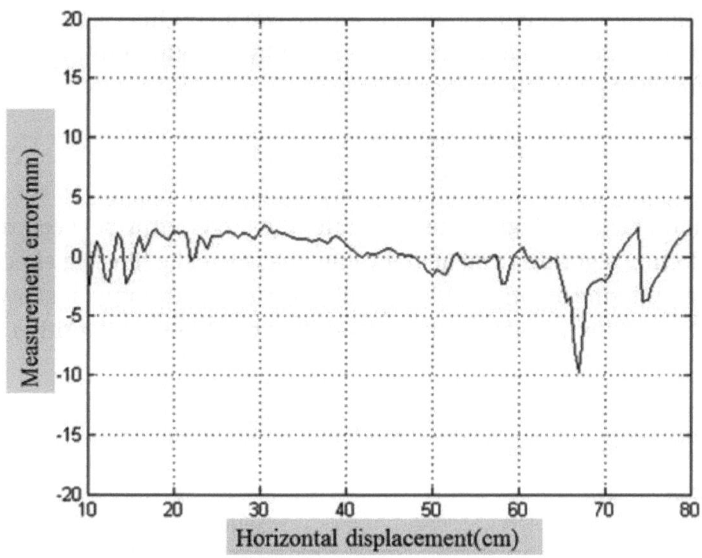

(a) Traditional Kalman filtering

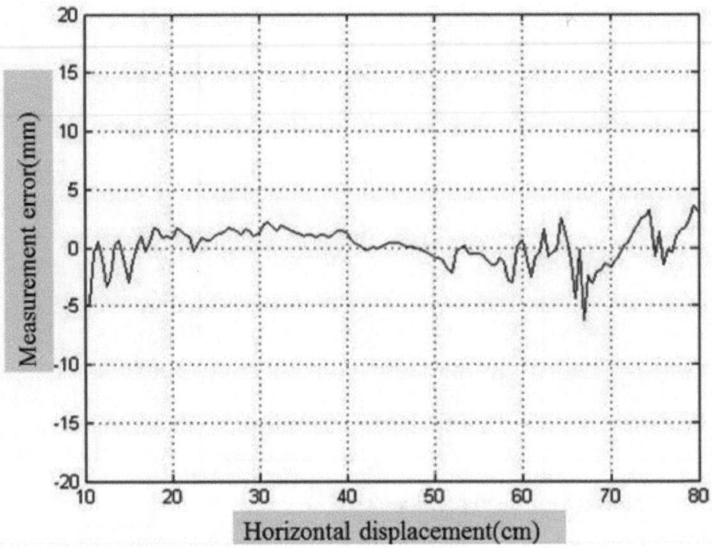

(b) Adaptive Kalman filtering

Fig. 6: Filtering error of gravels terrain

In order to check reliability of filtering model, compare two algorithms more clearly, this paper evaluated two Kalman filtering algorithm by adopting Error rate [8]. For calculating formula of error rate see equation (22).

$$\text{error rate} = \left|\frac{\text{Filtering estimation} - \text{actual value}}{\text{actual value}}\right| \times 100\% \tag{22}$$

According to equation (22), the higher the error rate, the larger error between result of a Kalman filtering algorithm and actual value, the worse the filtering effect, likewise, if a Kalman filtering algorithm has a result with lower error rate, it shows the algorithm has better filtering effect. Filtering result of two filtering algorithm is shown in Table 1.

Tab. 1: Filtering result of two filtering algorithm

type	traditional		adaptive	
	error rate	Max- error	error rate	Max- error
Hardwood terrain	7.45%	0.91	6.04%	0.52
Gravels terrain	8.27%	9.19	6.39%	5.89

According to Table 1, error rate of using adaptive Kalman filtering algorithm is lower than that of using traditional Kalman filtering algorithm, result after filtering is also more close to known terrain, comparing processing effect of local outliers, adaptive Kalman filtering algorithm can reduce error to about 60% that of traditional Kalman algorithm, with more significant smoothing effects.

6 Conclusion

Based on traditional Kalman filtering algorithm this paper proposes adaptive Kalman filtering algorithm, then simulate filtering effect of two algorithms through Matlab, and use them in ultrasonic detection result processing of hardwood and gravels terrain. By comparison, both algorithms can remove abnormal value in measurement, effect of adaptive Kalman filtering algorithm is better than that of traditional Kalman filtering algorithm, realizing optimization of traditional Kalman filtering algorithm, with better performance in estimation of terrain elevation. In the future, we will make further efforts to optimize Kalman filtering algorithm to further improve the filtering accuracy.

Reference

[1] W. Luo, F. Chen. Study on Error Compensation in Robot Ultrasonic Ranging Based on Wavelet Transformation [J].Coal Mine Machinery. 2007, 28(2):51–54.
[2] Y.H. Yang. Research on Automatic Control System of Cutting Height Based on Ultrasonic Sensor [D]. Northwest A&F University, 2007.
[3] Mohinder S. Grewal, Angus P.Andrews. Kalman filtering theory and practice [M].John Wiley &Sons, Inc.2001.
[4] Y. Zhang, T.Y. Zhou, J. Zhou. Research on Federated Kalman Filtering in Ranging of Highspeed Vehicle. Journal of Projectiles, Rockets, Missiles and Guidance, 2007, 27(1): 387–389.
[5] J.Z. Wang, C. Chen, Z.J. Wang. Range Adaptive strategy for Ultrasonic Distance Measurement Based on Kalman Filter [J]. Transactions of Beijing Institute of Technology, 2011, 31(3):283–288.
[6] J.H. Zhao, J.N. Liu. Beam Bathymetry and Image Data Processing [M]. Wuhan: Wuhan University Press, 2008.
[7] M.W. Zhu, Y.X. Li, X.Z. Pu. Test Signal Processing and Analysis [M]. Beijing: Beihang University Press, 2006.
[8] Y.T. Fei. Error Theory and Data Processing [M]. Beijing: Mechanical industry press, 2004.

Zhi-Xiong Su* and Jun-Min Yi

A Two-phase Heuristic Algorithm for No-wait Flow Shop Scheduling

Abstract: To solve the no-wait flow shop scheduling problem with make span criterion, a two-phase heuristic algorithm was presented. In the first stage, the Nawaz-Encore-Ham (NEH) algorithm was exploited to solve the original and reverse problem to find a better schedule as the initial solution. In the second stage, a swap-based local search was used to quickly improve the solution obtained. Last, the experimental results of benchmark instances indicate the effectiveness of the proposed algorithm. Compared with NEH, the average percentage relative deviation value by this method is decreased by 1.14%, and the average running time is just equal to 0.023 s.

Keywords: production scheduling, no-wait flow shop, heuristic algorithm, swap neighbourhood

1 Introduction

Production scheduling is a decision-making process that plays a critical role in manufacturing and service industries. It deals with the allocation of available production resources to tasks over given time periods, aiming at optimizing one or more objective. The no-wait flow shop scheduling problem is an important branch of 'zero buffers' scheduling problems. It arises due to the processing characteristics of certain products, where each process follows the previous one immediately. This type of problem widely exists in practical production systems, such as metals, electronics, chemicals, plastics and food-processing industries [1].

In this paper, we mainly focus on the no-wait flow shop problem with make span criterion. The minimization of the make span or the maximum completion time is the most common and popular measure of the performance. Using the three-field notation, m-machine no-wait flow shop problem can be denoted by $Fm \mid nwt \mid C_{\max}$ [2]. This kind of problem is strongly NP-hard for more than two machines [3]. Therefore, exact algorithms such as branch-and-bound and mixed-integer linear programming can optimally solve the small-size problems. For decades, much more

***Corresponding author: Zhi-Xiong Su,** School of Economics and Management, Xiamen University of Technology, Xiamen 361024, China. Email: z.su@163.com
Jun-Min Yi , School of Economics and Management, Xiamen University of Technology, Xiamen 361024, China. Email: yijunmin@xmut.edu.cn

DOI 10.1515/9783110540048-049,

effort has been devoted to searching high-quality solutions in a reasonable computational time by heuristics [4-7] and metaheuristics [1, 8-10]. The purpose of this work is to present a two-phase heuristic (TPH) algorithm for the $Fm|nwt|C_{\max}$ problem. The effectiveness of the proposed method is tested on 31 benchmark problems with different sizes.

2 Problem Formulation

The problem is defined as follows. There are *n* jobs to be processed on *m* machines to minimize the make span. In the no-wait flow shop, the processing of each job has to be continuous from start to end without interruptions either on or between machines. This means, when necessary, the start of a job on a given machine is delayed in order that the operation's completion coincides with the start of the next operation on the subsequent machine.

2.1 Notation

To introduce the model, the required notations are described as follows.
i Machine index, $i \in \{1,\cdots,m\}$
j Job index, $j \in \{1,\cdots,n\}$
k Position index, $k \in \{1,\cdots,n\}$
π $\pi = \{\pi_1,\pi_2,\cdots,\pi_n\}$ Is a feasible schedule
p_{ij} Processing time of job j on machine i
P_{ik} Processing time of job π_k on machine i
x_{jk} 1, if job j occupies position k, 0 otherwise
C_{ik} Completion time of job π_k on machine i
$C_{\max}$ Make span

2.2 Mathematical Model

Using the above notations, the $Fm|nwt|C_{\max}$ problem can be formulated as a mixed-integer linear programming.

$$\min C_{\max} \tag{1}$$

$$\text{s.t. } C_{\max} = C_{mn} \tag{2}$$

$$C_{ik} - P_{ik} \geq C_{i,k-1},\ \forall i,k \tag{3}$$

$$C_{i,k} = C_{i+1,k} - P_{i+1,k}, \ \forall i,k \tag{4}$$

$$P_{ik} = \sum_{j=1}^{n} x_{jk} p_{ij}, \ \forall i,k \tag{5}$$

$$\sum_{j=1}^{n} x_{jk} = 1, \ \forall k \tag{6}$$

$$\sum_{k=1}^{n} x_{jk} = 1, \ \forall j \tag{7}$$

$$x_{jk} \in \{0,1\}, \ \forall j,k \tag{8}$$

$$C_{ik} \geq 0, \ \forall i,k; \ C_{i0} = 0, \ \forall i; \ C_{0k} = 0, \ \forall k \tag{9}$$

Equation (1) is the objective function to minimize the makespan. Constraint (2) indicates that the completion time of the last job π_n on the last machine m is makespan. Constraint set (3) ensures that a job does not start on a machine until its predecessor has completed processing on that machine. Constraint set (4) ensures that the completion time of a job on a given machine must be equal to the starting time of the job on the next machine. Constraint set (5) shows that the relation between p_{ij} and P_{ik}. Constraint set (6) ensures that each position of schedule is assigned exactly once for all jobs. Constraint set (7) ensures that each job is assigned exactly once for all positions. Constraint set (8) and (9) define the domains of the decision variables.

3 Problem-Solving Strategies

3.1 Reversibility Property

It is well known that the regular flow shop scheduling problem has the reversibility property: the makespan does not change if the jobs traverse the flow shop in the opposite direction, in reverse order [11]. As a particular case of the regular flow shop problem, the no-wait flow shop problem also meets this property. By graph representation, it is easy to prove that the reverse problem is equivalent to its original problem. Therefore, the existing algorithms can be used directly to solve the reverse problem.

3.2 Solution Strategies

The NEH [12] was recognized as the highest performing heuristic for the permutation flow shop scheduling problem to minimize the makespan. On the basis of equivalence relation and the NEH algorithm, three strategies can be exploited to solve this problem:

(1) Forward scheduling-based NEH (FNEH): use the NEH to solve the original problem;
(2) Backward scheduling-based NEH (BNEH): use the NEH to solve the reverse problem;
(3) Forward-Backward scheduling-based NEH (FBNEH): combine FNEH and BNEH to obtain a better schedule.

4 The Proposed Algorithm

The proposed algorithm is a two-phase heuristic. In the first stage, FBNEH is used to generate an initial solution. In the second stage, a swap-based local search is exploited to improve the solution obtained.

4.1 Solution Representation

In the proposed algorithm, each solution is represented by a string of numbers consisting of a permutation of *n* jobs denoted by 1, 2... *n*. This permutation-based encoding is commonly used in most of the literature for flow shop scheduling problem. Suppose that one solution is represented by {4, 5, 3, 2, 1}, which means that the processing sequence is job 4, job 5, job 3, job 2 and job 1.

4.2 Makespan Calculation

For a given solution $\pi = \{\pi_1, \pi_2, \cdots, \pi_n\}$, the makespan can be calculated as follows [8]:

$$C_{\max} = \sum_{k=2}^{n} d_{\pi_{k-1},\pi_k} + \sum_{i=1}^{m} p_{i,\pi_n} \tag{10}$$

Where d_{π_{k-1},π_k} is the completion time distance from job π_{k-1} to job π_k that can be calculated in advance. Therefore, the computation of the makespan of a solution can be done in O $(n+m)$.

4.3 Initial Solution

In this paper, FBNEH will be used to obtain a better schedule as the initial solution. The NEH algorithm was implemented in matlab, as shown in Fig. 1.

```
function [Cmax, jobOrder]=NEH(P, D, TPT)
% P is the processing time matrix
% D is the delay matrix
% TPT=sum(P,2) is the total processing time
% computeCmax is a function to compute makespan
[nbjobs, nbmachines]=size(P);
[unused,index]=sort(TPT, 'descend');
if computeCmax(index(1:2), D, TPT)<=computeCmax(index(2:-1:1), D, TPT)
  jobOrder=index(1:2);
else
  jobOrder=index(2:-1:1);
end
for k=3:nbjobs
  Cmax_cmp=zeros(k,1);
  for nb=1:k
    Cmax_cmp(nb)=computeCmax([jobOrder(1:nb-1); index(k); jobOrder(nb:end)], D, TPT);
  end
  [minCmax, I]=min(Cmax_cmp);
  jobOrder=[jobOrder(1:I-1); index(k); jobOrder(I:end)];
end
Cmax=minCmax;
```

Fig. 1: Matlab code for NEH

4.4 Local Search

A swap-based local search procedure is applied to the initial solution to generate a new solution. There are mainly two neighbourhood exploration strategies: first improvement and best improvement. In this paper, we adopt the latter one. The local search algorithm was implemented in mat lab, as shown in Fig. 2.

```
function [gain, bestOrder, bestCmax]=swap(jobOrder, Cmax, D, TPT)
 % D is the delay matrix
% TPT=sum(P,2) is the total processing time
nbjobs=length(jobOrder);
gain=0;
for i=1:nbjobs-1
   for k=i+1:nbjobs
      newOrder=jobOrder;
      newOrder([k i]) =newOrder([i k]);
      newGain=Cmax-computeCmax(newOrder, D, TPT);
      if newGain>gain
         gain=newGain;
         swapRecord=[i k];
      end
   end
end
%%
if gain
   i=swapRecord(1);
   k=swapRecord(2);
   bestCmax=Cmax-gain;
   bestOrder=jobOrder;
   bestOrder([k i]) =bestOrder([i k]);
else
   bestCmax=Cmax;
   bestOrder=jobOrder;
end
```

Fig. 2: Matlab code for swap-based local search

5 Computational Experiments

5.1 Experimental Setup

The test problems used in the experiments are 31 benchmark problems from Carlier [13], Reeves [14] and Heller [15] with different sizes, which were also studied in the literature [1, 8, 9]. The comparison was performed using three algorithms: FNEH,

BNEH and FBNEH [8]. These four algorithms were implemented in Mat lab 8.3 on a PC with Intel core i3 3.4 GHz processor and 4GB memory. In this study, the performance of the proposed TPH algorithm is evaluated by two indices: (1) CPU time (ms), (2) percentage relative deviation (PRD). PRD can be calculated by the following formula:

$$PRD = (C_{TPH} - C_{ref})/C_{ref} \times 100 \tag{11}$$

where C_{TPH} is the makespan generated by the proposed TPH, whereas C_{ref} is the optimal value reported for the Car01 to Car08 instances and upper bounds for the Rec01 to Rec41, Hel01 and Hel02 instances by Rajendran [5].

5.2 Computational Results

The numerical comparisons of FNEH, BNEH, FBNEH and the proposed TPH algorithm are given in table 1. It can be seen from table 1 that the overall mean values of PRD is equal to -2.28, which is better than those generated by FNEH, BNEH and FBNEH. However, the difference in the overall mean values of PRD between FNEH, BNEH and FBNEH is very small. For the average computational time in which an algorithm finds the final solution, the TPH is slower than the other three algorithms, but the average time is just equal to 0.023 s.

Tab. 1: Comparison results

Instance	Cref	FNEH		BNEH		FBNEH		TPH	
		PRD	CPU(ms)	PRD	CPU(ms)	PRD	CPU(ms)	PRD	CPU(ms)
Car01	8142	0	6.54	0	6.99	0	10.52	0	15.01
Car02	8242	0.62	1.02	0.62	0.59	0.62	1.60	0.62	1.50
Car03	8866	1.05	0.50	1.05	0.87	1.05	0.90	0.27	2.89
Car04	9195	9.07	0.63	10.01	0.73	9.07	1.28	9.07	2.15
Car05	9159	6.62	0.41	6.62	0.51	6.62	0.80	0.32	2.72
Car06	9690	0	0.37	0	0.44	0	0.63	0	0.70
Car07	7705	2.57	0.33	2.57	0.40	2.57	0.44	0	0.62
Car08	9372	0.36	0.29	0.36	0.44	0.36	0.60	0	1.22
Rec01	1590	-1.32	1.39	-0.25	1.18	-1.32	2.18	-1.32	3.46
Rec03	1457	1.30	1.18	1.30	1.25	1.30	2.47	-3.29	8.96
Rec05	1637	-4.52	1.14	-4.52	1.94	-4.52	2.31	-4.52	3.23
Rec07	2119	3.21	1.15	3.21	1.16	3.21	2.21	-0.85	7.13
Rec09	2141	-1.31	1.27	-1.31	1.29	-1.31	2.28	-3.04	5.91

Rec11	1946	0.62	1.13	0.62	1.12	0.62	2.32	0.15	4.96
Rec13	2709	-0.15	1.15	-0.15	1.79	-0.15	2.43	-0.15	3.08
Rec15	2691	-4.46	1.12	-4.46	1.13	-4.46	2.19	-4.46	4.14
Rec17	2740	-4.27	1.13	-4.27	1.15	-4.27	2.29	-5.47	5.13
Rec19	3157	-5.54	2.80	-5.54	2.47	-5.54	4.79	-5.92	9.80
Rec21	3015	-1.46	2.68	-1.46	2.94	-1.46	4.82	-2.95	13.46
Rec23	3030	-5.48	3.29	-5.48	2.43	-5.48	4.71	-6.83	14.67
Rec25	3835	-2.27	3.24	-1.72	2.56	-2.27	4.80	-3.60	9.90
Rec27	3655	-2.49	2.39	-2.41	2.85	-2.49	4.77	-2.49	7.60
Rec29	3583	-4.16	2.69	-4.16	4.17	-4.16	4.86	-5.11	10.77
Rec31	4631	-2.33	6.74	-2.33	12.20	-2.33	13.27	-3.28	46.83
Rec33	4770	-0.46	6.80	-0.78	6.93	-0.78	13.46	-1.43	46.84
Rec35	4718	-3.03	6.86	-3.09	7.32	-3.09	13.51	-3.48	41.47
Rec37	8979	-6.75	16.03	-6.97	16.66	-6.97	32.08	-7.23	63.70
Rec39	9158	-2.31	16.35	-2.44	16.66	-2.44	33.18	-2.45	64.69
Rec41	9344	-3.70	18.74	-3.59	25.28	-3.70	32.91	-4.45	159.50
Hel01	780	-3.59	31.15	-3.85	34.45	-3.85	64.18	-3.97	130.34
Hel02	189	-1.06	1.18	-1.59	2.06	-1.59	2.36	-4.76	4.98
Mean		-1.14	4.57	-1.10	5.22	-1.19	8.75	-2.28	22.50

6 Conclusions

In this paper, we examined the $Fm|nwt|C_{\max}$ problem with makespan criterion and some properties. The problem is known to be strongly NP-hard for more than two machines, therefore most contributions to the problem have focused on finding approximate solutions in a relatively short time. In our work, a two-phase heuristic (TPH) was developed to solve this problem. To evaluate the performance of the TPH algorithm, it was tested on the well-known benchmark problems from Carlier, Reeves and Heller. Computational results show that the proposed TPH algorithm outperforms the other three algorithms in terms of the overall mean values of PRD.

Some important directions for future research as follows. First, the proposed algorithm can be applied to no-wait flow shop scheduling problems with different objective functions. Second, the application of the same heuristic technique to $Fm|nwt|C_{\max}$ in real-life situations such as steel production is worth further research. Third, the reversibility property and swap-based local search can be used to develop a metaheuristic for $Fm|nwt|C_{\max}$.

Acknowledgement: This work was financially supported by the National Natural Science Foundation of China (71371162), and Type a Project of Fujian Provincial Department of Education, China (JAS150422).

References

[1] Q. K. Pan, L. Wang, B. H. Zhao. "An improved iterated greedy algorithm for the no-wait flow shop scheduling problem with makespan criterion." The International Journal of Advanced Manufacturing Technology 38.7 (2008): 778–786.

[2] R. L. Graham, E. L. Lawler, J. K. Lenstra, A. R. Kan. "Optimization and approximation in deterministic sequencing and scheduling: a survey." Annals of discrete mathematics 5 (1979): 287–326.

[3] H. Röck. "The three-machine no-wait flow shop is NP-complete." Journal of the ACM (JACM) 31.2 (1984): 336–345.

[4] R. Gangadharan, C. Rajendran. "Heuristic algorithms for scheduling in the no-wait flowshop." International Journal of Production Economics 32.3 (1993): 285–290.

[5] C. Rajendran. "A no-wait flow shop scheduling heuristic to minimize makespan." Journal of the Operational Research Society 45.4 (1994): 472–478.

[6] X. Li, Q. Wang, C. Wu. "Heuristic for no-wait flow shops with makespan minimization." International Journal of Production Research 46.9 (2008): 2519–2530.

[7] D. Laha, U. K. Chakraborty. "A constructive heuristic for minimizing makespan in no-wait flow shop scheduling." The International Journal of Advanced Manufacturing Technology 41.1 (2009): 97–109.

[8] J. Grabowski, J. Pempera. "Some local search algorithms for no-wait flow-shop problem with makespan criterion." Computers & Operations Research 32.8 (2005): 2197–2212.

[9] S. W. Lin, K. C. Ying. "Optimization of makespan for no-wait flowshop scheduling problems using efficient matheuristics." Omega 64 (2016): 115–125.

[10] J. Y. Ding, S. Song, J. N. Gupta, R. Zhang, R. Chiong, C. Wu. "An improved iterated greedy algorithm with a Tabu-based reconstruction strategy for the no-wait flowshop scheduling problem." Applied Soft Computing 30 (2015): 604–613.

[11] M. Pinedo, Scheduling: theory, algorithms, and system, second ed., Prentice Hall, New Jersey, 2002.

[12] M. Nawaz, E. E. Enscore, I. Ham. "A heuristic algorithm for the m-machine, n-job flow-shop sequencing problem." Omega 11.1 (1983): 91–95.

[13] J. Carlier. "Ordonnancements constraints disjunctives." Revue française d'automatique, d'informatique ET de recherché operationally. Recherché operationally 12.4 (1978): 333–350.

[14] C. R. Reeves. "A genetic algorithm for flowshop sequencing." Computers & operations research 22.1 (1995): 5–13.

[15] J. Heller. "Some numerical experiments for an M× J flow shop and its decision-theoretical aspects." Operations Research 8.2 (1960): 178–184.

Zeng-Rong Zhao, Shi-Lei Cai, Tian-Xiang Li and Li-Wei Pang

The Design of Laser Liquid Level Measurement System Based on FPGA and MCU

Abstract: The paper studies on an intelligent, real-time laser liquid level measurement system which is based on FPGA as the core unit and AVR MCU as aided processing unit. It realizes to measure the liquid level according to the control signal and the information can be shown in field or in remote site or on the mobile phone at the same time. The experiment result shows that this system achieves to high measurement accuracy, which meets to the requirement of measuring the actual level of liquid petroleum, liquid glass etc. in industry.

Keywords: FPGA AVR MCU Laser liquid level measurement

1 Introduction

Liquid level is a very common index in industrial measurement system. The early liquid level measuring instruments required to contact with liquid level measurement directly, which has the disadvantages of slow measurement speed and large error, and not suit in these occasions with high temperature and high pressure that is inconvenient to contact with sensor directly such as metallurgy, petroleum and chemical industry. But the laser liquid level measurement system adapts the method of non-contact level measurement, which achieves the accurate measurement of liquid level by using laser's characteristics such as high precision, high speed, wide range and strong anti-interference ability to convey the level signal accurately. Through the subsequent information processing system, the level signal can be displayed in field, in remote control room or on the mobile phone, and the system can be controlled to measure or not which solves the problem of the non-contact measurement in long-distance [1].

Zeng-Rong Zhao, Career Technology College, Hebei Normal University College, Shijiazhuang, China. Email: zhaozzr@126.com
Shi-Lei Cai, Career Technology College, Hebei Normal University College, Shijiazhuang, China. Email: 2572136520@qq.com
Tian-Xiang Li, Career Technology College, Hebei Normal University College, Shijiazhuang, China. Email: 1256255903@qq.com
Li-Wei Pang, Career Technology College, Hebei Normal University College, Shijiazhuang, China. Email: 1842010669@qq.com

DOI 10.1515/9783110540048-050,

2 System Composition and Working Principle

The hardware components of laser liquid level measurement system is shown as Figure 1-1, which including FPGA core board, MCU core board, laser measuring range module, Bluetooth module, VGA interface, Android mobile phone, computer monitor and the measured liquid device.

In this system, FPGA is used as the main chip, MCU as the aided design to real-ize the complex logical circuit and drive the VGA display in high speed. Standard SPI interface is used to realize the communication between the MCU and FPGA. After the level information is converted and processed by the MCU which obtains from the laser ranging module, on the one hand the data is transferred to FPGA to calculate and drive VGA to display the liquid level data by the SPI bus, on the other hand, it establishes Bluetooth communication network to communicate with the mobile phone through the Bluetooth module, so the information of measured liquid level will be displayed on the mobile phone client via Bluetooth serial port. So the system realizes the intelligent liquid level monitoring and meets the requirement in petrochemical industry [3].

3 Major Modules and Technologies Related

3.1 Laser Ranging Module

In this design, the laser range sensor HJ-40A is adopted because of the high performance-price ratio. It uses the visible red light of 635 nm wave length, which belongs to the two level security laser and is not harmful to person, and its measuring range is 0.03–40 meters. Its detailed parameters are shown in Table 1.

Tab. 1: HJ-40A Parameters

Number	Name	Technology parameters
1	range	0.03~40m
2	Baud rate	19200bps
3	Power Consumption	<1mw
4	Measurement accuracy	±1mm

In theory, there are two main methods to calculate the distance by laser: phase method and pulse method.

Firstly, the principle of phase method is to calculate the phase difference between the received laser signal and the continuous emission signal. The formula is:

$$D = (c/2)\,(\Phi/2\pi f) \quad (1)$$

Where, D is the distance measured, c is the wave length, Φ is the phase difference and f is the laser modulation frequency.

By this measurement method, the distance D can be concluded from the formula as long as the phase difference between the received signal and the transmitted signal is obtained. But it has a drawback that the laser must be continuous, and also need the technology of difference frequency and modulation. Because it requires that the signal must be very accurate and strict, which makes it is not easy to achieve and is not adopted widely.

Secondly, the principle of laser ranging is that the laser is emitted out and reflected back when it meets the obstacle, the round trip time is obtained by a counter to calculate the distance. The formula is:

$$D = C*T/2 \quad (2)$$

Where, D is the distance measure, C is the light speed, T is the laser trip time. In this design, we adopted the pulse method to measure liquid level [4].

3.2 Bluetooth Serial Port Module

Bluetooth is a wireless communication technology standard, the wireless wave works at short baud and high frequency, the radio frequency range is from 2.4 to 2.485GHz.In a short range, data can be achieved interaction with mobile devices, such as the mobile phones, tablet PCs, desktop computers etc.

There are many Bluetooth serial modules in market; here the HC05 Bluetooth serial module is selected because of its good stability and compatibility. This module uses the TTL serial port, which is convenient to connect with MCU interface. Its parameters are listed in Table 2.

Tab. 2: Bluetooth module parameters

Number	Name	Technology parameters
1	Communication distance	10m
2	Baud rate	9600bps
3	Current	<5mA
4	Voltage	3.3V~3.6V

3.3 Micro Controller Unit Module

Micro controller unit (MCU) is mainly responsible for the communication between interfaces, after the serial port sends measurement command through the trigger button, the system begins to emit laser to access liquid level information, and this information is sent to the FPGA through the SPI bus and the Bluetooth module to process and display. The design flow is shown as in Figure 1.

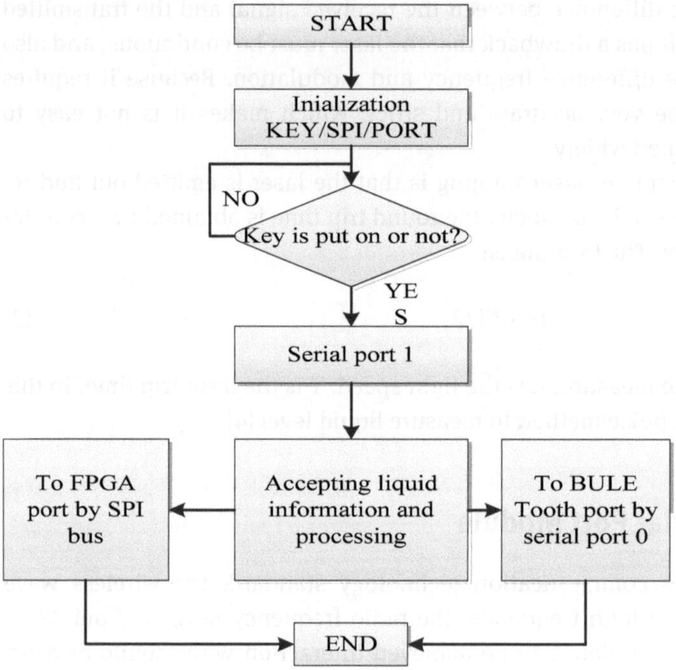

Fig. 1: Design flow of MCU module.

3.4 FPGA Module

Field programmable logic gate array (FPGA) comes from CPLD, GAL, PAL and other programmable device through technology innovation and progress, which can use VHDL or Verilog-HDL hardware programmable language to design the logic circuit to realize required function. VHDL language syntax is rigorous, and relatively close to the hardware, but the program is cumbersome. Otherwise Verilog-HDL is easier, which likes the C language. In this design, Verilog-HDL is used to program, ISE software and the hardware chip on the core board of Xilinx is used as the development platform. The design flow of FPGA module is shown as in Figure 2.

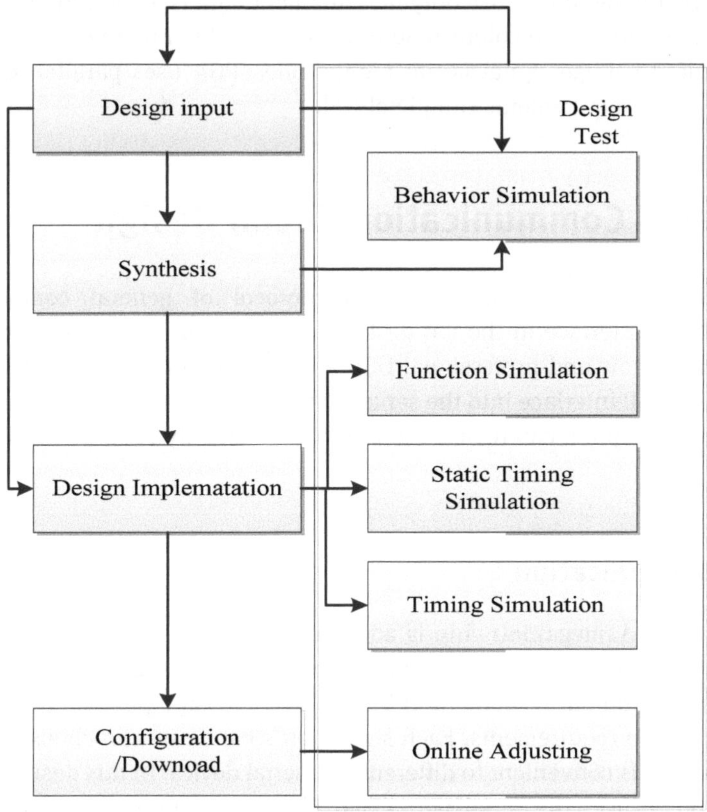

Fig. 2: Design flow of FPGA module.

3.5 VGA Display Realization

There are 15 pins in VGA interface, 9 pins are used in general. The driving principle of FPGA is shown as follows: a line synchronization signal is loaded at pin 13-HSYNC, while a field synchronizing signal at the pin 14-VSYNC, so the VGA signal begins to scan from the first pixel of the first row to the last line of the last pixel, and the scan cycle is repeated. However, if when a certain pixel is scanned, such as the VGA interface pins 1, 2, and 3, which corresponds to the RED, GREEN and BLUE respectively, if the appropriate analogy voltage is loaded, the screen will display different colours corresponding to the analogy voltage. But the FPGA output signals are digital signals, how to generate the analogy signal of 0-0.714V? In this design a simple and ingenious method is used to solve this problem which is a simple DAC circuit consisting of a resistor divider array to matching different analogy voltage, as shown in the figure 1–4. It is not accurate as dedicated DAC conversion chip, but meets the speed requirement because only measurement data is displayed only not the colour in this design, so the colour requirements is not high therefore, and the accuracy of required voltage signal is not high, while FPGA uses parallel mode, which meets the speed requirements completely [5].

4 Multi-Serial Communication Circuit Design

Serial communication is the communication protocol of general computer, equipped with RS232 interface in the old desktop computer. Because USB is more and more popularity now and professional USB serial communication chip can be used to convert the USB interface into the serial port of TTL directly, such as CH340, PL2303 chip, so the computer can realize serial transmission only equipped with the USB interface.

4.1 Serial Communication

In this design, AVR's ATmega2560 chip is adopted which contains 4 serial ports, serial port 0 contains pin 2 and pin 3, serial port 1 contains pin 45 and pin 46, serial port 2 contains pin 12 and pin 13, serial port 3 contains pin 63 and pin 64, and it can meet the system design requirements. Each serial port's baud rate can choose from 2400 to 115200, which is convenient to different rate serial device. In this design, the laser ranging module uses the serial port 1 and 19200 baud rate, and Bluetooth module uses the serial port 0 and 9600 baud rate. In the connection of laser ranging module and the microcontroller, series resistors are used to achieve the voltage conversion, so the power supply voltage is compatible from single-chip's 5V to laser ranging module's 3.3V.

4.2 MCU and FPGA Communication Design

SPI communication principle: Serial peripheral interface (SPI) is a kind of synchronous serial communication bus in short distance. SPI bus equipment direct communication works in full duplex mode, with high communication speed reaching to 1Mb/s in the ideal condition. Generally there are a host machine and one or more guest machine in SPI communication. Signals of frame data to wait writing and reading are generated by the host. If there is a few guest machines, the host machine will give the chip selected signal to choose one guest machine to response to the equipment' request.

The design of SPI communication circuit based on FPGA and MCU: There are many ways of communication between MCU and FPGA, such as serial communication, FSMC (flexible static memory controller) communication and SPI bus communication. The FSMC bus's speed is very fast to achieve 16MHz in write speed because of communication with the internal memory. But the FSMC interface pins are more, and its timing is complex. In contrast, SPI serial communication is more simple and quick, so it is adopted in the design. The diagram of communication hardware connection between FPGA and MCU is shown as in Figure1-4. Here microcontroller is the main equipment and FPGA is the slave device. After MCU obtains data's, they are transmitted to the FPGA through the SPI bus. In FPGA they are analysed and processed to display the liquid level information on screen through VGA interface.

Implementation of SPI communication program: SPI bus is a ring bus structure with data ports such as CS, CLK, MISO and MOSI. Under the CLK signal's controlling, bidirectional data's exchange is realized in two registers. Here SPI bus module is written by Verilog HDL in this design which realizes the real time communication between FPGA and SCM.

5 Conclusions

After the whole system is built, all programs are compiled and downloaded to the corresponding chip. When the start button is pressed, the liquid level measurement system begins to work: the laser ranging module emits and receives laser, MCU processes liquid level measurement information, on the one hand, the information is feedback to the mobile phone software to display the liquid level data via Bluetooth, on the other hand it is transmitted to the FPGA to process and drive the VGA interface to display the data on the remote screen. The mobile phone software results is shown as in Figure 4-1, computer display as in Figure 4-2, All the test data are shown in Table3.

Tab. 3: Measure Result

Times	Measured Liquid Level（mm）	Real Liquid Level（mm）	Error
1	5.05	5	1%
2	10.1	10	1%
3	15.06	15	0.4%
4	16.2	16	0.125%

According to the test data, we found that the accuracy of the liquid level measurement system is high with the error is low to 0.125 % in mm. This error meets the actual requirement of liquid measurement system in industry. And the system realizes the intelligent measurement of the display and control on the mobile phone or in the remote controller room. Through further improvement, it is easy to realize the liquid data's storage, statistics and comparison to the last data, which is more convenient to the management of the liquid level in the industrial production.

Acknowledgement: This paper is supported by Hebei science and technology project (13201711D) and Hebei science and technology project (11213501D).

References

[1] ZONG Jian, LI Bin, "Liquid level measuring system based on laser range finder without cooperant target"Journalof foreign electronic measurement technology, vol. 29(5), May 2010, pp. 82–86.

[2] SONG Nian-ning, LIU Huan and HOU Xiao-ning, "The elimination of laser liquid level measurement signal fluctuation based on the LMS adaptive noise cancellation" Journal of XI'an University of Technology, vol.28 (1), Sep 2012, pp. 90–93.

[3] Cai Li, "FMCW Radar Liquid Level Measuring System", Information&Communications, vol. 6(2), Sep 2012, pp. 14–15.

[4] JIA Ru-lei, LI Quan, GONG Hui, "Design of liquid level meter for portable railway tanker based on MCU," Journal of CHINA measurement, vol. 2016(10), Oct 2016, pp. 98–100.

[5] LIU Bin, LU Yong-Hua, and YANG Chao, Research on laser tracking measurement and control system based on PSD Mechanical Engineering, vol. 32(8), April 2015, pp. 1033–1037.

Na Lv, Zhan-Shou Chen and Jian-Qi Ma

On-line Monitoring Long Memory Parameter Change Point in FARIMA Processes

Abstract: In order to overcome the serious size distortion problem when online monitoring the long memory parameter change point in FARIMA(p, d, q) processes based on the critical values simulated byFARIMA$(0, d, 0)$ process, we propose a sieve bootstrap monitoring procedure. Simulations show that the critical values determined by the sieve bootstrap method can not only control the empirical size well below the test level, but also have satisfied test power.

Keywords: long memory, sieve bootstrap, change point monitoring, MOSUM test

1 Introduction

The research on the change point can be traced back to the 1950s, and it has gradually become a hotspot in statistics and econometrics since the 1970s, and last now. In recent years, the application research of the change point has also got the rapid development, and it is widely used in meteorological, financial, medical, economic, hydrological and other fields.

The detection of change point includes posteriori tests and on-line monitoring. The posteriori tests mainly analyse the observed historical data set, however, change point can appear at any given time and also new data arrives steadily it is necessary to detect a change in persistence as soon as possible in the application. On-line monitoring can find change point much earlier, which is beneficial to adjust decisions faster to reduce the loss. Starting with Bauer and Hackl [1] a strand of literature has emerged that studies on-line monitoring. Horvath et al. [2] studied the unconditional variance change point monitoring in conditionally heteroskedastic time series by CUSUM method, wave-type monitoring method, partial sum residual method and recursive residual method respectively. Berkes [3] researched on-line monitoring change point in GARCH (p, q) process. Chen [4] explored on-line monitoring parameter change in linear regression model. Chen [5-6] extended the kernel-

Na Lv, School of Mathematics and Statistics, Qinghai Normal University, Xining, China. Email: lvna___@yeah.net
Zhan-Shou Chen, School of Mathematics and Statistics, Qinghai Normal University, Xining, China. Email: chenzhanshou@126.com
Jian-Qi Ma, School of Mathematics and Statistics, Qinghai Normal University, Xining, China. Email: m18919486694_@163.com

DOI 10.1515/9783110540048-051,

weighted variance ratio method to the variance-infinite case and used it to monitor the stability of the heavy tailed random signals and the persistent change point from the stationary to the non-stationary respectively.

These studies mentioned above all considered short memory of time series model; in this paper we study online monitoring parameter change point of long memory time series. Due to plenty of actual data such as financial data is more suitable for long memory time series modelling, thus on-line monitoring parameter change point in long memory time series is also very important. Heine and Willet [7] constructed moving sums of cumulated residuals (abbreviated as MOSUM) statistics to monitor the autoregressive fractionally integrating moving-average process, namely theFARIMA (p, d, q) of persistent change point. However, when at least one of the autoregressive order p and the moving average order q is not equal to zero, if the critical value provided in the paper is continued using to monitor, the serious size distortion problem would occur. We use the sieve bootstrap method proposed by Buhrmann [8] to solve this problem. Chen [9][10][11] [12]monitored variance change, persistent change in long memory process and change from short to long memory by sieve bootstrap test respectively. Poskitt [12] studied the properties of the sieve bootstrap method and pointed out that the sieve bootstrap method is very useful in the analysis of the fractionally integrating process, but they only considered fixed sample data. We applied the sieve bootstrap method to the on-line monitoring parameter change point of long memory time series, and analysed the value of this method in online data.

The rest of the paper is organized as follows. Section 2 specifies the monitoring problem and shows all necessary assumptions and monitoring methods. In Section 3 we first compare the size, power and ARL performance of our monitoring procedures with literatures' by simulations. Section 4 concludes the paper.

2 Model Assumptions and Monitoring Methods

Consider theFARIMA(p, d, q) process

$$\Phi(B)(1-B)^d X_t = \Theta(B)\varepsilon_t \tag{1}$$

Where $\Phi(B)$ and$\Theta(B))$ are the AR- and MA- poly nominals respectively, ε_t are i.i.d random variables with mean zero and variance $\sigma_\varepsilon^2, t = 1,2, \dots, T$, B is the lag operator. The differencing parameter d is assumed to be$|d| < \frac{1}{2}$, which means X_t is always a stationary process.

Assumption 1 The first T samples are assumed to be non-polluting, i.e. the long memory parameter d in the model is not changed when observing the historical data set of size T

$$d_t = d_0, \quad t = 1,2,\dots,T. \tag{2}$$

Based on the first T observed samples (historical samples), the change points of the long memory parameter d are monitored from the $T+1$ new observed samples until the$[T\tau]$,$\tau > 1$, Where $[\cdot]$denotes the integer part of its argument. Namely, we want to test the following hypothesis test problem continuously

$$H_0:\ d_l = d_0,\ d_l = T+1, T+2, \dots [T\tau]$$

$$H_A:\ d_l = d_1, d_0 \neq d_1$$

To test the above hypothesis test, [7] proposed the following MOSUM monitoring function

$$MS_{T,h,d} = \max_{T+1\le k\le[T\tau]} \sigma^{-1} T^{-\frac{1}{2}-d} \left| \sum_{i=k-[Th]+1}^{k} x_i - \frac{[Th]}{T} \sum_{i=1}^{T} x_i \right| \tag{3}$$

Where $0 < h \le 1$ is a bandwidth parameter. In order to choose the appropriate boundary function, [7] found the boundary function $b(t) = z\sqrt{2\log_2'(m)}$ (where $m \in [0,1]$, z is a suitable scaling factor) is most effective under the long memory time series model compared with different boundary functions by simulating, so we still use this boundary function in this paper.

According to the mentioned monitoring function and boundary function, on-line monitoring parameter change point of long memory time series is carried out by the following method: for any given critical value $C(\alpha)$, the monitoring process is stopped when$MS_{T,h,d} > C(\alpha)b(k)$. It is assumed that the long memory parameter change point appears in the observed data, otherwise it will be monitoring until the maximum monitoring sample size appears.

Because σ_ε^2 is unknown in the practical problem, we use the estimator $\sigma_\varepsilon^2 = T^{-1}\sum_{i=1}^{T} x_i^2$, and the long memory parameter d can be estimated by extended Whittle method which is based on the first T non-contaminating samples and proposed by Abadir[13], and the critical value C (α) satisfies

$$P(MS_{T,h,d} > C(\alpha)b(k) \mid H_0) = \alpha.$$

The limit distribution of the monitoring function under the null hypothesis without change point is given and the consistency of the monitoring method under the alternative hypothesis is proved in [7]. And [7] proposed that the corresponding critical value $C(\alpha)$ could be determined by simulations. We list some critical values in Table 1. From model (1), we have

$$(1-B)^d X_t = \Phi^{-1}(B)\Theta(B)\varepsilon_t$$

Because$\Phi(B)$ and$\Theta(B)$ are assumed to have all roots outside the unit circle, so process$y_t = \Phi^{-1}(B)\Theta(B)\varepsilon_t$ still is a stationary short memory time series, i.e. $y_t \sim I(0)$, so The limit distribution of the monitoring function under the null hypothesis without change point. However, the simulations in the next section will point out that the critical value determined by [7] will present a serious size distortion problem when the autoregressive order p and the moving average order q in model (1) are not equal to 0. To overcome this problem, we propose the following sieve bootstrap test to determine the critical value$C(\alpha)$.

The basic idea of the bootstrap is to invest in computing resources to obtain a resampling based estimate of the control limit, instead of relying on the asymptotic distribution as an approximation. The traditional bootstrap method is quite useful as it overcomes the limitations of insufficient data size or unknown theoretical distribution, but this method has limitations when applied to dependent data such as long memory process. These barriers can be overcome thorough block bootstrap method or sieve bootstrap method. While the block bootstrap method has their advantages, there are some general drawbacks. For instance, the resampled blocks might not represent the actual behaviour of the time series under consideration. As a result of this, the bootstrapped series might have weaker dependency than that of original series. Sieve bootstrap can overcome such disadvantages. In this section we introduce the following fractional difference sieve bootstrap method to approximate the asymptotic critical value of statistic$MS_{T,h,d}$. The steps are constructed as follows:

Step 1: Estimate long memory parameter d, denote the estimate by $\hat{d}$

Step 2: Difference $X_1, \ldots X_t$ $\hat{d}$ times, denote the resulting series $Y_1, \ldots Y_t$ byY_t.

Step3: Fit a AR(pT) process on this process, i.e. $Y_t = a_1 Y_{t-1} + a_2 Y_{t-2} + \cdots + a_{pT} Y_{t-pT} + e_t$. Denote the residual of this operation by$\hat{e}^{\hat{d}}_{t,pT}$.

Step 4: Resample the fitted residual, we have$\{e_t^{\hat{d}^*}\}_{t=1}^{T}$.

Step 5: Calculate the statistic

$$\frac{MS^*{}_{T,h,d}}{b(k)} = \frac{\max_{T+1 \le k \le [T\tau]} \sigma^{-1} T^{-\frac{1}{2}-\hat{d}} \left| \sum_{i=k-[Th]+1}^{k} Y_t - \frac{[Th]}{T} \sum_{i=1}^{T} Y_t \right|}{b(k)}$$

Step 6: Repeat Steps 4-5 B times, take the $1-\alpha$ quantile of $\frac{MS^*{}_{T,h,d}}{b(k)}$ as critical value of significance levelα.

3 Simulations

In this section, we assess the finite sample performance of our monitoring procedures. All the simulations are implemented in the R. Table 1 lists critical value of monitoring function $MS_{T,h,d}$ which is obtained based on FARIMA$(0, d, 0)$ by

$T = 5000$, $\tau = 2$ generated long memory time series under 10000 cycles simulation, and $\alpha = 5\%$, bandwidth parameter h vary among{0.25, 0.5, 0.75}.

According to the critical value in Table 1, we set the historical sample$T = 200, 500$, monitoring sample $n = 2T$ to generate the long memory time series based on the modelFARIMA$(1, d, 0)$ and FARIMA$(0, d, 1)$ respectively. Table 2 and Table 3 show the empirical size resulted from the two data generation processes less than 1000 cycles, and from Table 2 and Table 3 we can see the empirical size is much higher than 5% of the test level. Although the size distortion decreases as the sample size increases, the test level is s is still not well controlled. This shows that when the historic al sample size is not large in the real problem, the critical value based onFARIMA$(0, d, 0)$ model cannot be directly used to monitor change point.

Table 4 and table 5 shows the empirical size resulted based on critical value determined by our sieve boot method, and the historical sample$T = 200, 500$, monitoring sample $n = 2T$, data generating process is based on the model FARIMA$(1, d, 0)$ and FARIMA$(0, d, 1)$ respectively. Compared with Table 2 and Table 3, we could see all the empirical sizes are controlled below the test level, but a bit too conservative. Some of the unlisted simulation results show that as the maximum monitoring sample size n increases gradually, empirical size will close to the selected test level.

Tab. 1: Critical values

d	$h = 0.25$	$h = 0.5$	$h = 0.75$
0.1	0.726	1.075	1.345
0.2	0.582	0.904	1.164
0.3	0.458	0.738	0.968
0.4	0.338	0.548	0.730

Tab. 2: Empirical sizes of [7] (%)

FARIMA$(1, d, 0)$, AR=0.7				FARIMA$(0, d, 1)$, MA=-0.5		
		h			h	
d	0.25	0.5	0.75	0.25	0.5	0.75
0.1	94.0	86.2	75.6	43.4	30.6	28.8
0.2	82.2	70.0	59.8	36.6	26.4	22.4
0.3	62.6	50.0	42.4	26.4	20.6	19.0
0.4	40.2	34.4	30.2	20.8	18.6	18.0

Tab. 3: Empirical sizes of [7] (%)

FARIMA$(1, d, 0)$, AR=0.7				FARIMA$(0, d, 1)$, MA=-0.5		
		h			h	
d	0.25	0.5	0.75	0.25	0.5	0.75
0.1	93.4	81	71.4	36.8	29.7	23.6
0.2	81.2	65.0	56.8	29.0	23.6	21.0
0.3	56.6	45.4	37.4	25.0	20.0	16.6
0.4	33.0	26.6	25.0	19.0	17.5	13.2

Tab. 4: Empirical sizes of sieve boot test (%)

FARIMA$(1, d, 0)$, AR=0.7				FARIMA$(0, d, 1)$, MA=-0.5		
		h			h	
d	0.25	0.5	0.75	0.25	0.5	0.75
0.1	2.8	3.6	2.8	1.8	1.8	2.6
0.2	1.1	0.8	1.4	0.6	0.6	0.8
0.3	1.0	0.8	0.8	0.4	0.4	0.2
0.4	0.8	0.8	1.2	0.6	0.4	0.6

Tab. 5: Empirical sizes of sieve boot test (%)

FARIMA$(1, d, 0)$, AR=0.7				FARIMA$(0, d, 1)$, MA=-0.5		
		h			h	
d	0.25	0.5	0.75	0.25	0.5	0.75
0.1	1.6	1.2	1.4	0.8	0.6	0.6
0.2	0.6	0.6	0.8	0.6	0.5	0.4
0.3	0.4	1.0	1.0	0.4	0.4	0.2
0.4	0.8	1.0	1.0	0.4	0.2	0.2

Tab. 6: Power (%) and ARL (see [.]) of sieve boot test (%)

	FARIMA$(1, d, 0)$, AR=0.7			FARIMA$(0, d, 1)$, MA=-0.5		
		h			h	
d	0.25	0.5	0.75	0.25	0.5	0.75
$0.1 \to 0.2$	15.2 [77.66]	12.8 [91.66]	11.0 [107.18]	11.4 [83.62]	10.6 [106.46]	9.8 [120.26]
$0.1 \to 0.3$	44.8 [84.13]	40.6 [97.25]	37.0 [108.37]	34.8 [93.95]	30.6 [105.82]	30.2 [118]
$0.1 \to 0.4$	80.0 [65.83]	74.8 [80.23]	70.2 [90.84]	75.4 [71.81]	69.6 [86.80]	63.4 [98.97]
$0.2 \to 0.3$	13.0 [114.16]	11.9 [107.18]	11.4 [115.95]	11.2 [105.53]	9.4 [111.38]	8.7 [121.97]
$0.2 \to 0.4$	45.8 [80.09]	44.0 [91.85]	42.2 [104.94]	39.4 [85.28]	35.6 [94.78]	40.4 [113.17]
$0.3 \to 0.4$	17.8 [86.13]	19.0 [102.68]	19.2 [112.22]	16.9 [96.98]	15.2 [108.13]	14.0 [127.66]

Tab. 7: Power (%) and ARL (see [.]) of sieve boot test (%)

	FARIMA$(1, d, 0)$, AR=0.7			FARIMA$(0, d, 1)$, MA=-0.5		
		h			h	
d	0.25	0.5	0.75	0.25	0.5	0.75
$0.1 \to 0.2$	13.6 [243.85]	12.4 [291.94]	11.0 [302.33]	10.2 [266.03]	8.8 [263.97]	5.6 [343.57]
$0.1 \to 0.3$	57.0 [216.32]	49.4 [266.32]	41.4 [278.43]	48.0 [252.19]	37 [281.44]	33.0 [331.59]
$0.1 \to 0.4$	90.2 [150.75]	83.2 [187.98]	76.6 [208.40]	87.4 [166.28]	79.6 [209.75]	71.5 [243.04]
$0.2 \to 0.3$	14.4 [231.97]	13.4 [269.39]	12 [286.54]	8.4 [278.19]	7.9 [308.21]	6.5 [299.75]
$0.2 \to 0.4$	59.4 [202.56]	55.4 [236.02]	47 [262.15]	43.4 [210.91]	40.8 [247.35]	35.5 [294.52]
$0.3 \to 0.4$	28.0 [212.85]	26.2 [250.75]	25.0 [294.04]	13.0 [211.7]	11.2 [270.95]	9.5 [332.53]

Now let the change point be$k^* = 0.5n$, compare the power and the average run length (ARL) when parameters change. The simulation conditions remain unchanged, Table 6 and Table 7 gives the power and ARL resulted from FARIMA$(1, d, 0)$ and FARIMA$(0, d, 1)$model data generation processes. Compare the change of parameter d in Table 6 and Table 7, it can be seen that when the difference between parameters d_1 and d_2 is larger, that is, the jumping degree is bigger, the power is higher. And the ARL decreases when the jumping degree increases, this is in agreement with the result obtained from the various change point monitoring methods in the literatures. Compare the bandwidth parameter$h$, we can see that the power decreases as h increases, and the ARL increases as h increases. When the bandwidth parameter is too large, the value of the power is too low, which may be due to the fact that the calculated sample size is too small to be stable when calculating the value of the monitoring statistic.

4 Conclusions

The change point monitoring method can not only analyse the change points in online data, but also can find change point earlier than the posteriori tests based on fixed samples in most cases, which is advantageous to adjust the decision in time to reduce the loss. For the serious size distortion problem of long memory time series long memory parameter change point using critical value determined directly by simulation method when the autoregressive order p and the moving average order q are not equal to 0. We proposed determine the critical value by sieve bootstrap resampling method to monitor and simulation results show that the test level can be well controlled and the power is higher as well.

Acknowledgement: This work was supported by the National Natural Science Foundation of China (No. 11661067), Natural Science Foundation of Qinghai Province (No. 2015-ZJ-717).

References

[1] Bauer P, Hackl, P. The use of MO sums for quality control [J]. Journal of the Royal Statistical Society, 1978, Series B, 57: 659–672.

[2] Horvath L, Kokosza P, ZHANG A. Monitoring constancy of variance in conditionally heteroskedastic time series [J]. Econometric Theory, 2006, 22(3): 373–402.

[3] Berkes I, Gombay E, Horváth L, et al. Sequential change-point detection in GARCH (p, q) models [J]. Econometric Theory, 2004, 20: 1140–1167.

[4] Chen Z, Tian Z. On-line monitoring parameter change in linear regression model [J].Systems Engineering -Theory &Practice, 2010,30(6): 1047–1054.(in Chinese)

[5] Chen Z, Tian Z. Online bootstrap monitoring of the stationarity for a class of heavy tailed random signals [J]. Control Theory& Applications, 2010, 27(7): 933–938. (in Chinese)

[6] Chen Z, Tian Z, Zhao C. Monitoring persistence change in infinite variance observations [J]. Journal of the Korean Statistical Society, 2012, 41: 61–73.

[7] Heinen F, Willert J. Monitoring a change in persistence of a long range dependent time series. http://hdl. handle.net/10419/73135

[8] Buhrmann P. Sieve bootstrap for time series [J]. Bernoulli, 1997, 3(2): 123–148.

[9] Chen Z, Tian Z, Xiao Q, Xing Y. Sieve bootstrap test for variance change in long memory time series [J]. International Joint Conference on Applied Mathematics, 2014, 41: 61–73.

[10] Chen Z, Tian Z, Xing Y. Sieve bootstrap monitoring persistence change in long memory process [J]. Statistics and Its Interface, 2016, 9: 37–45.

[11] Chen Z, Xing Y, Li F. Sieve bootstrap monitoring for change from short to long memory process [J]. Economic letters, 2016, 140: 53–56.

[12] Poskitt D S. Properties of the sieve Bootstrap for fractionally integrated and non-invertible processes [J]. Journal of Time Series Analysis, 2008, 29(2): 224–250.

[13] Abadir K M, W Distaso, L Giraitis. Nonstationary-Extended local Whittle estimation [J]. Journal of Econometrics, 1993(61): 821–856

Dun-Yuan Wang, Yong-Ping Zhang, Li-Mei Yao, Xian-Feng Xu and Jian-Guo Li

Detection System of Temperatures and Displacements for CNC Machine Tool Based On Lab VIEW

Abstract: The paper introduces a PXI system for detecting the temperature and displacement of CNC machine tool. The hardware of the system composes of high-performance embedded controller NI PXIe-8840, 9-slot chassis NI PXIe-1078 in which each slot accepts PXI Express module and X Series Data Acquisition Module NI PXIe-6341. Lab VIEW (Laboratory Virtual Instrument Engineering Workbench) software is used to process the acquired data from the digital potentiometer temperature sensor and grating displacement sensor according to CNC performance requirements. The developed system can be used to measure temperature, displacement, speed and acceleration accurately.

Keywords: Lab VIEW, digital potentiometer, grating, data acquisition

1 Introduction

Temperature and displacement are important test indicators of CNC machine tool, because the change in temperature during processing can lead to work piece's thermal deformation and the inaccuracy of the position may affect the accuracy of the work piece directly. In order to complete the test efficiently and flexibly, the NI virtual instrument technology is used. It consists of three components: the graphical programming software-Lab VIEW, modular I/O hardware and software hardware platform for integration. Lab VIEW is the most important part of NI virtual instrument technology, because communication function of GPIB, VXI, RS-232, RS-485 protocol hardware and data acquisition card is integrated and the function of in-

Dun-Yuan Wang, School of Electronic and Control Engineering, Chang' an University, Xian 710064, China. Email: 1213836557@qq.com
Yong-Ping Zhang, School of Electronic and Information Engineering, Ningbo University of Technology, Ningbo 315211, China. Email: ypz@nbut.edu.cn
Li-Mei Yao, GOLDCARD HIGH – TECHCO...LTD, Hangzhou 310018, China. Emil: 1621674790@qq.com
Xian-Feng Xu, School of Electronic and Control Engineering, Chang' an University, Xian 710064, China. Email: 3110259@qq.com
Jian-Guo Li, School of Electronic and Control Engineering, Chang' an University, Xian 710064, China. Email: 603757352@qq.com

DOI 10.1515/9783110540048-052,

strument panel control design, data analysis, instrument drive, I/O interface communication [1]can be achieved. The paper introduces a way to develop your own virtual instruments.

2 Temperature Detection

2.1 Uj33d-3 Digital Potentiometer

UJ33D-3 digital potentiometer uses advanced digital intelligent technology and traditional crafts, as shown in Fig.1. It can be used to read eight kinds of thermocouples' (K, E, J, S, T, B, R, and N) temperature values corresponding to the output or measured voltage value. Its DC voltage and current input/output range is 0-4999.0mv and 0-19.999mA.

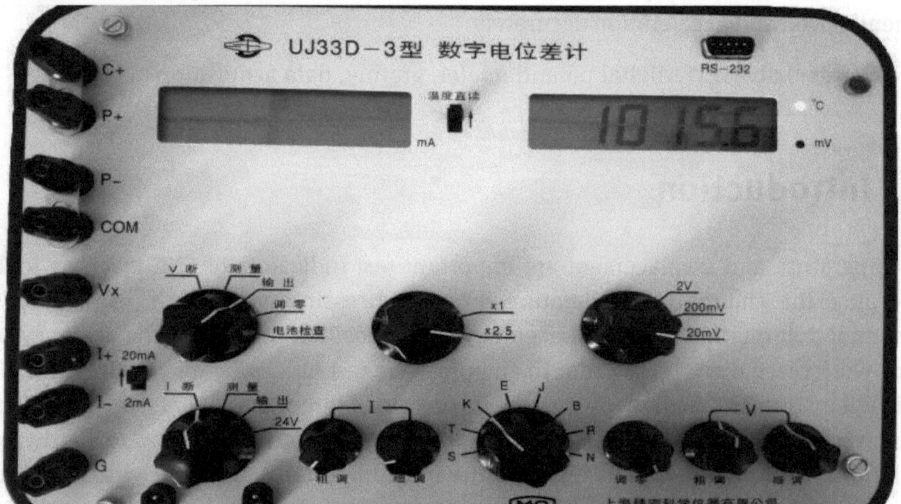

Fig. 1: UJ33D-3 digital potentiometer

RS-232 standard interface is used to communicate with the computer. RS232 parallel cable is unable to communicate with this device but RS232 cross cable can transmit data. The PC can be used to send commands to control the device's mode of data transmission, for example: the instrument can send data continuously to the PC by sending C or c. Its communication protocol is that baud rate is 9600 and data frame mode is 9 bits including 8 data bits and 1 stop bit without parity-proof bit.

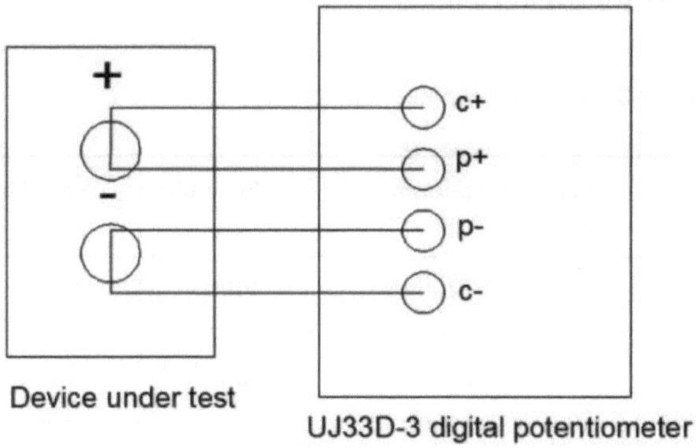

Fig. 2: Four-terminal output mode

Before using the device, it needs to be zeroed (set voltage function selector switch to "zero", and adjust the "zero potentiometer" so that the digital display is zero).In the 20mV and 50mV range, it needs to be preheated 5 to 10min so that the displayed voltage remains stable. In addition, as shown in Fig.2, by using four-terminal output mode, it can eliminate the voltage drop error generated by measuring wire when outputs small signal voltage. It can adjust the voltage to get the desired value through the coarse and fine adjustment. After setting the voltage value and thermocouple type, it can read the corresponding temperature value by opening the temperature read-only button up. The maximum test temperature is at 1700 °C and the temperature value will flash when the temperature value is out of range.

The main technical indicators as shown in Table 1:

Tab. 1: The main technical indicators of UJ33D-3 digital potentiometer

Graduation	Range	Basic error permissible limit	Resolution	Rated load
2V	0~1999.9mV	±(0.04%Ux+200μ V)	100μ V	2mA
200mV	0~199.99mV	±(0.04%Ux+20μ V)	10μ V	2mA
20mV	0~19.999mV	±(0.04%Ux+2μ V)	1μ V	2mA
2V×2.5	2000.0~4999.0mV	±(0.04%Ux+500μ V)	300μ V	2mA
200mV×2.5	200.0~499.90mV	±(0.04%Ux+50μ V)	30μ V	2mA
20mV×2.5	20.000~49.990mV	±(0.04%Ux+5μ V)	3μ V	2mA
(K)	0~1230.0°C	±(0.1%Tx+0.2°C)	0.1°C	
(E)	0~660.0°C	±(0.1%Tx+0.2°C)	0.1°C	
(J)	0~860.0°C	±(0.1%Tx+0.2°C)	0.1°C	
(S)	0~1768.0°C	±(0.1%Tx+1°C)	0.5°C	
(T)	0~380.0°C	±(0.1%Tx+0.2°C)	0.1°C	
(B)	300.0~1600.0°C	±(0.1%Tx+1°C)	0.5°C	
(R)	0~1600.0°C	±(0.1%Tx+1°C)	0.5°C	
(N)	0~1300.0°C	±(0.1%Tx+0.4°C)	0.2°C	

2.2 Test Plan

Use RS232 cross cable connect the digital potentiometer to the NI PXIe-8840's IOIOI interface properly. According to the type of temperature sensors, digital potentiometer in turn applies $0.1V_T$, $0.25V_T$, $0.5V_T$, $0.75V_T$ and V_T out voltage values to the input port of temperature detection system under test. According to (1), LabVIEW can be used to calculate the temperature of each test point indication error and determine whether this value is qualified by comparing with the allowable error. The maximum temperature indication error of all test points is taken as the measured value of the temperature indication accuracy of the system.

$$(T_C - T_S)/_{T_S}. \tag{1}$$

V_T— The temperature-sensor output signal value of the corresponding actual maximum used temperature, expressed in mV.

T_C—The corresponding temperature value read from the display unit of the system under test, expressed in°C.

T_S—The corresponding temperature value read from the display unit of the test system, expressed in°C.

2.3 Create the VI

After setting up the front panel, and call the function in the program block and connect the function with the input/output control according to the implemented function. As shown in Fig.4: initialize the serial port by using VISA Configure Serial Port VI[1], and connect the serial terminal of function with the corresponding serial input control; Enable Termination Char prepares the serial device to recognize Termination char; The role of Termination char is able to stop read operation when the Termination char is read from the serial device where the Termination char is set to the hexadecimal number 0xD for carriage return; Set the Message Based Settings:Send End Enable property (where the END flag is sent at the end of each write operation) through the Property Node; SARL End Out property: termination mode for serial transmission; Termination Char property: serial terminator to send; Call the For loop structure to connect the test number control to the count (N) terminal; Add a "Test: value change" event to the Event Structure and enable the test to start when you click the test control; VISA Flush I/O Buffer Function flushes the specified I/O buffer; VISA Write Function writes the command to the device specified by the VISA resource name; VISA Read Function reads the specified number of bytes from the device or interface specified by VISA resource name and returns the data in read buffer [1], and displays the read string on the digital potentiometer-display control and convert it to a numeric output via the Fractional/Exponential String To Number Function to a graphical display control; Call the functions of Numeric Function according to the test requirements logically; Case structure determine whether the results are qualified; Convert data to strings by using Number To Decimal String Function and Number To Fractional String Function, then the string will be displayed in the test table control through Build Array Function, and create a test table's local variable to save the last test data, then create the invoke node of the test table's reinitialize to default to clear the test table at the next test; Establish your own Word template: by entering title name and adding corresponding bookmarks in the new Word, generate the test Word that has the title name entered in the document, and select the inserted table to add the bookmark to save the data in the table, finally save the Word as a template file type, then double click on MS Office Report Express VI, and when the dialog box appears, enter the Word template path and select the custom report for Word from the Template Drop-down list box, and click OK to load the Word template into the function; VISA Close Function closes the device specified by VISA resource name; Simple Error Handler VI displays an error when an error occurs[1]; "Return to optional test main interface: value change" event structure realizes returning to the

'Test interface' when you click on return to optional main interface control, and make the test interface VI keep run and the front panel open; While Loop structure repeats the subroutine block; Flat Sequence Structure is executed in the order from left to right, and after the completion of each frame, the data is transferred to the next frame.

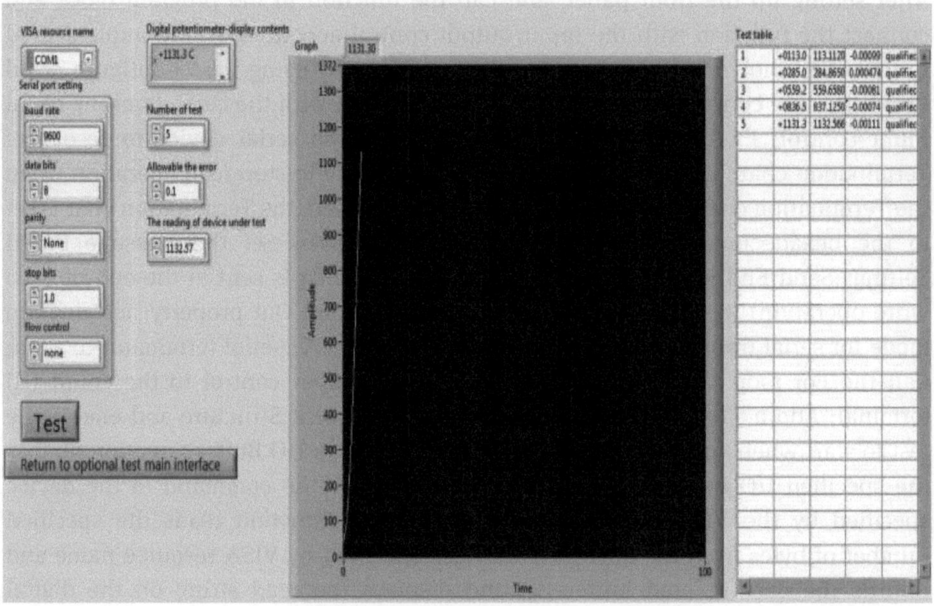

Fig. 3: Temperature control front panel

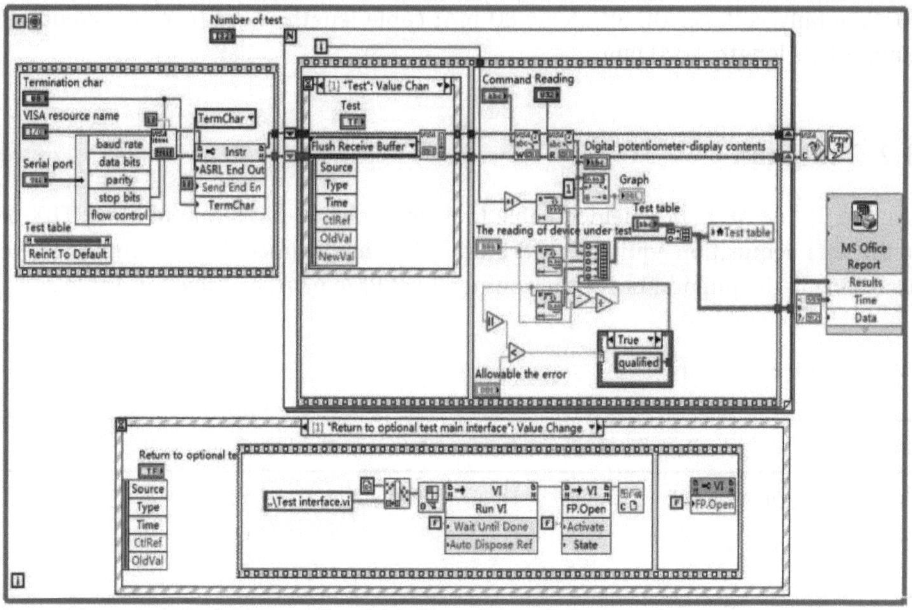

Fig. 4: Temperature control program

3 Displacement Detection

3.1 SINO KA-300 Grating

SINO KA-300 grating is a high precision displacement detection device formed by the light source, (1) scale grating, (2) cable, (3) indicating grating, (4) secure connection plate, photoelectric detection devices and other components. The scale grating and the indicating grating are installed on the machine's moving parts and fixed parts respectively. Its output is a square wave signal and there are three square wave signals: A phase, B phase, Z phase (A phase and B phase have the same cycle and the phase difference of 90°. A phase ahead of B phase represents the positive displacement; A phase lag B phase represents the negative displacement. Z signal can be used as a calibration signal to eliminate cumulative errors). The scale grating produces a count pulse when it moves a grid distance so grating can measure the relative displacement of the scale grating and the indicator grating by measuring the count pulses. Technical parameters of SINO KA-300 grating are resolution: 5μm (0.005 mm), grating pitch: 0.02mm, grating grid line: 50 lines/mm, reaction rate: 120 m/min (0.005 mm), operating temperature: 0–45 °C, storage temperature: –20–

70 °C, supply voltage: DC 5V ±5 %, 80 mA; cable length: 3.0m (can be extended), Main ruler's length: 1020 mm.

3.2 Test Plan

Use the SHC68-68-EPM Shielded cable to connect the 68-pin terminal block TBX-68 to the data acquisition equipment NI PXIe-6341. Then connect the grating to TBX-68 according to the pin number of the counter in NI PXIe-6341, as shown in Table 2:

Tab. 2: The wiring method

Grating pin	1	2	3	4	5	6	7	8	9
signal	Empty	0V	Empty	Empty	Empty	A	+5V	B	Z
Wiring color		Black				White	Orange	Red	Green
Counter pin		9				42	8	46	41

The grating is mounted on a displacement measuring apparatus; the digital input port of the system accepts the grating's output pulse and displacement curves will be displayed on the display unit after the system calculating the number of pulse. Performance requirements: measurement accuracy ≤0.01mm; measuring range (0-1000) mm. Based on the displacement and measured time during the displacement measurement, the linear velocity and acceleration of the related mechanical equipment are calculated. Curve reflects the dynamic response process of results. Speed detection performance requirements: accuracy≤0.1 %; measuring range (0–500) mm/s. Acceleration detection performance requirements: accuracy ≤1 %.

The grating drive device is installed on the displacement detection test device, and the grating drive device is connected with the tested control system properly. The motor drives moving parts toward the predetermined position in a straight line, and if the maximum displacement distance of the controlled moving parts is L(mm), the moving parts will be controlled to move L_S($L_S$may be 0.1 L, 0.3 L, 0.5 L, 0.7 L, 0.9 L)which is measured by the grating again. Each detection point will be tested 10 times, and the average of the absolute value of each detection point difference is calculated according to (2). The maximum difference among all detection points is used as the position detection accuracy index.

$$L_P = \sum_{n=1}^{10}(|L_{CN} - L_{S1}|)/10. \tag{2}$$

In the equation: L_P—the average of the absolute values of the difference in each detection point, expressed in mm;

L_{CN}—The position value read from the system under test, expressed in mm;

L_{S1}—The moving distance measured by the grating of moving parts, expressed in mm.

3.3 Create the VI

Create a Lab VIEW VI library under the test plan. First of all, create input and output controls that can control the test in the front panel, as shown in Fig.5:The counter control specifies the physical channel for task, because the NI PXIe-6341 data acquisition equipment is used in the third slot of the PXI chassis and counter 1 is used to count, so select PXI1Slot3/ctr1 in its drop-down list; A,B,Z input terminal controls specify the terminals of the A,B,Z signals, and select the corresponding pin in the drop-down list by looking at the pin number of the NI PXIe-6341;Z index enable Boolean control specifies whether to use the z signal; The cycle time(ms) control sets the time interval between two adjacent samples; The Displacement data control shows the amount of displacement; Digital Trigger Source control specifies the clock to be used for triggering, using 20MHz time base, where the trigger is a rising edge; The Start measurement button controls whether the measurement starts; Return to optional test main interface button controls whether to return to the Test interface VI; The Displacement waveform chart shows the displacement and displacement curve.

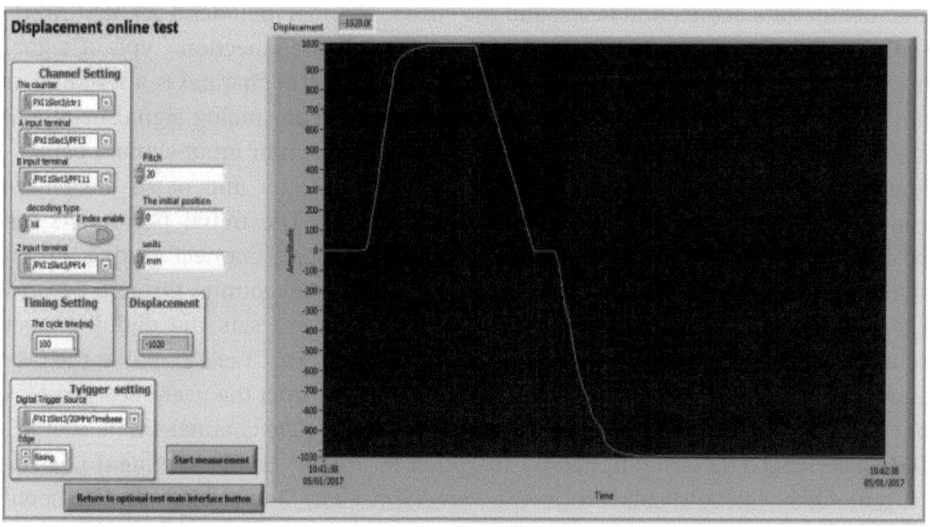

Fig. 5: Displacement Online test front panel

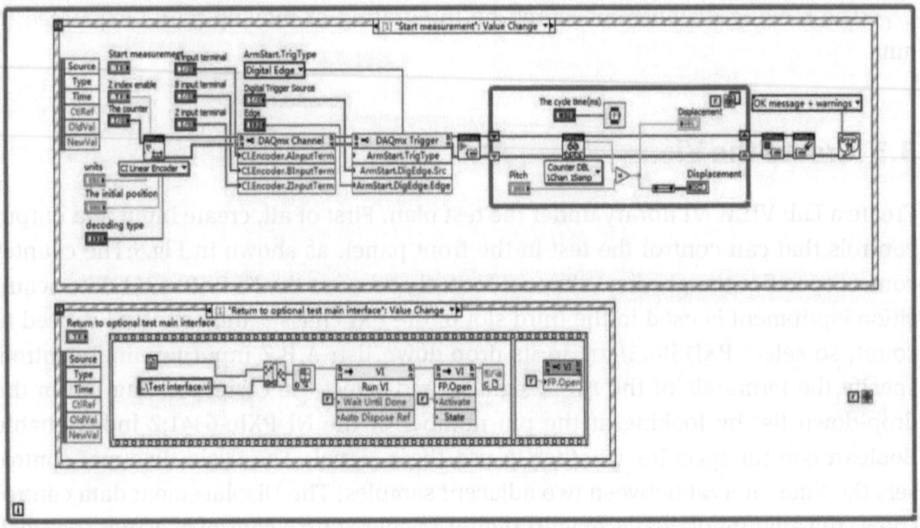

Fig. 6: Displacement online testing program

Call the function in the program block and connect the function with the input/output control according to the implemented function after setting up the front panel. As shown in Fig. 6: DAQmx Create Virtual Channel VI creates a virtual channel for data acquisition tasks, and the virtual channel includes a set of attribute settings, the name to assign, physical channel, input connection, type of signal measurement and scaling information. Where the physical channel is a terminal or pin of hardware device for measuring and generating an analog signal or digital signal [2], and the signal measurement type uses the counter input-signal-position-linear encoder. The decoding type specifies how to count and parse the pulses generated by the encoder on signal A and signal B. The initial position is the position of the grating when you begin the measurement; Connect grating output signal to the corresponding receiver of the virtual channel counter through DAQmx channel property node; DAQmx triggered property node sets the digital trigger source and trigger type for the virtual channel; DAQmx Start Task VI makes the task start the measurement. DAQmx Read VI reads samples from the user-specified task or virtual channel, and read type selection is counter-single channel-single sample-DBL; The read data is multiplied by the pitch to obtain the displacement [2]; The current time will be displayed on the waveform chart X-axis after displacement being processed into dynamic data by Merge Signals. DAQmx Stop Task VI returns the task to the state the task was in before the DAQmx Start Task VI ran; DAQmx Clear Task VI clears the task; Simple Error Handler VI displays an error when an error occurs[4]; "Start measurement: value change" event branch and "Return to optional test main interface: value change" event branch can't be placed within an event structure, then edit the “Start measurement: value change” event structures: uncheck “lock panel(defer processing of user action) until the event case completes”.

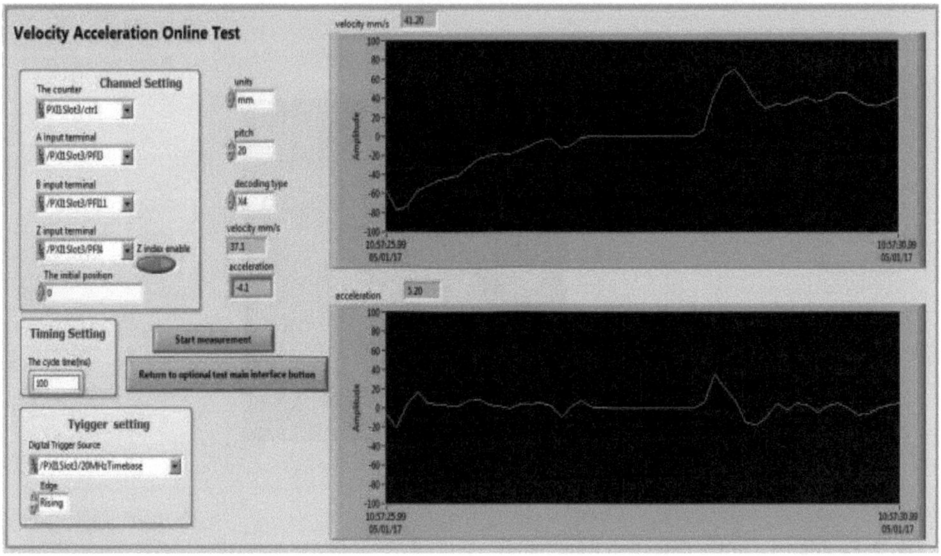

Fig. 7: Velocity Acceleration online test front panel

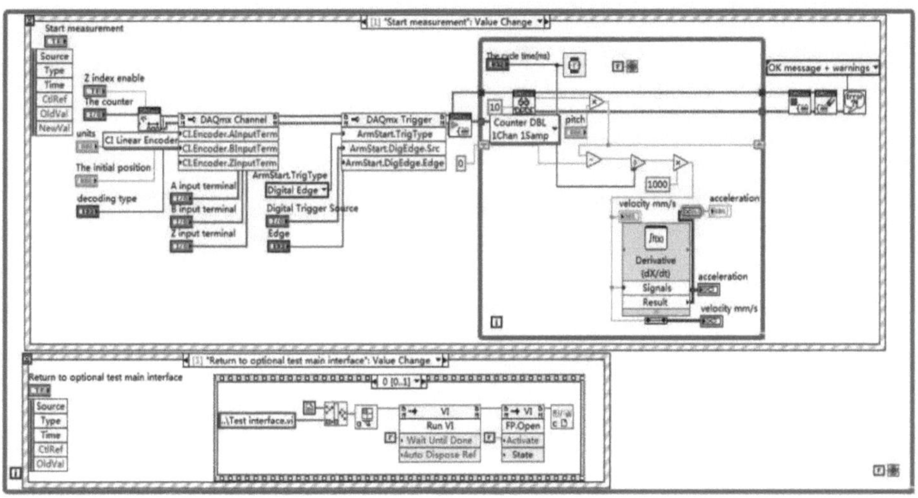

Fig. 8: Velocity Acceleration online test program

As shown in Fig.7 and Fig.8: subtract the displacement from the next two acquired acquisitions from the shift register and then divide by the time interval to get the velocity[5]; The time domain Math Express VI is used to derive the acceleration.

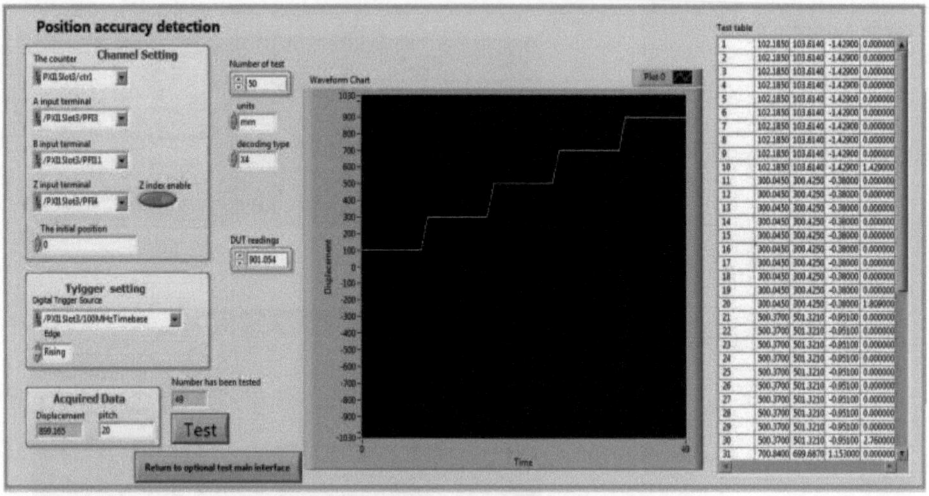

Fig. 9: Position accuracy detection front panel

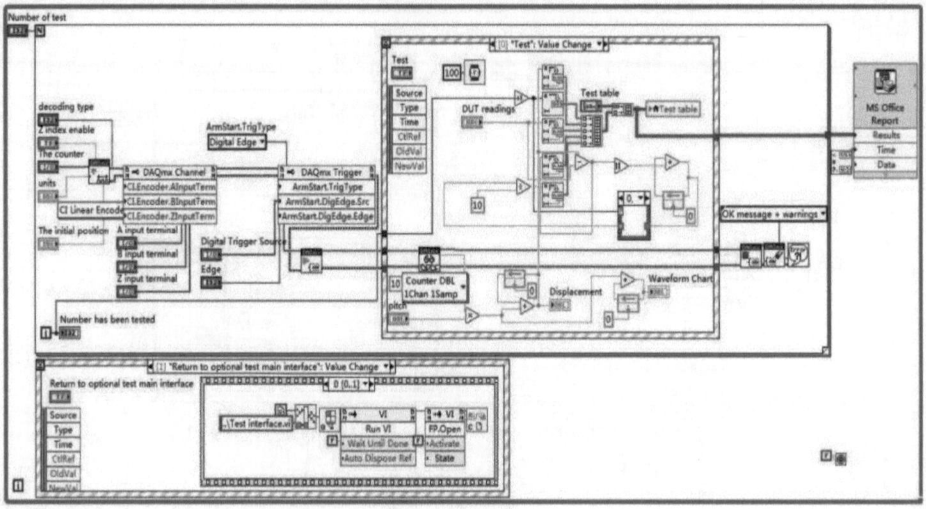

Fig. 10: Position accuracy detection program

As shown in Fig.9 and Fig.10: The feedback node is used to save the last run data and achieve the accumulation of displacement data; The case structure is used to determine whether the number of test points in each test point reaches 10, and calculate the average of the absolute value of each detection point's difference if the number reaches 10; The results will be displayed in the waveform chart and test table, and will generate Word document after the end of the measurement.

4 Conclusion

It is found that the data can be collected quickly and accurately by NI hardware and the data can be processed to meet the accuracy requirements through LabVIEW software. Because LabVIEW provides powerful functions library and the Express VI (based on configuration, no need to program), so you can write a program easily and quickly. Programs with the same functionality can be reused in different VIs, such as the program that counts grating's pulse of displacement, velocity, acceleration detection program[6]. If you change the hardware connection mode, just change the corresponding terminals on the front panel's input controls without changing your program. In addition, you can change the number of measurement as needed. LabVIEW provides modification function, so you can beautify your interface.

References

[1] National Instruments. LabVIEW help [S]. 2016.6.
[2] HAO Li, ZHAO Wei. Logical Relationship between MAX, DAQ Assistant and DAQmx VI [C], Research and Exploration in Laboratory, 2012.6(6): 83–85.
[3] Zhang Baihai, Chai Senchun, Jia Yuan yuan, Jiang Zemin, Peng Guangzheng. Sudy of Linear Encoder Signal Processing Methods in Data Acquisition Systems [C]. Machine Tool & Hydraulics, 2003(2): 118–119.
[4] Han Tao, Duan Shihui Design of Virtual Instrument Panel based on NI Data Acquisition Acquisition Card [J].ENGINEERING & TEST, 2016.12,56 (4): 60–62.
[5] FAN Xiao-lan, ZHAO Chun-feng. Design of displacement measuring platform based on LabVIEW [J], Manufacturing automation, 2012.8, and 34(8): 112–114.
[6] Liu Diantao, Ge Zongjiang, Wen Jun, Liu Chunfeng. The displacement sensor testing system based on LabVIEW [J], AUTOMOBILE APPLIED TECHNOLOGY, 2017, 1(1): 154–156.

Yan Wang and Xiang-Lei Yin

A Method of Diagnosing Boiler Four–Tube Leakage Rate

Abstract: Diagnosing boiler four–tube leakage rate is of great significance for finding leakage in time in the early leak, which is conducive to arrange maintenance, shorten the time of the maintenance and reduce the loss. Based on the linear relationship between the leakage rate and fault characteristic parameters, a new method of diagnosing leakage and leakage rate is proposed in this paper. The method is verified through the simulation experiment. Validation result shows that leakage can be identified correctly by the method, even if the leakage rate is low. Leakage rate can be diagnosed correctly, and the error is very small.

Keywords: Fault diagnosis, Leakage rate, Boiler, Fuzzy mathematics

1 Introduction

The boiler is the important equipment of thermal and power industry. Four–tube leakage is refers to pipes burst of the water wall, economizer, superheated and reheated, causing working medium leakage. Four–tube leakage has serious threat to the secure operation of the boiler. The characteristics of the leak is that when the leakage rate is low, the leak has no destruction and develops slowly, lasting a few days or weeks. When the leakage rate is higher, it turns into a destructive leak [1]. Therefore, in the four–tube leakage diagnosis, diagnosing leakage rate is of great significance for finding leakage in time in the early leak, which is conducive to arrange maintenance, shorten the time of the maintenance and reduce the loss.

At present the boiler four–tube leakage is diagnosed by the method of sound capture. This method is based on the analysis of the monitoring sound waves from furnace to judge whether the pipeline leak somewhere. However, the sound in the furnace is very complicated, which brings large difficulty to detect sound of leakage [2].What worse, in the early leakage, leakage rate is lesser; sound wave of leakage is difficult to be found. In this paper, a method of boiler four–tube leakage diagnosis is put forward based fuzzy mathematics. This method can diagnose four–tube leakage and calculate the leakage rate by monitoring the boiler operation parameters, making up for the inadequacy of sound capture method.

***Corresponding author: Yan Wang,** Zaozhuang University, No. 1 Beian Road, Zaozhuang, China. Email: yanhuadian@163.com

DOI 10.1515/9783110540048-053,

2 Theory Basis of Diagnosing Leakage Rate

Results of thermodynamic calculation of boiler four–tube leakage [3] point out that fault characteristic parameters, such as amount of exhaust smoke, exhaust smoke temperature and exhaust smoke enthalpy, increase with linear increase of the leakage rate. The conclusion is suitable for leakage, blocking and scale formation failure of thermal system, because they are all based on the law of mass conservation and energy conservation. Many other parameters of the boiler also have the linear relationship with the leakage rate. Diagnosis method proposed in this paper uses this conclusion as a theoretical basis.

3 Procedure and Workflow of Diagnosing Boiler Four–tube Leakage Rate

3.1 Selection of Fault and Fault Symptom

There are a lot of faults of four–tube leakage, to choose which fault as the research object, should be determined according to the type of boiler. At the same time, the fault of large possibility and serious harm should be listed as the research object. It should be pointed out that, the method proposed in this paper can't make accurate diagnosis to coordinates of leak location. This method only diagnoses that leakage exists in a certain area or certain equipment.

When fault characteristic parameter is chosen, attention should be paid to the relativity between parameters and the leakage fault. Moreover, it is necessary to make the chosen fault characteristic parameter easily distinguish the fault from other faults.

The selected faults constitute fault set. Fault set of which fault number is m can be expressed as $U = \{u_1, u_2 \ldots u_i \ldots u_m\}$, of which ui is fault *i*. In the same way, the selected faults symptom parameters constitute fault symptom set. Fault symptom set of which symptom number is n can be expressed as $Y = \{y_1, y_2 \ldots y_j \ldots y_n\}$, of which x_j is fault faults symptom parameters *j*.

3.2 Obtaining Fault Symptoms

Function of obtaining fault symptoms should adopt a continuous linear function, so as to avoid damaging linear relationship among fault characteristic parameter, fault symptoms and leakage rate. Assume that the maximum leakage rate diagnosed is w_m, and then leakage rate ranges of 0-w_m. Function of obtaining fault symptoms is expressed as the follow formula.

$$z=\begin{cases}1 & y\geq a\\ \dfrac{y-y_0}{a-y_0} & y_0<y<a\\ 0 & y=y_0\\ \dfrac{y-y_0}{y_0-b} & b<y<y_0\\ -1 & y\leq b\end{cases} \tag{1}$$

Where z is the symptoms value of a certain fault, y is transformed values of fault characteristic parameter, y_0 is normal value of y, a and b is the threshold value of transformed characteristic parameters when the leakage is w_m.

It should be pointed out that, for those symptoms parameters changing range of only increasing or reducing, formula (2) should also take unilateral value [4].

3.3 Establishment of Fault Knowledge Library

As mentioned above, fault characteristic parameters of a certain fault are: y_1, y_2... y_j...y_n, corresponding fault symptoms value z_1, z_2...z_j...z_n, then the feature vector of the fault can be got, as (z_1, z_2...z_j...z_n). Fault knowledge library **R** can be established based feature vectors of all faults (the number is m).

$$\mathbf{R}=\begin{pmatrix} z_{11} & z_{12} & \cdots & z_{1n}\\ z_{21} & z_{22} & & z_{2n}\\ \cdots & & z_{ij} & \cdots\\ z_{m1} & z_{m2} & \cdots & z_{mn}\end{pmatrix} \tag{2}$$

3.4 Fault Identification

The fault identification adopts the principle of maximum membership. Assume that the fault to be identified is fault f, and then fault feature vector of fault f is (z_{f1}, z_{f2}...z_{fj}...z_{fn}). Membership function is as following [5].

$$\mu(u_f,u_i)=\frac{\sum_{j=1}^{n} z_{fj}\cdot z_{ij}}{\sqrt{\sum_{j=1}^{n} z_{fj}^2}\cdot\sqrt{\sum_{j=1}^{n} z_{ij}^2}} \tag{3}$$

Where μ (u_f, u_i) is the membership degree of fault f the same as fault i.

If the maximum of membership degree of all faults is μ (u_f, u_c), fault c is the fault to be identified.

In theory, fault is able to identify accurately when leakage rate is very low, which cannot be achieved by many other membership functions.

3.5 Diagnosis of Leakage Rate

Leakage rate can be calculated according to the follow formula.

$$w_f = \sqrt{\frac{\sum_{j=1}^{n} z_{fj}^2}{\sum_{j=1}^{n} z_{cj}^2}} w_c \tag{4}$$

Where w_f is leakage rate of fault f (the fault to be identified), z_{cj} is symptoms value of fault c, w_c is leakage rate of fault c.

3.6 Workflow of Four–tube Leakage Rate Diagnosis

System of diagnosing boiler four–tube leakage rate is established based on simulation platform. The diagnosis system is installed on the simulation system or the DCS system of boiler to diagnose four - tube leakage rate or to do simulation experiments. The diagnosis system and workflow is shown in figure 1.

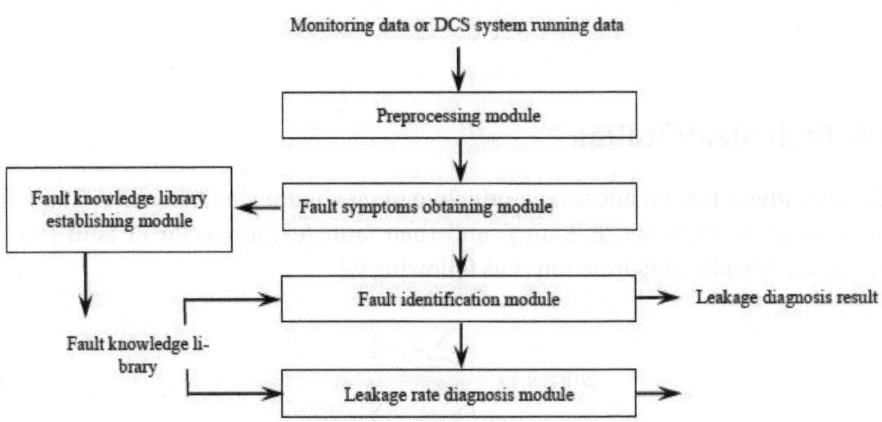

Fig. 1: The diagnosis system and workflow

4 Application Instance

4.1 Summary of the Boiler

Four - tube leakage was diagnosed based on the boiler of type SCL - 500 - 4.0/400. The main parameters of this boiler are shown in Table 1.

Tab. 1. Main parameters of the boiler of type SCL–500–4.0/400

Parameter	Parameter values
Rated Evaporation	41.6 t/h
Temperature of Main Steam	400°C
Pressure of Main Steam	4.0 MPa
Temperature of Feed Water	130°C
Pressure of Feed Water	6.0 MPa
Pressure of Drum	4.5 MPa
Temperature of Exhaust Smoke	190°C~210°C
Temperature of Air in Inlet of Air Pre–Heater	30°C
Flow of Primary Air	88400 m^3/h
Flow of Secondary Air	24316 m^3/h
Flow of Smoke	187500 m^3/h
Efficiency of Boiler	79.1 %

4.2 Diagnosis of leakage and leakage rate

For this boiler, 5 faults of steam and water pipe leakage and 14 corresponding characteristic parameters are selected, shown in Table 2 and Table 3 respectively.

Tab. 2. Faults Set of Boiler Steam and Water Pipe Leakage

symbol	Faults
u_1	Leakage of High Temperature Superheater
u_2	Leakage of Middle Temperature Superheater
u_3	Leakage of Low Temperature Superheater
u_4	Leakage of Economizer
u_5	Leakage of Water Wall

Tab. 3. Fault Symptom Set of Boiler Steam and Water Pipe Leakage

symbol	Fault Symptom
y_1	Water Level of Drum
y_2	Pressure of Drum
y_3	Temperature of Main Steam
y_4	Pressure of Main Steam
y_5	the Flow Difference between Main Steam and Feed Water
y_6	Furnace Pressure
y_7	Exhaust Smoke Temperature
y_8	Load of Unit
y_9	Electric Current of Induced Draft Fan
y_{10}	the Smoke Temperature Difference between Both Sides of High Temperature Superheater
y_{11}	the Smoke Temperature Difference between Both Sides of Middle Temperature Super-heater
y_{12}	the Smoke Temperature Difference between Both Sides of Low Temperature Superheater
y_{13}	the Smoke Temperature Difference between Both Sides of Economizer
y_{14}	the Smoke Temperature Difference between Both Sides of Water Wall

Simulation experiment for the fault in Table 2 was carried out based on simulation system above-mentioned, in which the leakage rate is 10%. The fault knowledge library is established by the method proposed in this paper. The fault knowledge library is shown as followed.

$$\mathbf{R}_{5\times14} = \begin{pmatrix} z_{11} & z_{12} & \cdots & z_{1n} \\ z_{21} & z_{22} & & z_{2n} \\ \cdots & & z_{ij} & \cdots \\ z_{m1} & z_{m2} & \cdots & z_{mn} \end{pmatrix}$$

$$= \begin{pmatrix} -0.100 & -0.801 & 1.000 & -0.673 & 0.594 & 0.450 & -0.012 & -0.487 & 0.986 & 0.404 & -0.200 & 0.149 & -0.094 & -0.049 \\ -0.940 & -1.000 & 0.000 & -0.846 & 0.843 & 0.721 & -0.115 & -0.860 & 0.991 & -0.282 & 0.500 & -0.125 & -0.356 & -0.153 \\ -0.342 & 0.119 & 0.000 & 0.087 & 0.776 & 0.537 & -0.134 & 0.087 & 0.112 & 0.087 & 0.309 & 0.915 & -0.810 & 0.500 \\ -0.500 & -0.077 & 0.078 & -0.031 & 0.829 & 0.856 & -0.134 & -0.142 & -0.760 & 0.000 & -0.407 & -0.111 & 0.347 & -0.077 \\ -0.866 & 1.000 & 0.000 & 0.995 & 0.920 & 0.623 & -0.964 & 0.715 & 0.134 & 0.469 & 0.500 & -0.193 & 0.463 & 0.966 \end{pmatrix}$$

Take leakage of middle temperature superheated as example, leakage rate is 5 % and 10 %. Fault feature vectors are calculated out by fault symptoms obtaining module, as follow:

u_{f1} (leakage rate is 5 %) (−0.453, −0.478, 0.000, −0.409, 0.410, 0.348, 0.000, −0.428, 0.491, −0.126, 0.239, −0.000,−0.147, −0.076)

u_{f2} (leakage rate is 10 %) (- 0.920, - 0.996, 0.000, - 0.813, 0.816, 0.706, - 0.100, - 0.836, 0.965, - 0.269, 0.489, - 0.113, - 0.287, - 0.136)

The two faults are identified by formula (3) and Leakage rate is diagnosed by formula (4), diagnosis results are shown in Table 4.

Tab. 4. Diagnosis Result of Leakage and Leakage Rate

Fault to be identified	Leakage Rate	u_1	u_2	u_3	u_4	u_5	Diagnosed Leakage Rate	Relative Error
u_{f1}	5%	0.447	0.626	0.225	0.150	−0.043	4.83%	3.4%
u_{f2}	10%	0.704	0.997	0.335	- 0.086	0.151	9.72%	2.8%

4.3 Analysis of Diagnosis Result

According to the principle of maximum membership, Table 4 shows that two faults are both identified as fault 2, which is consistent with actual fact. This shows that leakage can be correctly identified by the method, even if the leakage rate is low, which is conducive to find leakage in time.

Table 4 shows that leakage rate of two faults was diagnosed correctly, and the error is small, which is helpful to arrange maintenance and assess the loss.

5 Conclusions

1) Boiler four–tube leakage can be identified correctly by the method proposed in this paper.
2) When the leakage rate is low, leakage can be diagnosed correctly by the method proposed in this paper, which is conducive to find leakage in time, shorten the time of the maintenance and reduce the loss.
3) Leakage rate can be diagnosed correctly, and the error is small, which is helpful to arrange maintenance and assess the loss.

References

[1] H.W. Chen, S. Ding, and C. B. Wang, The "Four Tubes" Leakage Detection System for Power Plant Boiler Based on Acoustic Emission Technology, Process Automation Instrumentation, 27, 3, 42–44, 2006.

[2] L.S. An, G.S. Jiang, and S.L. Wang, K. Yang, A Study on The Trait Value of the Sound Fields Radiated from a Round Leaky Hole on the Heat Exchanger Tube Wall, Proceedings of the CSEE, 23, 5, 144–148, 2003.

[3] W.J. Yang, J.H. Zhou, and X.Y. Cao, Analyse on Heat Calculation of Boiler Tubes Leakage, Proceedings of the CSEE, 20, 8, 85–88, 2000.

[4] L.Y. Ma, B.S. Wang, J.Q. Gao, Study of Fault Fuzzy Knowledge Library of the Whole–Bypass U–Pipe High–Pressure Heater System and Its Fault Diagnosis with Neural Network, Power Engineering, 22, 1, 1615–1621, 2002.

[5] X.L. Yin, Research on Fuzzy Diagnosis Method of Boiler Steam and Water Pipe Leakage, Mechanical Engineering and Control Systems, Proceedings of the 2015 International Conference on Mechanical Engineering and Control Systems, 75–78, 2015.

Yu Ning, Xin-Li Xu*, Zhen Jiang and Bin-Yao Ning

Research on Fault Detection and Diagnosis for Small Unmanned Aerial Vehicle

Abstract: As the core of aircraft, the state of the flight control system can impact the survivability and mission effectiveness of the aircraft obviously. The fault diagnosis of the flight control system when it breaks down can provide some essential information for subsequent reconfigurable control, which can guarantee the system stability. At present the sensor and actuator faults are mostly studied and the damaged body of the small unmanned aerial vehicle (SUAV) considers very few. Some important methods and focus problems for fault diagnosis are firstly analysed and summarized. Then the research status of the air dynamics and fault detection and diagnosis (FDD) is introduced in the paper. Methods of the pneumatic parameters and properties on the SUAV and the key technologies of FDD are also analysed. At last the experiment of the SUAV is created by using the FDD subsystems. A specific plan and research methods are provided through the paper. Meanwhile, it provides the theory and research foundation for reconstruction control system.

Keywords: SUAV; flight control system; FDD; fault and damage; reconfigurable control

1 Introduction

The SUAV (internationally called portable UAV), which is shown as Fig1, has a simple structure, lightweight, high efficiency, flight speed and distance and cost advantages. So it is commonly used in the military and civil fields. The SUAV is the intricate system among multidisciplinary which is related to mathematics, mechanics, materials, energy, machinery, electronics, control, communication, navigation and computer, and so on. It exists in the obvious nonlinear and uncertainty during the working. Thus if any component failure of the SUAV is not found and removed in time, the huge economic loss even catastrophic consequences will be caused. For example, on October 14, 2011, china international general aviation conference was held in pushing, Shanxi Province. While the aerobatic flight was shown, the plane crashed and one pilot was injured and one was killed in the following explosion.

Yu Ning , Bin-Yao Ning, Departments of Automation, Shao Guan University, Shao guan, 512005, China. Email: nby6210@163.com.
***Corresponding author: Xin-Li Xu, Zhen Jiang** School of Mechatronics Engineering and Automation, Shanghai University, Shanghai, 200072, China. Email: lingyun308@shu.edu.cn.

DOI 10.1515/9783110540048-054,

The seven astronauts were killed in the Challenger space shuttle disaster in 1986. So it is the important thing that how to improve the reliability and safety of SUAV system and prevent and stop the occurrence and development [1, 2]. Therefore, the fault detection and diagnosis (FDD), as one of the most

Fig. 1: Structure diagram of fixing–wing SUAV

Important content about self–repairing technology is important to determine whether the failure occurs timely and accurately, position the location and type of failure, and determine the size and timing of failure [3–7]. Research on fault diagnosis is important to improve the safety and reliability of aircraft, survivability and combat effectiveness. And it has become one of the most popular research topics in comprehensive design of modern aircrafts. In this paper, based on a large number of literature references, fault diagnosis methods of flight control system are discussed comprehensively. And then the key technologies of the aerodynamic performance and FDD are put forward.

2 Problem Statement and Preliminaries

FDD has been widely researched because of its importance. To date, many fault diagnosis techniques have been suggested to deal with this problem by the researcher. And these methods can be divided into two categories according to the relationship with the model: one is based on the model and other has no direct links with the detailed model.

2.1 `FDD Method Based on the Model

The method is based on the modern control theory and optimization methods. First of all, the hardware redundancy is replaced by analytical redundancy based on the mathematical model of the system. Then, the residual signal can be got by comparing the actual output values and estimate values, which is calculated by the constructed observer. Next, analysis results are valuated based on the criterions or thresholds. According to the different generating forms of residual signals, these methods can be divided into state estimation, parameter estimation method and the equivalent spatial method.

2.1.1 State Estimation Diagnosis Method

State estimation method is firstly presented by Beard [8]. Numerous researches have been made under the theoretical foundation of Beard. And these studies are classified into branches: filtering method and observer methods. The kernel of this algorithm is that: (1) design the feedback gain matrix of observer in order to display the influences of one fault or faults in the residual signal in defined direction; (2) estimate the state based on the variations of residual signal, if there are faults in the system, the changes of residual signal will be stronger; vice versa, the changes of residual signal will be small under normal conditions.

It is obvious that the method based on state estimation has fine performances in practical application if we can get the accurate mathematical model of the system. Furthermore, the requirements of the input signal are not very strict and it is not necessary to keep the uninterrupted trigger signal all the time. However, with the development and non–linear characteristics of control systems, it is hard to build an accurate mathematical model. In addition, this method cannot be found the damage position and detect the damage degree clearly. These shortcomings limit the applications of state estimation methods.

2.1.2 Parameter Estimation Diagnosis Method

The basic principle of parameter estimation is to detect and analyse the faults based on the model parameters and its corresponding changes. It is not necessary to calculate the residual series. Generally speaking, aiming at reliable diagnostic results. This method is utilized cooperating with other state estimate method in the non–linear system. Currently, this method contains Kalman filter methods, a robust tracking filter, hybrid methods using parameter estimation and state estimation.

Comparing with the state estimation method, parameter estimation method is relatively worse in real time and the input signal on the system requirements more stringent. The uninterrupted trigger signal is crucial, which limits the practical application of the parameter estimation method. However, it can locate the damage position and detect the damage degree of sensors, actuators and controlled objected.

2.1.3 Equivalent Spatial Method

The essence of equivalent spatial method is the equivalent mathematical relationship between inputs and outputs of an analytical model. The purpose of fault detection and isolation can be completed by evaluating the inputs and outputs of system [9]. The primary means of equivalent spatial method is to set up "an equivalent vector and system" by excluding the input and the realization of the equivalent disturbance decoupling. And the special structures residuals are built through a weighting matrix.

Equivalent spatial method is a non–threshold approach and needs more redundant signal. The computation load will be increased significantly if the numbers of measurements are big. Therefore, this method is particularly suitable for measurements in a more redundant signals or low dimension system.

2.2 The Method Does Not Rely on the Model

The foundation of the techniques based on the model is in a position to build up the mathematical model. However, the system is becoming more and more complex with the automation of the system. It is not possible to get the accurate mathematical model. Therefore, diagnostic methods, which break the limitations of the system model, have got many attentions. Generally speaking, these methods are divided into two aspects: the method based signal processing and the methods based on experience knowledge.

2.2.1 Methods Based On Signal Processing

This this method is to analyse the frequency, amplitude, phase, and various features of output of the system based on the signal processing technique. These methods not only contains the conventional method, such as time domain analysis, FFT transform and spectrum analysis (Spectrum analysis) methods, but also includes wavelet transform (Wavelet transform), correlation analysis method, the signal modal estimation, information fusion, etc. [10–14]. Among them, the FDD based on wavelet transform is a new signal processing technique.

From above analysis, the advantages of based on signal processing method are not needed accurate mathematical model and have good performance on real time. Disadvantages of based on signal processing methods are summarized as follows:

–Depends on the results of signal processing
–Affected by signal noise
–Lack of fault information in the beginning
–Confined to specific signals and certain failures
–Difficult to expand or applied to different object

2.2.2 Methods Using Experience Knowledge

Currently, this area is a hot topic because of it does not need a quantitative mathematical model but take full advantage of expert diagnostic knowledge and diagnostic objects. All the methods in this area are divided into symptom–based approach and a qualitative model–based approach. The symptom–based approaches contain an expert system, fuzzy reasoning, neural networks, support vector machine, rough sets grounds occupied by the enemy, DS evidence theory, the gray system theory, etc. the methods based on qualitative model mainly include qualitative observer, knowledge observer, qualitative simulation, directed graph, fault trees. A few of typical methods are introduced in following context.

2.2.2.1 Methods Based on Artificial Neural Network

The artificial neural network has significant advantages over other methods of non–linear system because of its abilities in self–learning, high fault tolerance and parallel computing power. There are two kinds of working forms: one is off–line diagnosis. It is a learning process and uses separately. When the system fails, faulty information or phenomenon will be imputed to the neural network and the reasonable solution will be outputted after the self–organization, self–learning processes of neural network it. Fresh is on line diagnosis. The neural network is directly connected with the system, and fault information will be got automatically. That is to say. The processing of study and use is combined.

Neural network has the capability to simulate the continuous nonlinear function and to learn from the samples. Any knowledge of the rules can be stored in the form of a neural network through the example. It has good robustness and represents a new methodology to store the information in the distribution level. As for the specified problems, in order to achieve fault detection and diagnosis, faults reasons and resources can be deduced by the neural network and input signal [15–19].

2.2.2.2 Methods Based on Fuzzy Theory

Aiming at the uncertainty, un–accuracy and process noise of the process itself, and the fuzzy logic is proposed in fault diagnosis. It has noble superiorities in time-delay, time–varying and non–linear. The principles of fuzzy consist with the inherent human thought process. It is facilitated to deal with the qualitative knowledge. The concept of membership function is introduced in the processes of transforming the certain judge to define judge. For the application of fuzzy theory in fault diagnosis, there are three basic diagnostic thoughts [20–27]:

–Based on fuzzy relations and the synthesis algorithm

–Based on fuzzy knowledge processing technology

–Based on fuzzy clustering algorithm

2.2.2.3 Intelligent Diagnosis Based on Expert System

Fault diagnosis expert system is one of the most important studies in compelling diagnosis. One of the most widely used class of intelligent diagnosis system, mainly for those who are not difficult to build accurate mathematical models or mathematical models of complex systems. It uses the practice of experts in long–term experience accumulated knowledge base, and to design a computer program to simulate a human expert reasoning and decision–making process fault diagnosis. The system is constructed by the knowledge base, inference machine, integrated database, man–machine interface and explain the module parts. Knowledge base and the inference engine is the core of expert system [9].

Expert system–based fault diagnosis method to the use of expert knowledge and experience, without a mathematical model of the system and the diagnosis is easy to understand. It has been widely used. However, these methods have their shortcomings. First, knowledge acquisition is more difficult that is the main bottleneck. Secondly, the accuracy of diagnosis depends on the knowledge base of expertise and knowledge rich or low. Finally, when the rule is large, there is a match in the conflict in the reasoning process. Issues such as combinatorial explosion make the inference become slow and inefficient.

3 Aerodynamics for SUAV

3.1 Low Reynolds Numbers and Low Aspect Ratios

Low Reynolds is the important trait for SUAV, which has a great effect on the lift force character, drag force character and lift drag ratio character. For the miniature dimensions, the aspect ratio is expected to be very small. This result in the lift coefficient decrease, drag coefficient increase and easy air separation that make the pneumatic and motor performance degrades [28–30]. This is explored at home and abroad but mostly by experiments. For instance, aerodynamics of a series of flat, curving and rectangular thin wings and string Reynolds number from 60000 to 200000 are studied through the experiments in [31]. The aerodynamic forces and flow field is researched by experiments on low aspect ratios wings with different shapes and 70000 ~ 200000 series of Reynolds in [32] and in [33] the lift–to–drag ratio character in the different sharps wings is done the experimental research and so on. Their research findings are used in the design of SUAV structure.

3.2 Research Methods for Pneumatic Performance

In In SUAV aerodynamics model, the commonly used method is to set up a mathematical model, the CFD, wind tunnel test and aerodynamics identification based on flight test data, etc.[28–30,34,35].

Aerodynamic mathematical model is the analytical relation among the SUAV air power (lift, resistance, lateral force; pitch, yaw, roll pneumatic torque), flight motion state parameters (speed, angular velocity and Angle of attack, lateral spreads Angle, flight height, etc.) and control input (the elevator, aileron, rudder; various control the rudder deflection Angle, etc.). An accurate and reasonable boundary layer model is not obtained to SUAV with CFD method for low Reynolds number effect. So the existing software also can't solve its aerodynamic performance problems. Because of the small size, the wind tunnel tests are done and other institutions are used, which has all kinds of inevitable errors and affect the accuracy of the test data. Therefore it is just an effective and feasible method that kinematic equation is firstly built and then pneumatic parameter identification is done by flying experiments. Such as, aerodynamic characteristics with the low Reynolds number and low aspect ratios wings are studied through the numerical analysis, and the influence on the different aspect ratios and bending wings to aerodynamic characteristics is shown in [36]. In [37], dynamic equation are established for unmanned spacecraft, aerodynamic performance under low Reynolds number is researched by the orthographic hybrid mesh method and then lift coefficient , the resistance coefficient, lift–to–drag ratio in the different Reynolds number corresponding to different angle of attack is obtained. In [38], unsteady N–S equations are solved through the simple method. The flow field characteristics of the wings with a low Reynolds number and low aspect ratios are researched and the aerodynamic characteristics influence of laminar separation bubble of the front wing tips and the vortex is analysed. Aircraft system identification parameter model collection is built, and then multiple input multiple output system identification method was used to identify the unknown parameters, which shows that the algorithm has good stability and precision according to the simulation calculation and test flight results in [39].

Now the research on the dynamic characteristics of the SUAV with broken–wing is less done at home. In this paper, the dynamic equation is established; the dynamical parameters and performance are analysed on the normal condition. While the wing of SUAV is broken or lost. The change of the parameters has an effect on the SUAV system. So the method is used to for the dynamic model and the flight test combination, in order to study the aerodynamic characteristics with the aid of the nonlinear system parameter identification.

4 Key Technologies

Based on the present research situation and development of the discipline, the research contents and key technologies are put forward with damaged–wing SUAV, as shown in Fig.2, FDD for broken –wing SUAV

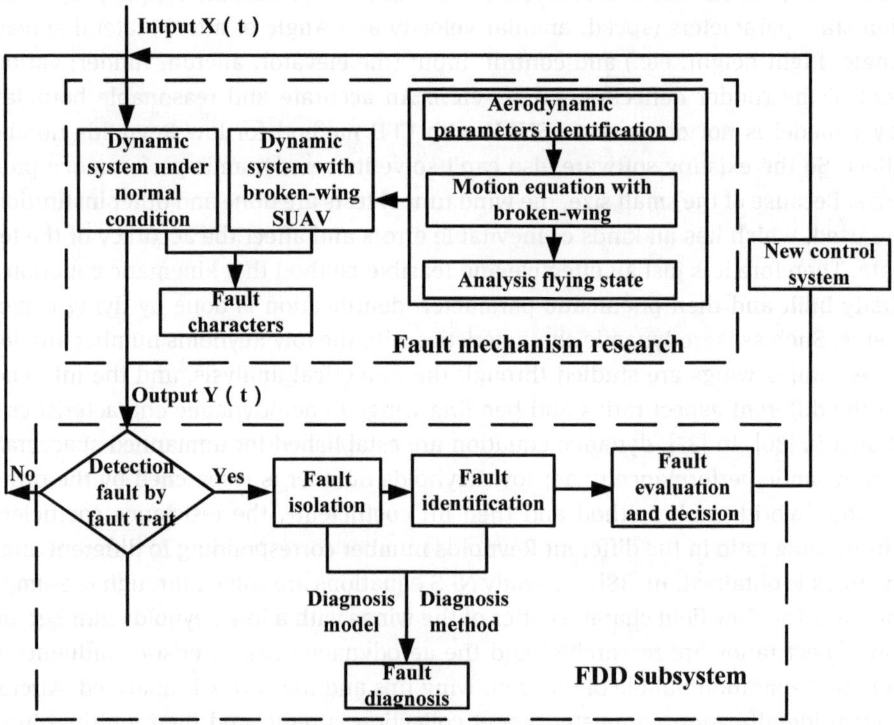

Fig. 2: Schematic diagram of FDD for broken–wing SUAV

Mainly include two parts. One part is called as the fault mechanism research. The research content is mainly that the motion equation is built for broke –wing SUAV, and then a new nonlinear system identification algorithm are explored. The aerodynamic characteristics are analysed in combination with SUAV flight test data analysis and pneumatic conditions to ensure the routine flight for broken–wing SUAV. The fault mechanism is studied and the motion state for broken–wing SUAV is simulated by using 3D software and virtual reality technology; the other part is FDD subsystem. The fault sensitive parameters are found through the movement equation or increase specified a device to obtain the diagnosis information and even gain the fault trait signal. By the abnormal state of the fault trait, a novel nonlinear fault pattern recognition and identification method is raised to establish a set of FDD subsystem, which can provide the conditions for the follow–up reconfigurable control system. The subsystem with strong robustness cannot be affected by noise. Interference, modelling error influence and also with the strong adaptive ability can improve itself to produce information such as external input change and structural changes. This is the key point for research.

Acknowledgement: This work is financially supported by the National Natural Science Foundations of China (No. 61175092 and 61433016). The authors would like to sincerely thank anonymous reviewers and editors for their constructive comments which helped us to improving this paper.

References

[1] Zhou Dong–Hua, Ye Yin–Zhong. Modern Fault Diagnosis and Fault–tolerant control. Beijing: Tsinghua University Press, 2000.

[2] Jiang Bin, Mao Ze–hui, Yang Hao, Zhang You–min. Fault Diagnosis and Fault Accommodation for Control Systems. Beijing: National Defense Industry Press, 2009.

[3] Yoshimura M, Frank P M, Ding X. Survey of robust residual generation and evaluation methods in observer–based fault detection systems[J]. Journal of Process Control, 7(6), 403–424, 1997.

[4] Patton R J, Frank P M, Clark R N. Issues of fault diagnosis for dynamic systems[C]. London: Springer–Verlag, 2000.

[5] Venkatasubramanian V, Rengaswamy R, Yin K, Kavuri S N. A review of process fault detection and diagnosis part I: quantitative model–based methods [J]. Computers and Chemical Engineering, 27(3), 293–311, 2003.

[6] Venkatasubramanian V, Rengaswamy R, Kavuri S N. A review of process fault detection and diagnosis part II: qualitative models and search strategies [J]. Computers and Chemical Engineering, 27(3), 313–326, 2003.

[7] Venkatasubramanian V, Rengaswamy R, Kavuri S N, Yin K. A review of process fault detection and diagnosis part III: process history based methods [J]. Computers and Chemical Engineering, 27(3), 327–346, 2003.

[8] R.V. Beard. Failure Accommodation in linear systems through self–reorganization[R]. Report MVT–71–1, Man Vehicle Lab, MIT, Cambridge, Massachusetts, 1971.

[9] Donghua Zhou, Yanyan Hu. Fault diagnosis techniques for dynamic systems [J]. Acta Automatic Sinica, 35(6), 748–758. (In Chinese) 2009.

[10] C. Kar, A.R. Mohanty. Vibration and current transient monitoring for gearbox fault detection using multiresolution fourier transform [J]. Journal of Sound and Vibration, 311(1–2), 109–132, 2008.

[11] H. Ye, G.–Z. Wang, S.X. Ding. A new parity space approach for fault detection based on stationary wavelet transform [J]. IEEE Transactions on Automatic Control, 49(2), 281–287, 2004.

[12] Shousong Hu, Chuan Zhou, Yuan Wang. A combination fault pattern recognition based on wavelet neural network [J]. Acta Automatica Sinica, 28(4): 540–543. (In Chinese) 2002.

[13] Xin Wen, Hongyue Zhang, Lu Zhou. Fault diagnosis and fault–tolerant control of control system [M]. Machinery Industry Press, (in Chinese) 2000.

[14] J.Q. Zhang, Y. Yan. A wavelet–based approach to abrupt fault detection and diagnosis of sensors [J]. IEEE Transactions on Instrumentation and Measurement, 50(5), 1389–1396, 2001.

[15] S. Rajakarunakaran, P. Venkumar, D.Devaraj, K.S.P.Rao. Artificial neural network approach for fault detection in rotary system [J]. Applied Soft Computing Journal, 8(1), 740–748, 2008.

[16] J.–D.Wu, P.–H.Chiang, Y.–W.Chang, Y.–J.Shiao. An expert system for fault diagnosis in internal combustion engines using probability neural network [J]. Expert Systems with Applications, 34(4): 2704–2713, 2008.

[17] H.Benitez–Perez, F. Cardenas–Flores, J.L. Ortega–Arjona, F. Garcia–Nocetti. Fault localization using neural networks and observers for autonomous elements [J]. Control and Intelligent Systems, 35(4): 319–328, 2007.

[18] Papadimitropoulos, G.A. Rovithakis, T.Parisini. Fault detection in mechanical systems with friction phenomena: An online neural approximation approach [J]. IEEE Transactions on Neural Networks, 18(4): 1067–1082., 2007.

[19] M.Michalik, M.Lsukowicz, W.Rebizant, S.–J.Lee, S.–H.Kang. New ANN–based algorithms for detecting HIFs in multigrounded MV networks [J]. IEEE Transactions on Power Delivery, 23(1): 58–66, 2008.

[20] Palma Luis Brito, Coito Fernando Vieira, Da Silva Rui Neves. Robust fault diagnosis approach using analytical and knowledge based techniques applied to a water tank system [J]. Engineering Intelligent Systems, 13(4): 237–244, 2005.

[21] Curry T, Collins E G Jr, Selekwa M. Robust fault detection using robust l–script 1 estimation and fuzzy logic[C]. Proceedings of the American Control Conference, Arlington, VA, (2): 1753~1758, 2001.

[22] Frank P M, Koeppen Seliger Birgit. New developments using AI in fault diagnosis [J]. Engineering Applications of Artificial Intelligence, 10(1): 3–14, 1997.

[23] Schneider H, Frank P M. Fuzzy logic based threshold adaption for fault detection in robots[C]. Proceedings of the IEEE Conference on Control Applications, Glasgow, UK, Aug, (2): 1127–1132, 1994.

[24] Schneider H, Frank P M. Observer–based supervision and fault detection in robots using nonlinear and fuzzy logic residual evaluation [J]. IEEE Transactions on Control Systems Technology, 4(3): 274–282, 1996.

[25] Liping Xu, Zhonghua Zhao, Yanhua Zhang. Navigation sensor fault detection strategies [J]. Journal of Shanghai Jiaotong University, 36(7): 966~969. (In Chinese), 2002.

[26] Lihua Hou, Xiaoyan Zhu, Xinyi Wang. Fuzzy inference algorithm used in FMS fault diagnosis and maintenance system [J]. Journal of Beijing Institute of Technology, 20(3): 281–285. (In Chinese), 2000.

[27] Bing Long, Xingwei Jiang, Zhengji Song. Telemetry data real–time monitoring and diagnostics based on multi–agent satellite [J]. Acta Aeronautica et Astronautica Sinica, 26(6): 726–732. (In Chinese), 2005.

[28] Zhu Bao–liu. Aerodynamics of pilotless plane. Beijing: Aviation Industry Press, 2006.
[29] Hu Yu–qun.Some Dynamic Problem in Micro Air Vehicles. Nanjing: Nanjing University of Aeronautics and Astronautics, 2002.
[30] Zheng Xiang–ming. Research on Nonlinear Flight Dynamics and Intelligent Flight Control of Micro Air Vehicles. Nanjing: Nanjing University of Aeronautics and Astronautics, 2008.
[31] Alain Pelletier and Thomas J.Mueller, Low Reynolds Number Aerodynamics of Low–Aspect–Ratio, Thin/Flat/Cambered–Plate Wings. Journal of Aircraft, Vol.37, No.5, 2000.
[32] Alain Pelletier and Thomas J.Mueller, Low Reynolds Number Aerodynamics of Low–Aspect–Ratio Wings, AIAA 99–3182.
[33] Gabriel E. Torres and Thomas J. Muller, Aerodynamics Characteristics of Low Aspect Ratio Wings at Low Reynolds Number, Fixed and Flapping Wing Aerodynamics for Micro Air Vehicle Applications, 115–141,2001
[34] LI Zhi–guo, ZHU Peng–cheng, LI Feng. Investigation on the low Reynolds number characteristics of the lift and drag of low aspect ratio wings. Experiments and Measurements in Fluid Mechanics, Vol.18, No.4, 2004.
[35] SONG Shu–heng. Numerical Simulation of Aerodynamic Characteristics for Micro Air Vehicles. Mianyang: China Aerodynamics Research and Development Center Graduate School, 2007.
[36] ZOU Xin–sheng.Research on the Nonlinear Parameter Identification and Robust Control of Aircraft. Beijing: Tsinghua University, 2006.
[37] YANG Ai–ming, WENG Pei–fen. Numerical analysis of low Reynolds number flows around thin wings for micro aerial vehicle. Acta Aerodynamica Sinica, 23(1), 57–67, 2005.
[38] BAI Yue, CAO Ping, GAO Qing–jia, SUN Qiang. Characterization of a multi rotor micro air vehicle and simulation of airfoil aerodynamics at low Reynolds–number. Acta Aerodynamica Sinica, 29(3), 325–329, 2011.
[39] LI Jian–hua, LI Feng. Low–Reynolds–Number numerical simulation of inverse Zimmerman wing. Acta Aerodynamica Sinica, 25(2), 220–225, 2007.

Yan-Bo Jia, Mei-Chun Chen, Wen-Bin Hu and Tie-Bing Liu*

Evaluation of the Uncertainty in the Determination of Arsenic in Cosmetics by Inductively Coupled Plasma-mass Spectrometry (ICP-MS)

Abstract: A mathematic model for evaluating the uncertainty of the determination procedure for arsenic in cosmetics by ICP-MS was established. The uncertainty sources for each of the parameters in the model were analysed. The uncertainties of type A (based on the calculation of the statistical distribution of the determination results) was evaluated. According to the international general method, the component uncertainties were synthesized and extended respectively to obtain the uncertainty evaluation. The results showed that the uncertainty of weighing process, the uncertainty of sample digestion solution constant volume, the uncertainty of standard material, the uncertainty of standard curve fitting, and the uncertainty of determination repeatability were main source of the uncertainty. After the evaluation of each uncertainty components, the main component is the uncertainty of the curve fitting introduced by least-square method; the second is the uncertainty of the standard material introduced.

Keywords: uncertainty, ICP-MS, arsenic, cosmetics

1 Introduction

Evaluation of uncertainty of measurement for chemical analysis and laboratory quality control is very important [1–6]. Uncertainty of measurement is composed of multiple components. Some of the component measured column with statistical distribution estimation are available, and characterization and the standard deviation. Other component is available based on experience or other information given probability distribution estimation, also can use the standard deviation.

Analysis of arsenic in cosmetics widely was paid attention. Recently there are some application methods such as atomic fluorescence spectrometry (AFS), inductively coupled plasma emission spectrometry (ICP-OES). Inductively coupled plas-

Yan-Bo Jia, Mei-Chun Chen, Wen-Bin Hu, Hangzhou Institute for Food and Drug Control, Hangzhou, Zhejiang, China, 310022. Email: 82998580@qq.com
***Corresponding author: Tie-Bing Liu,** Zhejiang University of Science and Technology, Hangzhou, Zhejiang, China, 310023. Email:tiebingliu@163.com.

DOI 10.1515/9783110540048-055,

ma mass spectrometry (ICP-MS) is a kind of high sensitivity, fast, high speed and efficient detection method, and has been widely used in many fields [7-13]. The results of uncertainty evaluation are particularly important. In this paper the uncertainty of as content in cosmetics by ICP-MS was studied. The results of the study have a certain guide for daily testing work in the laboratory.

2 Materials and Methods

2.1 Instruments, Reagents and Methods

2.1.1 Instruments and Reagents

ICP spectrometer (Agilent7700cx): United States Agilent company; MARSXpress microwave digestion: United States CEM company, with 40 digestion tubes; Electronic analytical balance (resolution 0.0001 g): mettler-toledo instrument (Shanghai) co., LTD. ; 5 mL pipette; 100 mL volumetric flask; 65% nitric acid (excellent level pure) : Germany MERK company products; Arsenic standard solution (1000 mg/L) : provided by the national research centre for certified reference materials; Germanium (Ge) (1000 mg/L) internal standard solution: provided by the national research centre for certified reference materials.

2.1.2 ICP-MS Analysis Parameters

Adjust instruments indicators according to the detection signal in the tuning solution. The indicators of the instrument sensitivity, mass, the resolution, oxide and double charge meet the measurement requirements. Main working parameters are RF power 1250 w, spray chamber temperature 2 °C, pump rate 0.1 r/s, carrier gas (argon) flow 1.0 L/min, Sampling depth 7.2 mm, oxide rate < 1.5 %, double charge rate< 3 %.The internal standard method was used to quantify with 1 mg/L germanium (Ge) as internal standard of arsenic.

2.2 Detection and Method

Weight 0.500 g sample in the digestion tank, add 5 mL nitric acid, stand 30 min, cover relief valve, the digestion tank was put into the microwave digestion system, set up the appropriate program of microwave digestion, catch acid according to relevant steps after complete digestion, remove the digestive solution to 25.0 mL test tube, and constant volume to the scale with the deionized water for determination. Determine ^{75}As concentration in digested solution in ^{72}Ge as online internal standard using inductively coupled plasma mass spectrometer, quantified after comparing with standard series solution.

3 The Results and Discussion

3.1 The Uncertainty Evaluation Mathematical Model

The mathematical model of arsenic content in the cosmetic as follows:

$$C_s = \frac{X_0 \times V}{W \times 1000}$$

C_s is arsenic content in the samples (mg/kg), X_0 is arsenic concentration in sampling solution (μg/L), V is the volume of digestion solution (mL), and W is sample weight (g)

3.2 Uncertainty Source Analysis

According to the detection method and mathematical model, the main sources of uncertainty components are as follows: the uncertainty of weighing process introduction, the uncertainty of sample digestion solution constant volume introduced; the uncertainty of standard material introduced (the uncertainty of arsenic standard solution, the uncertainty of standard series of solution preparation); the uncertainty of standard curve fitting introduction; the uncertainty of determination repeatability introduction.

3.2.1 Uncertainty Introduced In The Weighing Process $u_{rel}(M)$

The uncertainty introduced by the sample came mainly from the repeatability of the weighing changes and balance calibration. By balance calibration certificate the allowed error is 0.0001g, uniform probability distribution, take K=$\sqrt{3}$, and the standard uncertainty is

$$u(M)=\frac{0.0001}{\sqrt{3}} = 0.000058g$$

Sample average weight M is 0.5003 g, and the relative uncertainty from the sample weight is

$$u_{rel}(M) = \frac{u(M)}{M} = \frac{0.000058}{0.5003} = 0.000116$$

3.2.2 Uncertainty of Sample Digestion Solution Constant Volume Introduced $u_{rel}(V)$

Transfer sample digestion solution to volumetric flask of 25 mL. The uncertainty mainly includes the following parts: the calibration uncertainty and the uncertainty of temperature effect introduction. The uncertainty of the volumetric flask capacity allowance introduced is u_1 (V). On the basis of JJG196-2006 working glass container, the capacity allowance of a level volumetric flask of 25 mL is ±0.04 mL at 20 °C, uniform probability distribution, K is $\sqrt{3}$, and the standard uncertainty of volumetric flask capacity allowance was

$$u_1(V)\frac{0.04}{\sqrt{3}} = 0.0231\text{mL}$$

The uncertainty of temperature introduced is u_2 (V). As we all known, the expansion coefficient of water is 2.1×10^{-4}. The temperature difference is ± 4 °C, according to the confidence probability of 95 % (k = 1.96) the uncertainty of temperature introduced is

$$u_2(V) = \frac{25 \times 2.1 \times 10^{-4} \times 4}{1.96} = 0.0107\text{mL}$$

So the standard uncertainty produced by 25 mL volumetric flask was

$$u(V) = \sqrt{u_1(V)^2 + u_2(V)^2} = \sqrt{0.0231^2 + 0.0107^2} = 0.02546\text{mL}$$

Therefore, the relative uncertainty introduced by the constant volume of sample digestion solution was

$$u_{rel}(V) = \frac{u(V)}{V} = \frac{0.02546}{25} = 0.00102$$

3.2.3 Uncertainty of Standard Material Introduced $u_{rel}(S)$

1) Uncertainty of as standard solution $u_{rel}(As)$

According to the standard solution certificate (20 mg/L arsenic standard solution), the relative expanded uncertainty (including 2 factors) is 0.24 %, and the relative uncertainty is:

$$u_{rel}(As)=\frac{0.0024}{2}=0.0012$$

2) Uncertainty of standard series of solution preparation

Standard stock solution preparation: precise absorb arsenic (As) element standard solution (20 μg/mL) 1.0 mL into 100 mL volumetric flask, add 2 mL nitric acid, constant volume to the scale with deionized water, shake well, and arsenic (As) standard use solution content is 0.2 μg/mL. According to JJG196–2006 Working Glass Container, the capacity allowance of 1 mL pipette a grade is 0.01 mL. If the probability distribution is uniform, k $=\sqrt{3}$. The standard uncertainty of the capacity allowance of the pipette was:

$$u_1(1)=\frac{0.01}{\sqrt{3}}=0.00577\text{mL}$$

As we all known, the expansion coefficient of water is 2.1 x 10^{-4}. The temperature difference is ±4°C, according to the confidence probability of 95% (k = 1.96) the uncertainty of temperature introduced is

$$u_2(1)=\frac{1\times 2.1\times 10^{-4}\times 4}{1.96}=0.000429\text{mL}$$

So, the standard uncertainty introduced of 1 mL pipette was

$$u(1)=\sqrt{u_1(1)^2+u_2(1)^2}=\sqrt{0.00577^2+0.000429^2}=0.005786\text{mL}$$

Therefore in the preparation of arsenic standard stock solution the relative uncertainty introduced by 1 mL pipette was

$$u_{rel}(1)=\frac{u(1)}{1}=\frac{0.005786}{1}=0.005786\text{mL}$$

Standard series of solution preparation: from the above 0.2 μg /mL arsenic standard solution respectively absorbs 2 mL, 5 mL, 10 mL, 20 mL into 100 mL volumetric flask and constant volume with 2 % nitric acid solution. According to JJG196–2006 Working Glass Container, the capacity allowance of 10mLA grade pipet is ± 0.025 mL. If the probability distribution is uniform, k is $\sqrt{3}$. The standard uncertainty of the capacity allowance of a grade pipet of 10mL was:

$$u_1(10)\frac{0.025}{\sqrt{3}} = 0.01443\text{mL}$$

As we all known, the expansion coefficient of water is 2.1 x 10^{-4}. The temperature difference is ±4 °C, according to the confidence probability of 95% (k is 1.96) the uncertainty of temperature introduced is

$$u_2\ (10) = \frac{10\times 2.1\times 10^{-4}\times 4}{1.96} = 0.00429\text{mL}$$

So the standard uncertainty introduced of 10 mL pipette was

$$u(10) = \sqrt{u_1(10)^2 + u_2(10)^2} = \sqrt{0.01443^2 + 0.00429^2} = 0.01505\text{mL}$$

Therefore, in the preparation of arsenic standard series solution, the relative uncertainty introduced by 10 mL pipette was

$$u_{rel}(10) = \frac{u(10)}{10} = \frac{0.01505}{10} = 0.001505\text{mL}$$

The standard uncertainty of 100 mL volumetric flask was as follows. The uncertainty of capacity allowance of the volumetric flask is u_1 (V). On the basis of JJG196–2006 working glass container, at 20 °C the capacity allowance of 100 mL a level volumetric flask is ± 0.10 mL. If the probability distribution is uniform, k is $\sqrt{3}$. The standard uncertainty of the capacity allowance of the volumetric flask was

$$u_1(V) = \frac{0.10}{\sqrt{3}} = 0.0577\ \text{mL}$$

The uncertainty of temperature introduced is u_2 (V). As we all known, the expansion coefficient of water is 2.1 x 10^{-4}. Temperature difference is ± 4 °C, according to 95 % confidence probability of (k is 1.96), the standard uncertainty of introduced temperature was

$$u_2(V) = \frac{100\times 2.1\times 10^{-4}\times 4}{1.96} = 0.0428\text{mL}$$

So the standard uncertainty generated by the 100 mL volumetric flask was

$$u(V)=\sqrt{u_1(V)^2+u_2(V)^2}=\sqrt{0.0577^2+0.0428^2}=0.0718\text{mL}$$

Therefore, the relative uncertainty of constant volume introduced by sample digestion solution was

$$u_{rel}(V)=\frac{u(V)}{V}=\frac{0.0718}{100}=0.000718$$

To sum up, the uncertainty of standard substances was

$$u_{rel}(\text{standard})=\sqrt{u_{rel}(\text{As})^2+u_{rel}(1)^2+4u_{rel}(10)^2+5u_{rel}(100)^2}=0.00682$$

u_{rel}(As)is 0.0012, u_{rel}(1)is 0.005786, u_{rel}(10)is 0.001505,and u_{rel}(100)is 0.000718.

3.2.4 Uncertainty of Standard Curve Fitting $u_{rel}(Y)$

Standard series solution of five points was determined using online internal standard method, and each point was measured three times. Test results and calculation results were listed in Table 1. According to the data fitting, Y = 0.0106C + 0.0036, the slope (b) is 0.0106, the correlation coefficient is 0.9995.

Tab. 1: Test results of standard series solution

$C_{standard}$ 1)	Y_i 2)	Average	Curve simulation response values Y	$(Y_i-Y)^2$	$(C_{standard}-C_{standard}average)^2$
0.0000	0.0030	0.00315	0.0036	3.6×10^{-7}	219.4
	0.0037			1.0×10^{-7}	
	0.0027			8.1×10^{-7}	
4.00	0.0463	0.0464	0.0462	1.0×10^{-8}	116.64
	0.0460			4.0×10^{-8}	
	0.0469			4.9×10^{-7}	
10.00	0.1086	0.1091	0.1101	2.25×10^{-6}	
	0.1089			1.44×10^{-6}	
	0.1097			1.6×10^{-7}	
20.00	0.2189	0.2188	0.2166	5.29×10^{-6}	23.04
	0.2194			7.84×10^{-6}	
	0.2182			2.56×10^{-6}	
40.00	0.4292	0.42877	0.4296	1.6×10^{-7}	635.04
	0.4277			3.61×10^{-6}	
	0.4300			1.6×10^{-7}	
$\frac{\sum(Y_i-Y)^2}{p-2}$	-	-	-	8.4×10^{-6}	
$\sum(C_{standard\,i}-\overline{C_{standard}})^2$	-	-	-	-	3062.4

Note: 1) $C_{standard}$ is each point concentration of standard series solution (μg/L); 2) Y_i is the ratio of the corresponding value of the standard series solution for determined elements and the internal standard

$$u_{(bq)\,.\,rel}=\frac{S_Y}{b}\sqrt{\frac{1}{n}+\frac{1}{p}+\frac{(\overline{C_{standard}}-\overline{C_{sample}})^2}{\sum(C_{standard\,i}-\overline{C_{standard}})}}$$

In the type: b is the slope, S_Y is the residual standard deviation of the regression curve (residual standard deviation), p is the repeat number of the measurement sample (6 parallel samples, each sample was determined 3 times, p=6×3), n is the points number of the regression curve(n=5), C is the concentration average of test sampler, C_0 is the concentration average of the regression curve at various points, C_{0j} is the concentration of each standard solution.

The residual standard deviation of curve fitting was

$$S_Y = \sqrt{\frac{\sum (Y_i - Y)^2}{p-2}} = \sqrt{8.4 \times 10^{-6}} = 0.0029$$

$$C_{standard} = \frac{0+4+10+20+40}{5} = 14.8\mu g/L$$

The standard uncertainty of standard curve fitting was

$$u(Y) = \frac{S_Y}{b}\sqrt{\frac{1}{n} + \frac{1}{p} + \frac{(\overline{C_{standard}} - \overline{C_{sample}})^2}{\sum (C_{standard\ i} - \overline{C_{standard}})}}$$

$$= \frac{0.0029}{0.0106}\sqrt{\frac{1}{5} + \frac{1}{18} + \frac{(14.8 - 4.1484)^2}{3062.4}} = 0.0801\mu g/L$$

S_Y is the residual standard deviation of the fitted curve; b is the slope; n is the repeated number of the sample (n is 6); p is the repeated number of standard solution (5 points, each point was determined 3 times, p is15). $C_{standard}$ is arsenic concentration of the standard solution at various points; C_{sample} is arsenic concentration in digestion solution of the repeat measurement sample.

The relative uncertainty introduced by standard curve fitting was

$$u_{rel}(Y) = \frac{u(Y)}{C_{sample}} = \frac{0.0801}{4.1484} = 0.0193$$

3.2.5 Uncertainty of Determination Repeatability Introduction $u_{rel}(X)$

Sample 6 times measurement process and the results were shown in Table 2.

Weight 0.500g sample in digestion jar, add 5 mL nitric acid, put for 30 min, cover the relief valve, put the digestion tank into the microwave digestion system, set up the appropriate microwave digestion parameters according to relevant steps, after digestion completely catch the acid, remove the digestive solution to 25.0 mL test tube, scale with the deionized water for determination.

From Table 2, it showed that the sample standard deviation of the determination results were as follows:

$$s = \sqrt{\frac{\sum_{i}^{n} (X_i - \overline{X})^2}{n-1}} = 0.00663 mg/kg$$

C for solution concentration of the continuous measurement in six copies sample solution, each sample measured three times, C is 4.1484 μg/L, so the uncertainty standard of the average 6 determination replicates was as follow:

Tab. 2: Samples detemination resluts

Sample No.	Sample weight(M/g)	As concentration in digestion solution $C_{standard}$ /(μg/L)	Arsenic content in the samples X/(mg/kg)
1	0.5002	4.2212	0.2110
2	0.5003	4.1626	0.2080
3	0.5005	4.2658	0.2131
4	0.5004	4.0497	0.2023
5	0.5003	4.1656	0.2082
6	0.5002	4.0252	0.2012
Average Cstandard		4.1484	-
M average	0.5003	-	
X average		-	0.2073

3.3 Final Relative Synthesis Uncertainty of Standard Uncertainty

$$u_{rel}=\sqrt{u_{rel}(M)^2+u_{rel}(V)^2+u_{rel}(standard)^2+u_{rel}(Y)^2+u_{rel}(X)^2}=0.0205$$

$u_{rel}(M)$is 0.000116,$u_{rel}(V)$is 0.00102,$u_{rel}(S)$is 0.00682,$u_{rel}(Y)$is 0.0193, and $u_{rel}(X)$is 0.000461.

The average of arsenic content in the sample 6 replicates X is 0.2073 mg/kg, synthetic standard uncertainty is:

$$u=0.2073\times0.0205=0.00425 \text{ mg/kg}.$$

If extension factor k is 2, the expanded uncertainty U is that 0.00425 plus 2, the result is 0.008 mg/kg.

3.4 Results Expression

According to the above results of the uncertainty evaluation, the uncertainty evaluation result of determination on arsenic content by microwave digestion - inductively coupled plasma mass spectrometry in cosmetics should be represented as:

$$X= (0.207\pm0.008)\ mg/kg\ (k=2)$$

4 Conclusions

From our studies, mainly sources of the uncertainty for the determination on As content in cosmetics by ICP-MS was: the uncertainty of weighing process introduction, the uncertainty of sample digestion solution constant volume introduced, the uncertainty of standard material introduced (the uncertainty of arsenic standard solution, the uncertainty of standard series of solution preparation), the uncertainty of standard curve fitting introduction, the uncertainty of determination repeatability introduction. The uncertainty of the weighing process introduction is the least compared of the uncertainty caused by other factors, and the uncertainty of standard curve fitting introduction is the largest.

Acknowledgement: This research is financially supported by the National Quality Infrastructure of National Major Scientific Research Issues 2016YFF0201904-02 and the project of General Administration of Quality Supervision, Inspection and quarantine 2011QK144.

References

[1] ISO/IEC 17025, General Requirements for the Competence of Testing and Calibration [3] Laboratories. ISO, Geneva, Switzerland, 2005.
[2] Maroto A, Boque R, and Riu J, Trends in Analytical Chemistry, 18,577, 1999.
[3] Moser J, Wegscheider W, and Meisel T, Analytical and Bioanalytical Chemistry, 377, 97, 2003.
[4] Hund E, Massart D L, and Smeyers-Verbeke J, Analytica Chimica Acta, 480, 39, 2003.
[5] Luo Y, Gao S, and Longerich H P, Journal of Analytical Atomic Spectrometry, 22,122, 2007.
[6] XING Li, WANG Mei, and ZHAO Feng-min, Spectroscopy and Spectral Analysis, 34, 1,226, 2014.
[7] Beauchemin D, Analytical Chemistry, 82, 12, 4786, 2010.
[8] Llorent-Martinez E J, Fernandez de Cordova M L, and Ruiz-Medina A, Microchemical Journal, 102,33,2012.
[9] Zhu Y B, and Chiba K , Talanta, 90,57, 2012
[10] Chudzinska M, Debska A, and Baralkiewicz D, Accreditation and Quality Assurance, 17, 65, 2012.
[11] Yenisoy-Karakas, Food Chemistry, 132, 1555, 2012.
[12] Chen H, Fan C L, and Wang Z B, Analytical Methods, 5, 3291, 2013.
[13] Tanase I G, Popa D E, and Udristioiu G E, RSC Advances, 5, 11445, 2015.

Shu-Xin Wang

Experimental Study of Residual Stress State of Laser Induced Membrane System

Abstract: X–ray diffractometer (XRD) was used with three membrane–based samples under the system with different bonding strength and the residual stress was measured before and after laser shock with different energy. The influence of laser energy and the interface bonding strength on residual stress were analysed separately, the residual stress distribution on the back of the stainless steel film before and after the laser shock was studied, and the magnesium alloy substrate surface residual stress distribution was researched by experiment, the distribution of residual stresses (including residual principal stress and direction angles) in three directions is obtained.

Keywords: laser shock; membrane system; residual stress; experimental study

1 Introduction

Thin film technology is one of the important means to improve the performance of the material surface. By using a tiny amount of functional materials it endows the surface of parts and components with wear resistance, corrosion resistance, heat resistance, fatigue resistance, resistance to radiation and other special properties such as light, heat, electricity, magnetism. The bonding performance of membrane–based interface is the leading indicator which affects the quality of film–base system and the basic condition which determines the system's performance. To a large extent it determines the reliability and service life of the system. While laser shock wave loads film base system, the system responses dynamically, the membrane–based interface is under complex stress and ultimately produces residual stress field. The intensity of the laser parameters and the combination of membrane–based interface has a huge impact on the distribution of residual stress. Therefore, to study the distribution of the residual stress of membrane–based system induced by laser shock wave to further the study of the transmission mechanism of shock wave in membrane system is of great significance.

Shu-Xin Wang, Zhenjiang Watercraft College, Zhenjiang, Jiangsu, China. Email: doc.wsx@163.com

DOI 10.1515/9783110540048-056,

2 The Experimental Scheme

According to theoretical calculation and the measured values along 0 °, 45 ° and 90 ° directions, the principal stress and its direction angle on the surface of the measured point can be attained.

$$\tan 2\varphi = (\sigma_{0^\circ} + \sigma_{90^\circ} - 2\sigma_{45^\circ})/(\sigma_{0^\circ} - \sigma_{90^\circ})$$

$$\sigma_1 = (\sigma_{90^\circ} - \sigma_{0^\circ}\cot^2\varphi)/(1 - \cot^2\varphi)$$

$$\sigma_2 = (\sigma_{90^\circ} - \sigma_{0^\circ}\tan^2\varphi)/(1 - \tan^2\varphi)$$

σ_1 is the maximum principal stress, σ_2 is the minimum principal stress, φ is the principal stress direction angle, and $\sigma_{0^\circ}, \sigma_{45^\circ}, \sigma_{90^\circ}$ represents residual stress values along 0 °, 45 ° and 90 ° directions respectively.

The experiment adopts AZ31B magnesium alloy as the base body and 316 stainless steel foil as the coating. Three theoretical models of membrane base system are constructed by using water film, 502 glue and double–sided adhesive. The residual stress detection equipment is based on the stereoscopic XRD produced by Proto in Canada; MGR40P goniometer and special test software are equipped with as well.

Setting: X ray apparatus tube with a current of 25 mA, tube voltage of 30 kV, collimator tube diameterΦ0.5 mm, radiation with Cr–Ka characteristics, exposure time of 1 s, scanning step Angle of 0.1 °, angle bits value of 0 °, 25 °, 35 °, 45 °, crystal (122) for test plane, AZ31B magnesium alloy scan with a starting angle and an ending angle of 160 ° and 145 ° respectively, 316 stainless steel with a 140 °starting angle and a 125 °ending angle. Before the laser shock, residual stress is measured on the XRD upper and lower surfaces of AZ31B magnesium alloy and 316 stainless steel foil surface. Concretely the residual stress on the pre–shock spot centre and spot edge on the material surface is measured along 0 °, 45 ° and 90 ° directions each once. After the impact experiment is finished, the corresponding measurement will be done again in the same direction.

3 The Residual Stress Distribution on The Back of The Magnesium Alloy Substrate

3.1 The Effect of Laser Energy on Residual Stress

Table 1 lists the residual stress measured along 0 °, 45 ° and 90 ° three directions at the Centre of the impact spot and at the edge of the impact spot on the backside of membrane based systems, which based on the AZ31B magnesium alloy and used the double–sided adhesive as adhesive, Table 1 also lists the residual principal stress and direction angle by calculated.

Tab. 1: Residual stress distribution of the backside of sample C before and after LSP

Laser energy/mJ			$\sigma_{0°}$ /MPa	$\sigma_{45°}$ /MPa	$\sigma_{90°}$ /MPa	σ_1 /MPa	$\Delta\sigma_1$ /MPa	σ_2 /MPa	$\Delta\sigma_2$ /MPa	α_0 /°	$\Delta\alpha_0$ /°
400	centre	before	−7.6	2.4	−9.3	2.6	0.9	−19.5	3.9	−42.8	−0.1
		after	−5.4	3.5	−6.8	3.5		−15.6		−42.9	
	edge	before	−8.2	−1.3	−10.7	−1.3	−0.1	−17.6	0.2	−40.6	−0.8
		after	−8.4	−1.4	−10.4	−1.4		−17.4		−41.4	
600	centre	before	−13.2	−8.4	−15.7	−8.2	5.6	−20.7	5.4	−39.2	−3.3
		after	−8.4	−2.6	−9.5	−2.6		−15.3		−42.5	
	edge	before	−11.7	−5.4	−14.7	−5.2	0.8	−21.2	3.9	−39.6	9.8
		after	−7.6	−5.3	−14.1	−4.4		−17.3		−29.8	
800	centre	before	−3.7	7.4	−4.3	7.4	16.2	−15.3	19.8	−44.2	4.4
		after	15.8	23.4	12.4	23.6		4.6		−39.8	
	edge	before	−3.9	6.2	−4.2	6.7	4.3	−14.8	13.1	−44.6	15.7
		after	8.2	10.1	1.3	11.0		−1.7		−28.9	
1000	centre	before	−6.9	−10.8	−7.9	−4.0	29.3	−10.8	26.5	40.8	3.3
		after	20.7	15.9	20.4	25.3		15.7		44.1	
	edge	before	−9.3	−12.8	−8.5	−5.0	3.9	−12.8	18.5	−42.1	86.7
		after	2.4	−1.1	2.3	−1.1		−5.7		44.6	
								– 2.8			

The following findings can be achieved from Table 1:

(1) Under the action of pulsed laser, within the theory model base system using double–sided adhesive as adhesive, residual compressive stress values along the three directions on the back of the base, residual maximum principal stress

and residual minimum principal stress value generally become smaller; even the stress changes from the compressive stress state to the tensile stress state: the greater the laser energy, the greater the residual stress variation.
(2) When laser energy is small, it does not change significantly the residual stress on the boundary of the back impact spot; while with the increase of laser energy, the residual stress on the boundary of the back impact spot changes less quickly than that on the impact spot centre.

The analysis suggests that when the shock wave induced by pulsed laser is spread to the back of the substrate, the compression wave causes the material to experience plastic deformation and produces strain. When the laser energy is small, the corresponding induced shock wave pressure is small. Moreover in the process of transmission, the attenuation of pressure, the unloading of reflection wave and so on severely weakens the intensity of the shock wave. When the peak pressure is lower than the dynamic yield strength of the matrix or only slightly higher than it, no obvious plastic deformation will take place on the matrix surface; therefore residual compressive stress values along the three directions on the back of the base, residual maximum principal stress and residual minimum principal stress value change very little. When energy continues to increase and the intensity of the laser shock wave increases, the dynamic response of the induction material also becomes more intense, the plastic deformation of the material becomes greater and the greater the residual stress amplitude is. When the laser energy is small, say, 400 mJ, under the irradiation of laser energy, the residual stress on the impact edge remains almost the same. This is because the laser energy is in Gaussian distribution: when plastic deformation of the material happens in the centre of impact region, shock waves spread to the border area is still below the dynamic yield strength of the material, so the residual stress on the boundary area changes little. At the same time the dynamic response of the boundary is less intense than that in the impact spot area, resulting in different change rate of residual stress. The residual stress in the impact centre on the back changes faster.

3.2 The Effect of The Combined Strength of The Interface on Residual Stress

Since the variation regularity of residual primary stress and direction angle can reflect the variation characteristics of residual stress, three kinds of membrane model base systems are thus researched by taking into consideration the variation of the residual principal stress and the direction angle.

Figure 1 shows the variation of the residual principal stress in the Centre and on the boundary of the impact spot on the back of the substrate in the model of three membrane–based systems. $\Delta\sigma_1$ Is the maximum principal stress, $\Delta\sigma_2$, the minimum principal stress, and |$\Delta\alpha$ |, the growth of main stress angle. The following is what we can get from Fig.1:

(1) The maximum principal stress and the minimum principal stress increases the most with the water membrane as a binder followed by the gluing agent 502 and the double–sided adhesive.
(2) Before and after the laser shock, the main stress angle in the centre on the back of the material is basically unchanged, but the main stress angle on the shock edge is changed.

Analysis: Due to the weaker adhesion strength of the water film, the shock wave energy reaching the back of the matrix after overcoming the binding force of the interface is accordingly larger, so the dynamic response of the material is more intense, resulting in greater variation of principal stress. Because the laser is through focusing lens and the spot centre overlaps with the laser beam, the centre shock wave loading in the impact region is equivalent to direct impact and the shock wave thus loads vertically; meanwhile the reflection and transmission of the wave are also on a straight line and observe simple loading mechanism. The result is that the direction angle of the principal stress changes little. On the contrary, the shock wave loading on the boundary of the impact region is equivalent to oblique impact and transmitted wave and reflected wave interact complexly, resulting in greater change in the principal stress direction angle. With the increase of laser energy, direction angle changes greater.

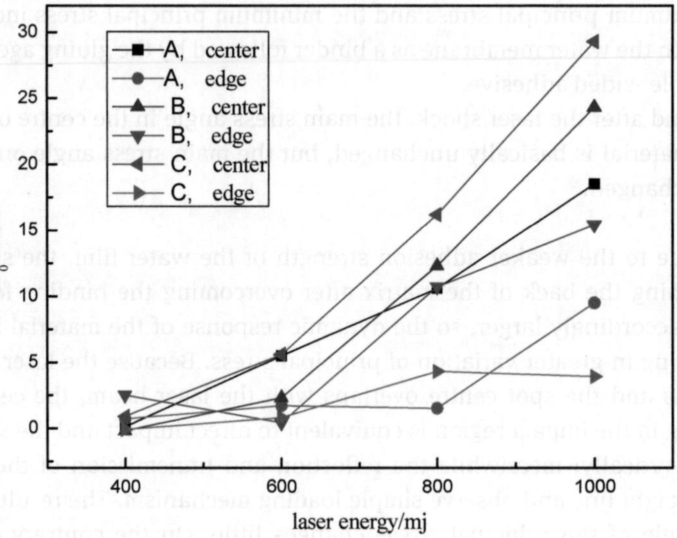

(a) $\Delta\sigma_1$

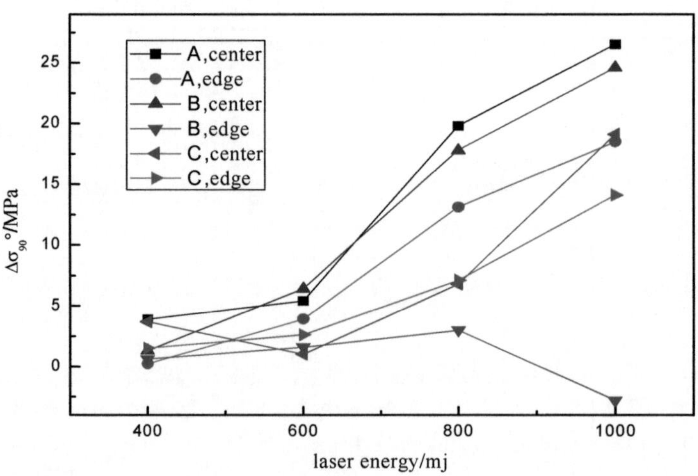

(b) $\Delta\sigma_2$

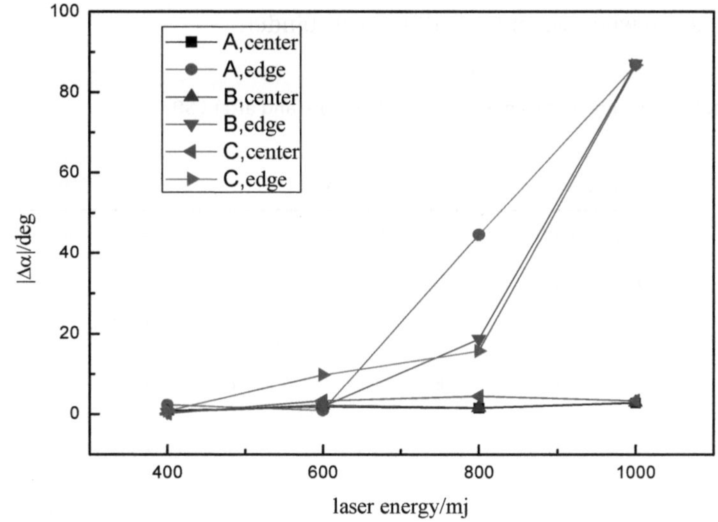

(c)|△α|

Fig. 1: Amplification of principal stress and its direction of substrate backside

4 Study of Residual Stress State Distribution in The Membrane System Interface

4.1 The Residual Stress State Distribution on The Back of The Stainless Steel Film

Table 2 lists the residual stress value along 0 °, 45 ° and 90 ° three directions as well as the variation near the light spectrum spot Centre on the back of 316 stainless steel foil before and after laser shock, figure 2 shows the residual principal stress increase rate on the back of stainless steel foil after laser shock. The following is what we can get from Fig.2

(1) under the different laser irradiation energy in the same membrane model base system, three directions of residual compressive stress value and the residual principal stress value is bigger, the more the laser energy, the greater variation of the residual stress.

(2) In the same laser irradiation energy, the maximum residual compressive stress variation when double–sided adhesive, the minimum residual compressive stress variation when water film adhesive as binder.

Tab. 2: Test result of residual stress of film backside before and after LSP

Sample	Laser energy/mJ	σ_{0°/MPa			σ_{45°/MPa			σ_{90°/MPa		
		before	after	$\Delta\sigma_{0^\circ}$	before	after	$\Delta\sigma_{45^\circ}$	before	after	$\Delta\sigma_{90^\circ}$
A	400	−31.2	−45.4	−14.2	−76.3	−86.5	−10.2	−80.3	−88.5	−8.2
	600	−44.5	−99.3	−54.8	−83.4	−125.1	−41.7	−76.1	−127.6	−51.5
	800	−24.3	−105.1	−80.8	−81.6	−148.9	−67.3	−80.4	−157.2	−78.8
	1000	−35.2	−143.1	−107.9	−74.7	−185.6	−110.9	−69.4	−170.4	−101.0
B	400	−52.3	−57.6	−5.3	−62.2	−66.3	−4.1	−70.7	−76.3	−5.6
	600	−48.4	−79.8	−31.4	−72.5	−103.8	−31.3	−81.3	−119.7	−38.4
	800	−61.4	−113.2	−51.8	−81.5	−139.3	−57.8	−74.3	−134.6	−60.3
	1000	−59.6	−140.9	−81.3	−77.5	−167.9	−90.4	−84.6	−173.6	−89.0
C	400	−26.5	−28.2	−1.8	−87.4	−91.6	−4.2	−64.2	−64.3	−0.1
	600	−31.4	−49.9	−18.5	−91.4	−120.6	−21.2	−77.7	−101.9	−24.2
	800	−44.5	−84.5	−40.0	−61.1	−97.7	−36.6	−80.5	−120.1	−39.6
	1000	−23.7	−81.3	−57.7	−80.6	−147.2	−66.6	−92.6	−163.8	−71.2

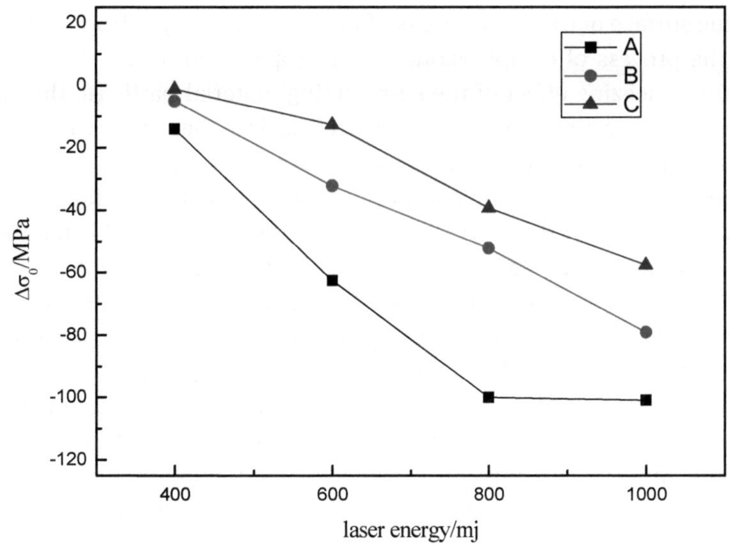

(a) $\Delta\sigma_1$

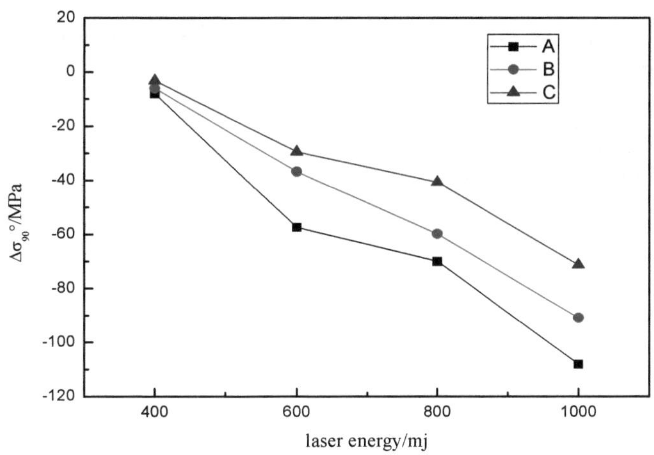

(b) $\Delta\sigma_2$

Fig. 2: Amplification of principal stress of film backside

The following findings can be achieved from figure 2, in the same membrane model base system under the irradiation of different laser energy, the laser shock wave affected on the surface of the stainless steel foil first, constantly extrude the material particles in the process of compressional wave propagation to the steel foil, combined with the squeezing effect of the surrounding material particles, the material compressive strain plastic deformation occurred at last. Because of the thin film thickness, caused the pressure strain plastic deformation on the back of stainless steel foil. The residual stress value and Residual principal stress value of the three directions were increased, the larger the laser energy, the more obvious the laser impact strengthening effect, so the change in residual stress value is greater. The difference of residual stress can be caused by the difference of interface binding performance under the same laser energy irradiation, when the laser–induced shock wave spreads to the interface junction, it's going to have a shock wave reflection, and the compression wave becomes a stretch wave. The interface is the most tightly bound when the double–sided adhesive is used as binder, the more work the stretching wave and the compression wave overcome the binding force consumption, the less the stretching wave that reflects to the back of the stainless steel foil, the worse the pressure strain induced by the discharge of laser shock, the maximum residual compressive stress value change is obtained at last. Because the water membrane is the minimum binding when it is a binder, the residual tensile wave unloading pressure strain works best, so the residual compressive stress change is the minimal

4.2 The Residual Stress State Distribution of Magnesium Alloy Substrate Surface

Figure 3 shows the increase of residual principal stress on the surface of AZ31B magnesium alloy after the laser impact.

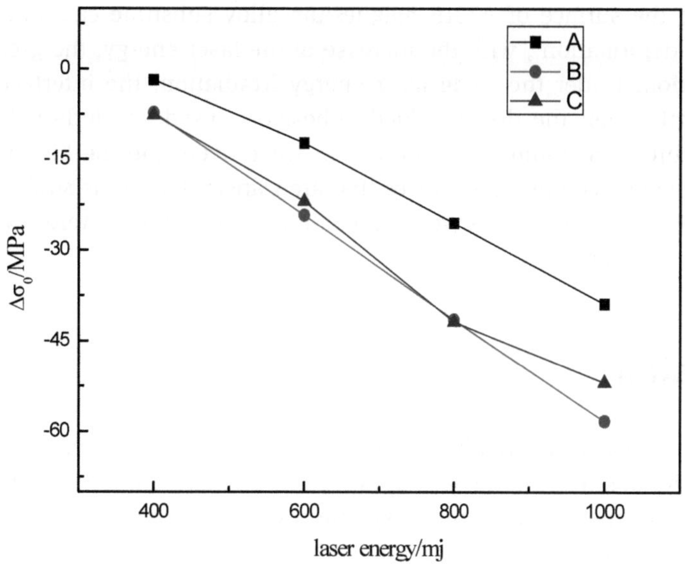

(a) $\Delta\sigma_1$

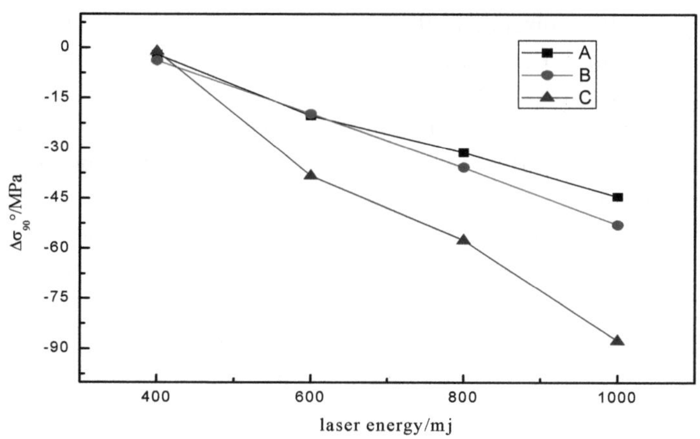

(b) $\Delta\sigma_2$

Fig. 3: Amplification of principal stress of substrate surface

The following findings can be achieved from figure 3, in the same membrane model base system under the irradiation of different laser energy, the laser shock wave transmission to the surface of AZ31B magnesium alloy substrate can still induce pressure strain deformation , with the increase of the laser energy, the greater the strain deformation. Under the same laser energy irradiation, the interface is the strongest strength when the double-sided adhesive is used as binder; the more work the stretching wave and the compression wave overcome the binding force consumption, the less compression waves that are transmitted to the surface of the matrix, the effect of deformation is worse, the minimum residual compressive stress change is obtained at last.

5 Conclusion

The residual stress was measured by XRD under different laser energy, the distribution rules of residual stress on the back of the matrix is analysed, the residual stress state distribution of the membrane base system was studied, the influence of laser energy and interface bonding strength on residual stress was analysed, two major conclusions have been drawn.

The residual compressive stress in the three directions on the back of the substrate under the theoretical model of the three membrane base systems is smaller, the residual maximum principal stress and residual minimum principal stress are also smaller, it is possible to change from compressive stress state to tensile stress state, the change of residual stress on the back of the boundary of the Light spot is not obvious than in the central area, the larger the laser energy, the greater the residual stress. The main stress Angle of the centre of the impact zone is basically unchanged, but the direction Angle of the main stress of the impact edge is larger. When the water membrane acts as a binder, the maximum principal stress and the minimum principal stress increasing maximum, the middle is 502 glue, the smallest is double-sided adhesive

In the membrane base system interface, the back of the film and the substrate surface are irradiated by different laser energy, the residual compressive stress and residual principal stress of the three directions are larger, The larger the laser energy, the greater the residual stress variation. Under the same laser energy irradiation, the change of residual compressive stress on the back of the thin film was the most significant when the Double-sided adhesive was used as binder, the residual compressive stress on the back of the film was the smallest when the water membrane was used as binder. However, the compressive stress on the surface of the matrix is the opposite of the stress on the back, the residual compressive stress change is the minimum when double-sided adhesive is binder, and the change of residual compressive stress is greatest when the water film is binder

When studying on dynamic static response of metal materials induced by laser shock wave, it contains many interdisciplinary techniques such as laser technology, thin film technology, table interface science, materials science, shockwave mechanics, stress strain detection technology, etc. There are still many problems to be solved, The key is how to establish the quantitative relation between the dynamic static response of metal material and the parameters of laser beam parameters(such as laser energy, pulse width, wavelength, loading mode, light spot etc), absorbing layer parameters and membrane material properties.

References

[1] Shuxin, Wang. Study on the Dynamic and Static Response Behaviour of Metal Material by the Laser Shock Wave [D]. Postdoctoral research work report, Jiangsu University, 2015(10).

[2] LI Zhi–yong, ZHU Wen–hui, CHENG Jing–yi et al. Experimental study on the shock wave induced by pulsed high power laser in aluminium target [J].Chinese J Lasers, 1997, 24(3): 259–262.

[3] Peyre P, Fabbro R, Merrien P, et al. Laser shock processing of aluminumalloys. Application to high cycle fatigue behaviour [J]. Materials Science and Engineering: A, 1996, 210(12): 102–113.

[4] Kouadria, Barrallier L.Texture Characterization of hexagonal metals: magnesium AZ91 alloy welded by laser processing [J]. Materials Science and Engineering, 2006, A429 (1):11–17.

[5] HU Yong–xiang. The numerical modeling and impact effect of laser shock processing technology [D]. Doctoral dissertation. Shanghai Jiao Tong University, 2008.

[6] Yongkang Zhang,Jian You,Jinzhong Lu et al. Effects of Laser Shock Processing on Stress Corrosion Cracking Susceptibility of AZ31B Magnesium Alloy [J]. Surface&Coatings Technology, 2010, 204:3947~3953.

[7] Chang Ye,Gary J. Cheng. Effects of Temperature on Laser Shock Induced Plastic Deformation: The Case of Copper [J]. Journal of Manufacturing Science and Engineering, 2010, Vol. 132(6).

[8] Mithila Achintha, David Nowell. Eigenstrain Modelling of Residual Stresses Generated by Laser Shock Peening [J]. Journal of Materials Processing Technology, 2011, 01(011).

Jian-Wei Ma, De-Ning Song, Xin Zhang and Zhen Liu

Contouring-Error Reduction by Combination of Tracking-Error Compensation and Cross-Coupled Control

Abstract: Complex curved surface parts are widely used in important fields such as aerospace, energy and power. In the machining process for the curved features of the parts, the contour-following error of the CNC (computer numerical control) system is increasingly significant due to the existence of servo delay for the feed-drive system and the dynamics mismatch between each axis. Then, the formed contouring error will directly affects the machining quality of the complex curved surface parts. To reduce the contour-following error, this paper presents a cross-coupled controller with tracking-error compensation. First, the tracking error for each feed axis is compensated in real time. On this basis, the numerical efficient contouring-error estimation method is utilized and the proportional cross-coupled controller is integrated with the tracking-error compensator. By using this method, the tracking error and contouring error can be synergistically reduced and the contour accuracy of the CNC system can be dramatically improved. Results of the verification tests demonstrate the feasibility of the presented approach, and the presented approach is significant for high-speed and high-precision machining of complex curved surface parts.

Keywords: contouring error, tracking error, complex parts, error compensation, cross-coupled control

1 Introduction

Contour-following tasks are widely applied in computer numerical control (CNC) machining of parts with complex curved surfaces [1, 2]. With the growing application of such complex parts with curved features in important fields such as aero-

Jian-Wei Ma, School of Mechanical Engineering, Dalian University of Technology, Dalian, China. Email:mjw2011@dlut.edu.cn
De-Ning Song, School of Mechanical Engineering, Dalian University of Technology, Dalian, China. Email:deningsong@163.com
Xin Zhang, School of Mechanical Engineering, Dalian University of Technology, Dalian, China. Email:992165798@qq.com
Zhen Liu, School of Mechanical Engineering, Dalian University of Technology, Dalian, China. Email:317149270@qq.com

DOI 10.1515/9783110540048-057,

space, energy and power, the contour-following accuracy becomes increasingly significant for ensuring the required profile precision of such parts. However, due to the existence of servo delay for the feed-drive system, the tracking error of each feed axis, defined as the lag value from the actual position to the desired position, is non-ignorable during high-feed-speed machining [3, 4]. As such, the contouring error, defined as the orthotropic distance between the actual motion positions to the desired contour, will be formed when following a curved machining toolpath of the complex parts, and this is unfavourable to the profile accuracy of such parts [5]. As a consequence, improvement of the contour-following accuracy of the CNC machine tools is of great significance for high-speed and high-precision machining of parts with curved surfaces.

The methods that can be utilized for improving contour accuracy can be mainly divided into two categories. The first category aims at reducing the tracking errors for every axes thus improving the contour accuracy in an indirect way, while the second one concentrates on reducing the contouring error directly.

In order to decrease the tracking errors of individual axes, Tomizuka [6] designed a zero phase error tracking controller (ZPETC) for CNC systems. Using the ZPETC, tracking error of each axis can be reduced by adding a reversed feed forward controller to the feed-drive system for cancelation of zeros and poles. As an improvement, Torfs [7] et al. extended the bandwidth of the ZPETC later. In addition, Zhang et al. [8] also improved the tracking accuracy by proposing of an iterative tracking-error pre-compensation approach. Although these methods can improve the tracking accuracy effectively, reduction of the tracking error for each single axis does not necessarily lead to the reduction of the contouring error for free-form curved contours, and vice versa [9].

To control the contouring error directly, Srinivasan et al. [10] designed a cross-coupled controller (CCC) for a biaxial motion system. In this method, the CCC was added among the two axes to harmonize their movements thus reducing the contouring error in spite of the tracking errors of the two axes. Later, a number of modified forms of CCC were widely researched and developed [11-14]. For these kinds of methods, the contouring error is first calculated and then used for contour control. Although computing of the contouring error for linear or circular paths in real time is effortless, it is a difficult or even an impossible task for free-form parametric curves because time-consuming algorithms should be used to calculate the point-to-curve distance [15]. In real application, estimation value of the contouring error is used as an alternative. However, it is hard to estimate the contouring error precisely when the tracking error of each single axis is relatively large, which may degrade the performance of the contour controllers.

Based on above analysis, it can be seen that more precise contouring-error control approaches remain in urgent need. This paper presents a contouring-error reduction method by combination of tracking-error compensation and cross-coupled control. The tracking error for each axis of the motion system is first compensated and the compensation value is derived according to the close-loop transfer function of the system. After tracking-error compensation, the tracking accuracy can be improved, which not only benefits the contour accuracy, but is also conductive to the improvement of contouring-error estimation precision. After the tracking-error compensation, the CCC is adopted for further direct improvement of the contour accuracy. Finally, performances of the presented contouring-error reduction approach are evaluated by simulation tests. By using this method, the tracking and contouring accuracy of the CNC systems can be simultaneously enhanced, which is significant for high-speed and high-precision machining of complex parts with curved features.

The reminder of this paper is organized as follows. The tracking-error compensation method is presented in Section 2. The cross-coupled control method for contouring-error control is introduced in Section 3. Section 4 illustrates the results of the verification tests. Conclusions are summarized in Section 5.

2 Compensation of The Tracking Error

For a positional close-loop feed-drive system with the model shown in Fig.1, the close-loop transfer function can be derived as:

$$G(s)=\frac{P(s)}{R(s)}=\frac{K_p}{Js^2+Bs+K_p} \tag{1}$$

Where$R(s)$ and $P(s)$ are respectively the desired reference and the actual output of the feed axis in s domain; K_p is the position-loop proportional gain; J and B are the equivalent rotational inertia and the equivalent damping factor of the feed-drive system, respectively.

By letting $\omega_n=\sqrt{K_p/J}$ and $\zeta=B/\left(2\sqrt{JK_p}\right)$, the transfer function can be expressed as a typical second-order system:

$$G(s)=\frac{\omega_n^2}{s^2+2\zeta\omega_n s+\omega_n^2} \tag{2}$$

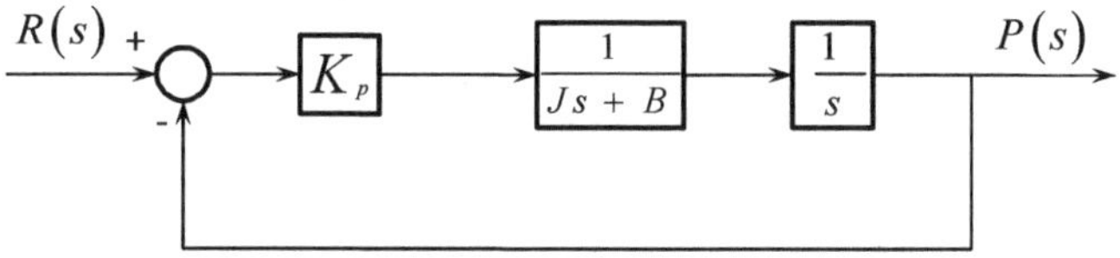

Fig. 1: Block diagram of the feed-drive system

In order to realize the tracking-error compensation, the actual output is expected to equal to the desired input reference, i.e. $P(s)$ =$R(s)$, when adding a compensation value $\Delta R(s)$ to the command reference. Accordingly, the following relationship is prospected to be satisfied:

$$\frac{R(s)}{R(s)+\Delta R(s)}=G(s) \tag{3}$$

By substituting Eq. (2) to Eq. (3), the following equation can be obtained:

$$\omega_n^2 \Delta R(s)=s^2 R(s)+2\xi\omega_n sR(s) \tag{4}$$

Thus, the tracking-error compensation value in time domain $\Delta r\ (t)$ can be derived according to Eq. (4) using reverse Laplace transformation as:

$$\Delta r(t)=\frac{1}{\omega_n^2}\ddot{r}(t)+\frac{2\xi}{\omega_n}\dot{r}(t) \tag{5}$$

Where (t) is the input command reference of the corresponding feed-drive system in time domain. As a result, the first and second order derivatives of $r\ (t)$ in Eq. (5) can be seen as the velocity $v\ (t)$ and acceleration $a\ (t)$ of the corresponding axis, respectively. Hence, the Eq. (5) can be rewritten as:

$$\Delta r(t)=\frac{1}{\omega_n^2}a(t)+\frac{2\xi}{\omega_n}v(t) \tag{6}$$

Thus, the tracking-error compensation value for each axis can be calculated using Eq. (6) according to the $v\ (t)$ and $a\ (t)$ obtained from the interpolator for the corresponding axis. By adding the compensation value to each axis in real time, the tracking error can be effectively reduced.

3 Cross-coupled Control with Tracking-Error Compensation

After tracking-error compensation, the contouring error can be easily estimated because the tracking errors for every axes are all small values. Therefore, the tangential approximation method is utilized here. Denote the command federates of x and y axes as v_x and v_y, respectively, thus the feed-direction vector of the desired contour can be expressed as $\mathbf{V}= [v_x, v_y]$. Consequently, the unit tangential vector of the desired path can be computed as:

$$\mathbf{T} = \frac{\mathbf{V}}{\|\mathbf{V}\|} \tag{7}$$

Denote the desired motion position and the actual motion position as **R** and **P**, respectively. According to the geometric relationship shown in Fig. 2, the estimated contouring-error vector $\hat{\boldsymbol{\varepsilon}}$ can be derived as:

$$\hat{\boldsymbol{\varepsilon}} = \mathbf{R} - \left((\mathbf{R} - \mathbf{P}) \cdot \mathbf{T} \right) \cdot \mathbf{T} - \mathbf{P} \tag{8}$$

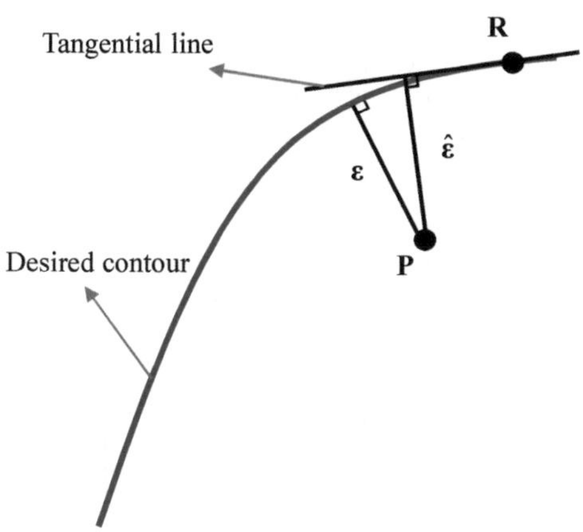

Fig. 2: Estimation of the contouring error

According to the estimated contouring error, the control values for the feed axes can be obtained as:

$$\begin{cases} \Delta C_x = K_c \hat{\boldsymbol{\varepsilon}}(1) \\ \Delta C_y = K_c \hat{\boldsymbol{\varepsilon}}(2) \end{cases} \tag{9}$$

Where K_c is the proportional cross-coupled gain; ΔC_x and ΔC_y are respectively the x-axis and y-axis cross-coupled compensation values. The model of the cross-coupled controller with tracking-error compensation is illustrated in Fig. 3. By combination of the tracking-error compensation and the cross-coupled control, the contour accuracy of the CNC systems can be effectively improved.

4 Verification Tests

In this section, simulation tests are conducted to verify the feasibility of the presented contouring-error reduction approach. A two-axis feed-drive model with the structure shown in Fig. 1 is established in MATLAB/Simulink, and the utilized parameters are illustrated in Table 1. An eight contour shown in Fig. 4 is employed here for the tests. The testing results are provided in Figs. 5-8. The maximum tracking error for x axis is decreased from the original 0.7041 mm to 0.2592 mm after CCC with tracking-error compensation, while that for y axis is decreased from 0.7500 mm to 0.2383 mm. The maximum contouring error before CCC is 0.0918 mm, while that after CCC and tracking-error compensation is reduced to 0.0156 mm. As a consequence, the feasibility of the presented approach is demonstrated.

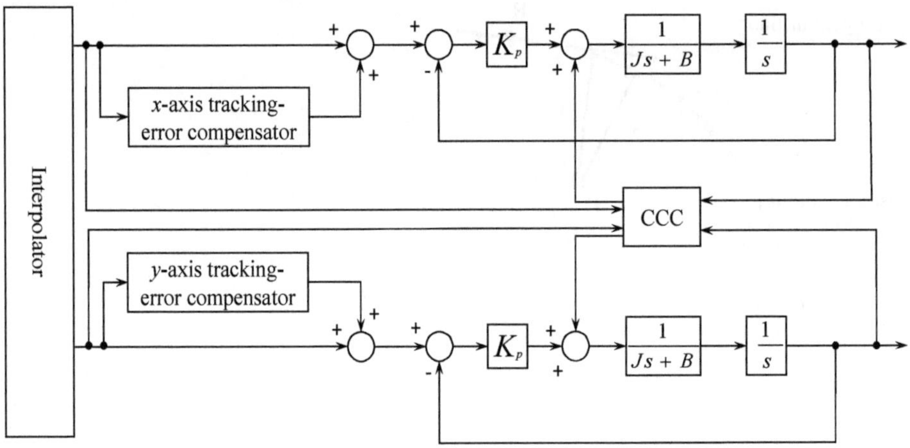

Fig. 3: Structure of the cross-coupled controller with tracking-error compensation

Tab. 1: Parameters of the Verification Tests

Parameter	x axis	y axis
K_p	75	80
J	0.01	0.01
B	1.0	1.2

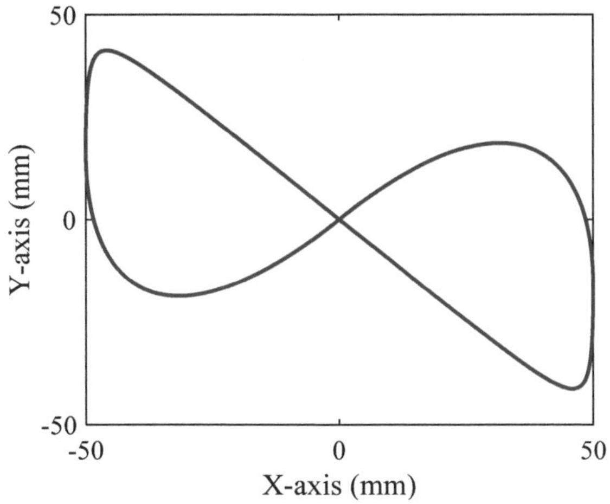

Fig. 4: The eight contour used for verification tests

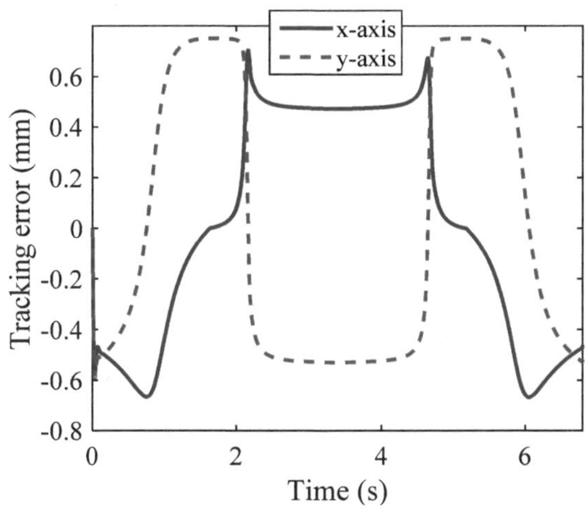

Fig. 5: Tracking errors for x and y axes before tracking-error compensation and CCC

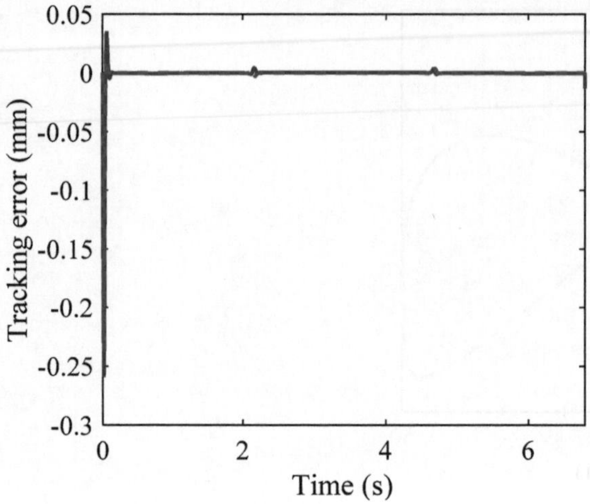

Fig. 6: Tracking errors for *x* and *y* axes after tracking-error compensation and CCC

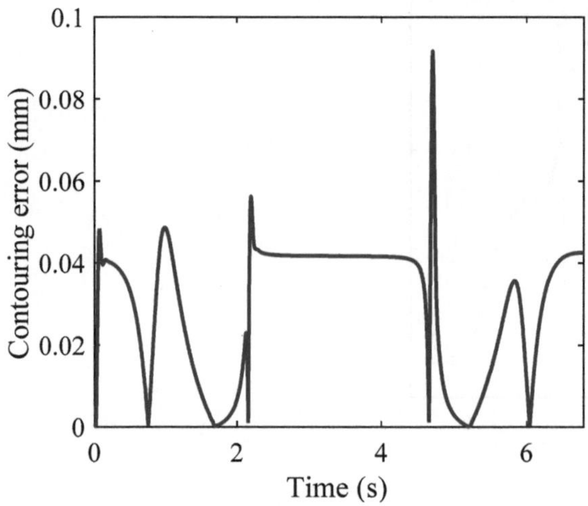

Fig. 7: Contouring error before tracking-error compensation and CCC

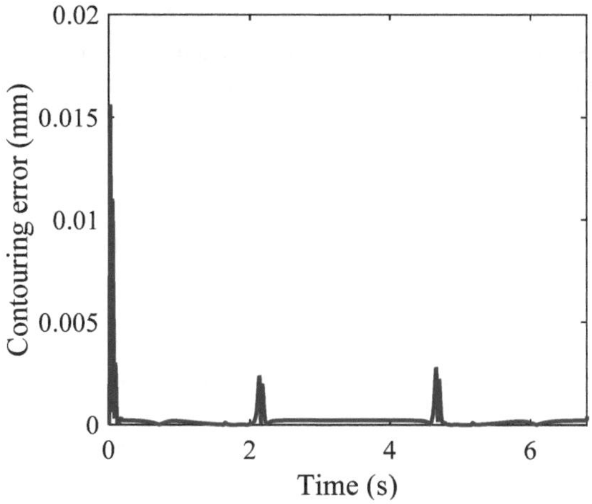

Fig. 8: Contouring error after tracking-error compensation and CCC

5 Conclusions

In this paper, a contouring-error reduction method by combination of tracking-error compensation and cross-coupled control is presented. First, the tracking-error compensation approach is given. Then, the cross-coupled controller with tracking-error compensator is provided. From the simulation results it can be seen that the presented method can simultaneously improve the tracking and contouring accuracy, which is significant for high-speed and high-precision machining of parts with curved features.

Acknowledgement: The project is supported by the National Natural Science Foundation of China (Nos. 51575087 and51675081), National Science and Technology Major Project of China (No. 2016ZX04001-002), Innovation Project for Supporting High-level Talent in Dalian (No. 2016RQ012), Science Fund for Creative Research Groups (No. 51621064) and the Fundamental Research Funds for the Central Universities (No. DUT17LAB13). The authors wish to thank the anonymous reviewers for their comments which led to improvements of this paper.

References

[1] El KM, and U. Naoki, "Discrete-time model predictive contouring control for biaxial feed drive systems and experimental verification, Mechatron" 2011, vol. 21, pp. 918–926.

[2] F. Huo, and A. N. Poo, "Precision contouring control of machine tools," Int. J. Adv. Manuf. Technol., 2013, vol. 64, pp. 319–333.

[3] T. Lie, and G. L. Robert, “Multiaxis contour control—the state of the art,” IEEE Trans. Control. Syst. Technol., 2013, vol. 21, pp. 1997–2010.
[4] J. Dong, T. Wang, B. and Li, Y. Ding, “Smooth federate planning for continuous short line tool path with contour error constraint,” Int. J. Mach. Tools Manuf., 2014, vol. 76, pp. 1–12.
[5] R. Mostafizur, S. Rudolf, and Y. Ian, “A new approach to contour error control in high speed machining,” Int. J. Mach. Tools Manuf., 2015, vol. 88, pp. 42–50.
[6] M. Tomizuka, “Zero phase error tracking algorithm for digital control,” J. Dyn. Syst. Meas. Control-Trans. ASME, 1987, vol. 109, pp. 65–68.
[7] D. Torfs, J. Deschutter, and J. Swevers, “Extended bandwidth error phase error tracking control of nonminimal phase systems,” J. Dyn. Syst. Meas. Control-Trans. ASME, 1992, vol. 114, pp. 347–351.
[8] D. Zhang, Y. Chen, and Y. Chen, “Iterative pre-compensation scheme of tracking error for contouring error reduction,” Int. J. Adv. Manuf. Technol., 2016, in press. DOI 10.1007/s00170-016–8735–9.
[9] R. Ramesh, M. A. Mannan, and A. N. Poo, “Tracking and contour error control in CNC servo systems,” Int. J. Mach. Tools Manuf., 2005, vol. 45, pp. 301–326.
[10] K. Srinivasan, and P. K. Kulkarni, “Cross-coupled control of biaxial feed drive servomechanisms,” J. Dyn. Syst. Meas. Control-Trans. ASME, 1990, vol. 112, pp. 225–232.
[11] S. S.Yeh, and P. L. Hsu, “Theory and applications of the robust cross-coupled control design,” J. Dyn. Syst. Meas. Control-Trans. ASME, 1999, vol. 121, pp. 524–530.
[12] W. Chen, D. Wang, Q. Geng, and C. Xia, “Robust adaptive cross-coupling position control of biaxial motion system,” Sci. China Tech. Sci., 2016, vol. 59, pp. 680–688.
[13] F. Huo, and A. N. Poo, “Improving contouring accuracy by using generalized cross-coupled control,” Int. J. Mach. Tools Manuf., 2012, vol. 63, pp. 49–57.
[14] D. Zhang, J. Yang, Y. Chen, and Y. Chen, “A two-layered cross coupling control scheme for a three-dimensional motion control system,” Int. J. Mach. Tools Manuf., 2015, vol. 98, pp. 12–20.
[15] H. Zhao, L. Zhu, and H. Ding, “Cross-coupled controller design for triaxial motion systems based on second-order contour error estimation,” Sci. China Tech. Sci., 2015, vol. 58, pp. 1209–1217.

Jian-Guo Li, Yong-Ping Zhang, Xian-Feng Xu, Yuan-Jiang Liao and Dun-Yuan Wang

The Multi-parameter Detection System of Industrial Controller

Abstract: The detection system of industrial controllers' temperature and angle using virtual instrument and LabVIEW as hardware and development software are introduced. This system adopts NI PXle-8840 controller through two slots No.3 and No.4 on NI PXle-1087 respectively using platinum thermistor acquisition card NI PXIe-4357 and universal card of data acquisition NI PXIe-6341 to make detections. One channel is to use platinum thermistor PR-18-2-100-1/4-6-E of temperature sensor through acquisition card of platinum thermistor to collect temperature signals which can be transmitted to NI PXle-8840 controller. The other channel is to use Omron E6HZ-CWZ6C rotary encoder to collate signals which are transmitted to NI PXle-8840 controller by NI PXIe-6341data acquisition card. Then, these signals are showed in the LabVIEW interface background and generated real-time data detection chart according to the data by temperature sensor and the writing data by manual. As a result, this detection system has not only plenty of features such as friendly man-machine interface, perceptual intuition, high precision, safety and reliability but also strong data visualization analysis, strong control ability of controlling instrument and the advantages of simple operation that is easy to complete daily sheet of watch keeper.

Keywords: LabVIEW, platinum thermistor acquisi-tion card NI PXIe-4357, platinum thermistor PR-18-2-100-1/4-6-E, NI PXIe-8840 controller, universal card of data acquisition NI PXle-6341, E6HZ-CWZ6C rotary encoder

Jian-Guo Li, College of Electronic and Controlled Engineering, Chang' an University, Xi'an, China. Email: 603757352@qq.com
Yong-Ping Zhang, College of Electronics and Information Engineering, Ningbo University of Technology, Ningbo, China. Email: ypz@nbut.edu.cn
Xian-Feng Xu, College of Electronic and Controlled Engineering, Chang' an University, Xi'an, China. Email: 3110259@qq.com
Yuan-Jiang Liao, College of Electronics and Information Engineering, Ningbo University of Technology, Ningbo, China. Email: 327290612@qq.com
Dun-Yuan Wang, College of Electronic and Controlled Engineering, Chang' an University, Xi'an, China. Email: 1213836557@qq.com

DOI 10.1515/9783110540048-058,

1 Introduction

The detection of industrial controllers' parameter is an important work in the field of industrial control, and having been always an important research topic on the properties of the detection of industrial controller [1]. With the development of the modern technology of industrial control, the detection technology of industrial controller is constantly also increasing at domestic and foreign, and having a wide range of applications. In particular, the detection of the original form of the sensor is more extensive [2].

The majority of the industrial controller detection systems are controlled by single-chip MCU, with a complex programming, complicated hardware circuit and inaccurate detection. This detection system of industrial controller is based on LabVIEW, PC, rotary encoder and temperature sensor. The system can detect temperature, angle, angular position and angular acceleration in real time, and a higher precision.

2 The Hardware Devices of Industrial Controller Detection System

Detection system consists of temperature sensors, rotary encoders, and PC and LabVIEW components. System block diagram is shown as Figure 1.

This system adopts the Virtual Instrument (Virtual Instrument, referred to as VI) hardware system, which is based on computer with Lab VIEW software, based on the test items and matched with the corresponding hardware module with function test as signal I/O interface, and accomplish the signal collection, measurement and conditioning [3]. According to the needs of users, designers can design and develop a virtual operator panel with good human-computer interaction ability that is the front panel of LabVIEW program. All the test functions are realized by LabVIEW and virtual instrument hardware, which is also a new type system of computer instrument.

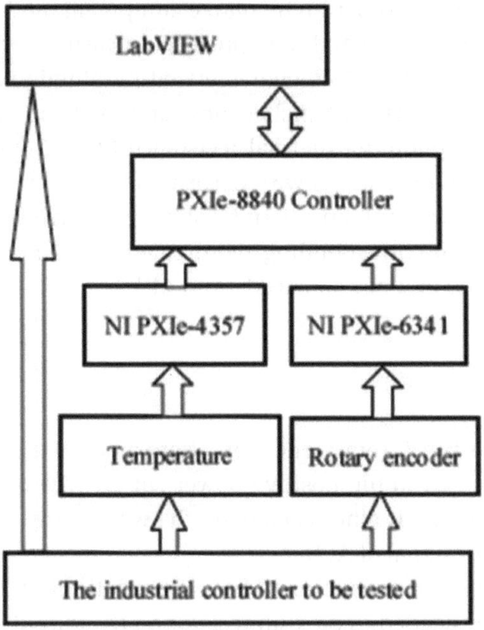

Fig. 1: Industrial controller multi-parameter detection system block diagram

Compared with traditional instrument, virtual instrument is enriched and enhanced the function of traditional instrument. Analyse, display, storage, printing and other management the collected signals, and focused on the computer for processing [4]. Virtual instrument hardware and software have developed an open industry standard, raise the repeatable utilization of resources, the function is easy to extend, standardized management, production, maintenance, and can save equipment purchase and maintenance costs [5]. Can be formed through the computer network of a complex and distributed test system for remote testing, monitoring, and diagnosis [6].

2.1 Temperature

Platinum thermal resistance PR–18–2–100–1/4–6–E is produced by the omega RTD thermal resistance sensors, RTD is one of the most accurate temperature sensors, it has high precision, long life, high resistivity and low heat capacity, and the corresponding advantages of fast speed, etc., and the resistance value has good linear relationship with temperature changes, but also provides the advantages of good stability and repeatability, RTD can also relatively avoid electrical noise, therefore, it is particularly suitable for temperature detection in complex industrial environments, especially around motors, engines and other high-voltage equipment, and is especially suitable for standard measuring device and high precision measurement. Platinum thermal resistance PR-18–2–100–1/4–6-E is characterized by aluminium head, 3 wires, using the PT100, 1/4 inch diameters, 6 inches long, and the temperature range -200°C to 600°C. Because the platinum thermal resistance PR-18-2-100-1/4-6-E is a standard three-wire configuration, the red wire Corresponding to the NI PXIe–4357 analogy signal EX +, the rest two black lines 1 and 2 of by respectively NI PXIe–4357 analogy signal AI+ and AI- (no corresponding requirements, 1, 2 black line can also be respective by NI PXIe-4357 analogy signal AI+ and AI-).

2.2 Rotary Encoder

This test system uses the Omron E6HZ-CWZ6C rotary encoder. There are five terminals, in which the brown wire is connected to the positive power supply, the blue wire is connected to the negative power supply, the black wire, the white wire and the orange wire is respectively the output signal of A, B, Z.

Because this is a type of NPN incremental encoder, and is an open collector output, when it is wired with NI PXIe–6341, needed to pick-up resistor on the output of the signal, and its resistance is in the range of $1k\Omega$ to $5k\Omega$. The schematic is shown as Figure 2.

2.3 Platinum RTD Acquisition Card NI PXI-4357

The NI PXIe-4357 has a typical accuracy of 0.09°C using 20 roads PT100 resistance temperature acquisition and supporting 2 wires, 3 wires and 4 wires RTD. There are 4 terminals in each analogy input channel, respectively, EX- said the excitation current return, EX+ said negative excitation current output, AI- said negative analogy input and AI+ said positive analogy input, and the three line of the thermal resistance we use three of them, respectively AI-, AI+ and EX+.

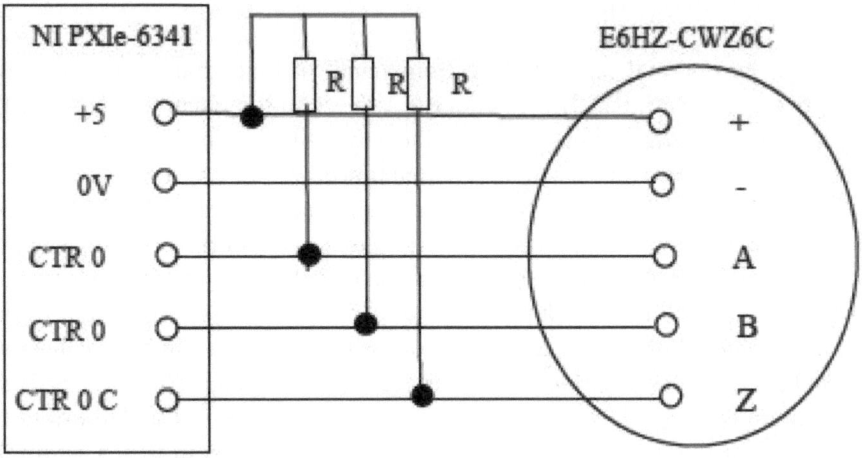

Fig. 2: Rotary encoder with a pull-up resistor with NI PXIe-6341 wiring diagram

2.4 General Acquisition Card NI PXIe-6341

NI PXIe-6341data acquisition card which contains 16 AI ports, 2 AO ports, 4 timers and 4 counters, and we only use the counter 0 in this system.

2.5 Controller NI PXIe-8840

The NI PXIe-8840 is a modular PC with a PXI Express 3U form-factor, the highest configuration of 2.6GHz quad-core processor. The controller is small, powerful, and expensive, often used in industrial technology, making it ideal for processor-intensive modular instrumentation and data acquisition applications.

3 The Program Design of Industrial Controller Detection System

LabVIEW programming is distinct from other programming languages, mainly graphical programming language divided into front panel and program block diagram, while the program is mainly completed in the program block diagram. The main task of the front panel is interface and data display in the process of running, for example, temperature, pressure, position and speed, and so on.

3.1 LabVIEW Program

Figure 3 is the optional test interface front panel of industrial controller multi-parameter detection system. This interface is a selection interface, that is set by the program to return to the optional test interface, the user must be used in the system of industrial controller when a self-selected interface.

Figure 4 is the optional test interface program block diagram of industrial controller multi-parameter detection system. While loop used in the program, the purpose of the test program interface can be managed so that in turn continues to jump to a temperature online detection and angular position and speed online testing program interface, but also allows detection of two interfaces in turn constantly return to the choice test interface. And then placed an event structure, respectively, angle position and speed online detection, temperature detection and return to the test interface for the three events added to the event structure, in which set up flat sequence structure with two frames in the event structure. In the first frame of the sequential structure, the purpose is to open to the vi of jump events, and then close the operation of the current interface in the second frame. Note that the sequence structure under every event in the event structure, create a name of the path function or relative path of the input name and the incident must be the event source corresponding to this test panel in the open when the corresponding test project , and will not be chaotic procedures.

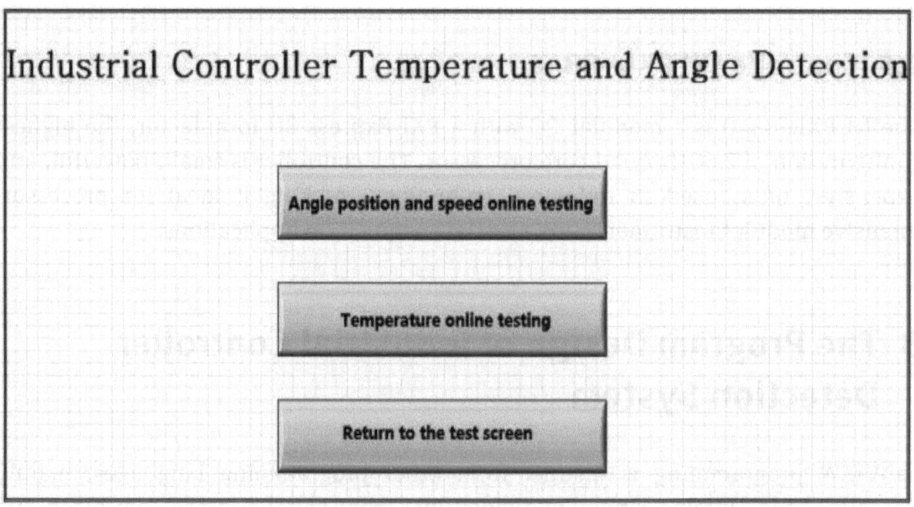

Fig. 3: The optional test interface front panel of industrial controller multi-parameter detection system

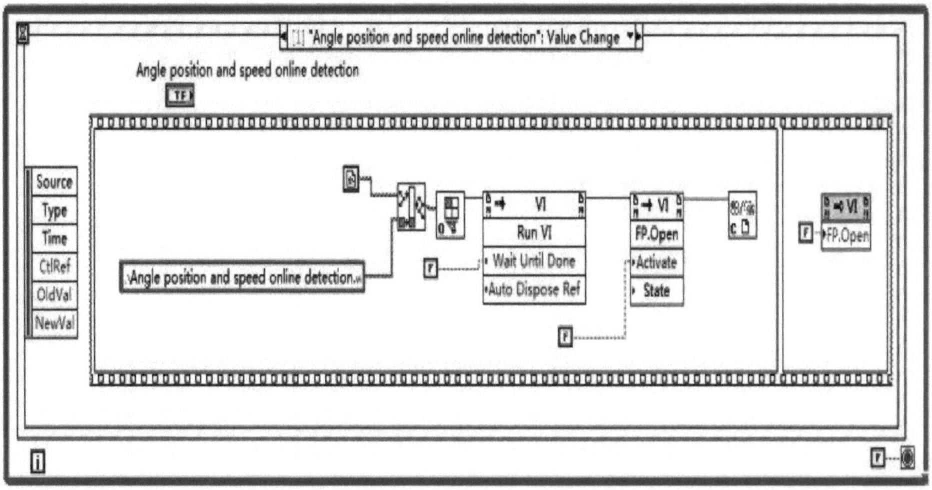

Fig. 4: The optional test interface program block diagram of industrial controller multi-parameter detection system

3.2 Design of Temperature Online Detection Program

Figure 5 is the front panel of real-time temperature detection system, which can display the current temperature value and its real-time curve, and finally saved in the form of a document on it. In the front panel, it has the basic information of the tester, the current environmental situation as well as the comparison of the test results with the device temperature.

Operation, firstly, the tester should be filled your personal information, equipment temperature 1, error requirement of temperature 1, equipment temperature 2 and error requirement of temperature 2 on the front panel. Then the program is running, clicking to start the measurement, it will detect the current real-time temperature, while producing the actual temperature 1 error value and actual temperature 2 error value. The temperature 1 and the temperature 2 will appear the real-time temperature by the temperature sensor detected. On the right side of Figure 5 showing the current real-time temperature value of the two forms of temperature value of the two forms of performance by the thermometer and the temperature curve, which form does tester like, can choose on the label above.

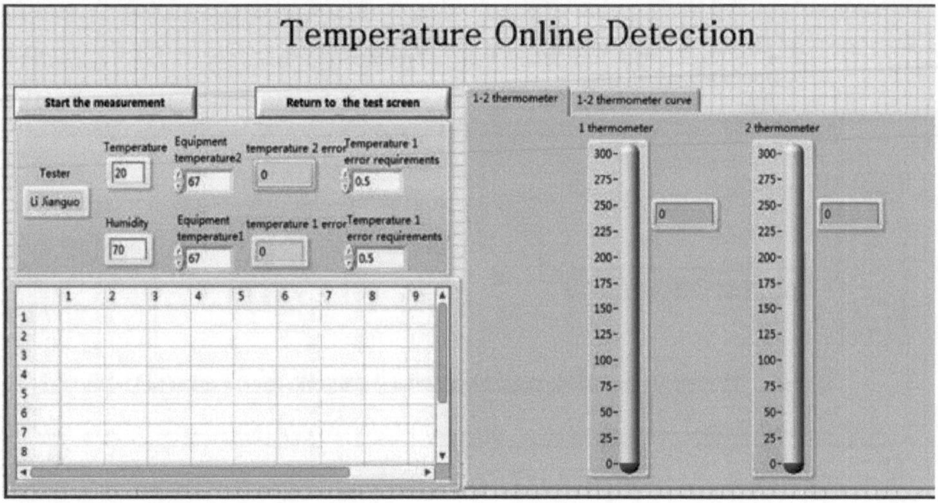

Fig. 5: The front panel of temperature online detection system

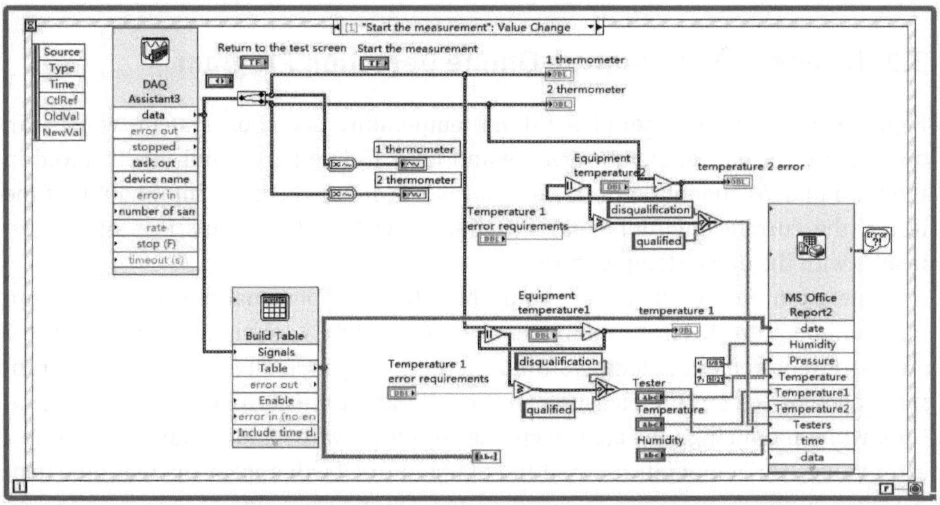

Fig. 6: The program block diagram of temperature online detection system

Figure 6 is the temperature online detection system program block diagram. Event 1 is the start of the measurement process, the data collected by sensor were being exported and split from the DAQ's port till temperature and curve being displayed, use the split temperature 1 and temperature 2 minus the equipment temperature 1 and equipment temperature 2 separately, then compare each result with corresponding precision standards of system that it was failed if greater than the temperature deviation standards of system, or it is qualified. And then through the creation of form, collected data to the form function from the creation of the form, so that real-time data will be displayed in the front panel of the table, and then create the table form's port directly connected to the side of the MS Office Report function's data, lastly, displayed in own set of word templates. The MS Office Report function is used for report display, while the corresponding system word template is located in the D: LabVIEW/software/installation path/LabVIEW 2016/templates/Report directory, and Excel and Word system templates in it. However, the report shows that the system is used from the word templates; you can copy the template from the system, modifying, and then saving to your own set directory below. Collected two ways' data by the two sensors minus the equipment temperature compared with the precision of the system requirements, which the results linked to MS Office Report function temperature 1 and 2, displayed in the final report. In the word document in which needed to write data and text inserted bookmarks respectively, the bookmark's names have date, time, testers, ambient temperature, humidity, pressure, data, temperature 1 and temperature 2 in the template of the system report. Note that when adding bookmarks to the data, add a bookmark to the area where the real-time temperature is displayed, and temperature 1 and temperature 2 are the result of comparing the test data with the system data, not temperature, and finally save as a document template (* .dot) format and added to the MS Office Report function inside.

Event 2 is a program that returns the selected test interface event, which sets a flat sequence structure with two frames. The main purpose of the program is running from the current program interface can jump back to the optional test interface in the first frame, that is, the temperature and speed test vi front panel, which is set in the second frame off the current test program in the program.

3.3 Design of Angle Position And Speed Online Detection Program

Figure 7 shows the front panel of angle position and speed online detection. The options for manual setting are channel settings, trigger settings, precision settings and cycle time settings. Which channel settings in the counter option, because our rotary encoder connected to the virtual instrument in the third slot of the counter 0. So here we choose PXI1Slot3/ctr0, rotary encoder A, B and Z three terminals are selected for PXI1Slot3/PFI8, PXI1Slot3/PFI10 and PFI9. Pulse per revolution and cycle times were set to 3600P/R and 100ms.The rest of the collected data is obtained by the position of the rotary encoder, the right angle waveform curve, tachometer curve and acceleration curve.

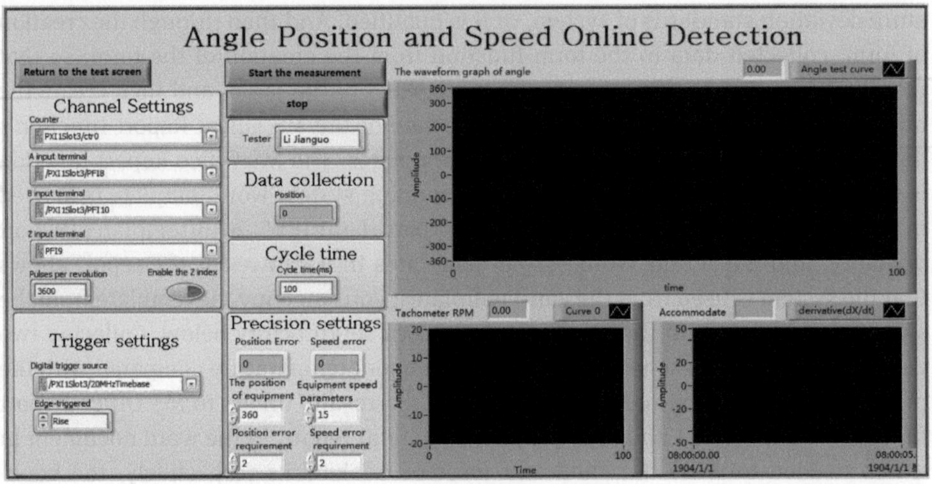

Fig. 7: The front panel of angle position and speed online detection

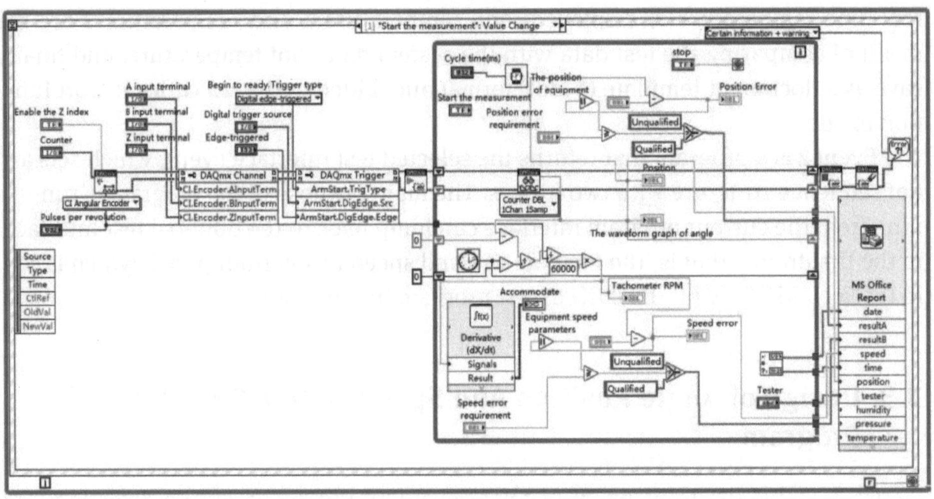

Fig. 8: The program block diagram of angle position and speed online detection

Figure 8 is the program block diagram of angle position and speed online detection. First of all, set a function of while loop, which allows the cycle of events inside each other. Next, put the events of start the measurement (angle position and speed online detection program) and return to the test screen added to the inside of the while loop structure.

At the beginning of the measurement procedure in the event 1, applied to an integrated function in LabVIEW, that is a DAQ assistant. It contains the data acquisition tasks in the sampling frequency, the number of sampling settings and other functions in this function, while the physical data channel this function can also be completed. In the program design, it can complete the result of the data acquisition, displaying and preserving and other functions. Consequently, we start creating a function of virtual channel by placing a DAQmx, added the corresponding task and the corresponding I/O type, the rest of the input parameters can be set in the front panel. We set up a DAQmx channel attribute property node function, in which three elements are added, namely, the input information of the three terminals A, B and Z of the rotary encoder and the corresponding input is set, and then the parameters are set on the front panel. Then set down a function of the DAQmx trigger property node and creates three inputs, namely, trigger type, trigger source and digital edge. Later beginning the task placed a DAQmx start task function, which makes the program in the running state that the measurement program is started from here. Let the real measure of the program carried out in the while loop, which is placed the function of DAQmx read, it is the data output of the rotary encoder angular position. Note that the angle position of rotary encoder regard as the next starting point measurement at the end of each measurement. So DAQmx read function of the data collected is real-time location of rotary encoder. Setting down a shift register, in which the initial value assigned to 0, so that every object begins, removed the results of last operating. However, the output of the shift register returns the result of the previous loop back to the input of the shift register. The consequences of one cycle minus initial value and divided by the time for taking to run. Here also used the shift register for recording time, the purpose of the same, clear the last cycle of running time, using the final time minus the start time as the running time. Therefore, the angular position difference is divided by the difference of next time what is every millisecond rotation angle, and because the encoder pulse per revolution to 3600P/R, and then the above obtained every millisecond turn angle to 3600 is the number of revolutions per millisecond, the last multiplied by 60000, into the standard unit, that is, turn the number of turns per minute, so the data collected here is the speed we require, and finally its demand to derive the acceleration.

Event 2 is to return to the settings of the self-selected test interface program. here we are the same a two-frame flat sequence structure as in the above event 2 measured temperature, which opened the main program of measuring temperature and speed in the first frame, then closed off the current operating procedures in the second frame.

4 Experiment

4.1 Experimental Introduction

The detection system of the industrial controller is mainly online detection for temperature, angle position, and angle velocity and angle acceleration. Temperature online detection, it can test simultaneously 8 channels, and can be extended to 16 channels reserved for more. The system uses two of them to detect, and one route is to test the controller under test, the other is placed in the air for indoor temperature detection. In the process of testing, both the device temperature 1 and the device temperature 2 of the front panel are filled with the self-temperature value and the same temperature error requirement value of the detected device. Finally, perform numerical and curve displays and generate reports. In the angular position and speed on-line detection, it is detected when the rotation of the rotating device on the controller under test is stopped just after one rotation. Finally, on the front panel, display the running status of the rotating device and the detection in real-time as well as generate detection reports.

4.2 Experiment Results

When the temperature of the industrial controller under test is detected on-line, the operator should fill in error requirements of temperature and temperature of the equipment on the front panel, then the program is running and test is started. When the difference between the real-time temperature and the temperature written manually is detected less than the accuracy of the system, the program will automatically make a judgment in the Word report which will generate a judgment result that is when the difference is greater than the error required, display as unqualified, otherwise qualified. The angular position and speed of on-line testing are just the same as the temperature online testing process. The following is a real-time report of multi-parameter detection for industrial controller, as shown in Figure 9.

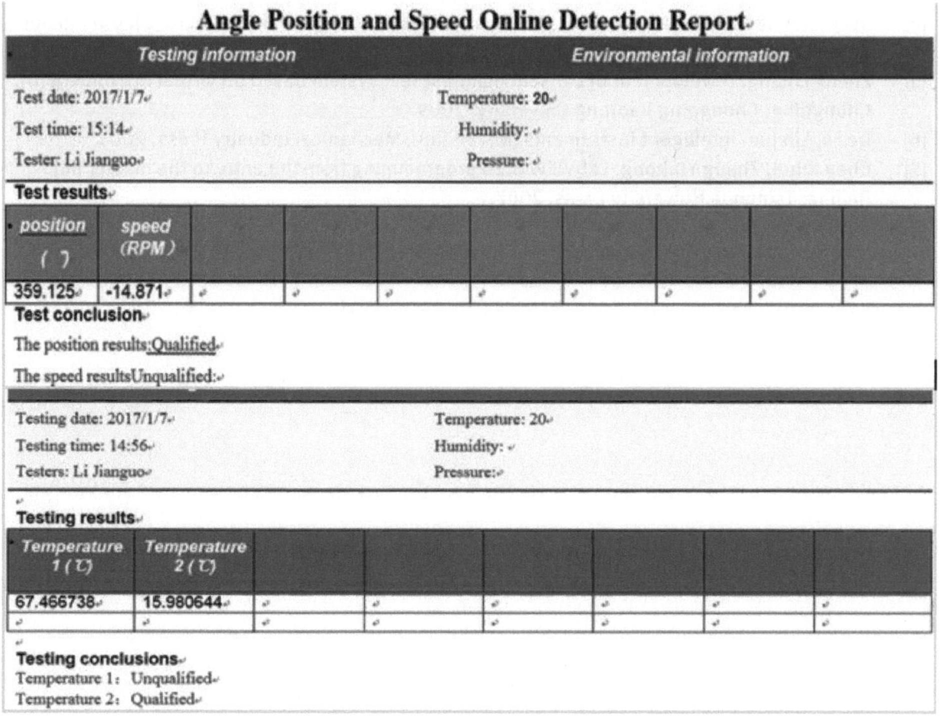

Angle Position and Speed Online Detection Report

Testing information	Environmental information
Test date: 2017/1/7	Temperature: 20
Test time: 15:14	Humidity:
Tester: Li Jianguo	Pressure:

Test results

position ( )	speed (RPM)								
359.125	-14.871								

Test conclusion

The position results:Qualified

The speed resultsUnqualified:

Testing date: 2017/1/7	Temperature: 20
Testing time: 14:56	Humidity:
Testers: Li Jianguo	Pressure:

Testing results

Temperature 1 (℃)	Temperature 2 (℃)						
67.466738	15.980644						

Testing conclusions

Temperature 1: Unqualified

Temperature 2: Qualified

Fig. 9: Industry controller multi-parameters online detection report

5 Conclusion

Through the industrial controller detection system can be seen that based on LabVIEW development test and measurement application development[7], programming is simple, but the use of different hardware equipment wiring will be different, such as the design of the encoder must be pulled up by a resistor. The system of virtual instrument detection is built by the sensors, PC and LabVIEW, and having great potential for later development.

References

[1] Zhang Wei, Liu Hongli. Design of temperature measurement and control system based on LabVIEW [J]. Foreign Electronic Component, 2008 (12): 19–21.

[2] Li Xiwen, Zhao Jian. Sensor and signal conditioning technology [M]. Xi'an University of Electronic Science and Technology Press, 2008.

[3] Liu Liying, Yang Wei. Virtual instrument design based on LabVIEW7.1 [J]. Journal of Heilongjiang Water College, 2004 (12): 113–114.

[4] Xu Min. Automatic test system for mobile phone based on virtual instrument [D]. Guangzhou: South China University of Technology, 2006.
[5] Zheng Lijiang. Development of car seat headrest test system based on virtual instrument [D]. Chongqing: Chongqing Jiaotong University, 2009.
[6] De Fu, Lin Jun. Intelligent instruments [M]. Beijing: Mechanical Industry Press, 2005.
[7] Chen Xihui, Zhang Yinhong. LabVIEW 8.20 programming from the entry to the master [M]. Beijing: Tsinghua University Press, 2007.

Zhi-Kai Cai, Jun-Nan Mi, Tong Wang, Xi Chen and Xiao-Min Lian

Mechatronic Dual-Redundancy Steering Gear System in Vehicle

Abstract: Mechatronic dual-redundancy steering gear system is an essential part in steer by wire system. It is not only the very important actuator in steering system, but also needed in autonomous driving car for its autonomous steering. A key point for mechatronic dual-redundancy steering gear system is its reliability and safety. To realize a kind of this system with high reliability and safety, the work on this paper proposed a kind of mechatronic dual-redundancy steering gear system, designed the mechanical architecture and controller architecture of the system, and built the redundancy management mechanism of mechatronic dual-redundancy steering gear system using redundancy theory. Simulation results verify the working ability and reliability of this system.

Keywords: mechatronic, redundancy, steer by wire, high reliability and safety

1 Introduction

Mechatronic dual-redundancy steering gear system is an essential part in steer by wire system in vehicle, which receives the steering angle information from steering wheel and executes steering operation using motors. Mechatronic dual-redundancy steering gear system is also needed in autonomous driving car for its autonomous steering. Concerning for the reliability and safety of steering, a redundant steer by wire system including a Mechatronic dual-redundancy steering gear system is came up with as vehicle steering system.

Many schools and companies have studied mechatronic dual-redundancy steering gear system while there are very less vehicles using it because the reliability and safety of the system is hard to guarantee. Yixin Yao and Brian Daugherty of Visteon

Zhi-Kai Cai, Lee Shau Kee Science and Technology Building A639, Tsinghua University, Beijing, China. Email: caizhikai292@126.com
Jun-Nan Mi, Lee Shau Kee Science and Technology Building A639, Tsinghua University, Beijing, China. Email: thumjn@163.com
Tong Wang, Lee Shau Kee Science and Technology Building A639, Tsinghua University, Beijing, China. Email: wangtong19929@foxmail.com
Xi Chen, Lee Shau Kee Science and Technology Building A639, Tsinghua University, Beijing, China. Email: chenxiholdon@163.com
Xiao-Min Lian, Automotive Research Institute 206, Tsinghua University, Beijing, China. Email: lianxm@tsinghua.edu.cn

DOI 10.1515/9783110540048-059,

Company [1] have proposed a structure of mechatronic dual-redundancy steering gear system by connecting two driving motor in an axis. They also did simulation study and got a good tracing performance and fault-tolerant ability. Dieter Heitzer and Alois Seewald in TRW Company [2] analyzed the mechatronic dual-redundancy steering gear system and proposed an electrical fault-tolerant structure of mechatronic dual-redundancy steering gear system and its redundant power supply system. They also proposed a HIL program of mechatronic dual-redundancy steering gear system. Yuichi Onoda and Yutaka Onuma [3] set up a kind of architecture which is based on mechatronic dual-redundancy steering gear system, and set up the model of steer by wire system. They did the assessment of the test vehicle simulation and got the steering features of steer by wire system in small steering angle. Bing Zheng and Cliff Altemare in Indiana's Purdue university [4] proposed a mechatronic dual-redundancy steering gear system with two motors and two controllers and Sohel Anwar and Lei Chen [5] optimized the two motor structure based on their study. And Mohammad Sharif-ul Hasan and Sohel Anwar [6] set up the sliding-mode observer and the method to long-run forecast and failure-tolerant control. And they used simulation to perfect system redundancy control. Also Isermann R, Schwarz R, Stolzl S [7] has their method of failure-tolerant in drive by wire system such as steer by wire system. Their works always have partial redundancy so the system may fail to work caused by one fault.

This paper proposes a structure of mechatronic dual-redundancy steering gear system with steer by wire technology in vehicle. The structure is based on redundancy theory and the system can reach once “failure/operation/warning” safety mechanism, which means that once a failure occurs in this system, the steering function is still operating as normal mode and the system will send warning information to let vehicle know that steering system has a fault.

2 The Structure of Mechatronic Dual-Redundancy Steering Gear System

Mechatronical dual-redundancy steering gear system has four modules, as shown in Fig. 1

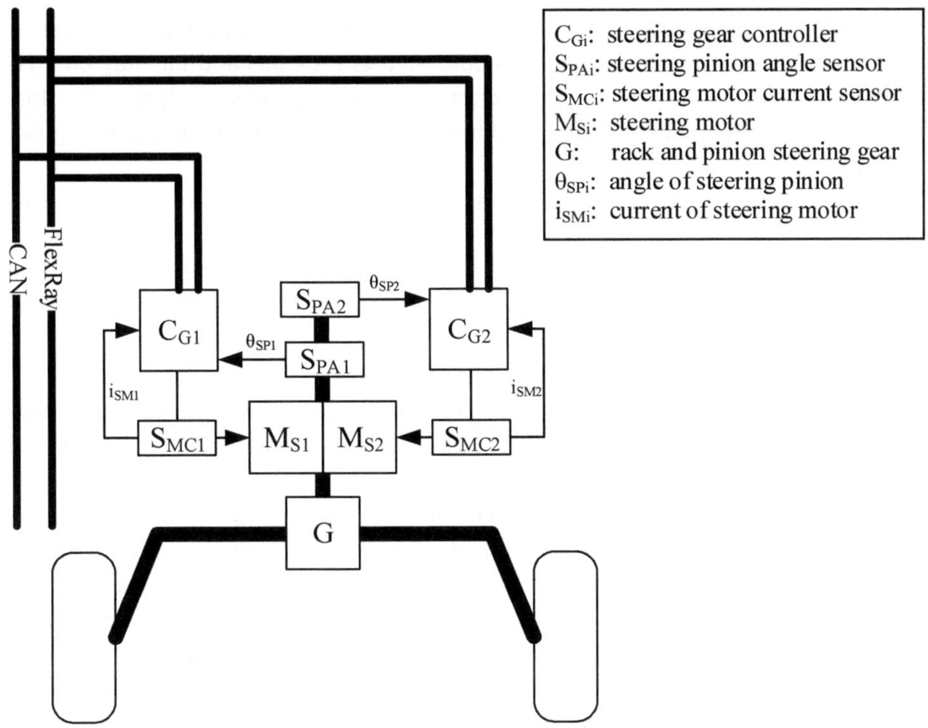

Fig. 1: the composition of mechatronic dual-redundancy steering gear system

In Fig. 1, mechatronic dual-redundancy steering gear system SG can be expressed as

$$SG=\{C, S, M, G\} \tag{1}$$

With elements
C = Redundant controllers set
S = Redundant sensors set
M= Redundant motor driving set
G = rack and pinion steering gear

In a mechatronic dual-redundancy steering gear system with high reliability safety, the system is formed with two steering motors, two controllers, two steering pinion angle sensors and two steering motor current sensors, which are shown in Fig. 2. In this system, the two steering motors M_{s1} and M_{s2} give the force to steer together and each motor is controlled by its controller. The same steering pinion angle information θ_{PA1} and θ_{PA2} is collected by steering pinion angle sensors S_{PA1} andS_{PA2}, and steering motor current information i_{SM1} and i_{SM2} is collected by steering motor current sensors S_{MC1} andS_{MC2}, then the information is transmitted to the corresponding controllers. The two controllers C_{G1} and C_{G2}communicate with each other by redundant communication network which is consisted of CAN bus and FlexRay bus. The network is also working for the communication between mechatronic dual-redundancy steering wheel system and mechatronic dual-redundancy steering gear system.

In this architecture, the two steering motors work together to steer in normal mode. One motor, one controller, one steering pinion angle sensor and one steering motor current sensor forms one motor driving channel and the others form the other motor driving channel. The information gotten from each kind of sensors in each channel is collected by both controllers and the same two pieces of information will be arbitrated to one which is correct for normal steering. But once one of the two controllers or motors or sensors breaks, there are some measures to detect the fault and it will immediately cut the broken motor driving channel and the system will use the motor which is in normal situation to steer.

The redundant motor driving set is consisted of two motors which are connected by gears and the two motors rotate together in normal mode. The output then is reduced by worm gear reducer and drives the rack and pinion steering gear. The steering pinion and the worm are in the same axle for steering. The gear ratio between two motors is 1:1 so the two motors can output the same amount of torque and have the same rotating speed. This is shown in Fig. 2 below.

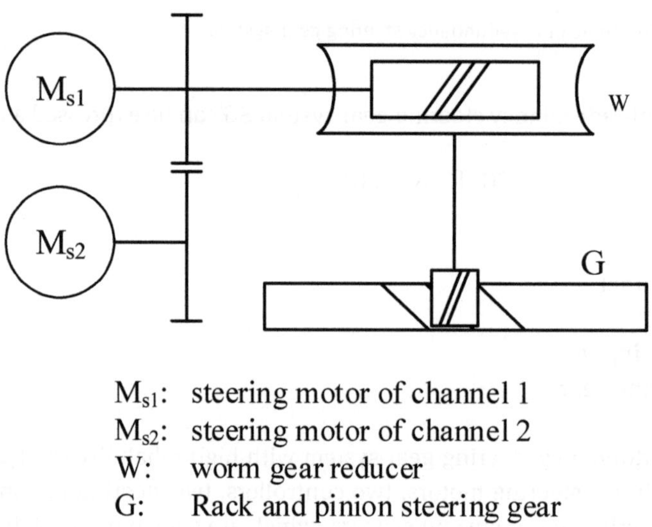

Fig. 2: The structure style of redundant motor driving set

3 The Controller System of Mechatronic Dual-Redundancy Steering Gear System

The function of mechatronic dual-redundancy steering gear control system is to receive the messages of sensors and steering wheel angles in the same time drive the steering motor to execute steering operation normally. Also redundancy management is needed in the control system so the system can reach 'failure/operation/warning' safety mechanism so as to insure its safety and reliability.

3.1 The System Architecture Of Mechatronic Dual-Redundancy Steering Gear Control System

Fig. 3 shows the controller architecture of mechatronic dual-redundancy steering gear control system. The controller mainly consists of steering gear upper sub-controller for upper decision, steering gear information transmission sub-controller for information transmission and steering motor sub-controller for controlling the steering motor. There are angle arbitration and current arbitration to arbitrate the angle and current values which the sensors get. Steering motor sub-controller executes the current close-loop to control the motor using PWM, and steering gear upper sub-controller executes the angle close-loop to let the steering pinion angle racks the steering wheel angle.

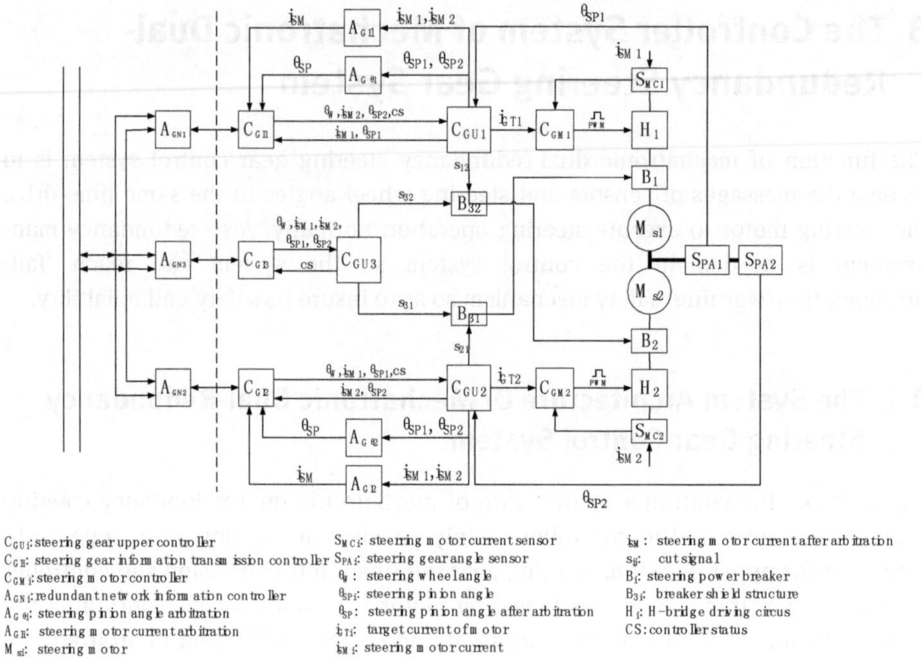

Fig. 3: system architecture of mechatronic dual-redundancy steering gear control system

Failure detection and failure diagnosis are both done in steering gear upper sub-controller and when failure occurs, both controllers are able to judge if the other is in a fault and use cut signal to execute channel isolation.

3.2 The Working Principle Of Mechatronic Dual-Redundancy Steering Gear Control System

In the mechatronic dual-redundancy steering gear system as shown in Fig. 3, the steering wheel angle message is transmitted from redundant network through redundant network information arbitration A_{GN1} and A_{GN2}. Then the message is gotten by steering gear upper sub-controller C_{GU1} and C_{GU2}through steering gear information transmission sub-controller C_{GI1} and C_{GI2}. Then the steering gear upper sub-controller calculates the target current of motor i_{GT1} and i_{GT2} to let the steering motor controllers C_{GM1} and C_{GM2} to execute the current close-loop of motors. So the steering motor controllers give the motor driving circus a proper PWM wave with a changed duty cycle so as to make the current of motor tracks the target current. The steering pinion angle sensors measure the angle θ_{SP1} and θ_{SP2} and the steering current sensors measure the current i_{SM1} and i_{SM2}, and transmit them to the controllers.

For one motor driving channel, it will receive the current and angle message of it, and will also receive the other channel's message through redundant network. Then both the two angle and the two current values will be arbitrated in each controller so the true angle θ_{SP} and the true current i_{SM} is gotten in each controller.

This is the normal working principle of mechatronic dual-redundancy steering gear system. And when failure occurs, both controllers have the ability to judge if the other channel is right or wrong and is able to cut the motor current channel of the other.

3.3 The Redundancy Management Of Mechatronic Dual-Redundancy Steering Gear System

The redundancy theory demands that the system should reach once "failure/operation/warning" safety mechanism that means even if there is a failure occurs, the system should keep operating and give a warning signal to upper system. So it needs failure diagnosis and failure isolation in mechatronic dual-redundancy steering gear system.

In the mechatronic dual-redundancy steering gear system, the failure diagnosis is working all the time to determine where the failure occurs. The failure diagnosis is mainly based on the angle and the current. The failure isolation mode of mechatronic dual-redundancy steering gear system is shown in Fig. 4.

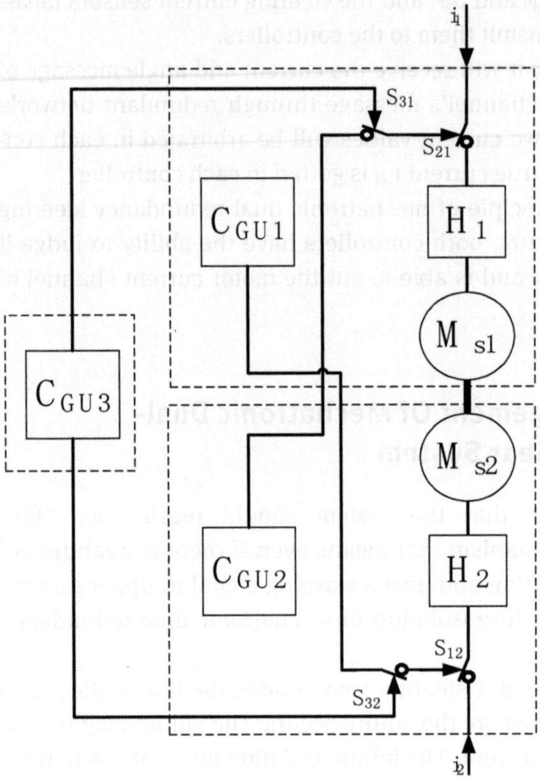

Fig. 4: The failure isolation mode of mechatronic dual-redundancy steering gear system

Fig. 4 shows the failure isolation mode of mechatronic dual-redundancy steering gear system. For example, when C_{GU1} is in a failure, the C_{GU2} will detect the failure and give the signal s_{21} to cut the motor current channel of M_{s1}. In the same time, C_{GU3} will also detect the failure and give the signal s_{32} to let the current channel of M_{s2} is always on so as to prevent C_{GU1} cut the current channel of M_{s2} wrongly. The logic circus of the failure isolation of mechatronic dual-redundancy steering gear system is shown in Fig. 5 and its true table is in Table. 1

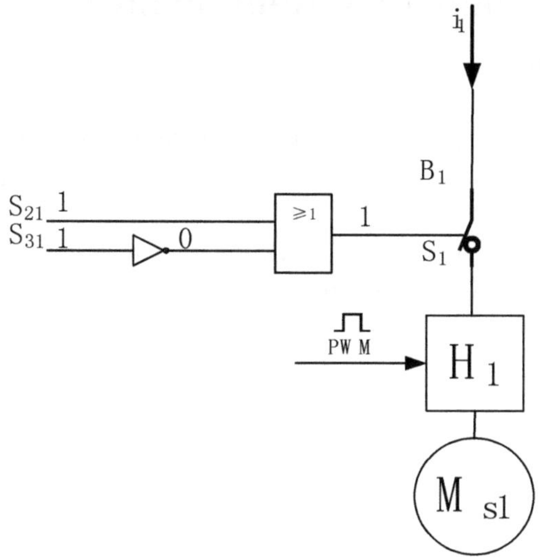

Fig. 5: The logic circus of the failure isolation of mechatronic dual-redundancy steering gear system

Tab. 1: True table of the failure isolation of mechatronic dual-redundancy steering gear system

S21	S31	S1
1	1	1
1	0	1
0	1	0
0	0	1

As shown in Fig. 5 and Table. 1. The on-off signal of current channel of M_{s1} s_1 is gotten from the logic circus of signal s_{21} ands_{31}, and$s_1 = s_{21} \cup \overline{s_{31}}$.

In normal mode, both s_{21} and s_{31} is equal to 1 and s_1 is 1, so the current channel of M_{s1} is on. When the channel of M_{s1} fails, s_{21} will be set to 0 so s_1 will turn to 0 and the current channel of M_{s1} is cut. And when the channel of M_{s2} fails, s_{31} will be set to 0 so s_1 will always be 0 no matter what s_{21} is. Then the wrong channel M_{s2} will never affect the normal channel M_{s1} so can make sure the system can work normally.

4 Simulation of Mechatronic Dual-Redundancy Steering Gear System

To verify the function and redundancy ability of mechatronic dual-redundancy steering gear system, this work uses system simulation by Matlab Simulink and Carsim.

In the simulation of the system, the two motors work together to follow the steering wheel angle. The steering wheel angle is simulated by actual test from steering wheel, which can be seen in Fig. 6. (a)

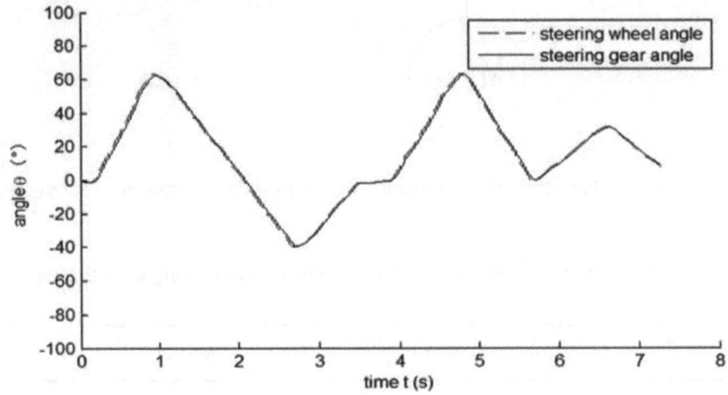

(a)

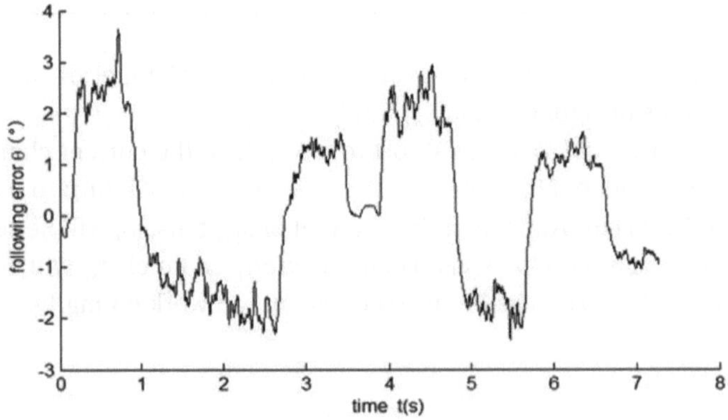

(b)

Fig. 6: steering angle tracking and tracking angle variation in normal mode

Fig. 6(a) shows that steering gear angle tracks steering wheel angle well by the two motors' working together. And Fig. 6(b) shows that the angle variation between steering gear angle and steering wheel angle is less than 4° every time, which is quite small. So it can be seen that the steering function plays well.

Matlab Simulink is also used to verify the redundancy ability of the system. In this test, steering wheel angle is different form that in two-motor mode. The simulation lasts almost 8 seconds and simulates a failure of one motor in the third second.

Fig. 7(a) shows steering angle tracking in this situation and it shows that the steering gear angle is tracking steering wheel angle as well as in two-motor mode after one failure occurs and there is only one motor working to track the steering wheel angle. In Fig. 7(b) it can be seen that in the third second there is little enlargement in the tracking angle variation between steering wheel angle and steering gear angle, which has no effect on the system working. And the tracking angle variation is less than 3° all the time, so the steering function is still normal by one motor steering.

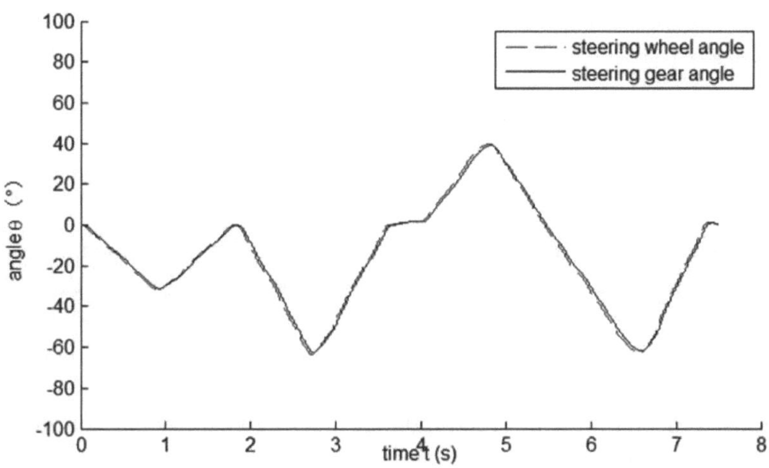

(a)

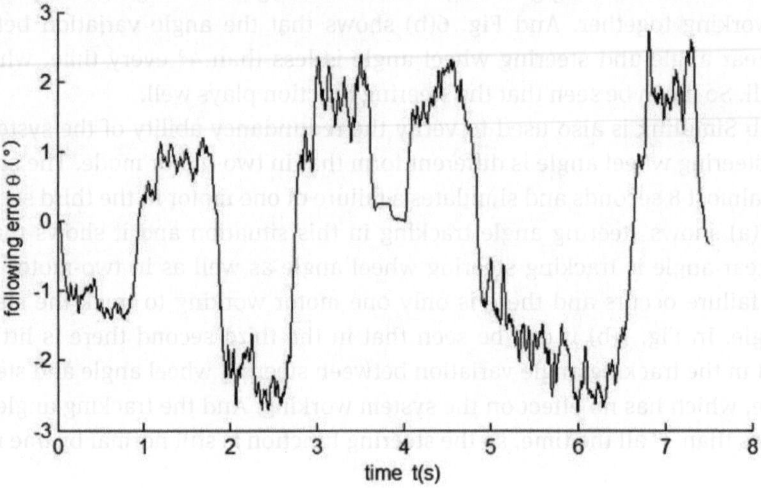

(b)

Fig. 7: steering angle tracking and tracking angle variation in one-failure situation

5 Conclusion

This paper presents a mechatronic dual-redundancy steering gear system, designs the mechanical architecture and controller architecture of it, and builds the redundancy management mechanism of mechatronic dual-redundancy steering gear system, including its failure diagnostician and failure isolation. The following conclusions are obtained:

(a) The proposed mechatronic dual-redundancy steering gear system not only can have good steering behavior but also can reach once 'failure/operation/warning' safety mechanism, having safer and more reliable steering ability.

(b) The proposed redundancy management strategy can realize failure diagnosis and failure isolation, so as to make the system keep working normally after one failure occurs.

(c) The proposed controller architecture of mechatronic dual-redundancy steering gear system can make the two motors in steering gear system work together in normal mode and change to one motor operating mode when one is failed, which makes the system work safely.

References

[1] Yao, Y. and B. Daugherty, Control method of dual motor-based steer-by-wire system. 2007, SAE Technical Paper.
[2] Heitzer, H., Development of a fault-tolerant steer-by-wire steering system. Auto Technology, 2003. 3(4): p. 56–60.
[3] Onoda, Y., et al., Design concept and advantages of steer-by-wire system. 2008, SAE Technical Paper. Zheng,
[4] Zheng, B., C. Altemare and S. Anwar. Fault tolerant steer-by-wire road wheel control system. In American Control Conference, 2005. Proceedings of the 2005. 2005: IEEE.
[5] Anwar, S. and L. Chen, an Analytical Redundancy-Based Fault Detection and Isolation Algorithm for a Road-Wheel Control Subsystem in a Steer-By-Wire System. IEEE Transactions on Vehicular Technology, 2007. 56(5): p. 2859–2869.
[6] Hasan, M.S. and S. Anwar, Sliding mode observer and long range prediction based fault tolerant control of a steer-by-wire equipped vehicle. 2008, SAE Technical Paper.
[7] Isermann R, Schwarz R, Stolzl S. Fault-tolerant drive-by-wire systems [J]. IEEE Control Systems, 2002, 22(5): 64–81.

Dao-Liang Xi, Xiao-Tong Li and Yu-Lin Zhang*

The Vector Control System Design of PMSM Based on DSP Motor Controller

Abstract: A drive vector controller for permanent magnet synchronous motor was researched and designed in this paper. At first, model of PMSM will be introduced and vector control principle and control strategy based on vector control are also determined. According to *dq* axis coefficient model of permanent magnet synchronous motor, i_d =0 vector control method is applied as the control strategy of permanent magnet synchronous motor to construct a permanent magnet synchronous motor controller model based on vector control. The digital signal processor (DSP) TMS320F28035 made by TI Company is adopted as core of the control system. The controller adopts internal A/D to sampling the voltage and current signal. And using incremental photoelectric encoder and the DSP with enhanced orthogonal coded pulse (EQEP) module measure speed and position. The experimental results show that the control system has the advantages of fast response and stable ccontrol effect.

Keywords: PMSM, vector controller, DSP, EQEP

1 Introduction

In the modern industrial production, the motor and its control system occupies a pivotal position. The development of motor control systems with higher operating accuracy, greater speed range, and shorter regulation times is a popular research area in the field of modern industrial control. The permanent magnet synchronous motor has become the mainstream motor in the industrial servo system because of its excellent characteristics. Therefore, it is more and more important to study the design of the control system of permanent magnet synchronous motor which can meet the requirements of modern industrial control.

The crucial technology of researching electric cars is how to make driving controller to work efficiently and steadily. This project uses TI's digital signal processor

Dao-Liang Xi, No. 336, West Road of Nan Xin Zhuang, Jinan 250022, Shandong, China. Email: ujn_xidl@qq.com
Xiao-Tong Li , No. 336, West Road of Nan Xin Zhuang, Jinan 250022, Shandong, China. Email: 648956098@qq.com
***Corresponding author: Yu-Lin Zhang,** No. 336, West Road of Nan Xin Zhuang, Jinan 250022, Shandong, China. Email: ise_zhangyl@ujn.edu.cn.

DOI 10.1515/9783110540048-060,

(DSP) TMS320F28035 chip as the control core of the system. It combines high-performance DSP cores and rich microcontroller peripheral functions to provide an ideal solution for motor control system applications. It simplifies the design of the system, improves the controller`s real-time processing capability and speed performance, enhances the system's reliability and integration, and is more perfect.

In the actual production process, it has much difference among the models of the drive motor and different motor solutions are in the research and demonstration process. PMSM has obvious advantages over other types of motors, such as permanent magnet synchronous generators because of its use of permanent magnets instead of the original excitation mechanism which causes the volume reduce.

Due to simple control theory of BLDC [2] motor and the structure of the similar as PMSM, controlling theory of BLDC motor should be mastered in the early stages. The PMSM mode is a nonlinear, multi-variable system with strong coupling [10]. The mathematical model of PMSM is established to understand the basic structure and operation principle of the motor on the basis of vector control. Then the PMSM motor controller running in the analysis, and uses the electronic design automatic (EDA) software design controller schematic. This paper presents design a small-power controller based on TMS320F28035 [4] which is used to verify the correctness of the control algorithm and the feasibility of driving the high power motor scheme.

2 Pmsm Mathematic Model

Permanent magnet synchronous motor is actually an AC motor; the stator operation is a phase difference of the three- phase alternating current, while the rotor is a permanent magnet. The vector control theory can help the AC motor including PMSM to be controlled like a DC motor. The most obvious advantage of this kind of motor is that the AC energy is provided by DC, so that the motor can be precisely controlled and the life of the brush is solved. Through the coordinate transformation, the potential of the armature in the stationary two-phase coordinate system can be decomposed into the straight axis potential and the cross axis. From a mathematical point of view, the coordinate transformation is the original set of variables in the equation with a new set of variables instead. Linear transformation refers to the linear relationship between the old and new variables.

The coordinate transformation used in the motor is a linear transformation [7]. Permanent magnet synchronous motor mathematical model of the establishment, we should first establish the motor reference space coordinates. As with the induction motor, the axis of the sinusoidal magnetic potential wave generated by the forward current flowing through the phase winding is defined as the axis of the phase winding and the A axis is used as the spatial reference coordinate of the ABC shaft [9]. Assuming that the positive direction of the induced electromotive force.

The positive direction of the current opposite; counter clock- wise direction for the rotation and electromagnetic torque in the positive direction, the positive direction of the load torque is the opposite. Figure 1 shows the ABC axis spatial reference coordinate system of the motor permanent magnet synchronous motor.

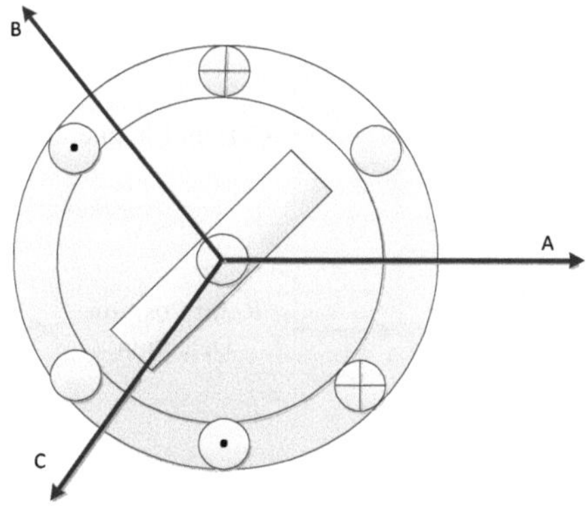

Fig. 1: ABC reference frame of PMSM

3 Hardware Design

The motor control system is given by figure 2 including DSP control circuit, three phase inverter, human-machine interface circuit and detection and protection circuit. The single chip DC/DC conversion integrated circuit Switch DPA made by Power Integrations Company will be adopted as power circuit. Because of the advantages of the TMS320F28035 in controlling motors (TMS320C2XX), this paper chooses TMS320F28035 as main control unit. The DSP control system can generate Corresponding PWM signals to make the motor work normally through sampled phase currents from the inverter [5].

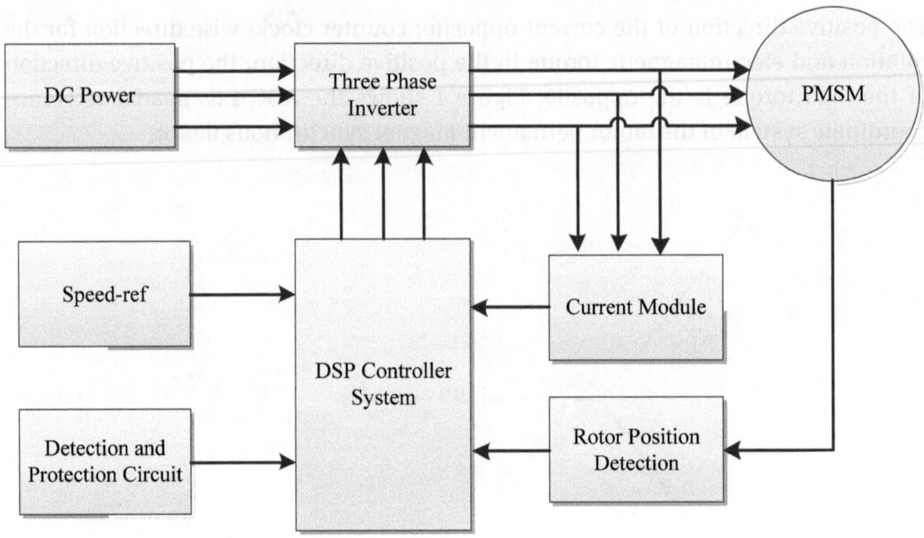

Fig. 2: Motor control system structure diagram

3.1 Power Circuit

Power master chip with DPA425 chip. The DPA425 is a highly integrated DC / DC power control chip designed by PI (Power Integration). It integrates a 200V high frequency power MOSFET and controls PWM control, operating frequency selection, input over voltage detection, programmable current limit, ON / OFF switch control, external clock synchronization, soft start and turn off auto restart, Thermal shutdown protection and other functions set in one. With very few external components can achieve many functions, not only to simplify the design, save space, and can reduce costs. DPA425 support forward and fly back mode, high frequency, chip package; if the external components and transformers using chip components and planar transformers, and the use of aluminum plate design, you can achieve modular design [1].

The main features of the DPA425 are: set 220V high voltage MOSFET and pulse width modulator with one, support forward and fly back mode; built-in automatic restart circuit for the output overload and open circuit protection. Through a pin can be set to 300KHZ or 400KHZ switching frequency and can be input low voltage shutdown protection. By monitoring the voltage drop across the MOSFET on-resistance R, you can achieve over-current protection, eliminating the need for external current detection circuit, the output using synchronous rectifier circuit, power conversion efficiency greater than 90%, with a soft start function [11].

DPA425 composition of the DC converter circuit shown in figure 3, using a single-ended fly back topology, the input voltage range of 36V~72V, the output voltage stabilized at 15V, the maximum output power of 45W, full load power efficiency of up to 80%, while Chip DPA425 operating frequency of 300KHZ. The work flow is as follows: DC voltage through the C1 filter ripple, and then all the way through the transformer T1 primary winding DPA425 internal power MOSFET drain, the other signal by R4, R5 step-down and sent to the DPA425 L feet, Provide start voltage for DPA425. The starting voltage charges the capacitor C3 through the C pin. When the charging reaches a certain threshold voltage, the internal control circuit is activated and the output PWM signal drives the internal power MOSFET to operate.

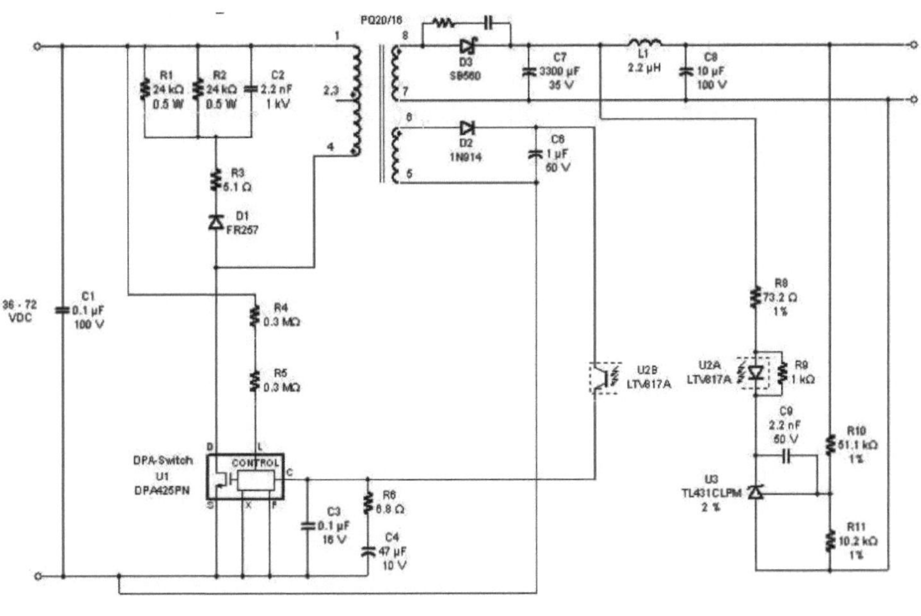

Fig. 3: Power circuit

3.2 Drive circuit and Three Phase circuit

The PWM current signal from DSP is very small so that cannot drive MOSFET tube directly [3]. The drive circuit must amplify current signal to drive more high power motor. Drive circuit is based on IR2127, which of A-phase is taken as an example shown in figure 4. In order to protect the driver chip will not be returned by the inverter circuit to burn the current, in the drive chip after the increase in the push-pull output circuit, which not only increases the drive capability, but also a good protection of the driver chip.

For the small and medium-power permanent magnet synchronous motor controller, the main circuit is generally used AC-DC-AC voltage converter, the inverter for the six-arm inverter bridge, six switch original anti-parallel freewheeling diode, according to a certain strategy control, Output three- phase AC. This main circuit topology has been proven, reliable performance. When the motor start or have high power output, phase current must be great, and current of MOSFET tubes are up to dozen amps in general. To acquire higher current out, MOSFET tubes must be placed in parallel. Therefore, body diode on the power MOSFET can be mainly to prevent the back-EMF to breakdown MOS tube. Here power MOSFET selects IXFA180, PWM amplifying uses IR2127. Three Phase circuit which of C-phase is taken as an example is shown in Fig 4. Two sampling resistor in the three phase circuit as shown in Fig-ure 4 is mainly to take sampling C-phase voltage to protect MOS tubes to be break-down. There is a sampling resistance to sense C-phase currents on below bridge arms. The purposes of sensing phase currents is to obtain new PWM duty cycles to drive motor to work as desired.

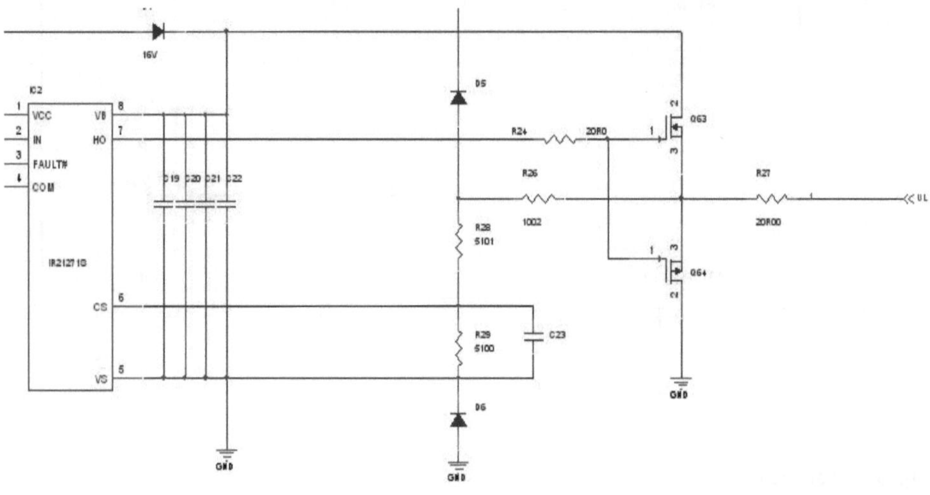

Fig. 4: Drive circuit

3.3 Phase Current Phase Voltage Detection Circuit

Phase current detection circuit is mainly through the detection of the corresponding voltage, according to Ohm's law in the program to achieve. The signal through the 0-3.3V limiter circuit input to the DSP chip integrated AD [8]. The detected phase current is mainly based on the reference speed to produce the corresponding PWM wave. Phase voltage detection principle and phase current detection principle is similar to its use is mainly to detect whether the phase voltage is too large, the protection of the circuit. The phase current phase voltage detection circuit is shown as figure.5

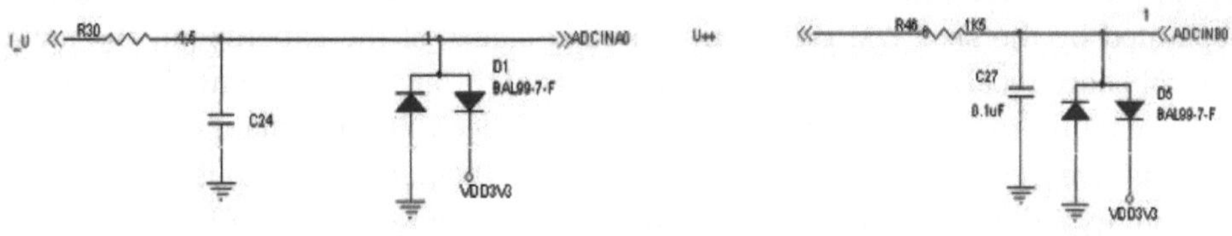

Fig. 5: Detection circuit

3.4 Chopper Circuit

As the name suggests chopping means cutting the current pulse peak, to prevent the motor in the start and braking when the current is too large burned MOS tube, so to adjust the current limit. The current chopper circuit first amplifies the measured three-phase current, and then compares the reference value and the input value through the comparator. Then the output signal is input to the DSP chip. The DSP chip outputs the PWM wave according to the signal output. The opening of the tube is closed to the purpose of limiting the current. The chopper circuit is shown as figure.6.

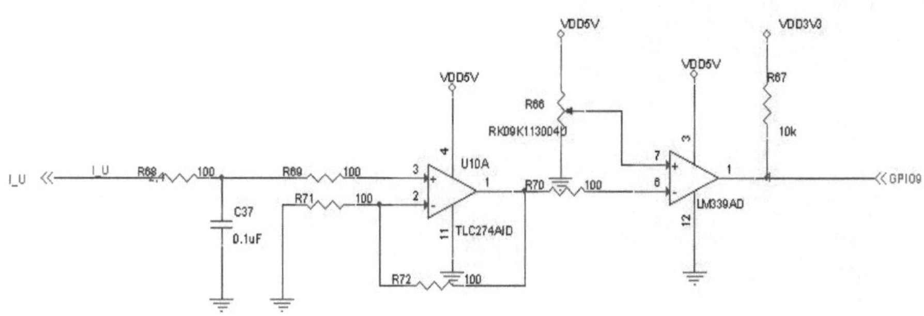

Fig. 6: Chopper circuit

4 Test Result Analysis

Because of the drive circuit and three phase inverter circuit are integrated to the main control board, to build an experimental platform, there still need the host computer and PMSM of 72W. In addition, the main circuit board there are other circuits such as the sampling circuit, communication circuit, photoelectric encoder circuit and the main control chip TMS320F28035 minimum system circuit. Driving signal of the three phase inverter is provided by the main control board which is separated by the six PWM signal. In the early stages of the experiment, the main control board receives a variety of sampling signal through SCI communication from host computer to control switch signal to adjust speed. But the main control board must receive the external voltage signal to achieve the purpose of adjusting speed in the end. Experiment platform assembled is given by figure.7. The software design of the motor control system needs MATLAB simulation to write the relevant code. Then the relevant code needs to be downloading to the TMS320F28035 with the Code composer studio (CCS) compiler. And the type of CCS compiler is XDS v100.There are mainly tow module chosen, for example debugged online or flash. After the relevant code is downloading to chip, the output of drive signals may be observed. The output waveform is monitored through an oscilloscope which is TDS 2012.

Fig. 7: The power circuit physical map

5 Conclusion

In this paper, based on the analysis of the characteristics of TI's high-speed digital signal processor (DSP) TMS320F28035, the permanent magnet synchronous motor servo control system based on the DSP is designed through its special development platform. The controller, power conversion unit and Position detection unit design, and on the basis of the design of a variety of protection circuits. The experimental results show that the control system is stable and reliable and can drive more powerful motor with a number of MOS tube in parallel manner.

Acknowledgement: First of all, I am most grateful to my mentor Professor Zhang Yulin, when I have a problem, for his rich experience and experience to give guidance on my paper. Zhang's attitude to the people and the open-minded mind deeply influenced me, in my future life and work in the moment to guide me in the right direction.

I am also very grateful to my classmates to give me the guidance and help in my diffcult times, as well as the care of my life, passed to me practical and serious and r igorous work attitude. Finally, I thank my alma mater who gives me this piece of land to let me continuing to recognize them, give me the spirit of food to help me step by step to a new life.

References

[1] Maozhen Cui, Changfan Zhang, and Jian Zhu. Realization of pmsm adjustable speed system based on sliding mode control. In Control and Decision Conference, pages 435–438, 2012.P.I. Chernousov, O.V. Golubev and A.L. Petelin, Metallurgist+, 55, 3–4, 242–250, 2011.

[2] B. Fahimi, Ali Emadi, and Raymond B. Sepe. A switched reluctance machine-based starter/alternator for more electric cars. IEEE Transac- tions on Energy Conversion, 19(1):116–124, 2004. P.I. Yugov, A.V. Sarychev and L.A. Baeva, Metallurgist+, 45, 9–10, 379–381, 2001.

[3] Dipl Ing Heiz. Position data and direction auto-determining system for a moving terrrestrial object, 1985.

[4] Wang Huan. The Research and Design of Permanent Magnet Syn- chronous Motor Vector Control System Based on DSP. PhD thesis, Dalian University of Technology, 2008.

[5] G. H. Jang, J. H. Park, and J. H. Chang. Position detection and start- up algorithm of a rotor in a sensorless bldc motor utilising inductance variation. IEE Proceedings - Electric Power Applications, 149(2):137–142, 2002.

[6] Ting Ting Liu, Tan Yu, Wu Gang, and Shu Mao Wang. Simulation of pmsm vector control system based on matlab/simulink. In 2009 international conference on measuring technology and mechatronics automation, pages 343–346, 2009.

[7] M. Marufuzzaman, M. B. I. Reaz, M. S. Rahman and M. A. M. Ali. Fpga implementation of an intelligent current dq pi controller for foc pmsm drive. In Computer Applications and Industrial Electronics (ICCAIE), 2010 International Conference on, pages 602–605, 2010.

[8] Peng Tao Mou, Jian Guo Song, and Qiang Song. Double closed-loop control system design of pmsm based on dsp. In International Confer- ence on Mechatronics, Materials, Chemistry and Computer Engineering, 2015.

[9] Arthur V. Radun. Design considerations for the switched reluctance motor. IEEE Transactions on Industry Applications, 31(5):1079–1087, 1995.

[10] L. Zhong, M. F. Rahman, W. Y. Hu, and K. W. Lim. Analysis of direct torque control in permanent magnet synchronous motor drives. IEEE Transactions on Power Electronics, 12(3):528–536, 1997.

[11] Yinghong Zhou, Xiaopei Feng, Siyuan Guo, and L. I. Zhizhong. Design of three-phase power charger based on tms320f28035. Electric Drive, 2016.

Kui Lu and Ju-Guang Qiang

Research on Improvement of Direct Torque Control System for Permanent Magnet Synchronous Motor

Abstract: In recent years, with the national policy to vigorously promote the permanent magnet material prices continue to reduce the permanent magnet synchronous motor to drive the development of new energy vehicles into the fast lane, the entire automotive industry progress with each passing day [1]. Direct torque control system because of the use of non-linear hysteresis regulator, the system response speed, often used as a vehicle permanent magnet synchronous motor speed control system, but the system also exists shortcomings such as torque ripple, in order to better to meet the actual needs of the vehicle conditions, to be improved to its research.

Keywords: permanent magnet synchronous motor, direct torque control system, torque ripple, improvement

1 Introduction

With the energy and environmental issues become increasingly serious, countries around the world are calling for energy-saving emission reduction, new energy vehicles also popular, which permanent magnet synchronous motor with its good speed characteristics and high energy conversion rate has become a new energy vehicle driver the first choice of the motor [2]. At present the vehicle permanent magnet synchronous motor speed control system is divided into two kinds, one is vector control; the other is the direct torque control. Which direct torque debugging system because the system response faster, can better meet the system's brake safety So the more popular people, but the traditional direct torque control system because the composition of the voltage vector table voltage vector is relatively small, and the size of the amplitude are fixed, the larger torque ripple, cannot be very good to meet the car driving process In the comfort and stability.

Therefore, based on the traditional direct torque control system, an improved scheme of torque ripple suppression is proposed by introducing vector control: a

Kui Lu, Anhui University of Science and Technology, Anhui, China232001, P.R. China. Email:klu@aust.edu.cn. Corresponding author.
Ju-Guang Qiang, Anhui University of Science and Technology, Anhui, China232001, P.R. China. Email: 397741219@qq.com

DOI 10.1515/9783110540048-061,

direct torque control system (SVM-DTC) based on vector control. The MATLAB simulation platform is used to analyze the improved and improved control system respectively.

2 Traditional Direct Torque Control System

Permanent magnet synchronous motor consists of base, stator core, stator winding, rotor core, permanent magnet, rotor shaft and other components. Through the rotary transformer to detect the motor rotor position information transmitted to the controller after the high-performance control.

The working principle of the motor is based on the interaction of the current in the stator windings and the interaction of the rotor magnetic field [3]. After determining the rotor information of the motor, the energization mode of the stator winding and the current value of each phase winding are controlled according to the information of the rotor so as to keep the stator current vector and the rotor synchronized.

It is difficult to analyze the mathematical model of the permanent magnet synchronous motor in the traditional three-phase stationary coordinate system, so it is usually analyzed by the coordinate transformation in the dq axis rotor coordinate system. The dynamic mathematical model of the dq rotor coordinate system is obtained as follows.

$$\begin{aligned} u_d &= R_1 i_d + p\psi_d - \omega\psi_q \\ u_q &= R_1 i_q + p\psi_q - \omega\psi_q \end{aligned} \tag{1}$$

Equation (1) is the stator voltage equation, where ω is the angular velocity of the rotor rotation is the differential operator.

$$\begin{aligned} \psi_d &= \psi_f + L_d i_d \\ \psi_q &= L_q i_q \end{aligned} \tag{2}$$

Since the PMSM rotor permanent magnet produces a sinusoidal distribution magnetic field, when the magnetic field is transformed into the rotor coordinate system, only the d-winding turns in the stator windings (ie, the term ψ_f in the stator winding, Chain amplitude), and with the q-winding without the chain (BLDMC is different).

$$T_e = 1.5 n_p (\psi_d i_q - \psi_q i_d) \tag{3}$$

The equation is the torque equation of the permanent magnet synchronous motor, where i_d and i_q are the stator current and the rotor current, n_p is the pole pair of the motor. From the formula, it can be seen that the stability of the electromagnetic torque must control the stator and rotor flux and.

$$T_e - T_1 = \frac{J}{n_p}\frac{d\omega}{dt} \tag{4}$$

Equation (4) is the equation of motion balance of the motor, which reflects the relationship between torque and load and speed, where T1 is the load torque.

The traditional direct torque control system used in this paper is a kind of circular flux trajectory control scheme proposed by Japanese scholar Takahashi, whose stator flux trajectory is quasi-circular. The stator current of the motor has a good sinusoidal degree only on the inverter switching capability has a greater demand. The scheme uses the method of querying the voltage vector table to control the stator flux and the torque of the motor [4]. By detecting the set electromagnetic torque and the stator flux linkage and detecting the actual value of the system, the hysteresis bang adjustment is carried out to obtain the control target ψQ and the TQ then selects the most suitable voltage vector from the voltage inverter according to the sector position where the stator flux vector is located, converts it into a control signal, controls the turn-on and turn-off of the IGBT. The working principle of the two-point magnetic hysteresis looped is shown below.

$$\psi Q = \begin{cases} 1 & : \ \psi_1 - \psi_1^* \geq \Delta\varepsilon \\ 0 & : \ \psi_1 - \psi_1^* \leq -\Delta\varepsilon \\ \text{cons tan t} & : \ \text{other} \end{cases} \tag{5}$$

According to the structure of the permanent magnet synchronous motor described above, the following simulation model of the traditional direct torque control system is established. It can be seen from Figure 1 that the traditional direct torque control system includes the motor body, the motor flux linkage and the torque observation unit, Flux chain sector determination unit, flux linkage and torque bang comparator unit and so on several parts[5]. The parameters of the system are shown in Table 1.

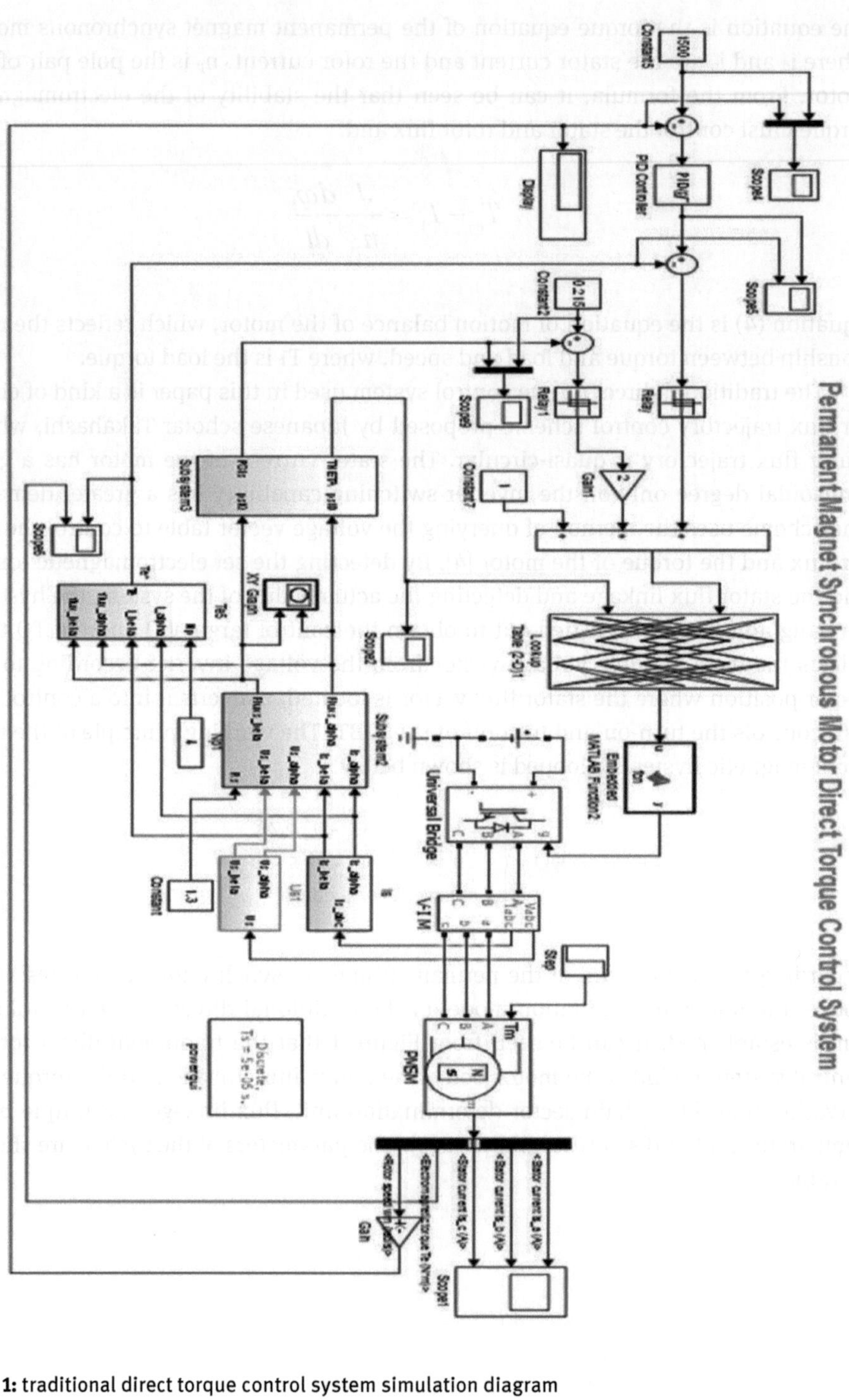

Fig. 1: traditional direct torque control system simulation diagram

Tab. 1: Simulation parameters

Parameter	Value	Parameter	Value
Stator winding	1.3Ω	Moment of inertia	0.0008kg.m^2
q axis inductance	8.5mh	Pole number	4
d axis inductance	8.5mh	DC voltage	310v
Permanent magnet flux linkage	0.174V·S	Rated voltage	220V

The torque and speed commands in the system are given by Simulink's Step module, and the specific parameters are shown in Figure 2 below.

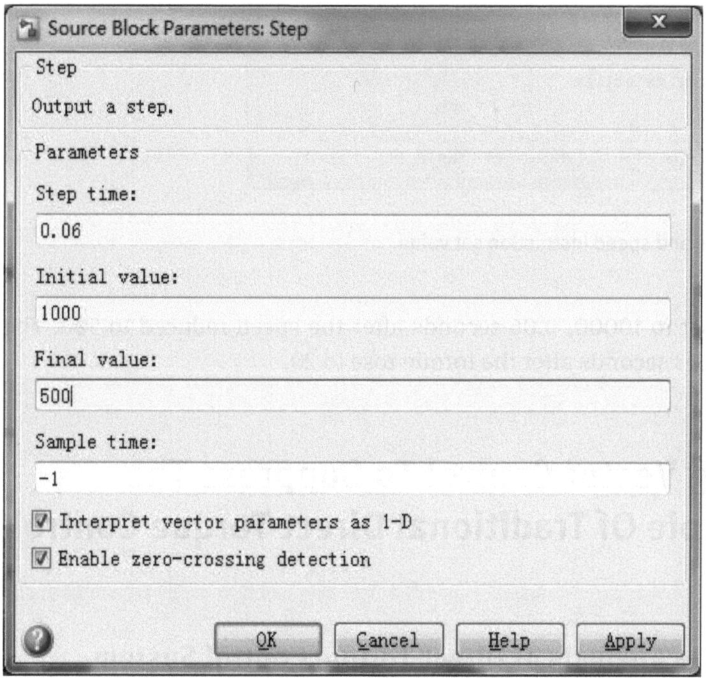

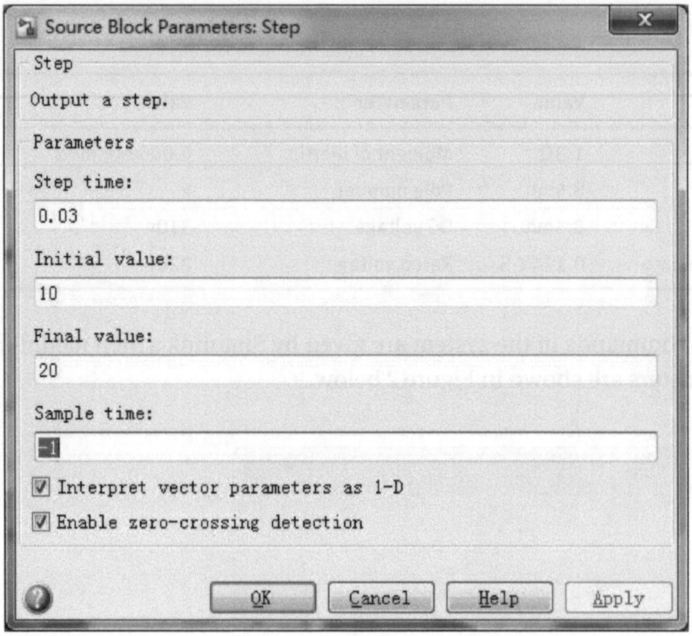

Fig. 2: PMSM load torque and speed instruction set value

The initial speed is set to 10000, 0.06 seconds after the speed reduced to 500. The initial torque is 10, 0.03 seconds after the torque rose to 20.

3 Introducing Vector Control To Suppress The Torque Ripple Of Traditional Direct Torque Control System

3.1 Torque Ripple Analysis of Direct Torque Control System

Since the voltage vector table in the conventional direct torque control is composed of only a small number of voltage vectors, the magnitude of the magnitude is constant; and after searching for a suitable voltage vector from the voltage vector table, the voltage vector The time TS on the motor. And the size of a change component ΔTe3 of the electromagnetic torque is difficult to control, which inevitably leads to a large torque ripple. Especially when the voltage required at low speed is low, the larger voltage vector obtained by querying the vector table will make the torque have a great change, which leads to a significant fluctuation, the control performance of the motor at low speed.

In the traditional direct torque control system, the required stator flux angular velocity requires several TSto achieve, and can only be some discontinuous value; and the magnetic field orientation vector control technology uses the mature SVPWM modulation technology, can The required stator flux angular velocity is generated in a TS, so that the torque equal to the load can be generated in a short time, so that the rotational speed can be stabilized and the generation of the torque ripple can be suppressed[6].

3.2 An Improved SVM-DTC Control System with SVPWM Modulation Technology

The usual SVPWM algorithm is to first determine the basic voltage space vector based on the spatial sector location of the U_g, as shown in Fig. 3 for the nonzero voltage vectors U_4 and U_6. Remember their respective role in the time were t_1 and t_2, then the following formula (5).

$$U_4 t_1 + U_6 t_2 = u_g t_g \tag{6}$$

And then calculate the value of t_1 and t_2, where U_{gm} is the amplitude of the U_g vector.

$$\begin{aligned} t_1 &= \frac{3}{2} t_g \left(\frac{u_{gm} \cos\theta}{U_d} - \frac{1}{\sqrt{3}} \frac{u_{gm} \sin\theta}{U_d} \right) \\ t_2 &= \sqrt{3} t_g \frac{u_{gm} \sin\theta}{U_d} \end{aligned} \tag{7}$$

In general, then the extra time $t_1 + t_2 \leq t_g$ can be evenly distributed over the two zero voltage vectors U_0 and U_7 because their effect does not affect the integration of the output voltage vector of the inverter.

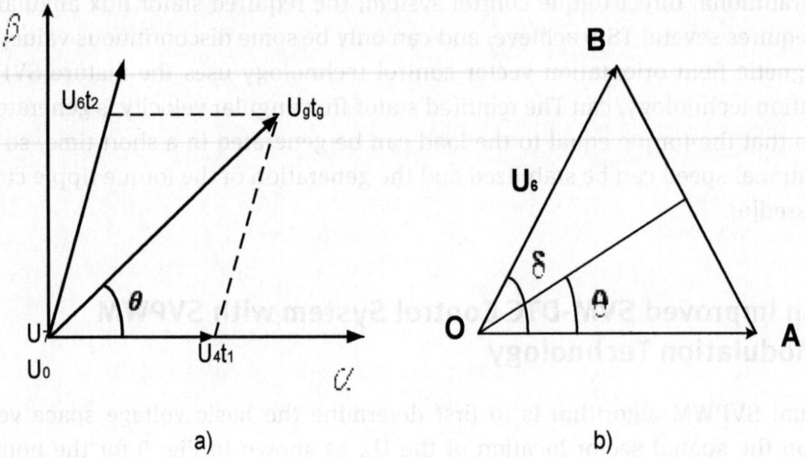

Schematic diagram of voltage vector synthesis (b) synthetic voltage vector anti-counterfeiting schematic

Fig. 3: Schematic diagram of space vector pulse width modulation

The improved direct torque control system SVM-DTC control system can be obtained by adding SVPWM modulation technology to the traditional direct torque control system.

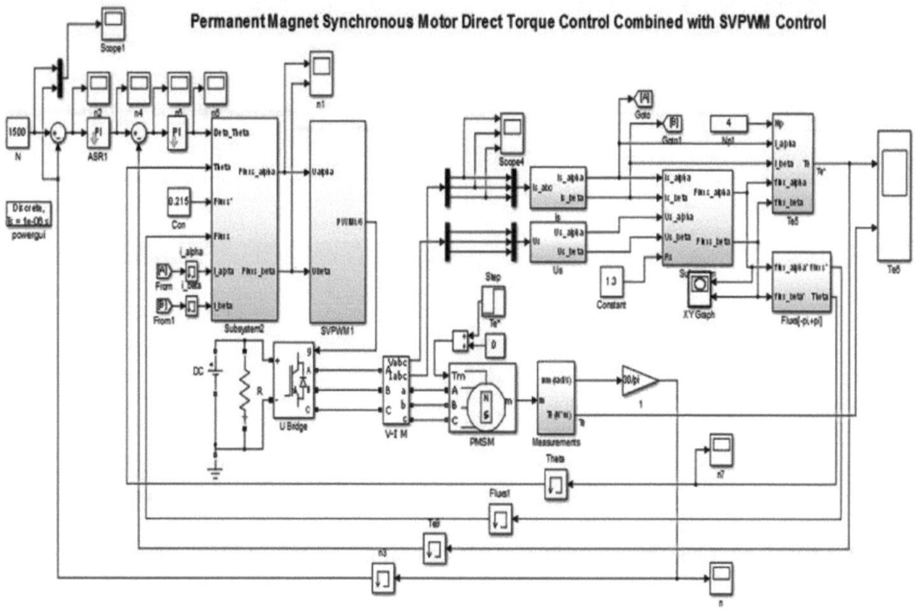

Fig. 4: Direct torque control system with SVPWM control technology

The system parameter setting is the same as the traditional direct torque control system of Fig. 4, and the simulation time of the system is set to 0.08 seconds. After the simulation run, the comparison between the conventional direct torque control system and the improved SVM-DTC control system is shown in Fig. 5 and Fig. 6. Figure 5 in the left for the improved speed and torque simulation waveform, the right side of the traditional direct torque control system simulation waveform. It can be clearly observed that the torque and speed of the improved direct torque control system are smaller and the system is more stable than the traditional direct torque control system, and the fastness of its response is still preserved.

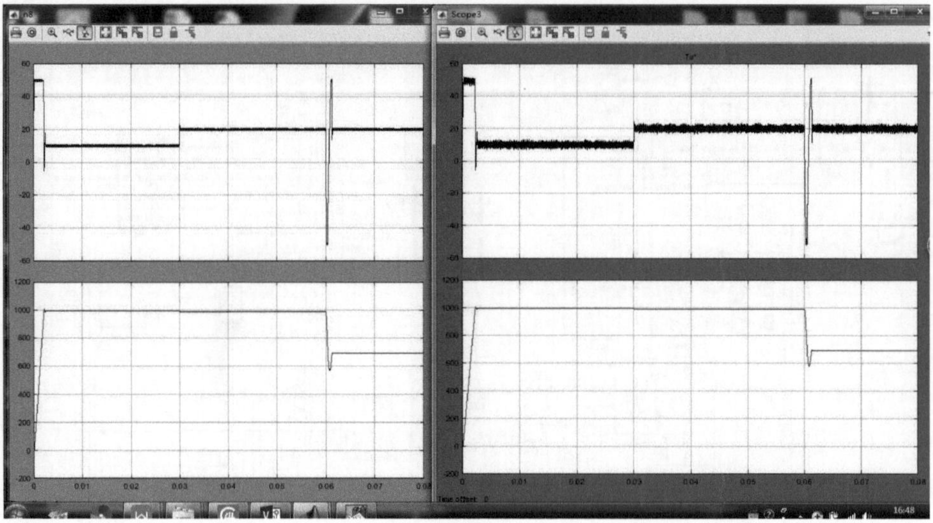

Fig. 5: SVM-DTC control system and traditional DTC control system torque speed simulation waveform comparison

Figure 6 for the two control system stator flux vector trajectory comparison chart, the left is the improved track, the right side of the traditional stator flux trajectory. It is found that the introduction of SVM-DTC control system can effectively improve the trajectory of the stator flux vector and make it more approach and circular.

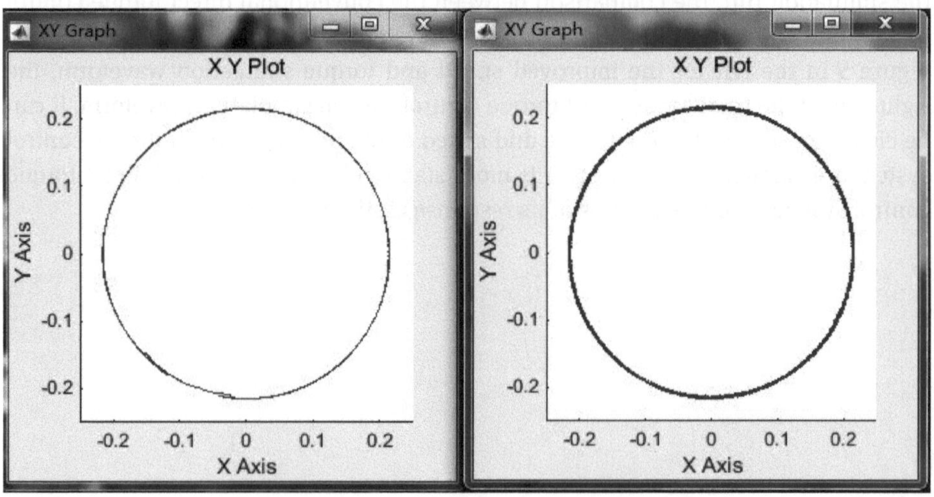

Fig. 6: Comparison of stator flux vector

4 Conclusion

Aiming at the problem of torque ripple in the traditional direct torque control system, SVPWM control technology is introduced to suppress the generation of torque ripple in combination with the principle of vector control. Through the simulation and modeling, the simulation model of SVM-DTC control system and the simulation model of the traditional direct torque control system are compared and analyzed. The improved system is successfully reduced by analyzing and comparing the improved SVM-DTC control system simulation model the torque ripple and the flux linkage flux error. It is proved that the improved scheme is successful and feasible.

References

[1] Bolognani S, Zordan M, Zigliotto M. Experimental fault-tolerant control of a PMSM drives [J]. IEEE Transactions on Industrial
[2] Electronics, 2000, 47(5):1134–1141.
[3] Moon H T, Kim H S, Youn M J. A discrete-time predictive current control for PMSM [J]. IEEE Transactions on Power
[4] Electronics, 2003, 18(1):464–472.
[5] Bolognani S, Oboe R, Zigliotto M. Sensorless full-digital PMSM Drive with EKF estimation of speed and rotor position [J].
[6] IEEE Transactions on Industrial Electronics, 1999, 46(1):184–191.
[7] Lu Q X, Dong X R. Application of DSP in Sensorless PMSM Control System [M]. 2009.
[8] GUO Ji-feng, GAO Cun-chen .Design of variable structure controller for uncertain time-delay singular system [J].Journal of
[9] Harbin Institute of Technology (New Series), 2009, 3(16):35–38.
[10] Andrzej Bartoszewicz Discrete-Time Quasi-Sliding-Mode Control Strategies [J]. IEEE TRANSACTIONS ON INDUSTRIAL
[11] ELECTRONICS, 1998, 4(45):78–81.

Xin-Li Xu, Yu Ning*, Bin-Yao Ning and Hai-Wen Liao

Cascade ADRC Based Robust Flight Control with System Potential Uncertainties

Abstract: This paper presents a novel robust control method for aircraft with potential system uncertainties such as parametric perturbations, external disturbances, and uncertainties due to unmodeled dynamics. A cascade active disturbance rejection controller (ADRC) is used to counteract the adverse effects against system uncertainties. This approach does not require the accurate mathematical model of the controlled system and ensures that the reference input value can be tracked rapidly and accurately. The stability and safety of the aircraft is much improved in the event of uncertain input. Finally, the simulation results are given to show the effectiveness and performance of the developed method.

Keywords: system uncertainties, parametric perturbations, unmodeled dynamics, robust control, cascade ADRC

1 Introduction

Over the past decades, due to the increasing demands for reliability and survivability in modern aircraft, robust control has been widely developed and used in the flight control system [1, 2]. A novel integral sliding mode control (ISMC) scheme was proposed for actuator outage [3]. For an uncertain non-linear system, this scheme consisted of adaptive control method and matrix full-rank factorization technique, and was synthesized with fuzzy logic systems to guarantee better robustness of the system against actuator faults. In [4], a robust algorithm based on MPC and H2 control theory was developed. In this work, the flight control law could follow independent parameters such as airspeed, sideslip angle, angle of attack, etc. [5] developed a new model reference adaptive control (MRAC) scheme for structural damage of aircraft, which used a discrete-time method to realize the flight control system. Several issues were addressed on damage aircraft, and the stability was also studied as well as asymptotic output tracking.

Moreover, a robust control method which consists of $H\infty$ and μ-synthesis control to optimize between the robustness and performance requirements was pre-

Xin-Li Xu, School of Mechatronics Engineering and Automation, Shanghai University, Shanghai, 200072, China. Email: lingyun308@shu.edu.cn
***Corresponding author: Yu Ning, Bin-Yao Ning, Hai-Wen Liao,** Departments of Automation, Shao Guan University, Shao guan, 512005, China. Email: nby6210@163.com.

DOI 10.1515/9783110540048-062,

sented for a multi-variable aircraft [6]. In order to alleviate the chattering phenomenon and obtain good stability, [7] presented a FTC scheme used the interval type-2 fuzzy logic and sliding mode control methodology for coaxial tri-rotor helicopter.

In [8], an adaptive global sliding mode control approach was proposed to address the problem of actuator malfunction of helicopter, and the quantum logic was used to increase the system accuracy. This work has a good identification performance and little time delay. Considering the wing flutter when actuator faults occur, a robust FTC was used to deal with actuator faults, parameter uncertainties and external disturbances. It consisted of neural network and H∞, and could suppress the wing flutter with better stability [9]. [10] Presented a novel body-centric modelling approach to a quad rotor UAV for the study of identification. By virtue of model parameters and data, acceleration tracking control strategy was investigated using a linear-quadratic (LQ) algorithm. Though many algorithms have been developed, the research of robust control is still a challenging with the complexity and nonlinearity of flight control system [11-21].

The rest of the paper is organized as follows. Section 2 is focused on the dynamic model of the aircraft, and the statement of faulty aircraft. Section 3 describes the fault tolerant control strategy. Simulation results are presented and discussed in Section 4. Finally, a brief conclusion and future work are given in Section 5.

2 Problem Statement and Preliminaries

Consider the flight control system that will be influenced by modelling error and external disturbance, as follows:

$$\begin{cases} \dot{x} = A(e)x(t) + B(e)u(t) + d(t) \\ y = Cx(t) \end{cases} \tag{1}$$

where $x(t) \in R^n$ and $y(t) \in R^n$ are respectively the state vector and output vector of the system, $u(t) \in R^m$ is the control input, $A(e) \in R^{n\times m}$ and $B(e) \in R^{n\times m}$ are the parameter matrices with model errors, $C \in R^{l\times n}$ is constant matrix, $d(t)$ is the lumped effect of uncertainties and disturbances.

Assumption 2.1 $A(e) = (1 \pm \varepsilon)A$ and $B(e) = (1 \pm \varepsilon)B$ are completely controllable ($rank(B) = m$, $0 \le \varepsilon \le 0.2$).

Assumption 2.2 $d(t)$ is bounded and unknown.

3 Design of Controller

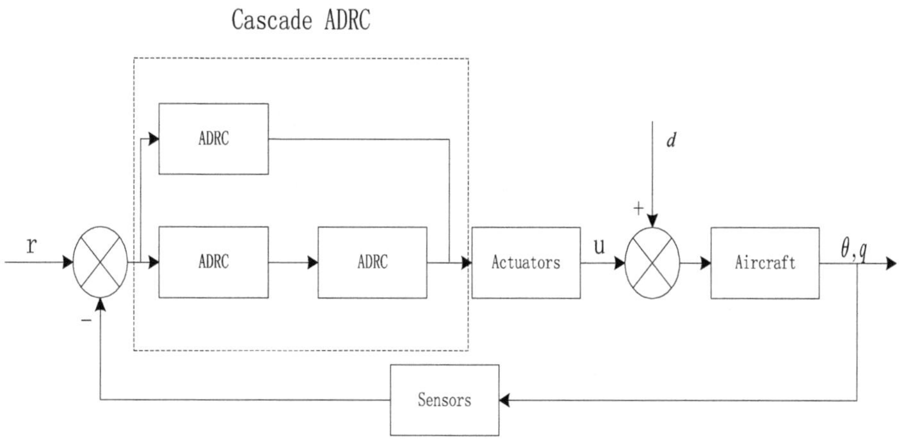

Fig. 1: The schematic of the proposed FTC

When the aircraft has unknown bounded perturbations which may be caused by disturbances or unmodeled dynamics, the presented controller guarantees the certain robustness for the perturbation or faults. That is to say, the controller is designed to ensure that the expression (1) is stable and has a good dynamic performance. Figure 1 shows the system block diagram of the Cascade ADRC method.

The Cascade ADRC has a strong ability for decoupling and is adaptable to disturbance as it is independent of the accurate mathematical model of the controlled plant. It uses the external disturbance and the internal uncertainty of the control system as a disturbance while compensating the total disturbance for tracking in real time.

When the actuator failure occurs, the circuit change can be estimated by ESO, and used to compensate for the system dynamically since it includes the actuator fault and external disturbance. Then the dynamic and steady-state performance of the circuit can be obtained by the nonlinear configuration of the state feedback.

Assuming that the control system can be approximate to a second order model under the effect of Cascade ADRC, it can be expressed as:

$$\begin{cases} \dot{x}_1 = x_2 \\ \dot{x}_2 = f + b_0 u \\ y = x_1 \end{cases} \tag{2}$$

Where u and y are the control signal and the output of the system respectively; b_0 is the parameters of Cascade ADRC.

Then we have that the following equation:

$$\begin{cases} \ddot{y} = f + b_0 u \\ f = g(t, y, \dot{y}, d) + (b - b_0)u \end{cases} \tag{3}$$

where b is the parameter of a controlled plant; g represents comprehensive characteristics combining with internal dynamics and external disturbances of the system; d is the external disturbances in the control loop; f is the extended state of the controlled plant, whose specific form of the expression is unknown.

ESO satisfies the following form:

$$\begin{cases} e = y - z_1 \\ \dot{z}_1 = z_2 + l_1(y - z_1) \\ \dot{z}_2 = z_3 + l_2(y - z_1) + b_0 u \\ \dot{z}_3 = l_3(y - z_1) \end{cases} \tag{4}$$

Where l_1, l_2 and l_3 are the parameters of observer which need to determine. z_1, z_2 and z_3 are the input variables of ESO. y, $\dot{y}$ and f will be tracked by z_1, z_2 and z_3 when the ESO in (4) is tuned properly.

From (4), the controller is defined as the third order Cascade ADRC. Therefore, the control law u_0 can be expressed as:

$$u_0 = b_1(r - z_1) - b_2 z_2 \tag{5}$$

Where r the reference is input variable of the system. b_1 and b_2 are the parameters of Cascade ADRC.

Then the control signal of Cascade ADRC can be expressed as:

$$u = (u_0 - z_3)/b_0 \tag{6}$$

Substituting (6) to (3), when the ESO in (4) is tuned properly and $z_3 \approx f$, the control system is converted to a double integral serial link:

$$\ddot{y} = f + (u_0 - z_3) \approx u_0 \tag{7}$$

Substituting (5) to (7) then the closed-loop expected dynamic equation of the system can be obtained:

$$\ddot{y}+b_2\dot{y}+b_1y=b_1r \tag{8}$$

From above, the parameters of the controller b_1, b_2, b_0 l_1, l_2 and l_3 have to be tuned.

The characteristic equation of (8) can be expressed, as:

$$D(s)=s^3+l_1s^2+l_2s+l_3 \tag{9}$$

Defining:

$$D(s)=(s+\omega_0)^3$$

Then l_1, l_2 and l_3 can be converted to the function of the observer bandwidth ω_0 :

$$\begin{cases} l_1=3\omega_0 \\ l_2=3\omega_0^{\ 2} \\ l_3=\omega_0^{\ 3} \end{cases} \tag{10}$$

Hence, we will obtain a good performance for Cascade ADRC by tuning b_1, b_2, b_0, ω_0.

4 Simulation

In this section, an example of an aircraft model is presented to demonstrate the feasibility of the proposed control method. The MATLAB software is used for a couple of numerical simulations. For comparison, the results under normal and uncertainty are presented simultaneously. The control parameters are used as the following: $b_0=0.8,0.8,-16$, $b_1=0.9,397,0.18$, $b_2=1.9,37,1$, $\omega_0=100,1,98$. The initial condition of the aircraft is that $M=0.53$, $h=13123ft$, where M is the Mach, h is the altitude. Then a longitudinal aircraft model can be described as:

$$\begin{bmatrix} \dot{q} \\ \dot{V} \\ \dot{\alpha} \\ \dot{\theta} \end{bmatrix} = \begin{bmatrix} -0.5818 & 0.0002 & -1.0280 & 0 \\ -0.1263 & -0.0062 & -3.6726 & -8.7912 \\ 1.0050 & -0.0006 & -0.5717 & 0 \\ -0.0050 & 0.0006 & 0.5818 & 0 \end{bmatrix} \begin{bmatrix} q \\ V \\ \alpha \\ \theta \end{bmatrix} + \begin{bmatrix} -1.5528 & 0.0152 \\ 0 & 1.3287 \\ -0.0378 & -0.0007 \\ 0.0378 & 0.0007 \end{bmatrix} \begin{bmatrix} \delta_e \\ \delta_T \end{bmatrix}$$

$$\begin{bmatrix} \theta \\ q \end{bmatrix} = \begin{bmatrix} 0 & 0 & 0 & 1 \\ 1 & 0 & 0 & 0 \end{bmatrix} \begin{bmatrix} q \\ V \\ \alpha \\ \theta \end{bmatrix}$$

where q denotes the pitch rate, V denotes the velocity, α denotes the attack angle, θ is the pitch angle, δ_e denotes the elevator deflection, and δ_T denotes the thrust.

Two scenarios are considered in the numerical simulations: a system uncertainty up to 30% and 60% of the nominal value is considered. In the first one, the disturbance $d_1 = 1.2\sin(0.9\pi t)$ and d_2 which is the white noise with intensity 0.3 are added into (1).

In Figure 2, the disturbances are injected $[15s, 20s]$ and $[30s, 35s]$ respectively, which may lead to transient of aircraft. Using the proposed method, the system stabilizes asymptotically quickly.

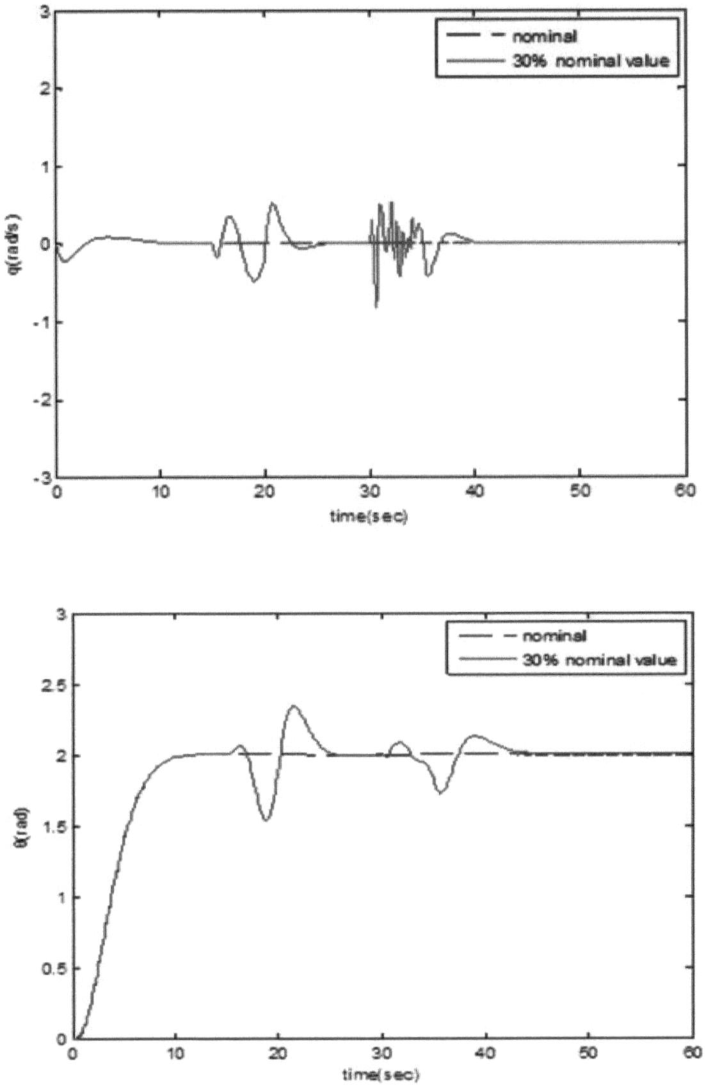

Fig. 2: The pitch angle and pitch rate with respect to loss of 30% nominal value

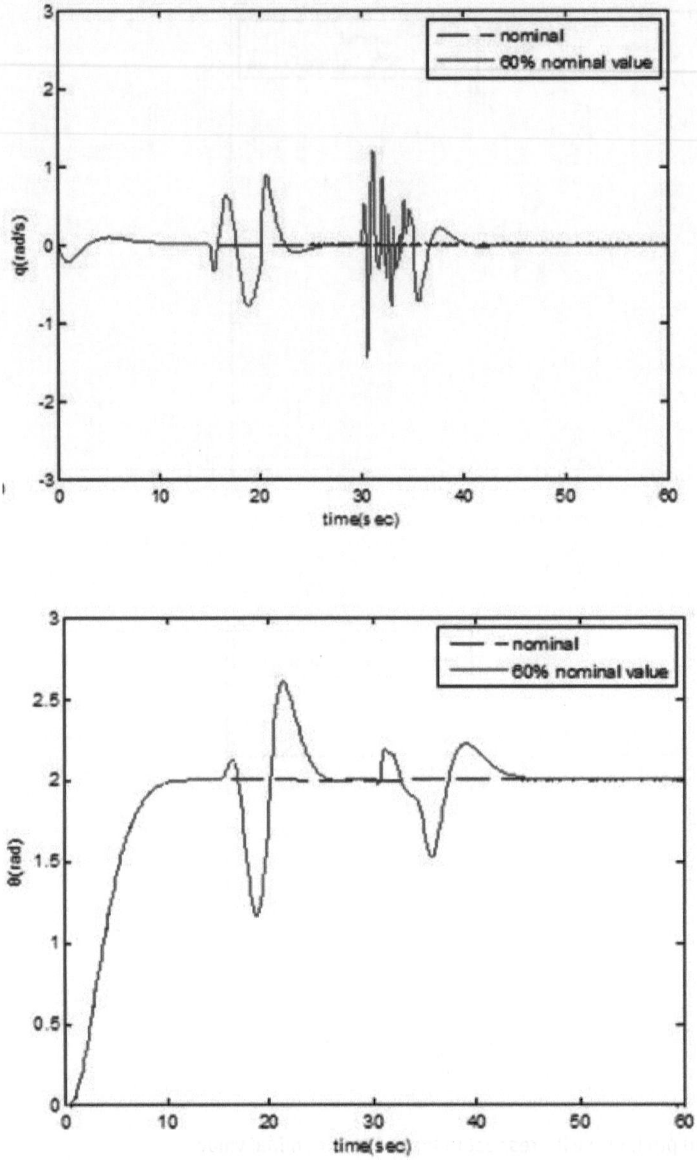

Fig. 3: The pitch angle and pitch rate with respect to loss of 60% nominal value

From Figure 3, a system uncertainty up to 60 % of the nominal value also occurs $[15s, 20s]$ and $[30s, 35s]$, which may lead to transient of pitch angle and pitch rate. The aircraft is manoeuvred by the developed approach back to the anticipated position after a transient volatile period.

Our future work is to apply the presented control method to an actual aircraft model with unexpected fast disturbance and strong uncertainties. The performance and effectiveness will be tested in and then it could be used in real aircrafts for a variety of tasks. The approach developed in this paper could also be used for hypersonic vehicle or missile control to overcome the system uncertainty, nonlinearity or parameter perturbation, improving both the tracking accuracy and the manoeuvrability.

Acknowledgement: This work is financially supported by the National Natural Science Foundations of China (No. 61175092 and 61433016) and by Guangdong college students' Special fund for cultivating scientific and technological innovation (No. pdjh2017b0456). The authors would like to sincerely thank anonymous reviewers and editors for their constructive comments which helped us to improving this paper.

References

[1] X. J. Wang, S. P. Wang, Z. W. Yang and C. Zhang, 'Active fault-tolerant control strategy of large civil aircraft under elevator failures', *Chinese Journal of Aeronautics*, Vol. 28, No. 6, 1568–1666, 2015.

[2] D. Adrien, N. Hassan, G. Laurent and P. Pascal, 'Actuators additive fault-tolerant control for combat aircraft', *IFAC*, Vol. 48, No. 21, 180–185, 2015.

[3] Y. H. Li, H.P. Ju and Y. Dan, 'Fuzzy logic systems-based integral sliding mode fault-tolerant control for a class of uncertain non-linear systems', *IET Control Theory and Applications*, Vol. 13, No. 3, 300–311, 2016.

[4] E. N. Hartley and J. M. Maciejowski, 'A longitudinal flight control law based on robust MPC and H2 methods to accommodate sensor loss in the reconfigure benchmark', *IFAC* , Vol. 48, No. 21, 1000–1005, 2015.

[5] J. X. Guo and G. Tao, 'A discrete-time multivariable MRAC scheme applied to a nonlinear aircraft model with structural damage', *Automatica*, Vol. 53, 43–52, 2015.

[6] H. A. Saif and A. Nabil, 'Aircraft robust flight tracking control against actuator efficiency faults', *2015 19th International Conference on System Theory, Control and Computing (ICSTCC)*, 901–906, 2015.

[7] Z. Samir, K. Kamel and S. Djamel, 'Fault tolerant control based on interval type-2 fuzzy sliding mode controller for coaxial trirotor aircraft', *ISA Transactions*, Vol. 59, 215–231, 2015.

[8] F. Y. Chen, R. Q. Jiang, C. Y. Wen and R. Su, 'Self-repairing control of a helicopter with input time delay via adaptive global sliding mode control and quantum logic', *Information Sciences*, Vol. 316,123–131, 2015.

[9] M. Z. Gao and G. P. Cai, 'Robust fault-tolerant control for wing flutter under actuator failure', *Chinese Journal of Aeronautics*, In press, 2016.

[10] T. A. Mohamad, M. B. Victor and H. William, 'Body-centric modelling, identification, and acceleration tracking control of a quadrotor UAV', International Journal of Modelling, Identification and Control, Vol. 24, No. 1, 29–41, 2015.

[11] C. Fikret and H. Chingiz, 'Reconfigurable control of an UAV against sensor/actuator failures', IFAC , Vol. 48, No. 9, 7–12, 2015.

[12] D. Z. Xu, B. Jiang, M. S. Qian and J. Zhao, 'Neural Networks-based Internal Model Optimal Control for Non-affine Nonlinear Systems with Control Inputs Constrain', ICIC Express Letters, Part B: Applications , Vol. 3, No. 6, 1341–1347, 2012.

[13] Intelligent travel information platform based on location base services to predict user travel behavior from user-generated GPS traces. International Journal of Computers and Applications, 1–14

[14] Memon I, Arain Q A, Memon M H, et al. Search me if you can: Multiple mix zones with location privacy protection for mapping services [J]. International Journal of Communication Systems, 2017.

[15] Mangi F A, Xiao S, Mallah G A, et al. Multiband Circular Polarizer Based on Fission Transmission of Linearly Polarized Wave for X-Band Applications [J]. Journal of Electrical & Computer Engineering, 2016, 2016:6.

[16] Memon I, Chen L, Majid A, et al. Travel Recommendation Using Geo-tagged Photos in Social Media for Tourist[C]// International Conference on Computer & Emerging Technologies March. 2014:1347–1362.

[17] Mangi F A, Xiao S, Jamro D A, et al. Manipulating Electromagnetic Wave Linear-to-Circular Polarization Conversion Transmitter Based on Periodic Strips Array[C]// International Conference on Information Science and Control Engineering. IEEE, 2016:1342–1345.

[18] Jamro D A, Hong J, Bah M H, et al. Triangular Antenna with Novel Techniques for RCS Reduction Applications [M]// Wireless Communications, Networking and Applications. Springer India, 2016.

[19] Mangi F A, Xiao S, Memon I, et al. Novel Design and Performance Analysis of Broadband Dual Layer Circular Polarizer Based on Frequency Selective Surface for 60 GHZ Application[M]// Wireless Communications, Networking and Applications. Springer India, 2016.

[20] Mangi F A, Xiao S, Memon I, et al. Single-layer Linear-to-Circular Polarization Transformer Based on Periodic Strips Array [C]. DEStech Transactions on Engineering and Technology Research, 2016.

[21] Memon I. Transmission Characteristics of Quarter Wave Plate Circular Polarizer Based on Frequency Selective Surface for High Frequency Applications [C]. International Conference on Computer Information Systems & Industrial Applications, 2015.

Yu Ning, Hou-Lin Liu, Ye Ye*, Quan-Xin Huang and Zu-Quen Su

A Digital Trigger of Thyristor Rectifier for High Voltage Formed Aluminium Foil Based on DSP

Abstract: The increasing availability of single-chip 32-bit digital signal controllers has made it possible to reconsider conventional hardware designs for a variety of triggering circuits. This paper developed a digital triggering controller of thyristor rectifier for high voltage formed aluminium foil based on TMS320F2812. The hardware and software design of the controller is investigated in details. This paper presented a grid frequency tracking and phase sequence adaptive method and steps. Experimental results show that the trigger controller has good performance due to simple hardware, high reliability.

Keywords: Phased rectifier; Phase sequence adaptive; Digital signal processor

1 Introduction

A thyristor rectifier and its controller are key equipment of high voltage formed aluminium foil production. Compared with the previous analogy integrated circuit chip, the 8-bit microcontroller trigger has been widely used in the industrial field for circuit structure simple, output waveform stability, high reliability and low temperature drift effects [1, 2]. With the advancement of science and technology, foil production urgently requires high efficiency DC power supply with low voltage ripple and high power factor. The operation speed of 8-bit microcontroller triggers is low. Its control accuracy is limited, and cannot do real-time calculation voltage and current and power factor and intelligent control algorithms.

This paper developed a 32-bit digital thyristor trigger control system of thyristor rectifier based on TMS320F2812 with grid frequency tracking, phase sequence adaptive, and has soft start, lack phase and over-current protection function.

Yu Ning, Departments of Automation, Shaoguan University, Shaoguan, 512005, China. Email: nby6210@163.com.
Hou-Lin Liu, Ruyuan Dongyangguang FormedFoil Co. Ltd., Shaoguan, China.
Ye Ye, Dhurakij Pundit University, Thailand, Email: 3736425@qq.com. Corresponding author.

DOI 10.1515/9783110540048-063,

2 Requirements for Trigger Control

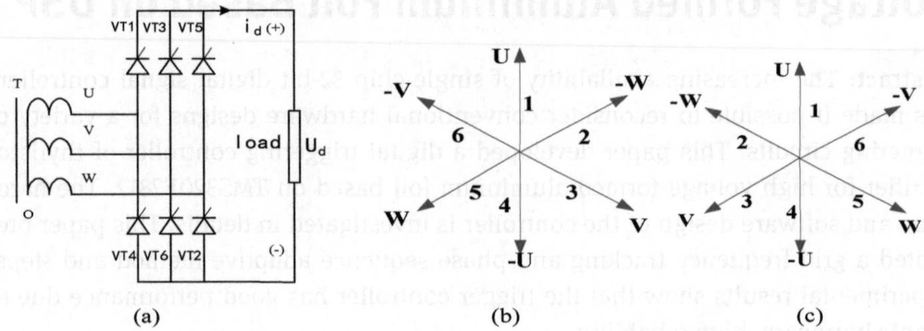

Fig. 1: relation between phase AC power and thyristor Conduction of thyristor rectifier (a) three-phase bridge rectifier (b) positive phase AC power (c) negative phase AC power

As seen from Fig. 1 (a), the three-phase bridge rectifier only conducting two thyristor at same time (e.g. VT1,VT2) to form a current loop, which requires generate dual narrow triggering pulse interval 600 sequentially according to the phase sequence of input AC power. Because it was not clear what the phase sequence of the AC power for free to join, so there may be positive or negative phase sequence shown in Fig. 1 (b), (c) below. Fig. 1 (a) provides conducting order of positive phase sequence, once connected to the negative phase sequence; the thyristor cannot be normal conduction. It is necessary to judge the correct phase sequence and determine the corresponding conduction order automatically. Meanwhile, in order to ensure precise control, there also will be required to adjust the pulse phase according to frequency fluctuation of the grid.

Thus, correct gating control of three-phase bridge rectifier is a nice feature of the adaptive phase sequence, automatic tracking grid frequency and dual narrow pulse. In addition, in order to ensure thyristor reliable and fast conduction, the trigger pulse circuit and the main circuit should be isolated; pulse waveform has a certain width and steepness [3].

3 Hardware Design

As showed Fig. 2, TMS320f2812 is a core of digital trigger control system; its peripheral circuit mainly consists of synchronization circuit, voltage current detection circuit, signal conditioning circuit and pulse power amplifier and isolation output circuit.

In addition, the SCI circuit of DSP2812 through an Ethernet port connected PC, which enables remote network control different from the previous stand-alone 8-bit microcontroller. The system runs according to receive command parameter from the host computer. And it feedback state parameters of rectifier such as phase shift angle, RMS AC voltage and current, power factor, the DC voltage and current data to host computer.

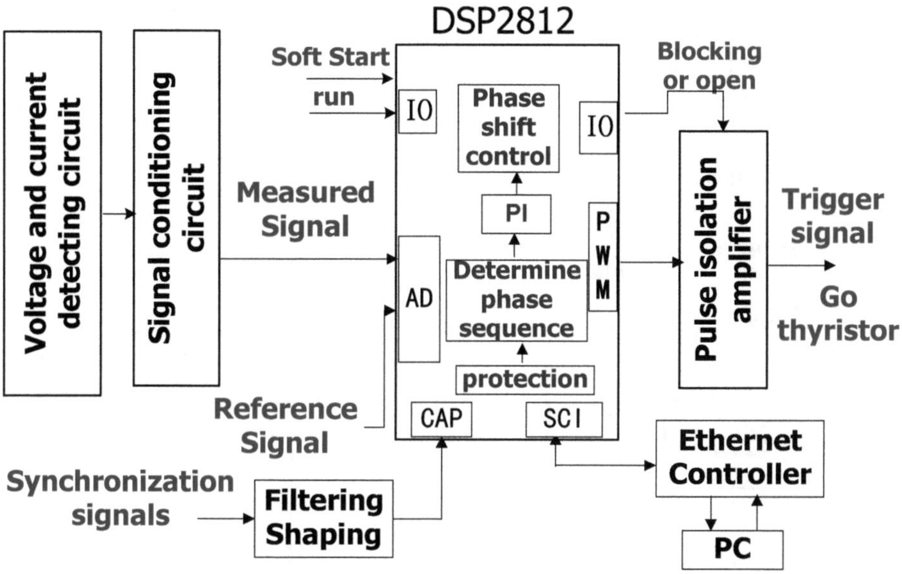

Fig. 2: DSP control system block diagram of phase-shift trigger

3.1 Synchronization Circuit

Synchronization control is a particular problem in thyristor phase shift rectifier [4, 5]. This paper adopted a dedicated synchronized transformer considering voltage zero crossing is not easy to make accurate for the voltage waveform of running rectifier will appear distortion. As saw from Fig.3,three transformer secondary line voltages filtered shaping into a square wave signal SUV, SUW, SVW access CAP1 ~ CAP3. A benefit of the line voltage zero crossing synchronization signal is natural commutation point (zero firing angle). Meanwhile, the three line voltage signal is also used as automatic phase sequence identification.

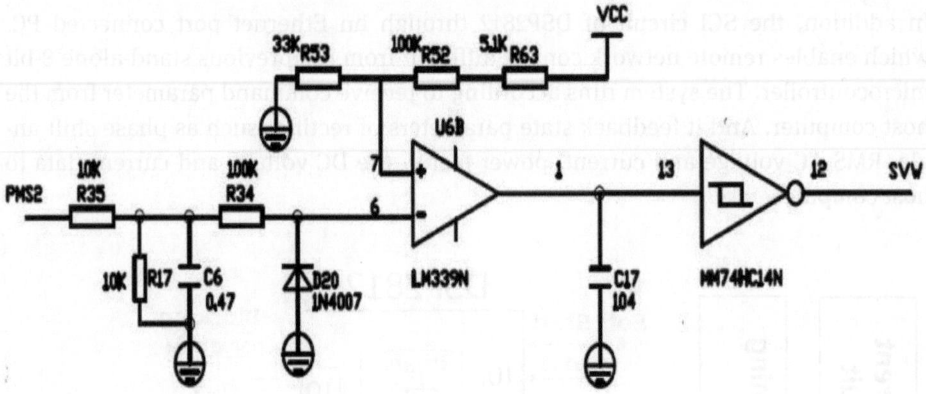

Fig. 3: Synchronization circuit diagram

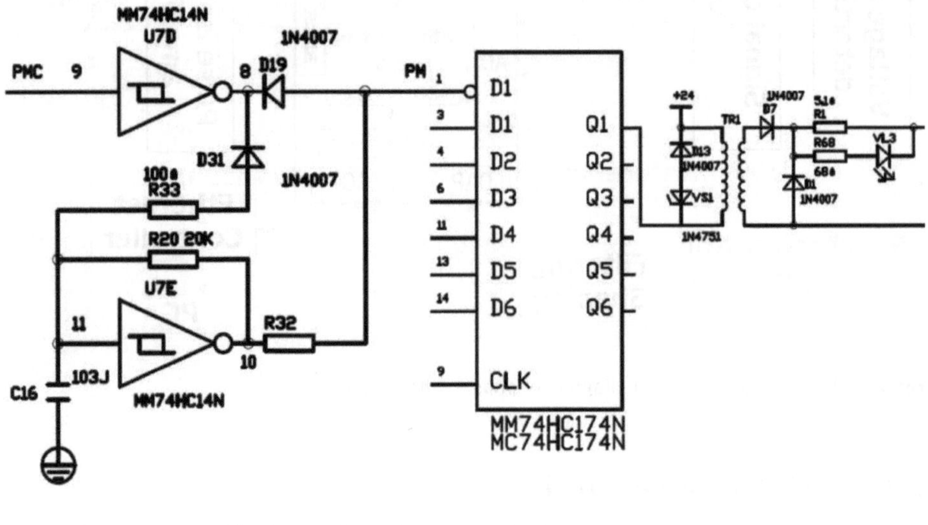

Fig. 4: Pulse power amplification and isolation output circuit

3.2 Pulse Power Amplification and Isolation Output Circuit

As saw from Fig.4, DSP pulse signal generated by the PWM pulse output port PMC latches MC74HC174N. It's a high frequency signal PM by DSP IO ports controlled. PM = 1 allowing pulse output. PM can also block trigger impulse when rectifier failure. On the left of Fig.4 is a high-frequency oscillator circuit, its role is to transform pulse signal into a high frequency pulse train. The frequency of the pulse train is determined by the resistor R and the capacitor C16. High frequency pulse train can reduce the power consumption of the pulse transformer and shrink the volume of the pulse transformer. Right in the Fig.4 are pulse power amplifier and transformer isolation circuit.

4 Software Programming

4.1 Trigger Control Method And Resource Allocation Dsp

A TMS320F2812 digital trigger is actually an analogy saw tooth synchronization trigger control principle. An incremental signal generated by the timer corresponds to the saw tooth signal compared with the control pulses generated phase shift control angle T_α. So trigger control is essentially timer control.

TMS320F2812 has three 32-bit CPU timers and four 16-bit event manager timers [6,7].As shown in Fig.5, capturing port CAP1 generate the CAP interrupt in synchronous square wave signal rise edge, as zero trigger angle and start up CPU timer, 2812 real-time computing grid voltage corresponding period T, 60 degrees corresponding time T_{60^0} and phase shift control angle corresponding time T_α, then start up timer T1, which is set to continuous incremental mode, T will be assigned to the period register, T_α to compare register T1, and generate triggering signal rise edge in pulse out port at time T_α, thus, the current device trigger pulse is formed. Generate triggering signal rises edge of the next device and to current device after time $T_{60^0} + T_\alpha$. It completes the trigger phase shift control according to the provisions of the order of the phase sequence triggering the remaining four thyristors in turn. In the next cycle, according to the real-time calculation of the period and phase control angle re assigned relevant registers, and then repeat the process, it realized the grid frequency tracking.

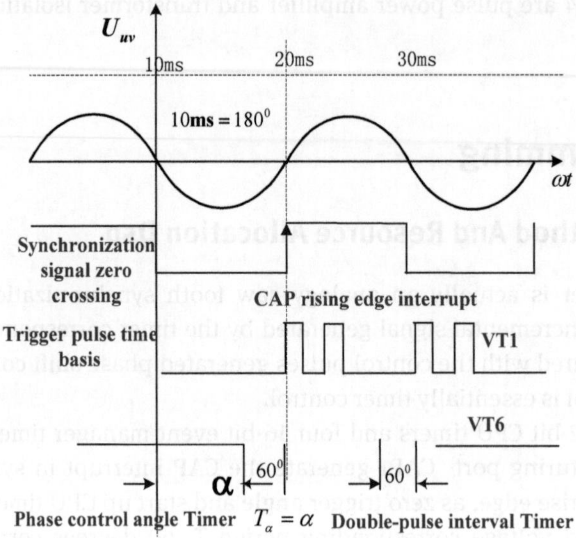

Fig. 5: Trigger control principle

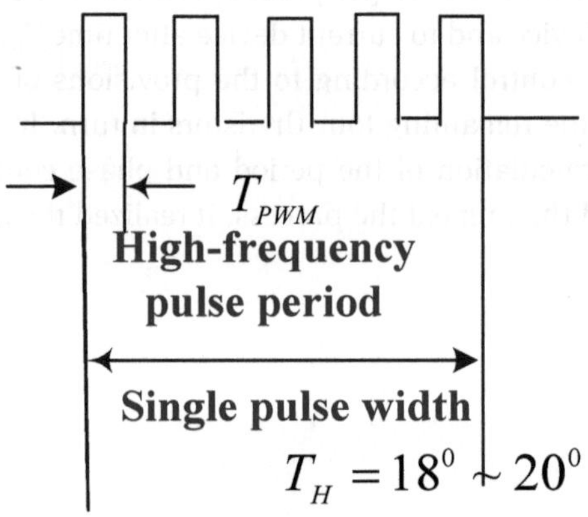

Fig. 6: High frequency single pulse Waveform

4.2 Phase Sequence Adaptive Algorithm

Phase sequences adaptive is automatically identified grid voltage phase sequence and determine trigger pulse order depending on the phase sequence, which can achieve a certain degree of free debugging. Therefore, it contains two questions: phase sequence judgment and phase sequence processing.

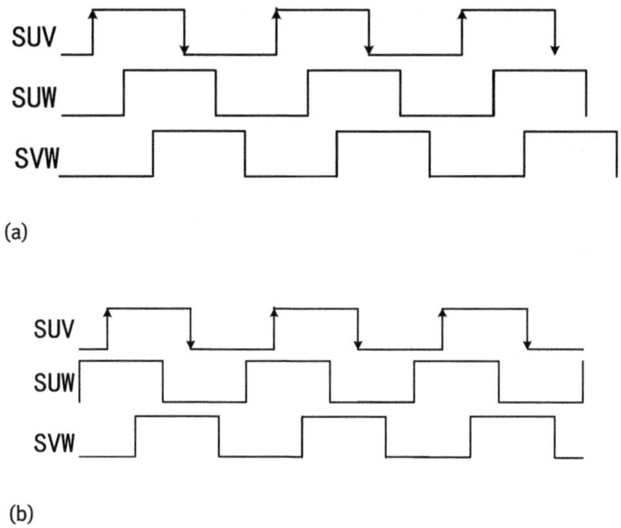

Fig. 7: Phase sequence determination principle (a) Line voltage positive phase sequence square wave (b) Line voltage negative phase sequence square wave

4.2.1 Phase Sequence Judgment

The line voltage signal SUV, SUW, SVW introduced 2812 three capture port, SUV, SUW, SVW three intervals 600, SUV set to capture the rising edge triggered interrupts, then the other two ports set to input , rising edges sequentially read SWV SUW value, if the two values are the same, i.e., with a "0" or "1" is the positive sequence (Figure 7 (a)), if difference is negative phase sequence, ie, a "0" and the other is "1." accordingly, relative to a variable assignment S "0" based on the positive and negative phase sequence and missing, "1", 0 represents the positive phase sequence , "1" represents the negative phase sequence, based on the value of S into the appropriate handler.

4.2.2 Phase Sequence Processing

Figure 1 (b), (c) can be known, the positive sequence dual pulse required to trigger the pulse sequence of (VT6, VT1) - (VT1, VT2) - (VT2, VT3) - (VT3, VT4) - (VT4, VT5) - (VT5, VT6) - (VT6, VT1), followed by cycle; negative double pulse sequence requires a trigger pulse sequence sent to (VT2, VT1) - (VT1, VT6) - (VT6, VT5) - (VT5, VT4) - (VT4, VT3) - (VT3, VT2) - (VT2, VT1), followed by cycle. The above sequence can do a loop to form a trigger pulse sequence has to meet the requirements of adaptive functions.

4.3 The Main Program Flow Chart

TMS320F2812 implement initial inspections procedures include overcorrect protection and phase sequence detection by software. After then, the system realize phase-shift trigger and voltage or current closed loop (PI)control according to external soft-start operation instructions and board given signal, Also issued instructions to block or open trigger signal according to the fault, to ensure the system safe and reliable. The main program block diagram is shown in Fig.8.

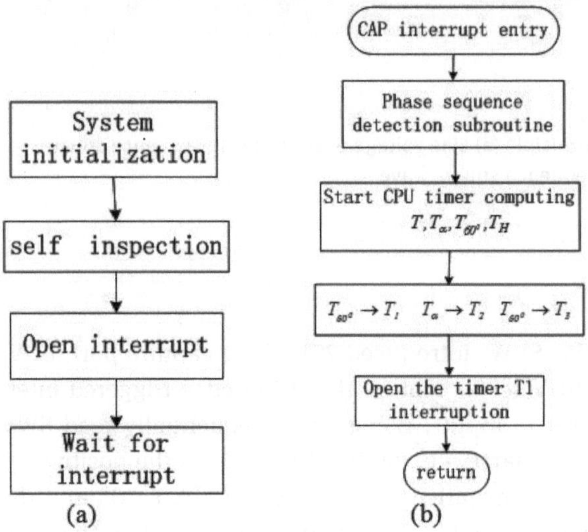

Fig. 8: The flow chart of software (a) main program (b) CAP interrupt

5 Simulation And Experimental Results

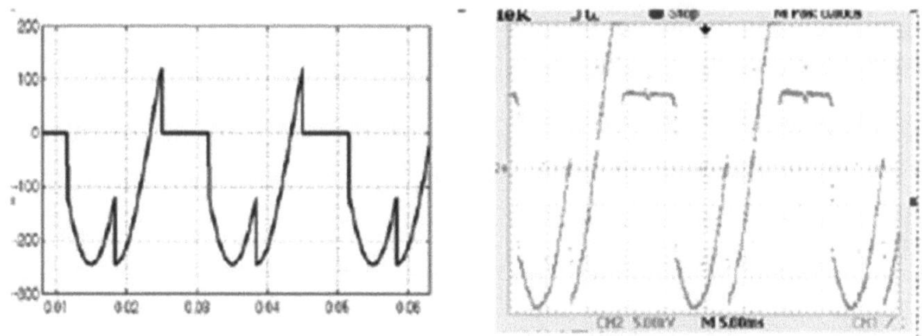

(a) Simulation waveform (b) experimental waveforms

Fig. 9: Rectified voltage waveform (α = 30⁰)

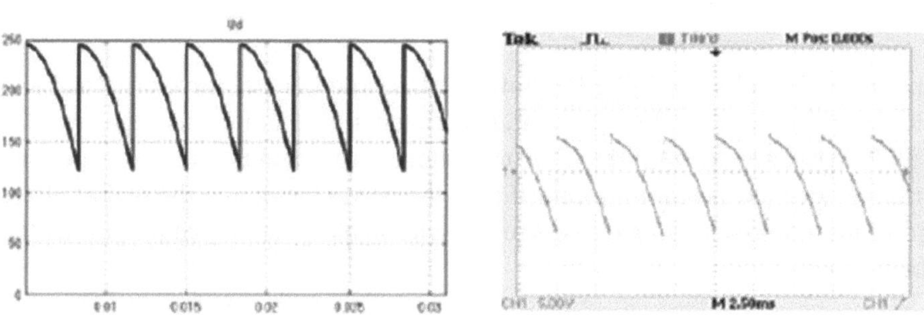

(a) Simulation waveform (b) experimental waveforms

Fig. 10: Thyristor terminal voltage waveform (α = 30⁰)

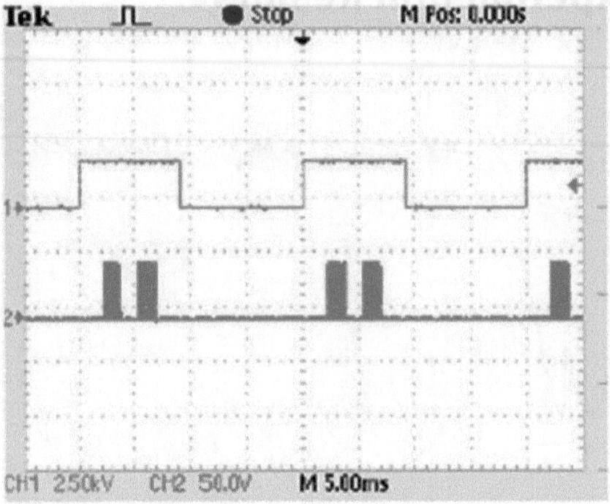

Fig. 11: Synchronization signal and the trigger pulse signal experiment waveform

6 Conclusions

Experimental results (Fig.9-Fig.11) show that the experimental waveforms and simulation waveform exactly. In this paper, experimental evaluation has been completed on DSP digital trigger, with the basic conditions for the industrial field experiment. Further work such as real-time detection of the power factor and the digital filtering algorithms required by direct access to the sync signal from the rectifier transformer is being developed.

Acknowledgement: This work was financially supported by the Shao guan innovation R & D Fund in 2017, and by Guangdong college students' Special fund for cultivating scientific and technological innovation (No. pdjh2017b0454 and No. pdjh2017b0456).

References

[1] F.P. Dawson.and R.Bonert,High Performance Single-Chip Gating Circuit for a Phase-Controlled Bridge, IEEE Transactionson Power Electronics, 41(4), 457–470,1994

[2] CHEN Zhong-xiao, LI Yan-fang, Three-phase bridge type all control of rectifying sequence phase automatic identification, Engineering and Technology (S-CET), 2012 Spring Congress on , 1(4),27–30,2012.

[3] Qiang Gao, An Hu, Na He, Weiming Ma, Liang Zhou, and Xinliang Zheng, A New Thyristor Triggering Scheme for Intermittent Pulse Rectifiers for a Large-Capacity Pulsed Power Supply, IEEE Trans. Plasma Science, 41(5), 1332–1339,2013.

[4] Z.-A.Wang, J.Huang, Power electronics technology (in Chinese), Mechanical Industry Press, Beijing, 2007.
[5] G.Zenginobuz, I. Cadirci, M. Ermis, and C. Barlak, Performance Optimization of Induction Motors During Voltage-Controlled Soft Starting, IEEE Trans. Energy Conversion, 19(2), 278–288, 2004.
[6] TMS320F2810, 2812 Digita signal processors data manual, Texas Instrument, 2002.
[7] TMS320F28x Event Manager (EV) peripheral reference guide, Texas Instrument, 2002.

Zhi-Min Li, Dong-Liang Wu and Men-Mei Li

Tcp Congestion Control Strategy Based on Multi-Level Priority Queue

Abstract: Seeing that the existing network congestion control methods do not consider that the asymmetry of network link and the transmission state of Acknowledgment (ACK) packets have influence on performance and stability of network, the article proposes the network congestion control strategy based on double-level priority queue (DPQCC) and describes the corresponding algorithm. To achieve greater ACK packets forwarding efficiency, the strategy sets up a high priority queue in the routing node for ACK packets that are saved temporarily and have not been retransmitted timely owing to link congestion, then retransmit ACK packets preferentially when the link is idle. Experimental results show that, compared with early congestion avoidance algorithm (RED), when using DPQCC algorithm, the packet loss rate is significantly lower, throughput has improved greatly, the end-to-end delay has reduced on the whole, and only when the network is under the severe congestion state, the end-to-end delay would slightly increase, but delay jitter is smaller, and network stability is enhanced.

Keywords: Network Congestion; The double-level priority queue; DPQCC algorithm; Network performance; ACK packet

1 Introduction

With the development of the Internet today, its success is attributed to the continuous improvement and strengthening of the protocols, in which the TCP/IP protocol plays an important role. TCP/IP has become the de facto standard protocol of the Internet. Whereas, with the growth of the WWW applications and a variety of multimedia information in the Internet, particularly booming the popularity of Internet + business idea, the weakness of the TCP/IP protocol emerges gradually. For example, fast information retrieval in self-organizing P2P network [1] and positioning algorithm for wireless sensor networks [2] should depend on the effective TCP con-

Zhi-Min Li, School of Information Technology, Hunan Mechanical & Electrical Polytechnic, Changsha, 410151, China. Email: 1446235363@qq.com
Dong-Liang Wu, School of Information Technology, Hunan Mechanical & Electrical Polytechnic, Changsha, 410151, China
Men-Mei Li, School of Computer Science and Technology, University of South China, Hengyang, 421001, China. Email: 2487932@qq.com

DOI 10.1515/9783110540048-064,

gestion control strategy.

The major reason of network congestion is the limited network resources and the imbalanced network traffic distribution; Congestion will not disappear with the increase of network capacity. Network's complexity, heterogeneity and dynamic make it difficult to design the congestion control strategy. So far, the congestion problem has not been solved well. Effective congestion control mechanism and queue management keep the network under good working condition. A lot of data packs without congestion control collapse the network. There are many TCP algorithms, such as addition increasing or decreasing multiplication based slow start, congestion avoidance, fast retransmit and fast recovery mechanism coupled with a simple FIFO queue management, which make Internet possible in a healthy run. But, with the increasing network bandwidth of all kinds of traffic on the Internet, the insufficient network bandwidth makes the current packet loss rate in the best effort delivery increase. Researches on the TCP layer's congestion control have very important significance for the network, because in the Internet, most of the work of congestion control is done by the TCP layer. Congestion control of the TCP layer can improve the utilization rate of network resources, which can avoid the collapse of the network.

With the rapid development and popularization of the Internet, all kinds of new network services have been put forward and applied; the degree of network link congestion is deepened increasingly with a sharp rise in network data traffic [3]. It is confirmed through the network measurement technology and network performance analysis technology that the fundamental reason of data packet loss, time delay, delay jitter, and a series of problems is the congestion of network link. At present, network congestion is the main factor for affecting network quality of service (QoS) and network performance [4]. In order to alleviate problems such as packet loss caused by the link congestion, a lot of research work is focused on the improvement on a router; setting up the buffer queue in the router is one of the solutions. The greater length of queue can save more data packets temporarily, reduce the number of packet loss, improve network throughput, at the same time, it will increase the queuing delay of data packets in the queue; On the contrary, the smaller length of queue cannot solve the problem of congestion effectively. Therefore, effective and stable queue congestion control method plays an important role in alleviating network congestion [5].

In this paper, we present an efficient TCP congestion control strategy based on multi-level priority queue that sets a high priority queue specially to store those ACK packets without being forwarded due to the link congestion. As corroborated by our experiments, the special high priority queue for ACK packets is crucial for solving the congestion problem. Furthermore, the multi-level priority queue offers the following key advantages over the TCP congestion control process:

Efficient ACK packets forwarding. The ACK packets are stored in a queue with higher priority, and can be forwarded firstly when the link is idle. So, the sender can know the transmit success at an early time. It can avoid many retransmissions in the network.

Significantly lower packet loss rate. The ACK packets and data packets, which have not been timely forwarded due to link congestion in the router, are stored respectively in two queues. In the idle link, the ACK packets can be forwarded firstly. So, the ACK packets can be stored and forwarded in priority especially under the congesting traffic. The number of the ACK packets the sender receives is the standard of the data transmission success. Therefore, it can lower the packet loss rate significantly.

Throughput has improved greatly. As we know that the success probability of data transmission has been improved greatly, less bandwidth will be consumed in our TCP congestion control strategy. So, the throughput of the network can also be improved greatly.

The rest of the paper is organized as follows. Section 2 discusses related work. Section 3 introduces the basic theory about the multi-level priority queue. We design a TCP congestion control strategy based on the multi-level priority queue in Section 4. Then, Section 5 describes the packet enqueuer and dequeuer algorithm in this multi-level priority queues. Section 6 shows the experimental results, such as the packet loss rate, the transmission delay, and the throughput. Section 7 shows conclusions as well as the future work. To summarize, this paper makes the following contributions:

A Double-level priority queue congestion control strategy. We present a double-level priority queue congestion control strategy to divide the ACK and data packets without being forwarded into two queues with different priorities. The ACK packets can be forwarded firstly when the link is idle.

A packet enqueuer and dequeuer algorithm. We propose a packet enqueuer and dequeuer algorithm according to the basic operations of the priority queue, which can store different kinds of packets. The packets in the higher-priority queue can be forwarded firstly.

Experimental evaluations. We present an experimental evaluation showing that our approach can achieve a better throughput with satisfactorily low loss rate in TCP congestion control as expected.

2 Related Research

Data is transferred in the network. When there are too many packets in a subnet, it will cause network performance decrease, which is called congestion as known. Network decreased performance is represented as the transmission increasing latency and network throughput reduction of the information packet. Fig. 1 shows that when the number of packets from the host port to the communication subnet is within its capacity range, they will all be sent to the destination (except for few packets not being transmitted correctly due to the transmission error). And the network load increases to a certain value, which we call “Knee”. If the network load exceeds its processing capacity, the communication subnet begins losing data packets and causes the deterioration of the situation. When the network load continues increasing to another value, the network throughput volume drops almost to zero. In this case, it cannot almost transfer any data, which is known as “Deadlock”.

The root cause of network congestion is that the network load has gone beyond the network resource capacity and processing capacity. It is also because that the sum of the demand for resources is greater than the sum of the available resources. In theory, if the resource is infinite, the congestion will not happen. But, it cannot happen in reality. The main factors leading to the congestion are the following:

Insufficient storage space. When many ports forward packets to the same output port, each packet needs to queue at the receiving port buffer. If the output does not have enough buffers, or the buffer is not enough, the packet will be discarded, especially for burst data stream.

Insufficient bandwidth capacity. High-speed data stream over a slow link can often produce the congestion. Only when the data transmission rate R is less than the channel capacity, the congestion can be avoided.

Slow Node processors. When the router receives a packet, it puts it into the caching of the CPU firstly, selects the routing path, and then forwards it to the appropriate node. Thus, the speed of the router processor is also an important cause of congestion.

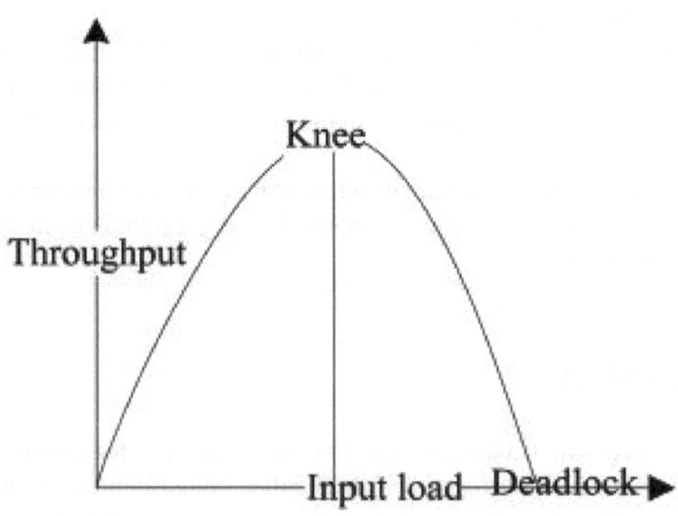

Fig. 1: The relationship between network load and throughput

Packets in network happen essentially suddenly, instead of constantly and continuingly. Separate TCP window mechanism [6, 7] is not able to control the network traffic effectively, or solve the network congestion problem. Effective router buffer queue management mechanism can reduce the queuing delay of packets and improve the transmission efficiency at a certain degree. At present, a lot of research work has been made on the network congestion problem. For example, random early congestion avoidance algorithm (RED) [8] proposed by Van Jacobson and Sally Floyed is among the first to put forward voluntary queue management strategy, which is applied to the packet switching network.

RED is an early drop technology. In the early congestion in the network, the portion of the packet data is discarded to improve the dynamic characteristics of the queue, which can increase the use of the network. Contrast to the lost tail algorithm, RED adds two mechanisms into the queue management:

(1) The packet discard does not occur under the buffer overflow of the queue. But a portion of packet data is discarded previously using probabilistic determination mechanism according to the queue length to prevent congestion from occurring.
(2) The queue length determining the probability of the packet loss is the average queue length, rather than the real-time queue length, which can absorb some of the short burst traffic to some extent.

RED queue management strategy based on the packet arrival or a fixed time interval calculates the average queue length $q_{ave}(t_n)$ using exponential weighted moving average (EWMA):

$$q_{ave}(t_{n+1}) = wq_{ave}(t_n) + (1 - w)q(t_n) \tag{1}$$

Where: w is filter coefficient, $q(t_n)$ is sampling instant queue length. Then, the probability of the packet loss/mark shall be determined according to the size of the average queue. When the average length of the queue is less than the minimum threshold value, indicating that no congestion occurs, no any arrival packets are treated; when the average length of the queue is greater than the maximum threshold, indicating that the network is congested, all incoming packets are discarded.

When the average length of the queue is between the minimum and maximum threshold, the probability P is marked in accordance with the following equation:

$$p = p_{max} \times \frac{q_{ave}(t) - q_{minth}}{q_{maxth} - q_{minth}} \tag{2}$$

Where: p_{max} is the maximum loss probability (also loss probability when the average length of the queue isq_{maxth}),q_{minth}, q_{maxth} are the minimum and maximum threshold value respectively. The process can be represented by Fig. 2.

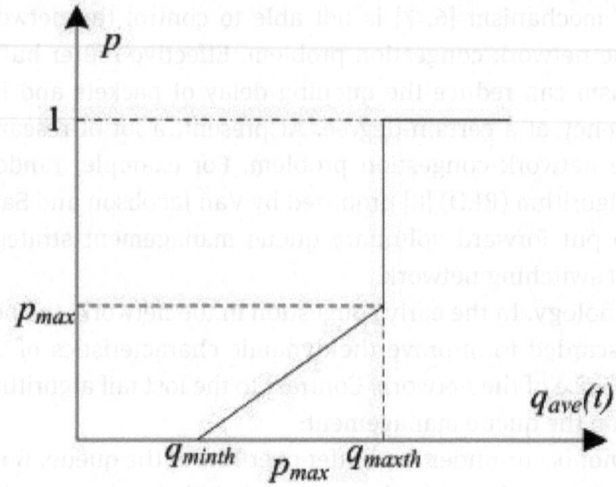

Fig. 2: RED LOSS/ MARK probability calculation Schematic diagram

RED algorithm overcomes the global synchronization issues, but it merely works in TCP flow, it is not able to deal with sudden flow effectively [9]; In order to reduce the influence of the parameters setting on algorithm, literature [10] puts forward adaptive random early discard algorithm (ARED), parameters' setting responds to changes in network traffic, but jitter of packet loss probability is bigger. Literature [11] proposes a new queue management algorithm based on the loss rate (LRED) algorithm, it achieves quick response by measuring the recent packet loss rate, which is regarded as the adaptive adjustment factor of queue length, but due to changes in the packet loss rate, the queue length is still jittery obviously.

BLUE [12] is an adaptive approach adjusting the drop probability, which determines a period with the random method. The mechanism updates the average queue length according to the number of data connections and traffic, when receiving the packet. The probability of being discarded is updated at the end of this period. If there is a greater packet drop, the drop probability increases; if the router cache is relatively free, drop probability is reduced. Contrast to RED, BLUE algorithm can obtain greater link utilization and lower packet loss rate.

Network system is a huge feedback control system [13], the complexity and dynamic of network itself, makes network control more disperse, therefore, the formulation of control strategy highly depends on part information [14, 15]. However, above classic congestion control algorithm does not consider the importance of information control in keeping network system stable. ACK packet's transmission delay or discard because of link congestion makes it impossible for the sender to get network state information timely and adjust the state of the network in time, which leaves the network in an unstable state, and seriously affects the service quality and holistic performance of the network [16, 17]. Therefore, it is important for improving network performance and maintaining network stability to improve the ACK packet forwarding priorities.

3 Multi-Level Priority Queues

In the multi-level priority queues, each queue has different priority. Before dispatching, firstly-arriving packets are classified, and then assigned to the queue with the appropriate priority. The same packets in the same queue are dispatched in accordance with the principle of first come first served. As for the different queues, the service objects are selected according to the priority of their queues.

Firstly, the queue with the highest priority is selected to be scheduled; if the queue with the highest priority is empty, the queue with the next highest priority is selected; when the queue with the second highest priority is empty, the queue with the lower priority is selected, and so on. Since the packets in the queue with higher priority are to be serviced firstly, their delay is relatively small and their throughput is relatively large. The basic scheduling principle is shown in Fig. 3.

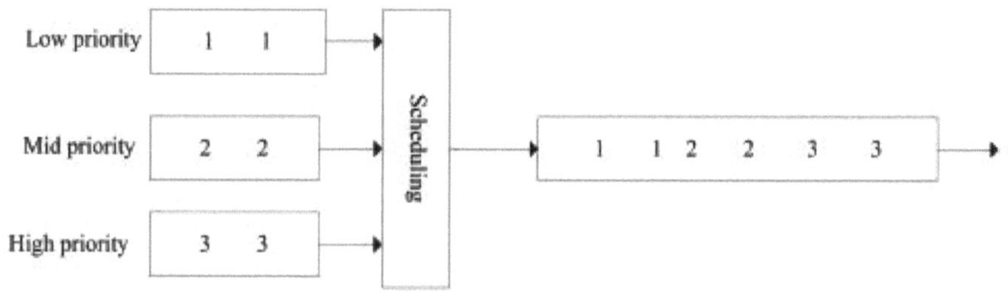

Fig. 3: Scheduling algorithm

Three operating containers are required to implement the priority queue as followings:
(1) Enqueuer puts objects into the container;
(2) Get Object returns the object in the container;
(3) Dequeuer deletes the object from the container.

An abstract class, called Priority Queue, is derived from the class Container. In addition to the inherited function, three functions are also included in the public interface.

```
Class Priority Queue: public virtual Container
{
Public:
virtual void Enqueuer(Object&)=0;
virtual object& Get Object() const=0;
virtual object& Dequeuer()=0;
};
```

Another way to implement priority queue is using the search tree. If multi-level priority queue is implemented with the balanced binary tree, the running time of all three operations is O (log n). Also, the functions provided by the search tree are more than what we need. We can find better ways to achieve the priority queue, among which the binary heap is one. The binary heap is a special binary tree with the ordered heap, which can be implemented with the array data structure. Its realization is based on the calculation of the array subscript, rather than the pointer manipulation. And, due to the use of the array, the overhead contained in the tree node pointer is essentially small. We can declare the concrete class, called Binary Heap, derived from the abstract class Priority Queue as following program code.

```
Class Binary Heap: public Priority Queue
{
Array<Object*>array;
Public:
Binary Heap(unsigned int);
~Binary Heap ();
//......
}
```

Herein, the implement of the Binary Heap class needs the member variable array only, which is declared as a pointer array instance pointing the class Object. This array stores the pointers which point to the objects contained in the binary tree. When the stack has n data items, the pointers, which point to these data item, occupy the n top position of the array.

4 The Double-Level Priority Queue Congestion Control Strategy (DPQCC)

In order to reduce the ACK packet loss rate, a queue will be set specially in the network routing node to save temporarily ACK packets which cannot be retransmitted timely because of link congestion; In order to reduce the ACK's queuing delay, the cache ACK queue priority can be improved, so that when the link is free, the ACK packet in the queue can be retransmitted preferentially, system structure is shown in Fig. 4.

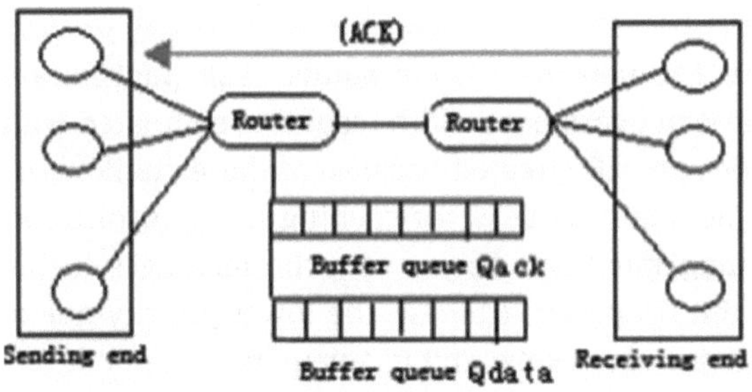

Fig. 4: Link Environment

Set the current queue length as q, in a single queue strategy, when the average queue length (q_avg) is less than the minimum threshold (q_min), the phenomenon of packet loss will not appear, packet queuing delay is the minimum at this time, it is:

$$t\_que = \frac{q}{Vout} \tag{3}$$

When q_avg > q_min, new arrival packet will not only increase the queuing delay, it might be discarded as well.

In the double-level priority queue strategy, set the current queue length of high priority queue Q_ack as q_ack, set the current queue length of low priority queue Q_data as q_data, then, the total length of the system's logical queue is: q_ = q_ack + q_data, Vout is the forwarding rate of router, Vin is the current access rate of router, now, the queuing delay of ACK message is:

$$t\_que' = \frac{q\_ack}{Vout} \tag{4}$$

Becasuse: q_ > q_ack,

Then: t_que > t_que,

When packet in the link continues to increase, the queuing delay of subsequent ACK message increases as well, but the packet loss rate will reduce significantly, only when the queue Q_ack and Q_data is not able to cache packet, the ACK packet is discarded.

5 The Packet Enqueuer and Dequeuer Algorithm

As the feedback information about the network system status, ACK packets are smaller, fewer, and take relatively less network bandwidth resources compared with the application packets. They can be specially treated because of these characteristics. The higher priority queue (Q_ack) can be used for caching the control information packets; and its queue management can ensure that the messages in the queue are timely forwarded using the Drop-Tail strategy. The lower priority queue (Q_data) is used for caching other types of packets; and its queue management can be using RED queue management strategy.

A. Packet Buffer Process

When the router queue receives a packet or the queue is not empty:

(1) Judge whether or not free (idle) the forwarding path is. If it is idle, retransmit the packet directly, turn to (5); otherwise, cache packets received, turn to (2).

(2) Obtain packet type from the packet header information, if the type of packet is "ack", it is proved that the packet is ACK packet, cache it to Q_ack, turn to (3); but if the packet type is not "ack", cache it to Qdata, then turn to (4).

(3) Calculate the current queue's length q, if q < q_ack_max, put packet into queue, turn to (1); otherwise, turn to (4).

(4) Calculate the average length of the current queue and packet loss rate, and deal with the new arrival packets by the current packet loss rate (follow the RED algorithm), turn to (1).

(5) Wait for the arrival of the packet.

B. The Packet Enqueuer Algorithms

When a router receives a packet, it will judge whether or not idle the forwarding path is, if it is idle, retransmit packet directly, otherwise, cache the packet to the queue. Packet enqueuer algorithm is described as follows:

Input: the received packet: Packet* p
Output: the packet enqueue or discarded
Algorithm Enqueue (Packet* p)
1. hrd_ip*p_type<-hdr_ip:: access(p);
2. If*strcmp(p_type,"ack") then

```
if((Q_ack->length+p->size)<q_max) then
    Q_ack->enque(p);
else
    q_avg<-calculate_avg(p);
    if (q_avg<q_min_threshold) then
        Qdata->enque(p);
    else
    if(q_min_threshold<q_avg<q_max_threshold) then
     calculate _p(q_avg,p);
    else
        drop(p);
else
 q_avg<-calculate_avg(p);
 if(q_avg<q_min_threshold) then
    Qdata->enque(p);
 else if(q_min_threshold<q_avg<q_max_threshold)
      then calculate _P(q_avg, p);
      else
     drop(p);
End
```

Among them, p_type is packet type; q max is the maximum length of queue; q_min_threshold is the minimum threshold of queue; q_max_threshold is the biggest threshold; q_avg is the average length of the current queue.

C. The Packet Dequeuer Algorithm

When the forwarding link is idle, remove the forwarding packets stored in the buffer, and then retransmit them. When choosing a forwarding packet, choose the control information packet preferentially. The packet dequeuer's algorithm is described as follows:

```
    Input: the link is ideal: ideal
Output: the dequeue Packet: P
Algorithm Packet* Dequeue()
Packet*p<-null;
p<-Q_ack->deque();
if(p!=null) then
   return p;
else{
   p<-Q_data->deque();
   if(p!=null) then
```

```
8.     return p;
9. }
10. return p;
11. End
```

Among them, the ideal is forwarding link path, p is the packet removed from the queue. It can achieve the fast retransmission of ACK message effectively through the algorithm, then, alleviate the network congestion, and maintain the stability of the network.

6 The Simulation Experiment

The simulation experiment use NS-2.34 [18], simulation experiment link topology is shown in Fig. 5. S in this picture is the sender, D is the receiving terminal. Data streams sent include data stream based on TCP and data stream based on UDP; The link's bandwidth between Routing node R1 and node R is 6 Mbps, transmission delay is 10 ms; There are two receiving terminals Dx and Dy added in the left of this picture, two sending nodes Sx and Sy added in the right of this picture, the purpose of setting the two links is to provide the background flow of the return link.

In order to verify the validity of the double-level queue DPQCC strategy, experiment based on single queue strategy will be done to compare with them.

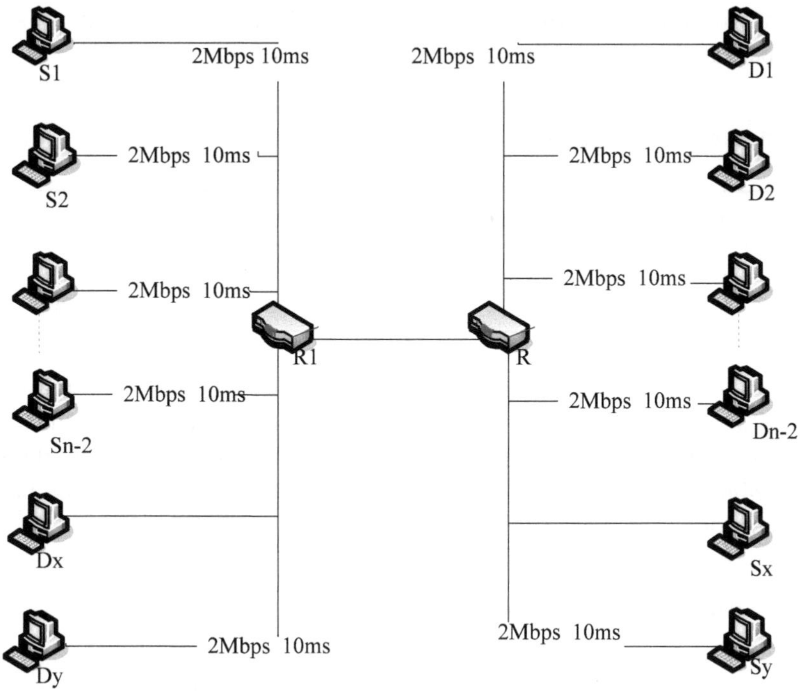

Fig. 5: ink topology

A. The Packet Loss Rate

Packet loss rate refers to the packet number discarded on the router and the percentage of the packets sent by the sender number.

In the same experimental environment, when using RED algorithm and DPQCC strategy, respectively count the number of packets received (en_count), the number of forwarding packets (f_count), the number of packets dropped (d_count), the number of packets received by the receiving terminal (r_count), and the relative packet loss rate (d) (10 groups of experiments, count experimental results every time, choose the average values of 10 groups of data).

The number of packets received by the router nodes (r_count), the number of forwarding packets (f_count), the number of packets dropped (d_count), and the relative packet loss rate (d) is given in Table 1 under the simulation test respectively based on RED algorithm and DPQCC strategy. As shown in Table 1: at the same simulation time, when using RED algorithm, the sender sends 26539 packets totally, 687 packets are discarded because of link congestion, the packet loss rate is 2.59%; However, when using DPQCC strategy, the sender sends 32086 packets totally, 555 packets are discarded because of link congestion, the packet loss rate is 1.73%.

Compared with the RED algorithm, when using DPQCC strategy, the sender sends more 5547 packets, packet loss rate was reduced by 0.86%.

Tab. 1: Statistical results of data packets

name	en_count	r_count	f_count	d_count	d
RED	26539	25852	25852	687	2.59%
DPQCC	32086	31531	31531	555	1.73%

Table 2 gives the situation of TCP packets' sending, receiving and missing under RED and DPQCC congestion control methods. As shown in the table, at the same simulation time, the distinction of packets' sending is large, compared with the RED algorithm, when using DPQCC, 245 TCP packets are discarded totally, the packet loss rate was reduced by 0.48%.

Tab. 2: Statistical results of TCP data packets

name	en_count	r_count	f_count	d_count	d
RED	15173	14897	14897	276	1.82%
DPQCC	18204	17959	17959	245	1.34%

Table 3 gives the situation of ACK packets' sending, receiving and missing in return link under RED and DPQCC methods. As shown in the table, at the same simulation time, the difference of ACK packets' transmission situation is obvious, compared with RED, when using DPQCC, ACK packet loss rate is reduced by 2.41%.

Table 3: Statistical results of ACK data packets

name	en_count	r_count	f_count	d_count	d
RED	4963	4748	4748	215	4.34%
DPQCC	6012	5896	5896	116	1.93%

It shows that the congestion control strategy based on the double-level priority queue can obviously reduce the packet loss rate of the network when the link environment is similar, the main reason can be summed up as the following two points:

(1) Using double queue mechanism expands the length of the buffer queue, absorbs more packets which has not been retransmitted timely, and then reduces the packet loss rate of network.
(2) ACK packets' priority forwarding and low packet loss rate makes ACK message send the current state of the network to the controller (the sender) faster and more completely, and then adjust to the current network status as soon as possible, so the entire network is in a relatively stable state.

B. The Transmission Delay

Under RED and DPQCC methods, the experimental results of TCP packets' end-to-end delay are shown in Fig. 6. Compared with RED, when using DPQCC, network at the initial stage (1 min to 2 min) cannot get to a stable state quickly, the time delay is jitter; After the initial stage, the network status tends to be stable, packet delay shaking is relatively small, packet end-to-end delay is between 30 ms and 42 ms; When the network is in a state of severe congestion (around 5 min, 7 min), delay shaking is bigger. Time delay increases because the buffer queue space increases, which make part of the data packets in the queue state for a long time, thus the queuing delay is increasing. But, on the whole, the TCP packet delay shaking is lesser, which indicates that the network stability is better.

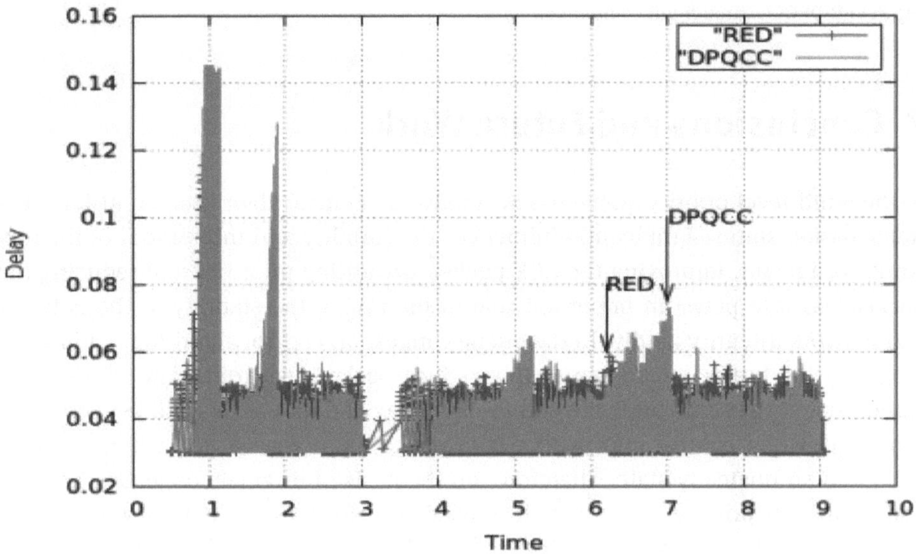

Fig. 6: RED-DPQCC delay comparison

C. The Throughput

According to Table 1, in the same network environment, when using RED, network average throughput is 24331 packets; however, when using DPQCC algorithm, the average throughput is 29482 packets. The difference is 5151 packets, and the latter's throughput is increased by 21.1% than the former's. The contrast of throughput change curve is shown in Fig. 7. It can be seen that: compared with RED, when us-ing DPQCC, network throughput is increased significantly.

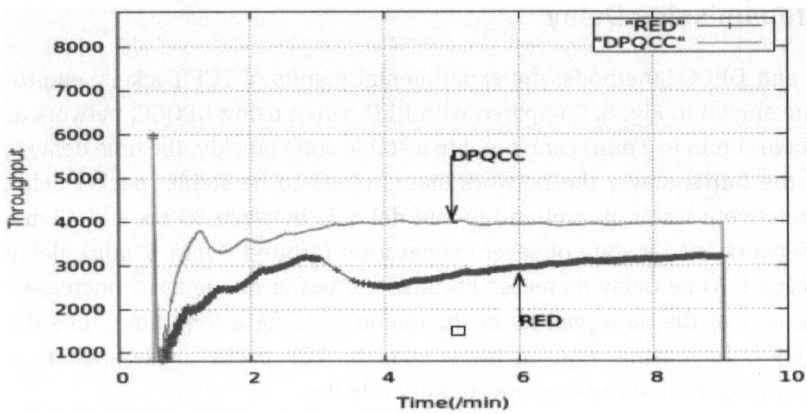

Fig. 7: RED-DPQCC throughput comparison

7 Conclusions and Future Work

In the multi-level priority queue based congestion control algorithm, an ACK packet transmission status significantly influences the stability and throughput of the network. As a result, improving the ACK packet forwarding priorities and reducing the packet loss rate plays an important role in improving the stability of the network and network integrity ability. In the queue congestion control algorithm, adding the length of the buffer queue can simply reduce packet discarding rate to a certain extent, but at the same time, the end-to-end delay of data packets will be increased, especially when control information data packet is increased, the whole network will be in an unsteady state. Therefore, on the basis of original congestion control, adding a high priority queue which is used to deal with the control information wholly, can achieve fast forwarding, reduce the time delay and reduce the packet loss rate.

In the future, we will adjust our TCP congestion control strategy dynamically during the data transition. Delay time factor is added into the forwarded packets.

Those packets with the longer delay time are put in higher-priority queue and can be forwarded earlier. This more fine-grained TCP congestion control strategy is expected to have lower packet loss rate and greater throughput.

Acknowledgement: This work was supported by Project 15C0487 of the Hunan Provincial Education Office Science Research of China.

References

[1] V. Carchiolo, M. Malgeri, G. Mangioni, V. Nicosia, "Fast Information Retrieval in a Self–Organising P2P Network," Journal of Computers (Taiwan), Vol. 2, No. 5, pp. 75–81, 2007.

[2] G.D. Stefano, A. Petricola, "A Distributed AOA Based Localization Algorithm for Wireless Sensor Networks," Journal of Computers (Taiwan), Vol. 3, No. 4, pp. 1–8, 2008.

[3] J.H. Yang, J.P. Wu, A Changing. Internet Measurement Theory and Application, the People's Posts and Telecommunications Press, Beijing, 2009.

[4] S.L. Zhang, "Research of End Network Congestion Control and Quality of Service," Zhejiang University, Hangzhou, 2004, 3.

[5] H.F. Feng, Y.T. Shu, "Study on Network Traffic Prediction Techniques," Wireless Communications, Networking and Mobile Computing Proceedings, Vol. 2, 995–998, 2005.

[6] G.Q. Guo, Y.X. Zhang, H.B. Wang, "IP traffic based on delay planning conditions," Computer Science, Vol. 3, pp. 49–50, 2002.

[7] S.S. Wang, Y. Hao, "Research on TCP Optimization Strategy of Application Delivery Network," IEEE, Vol. 11, pp. 14–17, 2011.

[8] S. Athuraliya, D.E. Lapsley, S. H. Low, "Random Early Marking for Internet Congestion Control," IEEE/ACM Transactions on Networking, Vol. 3, No. 15, pp. 48–53, 2001.

[9] X.F. Chen, "RED Strategy Based Congestion Control Algorithm," Dalian Maritime University, Dalian, 2008.

[10] S. Floyd, R. Gummadi, "Adaptive RED: An Algorithm for Increasing the Robustness of RED's Active Queue Management," AT&T Center for Internet Research at ICSI, 2001, 8.

[11] C.G. Wang, "LRED: A Robust and Responsive AQM Algorithm Using Packet Loss Ratio Measurement," IEEE Transactions on Parallel and Distributed Systems, pp. 29–43, 2007.

[12] W. Feng, D. Kandlur, D. Saha and K.G. Shin, "Blue: A New Class of Achieve Queue Management Algorithms," UM SCE-TR-387–99, 1999.

[13] Q. Yan, X.J. Hun, "Research of Active Queue Management Show Congestion Indication Mark," Computer Engineering and Applications, Vol. 48, No. 5, pp. 85–92, 2012.

[14] V. Misra, W.B. Gong, "Fluid based Analysis of a Network of AQM Routers Supporting TCP Flows with an Application to RED," SIGCOMM'00, Stockholm, Sweden, Vol. 4, pp. 113–224, 2000.

[15] B. Gao, M.S. Tan, Q. Chen, H. Zhao, "Research of ACK Packet Congestion Impact on the Network Performance," Network Security Technology & Application, Vol. 11, 2011.

[16] M. Borri, M. Casoni, M.L. Merani, "An Experimental Study on Congestion Control in Wireless and Wired Networks," ICC 2005, 2005 IEEE International Conference on Communications, Vol. 5, pp. 3570–3575, 2005.

[17] K. Hwangnam, J.C. Hou, "Network Calculus Based Simulation for TCP Congestion Control: Theorems, Implementation and Evaluation, "INFOCOM 2004, Twenty-third AnnualJoint Conference of the IEEE Computer and Communications Societies, 2004.

[18] The Network Simulator – ns2, http://www.isi.edu/nsnam/ns/.

Kui Lu, Run-Run Sun and Ting-Ting Dong

Power Control of Wireless Power Supply System Based on Fuzzy PID Control

Abstract: Wireless power supply technology because of non-contact power supply mode, has a good application prospect in the industry of instruments and meters. But in terms of mine electric locomotive adopted wireless power supply technology, loading condition changes and relative movement on power supply rail of mine electric locomotive will change the load and the coupling parameters of wireless power supply system. These changes of these two parameters will have great influence on the output power of the wireless power supply system. The above problems must be properly controlled so as to ensure the stability of the output power. Because the system contains a large number of non-linear elements, traditional quantitative analysis is too complex; and when the system encounters unexpected situations, the robustness of traditional control technique is poor. Therefore, the fuzzy PID controller is proposed to control the power output. Through validated, this program has stronger robustness and it can be applied in the wireless power supply system very well.

Keywords: fuzzy PID control, wireless power supply, robustness, instruments and meters

1 Introduction

Electrical equipment requires a stable input in order to ensure safe, stable, long-term operation [1].To stabilize the output of system power, it is necessary to take corresponding control measures to ensure the stability of the output power. Output power and the transmission efficiency of wireless power supply system have close relation with the coupling coefficient and its load. However, these two parameters can be changed easily [2]. If the charging device moves, it will result in changes of coupling parameters [3]. This paper adopts the non-isolated DC-DC control technique, because it is suitable for high power wireless power supply system and it

Kui Lu, Anhui University of Science and Technology, Anhui 232001, P.R.China. Email: klu@aust.edu.cn
Run-Run Sun, Anhui University of Science and Technology, Anhui 232001,P.R.China. Email: 1773486674@qq.com
Ting-Ting Dong, nhui University of Science and Technology, Anhui 232001,P.R.China.Email: 1198264511@qq.com

DOI 10.1515/9783110540048-065,

does not have the transformer usually used in isolated control technique that will increase the system loss and volume [4].To simplify the structure of system, the process of energy conversion and pressure transform, it employs BUCK-BOOST circuit because of the bigger range of output voltage and the post circuit protection utilized with the switch transistor. In order to stabilize output power. This paper uses the fuzzy PID controller to adjust the duty factor of BUCK-BOOST circuit [5].Because the system contains a large number of non-linear components, the system presents a very complicated non-linear dynamic behavior [6].The traditional PID controller is difficult to realize the rapid response of the system output, and the control of the overshoot amount is not very sufficient [7].Fuzzy PID control provides an efficient way for obtaining approximate and imprecise characteristics. When the traditional quantitative analysis is too complex or the input signal may only can described qualitatively and probably, fuzzy PID control technique is very effective for dealing with this problem[8].Finally, through proved by MATLAB simulation and physical experiment, this approach can be applied in stabilizing the output power of wireless power supply system very well.

2 Transmission of Wireless Power Supply System

When the primary and secondary coil stay in a resonant state, the system will have a high transmission efficiency. Therefore, the primary and secondary coil need access compensation capacitor circuit[8].This paper will study the transmission characteristics of wireless power supply system based on SS type[8].The topology of the system is shown in figure 1:

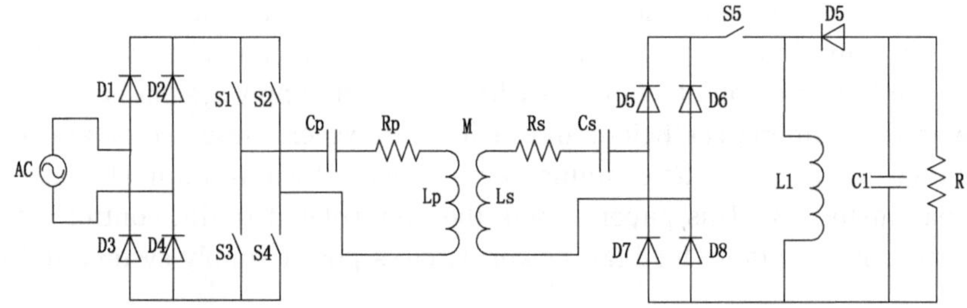

Fig. 1: Circuit of SS type wireless power supply system

AC rectified DC by rectifier bridge(D_1,D_2,D_3,D_4).Then through DC-AC inverter by inverter bridge(S_1,S_2,S_3,S_4),it will get the high frequency AC power. C_p and C_s are the compensation capacitors. R_p and R_s are internal resistance of the coils. L_p and L_s are the inductance of transmitter and receiver coils. M is the mutual inductance between coils. The receiver coil receives the energy. The rectifier bridge (D_5, $D6$, $D7$, $D8$) rectify AC into DC. Then it will control output power by the BUCK-BOOST circuit. Simplify the circuit as shown in figure 2.

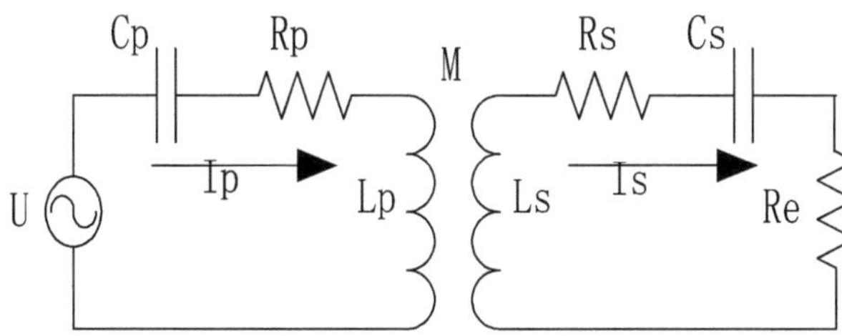

Fig. 2: Equivalent circuit of wireless power supply system

U is considered as the equivalent of high frequency AC. R_e is equivalent to after class resistance.

$$R_e = \frac{8}{K^2 \pi^2} R \tag{1}$$

K is a ratio of the BUCK-BOOST circuit's output and the input voltage. R is the load. The circuit equations of the transmitting coil and the receiving coil are listed by the KVL law:

$$\dot{U} = \left[R_P + j\left(\omega L_P - \frac{1}{\omega C_P} \right) \right] \dot{I}_P - j\omega M \dot{I}_S \tag{2}$$

$$0 = \left[R_S + R_e + j\left(\omega L_S - \frac{1}{\omega C_S} \right) \right] \dot{I}_S - j\omega M \dot{I}_P \tag{3}$$

$j\omega M\dot{I}_p$ is the induced electromotive force which transmitting coil reflect on the receiving coils. $-j\omega M\dot{I}_s$ is the induced electromotive force which the receiving coils reflect on transmitting coil. The reflected impedance Z_r of the receiving coil against the transmitting coil is:

$$Z_r = \frac{-j\omega M\dot{I}_S}{\dot{I}_P} = \frac{-j\omega M \dfrac{j\omega M\dot{I}_P}{Z_S}}{\dot{I}_P} = \frac{\omega^2 M^2}{Z_S} \tag{4}$$

Z_s is the total impedance of the receiving coil. When the receiving coil work in resonance:

$$\omega L_S = \frac{1}{\omega C_S}, Z_S = R_S + R_e \tag{5}$$

Through the above calculation, we can deduce the equation of the output power P_o:

$$P_o = \dot{I}_S^2 R_e = \frac{(\omega M)^2 \dot{I}_P^2}{(R_e + R_S)^2} R_e = \frac{(\omega M)^2 \dfrac{\dot{U}^2}{(R_P + Z_r)^2}}{(R_e + R_S)^2} R_e = \frac{(\omega M)^2 \dot{U}^2 R_e}{(R_P(R_S + R_e) + (\omega M)^2)^2} \tag{6}$$

Keep the input voltage constant. By the above equation derived, the P_o and transmission efficiency η will change with the M and R_e. Using MATLAB to draw when the R_e keep constant, the P_o and η change with the M, as shown in figure 3. Using MATLAB to draw when M is keep constant, the P_o and η change with the R_e, as shown in figure 4.

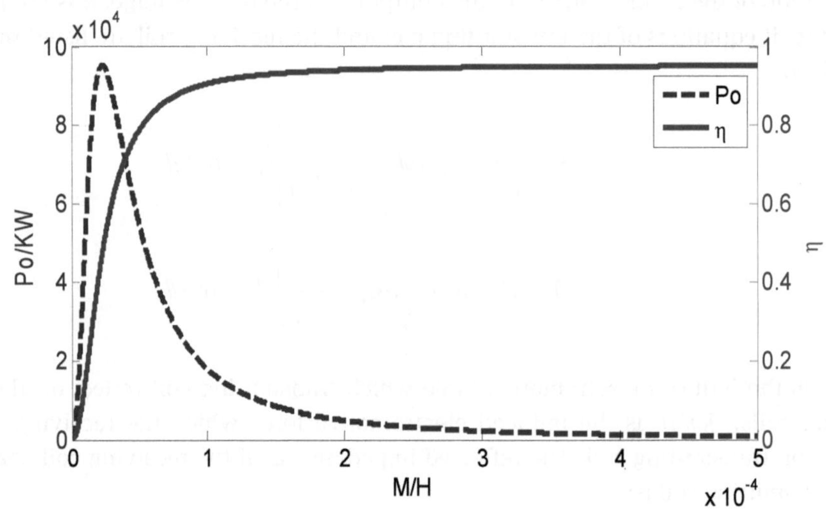

Fig. 3: Relation curve between P_o, η and M

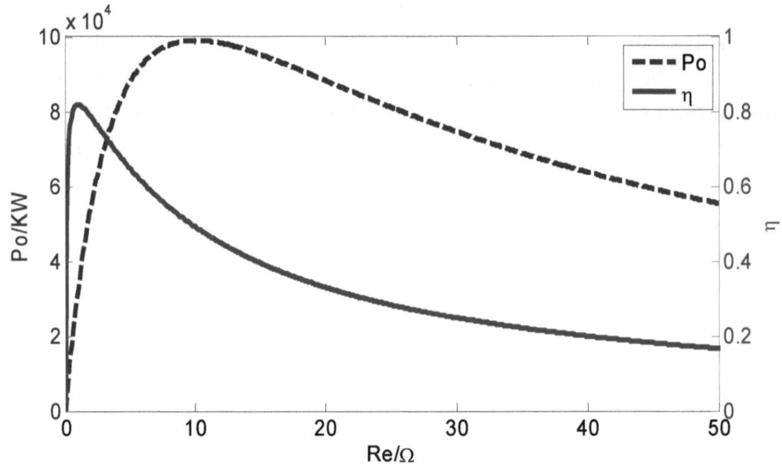

Fig. 4: Relation curve between P_o, η and R_e

From the above charts show, the output must be stabilized by the closed-loop control.

3 Impedance Simulation Results

Fuzzy PID controller compared with the traditional PID controller has less need of information, small overshoot, fast response, less error, etc., and it makes the controlled object has a good dynamic and static performance. Figure 5 for the schematic diagram:

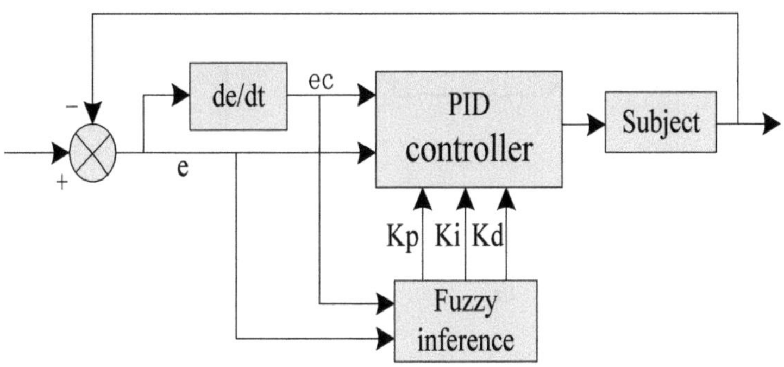

Fig. 5: Fuzzy PID control module

This paper adopts fuzzy PID control strategies. If the output current is constant, the output power will ensure constant because of the purely resistive load. The design idea is to compare the real-time current collected in this system with the set value, and take the differences as input samples. Controller output is the duty ratio *d* of BUCK-BOOST circuit. The constant current control of the system is realized by changing the duty ratio *d* of switch transistor.

In the MATLAB Simulink environment, a simulation model based on fuzzy PID controller and traditional PID controller is built respectively. Simulation parameters are set by theoretical calculation. The system requires 6A output current.

Simulation steps are: set the simulation time to 0.1s, and add controller in 0.01s.When time is 0.04s; the M is set from 30μH down to 20μH. When time is 0.07s, the R is set from 10Ω up to 30Ω. Simulation results show in figure 6 and 7.

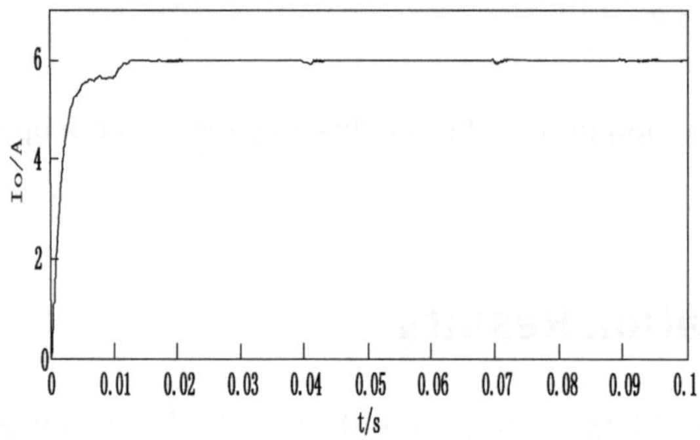

(a) Output current I_o

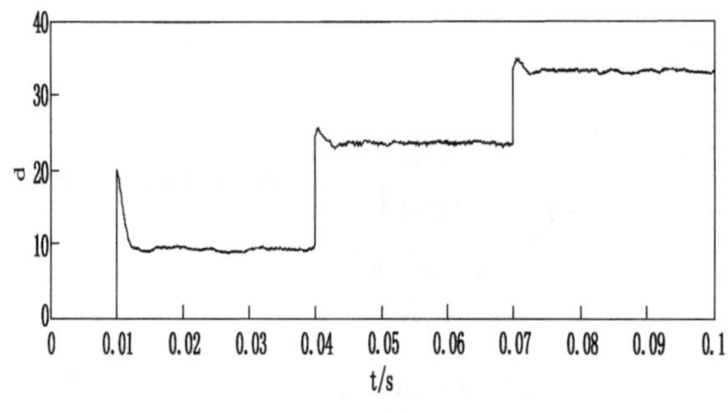

(b) Duty ratio *d*

Fig. 6: Waveforms of fuzzy PID control method

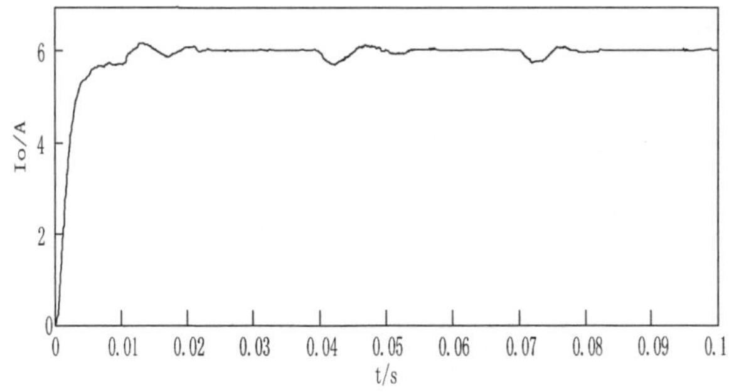

(a) Output current I_o

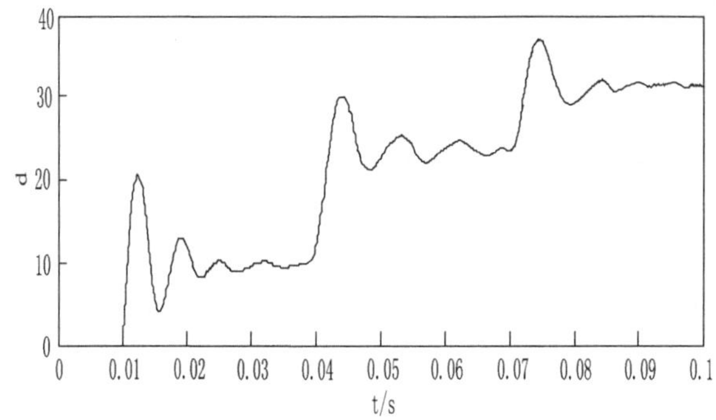

(b) Duty ratio d

Fig. 7: Waveforms of PID control method

The simulation waveforms show that after adding the controller, we can control the duty ratios for stabilizing output of the system. But by comparison, the fuzzy PID controller simulation waveforms significantly better than traditional PID controller simulation waveforms. The systems which use fuzzy PID control system have faster response, smaller overshoot and more stable output.

4 Experimental Results

According to the theoretical calculation, a simple physical model is built to verify that the output current remains constant when the system load changes and it proves the feasibility of using fuzzy PID control. Experiment distance between the coils is 10cm, as shown in figure 8.

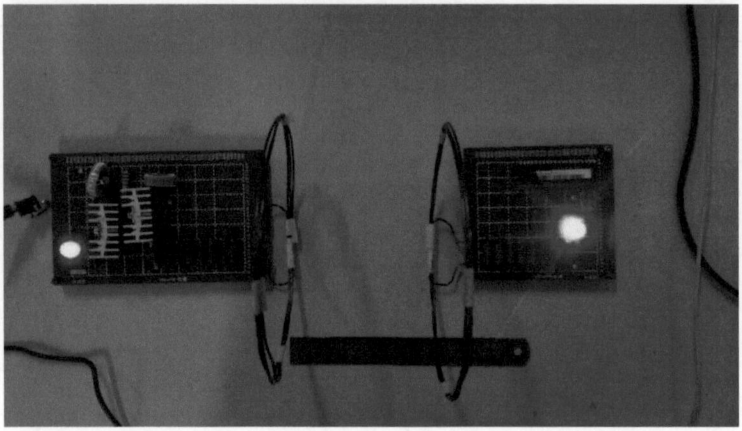

Fig. 8: Wireless power supply device based on fuzzy PID control

When the load is 10Ω, the output average voltage is 4.5V, as shown in figure 9. When the load is 30Ω, the output average voltage is 13.5V, as shown in figure 10.

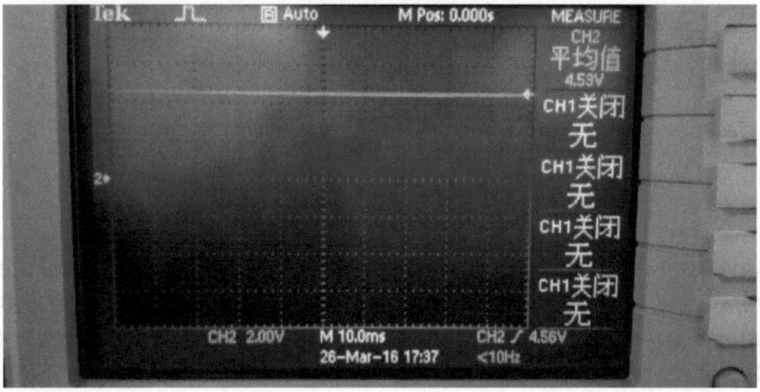

Fig. 9: Output voltage Uo(R=10Ω)

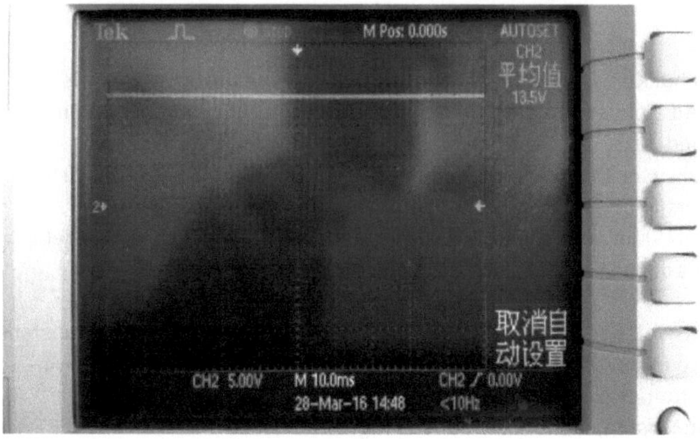

Fig. 10: Output voltage Uo(R=10Ω)

Measured experiments find that when the load change, the current remain at around 0.45A, only minimal change in the test is in the range of allowable error. The experimental results show that the fuzzy PID control strategy can stabilize the output power of the wireless power supply system.

5 Conclusions

Firstly, this article analyzes the transmission characteristics of the wireless power supply system. Through the calculative equations and the charts drawing by MATLAB, the output power P_o and the transmission efficiency η will change with the mutual inductance M and the load R. This conclusion proves that the output must be stabilized by the closed-loop control and the controller is necessary. Because the fuzzy PID controller utilized with the accurate mathematical model can get better results than the traditional control technique without relying on the system, the fuzzy PID controller which is more suitable for the wireless power supply system is adopted in this paper. Then by the illustrations of schematic diagram's design idea and the waveforms drawn based on the fuzzy PID controller and traditional PID controller using a simulation model in the environment of MATLAB, the results show that: the simulation model which based on fuzzy PID controller has more excellent performance and better than the traditional PID controller. Finally, this paper further verifies the feasibility of the scheme under the physical model. In this paper, the research of wireless power supply system power control has a certain reference value.

References

[1] SangCheol Moon, Bong-Chul Kim, Shin-Young Cho, et al, Analysis and design of a wireless power transfer system with an intermediate coil for high efficiency[C] ECCE Asia Down under, [S.l.]: IEEE, 1034–1040, and 2013.

[2] Mao Saijun, Research on key technologies for contactless inductive electric energy transfer system [D], Nanjing University of Aeronautics and Astronautics, Mao Saijun, 2006.

[3] Wu Ying, Yan Luguang, Xun Shangang, Stability analysis of the new contactless power delivery system[J], Proceedings of The Chinese Society for Electrical Engineering, 24(5),63–66, 2004.

[4] [Beh T C, Imura T, Kato M, et al, Basic study of improving efficiency of wireless power transfer via magnetic resonance coupling based on impedance matching[C], IEEE International Symposium on Industrial electronics, 2011–2016, 2010.

[5] Sun Yue, Wang Zhihui, Su Yugang,et al, Power transfer regulation mode for current fed CPT system[J], Journal of Chongqing University, 32(12), 1386–1391, 2009.

[6] Dai Xin, Huang Xiyue, Sun Yue, Study on discrete time mapping modeling and stability analysis for piecewise autonomous oscillation systems [J], Acta Automatica Sinica, 33(1), 72–77, 2007.

[7] Kiam Heong Ang Chong, G.Yun Li, PID control system analysis, design, and technology [J], Control Systems Technology, 13(4), 559–576, 2005.

[8] Petorv, M. Ganchev, I.Taneva, A Fuzzy PID control of nonlinear plants[J], Intelligent Systems, 26(1), 30–35,2002.

Jie Shen*, Qi Zhao, Xi-Chun Bao, Hai-Ming Zhou and Jun Li

Load Scheduling Based on Game Theory with Coupled Constraints

Abstract: In this paper, we study load scheduling problem with coupling constraint, and at the same time considering the game relation between the user and time coupling constraints, and build into accidentally and constraints of game problem. The algorithm is divided into two parts: the first dual decomposition, the original problem and then to the game after decoupling problem, Nash equilibrium is obtained by optimal reaction method. The numerical simulation results show that the proposed algorithm can effectively solve the load scheduling problem under the dual and constrained conditions, and through the load scheduling, we can smooth the fluctuation to ensure the stable operation of the energy internet.

Keywords: load schedule; Game theory; dual decomposition

1 Introduction

In an intelligent distribution network, users are connected to the power station's decompression transformer by wiring wires. In addition, the user through the smart meter is connected to the common communication facilities (such as neighbourhood network), and makes the power suppliers and users, and the users two-way communication is possible [1].

This ability enables the supply side and demand side to interact with each other through the exchange of electricity prices and requirements information, and the users can interact with the users through the requirements adjustment. These interactions can be easily done in an automated way through smart meters. When users are given enough incentive, they are willing to co-ordinate their use of electricity to achieve the trade-off between satisfaction and electricity. As shown in figure 1.

***Corresponding author: Jie Shen,** China Electrical Power Research Institute, Beijing P. R. China 100092. Email: shenjie74@163.com
Qi Zhao, Xi-Chun Bao, Hai-Ming Zhou and Jun Li, China Electrical Power Research Institute, Beijing P R. China 100092. Email: shenjie74@163.com

DOI 10.1515/9783110540048-066,

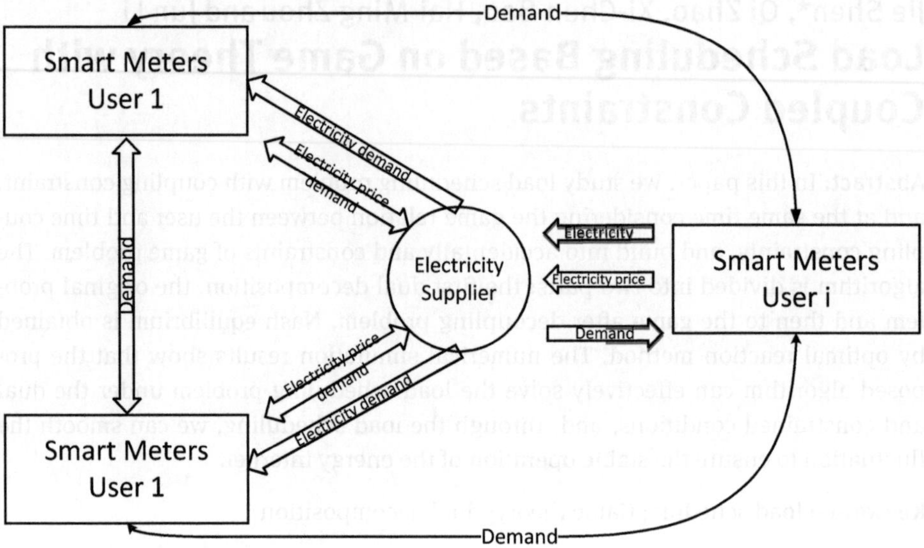

Fig. 1: smart distribution power network [1]

On load scheduling problem, where each user has a coupling constraints on time scales, which requires the cumulative power consumption is not less than the given threshold to complete every task (for example, electric cars need sufficient enough power to ensure the travel of second day),Thus, the problem has time coupling constraints. A comprehensive focus on the game relationship between users is coupled with time, and the load scheduling problem is built into the issue which is with coupled constraints. The original problem was decoupled by dual decomposition, and the Nash equilibrium was obtained by using the optimal response method. The optimal response method is calculated through the gradient projection, which does not reveal the private information of the supplier. In addition, the algorithm is extended to the online version to reduce the impact of the forecast error. The effectiveness of the theoretical analysis and the efficiency of the algorithm are proved by a large number of simulations. Numerical results show that the algorithm can effectively transfer peak load and decrease peak ratio, but also benefit the user. The scalability of the algorithm and its influence on its performance were also investigated.

2 System Model [2] and Problem Build

2.1 System Model

Consider a smart distribution grid made up of a power supplier and user sets N. The user's power requirements can be derived from historical data, which are main concern from user controllable household electrical appliances, such as electric cars, washing machines, dryers and so on(because uncontrolled home appliance, which unable to carry out electricity dispatching[2]).The next "electricity device" will refer specifically to household applicants. For this kind of electrical equipment, users only care about whether their tasks in a given time, which requires electrical equipment before the deadline for the accumulation of power consumption is not less than a certain threshold assumes that each user's smart meters is embedded with the load scheduler[3][4][5]. Its role is to control the various touch type with electrical switch and work cycle is divided into 2 sets (denoted as τ) of a day.

For the sake of statement, let's assume that each user operates only one device.

For each user $i \in N$, the load scheduling vector is defined as $x_i = \left[x_i^1, \ldots, x_i^t, \ldots x_i^T \right]$, where the scalar x_i^t represents the amount of power the user i is using in the time slot t, and the following is the bound:

$$\underline{x_i} \leqslant x_i^t \leqslant \overline{x_i}, \quad \forall t \in T \tag{1}$$

Where, $\underline{x_i}$, $\overline{x_i}$ is the minimum maximum electricity level for the user i . On the other hand, it also requires electricity equipment before the deadline for the accumulation of power consumption is not less than the given threshold to complete daily tasks, such as electric cars need sufficient enough power to ensure that the travel of second day, or need a dishwasher in the noon used tableware clean before dinner. Thus,

$$\sum_{t \in \tau} x_i^t \geqslant Y_i \tag{2}$$

Where, Y_i means that user i is required to complete the amount of electricity required for a given task, which is the threshold for the amount of electricity that the device accumulates. For example, electric cars need 16 KWH of electricity to make sure the next 40 miles a day.

2.2 Electrovalence Function

Use $C^t(\cdot)$ to indicate the cost function for power supply in time slot t; it's an increasing strictly convex function. Usually a secondary cost function such as

$C^t(l) + = a b l l^2$, where $a^t \quad > 0$, $b^t \quad > 0$ is a predefined parameter, which is different in different time slots. Consider the real-time tariff model $p^t(l^t)$; the price of electricity depends on the different times of the day and the total amount of electricity used at the time. Based on the price of electricity, the price of the user i in time slot t am [4]:

$$b^t = p^t(l^t) \cdot x_i^t$$

On the other hand, consider the profits of the suppliers, we can reasonably assume that

$$\sum_{i \in N} b^t = \kappa C^t(l^t)$$

Where, $k \geq 1$ is the benefit of the supplier Above all, the price of regulation and regulation is:

$$p^t(l^t) = \frac{\kappa C^t(l^t)}{l^t} = \kappa(a^t l^2 + b^t l)$$

Consider the real-time pricing model $p^t(l^t)$, where electricity prices depend on different times of the day and total electricity consumption at the time. Or, the price of electricity $p^t(l^t)$ is a time-varying function of total consumption:

$$p^t(l^t) = \kappa(a^t l^2 + b^t l) \tag{3}$$

2.3 Problem Structuring

The benefit of user i in time gap t am defined as:

$$W_i^t(x_i^t; p^t) = U_i^t(x_i^t) - p^t x_i^t$$

Where, the utility function $U_i^t(\cdot)$ indicates that the user i is satisfied with the use of the time slot$t$. Usually consider the quadratic utility function: such as,

$$U_i^t(x) = \begin{cases} \omega_i^t x - (\alpha / 2) x^2 & 0 \leq x < \omega_i^t / \alpha \\ \omega_i^t / (2\alpha) & x \geq \omega_i^t / \alpha \end{cases}$$

Where $\omega_i^t, \alpha > 0$ predefined parameters Different users are have different values in different times: and as [5]

$$U_i^t(x) = \begin{cases} -(x - y_i^t)^2 & 0 \le x < y_i^t \\ 0 & x \ge y_i^t \end{cases}$$

y_i^t Is a predefined parameter that represents the user's target power for the time gap t. User i has a day of income:

$$W_i(x_i; p) = \sum_{t \in \tau} W_i^t(x_i^t; p^t)$$

Where p is the price vector, so the load scheduling problem for each user i can be constructed:

$$\begin{aligned} \max_{x_i} \quad & W_i(x_i; p) \\ s.t. \quad & (1)(2) \end{aligned} \tag{4}$$

The problem of optimal optimization (4) is the problem of the bolt-on problem of coupling constraint, which is difficult to solve directly because of the time auxiliary constraint (2).Therefore, the game problem of unconstrained decoupling is first considered in section Then we introduce the dual decomposition to the original problem in section 3, so that the rest of the problem can be applied directly to the solution obtained in section 2.

2.4 The Game Of Coupling Constraint Is Not Considered

Because we don't consider time coupling constraints (3-23), maximizing the benefit of a user's day is equivalent to maximizing the benefits of each of the time gaps. So we can get rid of the sign t, and then the original problem (3-25) melt into, for each user i

$$\max_{x_i} \quad U_i(x_i) - p(l) \cdot x_i$$

$$s.t \ (1)$$

Because the price depends on how all user scheduling their electricity use, naturally into the game relation between the user and the user, needs, using relevant knowledge of game theory to solve.

2.4.1 Load Scheduling Game

The assumption is that all user actions are rational and individualistic, that is, each user wants to maximize his or her own gain. Therefore, the user's load scheduling problem can be modelled as a load scheduling game.

The concrete steps of the optimal response algorithm are summarized below:

Initialization: each user selects the policy as an initial value from his own policy set.

Optimal response: each user selects the best strategy based on other users' policies to maximize their own gain:

$$x_i^* = \arg \max_{x_i \in X_i} W_i(x_i, x_{-i}) \tag{5}$$

Note that if any user updates their own policies, new policies need to be told to other users;

(1)Repeat steps until all users stop updating their policies;

2.4.2 Gradient Projection Algorithm

The problem (5) can be solved directly by making the first partial derivative of the return function equal to zero, namely

$$\frac{\partial W_i(x_i, x_{-i})}{\partial x_i} = U_i'(x_i) - \kappa\left[a(l + x_i) + b\right] = 0$$

However, this approach might not work in practice, because each user needs to know the benefits of the power supplier and the cost of the power supply to solve the problem[8].And the information is private, and no power supplier is willing to divulge the information, the user may not have enough information to solve the problem. In view of this, the gradient projection method is designed to solve the problem (5) without disclosing the private information of the supplier. The detailed steps of the gradient projection algorithm are summarized as follows [8]:

(1) User i plus and all the user's power requirements, namely $l^k = x_i^k + \sum_{j \neq i \in N} x_j$ and inform the power supplier;

(2) After receiving all the user's total power demand, the power supplier calculates the price of the electricity p^k and feeds it to the user i;

(3) After receiving the electricity price information, user i first estimates the trend of price changes

$$p' = \begin{cases} \frac{p^k - p^{k-1}}{l^k - l^{k-1}} & k > 1 \\ 0 & k = 1 \end{cases}$$

The next iteration is then updated according to the gradient projection method, namely

$$x_i^{k+1} = \left\{ x_i^k + \gamma \left[U_i^t (x_i^k) - p^k - p' x_i^k \right] \right\}^{x_i}$$

Where$\gamma > 0$ is the step length, p is the projection of x onto the domain, k is the number of iterations.

This algorithm effectively protects the private information of the supplier, but the additional computation time and memory overhead are introduced because of the iterative steps [9].That is, each iteration requires time to compute the power requirements and price information, and memory is needed to store this information. However, because the method can quickly converge to the optimal solution, it can satisfy the real-time requirement of the smart grid.

The interaction between the supplier and the user, the user and the user is shown. Smart meters and two-way communication makes the supply side and demand side through the exchange of electricity and demand information for interaction, regulating interaction between the user and the user by demand [11][12].Through the gradient projection method, each user only needs a small amount of power to solve the problem, without the need for the private information of the supplier.

The interaction between the supplier and the user, the user and the user is shown in figure 3-18.Smart meters and two-way communication makes the supply side and demand side through the exchange of electricity and demand information for interaction, regulating interaction between the user and the user by demand. Through the gradient projection method, each user only needs a small amount of the price of electricity to solve the problem (5), without the private information of the supplier.

2.5 Game With Coupling Constraints

Considering the time coupling constraint (2), the original problem (4) is a game problem with complementary constraints. The original problem solver can be solved by dual decomposition, and the decoupling problem can be solved by the optimal response algorithm and the gradient projection algorithm. Due to the strong duality, the original problem is equivalent to the dual problem. Then we construct and solve the dual problem.

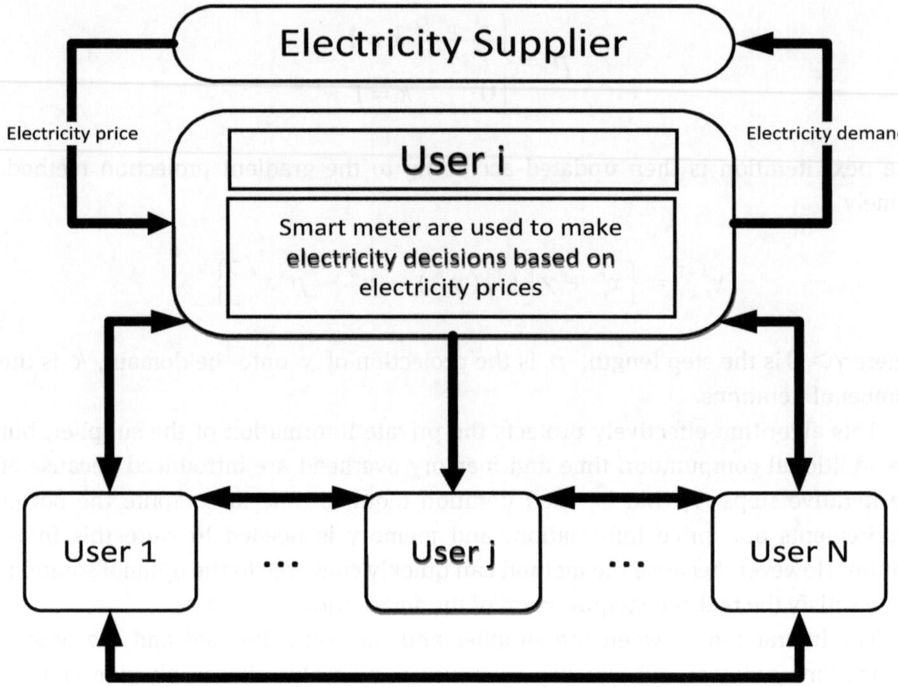

Fig. 1: The interaction between the supplier and the user, the user and the user

2.5.1 Two Level Iterative Algorithm

The dual problem can be solved by the gradient projection iteration and the Lagrange multiplier updating rule:

$$\lambda_i^{m+1} = \left[\lambda_i^m - \theta \left(\sum_{t \in T} x_i^t (\lambda^m) - Y_i \right) \right]^+$$

The two level iteration algorithm steps are as follows:

Each user by any cut beginning of Lagrange multiplier, sub problems by optimum reaction algorithm to get the Nash equilibrium of each game, as the power of the initial decision, namely

$$x_i(\lambda^1), \forall i \in N$$

(1) Each user updates his Lagrangian λ_i^{m+1} to the next iteration

(2) Based on the new Lagrange multiplier λ_i^{m+1}, the problem of the sub problem is given by the optimal response algorithm to the Nash equilibrium of each game, namely $x_i(\lambda^{m+1}), \forall i \in N$

(3) Note the best response to the optimal response algorithm, and the user selects the best policy:

$$x_i^{t*} = \arg \max_{x_i^t \in X_i} [W_i^t(x_i^t; p^t) + \lambda_i^{m+1} x_i^t] \quad \forall t \in \tau$$

The Lagrange multiplier is like a coordinated signal that takes the time to bind the bow into the game.

(4) Repeat steps (2)–(3) until each user no longer updates his Lagrange multiplier;

2.6 The Simulation Analysis

Consider a simple simulation of a smart distribution network consisting of a power supplier and three users. The cycle of the day is divided into T = 24 slots.

Figure 3 shows the algorithm performance when coupling constraints are not considered.

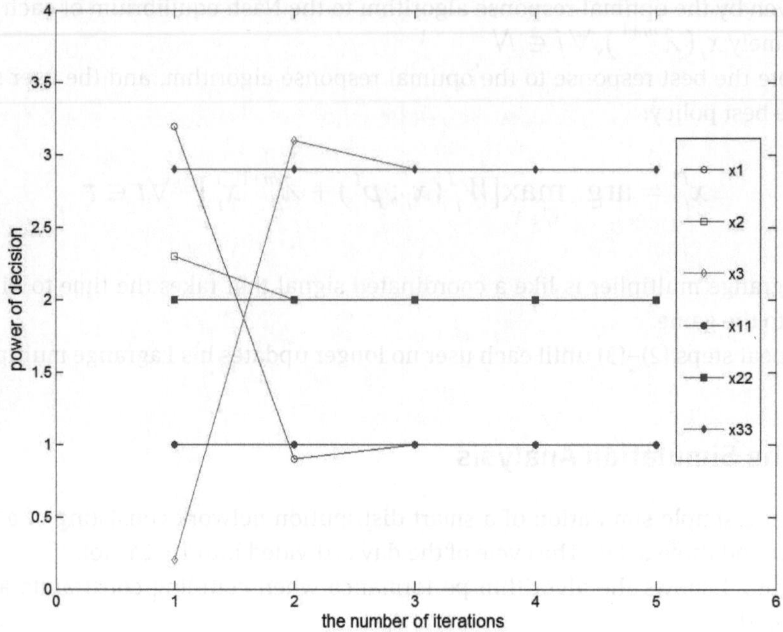

Fig. 2: The optimal projection algorithm for gradient projection

The graph shows that each user's electricity decision is able to converge to the optimal solution using the gradient projection algorithm. Moreover, the method converges fast, which can satisfy the real-time requirement of the smart grid.

Figure 4 shows the benefit of the users in the two-tier iteration algorithm with coupled constraints.

Figure 5 shows the transition curve of the Lagrange multiplier in the upper iteration with the number of iterations.

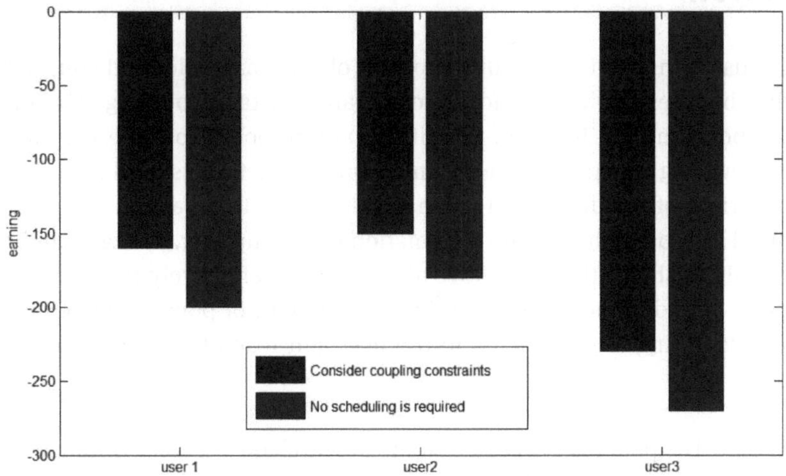

Fig. 3: Users in the two-tier iteration algorithm benefit

Figure 5 shows the transition curve of the Lagrange multiplier in the upper iteration with the number of iterations.

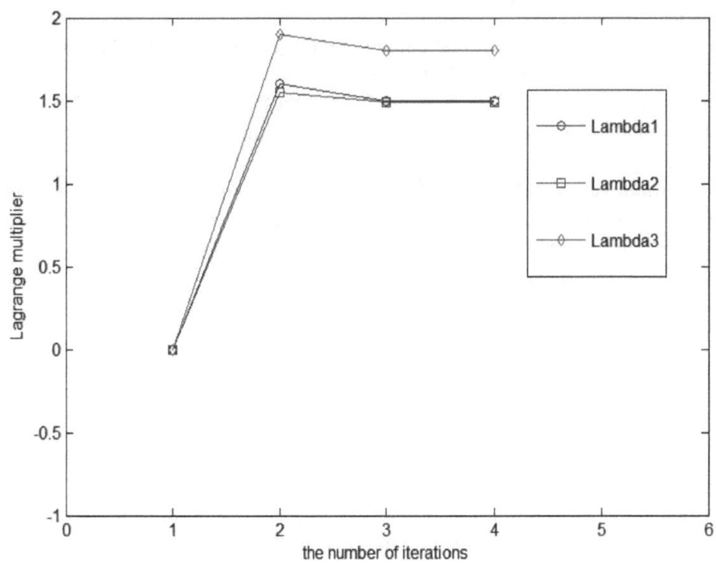

Fig. 4: Lagrange multiplier in the upper iteration

3 Conclusion

This article discusses the load scheduling problem of the smart grid, and considers the relationship between the users and the constraints of time coupling. Through the dual decomposition, two iteration algorithms are proposed to solve the game problem of the coupling constraint. The optimal response method is used to achieve the Nash equilibrium of the decoupled game problem, and the gradient projection algorithm is used to protect the private information of the supplier. A large number of numerical analysis shows that the above algorithm can effectively transfer peak load to non-peak period without affecting the total amount of power consumption per day. It can also bring more benefits to the user and reach the minimum peak ratio.

Acknowledgement: The paper is supported from "Research on Technical Framework of Energy Internet" project, which belongs to State Grid Corporation of China's research plan "Fundamental and Perspective Project". Thanks direction of Professor Hai-ming Zhou and discussion of colleagues engaged in the project.

Reference

[1] S Massoud Amin, Bruce F Wallenberg. Toward a smart grid: power delivery for 21st century [J].IEEE Power and Energy Magazine, 2005.3(5): 34–41.

[2] A-H Mohsen Ian-Rad, Vincent WS Wong, Juri Jatskevich, Robert Schobert. Optimal and autonomous incentive-based energy consumption scheduling algorithm for smart grid[C]//Proc. IEEE ISGT.2010: 1–6.

[3] Na Li, Lijun Chen, Steven H Low. Optimal demand response based on utility maximization in power networks[C]//Proc. IEEE PES-GM.2011: 1–8.

[4] W.I Wan Ishak, Tan Ming Yin and R.M. Hudzari (2011) "Development of Control Program for Plant Growth Parameter Analysis in Lowland Tropical Greenhouse", Journal of Applied Sciences, 11(3): 592–598.

[5] Muhammad Rizuwan Yahaya, Mohd Hudzari Hj Razali, Che Abdullah Abu Bakar, Wan Ishak Wan Ismail, Wan Musa Wan Muda, Nashriyah Mat and Abd. Jamil Zakaria (2013) " Automatic Alkaloid Removal System" Pakistan Journal of Biological Sciences .17(1), PP 141–145

[6] A-H Mohsenian-Rad, Alberto Leon-Garcia. Optimal residential load control with price prediction in real-time electricity pricing environments [J].IEEE Transactions on Smart Grid, 2010.1(2): 120–133.

[7] Pedram Samadi, Hamed Mohsenian-Rad, Robert Schober, Vincent WS Wong. Advanced demand side management for the future smart grid using mechanism design [J].IEEE Transactions on Smart Grid, 2012.3(3): 1170–1180.

[8] Deng ruilong. Smart grid use electrical information access and load scheduling research [D]. Zhejiang University, 2014.

[9] Zhang cong, Sun limin, Zhu zhibin. The generalized projection algorithm for extremely small problems [J]. Practice and Cognition in Mathematics, 2017, (2003): 259–264.

[10] Zhang Hua-jun, Zhao Jin, Luo Hui, Xie Xiang-peng. Based on the gradient projection method and the stochastic optimization algorithm of constrained optimization method [J]. Control and Decision, 2014, (10): 1777–1782.

[11] F. Farokhi, M. Cantoni and I. Shames. A game-theoretic approach to distributed scheduling of rigid demands on dynamical systems. 2016 Australian Control Conference (AuCC), 2016:147–152.

[12] W. Zhou, J. Wu, W. Zhong and S. Zou. Electricity Consumption Scheduling with Consumers' Comfort and Preference in Smart Grid [J].IEEE Chinese Journal of Electronics, 2016.25(6):1151–1158.

Guang Yang, Hong-Ru Ren, Zhong-Yuan Piao and Jian-Gang Qiao

Based On the Test Track Minimum Radius of Expressway Ramp

Abstract: Interchange is the throat motorway. It affects not only the capacity of the highway, but also traffic safety. Through the research on the form of single-track and double-lane transformation of freeway, the ramp model of single and double lane ramp transformation is constructed. On the basis of collecting the data of ramp characteristic points on the spot, the speed variation law of different vehicles on the ramp is analysed. According to the characteristics of the freeway, the minimum radius of the ramp on the driving conditions is obtained, and the design of the ramp is designed, which provides a theoretical basis for the design of ramp standard revision.

Keywords: Track test, ramp, minimum radius, traffic flow.

1 Introduction

Freeway ramp area is among the freeway accident-prone area, which subjects to terrain, cost and other factors. Linear ramp will always be biased index limit value, sometimes even more than or less than the limits. Hameed Aswad Mohammed analyzed the influence factors of traffic accidents by using accident analysis method ,and it showed that The number of lanes, the radius of the curve and the large longitudinal slope have great influence on traffic safety [1]. Ghazan Khan He found that the radius and length of the flat curve were the two factors that affected the safety. He also established the accident database of the 11427 flat curve section of the Wisconsin dry road network, and established the rural flat curve section Death and injury accident prediction model [2, 3]; Wilson He analyzes the accident rate on a sharp turn and a gentle turn to get a flat curve with a radius of less than 200 m. The accident rate on the corners is 4 to 5 times greater than the 900 m radius [4]; Mo Yang proposed structure the form of the ramp which meets meet the requirements of

Guang Yang , Jilin Traffic planning and Design Institute Changchun, China. Email: yg9898@126.com
Hong-Ru Ren, Hebei University of Technology Civil Engineering and Transportation Tianjin, China. Email:renhongru1991@qq.com
Zhong-Yuan Piao, Jilin Traffic planning and Design Institute Changchun, China. Email: 824220496@qq.com
Jian-Gang Qiao, Hebei University of Technology Civil Engineering and Transportation Tianjin, China. Email: qiaojg369@126.com

DOI 10.1515/9783110540048-067,

the parking distance curve minimum radius ratio to the requirements of the lateral force coefficient [5]; Research by Li Zhou Feng have shown that small radius plane curve at the outlet increases the likelihood of an accident, which provided that it had no significant effect on the accident in the convergence section and inlet section [6]; Shen Qiang believes that the construction of Diamond Interchange area should be analyzed to identify the horizon , and checked the main line linear index . It calculate the range of the radius of the main line of the main line through the special interchange to identify the line of sight, and provide the basis for the selection of the special interchange index and the choice of the exit position. [7,8]. Through the investigation and collection of vehicle ramp traffic organization forms were analyzed each ramp with the position and speed characteristics of different types of vehicles, articulated vehicles, aimed at cars with the vehicle running track test, which ensure safe driving conditions was in the off-ramp minimum radius.

2 Ramp Traffic Organization Analysis Form

According to the field survey of a highway, we found mainly single, two-lane ramp gradient in the form shown in Figure 1-3:

Fig. 1: Single lane entrance ramp connecting two-lane ramp form

Figure 1 uses the deceleration lane for the single-lane, two-lane ramp form, making travel faster than small cars can enter from the left side of the ramp with large vehicles and large cars are not changing lanes on the right side.

Fig. 2: Intermediate single-lane ramp connecting two-lane ramp form

Figure 2 is a single-lane ramp after the midpoint of the song becomes a two-lane ramp line ramp converted into a two-lane ramp acceleration lane having a functional, large-scale ramp right lane traveling by car, a small car accelerated into the left lane along the ramp mainline.

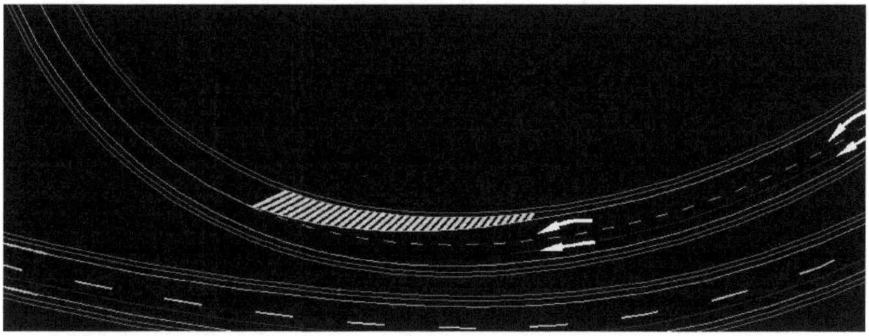

Fig. 3: Ramp midpoint of a two-lane ramp converted into a single-lane ramp form

Figure 3 is a two-lane ramp. You can pass through the middle of the ramp; the ramp will be converted into a single lane ramp, the right side of the vehicle lane into the left lane, the right side of the same set as the main hard shoulder for emergency parking. Vehicle lane change in advance into the acceleration lane can increase the safety of the main line.

When the deceleration lane is a single lane, the driver is accustomed to overtake on the left, so the entrance ramp connected by a single-lane ramp shift lane dual carriageway, traffic organization using the form on the left widened mainly; the terrain is limited short acceleration lane , it can widen the right of the lane from the ramp park midpoint of the curve, change to single-lane ramp traffic organization in the form of a two-lane ramp; the case of the acceleration lane to single lane, the vehicle is traveling in a two-lane ramp midpoint of the curve becomes traffic organization in the form of single-lane ramp to facilitate docking with the single-lane acceleration lane, lane will reduce the risk incorporated into the main line accidents.

3 Research On Operational Characteristics of Vehicles Ramp

Car on the ramp driving characteristics can be divided into: Split slowly down the process, uniform (or shift) and accelerated the process of merging with the process of moving. That left the main highway deceleration lane (shunt deceleration), into the ramp; the ramp (uniform or variable speed) driving; acceleration lane highway into the main line (confluence acceleration).

For some freeway interchange, the ramp is two-lane ramp, while shifting lanes is single-lane ramp, the split and form common single-lane ramp Frontage case two-lane ramp, at the confluence of the acceleration ramp run and there appear and two-lane ramp Cohesion single-lane ramp situation. Through Interchange on a branch point, gradual ramp segment, the midpoint of the song ramp, ramp confluence points to single-lane dual carriageway becomes a point of the vehicle speed data acquisition, will be divided into two axis model truck (2H), 3 axle truck (3H), 4 axis truck (4H), 5-axis truck (5H), 6-axis truck (6H), trucks (DH), small trucks (XH), minibuses (XK), the data finishing. We got Figures 4 and 5.

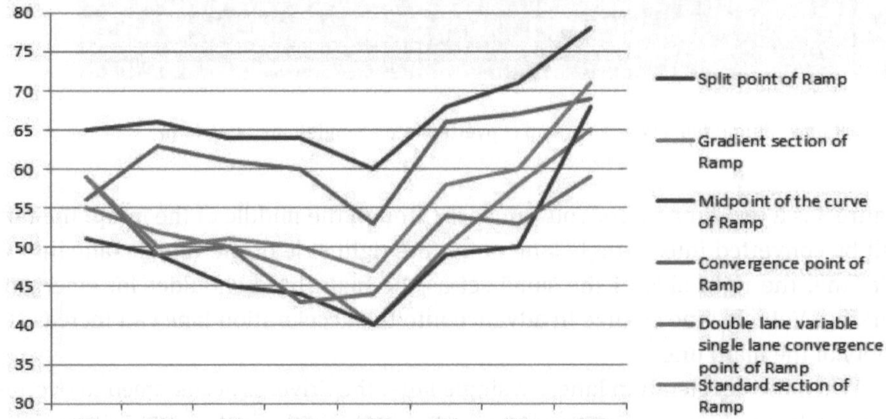

Fig. 4: Changchun southern interconnected ramp speed figure A

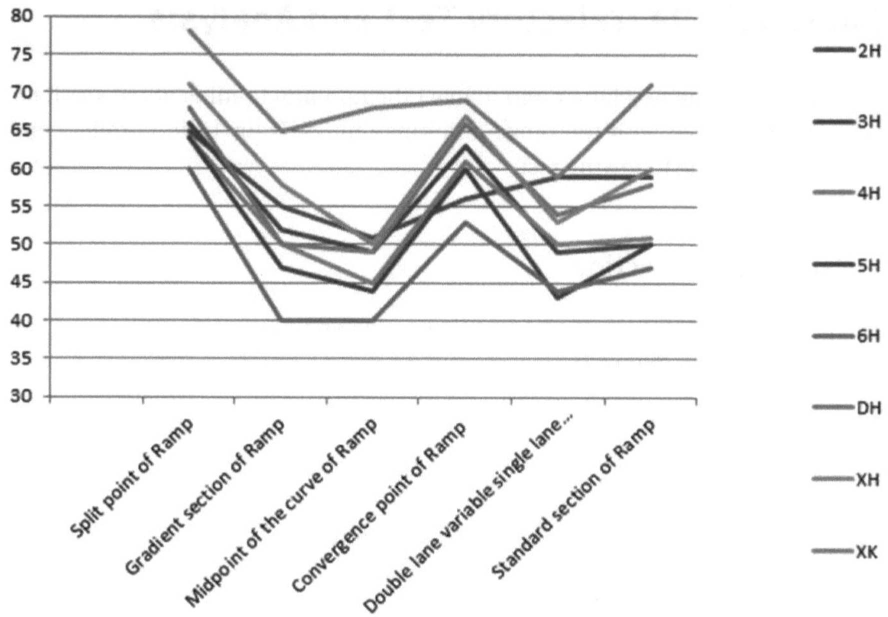

Fig. 5: Changchun southern ramp speed figure B

It can be seen from Figure 4 vehicle rules on the ramp; the vehicle from the deceleration lane diversion point has been decelerating until the vehicle speed linear ramp adaptation. Ramp with a gradient section gradually began to constant speed. After passing the midpoint, the vehicle started accelerating, but there would be a two-lane ramp to the single-lane ramp transition road when approaching the outlet and the vehicle would slow down again to converge on the lane. Finally enter the acceleration lane and started accelerating.

It can be seen from Fig 5. Large trucks with the increasing of the number of axes, the load of vehicle, the volume of the vehicle, the speed will slow down. From large trucks to small vans, it can be seen that speed of the passenger-car is the fastest, followed by the small truck, and speed of big truck is the slowest. In the operation of the ramp, the speed of the diversion point is the fastest, the speed of curved point is the slowest, two lane ramp to single lane ramp transform speed is lower than the standard section at the speed, not much difference with the transition section of the ramp, and the speed of the confluence point is also faster than the speed of the shunt.

4 Ramp vehicle trajectory Test and Analysis

In order to analyze the two-lane ramp to meet the minimum radius of the small passenger train in parallel with the hinge and hinge hinge Train Train parallel conditions permit, make the following test:

4.1 Parameter Selection

Vehicle parameters selected: minibus geometry shown in Figure 6, articulated vehicle geometry shown in Figure 7:

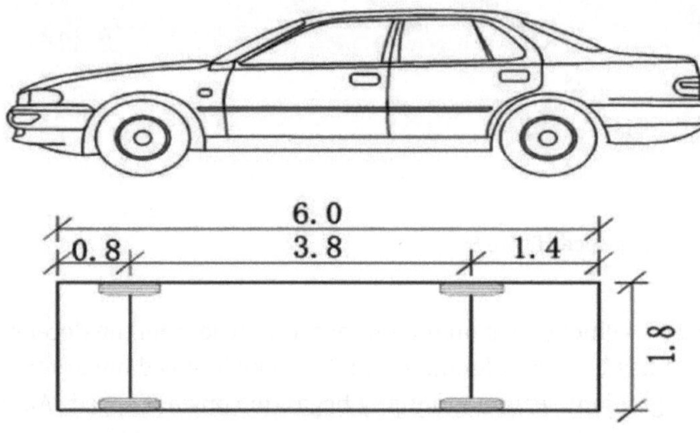

Fig. 6: Figure minibus geometry ometry

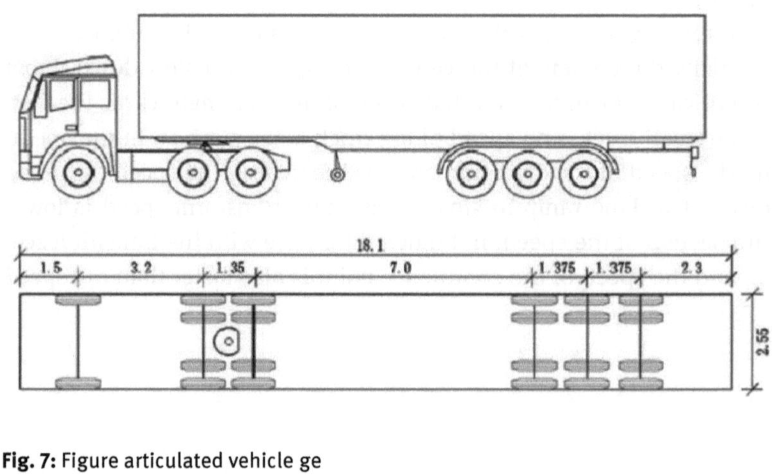

Fig. 7: Figure articulated vehicle ge

Road parameter selection: test road cross section shown in Figure 8

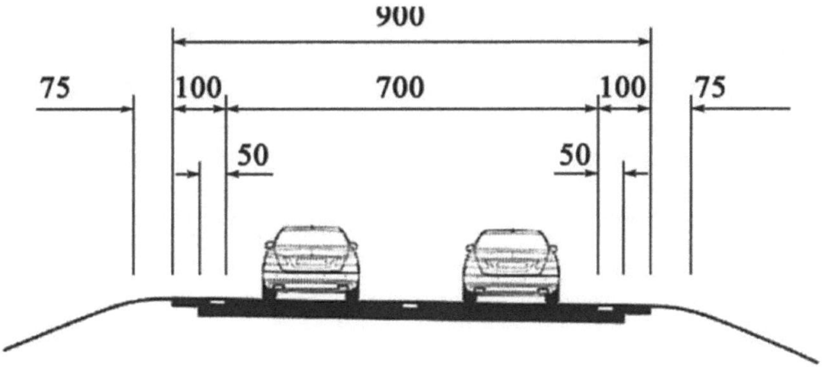

Fig. 8: Test cross-sectional view

Select other parameters: ramp design speed: 30 km/h ~ 80 km/h; ramp curve radius: Usually value of 20 m ~ 280 m, limit: 25 m ~ 230 m; safe distance: 0.25 m.

4.2 Vehicle Trajectory Simulations and Analysis of Least Squares Curve Radius

Condition 1: Ruoshi articulated train kept in the inside lane, and the distance from the vehicle lane is more than 0.25 m, the minimum horizontal curve radius of 140 m, as shown in figure 9.

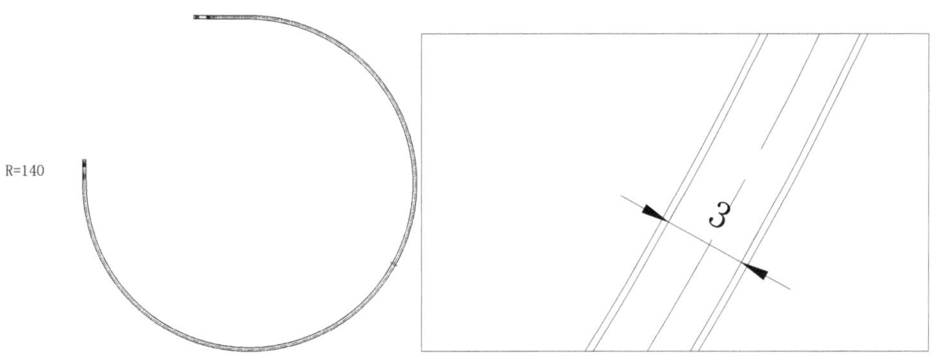

Fig. 9: Articulated train kept in the inside lane (lane edge line distance 0.25m)

Condition 2: Ruoshi articulated train kept in the inside lane, and the allowable distance from the vehicle lane is 0m, the minimum horizontal curve radius of 72 m, as shown in figure10.

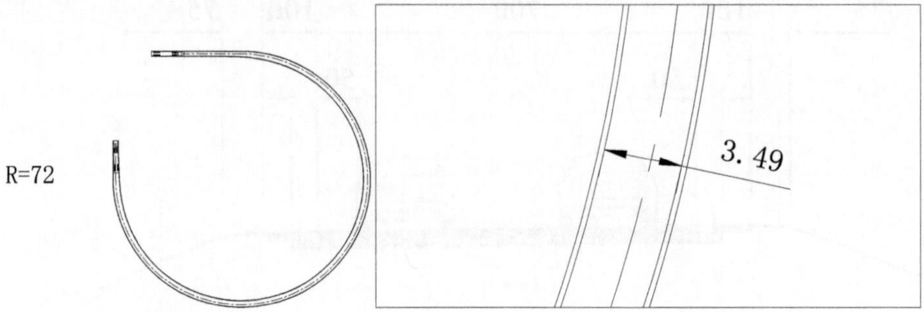

Fig. 10: Articulated trains (from the lane edge line 0m) in the lane

Condition 3: If the train remains occupied the adjacent lane, but the use of the hard shoulder width, the body maintains a safe distance from the edge of the hard shoulder width of 0.25 m, the minimum horizontal curve radius of 41 m, shown in figure11.

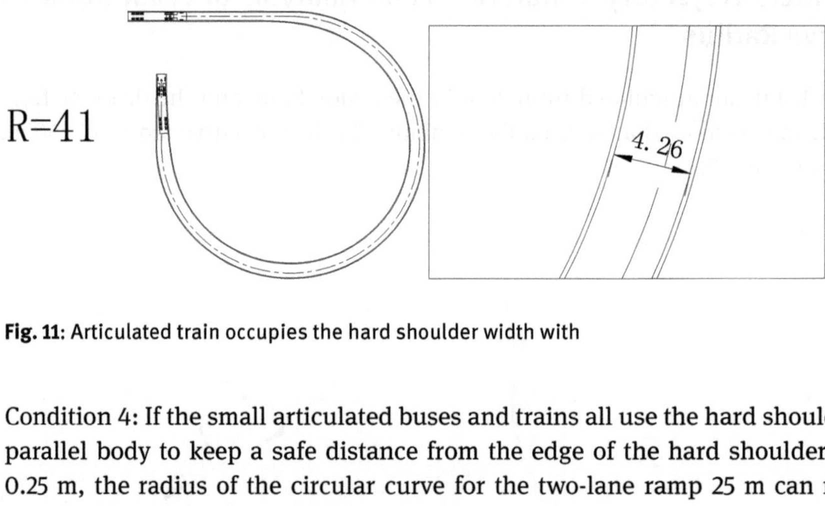

Fig. 11: Articulated train occupies the hard shoulder width with

Condition 4: If the small articulated buses and trains all use the hard shoulder width parallel body to keep a safe distance from the edge of the hard shoulder width of 0.25 m, the radius of the circular curve for the two-lane ramp 25 m can meet in a small passenger train in parallel with the hinge, parallel safety spacing between vehicles is 1.1m, shown in figure 12.

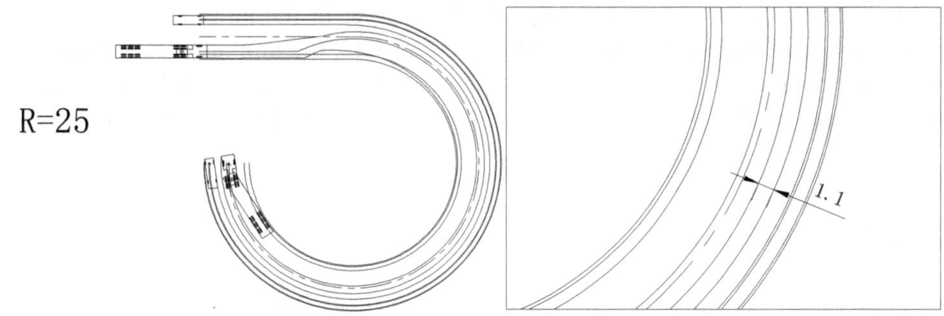

Fig. 12: Articulated trains and passenger cars in parallel with

Condition 5: If the vehicle maintain a safe distance from the edge of the hard shoulder width of 0.25 m, width 0.5 m between the vehicle remains safe, the two articulated train side by side in a two-lane ramp with, least squares curve radius required for the 43 m, shown in figure13.

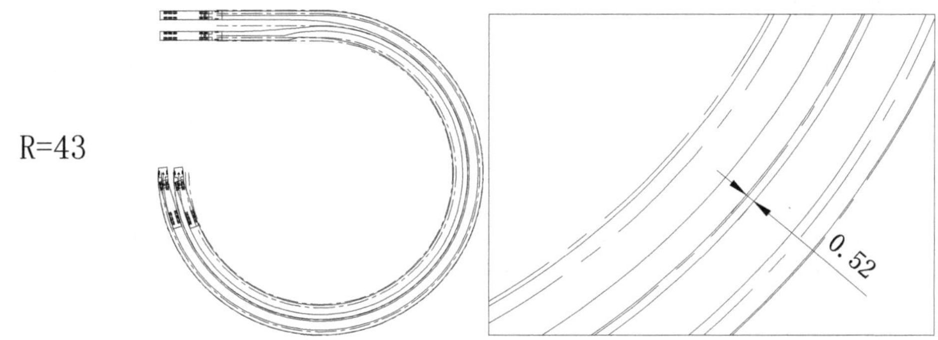

Fig. 13: Articulated trains and articulated parallel with the train

Tab. 1: The condition of one ~ five analysed

	Condition 1	Condition 2	Condition 3	Condition 4	Condition 5
Vehicle location	Articulated train kept in the inside lane (lane edge line distance 0.25m)	Articulated trains (from the lane edge line 0m) in the lane	Articulated train occupies the hard shoulder width with	Articulated trains and passenger cars in parallel with	Articulated train with the hinge parallel with the train
Radius (m)	140	72	41	25	43

Table 1 shows that a single vehicle gradually smaller in the inner lane ramp with radius, reduced safety, the safest radius of 140 meters, the minimum safe radius of 41 meters; when the two cars at the same time in parallel lane dual carriageway ramp minimum safety radius of 43 m, the limit minimum radius of 25 meters.

5 Summary

Through the investigation single-ramp and double-ramp forms on the freeway, we learned the main form of single and double lane ramp transform. Based on field speed data collection, the characteristics of running vehicles on the ramp speed are analyzed. In this paper, the ramp trajectory test of the passenger car and articulated vehicle is carried out, and the results are as follows: (1) the average value of the ramp curve radius is 72m, and the limit value is 43m. (2) If the passenger car is only allowed in the inner lane of traffic control, and the vehicle speed is effectively controlled, and the minimum radius of the ramp can be 25m.

References

[1] Mohammed H. The Influence of Road Geometric Design Elements on Highway Safety [J]. International Journal of Civil Engineering, 2013, 4(4): 146–162.

[2] Khan, Ghazan, Bill, Andrea R, Chitturi, Madhav, Noyce, David A. Horizontal Curves, Signs, and Safety [J].Transportation Research Record: Journal of the Transportation Research Board, 2012, 2279: 124–131.

[3] Khan, Ghazan, Bill, Andrea R, Chitturi, Madhav, Noyce, David A. Safety Evaluation of Horizontal Curves on Rural Undivided Roads. Transportation Research Record: Journal of the Transportation Research Board, 2013, 2386: 147–157.

[4] Wilson, T.D. Road Safety by Design [J].Journal of the Institute of Highway Engineers.1968, 15: 23–33.

[5] MO Yang. Study on Control Parameters for Ramp and Speed-change Lane of Urban Expressway [J]. Journal of Highway and Transportation Research and Development.2016, 33(5): 111–127.

[6] LI Zhoufeng, SUN Jian. AN Analysis of the impact of Geometry Design to the safety of Interchange Ramps of Urban Expressways [J].Journal of Transport Information and Safety.2015, 33(6): 89–95.

[7] SHEN Qiangru, YANG Shaowei, ZHAO Yifei, CAO Hui. Analysis on mainline alignment index in diamond interchange diverging [J].Journal of Harbin institute of technology.2015, 47(9): 90–94.

[8] SHEN Qiangru, YANG Shaowei, ZHAO Yifei, CAO Hui. Analysis on the main line of the main line in the export area of the special interchange based on the recognition sight distance [J]. Journal of Safety and Environment 2015, 15(3): 89–92.

Fu-Qing Li *, Zhao-Du Liu and Xiao-Mei Yue

Establishment of Key Components Testbed for Emergency Steering System Used in Heavy Duty Trucks

Abstract: Based on the working principle of emergency steering system of heavy-duty truck, the working principle and composition of hydraulic test-bed on key components of emergency steering system are introduced. The performance experiments of pump and valve are carried on. The results show that the emergency steering system performance of key components is capable to achieve the emergency steering of the vehicle.

Keywords: hydraulic system, emergency steering, data acquisition

1 Introduction

The heavy-duty vehicle steering load is usually very heavy, thus it is difficult to achieve anticipated steering relying on human manipulation alone. Therefore, heavy-duty vehicles are equipped with power steering systems. But when accidents happened such as engine flaming out or main power steering pump (hereinafter referred to as the main pump) failure, and when vehicle break down and get pulled, the power steering system failed to function. So the emergency steering system is usually installed based on the power steering system to achieve emergency steering [1, 2].

The emergency steering pump (hereinafter referred to as the emergency pump) and the emergency valve are key parts of emergency steering system, and their performance is directly related to whether the system function regularly. Therefore, it is necessary to establish a test-bed and carry on performance testing on the key components in the emergency steering system.

***Corresponding author: Fu-Qing Li**, School of Mechanical Engineering, Beijing Institute of Technology, Beijing, China. Email: lfqysb@bit.edu.cn.
Zhao-Du Liu, School of Mechanical Engineering, Beijing Institute of Technology, Beijing, China. Email: liuzhaod@bit.edu.cn
Xiao-Mei Yue, School of Mechanical Engineering, Beijing Institute of Technology, Beijing, China. Email: yuexiaomei@bit.edu.cn

DOI 10.1515/9783110540048-068,

2 The Development of The Hydraulic System Test-Bed

2.1 Working Principle of Emergency Steering System

With regard to heavy-duty vehicles equipped with emergency steering system, when the vehicle power steering system fails to work, the kinetic energy of wheels in driving state transfers to the power source of power-assisted steering. Through the emergency steering gear drive system, the wheel kinetic energy (the lower bound of system design speed is not less than 15 km/h) is introduced from the front axle of vehicle. Then growth regulation driven the emergency pump, and the emergency pump output pressure oil through the emergency valve into steering cylinder and achieve power steering.

According to the contrast between main pump flow rate and its own, the emergency valve performs the switch of oil-way automatically between the main pump and emergency pump. When the main pump works normally, the output flow is greater than the emergency valve setting flow, and steering oil pressure is mainly provided by the main pump, and emergency pump is in unloading state; when the main pump flow is below the emergency valve setting flow, the emergency pump is switched to steering oil-way automatically. Fig. 1 is schematic diagram of the working principle of emergency steering system.

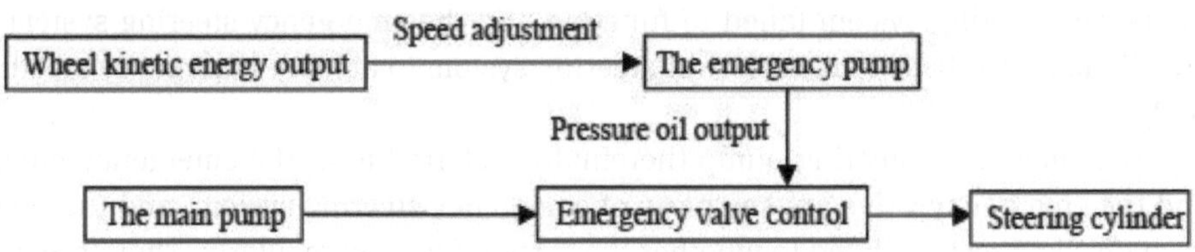

Fig. 1: Schematic diagram of the working principle of emergency steering system

2.2 Overview of The Emergency Steering System

The experimental table tests include the output characteristics of the emergency pump under different load conditions and the characteristics of emergency valve switching response under certain load conditions, the working principle and physical chart of the test bench for emergency steering system are shown in Figure 2.

The choice of the main pump, fluid medium and the diameter of the oil pipe in the experimental platform are consistent with the application of the real vehicle, and further guarantee the service function of the experimental platform. Based on the maximum output power of the main pump and emergency pump, the main motor and emergency motor are selected, and using the overflow valve as adjustable load [3]. First, start the main motor and emergency motor, and make sure the system load as the design load of the real vehicle emergency steering system by adjusting the relief valve. Second, after two motor working together for a period of time, close the main motor and test the switching characteristics of the emergency valve and the output characteristics of emergency pump under rated load. In the test of maximum output characteristics of emergency pump, using the branch valve to change the oil circuit by the system pressure will exceed the pressure limit of the emergency valve. A self-developed data acquisition system is established in the experimental test platform, and the data sampling period is 20ms. The operation control software is compiled by Basic Visual 6, with it we can obtain the functions of data acquisition, process display, data processing, control, and storage and output are realized.

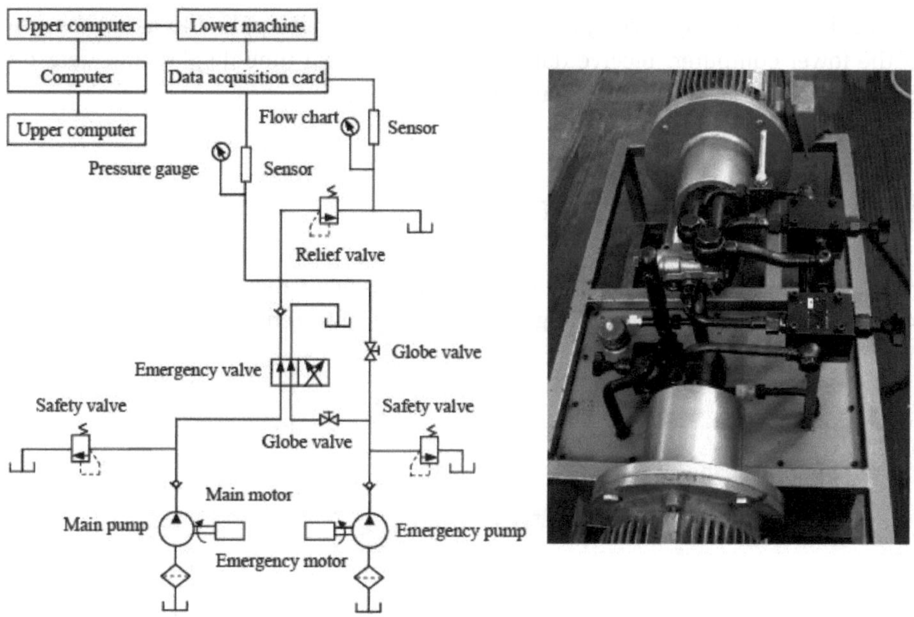

Fig. 2: Working principle and physical chart of the test bench for emergency steering system

2.3 Experimental Platform Data Acquisition Process

In the system of data acquisition, converting the physical quantity signal (pressure and flow signals) into electrical signal by using the sensor. In order to suppress the signal of the high frequency component of the interference noise signal, the signal should be filtered and limited, and then the signal is transmitted to the MCU (Microcontroller unit) by the interface circuit. And the single chip microcomputer is responsible for the upper computer for communications transmission [4, 5]. The concrete sampling flow is shown in Figure 3.

Fig. 3: Flow chart of experimental platform data acquisition

The lower computer of the acquisition system is MCU MC9S12DP256, and that is also the core of the data acquisition system. MCU MC9S12DP256 is mainly used for system initialization, serial interface to achieve control, signal acquisition and processing, and meet the needs of data communication and transmission with the host computer. The upper computer control procedures get written based on Visual Basic Microsoft 6.0. The upper computer control procedures is applied to assist the work of the lower computer, receive data, complete the real time display and data processing and complete the results summary and print.

3 Emergency Valve Test

Figure 4 shows the switch response of the emergency valve and the actual test results of the high - pressure working characteristics of the emergency pump.

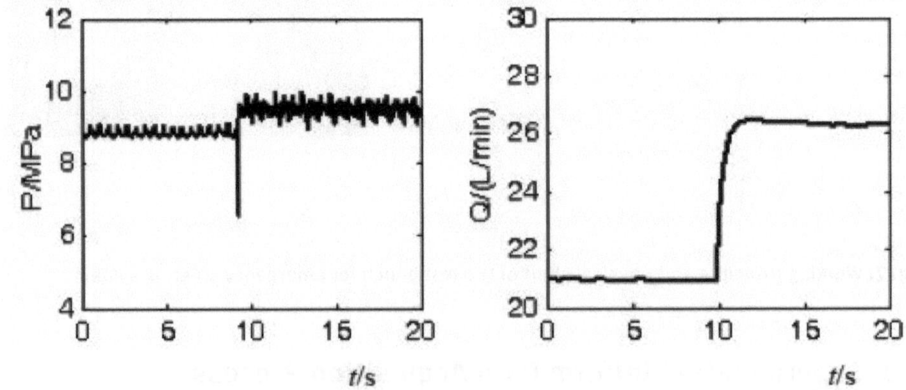

a) The switch response curve of the emergency valve

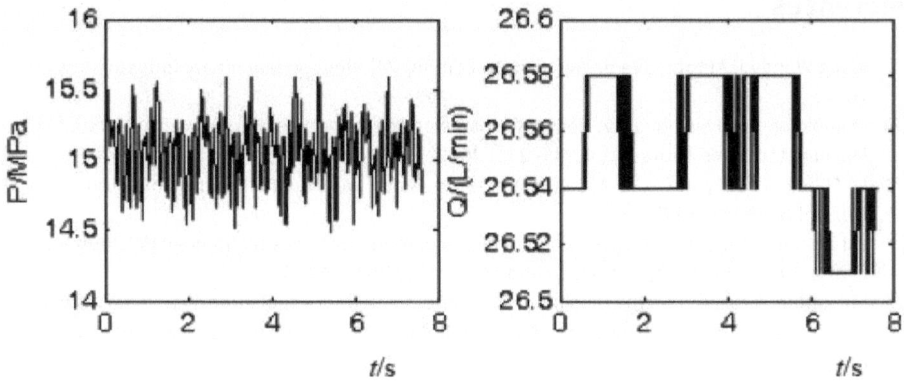

b) The characteristics of the emergency pump itself high pressure curve

Fig. 4: Test results of the valve performance of the emergency steering system

The measured results show that the load switching time under the emergency steering valve is 80ms rated in the vehicle, the pressure switch is 6.5MPa, can realize the fast handover and ensure that the required minimum power steering. Emergency steering pump load and its stability in the nominal pressure zone, pressure fluctuation is less than 3%, the flow fluctuation in about 0.1%, can provide plenty of power steering. This also shows the reliability of the real vehicle emergency steering system.

4 Conclusion

The test device can carry out the vehicle emergency steering performance test of key parts of the system in the emergency valve. The repeated experimental results showed that the testing principle is reliability, the test data is precise. The test device has functions of data storage and data output data under data post-processing. The automatic test has been realized in the test procedure, and the operation procedure is facility.

The performance test of the vehicle emergency steering performance test of key parts of the system show that, the Valve of hydraulic pump is rapidly switch able and it can provide stable steering power. This can make sure the vehicle could drive safely under the unexpected accident.

References

[1] Wang Wangyu. Automotive design (Fourth Edition) [M]. Beijing: Machinery Industry Press, 2004.
[2] Zhao Yong, Cao tree equality. Heavy off-road steering system emergency valve CAT study [J]. Machine tools and hydraulic, 6,235–236, 2002.
[3] Xu Fuling et al. Hydraulic and pneumatic transmission [M]. Wuhan: Central China University of Technology Press, 1997
[4] Liu Leshan. Principle and application of microcomputer interface technology [M]. Wuhan: Central China University of science and Technology Press, 1996.
[5] Shi Kaibin. Research on the development of integrated data acquisition system for automotive [D]. And ABS control logic ABS/ASR/ACC Beijing Institute of Technology, School of mechanical and Vehicular Engineering, 2003.

Zhe Du, Rui Guo and Zhi-Gang Hu*

Modeling and Testing of an ICPTs Used in Dynamic Balancing Device

Abstract: In order to balance the out-of-balance mass and reduce the harmful vibrations of motorized spindles, author developed a novel dynamic balancing device of standing-wave piezoelectric motor type with both simple and reliable structure. Since the whole balancing device is mounted on rotor of the spindle and rotating synchronously with it, this paper examines the possibility of using specially designed ICPTs to deliver electrical energy from stationary position to the balancing device with help of a novel contactless rotatable transformer.

Keywords: Contactless rottable transformer, ICPTs, Compensation, Ferrite material

1 Introduction

Today, majority of machine tool systems are equipped with motorized spindles to improve the machining efficiency and precision ,and one of the important consideration for speedup of spindles is the harmful vibration due to out-of-balance mass. Therefore, it is particularly necessary to balance these spindles at running state, as carefully as possible, to ensure smooth running. Up till now, various balancing systems have been developed to balance the out-of-balance mass in order to reduce the harmful vibrations [1-4]. Recently we also developed a novel dynamic balancing device of standing-wave piezoelectric motor type with both simple and reliable structures. This device can be used for effectively reducing the out-of-balance mass in a running motorized spindle to a relatively low level, and try not to influence the spindle's rotation characteristic in the meantime. Because the balance device has two sets of adjusting mechanism, two high-frequency AC power are required and need to be controlled with the electrical energy. Since the whole balancing device is mounted on rotor of the spindle and rotating in synchronous with it, it is evident that conventional electrical connection solutions such as trailing leads, rolling/sliding carbon/metal contacts can't meet the requirement of power transmission. In this paper, a specially designed inductively coupled power transfer system (ICPTs) is employed to deliver electrical energy from stationary position to the high-speed

Zhe Du, Henan University of Science and Technology, Luoyang 471003,China. duzhe@haust.edu.cn
Rui Guo, Henan University of Science and Technology, Luoyang 471003, China.guoruiyg@163.com
***Corresponding author: Zhi-Gang Hu,** Henan University of Science and Technology, Luoyang 471003, China. hu.robert@haust.edu.cn.

DOI 10.1515/9783110540048-069,

rotating balancing device with help of contactless rotatable transformer, which is the kernel of the whole ICPTs that can transmit electrical energy without any mechanical contact. So far a number of contactless rotatable transformers have been described in literatures [5-10].In contrast with these rotatable transformers generally consisting of a set of pot cores with a winding fitted into each of them, the one we investigated consists of two set of cores that can be thought of as an extended "C" core, with a winding running around the entire length of each of them in order to transmit two ways of high-frequency AC power alternately.

Due to the physical separation between the windings, and the availability of alternative magnetic-flux paths (air gap) rather than the ideal core path, the rotatable transformer exhibits relatively large leakage inductances and significantly reduced magnetizing inductance between the primary and the secondary, resulting in low gain and high loss for the ICPTs. Therefore, to solve the problem of poor efficiency of the ICPTs directly and effectively, a suitable impedance-matching circuit in the primary and secondary of the rotatable transformer making the compensated circuit operate at resonant frequency can be adopted in order to achieve the maximum system output power under the minimum system input VA rating.

In section 2 of this paper, the operation mechanism of the dynamic balancing device and the structure of the contactless rotatable transformer are introduced briefly. Section 3 presents the system modeling and compensation topology analysis of the ICPT system in detail. And then, section 4 determines the turns of the rotatable transformer windings based on previous section analysis, and derives inductance measurement of the contactless rotatable transformer. At last, the preliminary analysis on the experiment results of a prototype system and the conclusions are given in section 5 and section 6 respectively.

Nomenclature

Symbol	Explanation
L_e	Equivalent inductor of series subcircuit of the piezoelectric actuator
C_e	Equivalent capacity of series subcircuit of the piezoelectric actuator
R_e	Equivalent resistance of series subcircuit of the piezoelectric actuator
C_d	Clamping capacity of the piezoelectric actuator
f_p	Parallel resonance frequency of the piezoelectric actuator
f_s	Series resonance frequency of the piezoelectric actuator
ω_s	Series resonance angular frequency of the piezoelectric actuator
U_p	Primary input high frequency sinusoidal voltage
U_0	secondary output high frequency sinusoidal voltage
I_p	Primary input high frequency current
I_s	secondary output high frequency current

Symbol	Explanation
L_p	Primary self-inductance of the inner ring transformer
L_s	Secondary self-inductance of the inner ring transformer
M	mutual inductance of the inner ring transformer
R_p	Internal resistance of primary winding of the inner ring transformer
R_s	Internal resistance of secondary winding of the inner ring transformer
R_d	Equivalent resistance of the piezoelectric actuator
X_d	Equivalent reactance of the piezoelectric actuator
Z_s	Total impedance of the secondary circuit
C_p	Compensation capacity of the primary circuit
C_s	Compensation capacity of the secondary circuit
Z_r	Reflected impedance of the secondary circuit exerts on the primary
R_r	Reflected resistance of the secondary circuit exerts on the primary
X_r	Reflected reactance of the secondary circuit exerts on the primary
Z_p	Total impedance of the primary circuit
L_{diff}	Differential inductance
L_{cum}	Cumulative inductance

2 Contactless Power System Description

2.1 Operation Mechanism of the Dynamic Balancing Device

First of all, we want to briefly introduce the operation mechanism of the dynamic balancing device. Each device includes two sets of edge-driving piezoelectric actuators in the same form. A set of piezoelectric actuator consists of annular rotor and piezoelectric stator. The configuration of a set of piezoelectric actuator under consideration is shown in Fig.1 (A). An in-plane bending vibration B_{32} mode of the annular plate made of aluminium is used as the resonance mode of the stator vibrator. Fig.1 (B) shows polarization division arrangement of the piezoelectric ceramic disk for driving. The annular piezoelectric ceramic disk with six polarization partitions is bonded to the annular aluminium plate; both together make up a stator. Six projecting teeth are installed at the external circumference side of the annulus aluminium plate and located at certain positions. The rotor, with a thin frictional material bonded on its internal circumference side, is installed on the outside track of the stator. Through an appropriate preload force, the internal circumference side of the rotor and the external circumference side of the six teeth contact with each other. When AC power in specific frequencies is applied on the six electrodes of the stator, an ideal in–plane bending vibration mode will be excited as a result of the inverse piezoelectric effect. The teeth in contact with the rotor will vibrate in the radial and circumferential directions due to the vibration mode. Then a circumferential frictional force creates under the radial pressure between the teeth and the rotor, which will drive the rotor rotating along the circumferential direction. On account of the pre-installed counterweight mass on the rotor, the motion of the rotor can realize regulation balancing role.

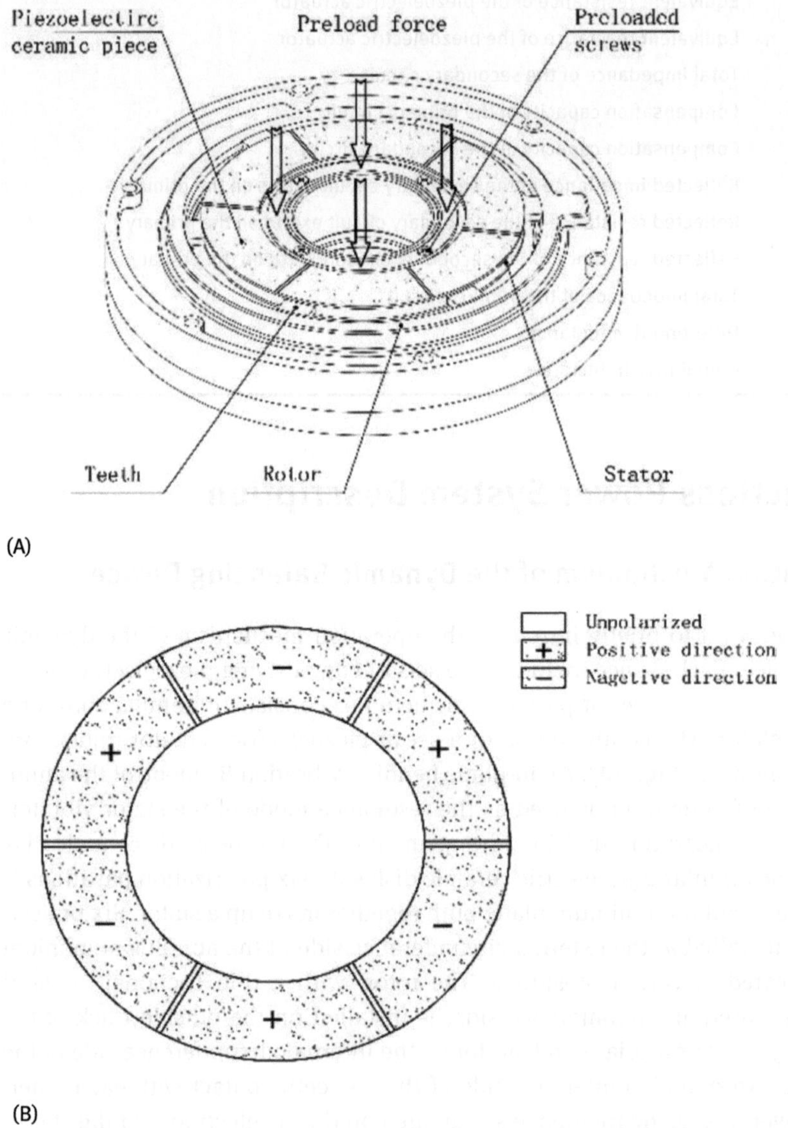

Fig. 1: Structure diagrams of the dynamic balancing device

The Electromechanical equivalent circuit of the piezoelectric actuator of the dynamic balancing device is similar as the well-known piezoelectric ultrasonic transducer, which is composed of a series sub circuit of equivalent inductor (L_e), equivalent capacity (C_e) and equivalent resistance (R_e) and a parallel branch of clamping capacity (C_d), as shown in Fig.2, which represents the electromechanical behaviour of the piezoelectric actuator [11, 12]. The C_d can be measured at the frequency far away from resonant frequency, which approximates to static capacity; the other values can be simply obtained from calculation of the impedance data.

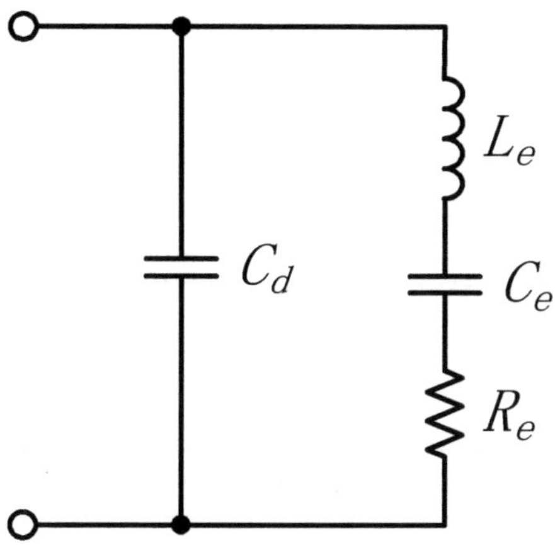

Fig. 2: The Electromechanical equivalent circuit of the piezoelectric actuator

The impedance characters of the piezoelectric actuator have been measured by a Frequency Response Analyser (*FRA*), as shown in Fig.3. According to the Fig.3, the various parameters can be obtained directly or further derived as shown in Tab.1:

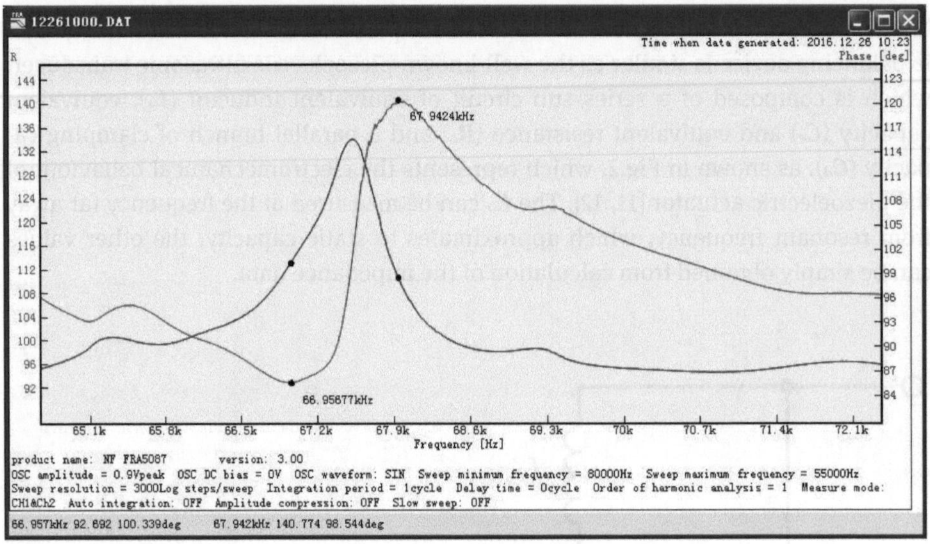

Fig. 3: Impedance characters of the piezoelectric stator at 55 kHz to 80 kHz frequency band

Tab. 1 : Parameters of the piezoelectric actuator

Parameters	**Acquisition method**	**Value**
C_d	Measured by capacitance meter	25nF
f_s	Obtained directly by Fig.3	66.96kHz
f_p	Obtained directly by Fig.3	67.94kHz
R_e	Obtained directly by Fig.3	115Ω
L_e	1/(4π2fs2Ce)	7.53mH
C_e	[(fp2/fs2)-1]Cd	0.75nF

2.2 Contactless Rotatable Transformer

As the introduction mentioned, the contactless rotatable transformers is the kernel of the whole ICPTs that can transmit electrical energy without any mechanical contact. The principle of transformer working on the electromagnetic induction is similar to the conventional transformer except that a large air gap between the primary and the secondary is arranged to enable the rotation of a part of the structure [8, 13].

A mechanical layout is proposed for the rotatable transformer as shown in Fig. 4. In order to transmit two high-frequency AC power, two sets of transformers are installed in a set of stepwise substrate. We call them the inner ring transformer and the outer ring transformer respectively. (The shape is designed in stepwise as to increase the air gap distance between the inner and outer ring transformer.)The stationary halves of the substrate are fixed on end cap of the spindle, and the rotatable one is mounted on the rotor of the spindle with dynamic balancing device. The primary core halves and windings are installed in the stationary substrate, and their counterparts, the secondary core and windings are installed in the rotary substrate.

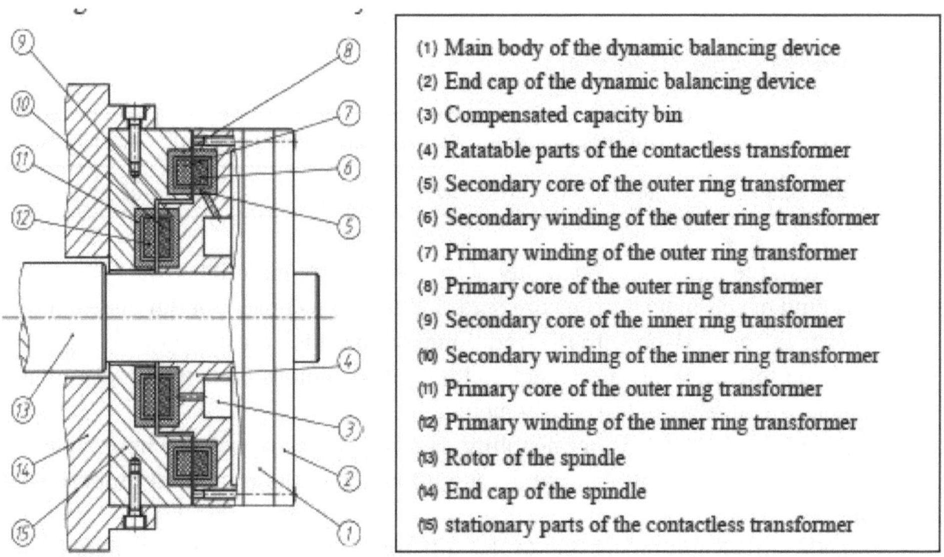

Fig. 4: Profile chart of the contactless rotatable transformer

Fig. 5: Photo of the contactless rotatable transformer without assembly

When transmitting high-frequency AC power, the conventional contactless transformer core is constructed in two ways, one is in use of high-temperature sintering ferrite cores [7, 8, 13], and the other is the use of air core (coreless) [14-16]. Based on the geometry of core is relatively small with thin wall in our study, high-temperature sintering ferrite core is prone to sintering deformation, and thus cannot meet the requirements of geometric accuracy. Furthermore, since the resonant transmission frequency of the system is relatively low, the air core will result in low transmission efficiency. Taking into account the two issues above, a room temperature vacuum curing of ferrite core material is proposed, and the formulation of this material is given by Tab.2.

Tab. 2: Ingredients of the core material

Ingredients	**Epoxy**	**dimethylbenzene**	**Curing agent**	**Ferrite powder**	**Iron powder**
weight ratio	10%	3.3%	10%	10%	66.7%

After real test, the relative permeability of this ferrite core material is 9.3.

This kind of self-made ferrite core material has been chosen for our research due to its merits in comparison with the traditional ones, which are as follows:

Good performance of later machinability. The dimensional tolerances of the post-processed cores meet the requirements;

Relatively high magnetic permeability. Coupling efficiency can be improved;

Relatively light weight, can reduce the influence of the rotational properties of the spindle;

Directly curing and processing in the substrate, this can avoid secondary bonding and improve the overall safety.

In help of this design, the windings of transformer in two circumferences can be separated and it can provide good path for the magnetic flux. Due to advantages of this structure, we will ignore the interference between the inner ring transformer and the outer one temporarily in the research work after, and take the inner transformer as the major study object.

Unlike the traditional method, during which the specific parameters of transformer windings were determined by reluctance method and/or empirical formula [31, 32], in our study, the parameters for transformer windings can be obtained according to the practical need of the ICPTs after the compensation topology research.

3 System Modeling and Compensation Topology Analysis

Among the literature proposed so far, a typical ICPT system consisting of converter, compensated circuits in the primary and secondary sides, inductive transformer and

DC converter is most commonly seen [17-23]. The structure of the ICPT system is shown in Fig.6.

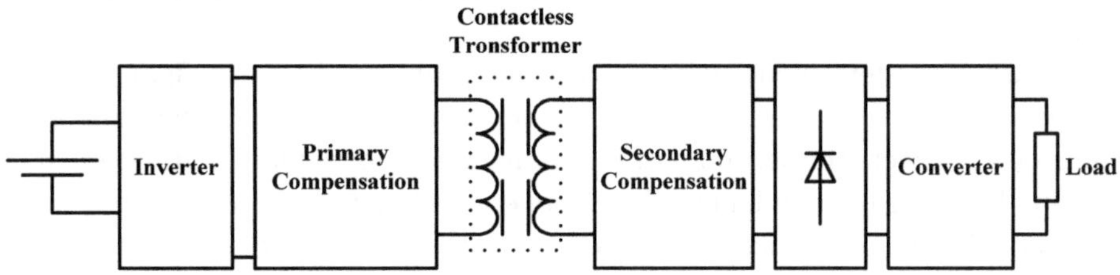

Fig. 6: Block diagram of a typical ICPTs

In this paper, the investigated ICPTs, as shown in Fig.7, employs a high frequency AC power to generate a standard sinusoidal voltage of Specific frequency. This voltage is fed to the rotatable transformer primary after primary compensation. The secondary transformer is connected to the special load--the dynamic balancing device directly to supply its power after secondary compensation. Due to the special load is driven by the high frequency sinusoidal voltage directly, so AC-to-DC converter is no longer need in secondary side of the ICPTs.

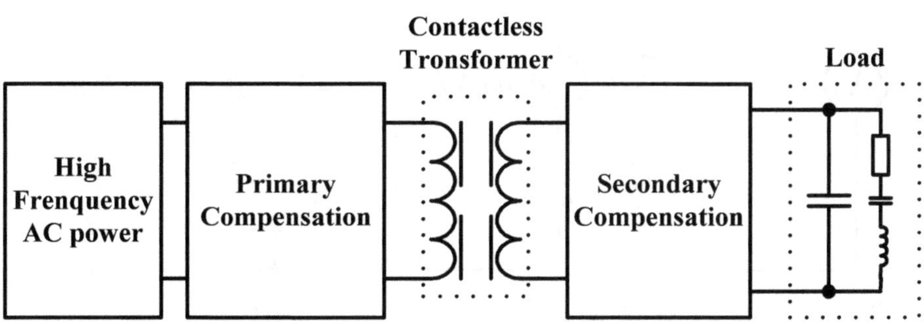

Fig. 7: Block diagram of the investigated ICPTs

To determine the characteristics of the ICPTs, the fundamental mode analysis (*FMA*) can be often used [24-26], which provides a much simpler and more intuitive analysis approach by modelling the resonant circuit in terms of the fundamental mode of the circuit waveforms, and thereby avoid mathematically complex from requiring a numerical solution to a large number of simultaneous transcendental equations. At the same time, mutual inductance modelling method [24, 25, 27] is adopted in this paper, to analyse the coupling effect of the contactless rotatable transformer. In the mutual inductance model, the induction voltage and reflected voltage are used to describe the coupling effect. The two concepts are expressed by mutual inductance. The reflected voltage shows the whole influence from secondary winding on primary winding. So it is not necessary to divide mutual inductance from leakage inductance.

The whole system model without compensation, as shown in fig.8, includes three parts. The first part is an AC power model that represents the high-frequency AC power. The second part is a transformer model that describes the relationships between the fundamental components related to the contactless transformer circuit. The resistance and inductance values of both sides are respectively R_p, L_p, R_s and L_s, and M is the mutual inductance. The inductive voltages in primary and secondary circuits are respectively $-j\omega_s MI_s$ and $j\omega_s MI_p$. Power transmission is realized during the process of inducing voltage mutually. The third part is the electromechanical equivalent model of the piezoelectric actuator, which describes the load characteristics of the ICPTs.

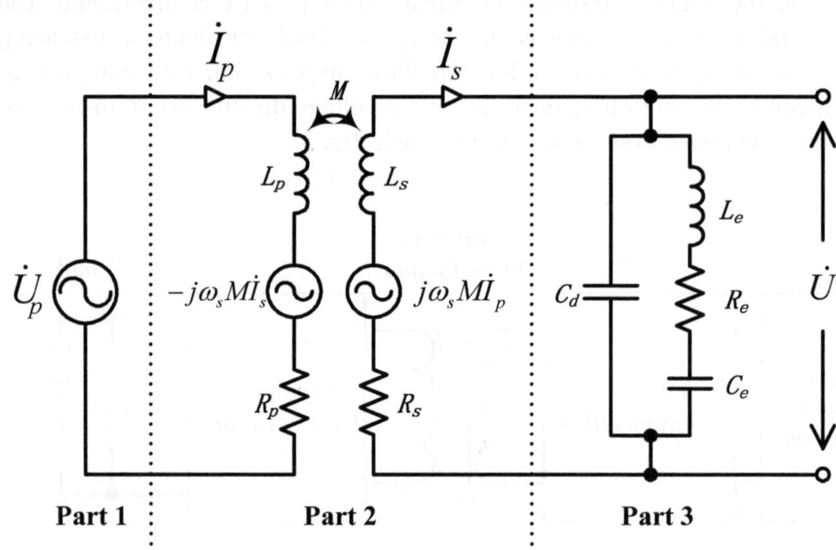

(A)

Fig. 8: Simplified circuit model of the ICPTs

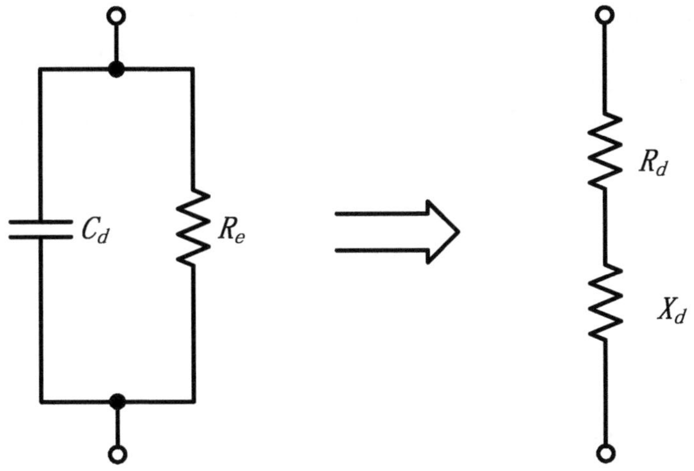

(B)

Fig. 9: circuit model of piezoelectric actuator

As the contactless rotating transformer has the characteristics of relative large leakage inductance and low magnetic inductance, and the system is highly reactive due to the high transmission frequency, the power transfer efficiency is poor. In order to improve the power transfer efficiency of the whole system, an efficacious approach using impedance matching to compensate the reactive power for the primary and secondary circuits respectively is necessary. The primary compensation can make the ICPTs ideally operating near zero phase angles between the primary input voltages and current at a locked frequency in order to minimize the VA rating of the supply, thus the power factor and supply quality can be meliorated. At the same time, the output power and transmission efficiency can be improved by the secondary compensation [24, 25, 27]. In further studies, the system circuit model and the compensation topology analysis will be implemented synchronously, and the process of analysis will be derived in reverse direction (from the secondary side to the primary side of the ICPTs).

In this paper, the nominal resonant transmission frequency of the ICPTs depends on the inherent frequency of the piezoelectric actuator. In order to produce the desired vibration mode and maximize efficiency of the piezoelectric actuator, its inherent frequency should be defined at series resonant frequency of a corresponding vibration mode. As mentioned above, the series resonance frequency (f_s) of the piezoelectric actuator can be obtained by real test. When the piezoelectric actuator is operated at this frequency, its present dynamic sub-circuit state of pure resistance and electromechanical equivalent circuit can be simplified, as shown in Fig.9 (A). In order to facilitate further analytical work, it can be continued to transform into the standard impedance model, as shown in Fig.9 (B).

The equivalent resistance (***R***$_d$) and reactance (X_d) of the piezoelectric actuator can be obtained as:

$$R_d = \frac{R_e}{1+\omega_s^2 C_d^2 R_e^2} \tag{1}$$

$$X_d = -\frac{\omega_s C_d R_e^2}{1+\omega_s^2 C_d^2 R_e^2} \tag{2}$$

Therefore, when the piezoelectric actuator operates in series resonant frequency, the system secondary circuit is equivalent to the Fig. 10:

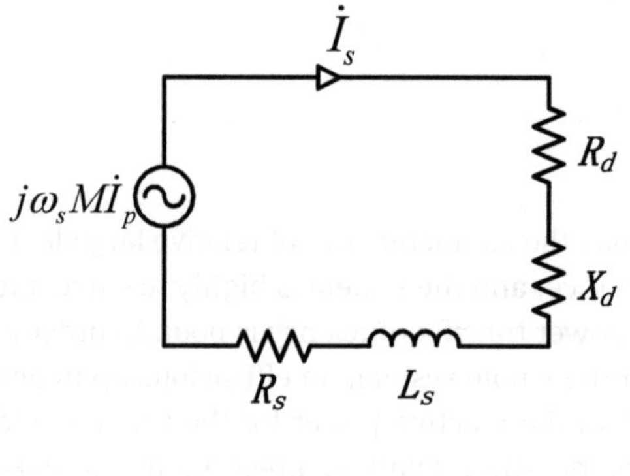

Fig. 10: Secondary circuit model without compensation

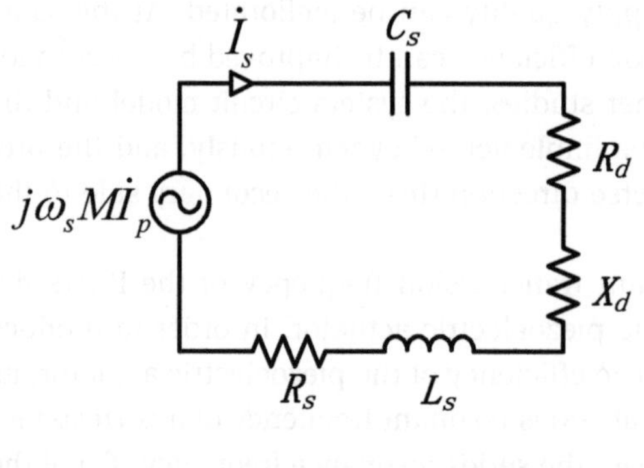

Fig. 11: Secondary circuit model with compensation

The expression of total impedance of the secondary circuit can be given by:

$$Z_s = R_s + \frac{R_e}{1+\omega_s^2 C_d^2 R_e^2} + j\omega_s (L_s - \frac{R_e^2 C_d}{1+\omega_s^2 C_d^2 R_e^2}) \tag{3}$$

Considering the above expression, it can be seen that the secondary circuit is highly reactive under high operational frequency. Thus if the secondary winding is connected directly to the piezoelectric actuator, reactive power loss of the secondary circuit must be greatly increased. Finally, it will become a reality that the power transfer capability cannot meet the requirements. For this reason, it is very necessary to employ secondary compensation to boost the power transfer capability from the primary to the secondary. As a result, there is no limit to the power transfer capability because the series compensated secondary is a voltage source in effect driving into the piezoelectric actuator at its series resonant frequency, and making the output voltage independent of the load and equal to the secondary open circuit voltage.

According to the circuit designed by us, the secondary circuit is inductive under no compensation, so $\boldsymbol{L_s}$ must be more than $R_e^2 C_d / (1+\omega_s^2 C_d^2 R_e^2)$. So far, the specific design parameter of the self-inductance of the secondary winding of the inner ring transformer has been determined, and can be complied in processing the transformer in next section.

Due to the designed secondary circuit is inductive, it is necessary to add compensation of series capacitance. The secondary compensation topology can be shown in Fig.11.

As a consequence, the expression of total impedance of the secondary circuit can be re-written in the form:

$$Z_s = R_s + \frac{R_e}{1+\omega_s^2 C_d^2 R_e^2} + j\omega_s (L_s - \frac{1}{\omega_s^2 C_s} - \frac{R_e^2 C_d}{1+\omega_s^2 C_d^2 R_e^2}) \tag{4}$$

In the above equations, the capacities of secondary side compensation capacitor can be selected in such a way that the imaginary part of $\boldsymbol{Z_s}$ equals zero. This means that the total impedance of the secondary circuit is purely resistive, the phase shift between secondary voltage and current, which supplies the piezoelectric actuator, is zero and no reactive power is drawn from the power. Hence the capacitor value for series compensation is as follows:

$$C_s = (\omega_s^2 L_s - \frac{\omega_s^2 R_e^2 C_d}{1+\omega_s^2 C_d^2 R_e^2})^{-1} \tag{5}$$

At the same time the final form of the expression of total impedance of the secondary circuit is:

$$Z_s = R_s + \frac{R_e}{1+\omega_s^2 C_d^2 R_e^2} \tag{6}$$

So far, the equivalent circuit of whole ICPTs is as follows:

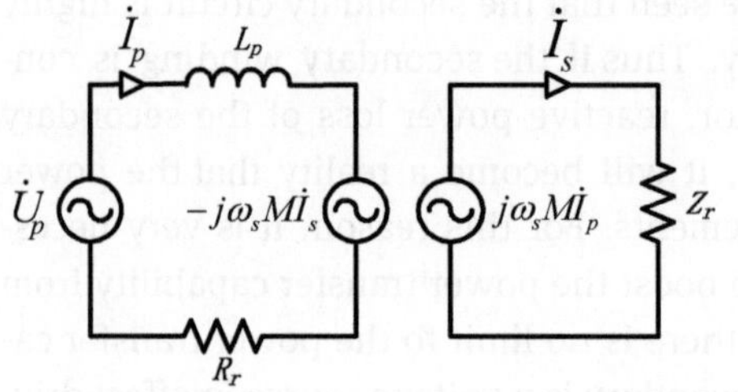

Fig. 12: Further simplified circuit model of the ICPTs

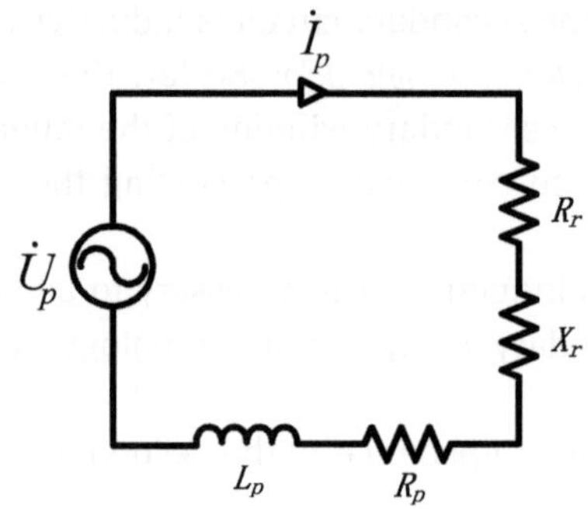

Fig. 13: Primary circuit model without compensation

If the positive directions of currents and voltages are as shown in the Fig.12, the following equations can be derived:

$$(R_p + j\omega_s L_p)\dot{I}_p - j\omega_s M\dot{I}_s = \dot{U}_p \tag{7}$$

$$Z_s\dot{I}_s - j\omega_s M\dot{I}_p = 0 \tag{8}$$

The above equations are the basic loop equations of the primary and the secondary circuit. Reflected impedance, $\boldsymbol{Z_r}$, is introduced to reflect the impact that the secondary circuit exerts on the primary. The reflected impedance shows the transmission performance of system energy. The complex power absorbed by reflected impedance is just the one of secondary system. $\boldsymbol{Z_r}$ can be tentatively assumed as follows:

$$Z_r = R_r + jX_r \tag{9}$$

Where $\boldsymbol{R_r}$ and $\boldsymbol{X_r}$ are reflected resistance and reflected reactance respectively. Furthermore, the equivalent circuit seen from the primary side can be derived, as shown in Fig. 13.

The basic loop equations of the primary can be re-written as follows:

$$(R_p + R_r + j(\omega_s L_p + X_r))\dot{I}_p = \dot{U}_p \tag{10}$$

We can reduce (7), (8), and (10) to

$$R_r = \frac{\omega_s^2 M^2}{Z_s} = \frac{\omega_s^2 M^2 (1 + \omega_s^2 C_d^2 R_e^2)}{R_e + R_s (1 + \omega_s^2 C_d^2 R_e^2)} \tag{11}$$

$$X_r = 0 \tag{12}$$

It can be seen that the imaginary part of $\boldsymbol{Z_r}$ equals zero, then the primary input impedance is purely resistive. The phase shift between input voltage and current is zero and no reactive power is drawn from the power amplifier. Summarizing, it may be concluded from the above equations that the secondary circuit impedance reflects to the primary side for purely resistive when the secondary circuit has been compensated for the purely resistive circuit. It is evident that primary reactive power is completely caused by self-induction of the primary winding. Therefore, the compensation of the primary circuit only offsets the primary self-inductance of the transformer, which produces reactive component. As to the rotatable transform, a concentrated primary winding often requires large primary currents; a parallel compensation plan is more attractive [28-30]. Hence a single capacitor at the terminals of the primary winding is used as shown in Fig. 14.

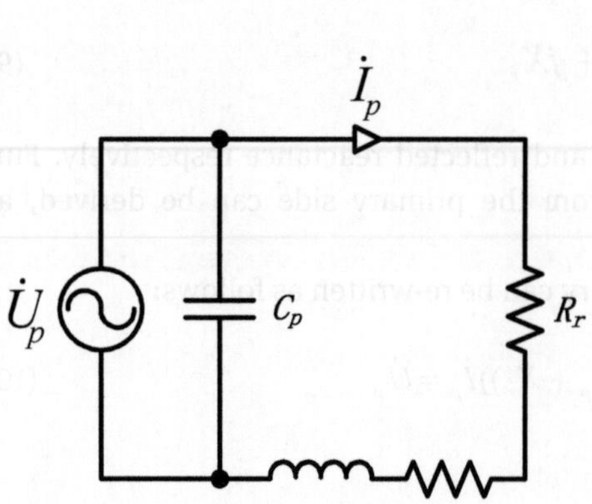

Fig. 14: Primary circuit model with compensation

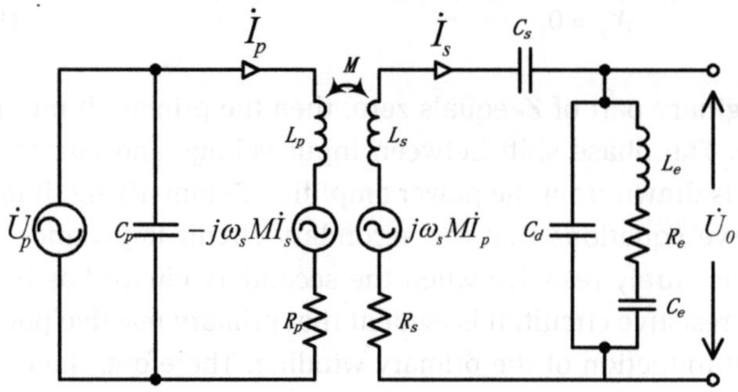

Fig. 15: Total circuit model of ICPTs with compensation

The expression of total impedance of the primary circuit with compensation is:

$$Z_p = \frac{(R_p+R_r)^3+\omega_s^2L_p^2(R_p+R_r)-j\omega_s\left[(R_p+R_r)^2+\omega_s^2L_p^2\right]\left\{\left[(R_p+R_r)^2+\omega_s^2L_p^2\right]C_p-L_p\right\}}{(R_p+R_r)^2+\omega_s^2\left\{\left[(R_p+R_r)^2+\omega_s^2L_p^2\right]C_p-L_p\right\}^2} \tag{13}$$

In the above equations, the capacities of the primary side compensation capacitor can be selected in such a way that the imaginary part of $\boldsymbol{Z_p}$ equals zero. This means that the total impedance of the primary circuit is purely resistive, and no reactive power is drawn from the power. Hence the capacitor value for parallel compensation is as follows:

$$C_p = \frac{L_p}{R_r^2+\omega_s^2L_p^2} \tag{14}$$

Finally, the total circuit model of the ICPTs with compensation can be drawn, and as shown in Fig.15.

4 Turns Choice Criterion and Inductance Measuring Method

Based on the previous section analysis, the self-inductance of the secondary winding of the inner ring transformer (***L_s***) must be more than $R_e^2 C_d / (1 + \omega_s^2 C_d^2 R_e^2)$, according the real test in **Section II (A)**, the following specifications are given: $C_d = 25nF$, $f_s = 67kHz$ and $R_e = 115\Omega$.

In this case, the self-inductance (Ls) must be more than 0.113 mH, and then an experimental prototype of the rotatable transformer has been developed with the following parameters. To simplify the analysis, the turn's ratio of the inner ring transformer is set to 1, tentatively.

Tab. 3: Winding parameters of the rotatable transformer

Parameter **(primary and secondary)**	**Inner ring transformer**	**Outer ring transformer**
turns	40	40
Diameter of wire (mm)	0.35	0.5

In this paper, the self-inductance of primary or secondary of the rotatable transformer can be directly measured when the secondary or the primary circuit is opened, respectively. A series-coupling test approach is adopted to measure the mutual inductances. The advantage of the series-coupling measurements over the traditional measurement approaches is that there is no problem of the winding resistance discussed before. As demonstrated below, the inductances measured in the series-coupling tests are in the algebraic addition and subtraction of inductances and are independent of winding resistances and parallel paths, thus facilitating direct measurement by the LCR meter [33].

The transformer can be setup into two different series-coupling measurement configurations that the secondary connections are reversed compared to each other. One of configurations is series opposing, yielding the differential inductance (***L_diff***), as shown in Fig. 16(A). The simplified expressions of the ***L_diff***, can be achieved as:

$$L_{diff} = L_p + L_s - 2M \tag{15}$$

In the cumulative measurement, as shown in Fig.16 (B), the cumulative inductance, L_{cum}, can be determined using the same test setup, except that the secondary connections are reversed. The inductance is given by

$$L_{cum} = L_p + L_s + 2M \tag{16}$$

According to (15) and (16), the general expression for the mutual inductance is given below:

$$M = (L_{cum} - L_{diff}) / 4 \tag{17}$$

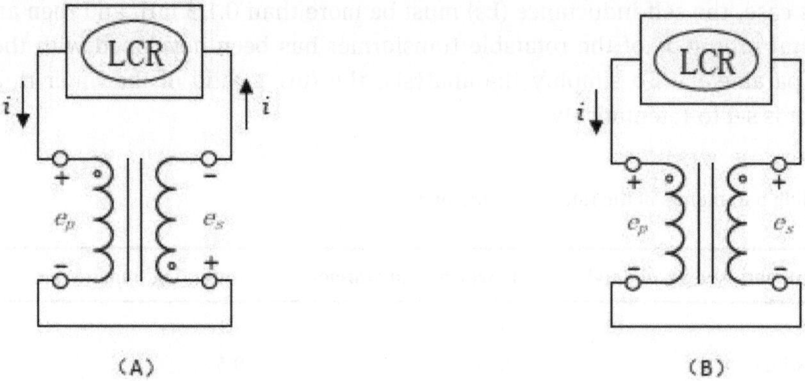

Fig. 16: Mutual inductance test setup

Equation (17) can be used to calculate M in the experimental section. In the case of different airgap distance, the measured data can be obtained, as shown by Tab 4 (LCR test frequency is fixed at 67 Hz).

Tab. 4 : Inductance measuring data for the rotatable transformer for different airgap distance

Air-gap	L_p	L_s	L_{cum}	L_{diff}	M
0.5	396.308	376.632	1301.85	245.073	264.194
1	382.87	363.56	1223.27	270.804	238.116
1.5	373.811	355.193	1169.91	289.12	220.197
2	365.646	347.464	1116.03	311.03	201.250
2.5	359.313	341.567	1071.49	330.846	185.161
3	353.517	335.885	1026.68	352.388	168.573
3.5	349.828	332.52	997.585	367.614	157.493
4	346.79	329.562	970.023	382.657	146.842
4.5	344.203	327.1	946.788	396.299	137.622
5	342.323	325.435	928.545	407.379	130.291
5.5	339.739	322.801	900.793	424.713	119.020
6	338.015	321.42	880.584	438.835	110.437

Through the above measured data, it is found that there is deviation in the self-inductance between the first and secondary windings due to deviation in process. Because it keeps steadily over ***19±1uH*** under different air gap, the deviation can be regarded as system error.

5 Experimental Evaluations of the ICPTs

To verify the real operation state of the ICPTs composed by the rotatable transformer, we established a set of experimental test bench, as bellow: Tektronix signal generator and RF power amplifier are used to construct the high-frequency AC power. Tektronix four ways oscilloscope is used as signal sampling device. The specific test parameters are stated here as the primary transformer is supplied by constant sine voltage power, voltage value of peak-peak is **100V** with frequency of 67kHz, air gap of the rotatable transformer is setup to 2mm, L_p=0.365646mH, L_s=0.347464mH, M=0.201250mH, the sampling resistors of the first and secondary current are both 1**Ω** of no-inductive precision resistors with power of 100W. The spindle speed is set for 3000 rpm, a slip ring installed on the rotor, is employed to deliver sampling signal from the rotating load (the balancing device) to stationary testing equipment's.

The first step is to do test without compensation in the inner transformer, and the equivalent test circuit is as following:

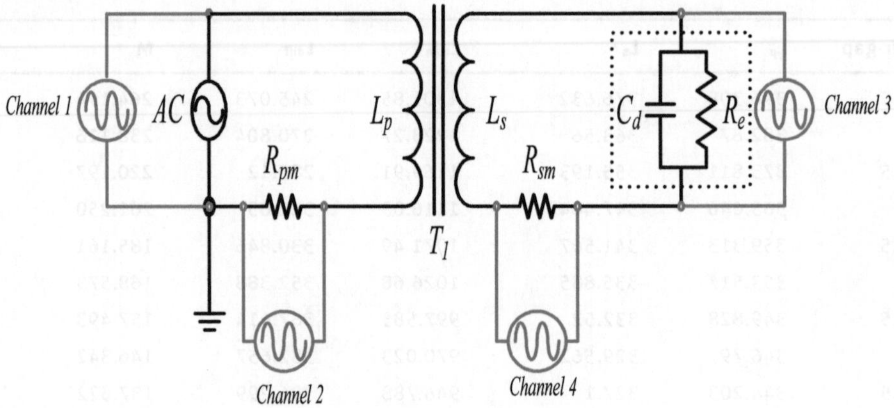

Fig. 17: Equivalent test circuit without compensation

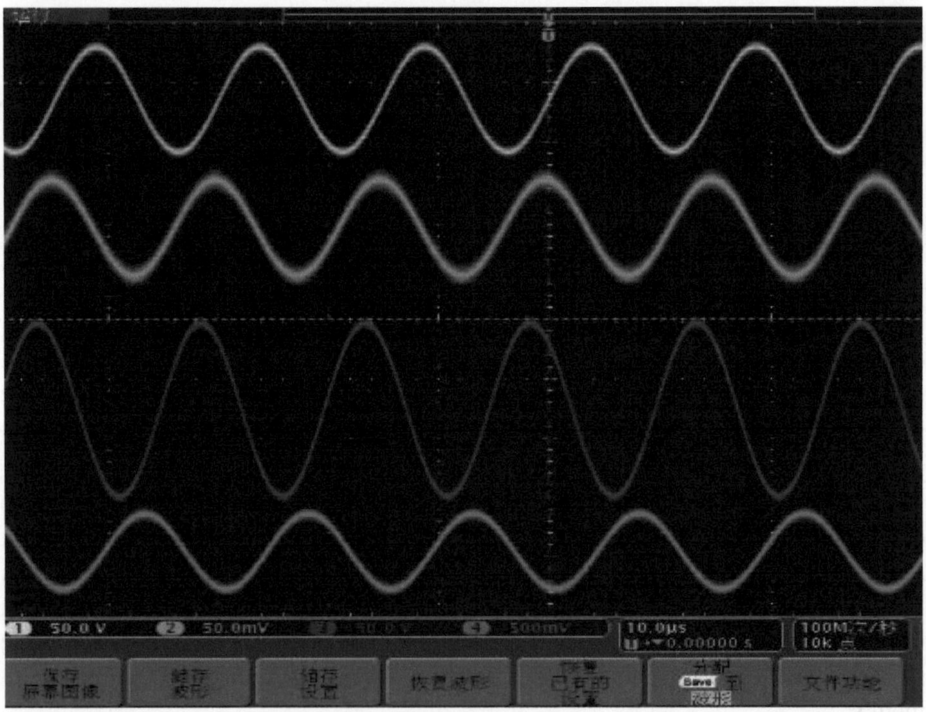

Fig. 18: Waveforms captured without compensation

In fig.17, AC power stands for the high-frequency AC power, transformer ***T1*** represents the inner transformer of the rotatable transformer, the part in dash line stands for the real piezoelectricity actuator, and $\boldsymbol{R_{pm}}$ and $\boldsymbol{R_{sm}}$ signify the sampling resistors of the primary and secondary side current respectively. In this case, the waveform captured from Tektronix oscilloscope which shows the test of actual voltage and current waveforms are as Fig.18. (Channel 1 to 4 show respectively the output voltage of the high-frequency AC power, the output current of the high-frequency AC power, the load voltage of the real piezoelectricity actuator and the load current of load).

The second step is to add the primary and secondary compensative capacities simultaneously in the above picture according to the compensation theory in the above analysis, and the experimental circuit is as below:

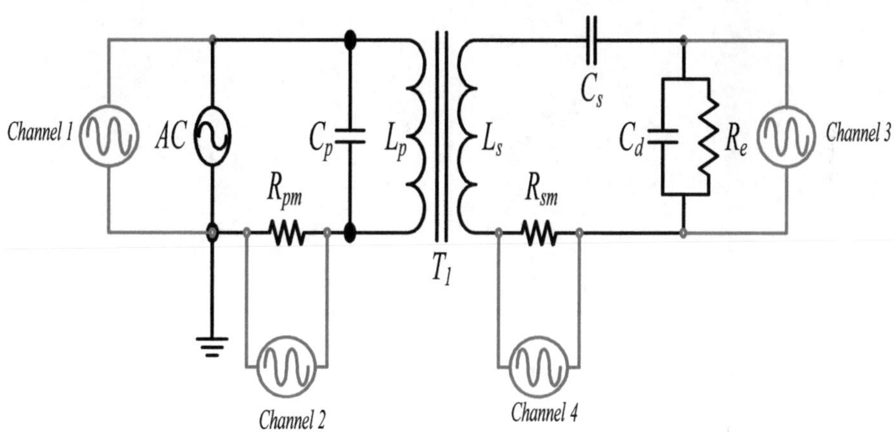

Fig. 19: Equivalent test circuit after compensation

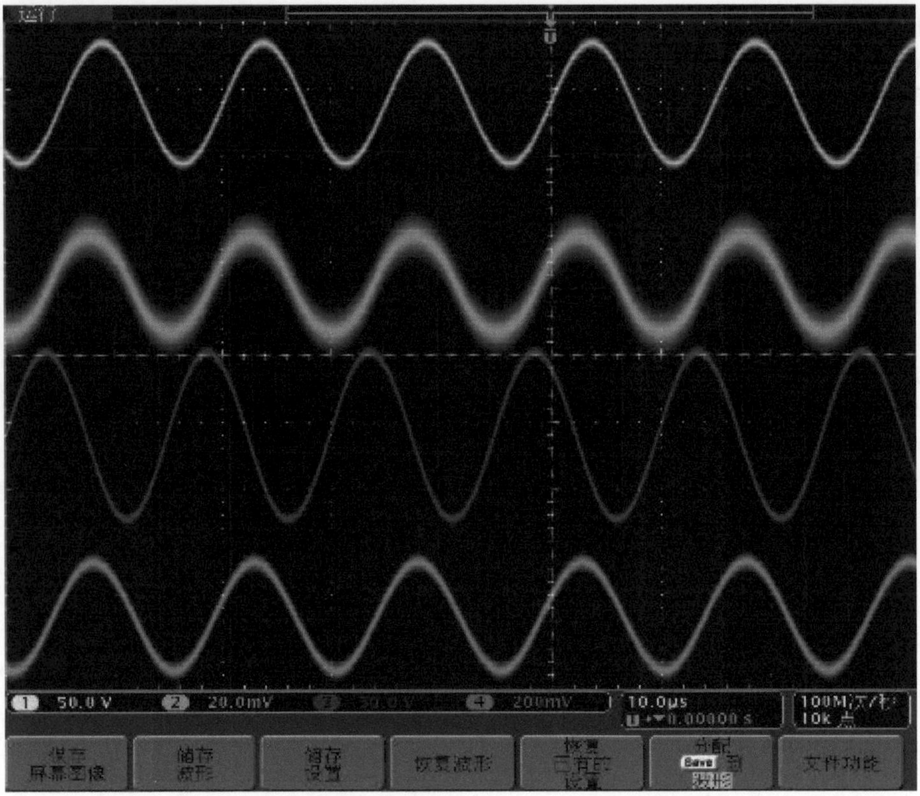

Fig. 20: Waveforms captured after compensation

The parameter of compensative capacitors in the figure is calculated as the compensation theory as:

$$\mathbf{C_p}=38.9\text{nF};\quad \mathbf{C_s}=1.75\text{nF}.$$

Because of the good high frequency characteristic of ceramic capacitor, four ceramic capacitors of 10nF/2kV are used in parallel as the first compensative capacitor, and secondary compensative capacitors are two ceramic capacitors of 1nF/2kV. And the parameter of actual compensative capacitors is:

$$\mathbf{C_p}=44\text{nF};\quad \mathbf{C_s}=2\text{nF}.$$

Similarly, the waveforms captured from Tektronix oscilloscope which shows the test of actual voltage and current waveforms are as shown in Fig.20.

The third step is to observe the interference from the outer transformer to the inner transformer. To configure the practical circuit for the outer transformer according to the compensative circuit of the inner one, it is different to replace the internal resistance of unloaded driving power with a resistor of $\boldsymbol{R_{p1}}$, as the following picture:

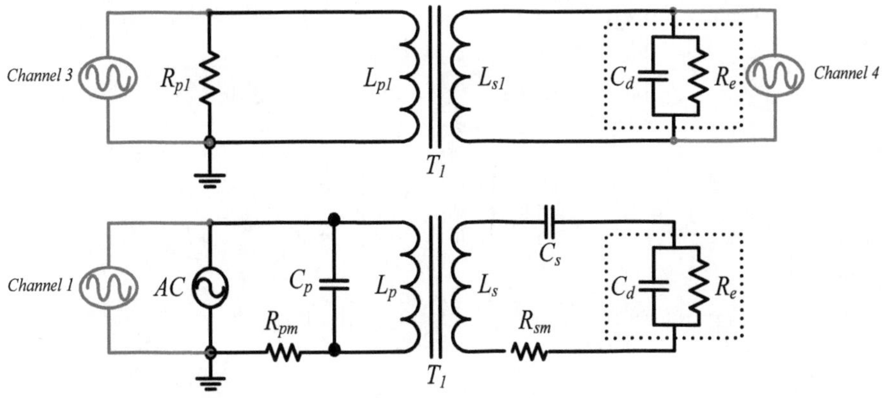

Fig. 21: Equivalent test circuit of the third step

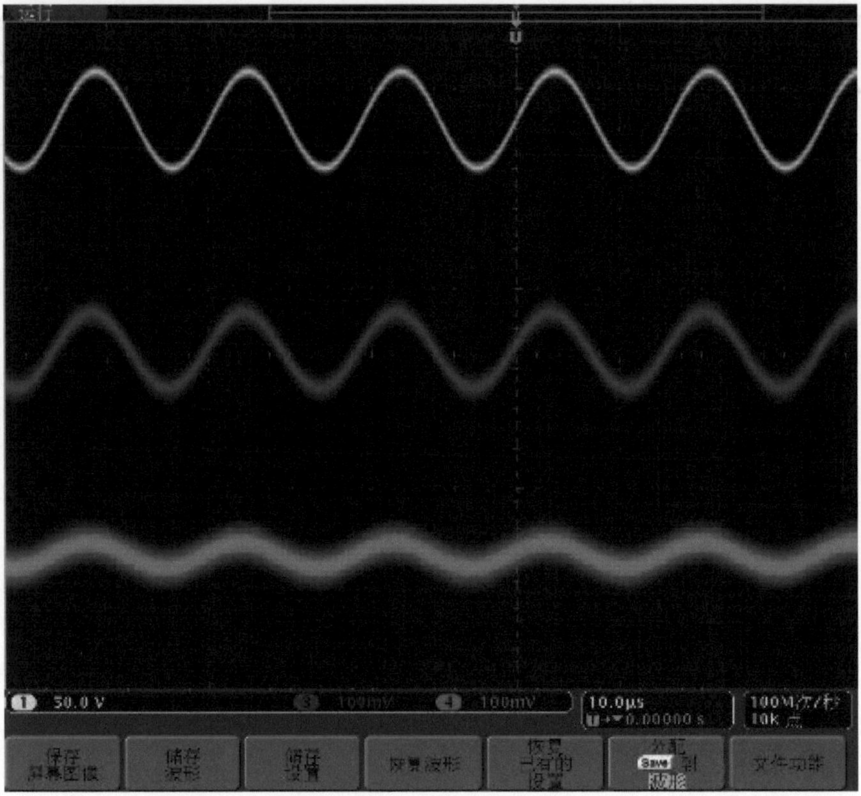

Fig. 22: waveforms captured of the third step

The waveforms captured from Tektronix oscilloscope which show the test of actual voltage waveforms are as Fig.22.

Experimental results drawn from the above experiments are as:

1 The rotatable transformer made of the self-made ferrite core can completely meet the requirements of the energy transfer and control of the dynamic balancing device.
2 Due to the application of the primary and secondary compensative circuit, operating at or near zero phase angle between the primary input voltage and the current is realized in the primary side circuit at first, and VA rating of the power amplifier can be reduced greatly, the result is that the peak-peak value of current of the supply power is reduced to 40mA from 100mA after compensation under the precondition that the first voltage keeps no change. Secondly, the reactive power of secondary circuit can be decreased greatly due to the presence of compensative capacitor. The peak-peak value of current of the load is reduced to 360mA from 700mA after compensation under the precondition that the driving voltage keeps no change.

3 Because the inner and outer ring transformers works alternately. It can be seen from Fig.22 that the inductive energy strength of the first and secondary windings of the outer ring transformer is very low (the inductive voltage of the primary winding is 140mV, and the one of secondary winding is less than 100mV) when the inner ring transformer is working. The dynamic balancing device driven by such energy cannot cause malfunction at all under such inductive energy strength.

6 Conclusion

This paper describes the implementation of ICPTs using a special stepwise contactless rotatable transformer to transfer high-frequency AC electric energy with power of about 10W to the balancing device which is mounted on rotor of the spindle and synchronous high-speed rotating with it. The rotatable transformer adopts special ferrite core material, which insures the machining accuracy and the whole strength of the transformer structure. This paper also makes detailed studies on the system models and compensation methods for the design and operation of the ICPTs operating at near zero phase angles between the primary input voltage and current at a locked frequency allowing maximum power transfer capability to be achieved. At last, contrast tests are applied to test the operation characteristic of the whole ICPTs under different operation state. The reliability of derivative result by theory is verified furthermore. By experiments, it is proved that this set of ICPTs can completely meet the requirements of the energy supply and control of the dynamic balance adjusting device.

Acknowledgement: This work was supported by the National Important Special Project under Grants [2015ZX04005001], Errors Measurement, Analysis and Compensation of CNC Machine Tools.

References

[1] Jongduk Moon, Bongsuk Kim, Soohun Lee, "Development of the active balancing device for high-speed spindle system using influence coefficients", International Journal of Machine Tools & Manufacture, Vol.46: pp. 978–987, 2006
[2] Michał Manka, Tadeusz Uhl, "Mechatronic design of fault detection isolation and restoration systems for rotating machineries" Mechanism and Machine Theory, Vol.44: pp. 1436–1449, 2009
[3] S. Das, M. C. Nighil, J. K. Dutt, H. Irretier, "Vibration control and stability analysis of rotor-shaft system with electromagnetic exciters", Mechanism and Machine Theory, Vol. 43: pp. 1295–1316, 2008

[4] S. W. Dyer, and J. Ni, "Adaptive Influence Coefficient Control of Single-Plane Active Balancing System for Rotating Machinery". Journal of Manufacturing Science and Engineering, Vol.123.1: pp. 291–298, 2001

[5] Esser and H. C. Skudelny "A New Approach to Power Supplies for Robots", IEEE Transactions on Industry Applications, Vol. 21.5: pp. 872–875, 1991

[6] Chengjun Liu, Ming Qi, Jibin Zou, "Modeling and Test of Contactless Transformer Used in Inductosyn", International Conference on Electrical Machines and Systems, Wuhan, P. R.China, 2008, pp. 4323–4327

[7] K. D. Papastergiou, D. E. Macpherson, and F. Fisher, "Contact-less transfer of energy by means of a rotating transformer," Proceedings of the IEEE International Symposium on Industrial Electronics, Dubrovnik, Croatia, 2005, pp. 1735–1740.

[8] K. D. Papastergiou and D. E. Macpherson, "An airborne radar power supply with contactless transfer of energy—Part I: Rotating Transformer", IEEE Transactions on Industrial Electronics, vol. 54(5) : pp. 2874–2884, 2007

[9] K. D. Papastergiou and D. E. Macpherson, "An airborne radar power supply with contactless transfer of energy—Part II: Converter design", IEEE Transactions on Industrial Electronics, vol. 54(5): pp. 2885–2893, 2007

[10] R. L. Steigerwald, C. F. Saj, G. A. Croff, "Analysis and design of a contactless rotary power transfer system", IEEE Power Electronics Specialists Conference, 32nd Annual, Vancouver, Canada, 2001, pp. 2125–2130

[11] Guidong Luan, Jinduo Zhang, Renqian Wang, Piezoelectric Transducers and Arrays, Peking University Press, 1990 (In Chinese)

[12] Shuyu Lin, Theory and design of ultrasonic transducers, Science Press, 2004 (In Chinese)

[13] J. Legranger, G. Friedrich, S. Vivier, J.C. Mipo, "Comparison of two optimal rotary transformer designs for highly constrained applications" IEEE International Electric Machines & Drives Conference, Antalya, Turkey, 2007, pp. 1546–1551

[14] M. Pern´ıa, I. C. Orille, J. A. Martinez, J. M. Ramos, J. A. Canal, F. Zacouto "Transcutaneous microvalve activation system using a coreless transformer" Sensors and Actuators A, Vol.136: pp. 313–320, 2007

[15] [15] C. Fernandez, O. Garcia, R. Prieto, J. A. Cobos, S. Gabriels, G. Van Der Borght, "Design issues of a core-less transformer for a contact-less application" 17th Annual IEEE Applied Power Electronics Conference and Exposition, Dallas, TX , USA , 2002. Vol.1: pp. 339–345

[16] J. G. Hayes, N. O'Donovan, M. G. Egan, T. O'Donnell "Inductance characterization of high-leakage transformers" IEEE Applied Power Electronics Conference and Exposition, Miami, Florida, USA, 2003. Vol.2: pp. 1150–1156

[17] D. A. G. Pedder, A. D. Brown and J. A. Skinner, "A contactless electrical energy transmission system", IEEE Transactions on Industrial Electronics, vol. 46: pp. 23–30, 1999

[18] Y. Jang, and M. M. Jovanovic, "A contactless electrical energy transmission system for portable-telephone battery chargers" , IEEE Transactions on Industrial Electronics, vol. 50, pp. 520–527, 2003

[19] Yuanhsin Chao, Jennjong Shieh, Chingtsai Pan and Weichih Shen, "A closed-form oriented compensator analysis for series-parallel loosely coupled inductive power transfer systems" , IEEE Power Electronics Specialists Conference, Orlando, FL, USA, 2007, pp.1215–1220

[20] G. B. Joung and B. H. Cho, "An energy transmission system for an artificial heart using leakage inductance compensation of transcutaneous transformer," IEEE Transactions on Power Electronics, vol.13 (6): pp. 1013–1022, 1998.

[21] C. Wang, G. A. Covic, O. H. Stielau, Power transfer capability and bifurcation phenomena of loosely coupled inductive power transfer systems, IEEE Transactions on Industrial Electronics, Vol. 51: pp. 148–157 , 2004

[22] R. Mecke, C. Rathge, "High frequency resonant inverter for contactless energy transmission over large airgap", Proceedings of the 25th Annual IEEE Power Electronics Specialist Conference, 2004, Vol.3: pp. 1737–1743
[23] H. Ayano, K. Yamamoto N. Hino, I. Yamato, "Highly efficient contactless electrical energy transmission system", Proceedings of the 28th Annual IEEE Industrial Electronics Society Conference, 2002, Vol.2: pp. 1364–1369
[24] Ying Wu, "Study on new contactless power supply system," Doctoral dissertation of Chinese academy of sciences, 2004.
[25] Ying Wu, Luguang Yan, Shangang Xu, "Modeling and Performance Analysis of the New Contactless Power Supply System" The Eighth International Conference on Electrical Machines and Systems, Nanjing, P.R.China, 2005,Vol. 3: 1983–1987
[26] Dengya Zhang, Wenxin Huang, Yumen Hu, "Study of a new rotor inductive power system", Power Supply Technologies and Application, Vol.8(10): pp. 61– 64, 2005 (In Chinese)
[27] M. Hao and W. Zhou, "Modeling a Current Source Push-Pull Resonant Converter for Loosely Coupled Power Transfer System," The 30th Annual Conference of the IEEE Industrial Electronics Society, Busan, Korea, 2004, Vol.2: pp.1024–1029
[28] H. Stielau, G. A. Covic, "Design of loosely coupled inductive power transfer systems", International Conference on Power System Technology, Perth, WA , Australia, 2000, Vol.1: pp. 85-90
[29] J. T. Boys, G. A. J. Elliott, and G. A. Covic, "An Appropriate Magnetic Coupling Co-Efficient for the Design and Comparison of ICPT Pickups," IEEE Transactions on Power Electronics, Vol. 22, Issue 1, pp. 333–335, January 2007
[30] H. Abe, H. Sakamoto and K. Harada, "A noncontact charger using a resonant converter with parallel capacitor of the secondary coil", IEEE Transactions on Industrial Applications, vol. 36(2), pp. 444–451, 2000.
[31] Wei Zhang, Qianhong Chen, S. C. Wong, C. K. Tse, Xinbo Ruan, "A novel transformer for contactless energy transmission systems", IEEE Energy Conversion Congress and Exposition, San Jose, CA, 2009, pp.3218–3224
[32] Houjian Zhan, Jiekang Wu, Nan Zhao, Cheng Jiang, Jiaxin Qi, "Parametric Design for Loosely Coupled Transformer in Contactless Inductive Power Transfer System" Modern Electric Power, Vol.26(1): pp.40– 44, 2009 (In Chinese)
[33] J. G. Hayes, N. O'Donovan, M. G. Egan, T. O'Donnell, "Inductance Characterization of High-Leakage Transformers", The 18th Annual IEEE Applied Power Electronics Conference and Exposition, 2003 vol.2: pp. 1150–1156

Gang Zhao*, Shuang Liu, Jia-Le Li, Cao Ji and Long-Gang Wu

Study on the Coordination Principle of Innovative Manufacturing Elements Based on Internet Plus

Abstract: Based on the related concepts of Internet plus,this paper researches on the Intelligent Manufacturing Collaborative Service System based on processing technology behaviour.The purpose is to enable the designer to realize close combination of product design and manufacturing, save product development time and improve the resource utilization rate and market flexibility of enterprises. Under this background, the following aspects of research have been mainly done in this paper:The concept of processing technology behavior and its hierarchical division are explained, and the relationship between the processing technology behavior at different levels is studied; The concept of Innovative Manufacturing Elements and system requirement are detailed, and the coordination principle is also studied; Based on the Internet plus scenarized demand, the feasibility and reliability of the coordination principle of Innovative Manufacturing Elements are validated with an example of an axle part.

Keywords: Internet plus, processing technology behavior, Innovative Manufacturing Element, Intelligent Manufacturing Collaborative Service System, scenarized manufacturing

1 Introduction

The concept of "Internet plus" which we refer to at present is different from the traditional academic concept. It has risen from a general academic concept to a new

***Corresponding author: Gang Zhao,** College of Mechanical and Electrical Engineering, Harbin Engineering University, Harbin, China.
Email: zhaogang@hrbeu.edu.cn.Corresponding author.
Shuang Liu, College of Mechanical and Electrical Engineering, Harbin Engineering University, Harbin, China. Email: CNliush@163.com.
Jia-Le Li, College of Mechanical and Electrical Engineering, Harbin Engineering University, Harbin, China. Email: lijiale engineer@163.com.
Cao Ji, College of Mechanical and Electrical Engineering, Harbin Engineering University, Harbin, China. Email: 554994327@qq.com.
Long-Gang Wu, College of Mechanical and Electrical Engineering, Harbin Engineering University, Harbin, China. Email: 734123745@qq.com.

DOI 10.1515/9783110540048-070,

concept of strategic importance to all sectors of the country's economy, politics, and so on [1]. The concept of Innovative Manufacturing Element is proposed in order to enable the mechanical manufacturing to adapt to the Internet plus. It is essentially a data structure of the software platform and it can be abstracted into a tetrahedral model, as shown in Fig. 1. The four sides of the model represent design, process, quality and economy respectively, which are also included in the mechanical manufacturing process. The coordination principle of Innovative Manufacturing Element is formulated according to the data information carried by this data structure, which is used to invoke the data structure rules. At home and abroad, the idea of internalizing processing manufacturing information into software platform has also been studied [2]. However, for most of the time only a single kind of information, such as price, delivery date, is internalized into the software platform. It only leads to the result that more mechanical price trading software has been spawned and it is impossible to include all kinds of parameters of the machine parts in the platform [3]. Therefore, a new idea that all information of parts in the mechanical assembly is internalized into software platform is introduced in this paper. Thus the specific part information can be coordinated through the Internet and the manpower coordination can be left out. It saves manpower and material resources, and greatly improves the efficiency of manufacturing.

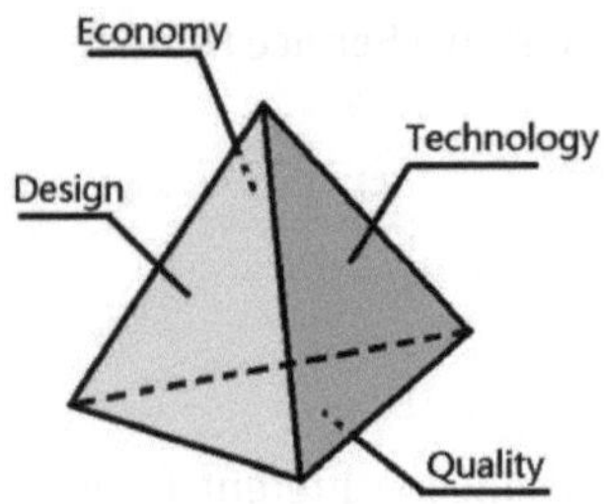

Fig. 1: The model of Innovative Manufacturing Element

2 Elaboration of Relevant Theories on Processing Technology Behavior

2.1 The Concept of Processing Technology Behavior

The essence of the whole product development process from the design to the finished product is that under the premise of making sure that the design requirements must match the manufacturing capabilities of the production facility and the producer can process the raw material through a single or combined process until the product meets the design requirements. The structure and appearance of the parts are essentially a series of geometric features formed by the processing technology and are combined in a certain order. Accordingly, two theorems about processing technology behavior are explained as under.

Theorem 1 Any part must be processed by a certain sequence of processing techniques, and parts that cannot be produced by the process don't exist.

Theorem 2 Any part can be described and expressed by using certain processing techniques.

Based on the above two theorems, the concept of processing technology behavior is given.

Processing technology behavior refers to the machining process through which parts are finished from raw material to final product with the constrains of specific design requirements, the manufacturing resources and capacity which belong to the subject of the processing technology behavior. Processing technologyy behavior contains all the parameter information of the corresponding processing characteristics such as machining method, geometry, dimension information, precision information, position information, production and processing resource information etc.

2.2 The Description Method of Processing Technology Behavior

Based on the understanding of the inherent requirements of processing technologyy behavior information processing, this paper argues that the information of processing technologyy behavior should be divided into different levels so that the intelligent goal of the system can be achieved.

According to the component assembly information of the product, the processing technology behavior is divided into different levels of Innovative Manufacturing Elements, parts, components, and products .And each level includes different processing technologyy behaviors. Cylinder turning, Keyway milling and Hole drilling etc. are included in Innovative Manufacturing Elements. Axles, Gears and Discs etc. are included in parts. Bearings, Couplings and Clutches etc. are included in components. Machine tools, Ships and Automobiles etc. are included in products.

The high-level machining characteristics can be shown in a certain combination of low-level processing technology behavior. Similarly, the processing technology behavior at the upper level can be split into multiple low-level ones, as shown in Fig. 2.

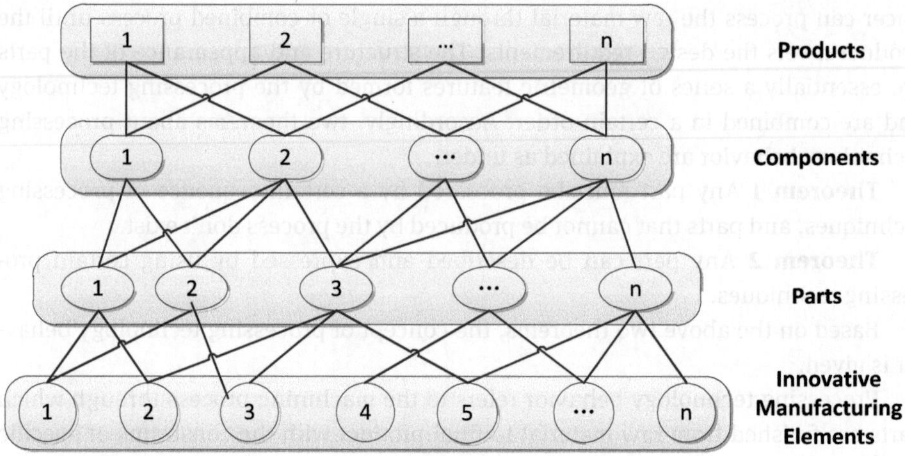

Fig. 2: Hierarchical relationship of processing technology behaviors

Each level of processing technology behavior has two coordinate systems which respectively are the positioning coordinate system that consists of positioning coordinates and positioning vectors, and the assembly coordinate system that consists of assembly coordinates and assembly vectors, as shown in Fig. 3. The positioning coordinate system is the coordinate system of the processing technology behavior itself, and the geometric characteristics of the processing technology behavior are mainly expressed in the positioning coordinate system. The establishment of the processing coordinate system is helpful to establish the processing method according to the current processing technology behavior. And it is especially beneficial to the generation of NC codes. Assembly coordinate system is the processing coordinate system of higher level process technology behavior. The position of the geometric features formed by the current processing technology behavior in the whole higher level behavior coordinate system is expressed in an assembly coordinate system. Assembly coordinate system is beneficial to the conversion of processing methods for multiple processing technology behaviors. Because the corresponding processing code of the same processing technology behavior in the same part's different location is not the same.

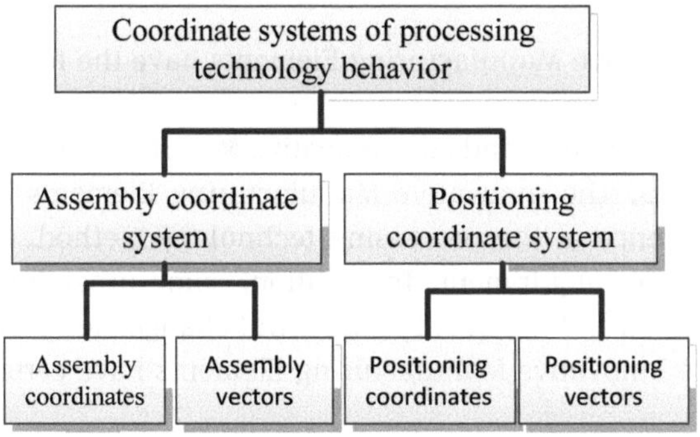

Fig. 3: Coordinate systems of processing technology behavior

Based on the division method and coordinate definition of the processing technology behavior above, the processing technology behavior can be described visually and accurately.

3 Research on Coordination Principle of Innovative Manufacturing Elements

3.1 The Concept of Innovative Manufacturing Element

Innovative Manufacturing Element refers to the smallest level of machining process that parts finish from raw material to final product with the constrains of specific design requirements, the manufacturing resources and capacity which belong to the subject of the processing technology behavior. It is the unit level of processing technology behavior.

Innovative Manufacturing Element is essentially a data structure from platform implementation point of view. Innovative Manufacturing Element can have all kinds of information about design and process, along with some data of quality and price are added to this data structure in order to make it have both quality and economic attributes. The quality and applicability of the final product can be ensured through setting design, process and economy reasonably.

3.2 The Characteristics of Innovative Manufacturing Elements

In the field of manufacturing, Innovative Manufacturing Elements have the following characteristics:

(1) Dynamic and diversity. There are many kinds of Innovative Manufacturing Elements for different applications. The Innovative Manufacturing Elements are different because of the difference of the processing technology method, the manufacturer and the production equipment. In addition, some other information, such as the current status and the quality grade, will change along with the processing service, so the Innovative Manufacturing Elements have certain dynamic characteristics.

(2) Heterogeneity and disreputability. The disreputability of Innovative Manufacturing Elements refers to the fact that Innovative Manufacturing Elements will not be confined to a particular region in Intelligent Manufacturing Collaborative Service System. These can be cross-region, cross-enterprise, cross-industry, and even cross-country. The universality and randomness of Innovative Manufacturing Elements' distribution result in the heterogeneity of understanding and description.

(3) Abstractness and similarity. In Intelligent Manufacturing Collaborative Service System, it is necessary to abstract the Innovative Manufacturing Elements into concrete model information, and this allows designers to invoke them in better way and thereby serve the users. Although Innovative Manufacturing Elements are different, they do have similarity in some attribute information, which brings convenience to the behavior description.

3.3 System Requirements for Innovative Manufacturing Elements

In order to achieve the goal of "guaranteeing quality and improving production efficiency", the Iintelligent Manufacturing Collaborative Service System puts forward "Two changes and Two properties" system requirements for Innovative Manufacturing Elements. "Two changes" means standardization and modularization; "Two properties" means reusability and composability.

3.3.1 Standardized Requirements for Innovative Manufacturing Elements

The essence of Iintelligent Manufacturing Collaborative Service System is a large-scale, heterogeneous, and resource efficiently-shared distributed networked collaborative manufacturing. The effective integration, management and cross invoking of heterogeneous Innovative Manufacturing Elements are the basis for the operation of InTelligent Manufacturing Collaborative Service System. This requires the Innovative Manufacturing Elements provided by the various providers in the manufacturing service system to use standardized description in order to shield the complexity of the underlying process, methods, and resources resulting from heterogeneity. After encapsulation, they are released to Intelligent Manufacturing Collaborative Service System in the form of service, and the manufacturing service platform can update the information in real time according to the actual situation of the Innovative Manufacturing Elements, and also provide interfaces for the operation of them [4].

3.3.2 Modular Requirements for Innovative Manufacturing Elements

The advantage of modularization is that the design platform has refined and classified various design elements which need to be used and they will finally be packaged into separate design modules for users to invoke [5].Compared with CAD system, the modularization mentioned in this paper is an upgrade from briefness to complexity and from a single line element to a specific part element. The modularization of the design platform should be focus on specific component design elements such as cylinder, chamfer, thread and threaded whole, etc. When a designer needs to design a column, what is needed to do is not to draw the cylinder itself, but to invoke the cylinder module directly. In that way, the repeated design process is omitted and the design is accomplished by invoking the module directly. The accuracy of design is ensured and the efficiency is improved.

3.3.3 Reusability Requirements for Innovative Manufacturing Elements

Reusability is the guarantee of modularization's vitality. The important function of reusability is that a module template can be used to implement the invoking of a module. For example, if a module is a threading module, reusability requires it to enable all types of threads to be invoked by setting parameters, specifying options and so on. Reusability allows the efficiency of modular invocation to produce a qualitative improvement and enables a module template to meet the invoking requirements of a type of processing technology behavior.

3.3.4 Composable Requirements for Innovative Manufacturing Elements

Due to the fact that the processing technology behaviors provided by different manufacturers are limited, in the face of complex manufacturing tasks, a single manufacturer may not fully meet the needs of it. At this point, different manufacturers' processing behaviors must be combined organically in order to accomplish the manufacturing task together[6].Composability requires invoked modules to perform a combination with each other in accordance with certain rules according to actual processing. Thus, all the Innovative Manufacturing Elements can be combined into a complete part and parts can be combined and then integrated into a kit or a whole machine [7].

3.4 Coordination Principle of Innovative Manufacturing Elements

Innovative Manufacturing Element, as a data structure, needs to be applied to software platforms. And the coordination principle of Innovative Manufacturing Elements is a rule which specifies the use of it. After a few years of research on Innovative Manufacturing Elements, we have developed a concrete idea of internalizing complex data structures into software platforms and the development of software platform has been initially completed. In the program code of software, various processing behaviors represented by Innovative Manufacturing Elements are defined as several C++ program classes. A member variable that identifies various information is added to the class. And methods for serializing member variables for sending information across the network are provided in order to realize the real-time coordination between the software manufacturing end and the client. Innovative Manufacturing Elements have been able to achieve real-time transmission of various processing information through the application of the methods described above. And then the real-time coordination of mechanical equipment processing can be realized to greatly improve the processing efficiency. And the processing is more intelligent and convenient with little manpower and few material resources.

4 Case Study of Innovative Manufacturing Elements Based on Internet Plus Scenarized Demand

Internet plus era makes the concept of “Scene”, which means a product or an application introduced by a business to meet a particular user’s needs, arise at the historic moment [8]. In the mobile Internet era, “Scene” is not a simple term, it is the reconstruction of human and business connections. The so-called “Scene” is actually a new experience, accompanied by the creation of a new scene. It is a new fashion, accompanied by new insights. It is a new way of life, a popular way of new scenes [9]. With the continuous expansion of user experience requirements, the idea of customized manufacturing has also been gradually emphasized in the modern manufacturing industry. The market is gradually changing from price orientation to scene orientation.

But today, many small and medium-sized manufacturing enterprises in China are still in traditional manufacturing mode, can't provide customized service for makers or other product design end. The experience is complicated, time-consuming and arduous from design to production, even leads to the failure to express the design thinking well. The demand for product manufacturing of different product design ends is different. The manufacturing demand for the same product design end when designs different products are different, too [10]. This kind of user demand, which varies with external factors or internal factors, is the scenarized demand. Based on this Internet plus scenarized demand, the concept of integration of design and manufacturing is put forward in this paper, and the goal of "Design is manufacturing, manufacturing is service and service is value" can be achieved through the combination and interaction of Innovative Manufacturing Elements. Personalized scenes can be created for the Internet model of the demand side and the supply side, the best matching way can be sorted out to provide service efficiently.

With the development of mass customization and network collaboration, according to the personalized demand of different design ends, the manufacturing enterprises of this platform need to receive many customized data from the Internet in real time and coordinate resources across the network, organize production and manage more kinds of data. Below is a simple axle part taken as an example. If a customer releases a specific axle part processing request through the platform, the platform decomposes it into several Innovative Manufacturing Elements by analyzing the parameters of the part, as shown in figures 4 and 5. The axle part shown in Fig. 5 is a collection of several Innovative Manufacturing Elements shown in Fig. 4 in a certain order. The part's Innovative Manufacturing Elements are three cylinder turnings, a chamfer turning, a key seat milling and a slot turning. The corresponding process characteristics are set up to become the technological characteristics of axle products by the interaction of Innovative Manufacturing Elements, thus a part is formed and ready for the next step.

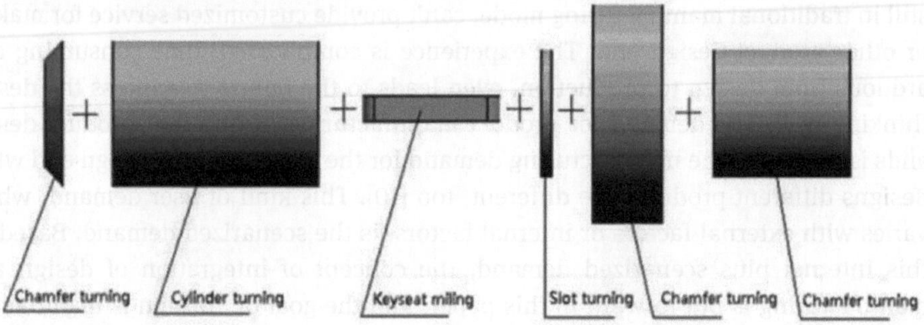

Fig. 4: Composite drawing of an axle part's Innovative Manufacturing Elements

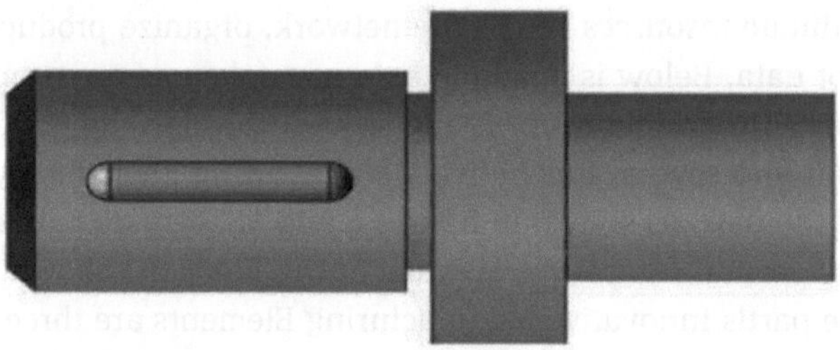

Fig. 5: Structural drawing of an axle part

5 Conclusion

Based on the related concepts of Internet plus, the hierarchical division of processing technology behavior is explained in this paper and thus the concept of Innovative Manufacturing Elements is proposed. The coordination function of Innovative Manufacturing Elements in the implementation process of the integration of design and manufacture is elaborated. The guiding significance of "Internet plus" scenarized demand for manufacturing is emphatically analyzed. The feasibility of the coordination principle of Innovative Manufacturing Elements is verified by the instantiation. And the system requirements of standardization, modularization, reusability and integration are satisfied. A solid foundation for realizing the goal of "Design is manufacturing, manufacturing is service and service is value" can be laid.

Acknowledgement: I would like to express my gratitude to all those who helped me during the writing of this thesis. My deepest gratitude goes first and foremost to Professor Zhao, my supervisor, for his constant encouragement and guidance. And second I would like to express my gratitude to the seniors in our laboratory who helped me work out my problems during the difficult course of the research. Last my thanks would go to my friends and my fellow teammates who gave me their help and time in listening to me.

References

[1] N. N. Fan, Research and Application for Technology Architecture of "Internet Smart Manufacturing", Suzhou University, 2016.
[2] S. Y. Cheng, Research on the Resources and Services Management Mode and its Prototype Development for Cloud Manufacturing, Zhejiang University, 2013.
[3] Z. L. Ye, G. N. Qi, X .J. Gu, Z. G. Bao, A cooperative process-management system based on the manufacturing grid, International Journal of Computer Integrated Manufacturing,20,2.244–253,2007.
[4] B. Jiang, Research on design method of parts based on manufacturing features, Harbin Engineering University, 2001.
[5] B. H. Li, Typical characteristics, technologies and applications of cloud manufacturing, Computer Integrated Manufacturing System, 18, 7, 1345–1356, 2012.
[6] Valiai, M. Houshmand, A collaborative and integrated platform to support distributed mabufacturing system using a service-oriented approach based on cloud computing paradigm, Robotics and Computer-Integrated Manufacturing,20,1,110–127,2013.
[7] J.H. Hao, S.D. Sun, et al. Manufacturing task granularity analysis of mechanical product supported by networked collaborative manufacturing, Computer Integrated Manufacturing Systems,12,3,451–457,2006.
[8] K. Schutzer, H. Schulz, Integrated design, process planning and automated NC programming on the basis of manufacturing features, Engineering Design,3,36–41,1996.
[9] T. X. Song, C.L. Zhang, Cloud manufacturing, service platform for small and medium enterprises, Computer Integrated Manufacturing Systems,19,5,1147–1154,2013.
[10] M. Wang, Research on the standard manufacturing services modeling for cloud manufacturing, Zhejiang University, 2016.

Xu-Yang Cao and Shan-Chao Liu*

Design of a Semi-active Heave Compensation System Combined Variable Frequency with Throttle

Abstract: In order to reduce the impact of wave to the ship crane, complete the lifting operations safely and efficiently, and with the view to decreasing the energy loss and increasing the efficiency of the existing system, a semi-active heave compensation system combined variable frequency with throttle is designed. By using the frequency conversion hydraulic technology and retaining valve control unit, the system can meet the requirements of position precision and energy consumption at the same time. Then the simulation models of the hydraulic system and the variable frequency motor are established respectively via the AMESim and Simulink. The compensation performance and energy saving effect of the system are analysed by joint simulation. The results show that 99.7% impact of the ship heave motion can be compensated. The energy consumption is only 25.6% of the valve-control active heave compensation system under the same operation conditions.

Keywords: heave compensation; variable frequency control; power matching; energy saving; joint simulation

1 Introduction

Cranes working on the sea is affected by the heave motion of the ship, which has a negative effect on hoisting. Heave compensation system is essential for the safe and efficient hoisting operation. According to the differences of energy source, the existing heave compensation system can be divided into passive system, active system and semi-active system [1, 2]. The energy consumed by the passive system stems from the heave motion of the ship, without additional power. Meanwhile, the passive system has disadvantages of large lag and low compensation precision. The active system is completely powered by the hydraulic power unit, so it has the characteristics of low hysteresis, high compensation precision, but huge energy consumption. The semi-active system, which combines the advantages of the above two

Xu-Yang Cao, School of Mechanical Engineering, Dalian University of Technology, Dalian, China. Email: saner@126.com

***Corresponding author: Shan-Chao Liu,** School of Mechanical Engineering, Dalian University of Technology, Dalian, China. Email: liushanchao_dlut@126.com

DOI 10.1515/9783110540048-071,

systems, can use the hydraulic power unit and the accumulator as power supply. And the compensation precision and system energy consumption are taken into account simultaneously.

Heave compensation technology has been comparatively mature in some western countries. In 1990s, the application of semi-active heave compensation system to the floating offshore drilling platform has been proposed in the United States [3]. Based on the secondary regulation of hydrostatic transmission, a new type of heave compensation system was designed and manufactured by Rexroth research. The compensation efficiency of this system can reach 95%, and 65% of the load gravity potential energy can be recycled [4]. In this field, some achievements have been made in recent years in China [5, 6]. Most of the researches have been focused on how to improve the compensation performance of heave compensation system. Comparatively, researches of energy conservation are less.

As a global power matching transmission mode, the variable frequency hydraulic speed regulation technology has been widely used in many fields because of its high efficiency and good energy saving effect. These advantages of the technology are more obvious in the case of large power and large flow fluctuation. For example, the world's first attempt to apply the variable frequency hydraulic technology to the hydraulic elevator was taken place in the Mitsubishi in Japan [7]. Another example is the application of variable frequency drive principle in injection moulding machine which was developed and tested by Helbig A [8].

In this paper, the frequency conversion technology is applied to heave compensation system and a new type of semi active heave compensation system is designed. Then the compensation and energy saving effectiveness of this system is verified by joint simulation. A new idea and method for the design and energy saving of heave compensation system are put forward.

2 System Principle Design

2.1 Analysis of the Ship and Compensation Motions

According to the relevant literature, the trajectory of ocean wave is a kind of nonlinear random curve [9]. It can be seen as a superposition of sine waves of different periods, amplitudes and phases. It can be roughly defined as

$$X_b = \sum_{n=1}^{N} \frac{H_n}{2} \sin\left(\frac{2\pi}{T_n} t + \theta_n\right) \tag{1}$$

As the wave height curve is generally similar to the shape of the sine curve, in order to facilitate the initial analysis and design, the wave motion can be approximated as a sinusoidal motion, which can be described by the following equations

$$x_b = \frac{H}{2}\sin\left(\frac{2\pi}{T}t\right) \tag{2}$$

The frequency of ship's heave motion is the same as that of wave motion, and the amplitude is smaller. The equation of the ship's heave motion is

$$x_c = \frac{\mu H}{2}\sin\left(\frac{2\pi}{T}t\right) \tag{3}$$

When the pulley block magnification is 1, in order to make the load in the suspended state is not affected by the heave motion of the ship, the compensation motion should content the following equation

$$x_g = -\frac{\mu H}{4}\sin\left(\frac{2\pi}{T}t\right) \tag{4}$$

In the above equations, T is the wave period, H is the wave height, the θ is the initial phase angle, and the μ is the ratio of the ship's heave displacement to the wave height.

2.2 Principle of the Variable Frequency Hydraulic Speed Regulation

Variable frequency hydraulic speed regulation technology is a kind of global energy-saving driving mode, which combines the frequency conversion technology and hydraulic transmission [10]. AC variable frequency motor drives hydraulic quantitative pump to output flow required by the system. Then the motor can be always in high efficiency. The output flow of hydraulic pump is defined as

$$Q_V = V_P n \eta_v = \frac{60 V_P \eta_v f(1-s)}{p} \tag{5}$$

Where V_P, n and η_v are hydraulic pump displacement, input speed and volume efficiency. f, s and p are the motor's power frequency, slip ratio and the number of poles.

Variable frequency hydraulic speed regulation system has no overflow loss and throttling loss, so it has higher power efficiency. Compared with the traditional variable displacement speed control mode, it also has the advantages of simple structure, low noise, large speed range, convenient control and better energy saving, etc. Meanwhile, there are some disadvantages such as slow dynamic response, poor low speed characteristics, and low speed accuracy. The vector control inverter is a good solution for these problems. Vector control simulates asynchronous motor to DC motor by coordinate transformation, and realizes the decoupling control of flux and torque, so that AC motor can achieve the control effect of DC motor.

2.3 System Principle Design

The semi-active heave compensation system combined variable frequency with throttle, as shown in Fig. 1, consists of passive compensation part and active compensation part. The passive compensation part formed by cavity C and an accumulator 13, which is equivalent to a hydraulic spring, bears most of the weight of the load. The active compensation part is comprised of cavity D, cavity E, servo valve 8, inverter 3, motor 4, pump 5, etc. Its function is to overcome the load inertia and friction. The Ship heave displacement signal x1 and piston rod displacement signal x2 are both detected by the sensor and used as the input of the controller. The combined action of solenoid directional valve 12, servo valve 8 and inverter 3, which are controlled and coordinated by the controller, makes the piston rod complete the compensation accurately.

This system not only retains the advantages of the valve control system, such as high control precision, fast response, but also allows full play to the characteristics of variable frequency hydraulic speed control system, which contains large speed range and good energy-saving. On the one hand, the difference between the pump outlet pressure and the working pressure of the hydraulic cylinder is kept constant under the control of the solenoid directional valve 12 and the pressure compensation valve 6. Thus, the flow into the actuator is only determined by the servo valve 8 spool size and the servo valve 8 is driven by the control signal to realize the accurate and fast control of the cylinder 9. On the other hand, in order to achieve energy saving and high efficiency, a variety of measures are applied to the system. The energy consumption of active compensation part is greatly reduced because the passive compensation part supports the load. The solenoid valve 12 and the pressure compensation valve 6 can be used to fulfil the pressure matching of the system. The real-time required flow of the system is calculated according to the servo valve control signal. Then the inverter motor according to the results of this calculation to adjust the output flow of the pump is adjusted by the inverter motor to achieve the system flow matching.

The working process of the system is divided into ascending compensation and descending compensation. When the controller 11 receives the rising signal of the ship, it drives the servo valve 8 to move up. The hydraulic oil flows from cavity D pass the solenoid valve 12 into the spring chamber of pressure compensation valve 6. At the same time, the control signal is sent out by analysing the displacement signals x1 and x2. Servo valve 8 moves toward right and spool opening size is controlled. Frequency conversion motor 4 and constant pump 5 provide enough flow for the system. The high pressure oil enters the cavity D through the servo valve 8 to push the piston rod to retract. Wire rope is released to achieve load displacement compensation. The process of descending compensation is similar and it's not described in this article.

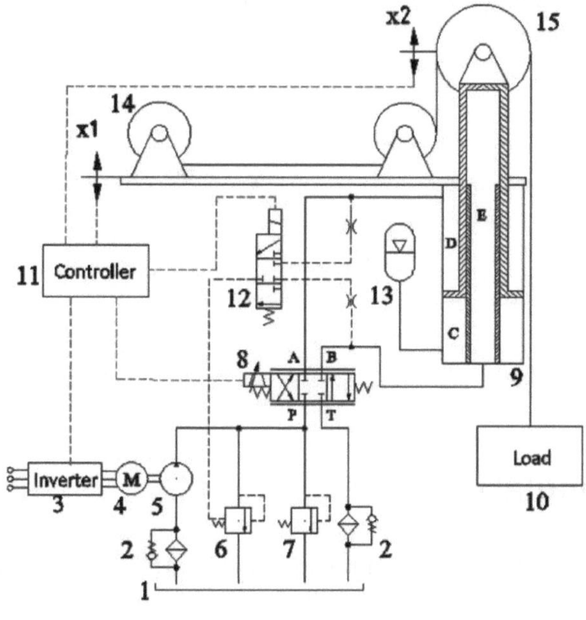

1. Tank 2. Filter 3. Inverter 4. Motor 5. Hydraulic quantitative pump 6. Pressure compensation valve 7. Relief valve 8. Servo valve 9. Compound hydraulic cylinder 10. Load 11. Controller 12. Solenoid valve 13. Accumulator 14. Winch 15. Movable pulley

Fig. 1: Schematic diagram of hydraulic system

2.4 Controller Design

The heave compensation control mode can be divided into displacement compensation, velocity compensation and tension compensation according to the purpose of compensation. A control method combining velocity compensation and displacement compensation is designed, as shown in Fig. 2, and it enables the system to compensate random waves efficiently. According to Eq. 3 and Eq. 4, the relative velocity of the piston rod and the cylinder body should be 1/2 of the ship's heave velocity in opposite direction. The expected velocity is used as the instruction signal of the speed controller which is open loop. In this way, the hydraulic cylinder can respond in time to make the compensation action. The difference between the ship heave displacement and the relative displacement of piston rod and cylinder body is used as the input of the displacement controller to form a closed loop control. Then, the cumulative error of velocity compensation is eliminated and the accuracy and stability of the system are guaranteed.

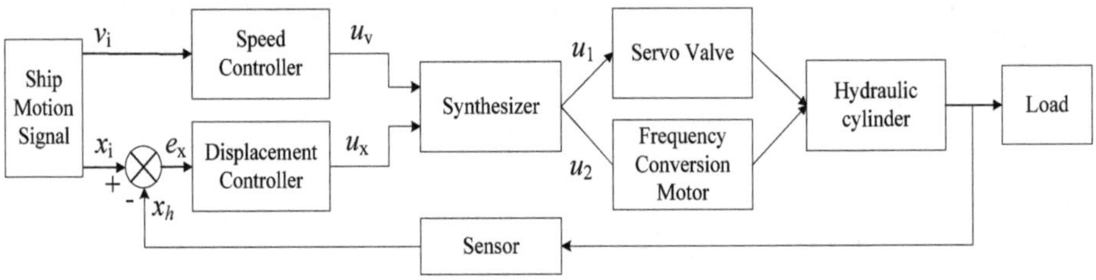

Fig. 2: Block diagram of the control principle

3 Modeling and Simulation

There is an independent model of the speed controlled motor in AMESim, but the model is too idealistic, which will reduce the reliability of the simulation results. Therefore, after establishing AMESim model of hydraulic system and Simulink model of vector variable frequency motor, the whole system is analysed by using the joint simulation, which can give full play to the advantages of two kinds of software in their respective fields.

3.1 Hydraulic System Modeling

According to the schematic diagram, the hydraulic system simulation model, as shown in Fig. 3(a), is established. During the modelling, the influence of pressure on the viscosity and elasticity of oil is not considered; the elasticity of the steel wire rope is neglected and efficiencies of hydraulic cylinders and pump is not counted.

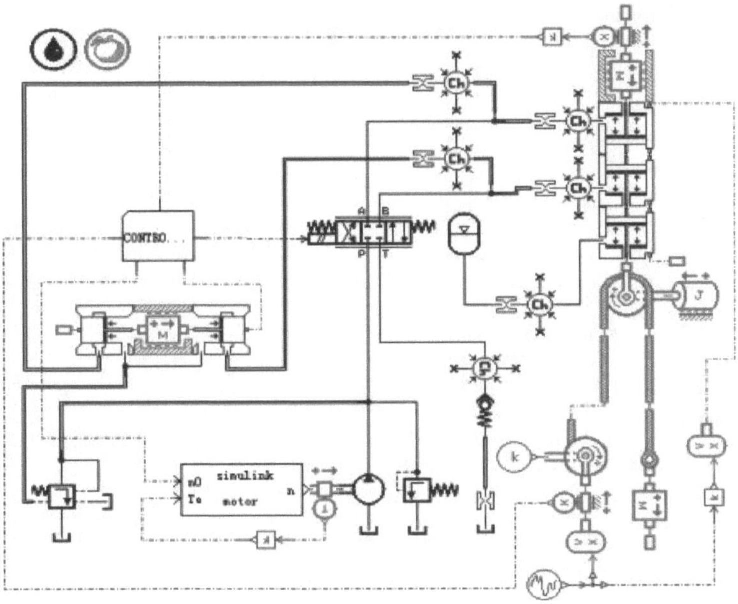

(a)

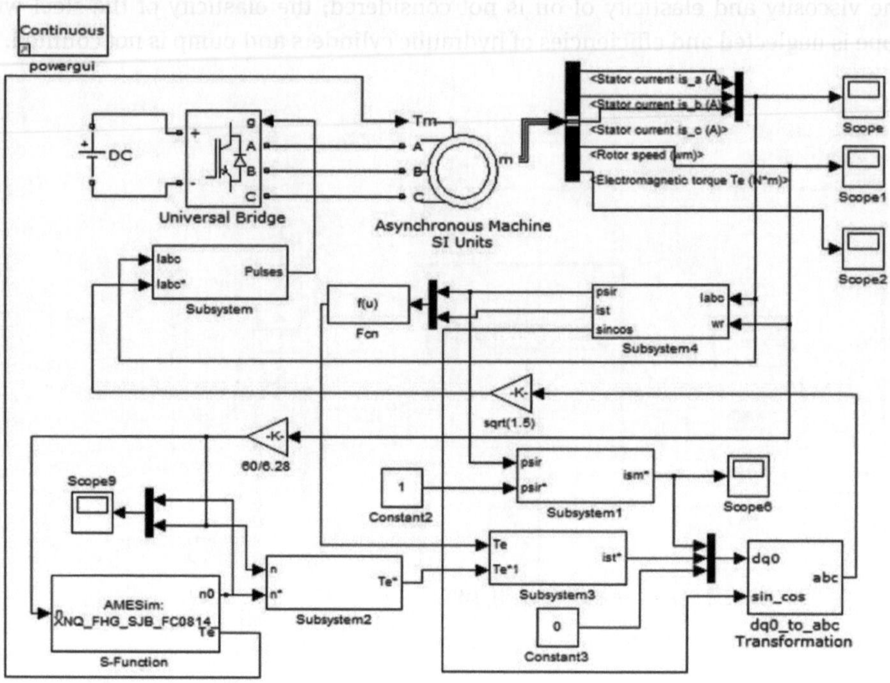

(b)

Fig. 3: (a): The hydraulic system simulation model. (b) The vector frequency conversion motor simulation model

3.2 Vector Variable Frequency Motor Modeling

The simulation model of vector frequency conversion motor is shown in Fig. 3(b). The model is composed of motor module, inverter, DC power supply, motor measurement unit, hysteresis pulse generator, flux observer, coordinate conversion, flux regulator, torque regulator, speed regulator and other sub modules. Coordinate conversion module can be established according to the mathematical model, or replaced by the ready-made modules, which are abc_dq0, Transformation and dq0_abc Transformation. Flux regulator, torque regulator and speed regulator are modelled as the PI regulator.

4 Simulation Analysis

Ships equipped with this heave compensation system should be able to work under degree 4 sea situations, which mean the maximum compensation height is 3 meters and the minimum compensation period is 5 seconds. A load of 10 tons is chosen for simulation. The main parameters of the simulation model are calculated, as shown in Table 1.

Tab. 1: Main parameters of amesim simulation model

title	unit	value	title	unit	value
Load quality	m/t	10	Pump displacement	ml/r	80
Passive cavity working area	mm2	11684	Rated flow of servo valve (Δp=30bar)	L/min	200
Cavity D working area	mm2	8858	Accumulator volume	L	200
Cavity E working area	mm2	8858	Accumulator precharge pressure	MPa	17.1
Maximum stroke of compound cylinder	mm	500	Accumulator initial pressure	MPa	17.5

4.1 Analysis of Harmonic Input

The harmonic signal generated by multiple sinusoidal signals is used as the input of the system to simulate the heave motion of the ship under the action of random waves. These sinusoidal signals' amplitudes range from 0.05 to 0.25 meter and their periods range from 5 to 15 seconds. The displacement curves of ship and load are shown in Fig. 4(a). From this figure, the ship heave displacement amplitude is 0.8 meter, and Load displacement amplitude is 2.5 millimetres. The displacement compensation efficiency of the system is 99.7%, which is very ideal.

Accumulator volume and pressure curves are shown in Fig. 4(b). The accumulator pressure fluctuates with the changes of ship's heave and the accumulator volume is changed reversely. When the ship rises up, the piston of the composite cylinder moves down, the oil in the passive chamber is pressed into the accumulator, the pressure of the accumulator is increased and the volume is reduced. When the ship descends down, the piston of the compound cylinder moves up, the oil are entered the passive cavity from the accumulator, the pressure of the accumulator is reduced, and the volume is increased.

According to the pressure change curve of the servo valve, as shown in Fig. 4(c), port P pressure is always 3MPa higher than the pressure of port A or port B. In other words, the pump output pressure is well matched with the load pressure. The maximum pressure of the port P of the servo valve is 9MPa and the pressure variation range is about 3MPa, which are very beneficial to stability of the hydraulic system. The cause of the lower system pressure and smaller pressure fluctuation range are that the high pressure oil provided by the pump is only used to overcome the inertia and friction of the system.

The flow tracking curve of system is shown in Fig. 4(d) and the speed tracking curve of vector motor is shown in Fig. 4(e). It can be seen that the vector control motor meets the speed control requirements of the system and drives the pump to provide just enough flow, though there is a little lag between the output speed and the desired speed. The average flow rate of overflow is 11.9L/min, while the output flow rate of the pump is about 56.6L/min, which ensures a high efficiency of the system. There are two main reasons for the overflow loss. The first reason is that the output flow of the pump is set to be slightly larger than that required by the hydraulic cylinder. The other reason is that the motor speed should not be lower than the minimum speed, which is 300r/min in this system, to ensure the normal operation of the pump. When the required flow rate is less than 24L/min, the pump is driven by the motor at the minimum speed and the system is only controlled by the valve.

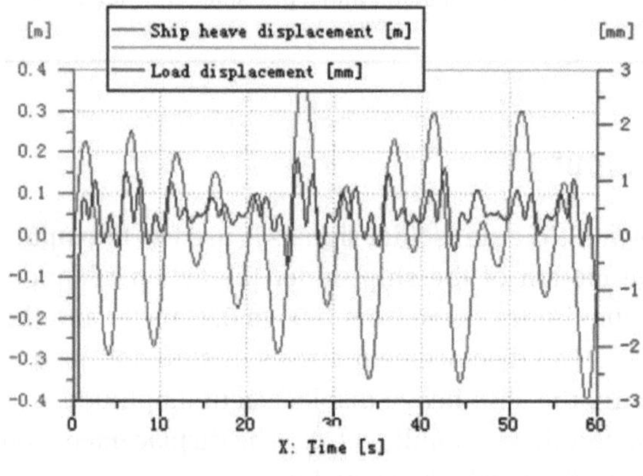

(a)

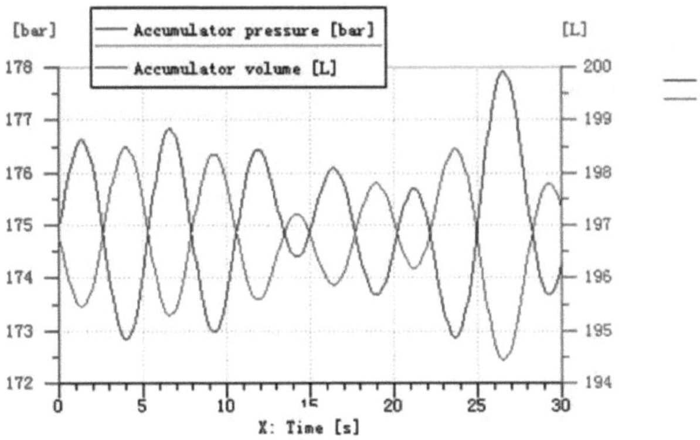
[bar]
Accumulator pressure [bar]
Accumulator volume [L]
[L]
178
177
176
175
174
173
172
200
199
198
197
196
195
194
0
5
10
20
25
30
X: Time [s]

(b)

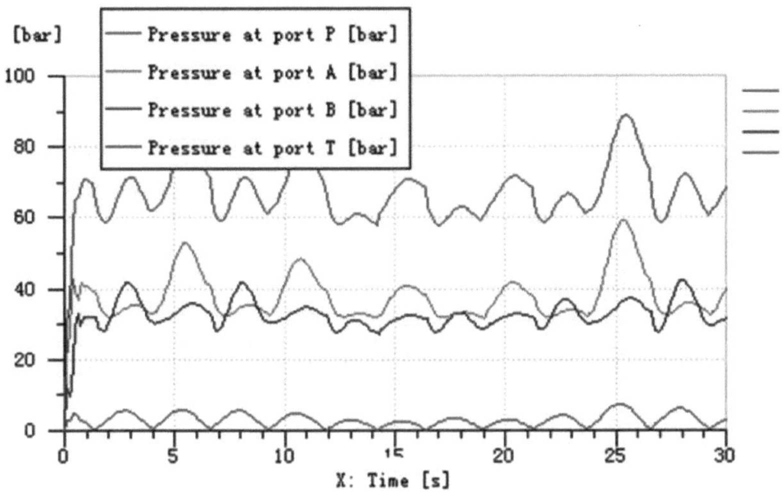
[bar]
Pressure at port P [bar]
Pressure at port A [bar]
Pressure at port B [bar]
Pressure at port T [bar]
100
80
60
40
20
0
0
5
10
20
25
30
X: Time [s]

(c)

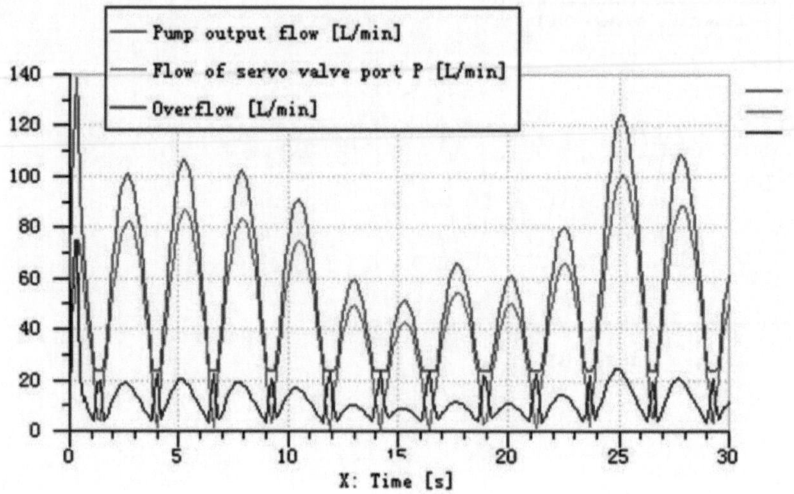

(d)

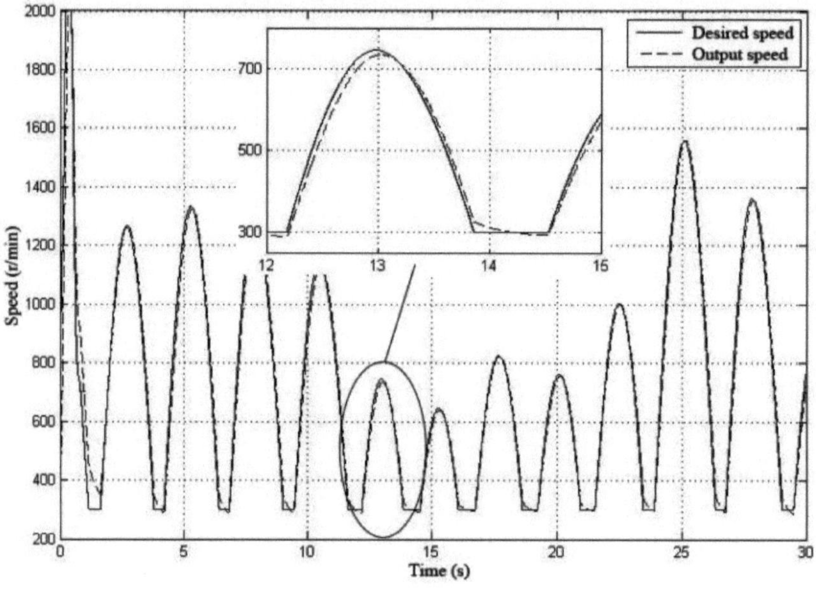

(e)

Fig. 4: (a) Displacement curves of ship and load. (b) Curves of accumulator volume and pressure. (c) Pressure fluctuation of the servo valve oil ports. (d) Flow tracking curve of system. (e) Speed tracking curve of vector motor

4.2 Comparison of System Energy Consumption

In order to compare the energy consumption of the above system and the existing active and semi-active heave compensation system, the simulation models of the other three systems are established: valve-control active system (Fig. 5(a)), active system combined variable frequency with throttle (Fig. 5(b)) and valve-control semi-active system (Fig. 5(c)).

The sine wave signal, whose amplitude is 0.45 meter and period is 5 seconds, is used as a unified simulation input signal. Simulation time is set to 30 seconds. The energy consumption of the system can be calculated by the equation

$$E = \int p_p q_p \mathrm{d}t \tag{6}$$

Where the pump output pressure is p_p and the pump output flow is q_p. The energy consumptions of the four systems can be easily obtained by using the post-processing function of AMESim. The results of the initial 5 seconds were ignored to eliminate the impact of the instability at the beginning of the simulation on the analysis. Comparison data are shown in Table 2.

As shown in Table 2, the system designed in this paper has an obvious advantage in energy saving. Its energy consumption is only 25.6% of that of the valve-control active system, or 45% of that of the active system combined variable frequency with throttle, or 53.1% of that of the valve-control semi-active system.

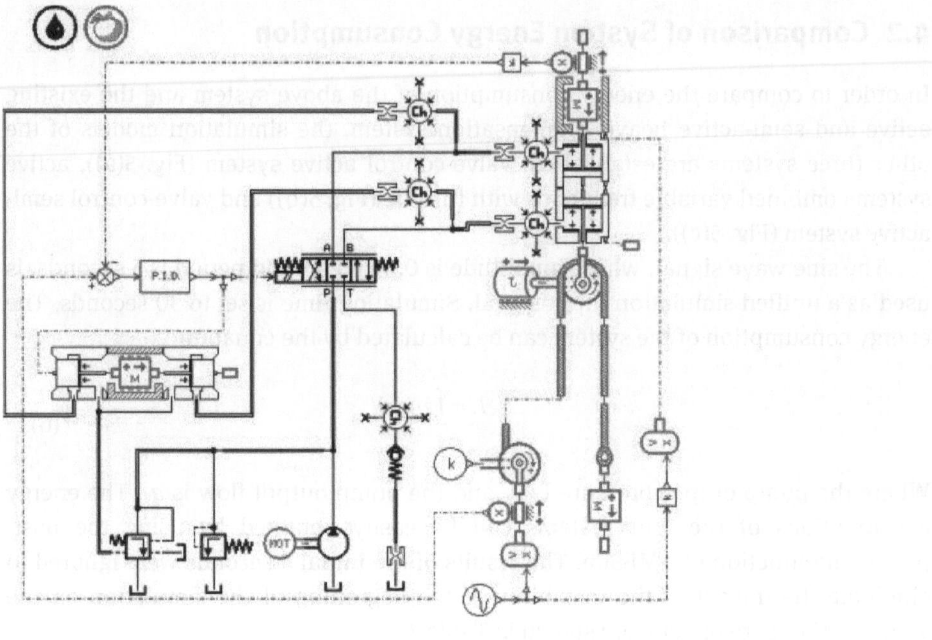

(a)

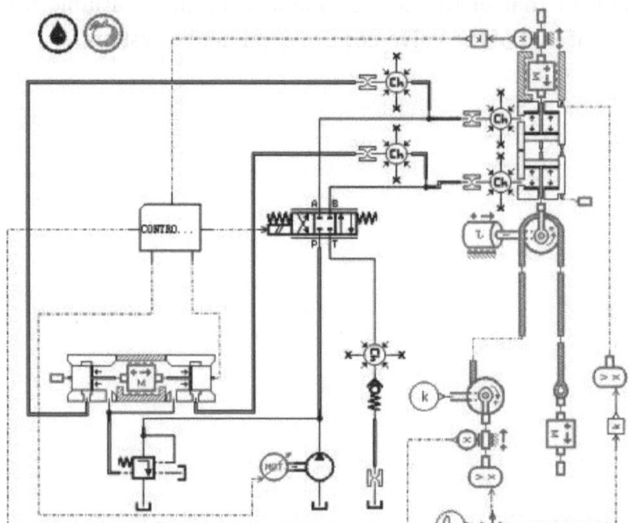

(b)

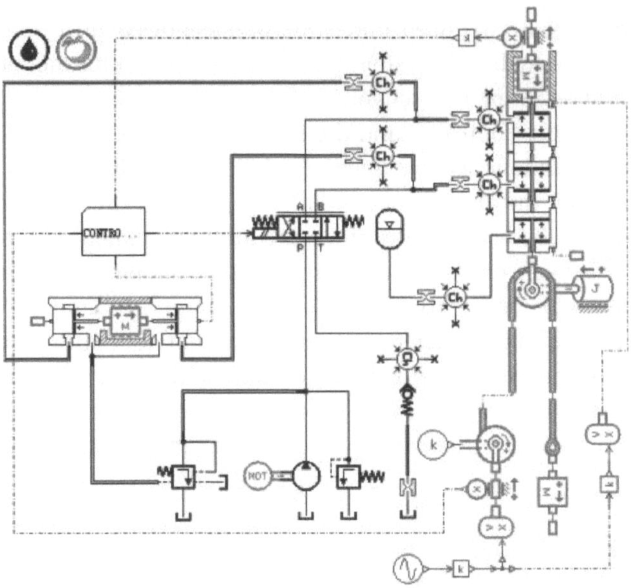

(c)

Fig. 5: (a) The simulation model of valve-control active system. (b) The simulation model of active system combined variable frequency with throttle. (c) The simulation model of valve-control semi-active system

Tab. 2: Comparison of energy consumption of four systems

system name	energy consumption in 5 to 30 seconds (kJ)	average energy consumption of a cycle (kJ)
valve-control active system	1809.5	361.9
active system combined variable frequency with throttle	1033.1	206.6
valve-control semi-active system	873.5	174.7
semi-active system combined variable frequency with throttle	464.2	92.8

5 Conclusion

A semi active heave compensation system is designed by combining the existing heave compensation system with the variable frequency hydraulic technology. The system energy consumption is greatly reduced under the premise of ensuring the high compensation precision. The hydraulic system model and vector motor model are established by AMESim and Simulink respectively.

Under the precise control of the controller and the servo valve, the compensation efficiency of the system for random heave motion of ship is increased to 99.7%. The weight of the load is born by the "hydraulic spring" composed of the passive cavity of composite cylinder and the accumulator. Pressure and flow matching are achieved simultaneously by active compensation part and the efficiency of the whole system, including the motor, is improved. The energy saving advantage of this system is verified by comparing the energy consumptions of four different heave compensation systems. When the experimental conditions are ripe, a prototype should be built according to this paper to verify the correctness of the simulation results and to lay the foundation for the batch production of the corresponding products.

Acknowledgement: The work was funded by the National Natural Science Foundation of China (approve number: 51475068) and supported by the National Science and Technology Support Program (approve number: 2015BAF06B01).

References

[1] U. A. Korde, Active heave compensation on drill-ships in irregular waves, Ocean Engineering, 7, 25, 541–561, 1998.
[2] K. D. Do and J. Pan, Nonlinear control of an active heaves compensation system, Ocean Engineering, 5, 35. 558–571, 2008.
[3] L. R. Robichaux and J. T. Hatleskog, Semi-active heave compensation system for marine vessels, U.S. Patent, No. 5,209,302, 11, May, 1993.
[4] BOSCH Rexroth, Heave compensation, unpublished.
[5] D. Zhang, J. Wu, and F. Lu, Analysis on Heave Compensation System of Ship-mounted Crane, Mechanical Science and Technology for Aerospace Engineering, 02, 265–269, 2012.
[6] W. Quan, A. Zhang, and Z. Zhang, Modeling and Simulation for Semi-active Heave Compensation System of Composite-cylinder Type, Machine Tool & Hydraulics, 1, 41, 137–141, 2013.
[7] T. Yamamoto, Apparatus for controlling a hydraulic elevator, U.S. Patent, No. 4,593,792, 10, Jun, 1986.
[8] Helbig, A. Injection moulding machine with electric-hydrostatic drives, Proc. of the 3. International Fluid Power Conference (3. IFK), 1, 67–82, Mar. 2002.
[9] J. Ma, J. Tian, and F. Peng, Numerical simulation and simulation of ocean waves, Journal of Huazhong University of Science and Technology, 04, 63–65, 2000.
[10] Y TANAKA, T Machda, and K Nakano, Speed and displacement control of pump system for energy saving, Journal of fluid control, 5, 17, 78–81, 1998.

Ze Li

Design of Auxiliary Arm for High Voltage Live Working Robot

Abstract: In this paper, the mechanical structure and control system of the auxiliary arm of the high voltage live working robot are designed and analysed. The 3D model was established by Solid works and the finite element analysis was carried out by ANSYS software to improve the performance and parameters. The motion control algorithm and motion control flow of Auxiliary arm is studied by mathematical modelling.

Keywords: working, auxiliary arm, simulation design, 3D modelling, control system

1 Introductions

China's artificial live operation began in the 1950s; there are great differences in the level of work carried out in various regions. For overhead lines and rod-type installation of a wide range of the scene, there are two or more multi-loop juxtaposition and other arrangements, for example cross-bar, equipment installation rods, therefore the live safety risk is high, limiting the promotion of artificial live operation technology. At present, live robot work content in China is relatively simple, cannot cover the entire distribution network electrification project. Robot cannot carry out operations in a complex line environment, and there is a gap with the actual production of distribution network for live working [1]. In order to solve the above problems, this paper introduces the design of the auxiliary arm system of high voltage live operation robot, which can carry out the live operation project in the complex line environment, effectively replace the manual operation, and improve the safety of the operation.

High load auxiliary arm has always been one of the hotspots in the field of robot research, and the international industrial robot load ratio below 1:10. In the field of live electricity distribution, limited by the live operating environment and the impact of insulated arm car, the overall weight of the table is generally not more than 500Kg. According to the research experience, the auxiliary arm should be limited to 100Kg. The auxiliary arm developed in this paper adopts the new type of material, and special structural design to ensure the rigidity of the system, the accuracy and stability of the end control, and meanwhile accomplish the high load ratio.

Ze Li, North China Electric Power University (Baoding), NCEPU, Baoding, China.
Email:jenny_lize@163.com

DOI 10.1515/9783110540048-072,

2 Mechanical Part

2.1 Overall Design

High-voltage live operation robot auxiliary arm consists mainly the mechanical part and control part. The mechanical part mainly includes the base, the waist turning part, the arm part, the wrist joint, the gripper part, the displacement sensor and so on. The mechanical part of the auxiliary arm is designed to follow four basic principles. Reliability principle: the requirements of the product under the conditions required to complete the functional requirements, and to ensure the stability of the system. Moment of inertia principle: too large or too small will have a greater impact on the auxiliary arm, when the inertia is too large, the auxiliary arm response time will be correspondingly longer, affecting the speed and accuracy of the auxiliary arm to complete the task; When the auxiliary arm of the moment of inertia is too small, the auxiliary arm will be vulnerable to external interference or vibration itself, control accuracy is also affected. Therefore, in the design process through a reasonable selection of the material, and reasonable design the size of various parts and structure to control the moment of inertia. Strength design principles: In the design process using mechanical calculation, computer simulation analysis and other methods to ensure that the material in normal work and the strength of parts are not damaged. Stiffness design principles: In the elastic range, the stiffness coefficient of the auxiliary arm is proportional to the load and displacement of the auxiliary arm. In the design process of the auxiliary arm, the deformation must be controlled strictly.

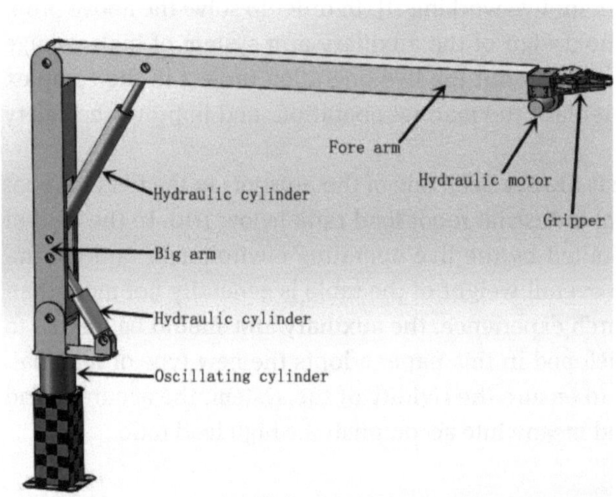

Fig. 1: Auxiliary arm

The overall structure of the auxiliary arm shown in Fig.1 mainly includes the swing cylinder, big arm pitch hydraulic cylinder, forearm pitch hydraulic cylinder, wrist swing hydraulic motor, hand claw opening and closing piston rod and other components.

2.2 Design Calculation

The auxiliary arm is 50 kg, and the end of the auxiliary arm is 100 kg; the total length between the big arm and the forearm joints to the base center line of the limit distance is 2800 mm, the size of the gripper is 300 mm; as shown in Fig. 2. The formula for calculating the moment of inertia based on the parallel axis theorem is:

$$I_G = \frac{1}{12} m L_2^2$$

$$I = I_G + m L_1^2$$

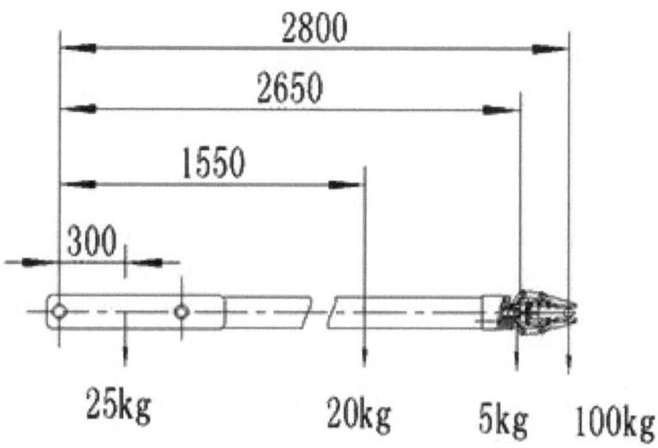

Fig. 2: Mass distribution of each axis

According to the mass distribution of each axis shown in Fig. 2, the first moment of inertia is calculated as:

$$I = \sum (I_G + mL_1^2) | \sum mL_1^2 = 100 | 2.8^2 + 5 | 2.65^2 + 20 | 1.55^2 + 25 | 0.3^2$$
$$= 869.4 (kg \cdot m^2)$$

The front end of the auxiliary arm rotates at 200 cm/s speed, and the rotation speed of the shaft is:

$$\omega = \frac{L}{R} = \frac{20}{280} = 0.07\left(\frac{rad}{s}\right)$$

From ω = 0 to ω = 0.07 rad/s time is 0.1 s, then the first axis of the swing cylinder torque:

$$\mathrm{T} = \mathrm{I}\dot{\omega} = 869.4 \times \frac{0.07 - 0}{0.1} = 608.6(N \cdot m)$$

The base angular velocity is:

$$\omega = \frac{0.07}{2\pi} \times 60|0.7\left(\frac{r}{min}\right)$$

Arm swing joint load calculation: According to swing cylinder cantilever installation to bear the bending moment calculation formula is as follows:

$$\mathrm{M} = \sum FL = 100 * 9.8 * 2.8 + 5 * 9.8 * 2.65 + 20 * 9.8 * 1.55 + 25 * 9.8 * 0.3 = 3251(N.m)$$

Swing cylinder to bear the bending moment: 3251N·m.

According to the torque balance shown in Fig. 3, the two kinds of extreme cases to calculate the hydraulic cylinder thrust F. In the first case, the hydraulic cylinder thrust F = 23161N is calculated as shown in Fig.3 (a). In the second case, the hydraulic cylinder thrust F = 20539N is calculated as shown in Fig.3 (b). So the big arm pitch cylinder maximum thrust 23161N.

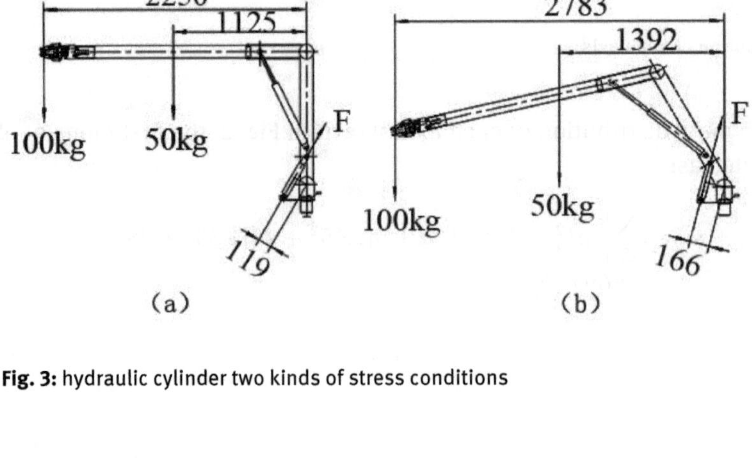

Fig. 3: hydraulic cylinder two kinds of stress conditions

Big arm selection HSG standard hydraulic cylinder, the connection for the earrings with joint bearing connection; Bore D = Ø50, d = Ø28; Travel = 100; Installation distance L4 = 280 + S = 380. According to the calculation of the boom, the forearm force is better than the arm, forearm pitch cylinder parameters, Bore D = Ø50, d = Ø28; Travel = 400; Installation distance L4 = 280 + S = 680.

2.3 Part Design

According to the design of the auxiliary arm, the design of the various parts of the auxiliary arm, the two-dimensional map, and the assembly of different components, is shown in Fig.4.

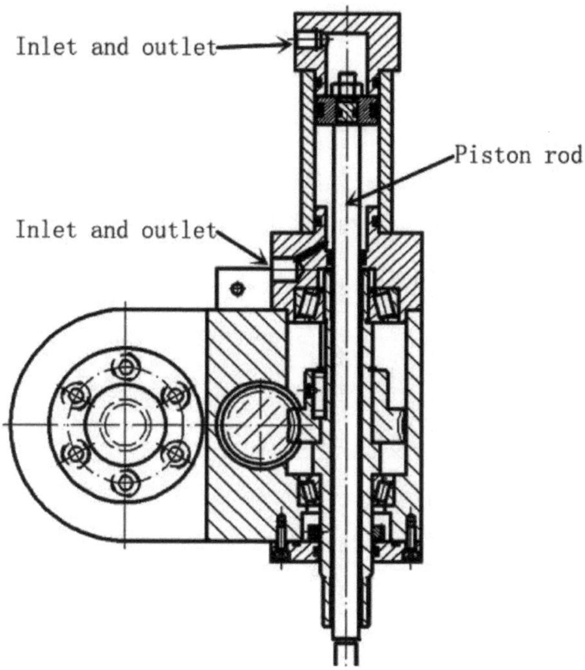

Fig. 4: Auxiliary arm gripper

The piston rod moving up and down, driving the gripper to grasp the action is moved by the hydraulic pressure, to complete the crawling action. Piston rod force is shown in Fig. 5.

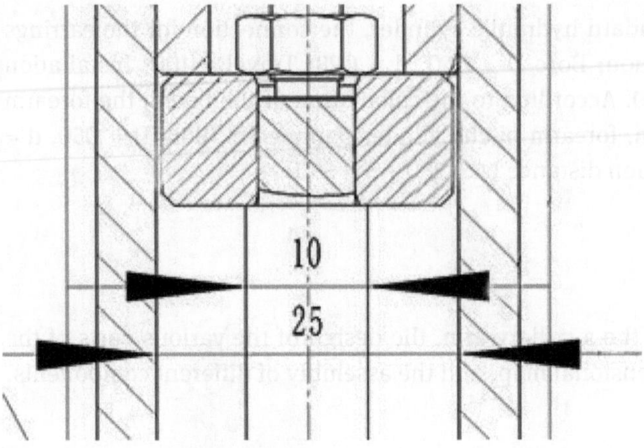

Fig. 5: Piston rod force diagram

Piston rod force area:

$$\begin{aligned} S &= \pi \times R^2 - \pi \times r^2 \\ &= \pi \times \left(12.5 \times 10^{-3}\right)^2 - \pi \times \left(5 \times 10^{-3}\right)^2 \\ &= \pi \times 131.25 \times 10^{-6} \left(m^2\right) \end{aligned}$$

The force of the piston rod is:

$$\begin{aligned} F &= P \times S \\ &= 20 \times 10^6 \times \pi \times 131.25 \times 10^{-6} \\ &= 8242.5 \,(N) \end{aligned}$$

2.4 3D Modelling

The 3D models are designed according to the 2D part drawing. The three-dimensional model is formed by assembling the three-dimensional parts, and the three-dimensional model of the parts is shown in Fig. 6.

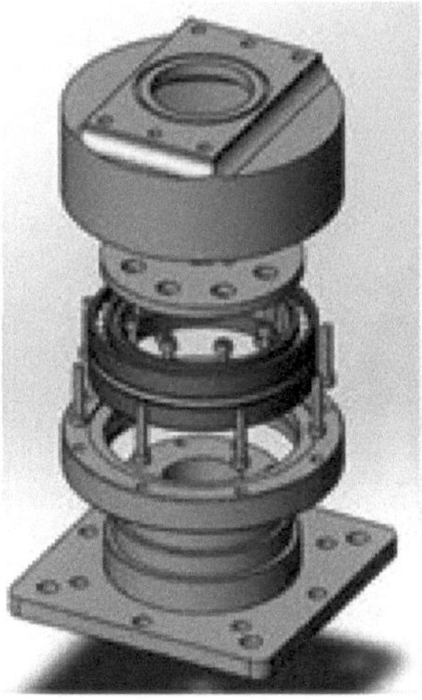

Fig. 6: Part of the three-dimensional model

2.5 Simulation Analysis

For the convenience of analysis, the three-dimensional models are imported into the ANSYS software, and in the hand set 1000N force. The following is the deformation, stress and strain results of the auxiliary arm under this load, as shown in Fig. 7, Fig. 8, and Fig. 9. Calculated by computer analysis, the maximum stress of 99 Mpa and strain is also small, within the safe range. However, the forearm deformation is large, reaching 45.54 mm, which is quite large for the auxiliary arm. Therefore we still need to take measures to reduce the amount of deformation [2].

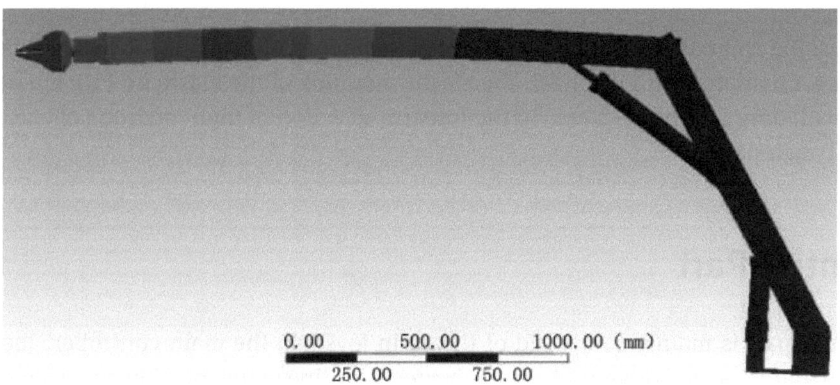

Fig. 7: Deformation

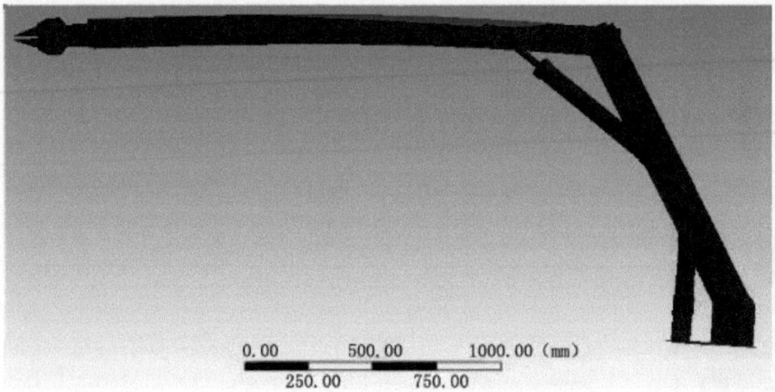

Fig. 8: Strain

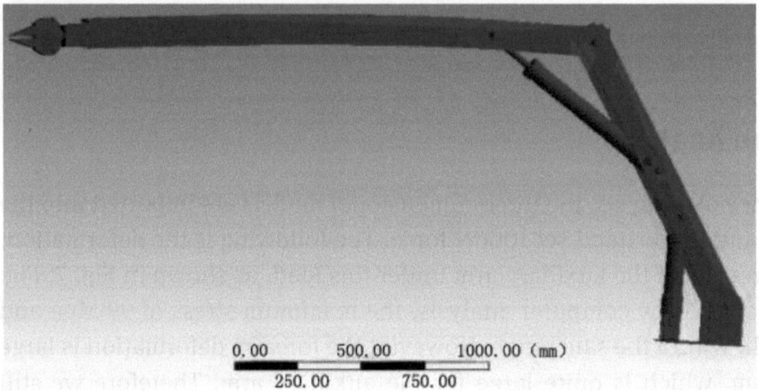

Fig. 9: Stress

After brushing comparison, high modulus of glass fiber reinforced plastic material is the choice. Compared with the steel, the elastic modulus of the elasticity FRP material is small, only 1/10 of the steel. So the forearm selection of high modulus of elasticity FRP material.

3 Control Part

The control part is mainly composed of the main joystick, the main controller, the electro-hydraulic servo valve and the high precision linear sensor. Control system design mainly uses two basic principles. One is the principle of design to meet the production of various control requirements, and another is the process design to meet the electrical control device itself, manufacturing, use and maintenance needs

[3]. The rational design determines the rationality and advanced nature of the production machinery and equipment. The process design determines whether the electrical control system has the characteristics of production feasibility, economy, appearance, easy maintenance and so on. Therefore the design of control system should consider two aspects [4].

3.1 Main Operating Lever

The main joystick selection is shown in Fig.10. The joystick has X, Y two degrees of freedom of movement, and has X, Y direction of the motion control signal output. The data signal is 0/1 switch and -10V~10V analogy voltage. S1 button controls the arm rotation and forearm rotation, S2 button is used to select whether the single axis or multi-axis linkage, S3 button controls hand rotation and the base rotation, and S4 button controls gripper open and hand closed.

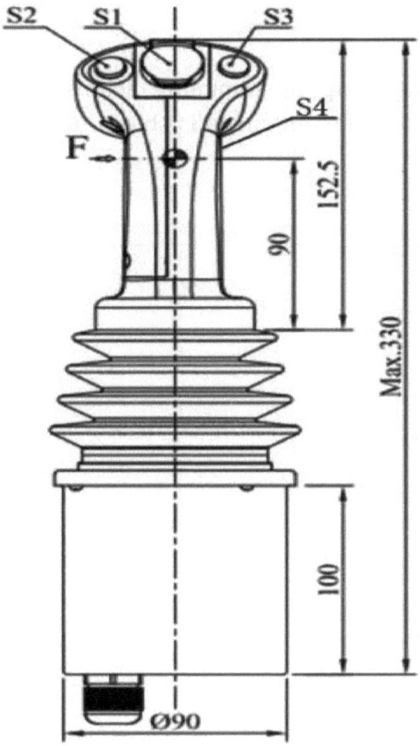

Fig. 10: Main operating lever

3.2 Main Controller

The main controller selection Keith PLC, using Keith PLC's design program includes: 1 Keith PLC KV-N32DT-type main controller, a four-way analogy input expansion module KV-NC4AD and three two Road analogy output expansion module KV-NC2DA. The system composition is shown in Fig.11.

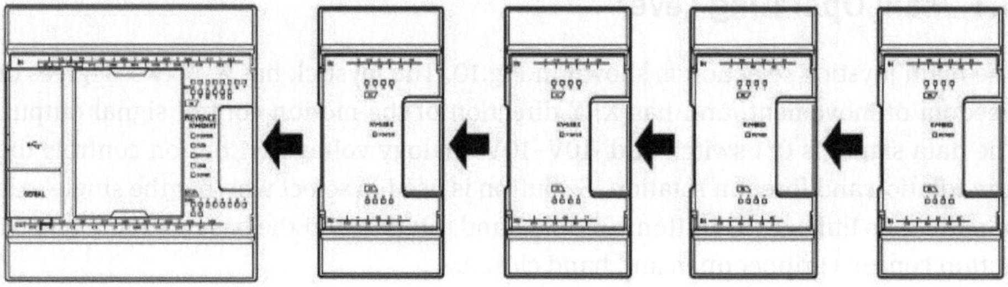

Fig. 11: Keith PLC controller system

There are two ways to program PLC, C language script programming mode and ladder programming mode, C language script programming mode is shown in Fig.12, and ladder programming is shown in Fig.13.

```
00005 [+] IF CR2006=ON  THEN
              R501=ON
              R502=ON
          ELSE
              R501=OFF
              R502=OFF
          END IF
00021 [+] DM1000=sin(DM1004/12.5)*cos(DM1006/13.5)
```

Fig. 12: Keith PLC scripting C language programming method

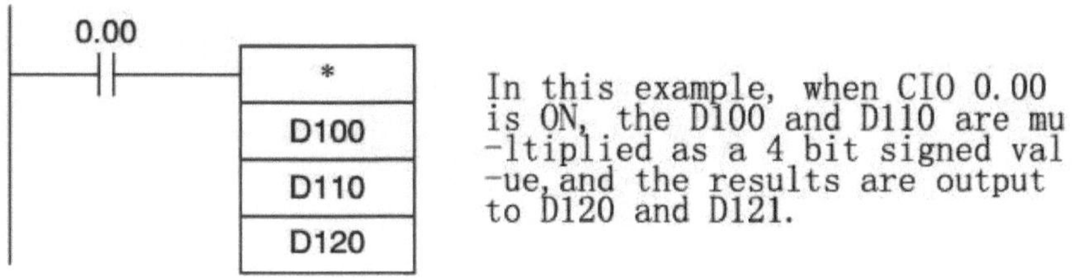

Fig. 13: Keith PLC programming method of ladder diagram

3.3 Straight Line Sensor

Linear sensor selection is the magnetostrictive displacement sensor, using magnetostrictive principle; through two different magnetic fields intersect to produce a strain pulse signal to accurately measure the location. The measuring element is a waveguide, and the sensing element in the waveguide is made of a special magnetostrictive material. The measurement process is a current pulse generated in the electronic chamber of the sensor which is transmitted in the waveguide to produce a circumferential magnetic field outside. When the magnetic field and the magnetic field which is arranged on the guide tube are used as the magnetic field of the position to change, due to the role of magnetostriction, the waveguide will produce a strain mechanical wave pulse signal, the strain mechanical wave pulse signal to a fixed sound speed transmission, and was quickly detected by the electronic room.

Because the propagation time of the strain mechanical pulse signal in the waveguide is proportional to the distance between the active magnetic ring and the electronic chamber. By measuring the time, the distance can be determined with high accuracy. Since the output signal is a true absolute value, not a proportional or amplified signal, there is no signal drift or variable value, and there is no need for periodic re-marking.

Magnetostrictive displacement sensor is a high precision, long stroke, absolute position measurement of the displacement sensor. It uses the internal non-contact measurement method, has long service life, strong environmental adaptability, high reliability, good security, fast response, high repeatability, good compatibility, strong anti-interference ability and convenient system automation work. In a harsh industrial environment, it can work properly because of its fully enclosed structure. The magnetostrictive displacement sensor is shown in Fig.14.

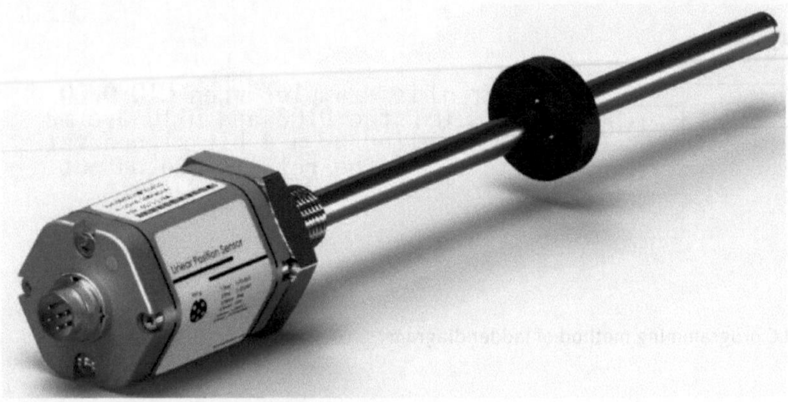

Fig. 14: Magnetostrictive displacement sensor

3.4 Control the Process

The control flow chart is shown in Fig.15. The output data of the main control rod and the output of analogy data are used as input to the PLC controller. The motion data of the manipulator joint is fed back to the PLC controller by a high precision linear sensor. After entering the PLC controller, two algorithms are used to calculate the driving data of the servo valve. The PLC controller controls the movement of the hydraulic actuator by controlling the flow of the servo valve.

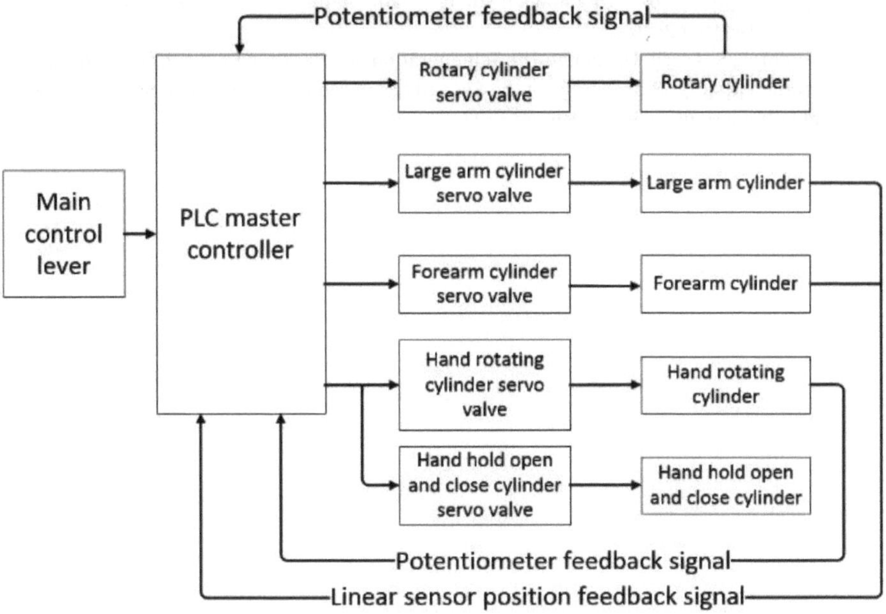

Fig. 15: Control flow chart

4 Conclusion

This paper introduces the mechanical design and control system of the auxiliary arm of the high voltage live working in the distribution network, designs the overall structure of the auxiliary arm, carries out the 3D modelling and the simulation analysis, and carries out the control part selection, the control flow and the algorithm design[5]. The robot can complete the complex line environment under the live operation of the project to meet the 10kV following the daily distribution network maintenance and repair operation needs.

References

[1] Qi Hui Design and Simulation of mechanical arm with high voltage operation insulation, Carpi2016, pp.22–24, October 2016.

[2] LI Jian, LU Shou-yin, LI Yun-Chang, ET. "The practical research of robot on live repair working for 10kV distribution lines," Manufacturing Automation. Robot technology, vol. 34, pp. 90–94, Nov. 2011.

[3] ZHAO Yu-Liang, ZHANG Qi, QI Hui, ET. "Design of a non-force feedback hydraulic manipulator for live working", Machinery Design &Manufacture, manipulator, vol. 9, pp. 117–119, Sep, 2012.

[4] Wang Hai-lei Design of High Voltage Insulated Manipulator Control System, Carpi2016, pp.16–18, October 2016.
[5] A Barrientos, R. Aracil, O. Luengo The Teleoperation of a Backhoe Excavator through Force Measurements[C]// Proceedings of the Second International Workshop on Service and Personal Robots, Genova, Italy, October 1997.

Yu Ning, Hou-Lin Liu, Zi-Pei Zhang*, Quan-Xin Huang and Zu-Quen Su

Investigation on Integrated Digital Platform of High Power Converter with Power Factor Correction for Formed Aluminium Foil

Abstract: Aluminum foil industry is the high energy-consuming industry. In order to realize energy saving, green and environmental protection, this paper presents an integrated topology for low-voltage and high-current aluminum forming foil power supply. This is characterized by the use of pre-PFC circuit to improve power factor, to provide a stable DC voltage on phase-shifted full-bridge soft-switching DC/DC converter to reduce design difficult and synchronous rectification to improve the low-voltage high-current power supply system power density and efficiency. This paper focuses on the design of the 28027-based digital control system and its software system. It uses SPI to implement communications of two 28027 to enhance the control performance of the system. The 650W prototype demonstrates advantages of digital control. Test data are shown that the digital development platform has the advantages of high efficiency and high-power factor.

Keywords: digital control, DSC, PFC, average current control, phase-shift full-bridge

1 Introduction

With the rapid development of many emerging fields such as electric drive and new energy automobile electronic technology, increasing demands into the aluminum foil. For formed aluminum foil production is energy intensive industries, in order to reduce the production cost, enhance the competitiveness of their products, it is increasing demands that to develop low voltage large current of high power dc power supply with high reliability, high efficiency and high power factor.

Traditional thyristor rectifier is a low efficiency low-frequency rectifier with large loss, so, high power efficiency high power factor of high-frequency soft switch

Yu Ning, Quan-Xin Huang, Zu-Quen Su, Departments of Automation, Shaoguan University, Shaoguan, China. Email: nby6210@163.com
Hou-Lin Liu, Ruyuan Dongyangguang Formed Foil Co. Ltd. Shaoguan, China.
***Corresponding author: Zi-Pei Zhang,** College of Automation, Guangdong University of Technology, Guangzhou, China. Email: 891330772@qq.com,

DOI 10.1515/9783110540048-073,

power becomes a development trend. Since its invention in the 1980s, the conventional phase-shifted full-bridge (PSFB) dc-dc converter is extensively used for high-power applications [1-6]. This power supply adopts generally multi-modules parallel redundant technology. Some of the literature in-depth analysis of the PSFB circuit theory, topology analysis and steady-state and dynamic performance [7-13]. However, its performance is affected mostly for the analog control and passive power factor control. The digital control for PSFB converters and related issues has been researched in recent years [14-18]. In this paper, the digital development platform of the power supply module is developed, which highly integrated DSP-based realizations feature adaptive two-phase interleaved power factor correction front end, zero voltage switching (ZVS) and synchronous rectification schemes. It can effectively reduce development costs of the power supply and speed up product upgrading.

Details for generating complex gate drive waveforms, required by these control schemes, and intelligent timing control, to optimize system performance under changing operating conditions, are provided.

2 Main Circuit Topology

The input AC line is passed through an EMI filter, and then through a bridge rectifier. The next stage is a boost converter that boosts the AC line voltage, at any given instant, to the DC bus voltage. Since the boost converter output is discontinuous, a capacitor acts as an energy reservoir that reduces the output ripple.

A phase shift full bridge (PSFB) inverter is core of the circuit, which consists of four MOSFET switches that form a full-bridge on the primary side of the isolation high- frequency transformer and MOSFET switches for synchronous rectification (SR) on the secondary side. This circuit allows all the switching devices to switch with ZVS resulting in lower switching losses and an efficient converter. In this work, ZVS for switches in the one leg of the full bridge and zero or low voltage or valley switching for switches in the other leg is achieved across the complete load range, by changing dead-times for primary side switches based on load conditions.

For such an isolated topology, signal rectification is required on the secondary side. For systems with low output voltage or high output current ratings, implementing synchronous rectification instead of diode rectification achieves the best possible performance by avoiding diode rectification losses. In this work, current double synchronous rectification is implemented on the secondary side with different switching schemes to achieve optimum performance under varying load conditions.

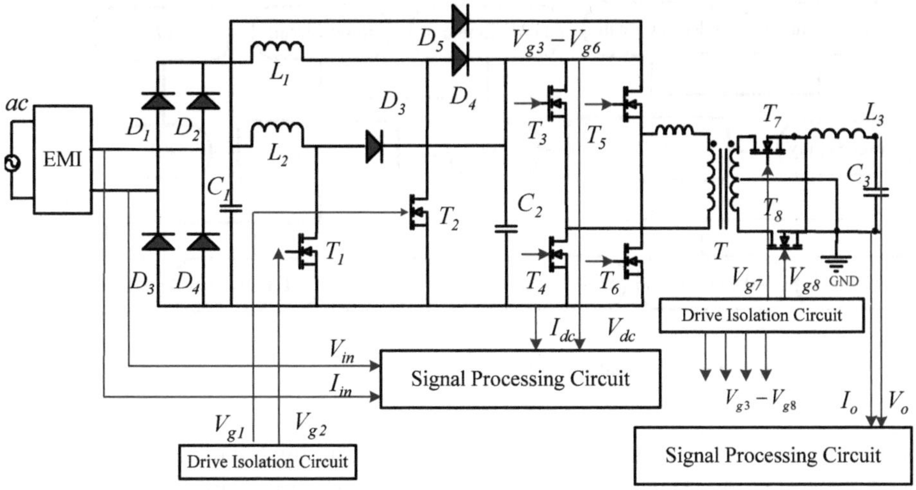

Fig. 1: Block diagram of overall main circuit topology

Fig.1 shows a simplified circuit of a phase shifted full bridge. MOSFET switches T3, T4, T5 and T6 form the full-bridge on the primary side of the transformer T1. T3and T4 is switched at 50% duty and 180^0 out of phase with each other. Similarly, T5 and T6 are switched at 50% duty and 180^0 out of phase with each other. The PWM switching signals for leg T5-T6 of the full bridge are phase shifted with respect to those for leg T3-T4. The amount of this phase shift decides the amount of overlap between diagonal switches, which in turn decides the amount of energy transferred. T7, T8 provide the current double rectification on the secondary, while L3 and C3 form the output filter. The inductor LR provides assistance to the transformer leakage inductance for the resonance operation with MOSFET capacitance and facilitates ZVS.

In figure 1, V_{in}, I_{in} is instantaneous value of the input ac voltage and current, V_{dc}, I_{dc} is instantaneous value of the power factor correction circuit, V_{dc}, I_{dc} is instantaneous value of the phase shifting full bridge circuit and Synchronous rectification circuit, V_{g1}, V_{g2} is gate signal of T_1, T_2 and so on.

3 Software Structure Of Digital Controller

System overall scheme of digital controller is showed in figure 2. The circuit of the two-stage scheme such as inter-leaved PFC and PSFB has its own control circuit. Its communication is by SPI of 28027.

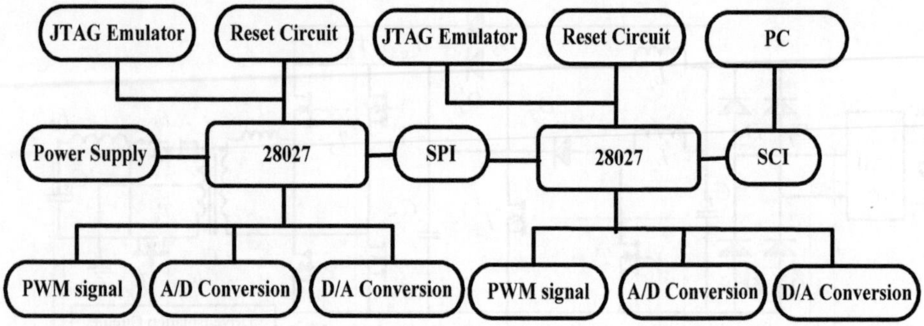

Fig. 2: System architecture of digital controller

The overall scheme of the software design as shown in Fig.2, which is mainly composed of three parts: (1) leading power factor compensation circuit of digital control algorithm; (2) DC/DC circuit of digital control algorithm; (3) man-machine interface.

28027 mainly complete control of the leading power factor compensation circuit, realize the input current tracking the input voltage, based on the average current control implementation. Second 28027 mainly complete DC/DC circuit of digital control; this paper adopts the phase shifting control. Man-machine interface can implement the system of condition monitoring, fault detection and real-time parameter adjustment, fully embodies the advantages of digital power based on the C#. We take the PFC scheme so inevitably involves the communication between the two levels before and after parts, moreover also includes protection, D/A program such as auxiliary module.

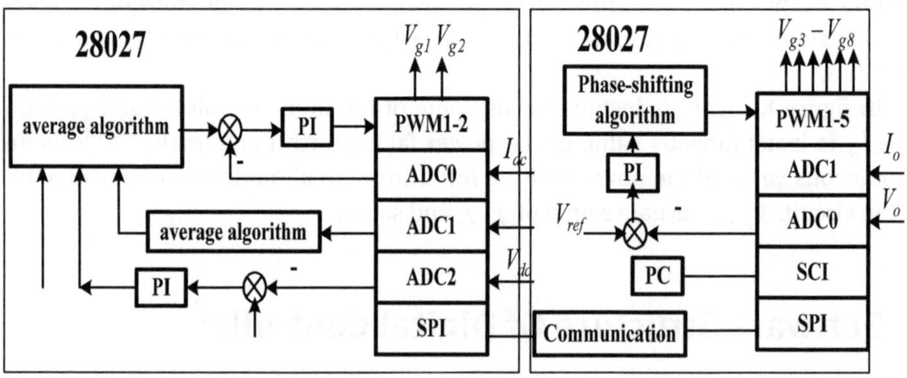

Fig. 3: Software structure diagram

The CCS project for C2000 ILPFC mostly makes use of the "C-background/ASM-ISR" framework. The main fast ISR (100 kHz) runs in assembly environment. A slower ISR (10 kHz) is also run from C environment. This slow ISR is made interruptible by the fast ISR.

The CCS project uses C-code as the main supporting program for the application, and is responsible for all system management tasks, monitoring, decision making, intelligence, and host interaction. The assembly code is strictly limited to the fast Interrupt Service Routine (ISR), which runs all the critical control code. Typically this includes reading ADC values, input line cycle polarity detect, sensed line volt rectification, control calculations, and PWM updates. The slower ISR in the C environment calculates the RMS input voltage, RMS input current, and RMS input power and frequency of the input line voltage. Fig.4 depicts the general software flow for this project.

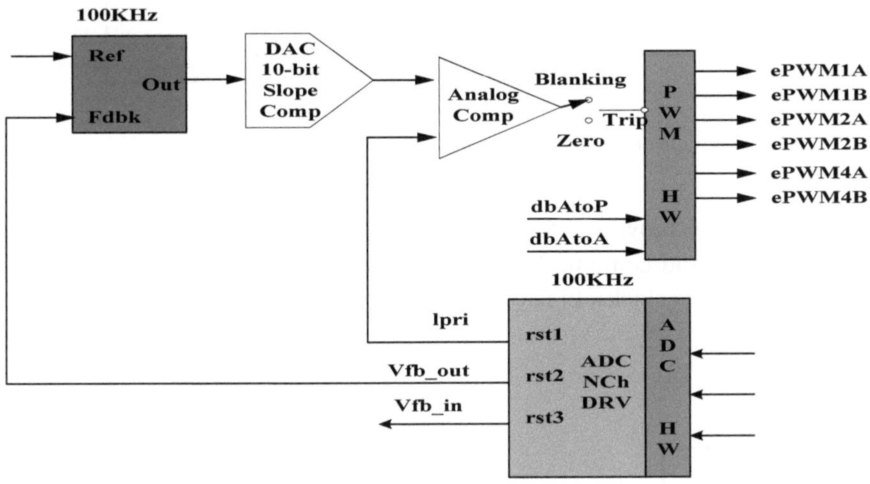

Fig. 4: PCMC structure diagram

The key framework C files used in this project are: interleaved PFC-Main. C-this file is used to initialize, run, and manage the application. Interleaved PFC-De v/nit F2802x. c. This file is used for 2802x controller initialization. A

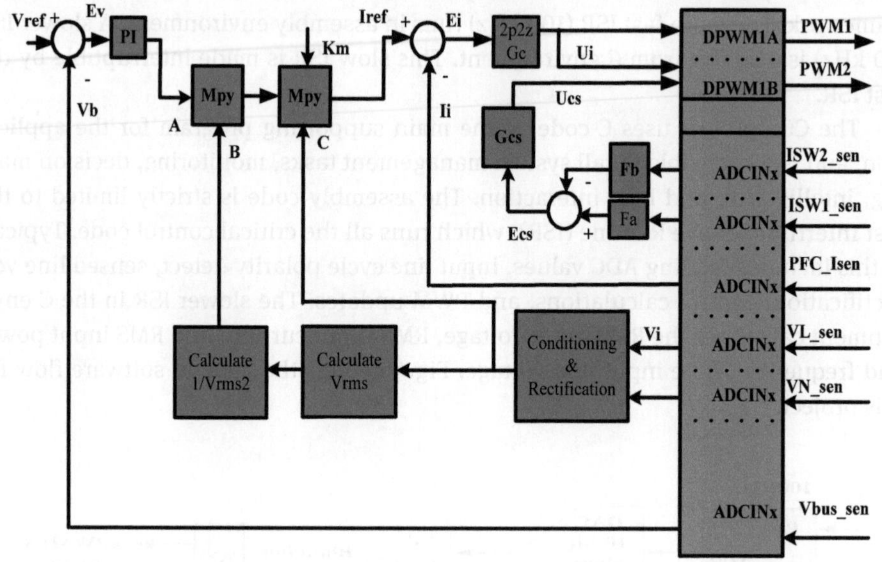

Fig. 5: Interleaved PFC Converter Control Using the C2000 Microcontroller

2802x control card is provided with the IL PFC EVM. This file is responsible for a one time initialization and configuration of the F280xx device, and includes functions such as setting up the clocks, PLL, GPIO, etc.

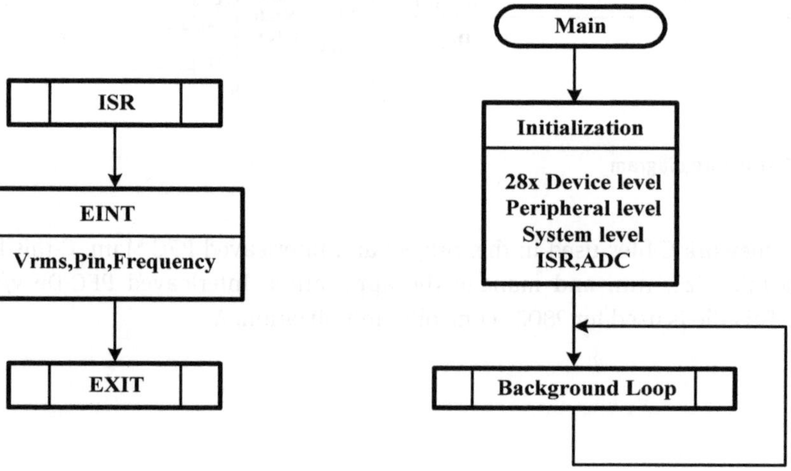

Fig. 6: IL PFC Software Flow Diagram

As mentioned in Section 1, the IL PFC system is controlled by two feedback loops. The outer voltage loop regulates the DC bus voltage, while a faster inner current loop shapes the input current in order to maintain a high input power factor. Fig.6 also shows the rate at which the software modules are executed. For example, the current controller is executed at a rate of 100 kHz (half of the PWM switching frequency) while the voltage controller is executed at 50 kHz rate.

4 Experimental Results

The experimental results of one FB dc-dc converter are presented in Fig.7-Fig.11. It should be mentioned that feedback loop is disabled intentionally to test circuit with 50% duty cycle. Fig.7 illustrates the Gate signals of T3 to T6 in FB dc-dc converter. Fig.8-Fig.10 illustrates voltage waveforms of power switch in FB dc-dc converters at the zero voltage. Fig.11 illustrates the waveforms of input voltage and current of dc-dc converter.

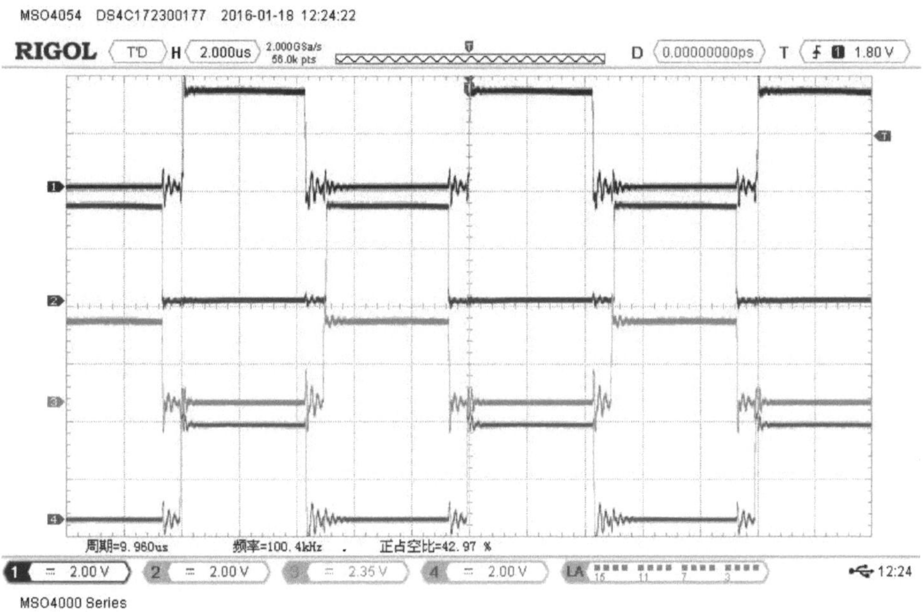

Fig. 7: Gate signals of T3 to T6 in FB dc-dc converter

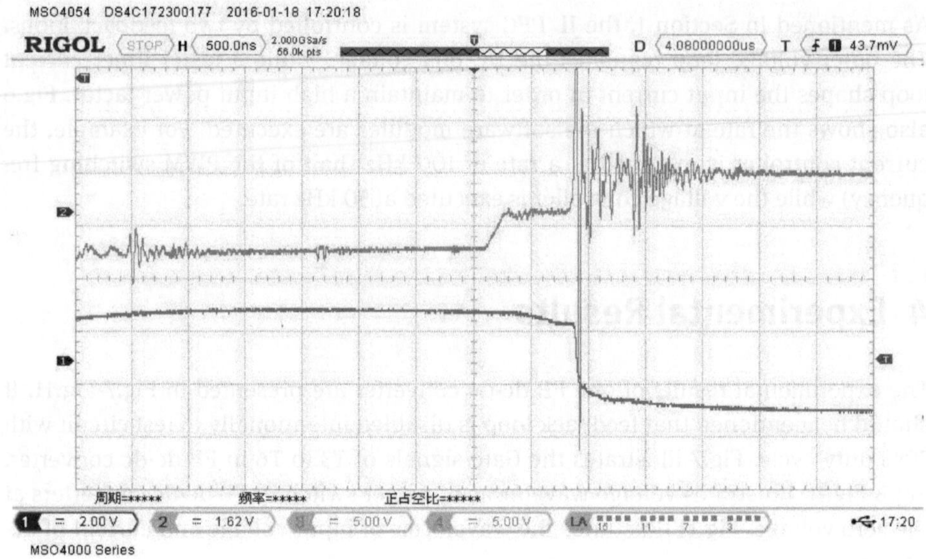

Fig. 8: Voltage of power switches in FB dc-dc converters opening at the zero voltage

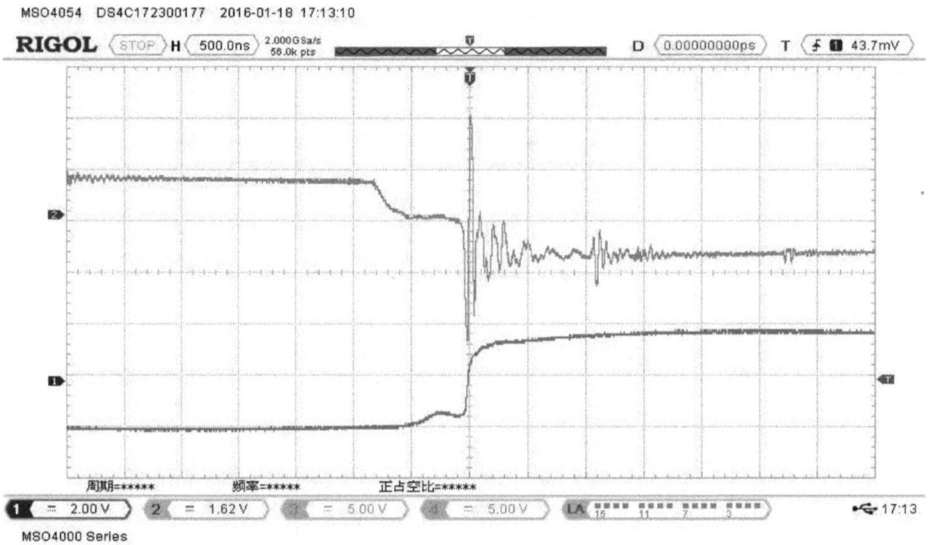

Fig. 9: Voltage of power switches in FB dc-dc converters turn off at zero voltage

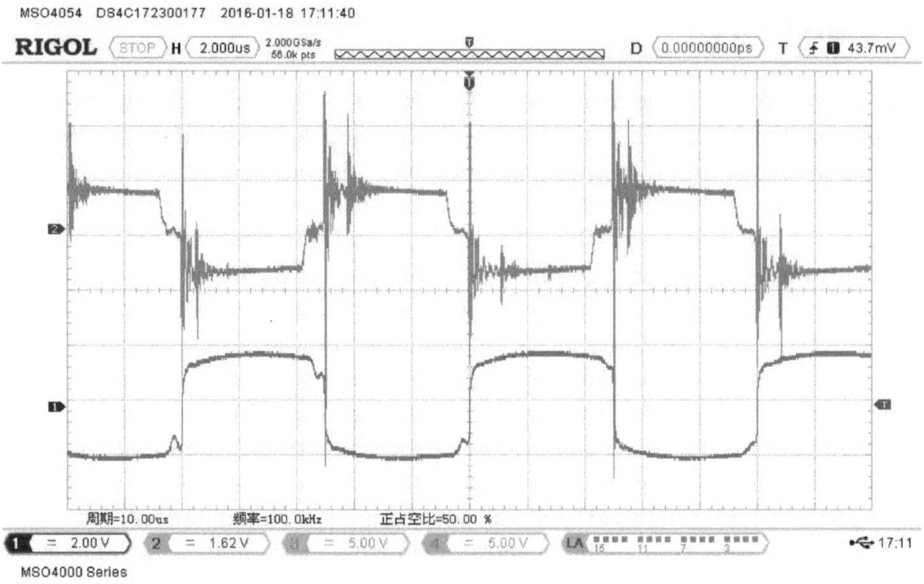

Fig. 10: Voltage of power switches in FB dc-dc converters

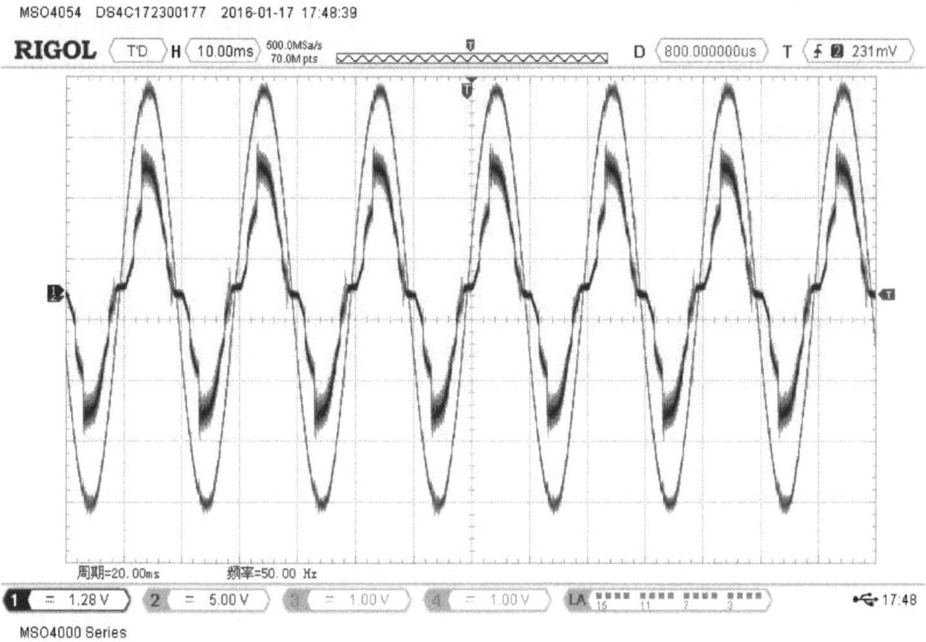

Fig. 11: Input voltage and current of power supply

5 Conclusion

This paper focuses on the digital control design of low-voltage high-current power supply for aluminium foil. In the power converter topology converter after the general requirements with high-speed high-precision AD conversion module, is mainly responsible for the output voltage and current digital sampling and processing. Pre-controller generally requires a high frequency, to meet the complexity of the computational complexity of the algorithm, to be responsible for the implementation of various types of control loop algorithm and the final output PWM control signal. This article uses two 28027 to achieve before and after the level of serial communication between. After the controller is configured as the host, the former controller is configured as a slave, which realizes the integrated control of the high efficiency and high power factor power supply. The experimental results show that the digital control method and software design are feasible. It provides a powerful development platform for the high-power digital power supply of aluminium foil.

Our future work is to apply the integrated digital platform to actual aluminium forming foil large power supply. The performance and effectiveness will be tested. The approach developed in this paper could also be utilized.

Acknowledgement: This work is supported by the Shao guan innovation R & D Fund in 2017,and by Guangdong college students' innovative entrepreneurial training plan(No. 201510576033), and by Guangdong college students' Special fund for cultivating scientific and technological innovation (No. pdjh2017b0454 and No. pdjh2017b0456).

References

[1] X. Wu, G. Hua, J. Zhang , Z. Qian, A New Current-Driven Synchronous Rectifier for Series-Parallel Resonant(LLC) DC-DC Converter, IEEE Transactions on Industrial Electronics[J], 58 (1) :289–297,2011.

[2] M. Yu, D. Sha, X. Liao, Hybrid phase shifted full bridge and LLC half bridge DC-DC converter for low-voltage and high-current output applications, IET Power Electronics[J], 7 (7) :1832–1840,2014.

[3] Z. Sun,F. Peng,P. Magne,A Emadi A Phase Shifted Full Bridge Converter with ZCS Synchronous Rectifier for Auxiliary Power Units[C], Conference of the IEEE Industrial Electronics Society:2945–2951, 2015.

[4] H. Cho, K. M. Cho, J. W. Kim, and Ci. W. Moon, a New Phase-Shifted Full-Bridge Converter with Maximum Duty Operation for Server Power System[C], IEEE International Conference on Power Electronics & Ecce Asia, 26 (12):1278–1285, 2011.

[5] A Mangi, S Xiao, GA Mallah, DA Jamro, I Memon, Multiband Circular Polarizer Based on Fission Transmission of Linearly Polarized Wave for X-Band Applications. Journal of Electrical and Computer Engineering, 2016.

[6] FA Mangi,S Xiao, I Memon, DA Jamro, GF Kakepoto, Single-layer Linear-to-Circular Polarization Transformer Based on Periodic Strips Array. DEStech Transactions on Engineering and Technology Research, 2016.
[7] Wu, X., Zhang, J., Xie, X., Qian, Z.,Analysis and optimal design considerations for an improved full bridge ZVS DC–DC converter with high efficiency[J], IEEE Trans. Power Electron., 21, (5),1225–1234, 2006.
[8] B. R. Lip, C. L. Huang, and J. F. Wan, Analysis, design, and implementation of a parallel ZVS converter[J]. IEEE Trans. Ind. Electron., 55(4), 1586–1594., 2008.
[9] Chen, W., Ruan, X., Zhang, R., A novel zero-voltage-switching PWM full bridge converter [J], IEEE Trans. Power Electron., 23, (2), 793–801, 2008.
[10] D. Wang, X. He, and J. Shi, Design and analysis of an interleaved flyback-forward boost converter with the current autobalance characteristic [J].IEEE Trans. Power Electron., 25(2), pp. 489-498. 2010.
[11] I.H.Cho, H.W.Seong, S.M.Jung, J.S.Park, G. W. Moon, and M.J. Youn, Implementation of digitally controlled phase shift full bridge converter for server power supply[C], in Proc. Energy Conveys. Congr. Expo. 802–809, 2010.
[12] Chen, Z., Liu, S., Shi, L., Improved zero-voltage-switching pulse width modulation full bridge converter with self-regulating auxiliary current [J], IET Power Electron., 6, (2), 287–296, 2013.
[13] G.N.B. Yadav, N.L. Nagayama, and An Active Soft Switched Phase-Shifted Full-Bridge DC-DC Converter: Analysis, Modelling, Design, and Implementation, IEEE Transactions on Power Electronics, 29 (9):4538–4550, 2014.
[14] S.Saggini,M.Ghioni, A.Geraci,An Innovative Digital Control Architecture for Low-Voltage,High-Current DC-DC Converters With Tight Voltage Regulation, IEEE Transactions on Power Electronics, 19 (1) :210–218,2004.
[15] S. Zheng, D. Czarkowski, Modeling and Digital Control of a Phase-Controlled Series-Parallel Resonant Converter [J], IEEE Transactions on Industrial Electronics, 54 (2):707–715, 2007.
[16] J.H. Cho, H.W. Seong, S.M. Jung, J.S. Park ,Implementation of Digitally Controlled Phase Shift Full Bridge Converter for Server Power Supply[C], Energy Conversion Congress & Exposition :802–809,2010
[17] A.K. Sadigh, V. Dargahi, K. Corzine, DSP-Based Digital Control of a Set of Phase-Shifted Full-Bridge DC-DC Converters[C], Power Systems Conference:1–6, 2015.
[18] W. Yang, P. Geng, Y. Wang, P. Zhang, Z. Xu, The effect of dead-time to ZVS phase-shifted full-bridge converter, International Conference on Electrical Machines & Systems[C] :578–580,2016.

Ping Zhou

Thermal-mechanical Coupling Analysis of Screw Discharge Machine Based on ANSYS

Abstract: This paper establishes the geometric model of screw discharge machine, and analyses its temperature field in ANSYS. In considering the boundary conditions of radiating heat transferring, heat transferring and high temperature gas, it conducts the stress analysis of the materials of screw discharge machine under high temperature, sets up the thermal-mechanical coupling model, shows the details about the temperature field coupling with stress field, and obtains specific distribution of the stress field.

Keywords: Screw discharge machine, ANSYS, Thermal-mechanical Coupling, Temperature field

1 Introduction

Rotary hearth furnace craftwork is a new technology of iron-making in the iron and steel production. Because of its advantages of no fuel preparation, no processing of raw materials and environmental protection, it has been widely applied. As the key equipment of rotary hearth furnace, the main function of screw discharge machine is to fast deliver metalized pellets generated by means of screw rotation to cylinder cooling machine. Because screw discharge machine is located in the area where there are metallic pellets whose temperature is about 1100°C, the material property decreases with the increase of temperature, which makes a spiral duct bent and the deformed blades burned. Screw discharge machine must be cooled in order to ensure the property. The material deformation and stress of the screw discharge machine is the important factors affecting the operating performance when it is cooled [1]. It is of important significance for the design of screw discharge machine to research cooling temperature field distribution and mechanical properties by using ANSYS.

Ping Zhou, School of Arts and Sciences, Jiangxi University of Engineering, Xinyu Jiangxi 338000, China. Email: 2865181276@qq.com

DOI 10.1515/9783110540048-074,

2 Establishment of Temperature Field Model

This paper takes screw discharge machine used by rotary hearth furnace in a steel company as the simulation object. The interior of the rotary hearth furnace has a temperature of 1100°C when the metal pellets is send out of the rotary hearth furnace by the screw discharge machine. Screw discharge machine adopts the method of water cooling, and the cooling water is sent into the internal spiral duct through the shaft and drained out by the drive shaft through spiral channels of internal and external aqueducts. The structure diagram of screw discharge machine is shown as Fig. 1.

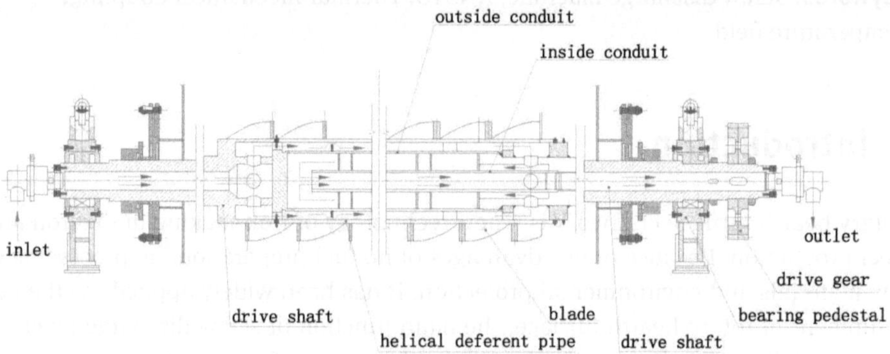

Fig. 1: Structure Diagram of Screw Discharge Machine

2.1 Assumptions

The temperature field model of screw discharge machine is set up with the following assumptions: (1) The parts of screw discharge machine have the same material properties; (2) The heat exchange between air and furnace burden is constant; (3) The thermal deformation produced by the deformation of screw discharge machine is ignored.

2.2 Establishment of Finite Element Three-dimensional Model

According to the actual structure of the screw discharge machine, the 3D geometric model is created, as shown in Fig. 2. When modeling, we should take it into consideration whether the transmission shaft, leaf, spiral duct, as well as water pipe are deformed under the high temperature and thermal radiation, but the drive gear and bearing are not considered in the model.

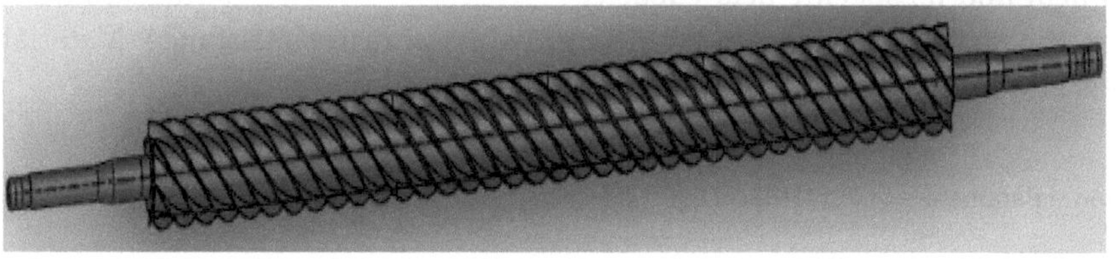

Fig. 2: Geometric Model of Screw Discharge Machine

Screw discharge machine is surrounded by high temperature gas in the rotary hearth furnace, so an area should be set for the analysis of surrounding flowing air field when we establish geometric model of screw discharge machine [2], which is shown as Fig. 3.

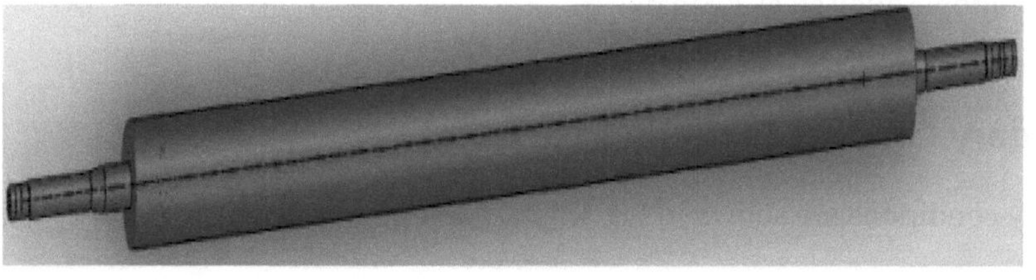

Fig. 3: Geometric Model of Screw Discharge Machine after considering the outside Air Field

As the model of screw discharge machine is about plane symmetry of central axis, it can go to a half when researching it. Thermal analysis is conducted by using a three-dimensional unit SHELL57 which is capable of the plane heat conduction. The finite element model of grid meshing is shown as Fig.4. The average cell size is 8.5 mm; the total number of cells is 2.975 million cells.

Fig. 4: Results of Grid Meshing

2.3 Parameter Determination

The material of blade in screw discharge machine is steel 0Cr25Ni20 which is resistant to heat and its specific heat capacity $C = 500J/(Kg.°C)$. The carbide is welded on the surface of blade so as to improve its abrasive resistance. Material parameters at different temperatures are shown in Table1.

Tab. 1: Material Parameters of 0Cr25Ni20

Temperature(°C)	Yield strength (MPa)	Tensile strength (MPa)
25	314	624
93	286	575
204	254	533

The material of spiral pipe is 0Cr18Ni9, the specific heat capacity $C = 500J/(Kg.°C)$. The material of the aqueduct is carbon steel Q235, and its rotating shaft adopts alloy steel 20MnMoNb.

2.4 Boundary Conditions Setting

The boundary conditions are shown in Fig.5.

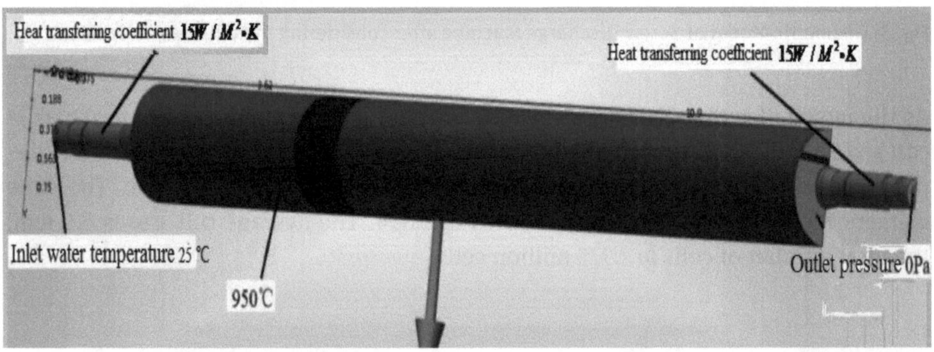

Fig. 5: Boundary conditions

The half model of screw discharge machine is set symmetrically, whose inlet water temperature on the left side is 25°C, and the water flow velocity is $3m/s$, and whose outlet pressure on the right side is 0 pa. The temperature of the outside screw is defined as 950°C, and radiate to the whole screw feeder. On the surface of the shaft on the outside of rotary hearth furnace is set heat transferring coefficient of $15W / M^2 \cdot K$, and ambient temperature of 35°C.

3 Results of Calculation of Temperature Field Model

The results of temperature field model of screw discharge machine calculation using ANSYS is shown as Fig.6-7. The water outlet temperature is 51.8584°C and the inlet water rate is $84.78 t / h$. The outside average temperature of spiral duct is 78.6°C, and the outside temperature of blade is 847°C. Temperature of blade changes greatly from the roots to the top. As can be seen from the Fig.7, cooling section changes at the position of the shaft contact with spiral duct, and the temperature field changes sharply.

Fig. 6: Cloud Picture of Temperature Distribution

Fig. 7: Profile Picture of Temperature Field Distribution

4 Establishment of Thermal - mechanical Coupling Model

4.1 Working Principle and Mechanical Model of Screw Discharge Machine

The drive mode of screw discharge machine adopts variable frequency motor which transmits torque to drive shaft through the reduction gears. The speed is assumed as $2\ r/min$ in the model and transport capacity as $110\ t/h$. Its drive torque is 95500 $N \cdot M$ during movement. One end of axle bearing is fixed, while the other can move around. Therefore, under the condition of high temperature, the axial thrust caused by the thermal expansion is not considered.

4.2 Establishment of 3D Model and Grid

Mechanics analysis of grid model is shown in Fig.8, whose cell size is 25 mm, node number is 1.168 million, and cell number is 641200.

Fig. 8: Grid Model

4.3 Boundary Conditions for Thermal Analysis

Temperature boundary conditions are set as shown in Fig.9, A shown in figure on the left lateral stands for environment value of 950°C, B and C stand for the two side of adiabatic surfaces in the rotary hearth furnace, D is the thermal coefficient value (0.95) of spiral surface under the temperature of 950°C, E is surface radiation coefficient value of 950°C, L, F, G, H, I, J, and so on are boundary conditions of water channel average temperature.

Fig. 9: Temperature Boundary Setting

Structural analysis of boundary conditions is shown in Fig.10. A showed on the left side in the diagram is the acceleration of gravity, B is fixed support, C is frictionless supporting bearings, and D is torque acting on the blade root, whose value is 95500 $N \cdot M$.

Fig. 10: Structural Analysis of Boundary Conditions

5 Results of Calculation of Thermal - mechanical Coupling Model

The cloud picture of stress distribution after thermal - mechanical coupling is shown in Fig.11.

Fig. 11: Cloud Picture of Stress Distribution

The cloud picture of deformation distribution after thermal - mechanical coupling is shown in Fig.12.

Fig. 12: Cloud Picture of Deformation Distribution

From results of ANSYS analysis, under the condition of water cooled, temperature of the blade varies from 78.5°C at the root to 847°C at the top. There is large temperature difference and large thermal stress on the blade, about 2290Mpa stress in blade's edges, and about 250Mpa stress at the bottom of the blade cylinder. Material mechanics properties of heat resistant steel 0cr25ni20 of blade can basically meet the requirements when the temperature is 78.5°C, but blade can't work continuously under the condition of high temperature considering the fatigue mechanical properties of materials.

6 Conclusion

This paper establishes thermal - mechanical coupling model of screw discharge machine by using ANSYS software, calculates the mechanical properties under high temperature, and obtains the weak spot of material properties of a screw discharging machine under the condition of high temperature. According to the working characteristics of the screw discharging machine, some of the heat-resistant material suitable for high temperature is applied to the key parts, and the parameters of material mechanics performance is given, which is guidance to the design of screw discharge machine.

Acknowledgement: Project supported by Science and Technology Research Project of Jiangxi Provincial Department of Education (GJJ161294).

References

[1] Wang Hualin, Zhou Xiaoping, LU Dacheng. Design analysis of screw discharge machine. Industrial Furnace. Vol. 38 (2016) No. 6, p. 28–29.

[2] Yin Zhongjun, Chen Bing, Sun Jie, Zhang Lianwan. FEA of temperature field of blade of screw conveyor. Mining & Processing Equipment. Vol. 38 (2010) No. 13, p. 70–73.

[3] Wang Jiangang. Design of water cooled discharge screw for dr rotary hearth furnace. Sintering and Pelletizing. Vol. 22 (2016) No. 2, p. 36–38.

[4] Li Hang. The thermal-mechanical coupling analysis of excavator bucket based on ABAQUS. Machine Design and Manufacturing Engineering. Vol. 42 (2013) No. 4, p. 14–18.

[5] Gao JQ, Lee SC, Ai XL, et al. An FFT-based transient flash temperature model for general three-dimensional rough surface contacts. ASME Journal of Tribology, 2000, 122:519–523.

[6] Colin F, Floquet A, Play D. Thermal contact simulation in 2-D and 3-D mechanisms. ASME Journal of Tribology, 1988, 110: 247–252.

[7] Sun Qiang, Nuo Lichun. Steel column dynamic analysis under heat-coupled force. Journal of University of Science and Technology of China. Vol. 39 (2009) No. 6, p. 644–648.

Wen Zhang, Hong Hu* and Dongik Shin

Improved XY-axis Calibration Based on Bilinear Interpolation

Abstract: Based on bilinear interpolation, this paper presents an improved calibration method for machine vision systems. An experiment is conducted that involves the construction of a model between the image plane of the camera and the world plane of the object, which provides a relationship of the planes' coordinate systems. An improved distortion correction method is then applied to improve accuracy. The results of the experiment demonstrate the advantages of this method compared to traditional methods.

Keywords: calibration, distortion correction, bilinear interpolation, machine vision

1 Introduction

Although machine vision has become well established, the images obtained are distorted due to the poor structure and installation of cameras. Camera calibration is therefore an important issue in machine vision. Camera calibration is widely considered as the process of estimating the parameters of a pinhole camera model, which approximates the actual camera and then achieves the relationship between image and world points [1]. CCD cameras play an important role in modern machine vision systems because the precision of CCD camera calibration will strongly influence the overall performance of the system [2]. Therefore, to improve the precision of machine vision systems, the precise calibration of the camera must firstly be ensured.

Currently, there are two types of calibration targets used widely in machine vision: the first target is the two-dimensional square matrix [3], where a linear 3×4 matrix is applied that denotes the perspective transformation; and the second target is chessboard, which applies the non-linear least square method [4] and the deviation pie correction method [5] to correct both parameters of the distortion and the camera itself. Because the second type is applied to compute distortion parameters

Wen Zhang, Harbin Institute of Technology Shenzhen Graduate School, Shenzhen, China. Email: honghu@hit.edu.cn
***Corresponding author: Hong Hu,** Harbin Institute of Technology Shenzhen Graduate School, Shenzhen, China. Email: honghu@hit.edu.cn
Dongik Shin, Shenzhen Axxon Automation Con., Ltd., Shenzhen, China. Email: sdongik@hotmail.com

DOI 10.1515/9783110540048-075,

and camera parameters at the same time, the complexity and running time of calculation is increased, both of which researchers are constantly trying to decrease. Reference [6] proposed adaptive distortion correction to improve the calibration speed without loss of the predicted accuracy, while reference [7] only models the distortion parameters without the inner and outer parameters of the camera so that the calibration speed is substantially improved.

Considering that most calibration approaches use detected corners on the chessboard, it is imprecise to correct a certain point located inside the grids because if that point is located exactly at the corners of the chessboard, it can be corrected by corrected coefficients directly. This paper therefore introduces bilinear interpolation in the direction of X-axis and Y-axis respectively to improve the precision of the calibration on these feature points located inside the grids after finishing the camera calibration.

2 Distortion Correction Model

2.1 Imaging Model

The pinhole camera that is the object of this study is shown as Fig. 1. When using a pinhole camera, the geometric mapping from 3D to 2D is called a perspective projection [8]. O-XYZ is the 3D world coordinate system in which the origin O is regarded as the center of calibration board and any feature point P is represented three-dimensionally by X, Y, Z; o-uv is the image pixel coordinate system where the plane (u, v) being perpendicular to the optical axis is just the image plane of the camera (in pixels) and the origin o is in the left upper corner. Given a camera with an optical axis collinear to the Z-axis with all objects located at the plane $Z=Z_0$, this study aims to find the transformation relationship between the image plane (u, v) and the world plane (X,Y).

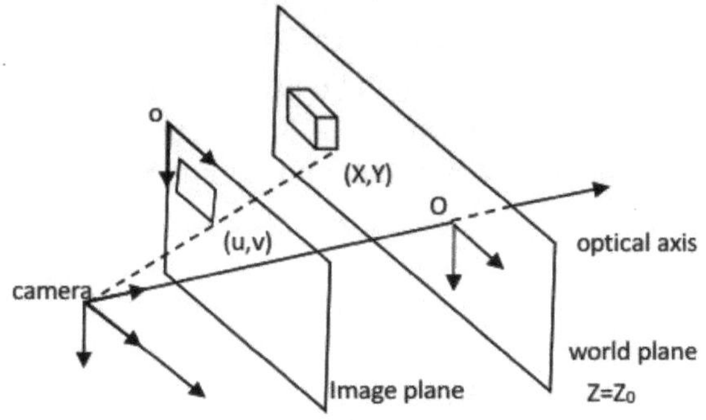

Fig. 1: Imaging model with image and world plane

2.2 Bilinear Interpolation

In mathematics, linear interpolation is a method to construct new data points with a set of known data points for curve fitting, which applies linear polynomials [9]. If two points are given by the coordinates (x_0,y_0) and (x_1,y_1), the linear interpolation is the straight line between them, which is shown as Fig. 2. For a value x in the interval $(x_0, x1)$, the value y along the straight line is regarded as being given from:

$$\frac{y-y_0}{x-x_0}=\frac{y_1-y_0}{x_1-x_0}. \tag{1}$$

We will obtain y by solving (1) with unknown value at x,

$$y=y_0+(x-x_0)\frac{y_1-y_0}{x_1-x_0}, \tag{2}$$

Which is just the formula for linear interpolation in the interval (x_0, x_1).

The formula can also be understood as a weighted average both in the x and y direction. The weights are inversely related to the distance from the end points to the unknown point. That is to say, the closer point has more influence than the father point. They are therefore supposed to be $(x-x_0)/(x_1-x_0)$ and $(x_1-x)/(x_1-x_0)$, which are normalized distances between the unknown point and each of the end points, and because they sum to 1 it yields the formula for linear interpolation as:

$$y=y_0(1-\frac{x-x_0}{x_1-x_0})+y_1(1-\frac{x_1-x}{x_1-x_0})=y_0(\frac{x_1-x}{x_1-x_0})+y_1(\frac{x-x_0}{x_1-x_0}) \tag{3}$$

Bilinear interpolation is actually an extension of linear interpolation with interpolating functions of two variables (i.e., x and y) on a rectilinear 2D grid [10]. The key idea in bilinear interpolation is to perform linear interpolation first in one direction, and then again in the other direction, which is shown as Fig. 3. Although each step is linear in the sampled position, as a whole it is not linear but rather quadratic in the sample location.

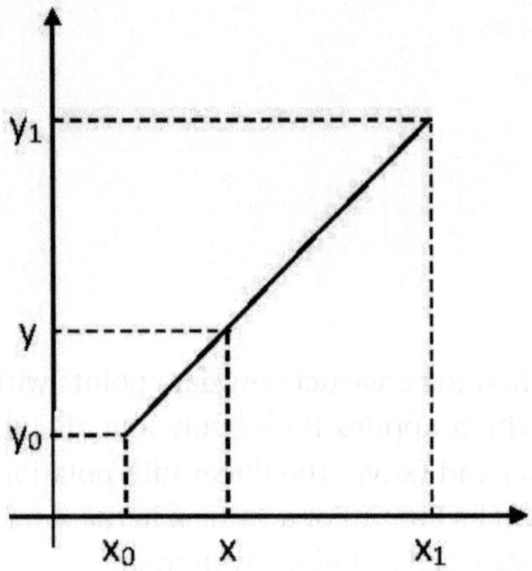

Fig. 2: Linear interpolation

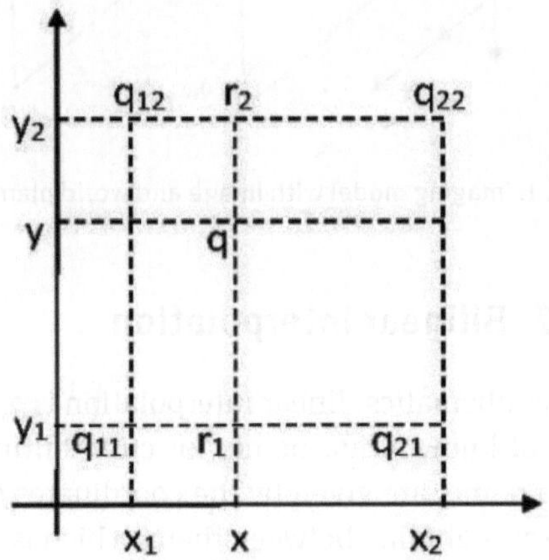

Fig. 3: Bilinear interpolation

What we want to find out is the value of an unknown function f at one certain point (x, y) according to values of this function f at the four points $q_{11}=(x_1,y_1)$, $q_{12}=(x_1,y_2)$, $q_{21}=(x_2,y_1)$, $q_{22}=(x_2,y_2)$. Through bilinear interpolation, first linear interpolation is applied in the x-direction:

$$f(x,y_1)\approx\frac{x_2-x}{x_2-x_1}f(q_{11})+\frac{x-x_1}{x_2-x_1}f(q_{21})$$
$$f(x,y_2)\approx\frac{x_2-x}{x_2-x_1}f(q_{12})+\frac{x-x_1}{x_2-x_1}f(q_{22}) \tag{4}$$

Then proceed with interpolation in the y-direction to obtain the desired estimate:

$$\begin{aligned}f(x,y)&\approx\frac{y_2-y}{y_2-y_1}f(x,y_1)+\frac{y-y_1}{y_2-y_1}f(x,y_2)\\&=\frac{y_2-y}{y_2-y_1}\left(\frac{x_2-x}{x_2-x_1}f(q_{11})+\frac{x-x_1}{x_2-x_1}f(q_{21})\right)+\frac{y-y_1}{y_2-y_1}\left(\frac{x_2-x}{x_2-x_1}f(q_{12})+\frac{x-x_1}{x_2-x_1}f(q_{22})\right).\end{aligned} \tag{5}$$

The same result will be arrived at if the interpolation is done first along the y-direction and then along the x-direction.

Similarly, let (5) be understood as a weighted average in x and y directions and the weights are $t=(x-x_1)/(x_2-x_1)$ and $s=(y-y_1)/(y_2-y_1)$, so that the bilinear interpolation formula is considered as:

$$f(x,y)=(1-s)((1-t)f(q_{11})+tf(q_{21}))+s((1-t)f(q_{12})+tf(q_{22})). \tag{6}$$

2.3 Distortion Correction Model

As shown in Fig. 4, the whole distortion correction model is based on bilinear interpolation. If we know the world coordinate (X, Y) of a certain point P, we can apply the calibration method to find out its exact image coordinate (u, v). When we get the world coordinates of four vertexes P_{11}, P_{12}, P_{21}, P_{22}, which are (X_1,Y_1), (X_1,Y_2), (X_2,Y_1) (X_2,Y_2) of the grid where (X,Y) is located, we can apply feature detection to obtain the corresponding image coordinates (u_{11},v_{11}), (u_{12},v_{12}), (u_{21},v_{21}), (u_{22},v_{22}). On the basis of the bilinear interpolation formula, suppose that u-coordinate and v-coordinate in the image coordinate system are regarded as the unknown function u (P) and v (P) respectively, and then the distortion correction model is shown as:

$$\begin{aligned} u(P)&=(1-s)((1-t)u(P_{11})+tu(P_{12}))+s((1-t)u(P_{21})+tu(P_{22})) \\ v(P)&=(1-s)((1-t)v(P_{11})+tv(P_{12}))+s((1-t)v(P_{21})+tv(P_{22})). \end{aligned} \tag{7}$$

That is to say:

$$\begin{aligned} u&=(1-s)((1-t)u_{11}+tu_{12})+s((1-t)u_{21}+tu_{22}) \\ v&=(1-s)((1-t)v_{11}+tv_{12})+s((1-t)v_{21}+tv_{22}), \end{aligned} \tag{8}$$

Where $t=(X-X_1)/(X_2-X_1)$ and $s=(Y-Y_1)/(Y_2-Y_1)$. Therefore, all we need to do is to put world coordinates of P_{11}, P_{12}, P_{21}, P_{22} into the correction model and then (u, v) is just the image coordinate for the point p.

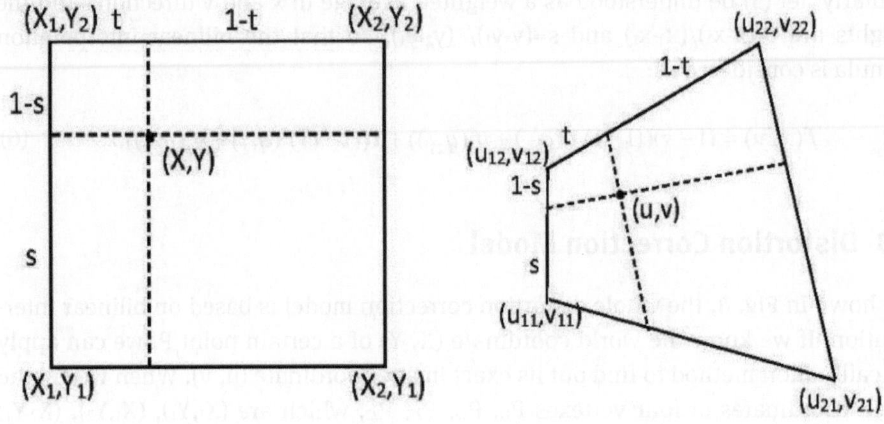

(a)World coordinate (b) Image coordinate

Fig. 4: Distortion correction model

Conversely, if the world coordinate (X, Y) of a certain point is known, its exact image coordinate can also be calculated through inverse bilinear interpolation. If the image and world coordinates of four vertexes of the grid where the calculated point location is known, we can also solve the problem from the bilinear interpolation formula. Different from computing the image coordinate (u, v), if we want to solve this equation to find the world coordinate (X, Y) with known (u, v), we first reorganize these equations to be:

$$\begin{aligned}(u_{11}+u_{22}-u_{12}-u_{21})st+(u_{21}-u_{11})s+(u_{12}-u_{11})t=u-u_{11}\\(v_{11}+v_{22}-v_{12}-v_{21})st+(v_{21}-v_{11})s+(v_{12}-v_{11})t=v-v_{11},\end{aligned} \tag{9}$$

Then the value s and t can be solved from this, so the world coordinate of the unknown point (X, Y) will be found as:

$$\begin{aligned}X=X_1+t(X_2-X_1)\\Y=Y_1+s(Y_2-Y_1).\end{aligned} \tag{10}$$

3 Distortion Correction Method

3.1 Set the Grid System

Prepare a large chessboard with known dimensions whose vertexes can be represented as a certain set V= {(X_0, Y_0), (X_1, Y_1})... (X_n, Y_n)}, and take a photo of the chessboard using the camera that needs to be calibrated. Apply corner detection to this photograph in order to get the image coordinates corresponding to the vertexes on the chessboard, which can be represented as a set v= {(u_0, v_0), (u_1, v_1)... (u_n, v_n)}. All the calibration is based on these two coordinate sets, and all values to be used to interpolate are from them. The image grid system and the world coordinate system are shown in Fig. 5.

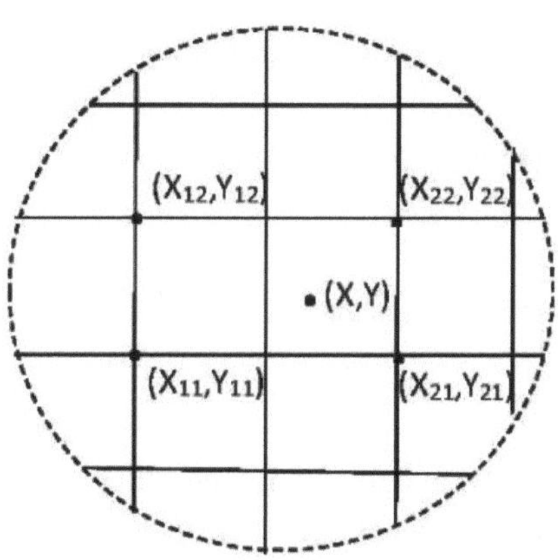

(a) World coordinates

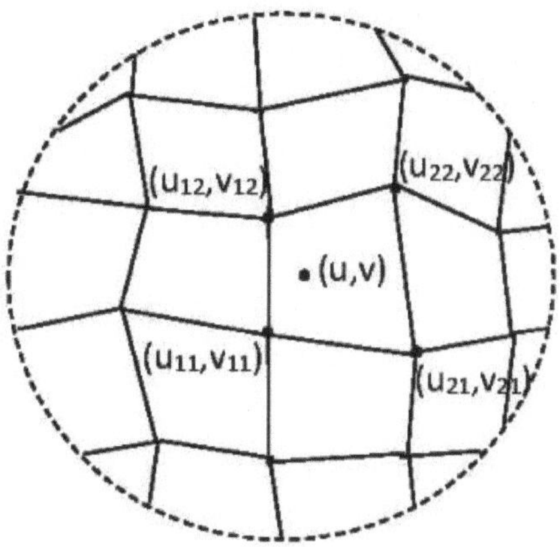

(b) Image coordinate

Fig. 5: Coordinate system

3.2 World-to-Image Coordinate System Calibration

If we know one certain point P=(X, Y) in the world coordinate system, we can find out the exact coordinate (u, v) in the image coordinate system. Fig. 5 shows the principle of CCD camera lens distortion correction by bilinear interpolation where the image coordinate is to be computed based on its known world coordinate.

Thanks to the standard grid in chessboard, it is convenient to determine the exact grid location of P. As shown in Fig. 5(a), we just need to compare X and Y with the points from set V to find out four vertexes P_1= (X_i, Y_j), P_2= (X_i, Y_{j+1}), P_3= (X_{i+1}, Y_j),

$P_4 = (X_{i+1}, Y_{j+1})$, which satisfy P_1, P_2, P3, $P_4 \in V$ and $X_i<X<X_{i+1}$, $Y_j<Y<Y_{j+1}$. According to the prepared grid system, their corresponding image coordinates can be obtained from set v as (u_{11}, v_{11}), (u_{12}, v_{12}), (u_{21}, v_{21}), (u_{22}, v_{22}). By putting them into the bilinear interpolation formula, the image coordinate (u, v) can be easily solved.

3.3 Image-to-World Coordinate System Calibration

Unlike the world-to-image coordinate system calibration, if we only know the image coordinate of a certain point and the prepared grid system, there is no way to find the exact grid location of that point by simply comparing. Because of the lens distortion, shown in Fig. 5(b), which causes the u and v axis not to be straight and perpendicular, it is not possible to compare values in the u and v directions to confirm the four vertexes of the grid for the calculated point. The algorithm therefore calculates the cross product to confirm the location of the grid.

Suppose the grid system in the image plane is as shown as Fig. 6. The main idea of cross product to confirm the

Location is that we search each grid in the image plane to calculate cross product between vector $q_{11}q_{21}$ and vector $q_{11}q$:

$$\overrightarrow{q_{11}q_{21}} \times \overrightarrow{q_{11}q} = \left|\overrightarrow{q_{11}q_{21}}\right| \cdot \left|\overrightarrow{q_{11}q}\right| \cdot \sin < \overrightarrow{q_{11}q_{21}}, \overrightarrow{q_{11}q} > . \tag{11}$$

If this cross product is larger than the value 0, vector $q_{11}q$ is located within 180 degrees anti-clockwise of vector $q_{11}q_{21}$; otherwise it is beyond 180 degree. Once we have calculated the cross products at all four vertexes $q_{21}q_{22}\times q_{21}q$, $q_{22}q_{12}\times q_{22}q$, $q_{12}q_{11}\times q_{12}q$ and they are all larger than 0, it is proved that the target point is located inside this grid. We can then get image coordinates of these four vertexes (u_{11}, v_{11}), (u_{12}, v_{12}), (u_{21}, v_{21}), (u_{22}, v_{22}) and their corresponding world coordinates.

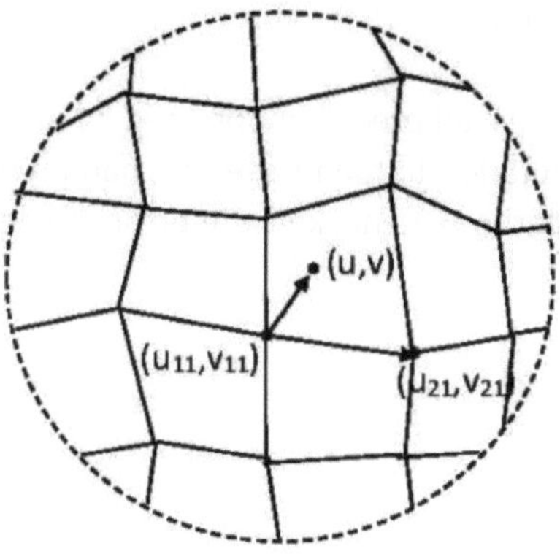

Fig. 6: Cross product

Once we have obtained four vertexes, inverse bilinear interpolation is applied to solve the problem. What we need to do is put the image coordinates of these four vertexes into the formula to find (u, v) and finish the calibration.

4 Experiments

The current most common calibration method is that based on a planar template, which was proposed by professor Zhang [11]. The method proposed in this paper improves upon that method by introducing a two-step calibration approach by first obtaining a set of photographs taken at different orientations of the planar template and then solving the intrinsic and extrinsic parameter matrix of the camera. An experiment on this improved camera calibration method was conducted in order to prove that it works.

The camera used for calibration was a CCD camera. The model plane we used was a pattern of 8×8 squares, which is to say 20 inner corners with each square 24mm × 24mm. This prepared chessboard was precisely printed. During this experiment, eight images of this chessboard were first taken in accordance with Zhang's calibration method to obtain the intrinsic and extrinsic parameter matrix and also the distortion parameters. Secondly, we applied the calibration results to 22 points in the image that had been prepared for calibration with world and image coordinates already known, as shown in Fig. 7. Using the known image coordinates of each point, the corresponding world coordinates were calculated by Zhang's method. Suppose (u_o, v_o) is the actual value for the image coordinate and (u_c, v_c) represents the calculated value obtained by parameters of the camera calibration, so the error of each point is calculated by:

$$e = (u_c - u_0)^2 + (v_c - v_0)^2. \tag{12}$$

The results are shown in Table 1. To prove the increased precision obtained by using the proposed method, we applied bilinear interpolation to the image that had previously been calibrated by Zhang's method. Similarly, we checked the error of each point and the results are shown in Table 1 and Fig. 8.

Fig. 7: Calibrated image

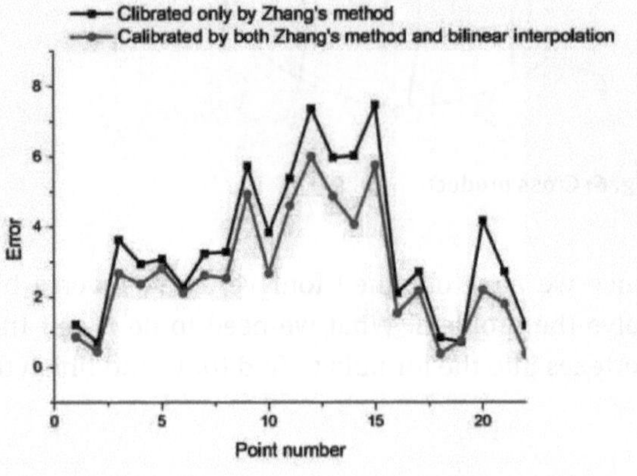

Fig. 8: Comparison of two calibration methods

As can be seen from Table 1, after calibration by the proposed method the average error of these points is clearly smaller than the calibration error incurred using Zhang's method alone. So we can see from the results in this experiment that the method based on bilinear interpolation can increase the accuracy in the transformation between image and world coordinate systems.

Tab. 1: Comparison of calibration errors

Calibration procedure	Calibration variance
Calibrated only by Zhang's method	3.480
Calibrated by both Zhang's method and bilinear interpolation	2.658

5 Conclusion

This paper introduced bilinear interpolation to correct camera distortion and proposed improved algorithms for the world-to-image coordinate system for calibration and the image-to-world coordinate system for calibration. Results from the experiments show that calibration based on bilinear interpolation can be considered as a new calibration strategy for increasing the precision of XY-calibration.

Acknowledgement: The authors gratefully acknowledge the support provided by the National Nature Science Foundation of China under NSFC Grant 51375118, as well as the support from Shenzhen Government Fund JCYJ20160330161701343.

References

[1] Pan Feng and Wang Xuanyin, "New versatile camera calibration technique based on linear rectification," in Chinese Journal of Mechanical Engineering, vol. 17, No. 4, 2004.
[2] Qiu N L and Ma S D, "The nonparametric approach for camera calibration," in IEEE Proceedings of the Fifth Internal Conference on Computer Vision, 1995, pp. 50–55.
[3] Ma Song-de and Zhang Zheng-you, "Computer vision elements of computation theory and algorithms," Beijing[M], science press, 1998, pp. 52–71.Ma Song-de and Zhang Zheng-you, "Computer vision elements of computation theory and algorithms," Beijing[M], science press, 1998, pp. 52–71.
[4] Janne H, "Geometric camera calibration using circular control points," in IEEE Transactions on Pattern Analysis and Machine Intelligence, vol. 22, 2000, pp. 1330–1334.
[5] Alan M. Mclvor, "Nonlinear calibration of a laser stripe profiler," in Optical Engineering, vol. 41, 2002, pp. 205–212.
[6] Chiung-Wei Huang and Chang-Min Chou, "Adaptive calibration method for camera distortion," in Journal of Vibration and Control, vol. 19, 2012, pp. 86–93.
[7] WANG Shuo, et al, "An improved adaptive correction method for camera distortion," Proceedings of the 33rd Chinese Control Conference, July 28–30, 2014, Nanjing, China.
[8] YB Leng, et al, "The beam-based calibration of an X-ray pinhole camera at SSRF," in Chinese Physics C, vol. 36, 2012, pp. 80–83.
[9] M Zeng, et al, "A new linear interpolation algorithm," International Conference on Computer Engineering, 2016.
[10] Rukundo O, et al, "Optimization of bilinear interpolation based on ant colony algorithm," in Electrical Engineering, vol.137, 2012, pp. 571–580.
[11] P Lu, et al, "Camera calibration implementation based on Zhang Zheng you Plane Method," Springer Berlin Heidelberg, 2016.

Chun-Run Li, Jian Liu* and Hua-Qing Yang

Research of E36 Steel Welding Process for Jinzhou 9-3 Offshore Platform

Abstract: In this paper, the welding process of E36 high strength steel involved in Jinzhou 9-3 offshore platform was studied. The mechanical properties such as tensile, bending, notch impact and surface hardness were tested and analysed with the Welding specimens. The residual stress distribution of the weld after welding was tested by X-ray diffraction, and the distribution rules of residual stress in the weld and nearby area were obtained. The test results showed that the performance of the welded joints of the high strength steel E36 met the requirements of the AWS D1.1 standard and the welding process was practicable.

Keywords: High strength Steel E36 Welding Process Residual Stress
List 3-5 keywords covered in your paper, which will also be used by the publisher to produce a keyword index.

1 Introduction

The development of China's offshore oil has brought about the ever-expanding demand for steel for offshore platforms. E36 steel is the main structural steel for marine engineering in China. It has been widely used in hull structure and offshore platform construction with its high strength, high toughness, good fatigue resistance, anti-layer tearing and good weldability and seawater corrosion resistance and other excellent performances. E36 high strength steel was widely used in the Jinzhou 9-3 offshore platform, the performance of welding joints would directly affect the quality of the entire offshore platform. Therefore, in this paper, the E36 high-strength steel welding process was systematically studied combined with Jinzhou 9-3 offshore platform construction project and the joints mechanical performance and residual stress was tested in order to ensure the quality of platform structure and provide technical guidance of safety service.

Chun-Run Li, CNPC Research Institute of Engineering Technology, No.40 Jing tang Road, Tian jin, China. Email: licr@cnpc.com.cn
Jian Liu, Key Laboratory of Offshore Engineering of CNPC No.40 Jing tang Road, Tian jin, China. Email: liujian3@cnpc.com.cn. Corresponding author
Hua-Qing Yang, CNPC Research Institute of Engineering Technology, No.40 Jing tang Road, Tian jin, China. Email: yanghq1@cnpc.com.cn

DOI 10.1515/9783110540048-076,

2 Welding Process Experiment

2.1 Weldability Analysis of E36

E36 steel is a low alloy high strength structural steel containing trace alloy elements Nb and V [1]. It has shown in studies that trace alloying elements Nb and V can strengthen and toughen the steel. The chemical composition and mechanical properties of E36 steel were respectively shown in Table 1 and Table 2.

Tab. 1: Chemical composition of E36 steel (× 0.001%)

C	Si	Mn	P	S	Cu	Al	Cr	Ni	Ti	Mo	Nb	V
160	300	1440	12	6	150	32	40	40	2	21	2	34

Tab. 2: Mechanical properties of E36 steel

Yield strengt/ σs	tensile strength/σb	Elongation/δ	Impact energy/-40°C, J		
375N/mm2	540N/mm2	31%	269	275	200

According to the carbon equivalent formula of the International Welding Society (IIW) and the International Classification Society (IACS):

$$Ceq=C+Mn/6+ (Cr+Mo+V)/5+ (Ni+Cu)/15\ (\%) \quad (2)$$

It can be seen that the E36 had a high carbon equivalent, a high rigidity, a poor weldability and a high sensitivity to welding hydrogen-induced cracks. Therefore, it was necessary to preheat when the ambient temperature was not high.

The principle of the groove is to ensure that the full penetration, but also to minimize the amount of welding, and strive to make the joint deformation to a minimum. According to Jinzhou 9-3 project actual situation, taking into account the tube node of various connections in the platform construction process, the groove type was designed as shown in Figure 1.

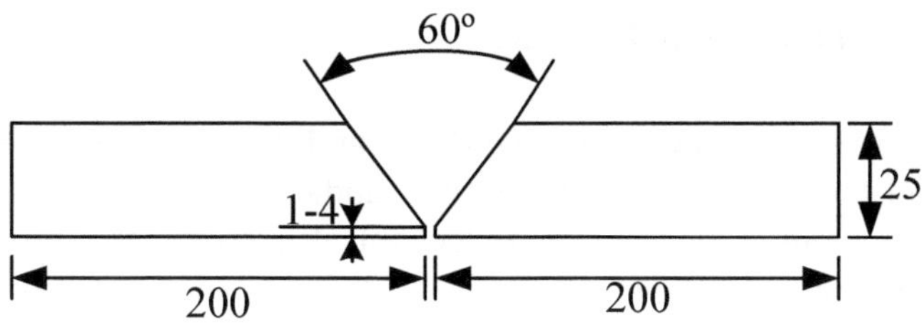

Fig. 1: Groove type

In order to ensure the strength, toughness, plasticity and other mechanical properties of the welded joint to match the base metal, the choice of welding materials should be based on the mechanical properties of the base metal, rather than from the chemical composition of the base material composition and the same choice of welding materials. In this study, LB-52U primer was used, CHE58-1 was filled and TWE-71Ni was filled.

2.2 Welding Methods and Process Specifications

Welding process test was done with a test plate whose size was 400mm × 400mm × 40mm, using manual arc welding (SMAW) for primer and flux-cored wire gas shielded welding (FCAW-G) for covering [2].

The range of preheating temperature before welding should be 100 ~ 150°C, and the temperature should be between 100 ~ 200°C. The preheating range is not less than the maximum thickness of the weld in the direction of the weld bead, but at least 75mm. When the temperature of base metal is lower than 0°C, the preheating temperature is at least 210°C. The heating method must ensure that the base material can be preheated thoroughly. It should be cleaned by polishing or other suitable means at least 15 mm on both sides of the welding area. The surface of the weld must be cleaned by brush and angle polisher when a weld layer is finished. Pay attention to using electrode reasonably, to prevent the edge of the layers of non-fusion, undercut, porosity and other defects.

According to the above method and specification, the welding technology of 6G position of the plate specimen and the 6G position of the tube specimen were carried out respectively. After a series of welding process parameters, the typical welding process parameters were obtained as shown in Table 3.

3 Mechanical Performance Test

After all the welding of test specimen was finished, X-ray and ultrasonic tests were carried out, the results of the weld inspection met the standard requirements, and then the mechanical tests were carried out. The mechanical property tests of the welding process evaluation included tensile, bending, notched impact and surface hardness [3]. The tensile and bending tests were done in the SHT5106-P microcomputer control current servo universal testing machine and the impact tests was done in the ZBC2602-C low temperature impact testing machine. Weld joint test specimen, processing and test methods were according to AWS D1.1 standard [4].

Tab. 3: Weld parameters

weld layer	weld method	diameter(mm)	brand name	type &polarity	current (A)	voltage(V)	arc type	arc flow rate(L/min)	weld speed (mm/min)	heat input (KJ/mm)
Root	SMAW	3.2	LB-52U	DC(+)	80-120	20-23	-	-	50-100	1.2-2.5
Fill(or)	SMAW	3.2	CHE58-1	DC(+)	100-130	20-25	-	-	50-160	1.2-2.0
Fill(or)	SMAW	4.0	CHE58-1	DC(+)	140-180	22-26	-	-	50-160	1.2-2.0
Fill	FCAW-G	1.2	TWE-711Ni	DC(+)	150-200	20-25	CO2	20-25	160-280	0.9-1.6
Cap	FCAW-G	1.2	TWE-711Ni	DC(+)	150-200	20-25	CO2	20-25	180-320	0.8-1.4

3.1 Bending Test

The bending test was carried out with a bending angle of 180 °according to the requirements of the AWS D1.1 standard. The test results were shown in Table 4.

Tab. 4: Results of welded joint bending test

Specimen source	Specimen No	Bending angle	Test results
Plate	1	180°	a 1.2mm crack near the root
	2		No crack
	3		No crack
	4		No crack
Pipeline	5		a 0.8mm crack in the root
	6		No crack
	7		No crack
	8		No crack

The test results showed that the bending test results of the specimen are in accordance with the provisions of the AWS D1.1 standard. Which was that the length of the single crack was less than 3mm and the total length of the crack was less than 10mm.

3.2 Tensile Test

The results of the tensile test of the welded joint were shown in Table 5. They showed that the tensile test results of the welded joints met the requirements of the AWS D1.1 standard.

Tab. 5: Welded joint tensile test results

Specimen source	Specimen No	Specimen size width×thickness(mm)		Breaking power Pmax (kN)	Tensile strength σb(MPa)	Reduction of area Ψ(%)	Breaking section
		Before test	After test				
Plate	1	19.57×29.45	13.83×20.26	288	489	51.4	Base metal
	2	19.85×29.25	12.90×19.56	281	487	56.5	Base metal
Pipe	3	20.25×25.30	12.20×13.94	271	520	66.8	Base metal
	4	20.15×24.46	11.98×13.76	269	528	66.6	Base metal

3.3 Impact Test

The positions of the impact specimens were in the center of the weld, 1mm away from the fusion zone and 5mm from the fusion zone. The surface of the specimens was perpendicular to the weld surface. The impact test temperature was -40°C [5]. The test results were showed in Table 6 and table 7.

Tab. 6: Impact test results of plate

Specimen source	Specimen size (mm)	Test temperature	Test area	Specimen No	Impact absorbing energy (J)	Average (J)
plate	10×10×55	-40oC	weld	1	33.72	78.54
				2	81.01	
				3	120.89	
			fusion-line+1mm	1	67.04	48.68
				2	33.12	
				3	45.89	
			fusion-line+5mm	1	45.04	73.17
				2	89.29	
				3	85.20	

Tab. 7: Impact test results of pipe

Specimen source	Specimen size (mm)	Test temperature	Test area	Specimen No	Impact absorbing energy (J)	Average (J)
Pipe	10×10×55	-40oC	weld	1	68.56	69.41
				2	41.02	
				3	98.65	
			fusion-line+1mm	1	143.18	119.11
				2	170.32	
				3	43.85	
			fusion-line+5mm	1	71.20	91.03
				2	45.90	
				3	156.00	

The test results met the requirements of the AWS D1.1 standard.

3.4 Hardness Test

Two plates and two pipes were taken to test for HV10 values. Test point position of butt joint hardness was showed in Figure 2, test results of the board sample Vickers hardness were showed in Table 8.

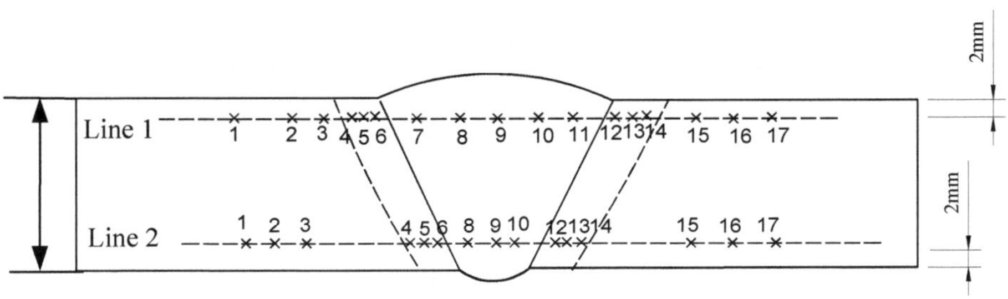

Fig. 2: Hardness test point of butt joint

Tab. 8: Butt joint Vickers hardness test results (HV10)

Specimen No	Test point	area	Line1	Line2
Plate 2	1	Base metal	168	202
	2		171	188
	3		172	172
	4	HAZ	239	199
	5		205	211
	6		198	221
	7	weld	188	
	8		186	211
	9		185	219
	10		187	218
	11		187	
	12	HAZ	214	210
	13		219	208
	14		232	211
	15	Base matel	171	179
	16		169	177
	17		173	175

Test results showed that HV10 values were less than 280, which met the requirements of AWS D1.1 standard.

3.5 Macro Corrosion test

Macro corrosion photograph of welded joint specimen was shown in Figure 3. The results of the macro corrosion test were shown in table 9.

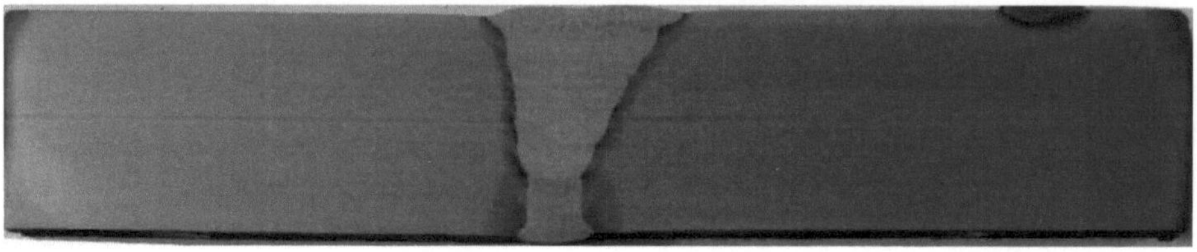

Fig. 3: Macro corrosion photograph of the specimen No. 3

Tab. 9: Macro Corrosion test results

Joint type	Specimen No	Test result
Butt joint	1	No defect
Butt joint	2	No defect
Butt joint	3	No defect
Butt joint	4	No defect

4 Residual Stress Test

After the welding of the test specimen, the residual stress of the specimen was tested by X - ray diffraction. The equipment was a Japanese Science X-ray stress meter (MSF-3M). As the specimen is symmetrical about the centreline and the weld centre, only a quarter of the position was measured. The test results were shown in Fig. 4 and Fig.5.

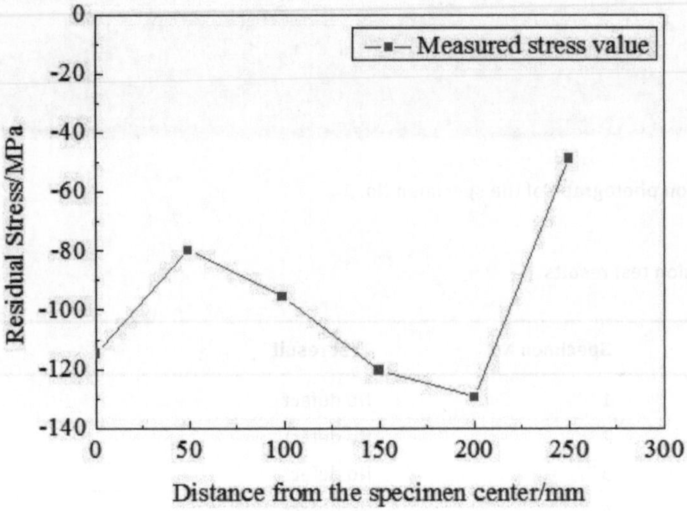

Fig. 4: residual stress in the direction of the weld

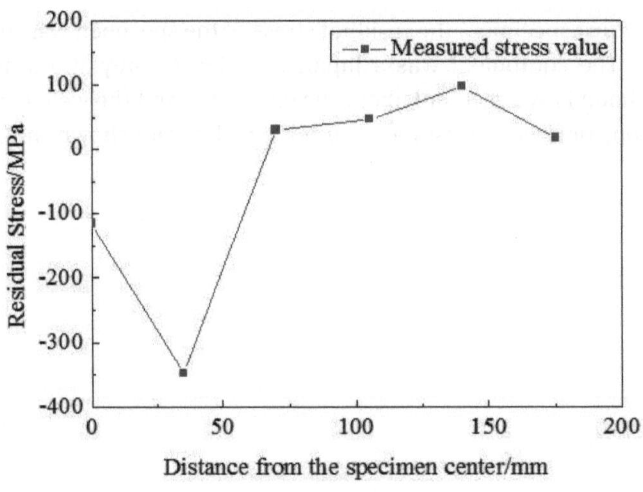

Fig. 5: residual stress perpendicular to the weld direction

It can be seen from the figure, whether in the direction of the weld, or perpendicular to the weld direction, the longitudinal residual stress in the vicinity of the weld was compressive stress. In the position away from the weld, it maintained compressive stress along the weld direction of the stress, and it became tensile stress perpendicular to the weld direction. It is known that the compressive stress is beneficial to improve the fatigue performance and corrosion resistance of the weld. Therefore, the compressive stress distribution in the weld and the surrounding area was beneficial to improve the fatigue performance and stress corrosion resistance of the structure.

5 Conclusion

(1) The welding process was successfully used for the Jinzhou 9-3 offshore platform construction, and the welding quality was reliable.
(2) The test results of welded joints' Mechanical properties showed that the performance of the joints met the relevant requirements of the AWS D1.1 standard.
(3) It is shown that a compressive stress distribution was in the weld and the vicinal area by the welding process, which was conducive to improving the fatigue performance and stress corrosion resistance of Jinzhou 9-3 offshore platform.

References

[1] Liu Nianfu, He Kuangnian, etc., Trial-Production of Offshore Platform Plate in Grade E36, Steel, Vol 46, No. 5, May 2001, 97–100.
[2] Liu Guangyun, Ren Aizhen et al., Welding technology research on EH36 steel, Welding Technology, Vol. 36, No. 2, April 2007, 35–36
[3] Wang Jianli, Welding process assessment of high strength steel, Yunnan Water Power, Vol 23, No. 2, 2007, 111–114
[4] Structural Welding Code AWS D1.1, 21st Edition, American Welding society, 2008
[5] Qu Zhanyuan, Influence of Heat Input on Low Temperature Toughness of Submerged Arc Welding Seam on Marine E36 Steel, Development and Application of Materials, 2010 04, 7–9

Wei-Guo Gao, Tian-Xiang Wang, Shi-Guo Xu*, Guang-Yu Su and Xu Wang

Structural Property and Stratified Sampling Technique of Reservoir Sediment

Abstract: Sediment pollution becomes an important factor affecting the ecological environment of reservoir after many years of reservoir operation. Taking representative sediment samples is the basis of related research. The investigation shows that there are four main types of sediment collection: digging, grasping, gravity sampling and vibration sampling. Through the investigation and analysis of the sediments of the typical reservoirs, it is found that the sediments are imported by exogenous pollution and gradually deposited in the still water environment, which have obvious stratification and timing property after years of accumulation. The existing sampling equipment's have their own application conditions, but they cannot fully apply to the soft and stratified sediment sampling of reservoirs. By investigating the structural property of reservoir sediments and analysing the key techniques of sampling, this paper developed a sampling device suitable for the sediment collection, which has great application value.

Keywords: reservoirs, sediments, property, fidelity, sampling

1 Introduction

As an important water conservancy project, the reservoir has many functions such as flood protection, water supply and so on [1]. With the further development of urbanization in China, the problem of water supply is becoming increasingly prominent. However, the operation of the reservoir is affected by the double effect of "artificial-natural" regulation; the sediment pollution problem gradually emerged after several decades of operation [2]. For example, Tang he Reservoir is affected by

Wei-Guo Gao , School of Hydraulic Engineering. Dalian University of Technology, Ganjingzi District, Dalian City, Liaoning Province, P.R.C. Email: gaoweiguo@mail.dlut.edu.cn
Tian-Xiang Wang, No.2 Linggong Road, Ganjingzi District, Dalian City, Liaoning Province, P.R.C. Email: tianxiang@mail. dlut.edu.cn
Shi-Guo Xu, No.2 Linggong Road, Ganjingzi District, Dalian City, Liaoning Province, P.R.C. Email: sgxu@dlut.edu.cn
Guang-Yu Su, No.2 Linggong Road, Ganjingzi District, Dalian City, Liaoning Province, P.R.C. Email: sugy@mail.dlut.edu.cn
Xu Wang, No.2 Linggong Road, Ganjingzi District, Dalian City, Liaoning Province, P.R.C. Email: xuwang@mail.dlut.edu.cn

DOI 10.1515/9783110540048-077,

the release of sediment pollution and as a result, the hydrogen sulfide pollution incident broke out [3]. Sediment pollution becomes a hot topic in current research, and the collection of representative samples is the basis of related research. So the development of accurate and efficient sediment sampling equipment is of great significance.

There are four main types of sediment collection equipment now: digging, grasping, gravity sampling and vibration sampling. It's very easy to use a shovel or other digging tools in shallow water. Three other types can be used in deep water. The grasping sampler collects the surface sediments by closing the grab at the moment of touching the sediment. This method can only collect sediments from the surface layer [4].The gravity sampler which relies on its own gravity to inset the sampling tubes into the sediments, which destroys the original physical structure of the surface sediments during the equipment's falling [5]. The Danish KC-Denmark company HAPS Sediment Columnar Sampler is a sampler with a vibrating motor and the total weight of the sampler reaches 168Kg. So in the actual sampling, there must have large ships and winch. And the use of this equipment will damage the stability of the sediment and the sampling depth is limited. So that it is not suitable for the investigation and sampling of reservoir sediments in inland areas.

Based on the investigation of the structural property of reservoir sediments, this paper analyses the key techniques of reservoir sediment collection, and develops a kind of reservoir sediments sampler, which provides an effective sampling tool for the collection of reservoir sediments.

2 Structure Property of Reservoir Sediments

After the dam was built, the natural runoff characteristics of the river were destroyed, the flow velocity in the reservoir area was reduced, and the sediment capacity was weakened. A certain degree of siltation always occurs in the reservoir after years of operation. There are many factors affecting the distribution and form of siltation likestorage volume, sediment concentration, sediment composition, reservoir dispatching. In the change backwater area, flush and silting staggered and the sediment is stratified.And the stratification phenomenon of sediment is becoming more obvious after many years because of the different quantity and species of flood-carrying substances by different floods. The concept map of reservoir sediment distribution is shown in Figure1.

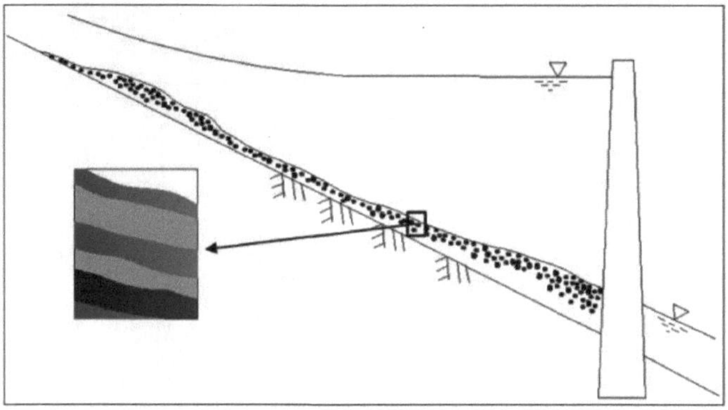

Fig. 1: Reservoir sediments distribution concept map

The comprehensive master of the structure state of reservoir sediments is the basis of collecting sediment samples, analyzing sediment properties, and then studying the countermeasures of sediment management. In order to make the research in detail, this study chooses the Biliuhe Reservoir in Dalian as an object to study and analyse its actual state of sediments. Since its completion in 1984, Biliuhe Reservoir has been running for 32 years. It is a large-scale reservoir with comprehensive utilization of flood protection, power generation, fish farming, irrigation and tourism [6]. After several decades of operation, there has been a certain degree of deposition. This paper focuses on the sediment status in the entrance section and the dam front area of Biliuhe Reservoir.

2.1 Sediment Property in the Entrance Section

This paper investigated the sediment property of three entrance sections of Guiyunhua, Dabao and Zhongling in the period of the dry season. The actual sedimentation status and morphology of the reservoirs were analyzed comprehensively, and samples of sediments at each point were collected. The field investigation (shown in Figure 2) shows that the sediments show a distinct stratification in the vertical direction.

a .The sediment in Guiyunhua section b .The sediment in Dabao section c .The sediment in Zhongling section

Fig. 2: Sediment stratification in the survey area

As shown in Figure 2, the thickness, particle size, color of the soils of different depths are definitely not same and the sediment stratification characteristic is obvious. Surface sediments in some areas contain a lot of straw, hay and other substances because they don't decay into muddy sediments completely due to the short deposition time. In addition, the sediments are often entrapped with refractory substances such as plastic bags, and we found that the date of production attached to the food packaging bag corresponds to the flood event for a particular year. It is concluded that the formation of sediments is related to the type of material entering the reservoir, hydraulic handling and other factors, and a flood often carries large amounts of straw and garbage into the reservoir [7]. The different kinds of substances carried by the floods make the sediment form a distinct stratification phenomenon, and the amount of silt, the amount of organic matter and the like carried by different levels of floods are obviously different, and sediments at different depths reflect the hydrological events, pollution characteristics and watershed characteristics of reservoirs in different years.

2.2 Sediments Structural Property in front of the Dam

In this paper, we use the grasping and gravity type sampler to investigate the reservoir sediment based on the "water environment monitoring specification"(SL219-2013) and portable GPS. The sediment status is shown in Figure3.

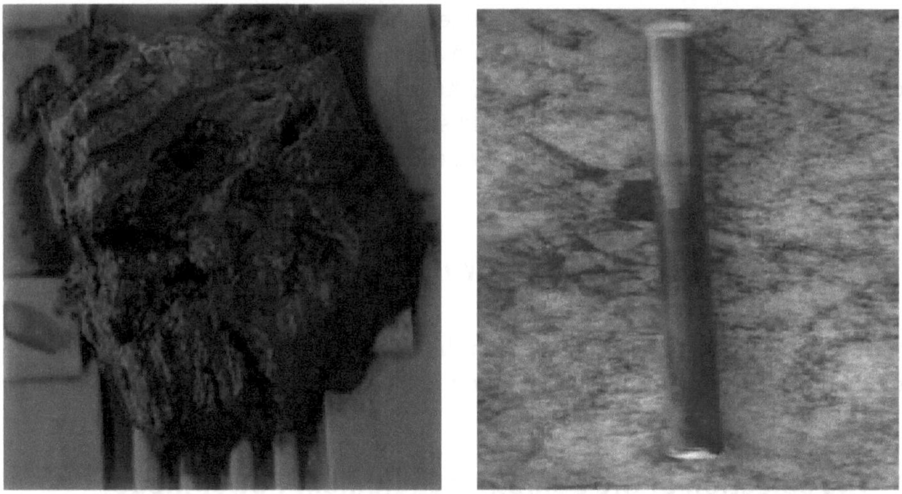

a .The surface sediment b. The columnar sediment

Fig. 3: Sediment samples from Biliuhe reservoir

After more than 30 years of reservoir operation, the reservoir sediments have a deep black color that indicated that the sediment pollution is serious [8]. At the same time, it was observed that the surface sediments of the reservoir were very sensitive to disturb by means of underwater camera equipment. Indoor monitoring also indicated that the sediment water content reached 180% and the carrying capacity was low. Reservoir sediments are perennial accumulated under the water, so it is different from the runoff erosion and ocean current disturbance of ocean. In addition to the flood season, reservoir in most of the time is in a quiescent state, which causes the sediment to be softer. The survey found that the sediments grain size in the vertical direction showed a fine-coarse-fine distribution characteristic, and similar phenomena were found in the study of Lianhuan Lake [9]. The columnar sediments show different ribbon distribution related to the physicochemical properties of the sediments. The distribution of TN, TP and organic mattern the sediments of Tai Lake have an obvious difference in the vertical distribution: the surface sediment content is high and the content of the bottom layer is low [10]. Besides, the particle size, nutrition, heavy metals and other physical and chemical indicators, even microbial all have an obvious stratification phenomenon in the vertical direction.The study of Tai Lake shows that the microbial community hannon-Weiner index in the upper sediments is higher than that the lower layer[11]. These existing studies have validated the stratification characteristic of sediments by detecting physical and chemical indicators, so it is necessary to ensure the sediments stratification characteristic are not destroyed when collecting the sediment.

Above all, it is found that the surface sediments are relatively soft and vulnerable to disturbed, and the sediments stratification features are obvious. In order to accurately grasp the state of reservoir sediments and analyze the sediments physicochemical properties, the susceptibility and stratification characteristics of sediment must be taken into consideration when we design the collecting equipment.

3 Key Techniques of Reservoir Sediment Sampling

According to the stratification and susceptibility characteristics of the sediment, we think that the following techniques should be solved.

3.1 The selection of the Sampling Equipment Power Mode

Sampling device should be able to collect sufficient depth of sediment sample in powerful mode. It can not only provide the power forcollecting deep sediment, but also not destroy the original layer of the sediment. It is very important to obtain the sedimentary characteristics of reservoirs through the deep sediment samples. The power of traditional sample collectionis mainly by the weight of the sampler and the momentum of the descent, and the length of the collected sample depends on the mass and weight of the counterweight. Not only we cannot get sediments samples with enough depth, but also there are many problems: the inclination of the sampling tube and lack of space utilization of the sampling tube. How to ensure the collection of sufficient depth of sediment and reduce the disturbance of sediment samples, and select a reasonable sampling power mode is the key points to deep sediment collection.

3.2 The Design of the Sampling Tube Structure

When collecting sediments, it is necessary to minimize the disturbance to the sediments. Material, shape and size of the sampling tubes are the key factors that influence the sample disturbance in the sampling process. The surface sediments of reservoirs are soft, which causes a problem that all existing gravity dependent methods have to face with that because of the existence of a great friction between the inner wall and the outer wall of the sampling tube, it would be easy to produce "pile effect" phenomenon [12]. Due to the effect of the "pile effect", the length of the collected sample is shorter than the sampling depth, also the sediment is compressed. Deeper deposits cannot enter the sampling tube, so the original state of the sediments is destroyed. Failure to collect microscopic sediments directly causes the error of the study of sediment pollution. Therefore, it is necessary to reasonably design the form and parameters of the sampling tube, so that the collected samples can have better representation.

3.3 The Design of the Sealing Structure

After collecting the sediment by sampler, it is necessary to seal the upper and lower part of the sampling tube to avoid the deposition of the sediment from the tube and to reduce the disturbance to the sediment during the lifting of the sampler. The bottom of existing gravity-type sediment samplers adopts "petal-type" closure device to block the sampling tube. This device increases the disturbance to the sediment during the sampling process. Moreover, if the collected sediment samples are too soft, they would be easily to leak and finally result in sample loss. The upper drainage device adopts the rubber buoyancy valve to achieve the drainage and sealing on the top of the sampling tube, but the sealing effect is limited. Because the vacuum on the upper part of the sampling tube cannot be maintained, the loss of the bottom sample would be aggravated. Therefore, the reasonable plugging and the design of the upper sealing cover is another key point of the sampler design.

Based on the above analysis, the selection of the sampling equipment power mode, the design of the sampling tube structure and the sealing structure are related to the fidelity and representative of the sediment samples. The above problems should be taken into account to make sure the sampler has a small disturbance to the sediment during the collection process and has ability to collect sufficient depth and representative stratified sediment samples.

4 Development of Stratification Sampling Equipment

Aiming at the soft texture, easily disturbed and obvious stratification characteristics of reservoir sediment, and combined with analysis of the key issues of sampling, this paper designs a kind of stratification sediment sample collector.

4.1 Sampler Structure Design

The stratification sediment sampler mainly consists of vibration mechanism, drainage sealer, sampling tube, pipe clamp, plug mechanism, cable, plug cable and vibrator cable. The overall structure is shown in Figure 4. Under the influence of the vibration of the sampling tube driven by the vibration mechanism, sediments around the sampling tube wall will be "liquefied", and then the sampling tube sinks to actualize the collection of stratification sediment samples. The drainage sealer mechanism achieves the drainage and sealing of the upper part of the sampling tube, while the plug mechanism realizes the block of the lower part of the sampling tube. Under the cooperation of these two mechanisms, the sampler successfully avoids the loss of sample and reduces the disturbance to the collected sediments.

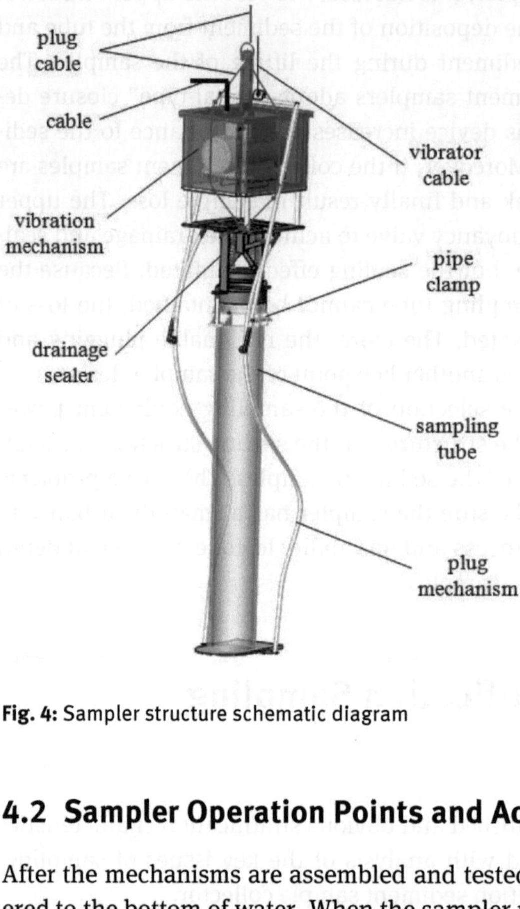

Fig. 4: Sampler structure schematic diagram

4.2 Sampler Operation Points and Action Process

After the mechanisms are assembled and tested, the sampler should be slowly lowered to the bottom of water. When the sampler reaches the bottom of the water, turn on the power to make the vibration mechanism work. Under the action of the vibration, sediments around the sampling tube would be “liquefied”, so the sampling tube can enter the sediment to carry out a sample collection at a uniform rate.

When sampling is complete, turn off the DC vibration motor and continue to release the vibrator cable to disengage the cable from the hook. Then pull the vibrator cable, the pulling force of the cable will act on the top rod of the drainage sealing device, and the top rod pushes the seal cap move downward to seal the top of the sampling tube. Release the plug mechanism cable when the sampler is completely removed from the sediment, at this point the plug mechanism flips downward under its own gravity to block the sampling tube. Pulling the vibrator cable to put the device out of the water, thus a sample is collected.

The sampler is easy to operate and can be easily and quickly sampled at a specific sampling point.

4.3 Sampler Application Practice

To examine the sampling effect of this sampler, this study applies this sampler to the actual sediment sampling. The results of the sampling test in Biliuhe reservoir are shown in Figure 5.

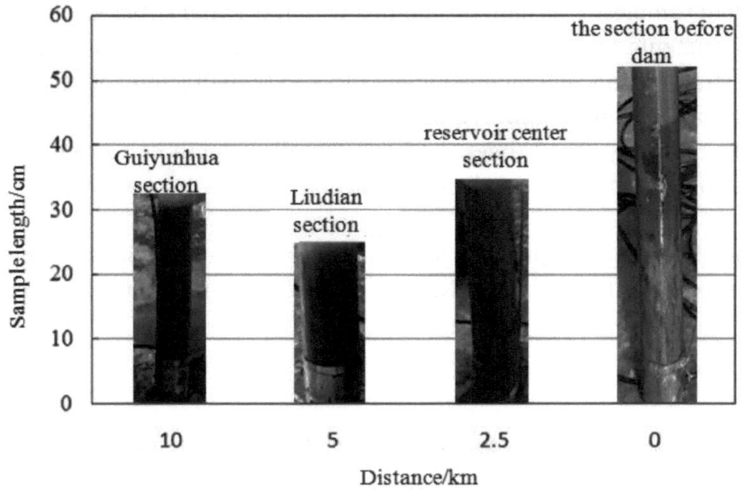

Fig. 5: Sediment samples of Biliuhe reservoir

The length of sediment samples of Guyunhua section, Liudian section, the reservoir center section and the section before dam is 32cm, 25cm, 34cm and 52cm. Due to the reduction of flow rate in the Guiyunhua section, which provides favorable conditions for the deposition of materials through input runoff, the mount of sediment deposition is huge. And then from Liudian section, as the decreases of distance from sampling section to the dam, the length of sediment obtained gradually increased.

Compared with the bulky gravity sampler, the total weight of the sampler is only ten kilograms, which makes it easy to carry and operate the success rate of sampling and sampling efficiency is greatly improved. The sampler can adapt to different water depth of the field sampling requirements, reduce the labor intensity and improve possibility of sampling success. The vibrator mechanism of the sampler is powered by DC 24V lithium battery; it is more suitable for field sampling than AC vibratory samplers. The application in practice shows that samples collected by the sampler are obviously stratified and the degree of fidelity is high. The problems of low success rate, complex operation and large disturbance of sediment are effectively solved.

5 Conclusion

Field investigation and literature analysis show that reservoir sediments have distinct temporal and stratification characteristics, and representative sediment stratification sample collection is the basis of subsequent studies. The existing sampling equipment has its own characteristics, but cannot be fully applied to the collection of reservoir sediment samples. This paper combined with the actual structural property of reservoir sediments and comprehensive analysis of the key issues of sampling, a vibration stratification sediment sampler with good fidelity and convenient operation was developed. Compared with other types of equipment, the sampler can raise the sampling efficiency by 10 times, and the sampling depth can reach 100 meters, which has good scientific research and application value.

Acknowledgement: The authors would like to thank the anonymous reviewers. This work was supported by the National Natural Science Foundation of China (51327004).

References

[1] Z.H. Sun, L. Zhang, Z. D. Duan. Quantity and distribution of reservoir engineering in China[J]. China Water Resources, 07: 10–11, 2013.
[2] F. Zhang, D. Wu. Genetic analysis and Countermeasures of abnormal water quality in partial water of Tanghe Reservoir [J]. Agriculture and Technology, 02: 67–69, 2010.
[3] Z.Z. Zhou, T.L Huang, Y. Li, et al. Sediment pollution characteristics and in situ control in a deep drinking water reservoir [J]. Journal of Environmental Sciences, 2016.
[4] G. Y. Li, L. P. Li. Process design and process guarantee in the production process of the surface series mud sampler [J]. Marine Technology, 04: 58–63, 2000.
[5] R. Ruan. Gravity sampling technology and its development [A]. Surveying and Mapping Press. "Survey Bulletin" Surveying and Mapping Science Frontier Technical Forum Summary [C]. Surveying and Mapping Press: 5, 2008.
[6] S.Y. Zhang, J. Sun. Investigation and analysis of water quality in the Biliuhe Reservoir [J]. Modern Agricultural Science and Technology, 07: 278–285, 2012.
[7] Y. Zhao, C.H. Hu, Strategies for managing reservoir sedimentation [J]. International Journal of Sediment Research, 04: 369–384, 2009.
[8] L.J. Zhang, G. Wang. Study on heavy metal pollution characteristics of sediments in Li village, Jiaozhou bay [J]. Journal of Shandong University of Technology (NATURAL SCIENCE EDITION), 2003, 01:8–14.

[9] L. Sun. Study on the characteristic of sediment size and change of magnetic susceptibility and significance of sediments I in Lianhuan Lake[J]; Journal of Harbin Normal University (Natural Science Edition)
[10] J. Li, X.C. Jin. Physicochemical properties of sediments and vertical changes of phosphorus in lakes of different tropic levels in Tai Lake [J]. Acta Scientiae Scientific Research, 04: 64–69, 2007.
[11] X.Q. Zhao. Vertical distribution characteristics of microbial diversity in sediments of Tai Lake [J].Journal of Geography, 06: 177–184, 2008.
[12] L. Zhu. Study on fidelity technique of deep-sea sediment fidelity sampler [D]. Zhejiang University, 2005.

Han Weng, Guo-Jin Chen and Hui-Ying Li*

Finite Element Analysis and Optimum Design of a Video Llaryngoscope

Abstract: With systemic analyseis and comparison on the current products and clinical feedback, this study aims to develop a new type video laryngoscope which possessing new structure, material and convenient for clinical application. The three-dimensional model was imported into ANSYS software for FEM analysis after design. Results suggest that the life of the optimized handle increase about 5 times compared with the traditional laryngoscope. Besides, the stress and strain at the edge of the laryngoscope lens are larger than those of the surroundings, and are vulnerable to fatigue fracture. Therefore, we recommend that doctors are better to use it as a disposable one.

Keywords: video laryngoscope, FEM analysis, fracture

1 Introduction

Early laryngoscope was invented for larynx examination and laryngeal operation. With the development of medical, laryngoscopy is not only used in laryngeal science, but also widely used in other clinical fields. From the initial indirect laryngoscope, until now the video laryngoscope, it probably experienced a hundred and sixty years of development. Although it has a high success rate on modern clinical, there are many defects that cannot be ignored. For example, the Glide Scope video laryngoscope produced by Verathon, which is hard to hold as the structure, does not match the structure of hand [1]. Mc Grath video laryngoscope manufactured by Medtronic is too expensive to use [2]. The Yian video laryngoscope produced by Tianjin Mai Dian Medical Electronic Technology company is relatively not such expensive but easy to fracture with the poor material [3]. In view of the above defects, this study improved a new type of video laryngoscope, which provides a more practical and extensible tracheal intubation assistant for clinical.

Han Weng, Hangzhou Dianzi University, Hangzhou, China. Email: 649648181@ qq.com
Guo-Jin Chen, Hangzhou Dianzi University, Hangzhou, China. Email: chenguojin@163.com
***Corresponding author: Hui-Ying Li,** Hangzhou Dianzi University, Hangzhou, China. Email: 775210780@ qq.com. Corresponding author.

DOI 10.1515/9783110540048-078,

2 Mechanical Structure Design of the Video Laryngoscope

The handle of tradition laryngoscope is straight, as shown in Figure 1. It is easy to slide and need a larger upward force when operation. In addition, the radian of laryngoscope lens is not well in line with the ergonomics, so this study mainly analyzes the handle and laryngoscope lens.

The new structure is shown in Figure 2, for convenience we set the handle and elbow as part 1 and 2 respectively, and the laryngoscope lens is set as part 3.

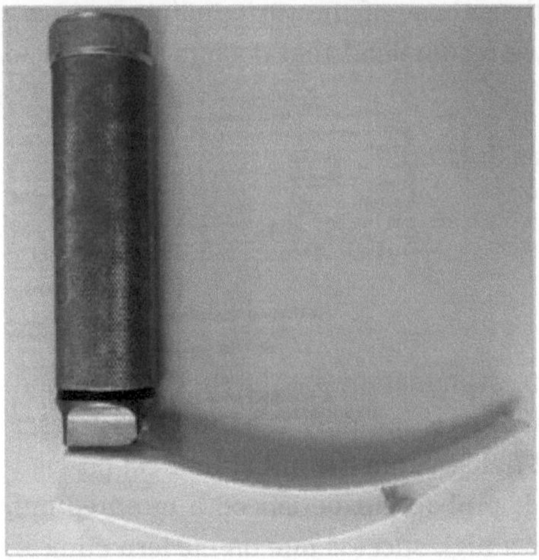

Fig. 1: Handle of tradition laryngoscope

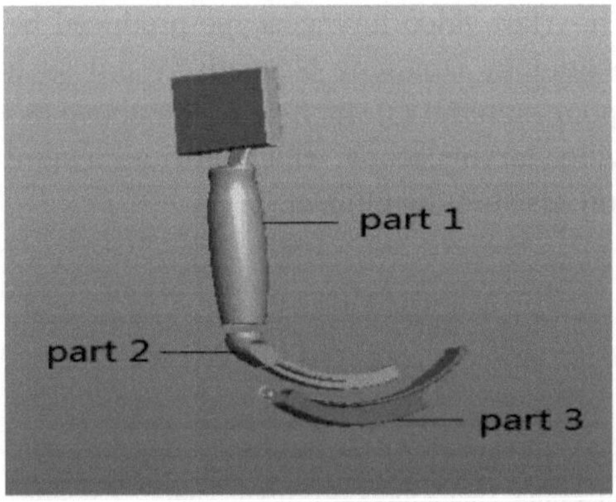

Fig. 2: Mechanical structure of the video laryngoscope

2.1 Improvement of Handle

2.1.1 Bending Curvature

As the most indispensable component of the video laryngoscope, the radian of part 1 must be reasonable. For patients with limited neck activity or thoracic obese, when intubated it needs a larger force to explore the glimpse, so its structure is better to design with a certain curvature to reduce the force [4]. We considered the physical structure of hand combined with aesthetics, finally set the handle curvature is 5 degree.

2.1.2 Material

In order to ensure the strength, the general material used in clinical is stainless steel, but there is no doubt that it increased weight. Considering the upward force (0.4-1.5KG) when using, we choose the aluminum alloy instead, which has the same mechanical strength. And part from this, its light weight is convenient for doctors to operation. The material properties of Aluminum alloy are: 2770kg/m^3 of density, 0.33 of Poisson Ratio, 280MPa of yield stress.

2.2 The Design of Laryngoscope Lens

Material is the key to design a laryngoscope lens, it not only need high-definition but also need a better strength. At present ABS is widely used to make a laryngoscope lens, in order to protect the patient's teeth and improve the imaging results, we choose PC instead. It has a higher impact resistance, toughness and transparency. The material properties of PC are: 1220kg/m^3 of density, 0.39 of Poisson Ratio, 66MPa of yield stress, and 106MPa of bending strength.

3 Mechanics Modeling and Analysis

For the purpose of proving the rationality of the structure and analyze the force relations between the joint points of each component, the three-dimensional mechanical model was imported into ANSYS software for Finite Element Analysis (FEM). According to the anti-fatigue test requirements, the model was simplified, the loading of external forces and boundary conditions are equivalent to loading a moment on the laryngoscope, the simplified model was then for stress cloud calculation. The maximum stress and dangerous nodes of the model were obtained after analysis. Based on the mathematical formula of the S-N curve and Goodman curve, finally we calculated the fatigue life (N).

3.1 Statics Model

Part 1 can be simplified without considering the bending deformation of part 2 and part 3 during stress. The length of part 2 is 13cm with the force surface of 2.25cm^2, the length of part 3 is 15cm, and the average thickness of head is 3mm. When operation, part 3 is fitted to part 2, then both of them are put into the throat of the patient, the laryngoscope lens was caught in the throat of the person, and the handle was lifted up to observe the vocal and glottis through the camera.

The laryngoscope in human body mainly undergoes two types of deformation: stretch and bend. Biomechanical research has reported that bend plays a dominant role during operation. The external force and the boundary condition can be considered completely loaded on part 3. According to the force of the video laryngoscope in human throat, it is equivalent to apply a vertical upward force to the top surface (A) of part 2 for finite element analysis, and then calculate the structure strength, the equivalent loading method are shown in Figure 3. It is reported that the upward force is between 0.5-1.4KG, so force 5N, 10N and 14N were used for calculation, and the head of part 3 (B) was selected as the fixed fulcrum.

3.2 Finite Element Fatigue Analysis and Results

3.2.1 The Finite Element Mesh Model

The three-dimensional model was imported into Workbench statics analysis module of ANSYS. Regardless of the relative motion between part 2 and part 3, the contact constraint of the mating position was simplified. The form-a-new-part method was used to make the grid have a common node on the interface. When the grid is divided, the global grid size is automatically set to 2.5mm. From the theoretical point of view, the position where the stress and strain are relatively high appears on the part 3, so the subdivision size of grid is set to 1.2mm, and the whole model was divided into 84292 grid cells, as shown in figure 4.

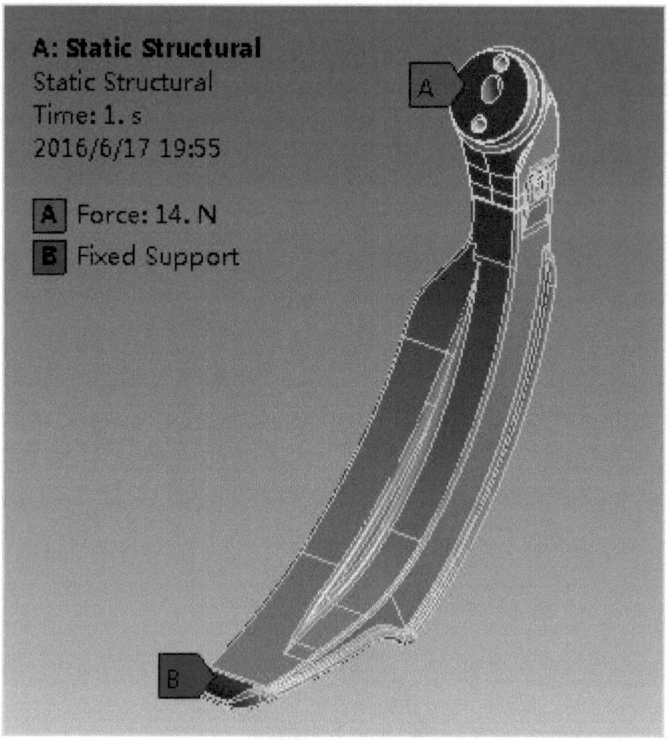

Fig. 3: Equivalent loading method

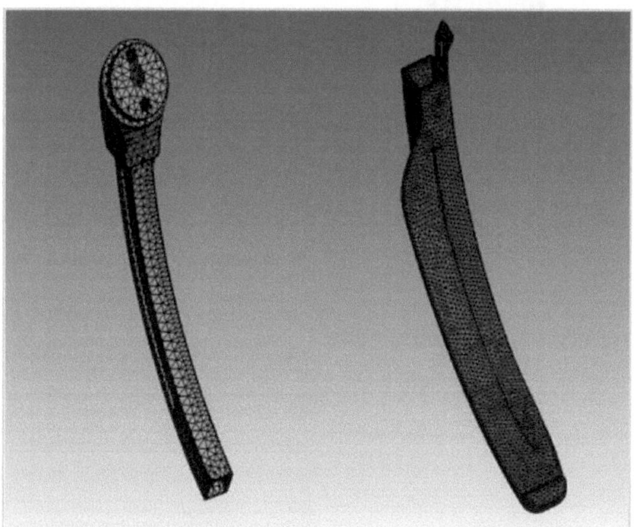

Fig. 4: Mesh model

3.2.2 FEM Results

After the grid is generated, the load of 5N, 10N and 14N is applied to the model stress surface in the current analysis, the calculated stress clouds for the 3 kinds of force were shown in Figure 5 ~ 7 below.

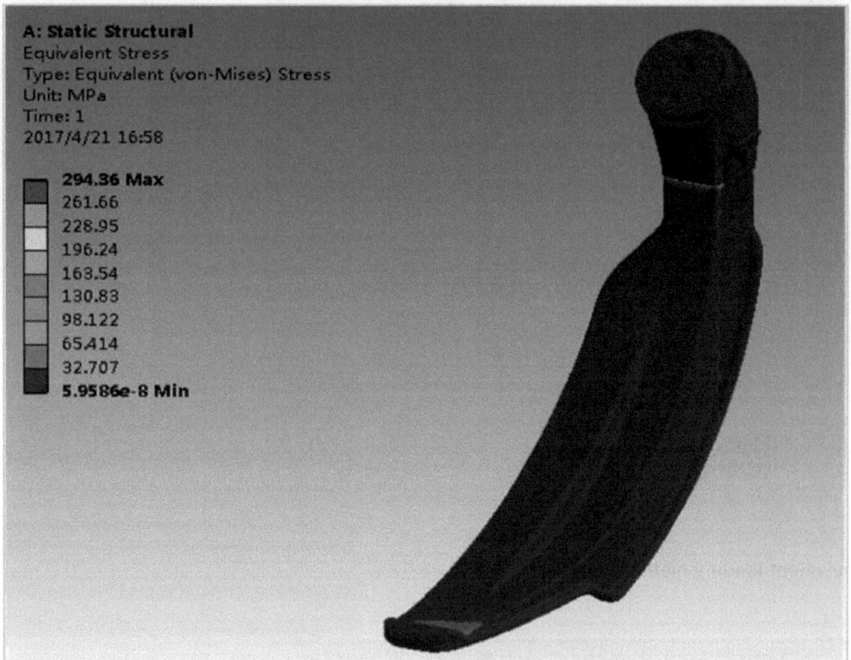

Fig. 5: Stress cloud for load of 5N

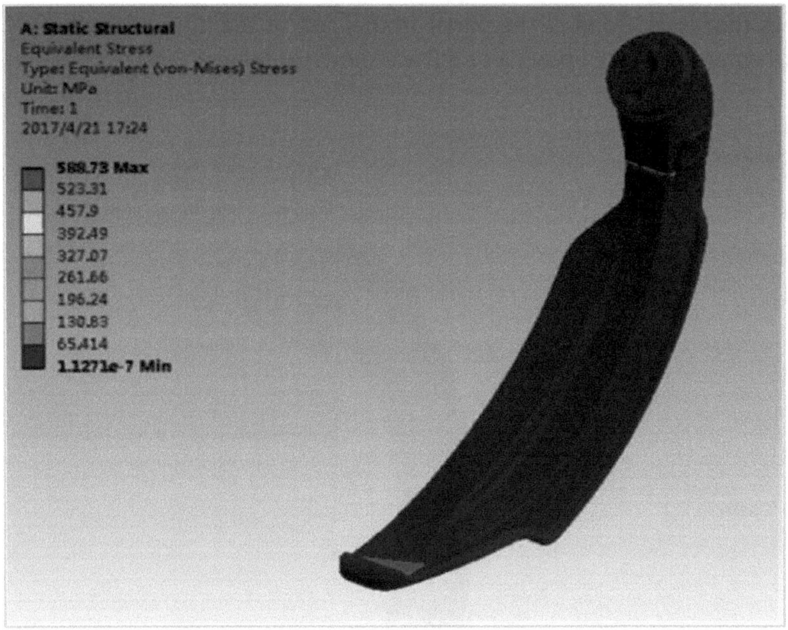

Fig. 6: Stress cloud for load of 10N

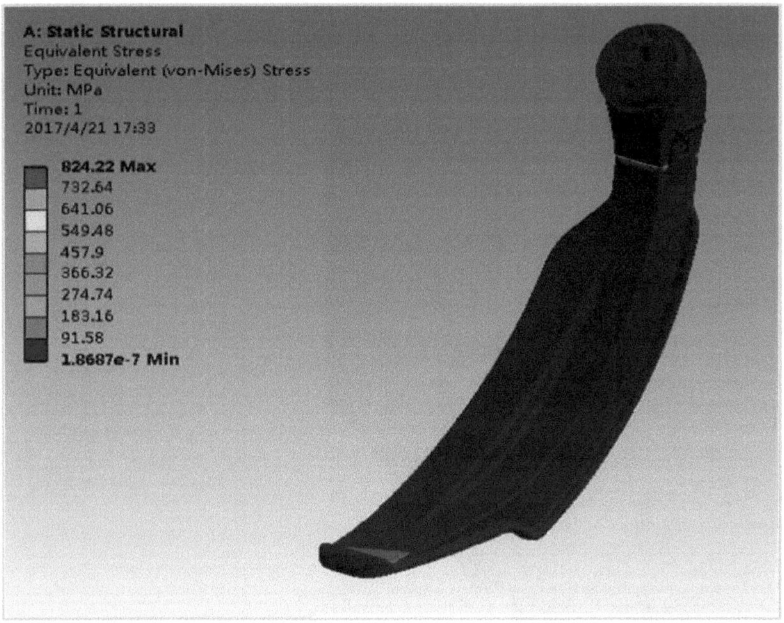

Fig. 7: Stress cloud for load of 14N

Figure 8 shows the stress cloud of the part 2 in the case of the ultimate force 14N, and the partial enlarged view of parts A and B are shown in Figures 9-10.

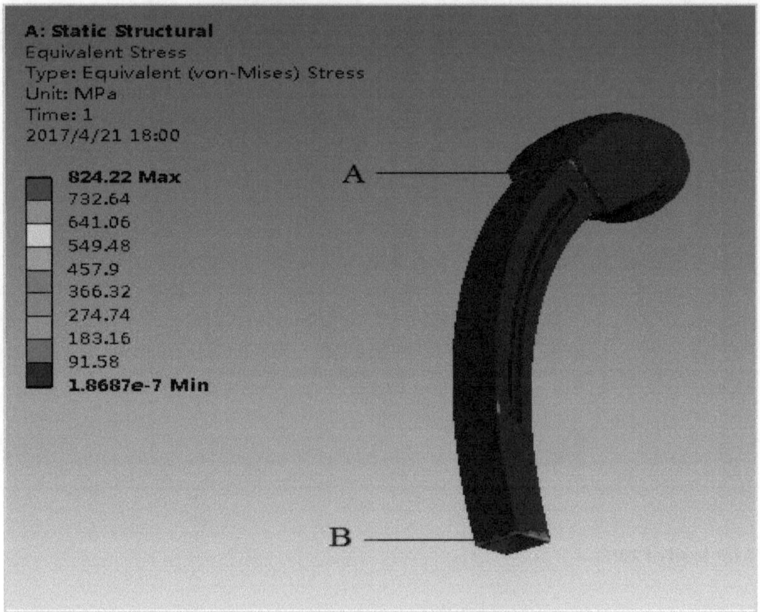

Fig. 8: Stress cloud of part 2 on 14N

Fig. 9: Enlarged view of A

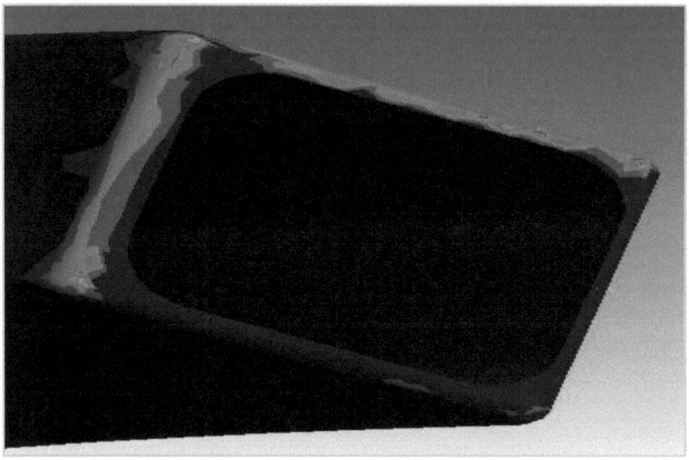

Fig. 10: Enlarged view of B

From figure 5 to figure 7, we can see that the stress under the three loads is mainly distributed at the end of the handle and the junction of the fixed boundary and the deformable grid of part 3. The maximum stress on part 3 is 91.58MPa, which is less than but close to its bending strength of 106MPa. Most of the parts have a small stress, less than about 2% elastic deformation. Figure 9-10 indicate the maximum stress value is 824.22MPa, which appears in the B of the part 2, but only a very small number of nodes' stress beyond the limitation, stress in most part is about 100MPa, no more than the yield strength of Aluminum alloy, so in the case of limit force both of the handle and laryngoscope lens meet the material permit situation [5].

3.3 Solution Fatigue Life

The classic stress fatigue theory relates the stress (S) with the fatigue life (N) by the S-N curve formula:

$$S^m N = C \tag{1}$$

Where m and C are the parameters associated with material properties, stress ratio, and the corresponding loading method.

m can be calculated by taking two points of the high cycle fatigue S-N curve for Aluminum alloy [6], as shown in figure 11, using the following formula:

$$m = \frac{\log_{10}\left(N1/N2\right)}{\log_{10}\left(S2/S1\right)} \tag{2}$$

And C is calculated with:

$$C = N1 * S1^{m} \tag{3}$$

Where (N1, S1) and (N2, S2) are any of two points on the S-N curve.

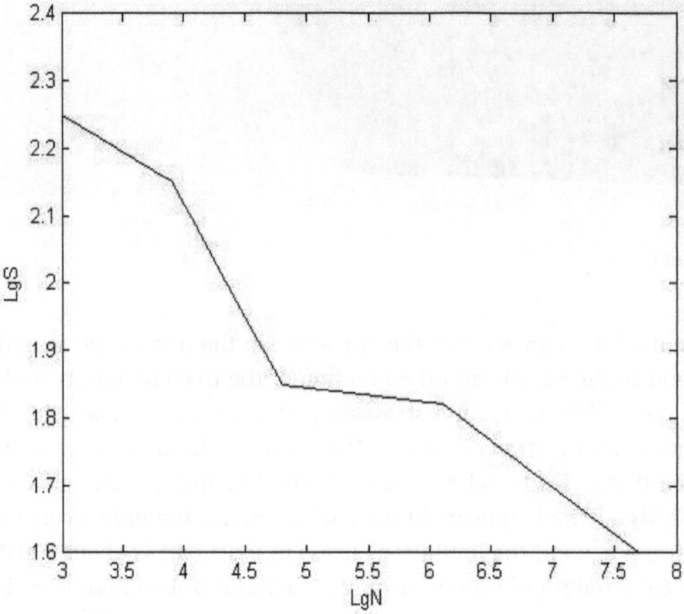

Fig. 11: High cycle fatigue S-N curve for Aluminum alloy

Goodman Curve follows the following relationship:

$$\frac{S_a}{S_{-1}} + \frac{S_m}{S_u} = 1 \tag{4}$$

Where S_a is the average stress, S_m the stress amplitude, S_{-1} the stress cycling characteristics of the stress at cycle for symmetry, S_u the material fatigue limit.

Combining the S-N curve and Goodman curve, the circle life N of part 2 can be calculated by the following steps:

$$S_u = 0.6 \times S_s \tag{5}$$

$$S_a = \frac{(S_{max} - S_{min})}{2} \tag{6}$$

$$S_m = \frac{(S_{max} + S_{min})}{2} \tag{7}$$

$$S\_a = \frac{S_a}{(1 - S_m / S_u)} \tag{8}$$

$$N = \frac{C}{(S\_a)^m} \tag{9}$$

where S_{max} is the maximum work stress of the plate under loads, S_{min} the minimum work stress under load, Ss the tensile strength for Aluminium alloy, S_a the cyclic stress corresponding to the stress ratio R =-1.

The fatigue circle life N was calculated for the part 2 under three loads. The results were listed in Table 1.

Tab. 1: The fatigue life of part 2 under three kinds of load (thousand times)

Load(N)	times
5	12590
10	800
14	180

4 Conclusion

On the basis of using environment and habit, this study optimized the main components of the video laryngoscope, and then the FEM analysis of the simplified model was carried out in the current study. The FEM results shows that the most fragile part under three kinds of load is at the end of the handle and the edge of the laryngoscope, for the handle part it turned out that the life increased 5 times compared with before. As for the laryngoscope lens, the maximum stress is 91.58MPa, which is close to its bending strength of 106MPa. So we recommend that doctors are better to use it as a disposable one.

Acknowledgement: The work was supported by Key Scientific Research Project of Zhejiang Province; Zhejiang Hisern Medical Devices Inc; National Natural Science Foundation of China (Grant No.51675148).

References

[1] Guang Xu. Research on Portable laryngoscope based on Embedded System [D]. Zhe Jiang University, 2014:3–5.

[2] Yi Cheng, Fushan Xue et al. Recent characteristic of newly developed intubation devices [J]. International Journal of Anesthesiology and Resuscitation, 2013, 34(11):1003–1008.

[3] Gang Tan, Jie Yi. Comparison between two portable video laryngoscopes and Macintosh laryngoscope in airway intubation under general anesthesia [J]. Journal of Navy Medicine, 2013, 34(5):334–336.

[4] JD Walker. Posture used by anaesthetists during laryngoscopy [J]. British Journal of Anaesthesia, 2002, 89(5):772–774.

[5] Ning Zhang. Finite Element Analysis and Optimum Design of Aluminum Alloy Wheel's Forced State [D]. Chong qing University, 2010, 4.

[6] Aye Thant Htoo, Yukio Miyashita, Yuichi Otuska, Yoshiharu Mutoh, Shigeo Sakurai. Variation of local stress ratio and its effect on notch fatigue behavior of 2024–T4 aluminum alloy [J]. International Journal of Fatigue, 2016, 88:19–28

Xin-Yao Hu, Shu-Jian Zhong and Xing-Da Qu

An Instrumented Glove for Real Time Feedback on Hand Gripping Pressure and Bending a Pilot Study

Abstract: Previous research has shown that prolonged hand grips, forceful grips, or grips with awkward postures are associated with many kinds of upper limb musculoskeletal disorders (MSDs). For many years, different preventive measures have been taken to reduce the risk of upper limb MSDs. All these preventive measures require the hand tool or the task to be carefully analyzed. This study presented an instrumented glove based on the Force Sensitive Resistors (FSRs) and bending sensors that can provide real time feedback on hand gripping pressure and bending. This paper introduced the system design of this instrumented glove. And a pilot study with a camera gripping testing was followed. The results showed that the instrumented glove can provide quantitative data for real time analysis for different hand gripping tasks. Thus, this instrumented glove can be used for ergonomic analysis of hand tools or tasks. It can be implemented to help minimize risk of repetitive strain injuries of upper limbs in the future.

Keywords: hand gripping, instrumented glove, musculoskeletal disorders (MSDs), FSR, and bending sensor

1 Introduction

Gripping is a common task our hand performed in many occupational and home settings. Especially, with the recent technology booming, hand-held devices, such as cell phones, digital cameras, and gaming devices (e.g. joystick) have been widely used in our daily lives [1].Gripping objects like these often requires the muscles from fingers, hands or forearms to function repeatedly. And this might increase the risk of repetitive strain injuries (also known as cumulative trauma disorders). Previous research has shown that prolonged grips, forceful grips, or grips with awkward postures are associated with many kinds of upper limb musculoskeletal disorders (MSDs), such as "trigger finger" and carpal tunnel syndrome (CTS) [2].

For many years, different preventive measures have been taken to reduce the risk of suffering upper limb MSDs [3]. In general, those preventive measures can be

Xin-Yao Hu, Shu-Jian Zhong , Xing-Da Qu ,Institute of Human Factors and Ergonomics, Shenzhen University, Shenzhen, China. Email: huxinyao@szu.edu.cn

DOI 10.1515/9783110540048-079,

categorized as three strategies. First, quantifying the risk factors of MSDs, that can provide a scientific basis regarding to the tools or tasks that might directly lead to cumulative trauma disorders. Second, ergonomic design changes, such as improving the design of hand tools or computer mouse, can reduce the case of awkward postures. Third, work place practice modification, such as involving frequent work break or job rotation, can also minimize the risk of cumulative trauma or injuries [3].

All these preventive measures require the hand tool or the task to be carefully analyzed. Previous analysis approach was qualitative. For instance, Ducharmeet al. [4] analyzed the tool grips at the US Air Force. They found that those tool grips were too wide for female works that often required two hands to operate and suggested that ergonomic changes need to be made. The qualitative analysis help certain the necessity of making ergonomic or work place practice changes. It is often lack of quantitative data to support a more detailed modification. In other words, a quantitative analysis will be much more useful, in terms of offering the insight about how to make changes to prevent upper limb MSDs.

For hand gripping task in particular, quantitative analysis can be achieved by providing the real time feedback on hand pressure and hand posture. There are a few commercially available equipment to do this. For instance, the Grip™ system made by Tekscan, Inc. (MA, USA), and the pliance® hand sensor made by Novel Inc. (CA, USA), both can measure the interface pressure for human hand and finger gripping applications. However, those devices are still quite expensive. Therefore, they are not widely used in occupational settings. Besides, these devices cannot provide information regarding to the hand postures. In fact, in addition to the pressure, awkward postures affect the amount of force that a worker can produce.

This paper proposed an augmented glove that can provide real time feedback on hand gripping pressure and figure bending. The system integrated multiple force sensitive sensors (FSRs) and bending sensors that can simultaneous capture pressure and bending signals. With a customized graphic user interface (GUI), this system can have the information displayed in PC end.

2 System Design

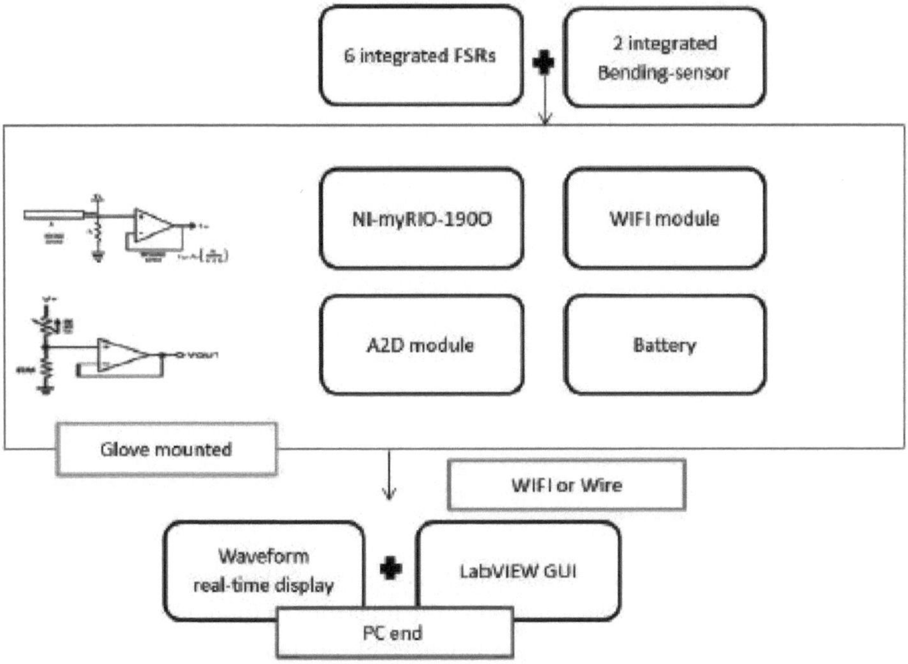

Fig. 1: System overview of the instrumented glove

The system overview is depicted in Fig. 1. The instrumented glove system integrated 6 FSRs (FSR402, Interlink Electronics, CA, US) and 2 bend-sensors (FlexSensor, Spectra Symbol, US) (Fig. 2), a glove mounted unit, and a PC end graphic user interface (GUI) (Fig. 3).

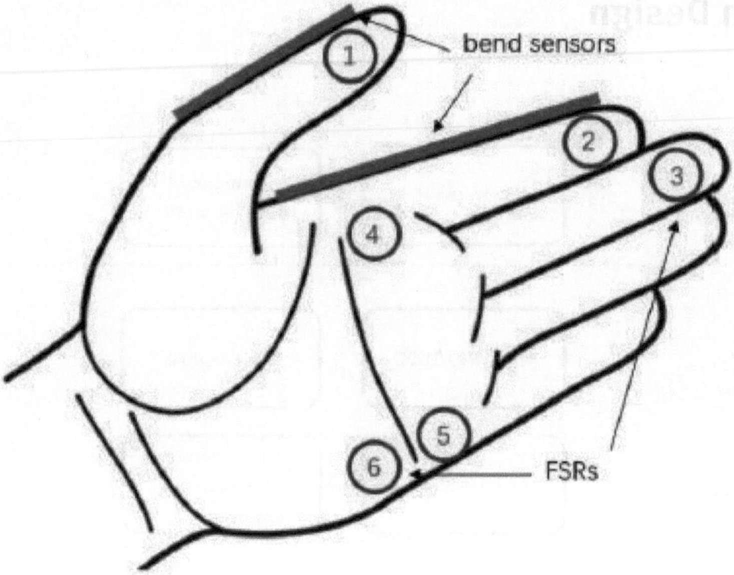

Fig. 2: Sensors Layout

The layout of the 6 FSRs and 2 bend-sensors is depicted in Fig. 2. Six FSRs were strategically placed: Sensor 1-3 were placed at the fingertips of thumb, index finger and middle finger, respectively. Sensor 4 was placed at the hand palm closed to the end of the index finger. Finger 5 and 6 were placed at the lateral edge of the hand palm. This sensor layout is similar to what has been suggested by Bergamasco et al. [5], which cover the important hand pressure distribution area of hand during gripping. For each FSR sensor, it has a sensing area made of fiberglass resin (12.7 mm in diameter), which was attached to a base with 18.1 mm diameter. This design helps to allow the force applied to be evenly distributed over its surface and decrease the possibility of saturation.

The bend-sensors were attached to thumb and index finger (Fig. 2). One of the bend-sensor is known as the 4.5 (inches)' type, with the length of 112mm (active length: 95 mm). Another one is 2.2 (inches)' type, with the length of 74mm long (active length: 55.37mm).

Above mentioned sensors were integrated and connected to the glove mounted unit (Fig. 1). This unit consists of a micro controller (myRIO-1900, National Instrument, US), a Voltage-to-Current (V2C) converting module and a WIFI module and a battery. The V2C converting circuit converts the FSR and bending sensor resistance value to an inverse voltage output, which is subsequently converted into readable voltage output through a 12-bit analog-2-digital (A2D) module. These data were transmitted wirelessly by a WIFI module to the PC end receiver. The sampling frequency was set at 50 Hz. The hand pressure and bending information was displayed in a GUI (as shown in Fig. 3) which is designed by a customized Lab VIEW script.

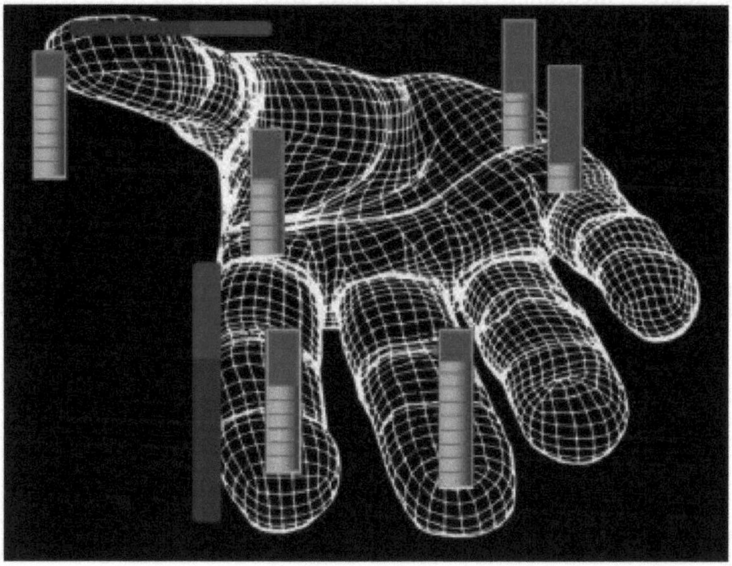

Fig. 3: The GUI shows the hand pressure and bending information in real time

3 System Evaluation

3.1 Sensor Calibration

Previous research has shown thatthe force-to-voltage response behaviour could vary among different FSRs [6].This is probably due to manufacturing technique flaws [7]. Therefore, the reliability and accuracy of FSR rely much on the calibrations. In the presenting study, a series of weights, ranging from 100 mg to 800 mg were used to calibrate the FSRs. Each time, a weight was placed at the centre of FSR in random sequence. And the A2D output (i.e. voltage) recorded from each FSR during the calibration was plotted against the weight. A least-squares-fitting curve was plotted for each FSR, as can be seen in Fig. 4. A fifth-degree polynomial function is then obtained from a multiple regression approach to establish the relationship between the voltage (V) and the force (N) applied. The model coefficients were as shown in Eq. 1.

$$f(x) = 0.001417 * x^5 - 0.03566 * x^4 + 0.3473 * x^3 - 1.655 * x^2 + 4.069 * x + 0.02455 \quad (1)$$

Similarly, the force-to-voltage response behaviour of the bend-sensor is different for the 4.5' bend-sensor and 2.2' bending sensor. And due to sensor resolution, its bending characteristics were classified into three categories, i. e. small bending (bending angles from 1 to 60 degree), medium bending (bending angles from 61 to 120 degree) and large bending (bending angles from 121 to 180 degree). Experimental testing has been carried out to determine the sensor output (in voltage) that corresponded to the range of the bending angles. The results are shown in Table 1.

Tab. 1: bend-sensor calibration: the angle range and corresponding A2Coutput

Sensor Output (V)	No bending (0°)	Small bending (1°-60°)	Medium bending (61°-120°)	Large bending (121°-180°)
4.5' type	3	1.76-2.99	1.1-1.75	0.68-1.74
2.2' type	3	1.76-2.99	1.25-1.75	0.55-1.24

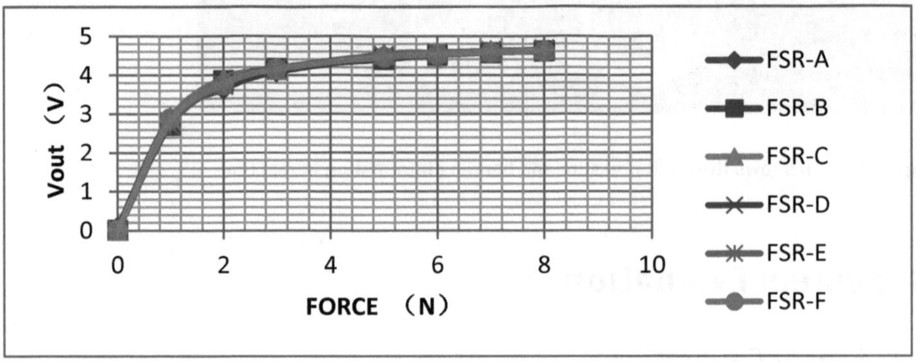

Fig. 4: The calibration of FSRs

3.2 Hand Gripping Test

As a pilot study, one young male volunteer was involved in the hand gripping test. To demonstrate that the instrumented glove can provide real time feedback on the band pressure and bending, a camera gripping task was designed. The volunteer was asked to hold different cameras (DCs) in his own comfortable way. And he need to perform the task of "pressing the camera shutter for three times, consecutively". Two cameras were selected for this pilot study. As can be seen in Fig. 5, Camera A is small DC, and Camera B is a relatively large digital single-lens reflex camera (DSLR).

Fig. 5: Hand gripping test with two digital cameras (Camera A at the left, Camera B at the right)

The testing results are shown in Fig. 6. When gripping Camera A, only FSR 1, 2 and 4 had readings, indicating that the hand pressure was applied only at the thumb and index finger and a small part near the end of the index finger (FSR 4). It is clear that there was limited support from the hand palm when gripping camera A. In other words, this kind of posture might compromise the holding stability, which means the thumb and index finger need additional work to compensate the stability. That might increase the risk of repetitive strain injuries for the thumb and index finger when holding the camera for a long period of time. On the other hand, when gripping Camera B, all the FSRs had similar readings except Sensor5 (which is located at the lateral edge of the hand palm), indicating that pressure was distributed quite evenly across the hand palm and finger. Thus, this kind of hand posture is preferable. Moreover, Despite the Camera B is heavier than Camera A, the results show that the hand pressure is almost similar to grip each camera. Again, this might suggest that the gripping posture of camera B helped the pressure to be distributed more evenly.

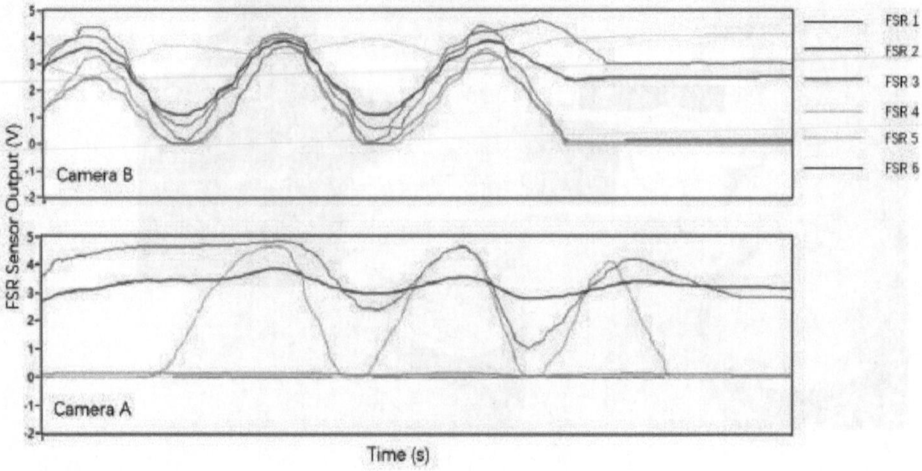

Fig. 6: Hand gripping test results for two digital cameras

4 Conclusion and Future Work

The presenting study proposed a novel approach for measuring hand pressure and bending by a wearable device based on low cost FSRs and bending sensors. The results showed that overall, the data are sufficient for analysing the hand pressure and posture. This study contributes to the current research by introducing a low-cost solution to measure the hand pressure and bending during gripping task. Thus, it can benefit the future clinical application to allow the hand tools and tasks to be analysed in occupational settings.

As a pilot study, this work has some limitations and future work needs to be done. First, only one volunteer was involved in the experiment study, and only one trial of data was analysed. Therefore, the result of this study was limited by its generality. Future studies need to done with more participant. Second, only preliminary analysis was done in the presenting study. Future work is granted for more detailed quantitative analysis.

Acknowledgement: This work was supported in part by Natural Science Foundation of Guangdong Province under Grant number [2015A030313553] and [2016A030310068]; Science, Technology and Innovation Committee of Shenzhen City under Grant number [JCYJ20160422145322758] and [JCYJ20150525092940994].

References

[1] L. Rainie, Pew Research Centre, 1–5, 2010.
[2] S. H. Snook, D. R. Vaillancourt, V. M. Ciriello and B. S. Webster, Ergonomics, 38, 7, 1488–1507, 1995.
[3] T. J. Armstrong, Ergonomics and cumulative trauma disorders, Hand Clinics, 2, 553–565, 1992.
[4] R. Chaillet, Hand Pain and Impairment, 4th. ed. (Philadelphia : FA Davis Company), 2005
[5] M. Bergamasco, IEEE International Workshop on Robot and Human Communication, 1992.
[6] P. S. Dyer and S. J. M. Bamberg, IEEE-EMBC, 6805–6809. 2011.
[7] J. A. Flores and A. Velasquez, IEEE ANDESCON, 10, 1–6. 2010.

Li-Gang Zhao*, Tao Ruan, Jun-Zheng Ding and Ai-Sheng Wu

Optimization of Crank Mechanism of Multifunctional Nursing Bed Based on ADAMS

Abstract: The crank mechanism of multifunctional nursing bed is an important institution to assist patients exercising their legs. In order to optimize the function of the crank mechanism and improve the quality of rehabilitation, the kinematic relationship of the crank mechanism has been studied and the virtual prototype of the crank mechanism is established by using the design variables instead of the coordinates' value based on ADAMS [1, 2]. The influence of related design variables on the angle acceleration of thigh plate is studied by virtual prototyping. The sensitivity of related design variables to the angle acceleration of the thigh plate is obtained and those variables with larger sensitivity are optimized to be the best value. The maximum value of the angular acceleration is reduced and the curve is gentler. The mechanism is running more smoothly and the crank mechanism has been optimized.

Keywords: ADAMS; nursing bed; crank mechanism; structure optimization

1 Introduction

China has entered the era of aging under the background of global aging. With the increasing age, the vitality of people and the resistance to disease will be weakened. Especially in the older stage people are more susceptible to chronic diseases. According to statistics, about 79.1% of the elderly in our country suffer from one or more chronic diseases. Treatment of chronic diseases is not only long and slow. The chronic makes the majority of patients pain in the physical and psychological and

---

***Corresponding author: Li-Gang Zhao,** Jiangsu Provincial Key Laboratory of Advanced Manufacturing for Marine Mechanical Equipment, Jiangsu University of Science and Technology, Zhenjiang, China. Email: zhaoligang@just.edu.cn.
Tao Ruan, Jiangsu Provincial Key Laboratory of Advanced Manufacturing for Marine Mechanical Equipment, Jiangsu University of Science and Technology, Zhenjiang, China. Email:hnruantao@163.com.
Jun-Zheng Ding, Jiangsu Provincial Key Laboratory of Advanced Manufacturing for Marine Mechanical Equipment, Jiangsu University of Science and Technology, Zhenjiang, China. Email:m18252580230@163.com.
Ai-Sheng Wu, Jiangsu Provincial Key Laboratory of Advanced Manufacturing for Marine Mechanical Equipment, Jiangsu University of Science and Technology, Zhenjiang, China. Email:22871120@qq.com.

DOI 10.1515/9783110540048-080,

the difficulty is improved for family of patients and medical staff to care. The social resources and family resources for the care of the aging population are tense with the process of aging [3]. Multifunctional nursing bed has largely relieved the adverse effects of caring aging population which are caused by lacking of nursing resources. Multifunctional nursing bed can reduce the burden of family, medical staffs and improve the quality of care [4]. Stroke is a common chronic disease in the elderly. Legs and arms of stroke patients are more affected. They cannot walk normally and need regular exercise to improve the rehabilitation effect. The cost of helping the patients to exercise muscles of leg is high. The care bed with the crank mechanism is created under this background. The crank mechanism of multifunctional nursing bed is an important institution which can assist patients to exercise their legs. It can help stroke, hemiplegic and postoperative patients to do daily rehabilitation exercise [5, 6]. In order to improve the quality of care, this paper has studied the kinematic of the crank mechanism and established the parameterized model to research and optimizes the mechanism.

2 Optimization of Crank Mechanism

2.1 Research on Kinematics

The crank mechanism includes an electric putter L_3, a thigh plate L_{OA}, a calf plate L_{AB} and a support plate L_{CD}. The thigh and calf plates are connected by hinges. Calf plate always moves in horizontal direction and the angle of the thigh plate is 0°~30°. The chart of the crank mechanism is shown in Fig.1.

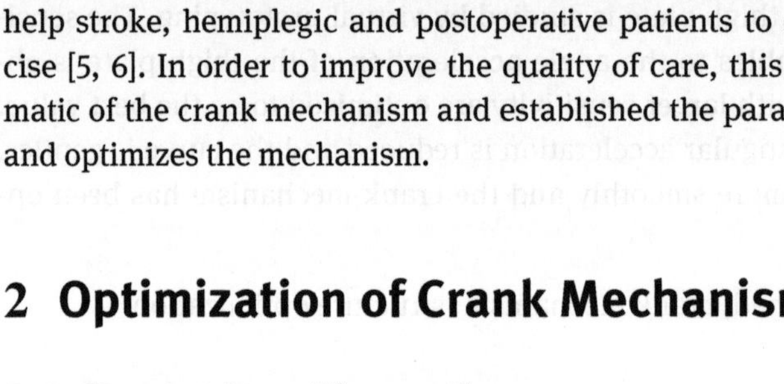

Fig. 1: Chart of the crank mechanism

Assuming that the articulation point of the thigh plate and the bed frame is the coordinate origin, and the coordinate is (0, 0). Assuming that the coordinate of point A is (-400, 0), point C is (-414, 0), point D is (-760, 0), point E is (27, -101) and point F is

(-799, -58) taking into account the angle of thigh plate, the stroke of electric putter, bed space and other factors. According to the vector equation analysis method, the vector equation is listed by ΔOEF:

$$\vec{L_1} = \vec{L_2} + \vec{L_3} \tag{1}$$

Assuming that $\varphi = 180 + \theta - \alpha$

$$\begin{cases} L_1 \cos\varphi = L_2 \cos(180+\beta) + L_3 \cos(360-\delta) \\ L_1 \sin\varphi = L_2 \sin(180+\beta) + L_3 \sin(360-\delta) \end{cases} \tag{2}$$

The equation (2) is subjected to a second derivative of t:

$$\begin{cases} -\alpha_1 L_1 \sin\varphi - \omega_1{}^2 L_1 \cos\varphi = \ddot{L_3} \cos\delta + \omega_3 \dot{L_3} \sin\delta + \alpha_3 L_3 \sin\delta - \omega_3{}^2 L_3 \cos\delta \\ \alpha_1 L_1 \cos\varphi - \omega_1{}^2 L_1 \sin\varphi = -\ddot{L_3} \sin\delta + \omega_3 \dot{L_3} \cos\delta + \alpha_3 L_3 \cos\delta + \omega_3{}^2 L_3 \sin\delta \end{cases} \tag{3}$$

To eliminate α_3 in equation (3) [7]:

$$\alpha_1 = \frac{\left(\omega_1{}^2 L_1 \sin\varphi - \ddot{L_3} \sin\delta + \omega_3 \dot{L_3} \cos\delta + \omega_3{}^2 L_3 \sin\delta\right)}{L_1 \cos\varphi}$$

$$+ \frac{\cos\delta\left[\left(\omega_3{}^2 L - \ddot{L_3}\right)\cos(\delta+\varphi) - \omega_3 \dot{L_3} \sin(\delta+\varphi) - \omega_1{}^2 L_1\right]}{L_1 \cos\varphi \sin(\delta+\varphi)} \tag{4}$$

α_1 is the angle acceleration of thigh plate and we can get the data of α_1 after substituting known data.

2.2 Parametric Model

We can see from Fig.1 that the thigh plate and calf plate are implemented components. The thigh plate moves around the point O and calf moves in horizontal movement. The thigh plate drives the calf and determines the condition of calf. So the optimization of the whole crank mechanism can be realized if the size of the thigh plate mechanism is optimized.

The mechanism of thigh plate includes four hinge points O (X1, Y1), E (X2, Y2), F (X3, Y3), A (X4, Y4); Point A has no effect on the angle acceleration of thigh plate. Therefore, we just need to determine the best location of point O, point E and point F. We can use DV_1~DV_6 instead of the coordinates of the three hinge points. The correspondence of design variables and coordinates value is shown in table 1.

Tab. 1: Correspondence of design variables and coordinates value

design variable	coordinate value	initial value
DV_1	X_1	0.00
DV_2	Y_1	0.00
DV_3	X_2	27.00
DV_4	Y_2	-101.00
DV_5	X_3	-799.00
DV_6	Y_3	-58.00

To create some hinge points in ADAMS and replace theirs coordinates with the corresponding variables. The electric putter, thigh plate, calf plate, support plate and other components have been established based on the hinge points. The driver and motion pair are added to the components to complete the parametric model, as shown in Fig.2.

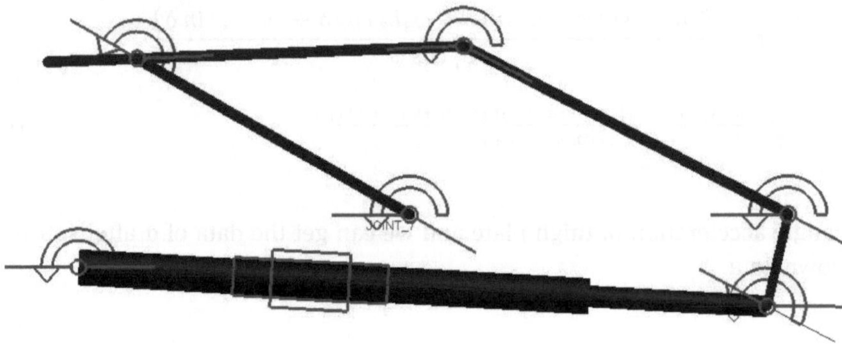

Fig. 2: Parametric model of crank mechanism in ADAMS

According to the parametric model established in Fig. 2, the curve of angle acceleration of thigh plate before optimization can be obtained, as shown in Fig.3.

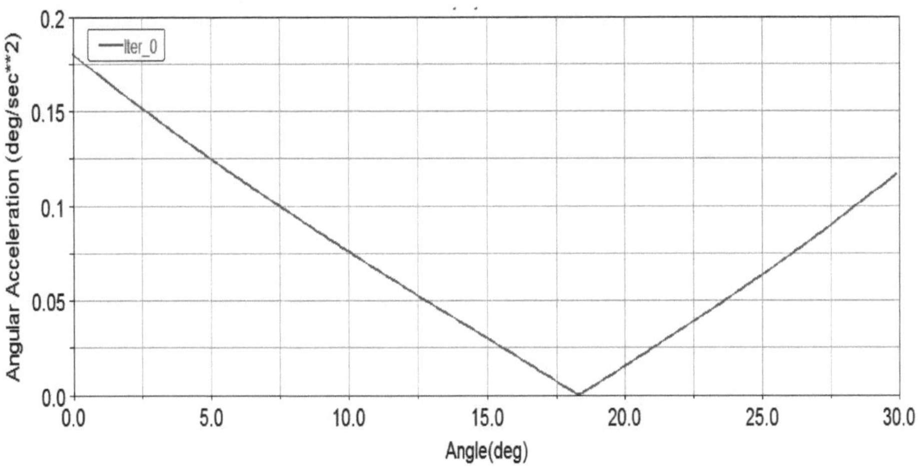

Fig. 3: Curve of angle acceleration of the thigh plate before optimization

2.3 Research on Design

The research on design mainly studies the changing trend of performance caused by different values of design variables. We can get the sensitivity of the design variables to the performance of virtual prototype by research on design. The sensitivity is that the change of the virtual prototype performance which is caused by the change of the unit value of the design variables [8]. The size of the sensitivity reflects the size of the design variables affecting the virtual prototyping performance. First, we can study the design variables before optimization to analyse the sensitivity of variables. Then, we optimize those variables with large value sensitivity so that we can improve the efficiency of optimization and ensure the accuracy of the results.

The research on design mainly includes three elements: objective function, design variables, level of variables. Firstly, the DV_1 is studied. The objective function of this research is the maximum value of angle acceleration of thigh plate. The design variable is DV_1 and the level of variable is 5. During the research, the curve of angle acceleration of the thigh plate with DV_1 is shown in Fig. 4, and the sensitivity report is shown in Table 2.

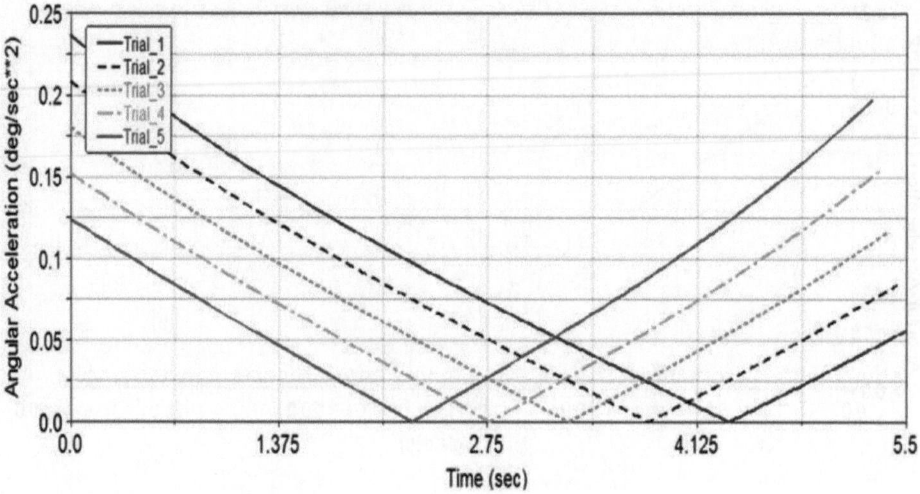

Fig. 4: Curve of the angle acceleration of the thigh plate with DV_1

Tab. 2: Sensitivity report of DV_1

Number	DV_1	Maximum value of angular acceleration (deg/s²)	Sensitivity (deg/s²)/mm
1	-10.000	0.23690	-0.005670
2	-5.000	0.20857	-0.005662
3	0.000	0.18030	-0.005413
4	5.000	0.15444	0.001707
5	10.000	0.19737	0.008586

It can be seen from Fig. 4 and Table 2 that the curve of angle acceleration of the optimization of the thigh plate is changed with the value of DV_1. The maximum of angular acceleration occurs at the start or the end. First, the maximum of angular acceleration reduced and then increased when the value of DV_1 from small to large. When DV_1 is 0, the maximum of angular acceleration of the thigh plate is 0.18030 (deg/s²) and the sensitivity is -0.005413 (deg/s²) / mm. The remaining design variables are studied and the research report of variables is shown in table 3.

Tab. 3: Research report of design variables

Design variable	Initial value	Coordinate value	Sensitivity $(deg/s^2)/mm$
DV_1	0.000	X_1	-0.005413
DV_2	0.000	Y_1	-0.005193
DV_3	27.000	X_2	0.005609
DV_4	-101.000	Y_2	0.004371
DV_5	-799.000	X_3	3.60e-005
DV_6	-58.000	Y_3	0.000857

We can get the sensitivity of every design variable from the table 3. The sensitivity of DV_3 is maximum and the value is 0.005609 (deg/s^2) / mm. The sensitivity of DV_5 is minimal and the value is 3.60e-005 (deg/s^2) / mm. Since the sensitivity of DV_1, DV_2, DV_3, DV_4 and DV_6 are higher than DV_5 by 1 ~ 2 orders of magnitude, DV_5 can be ignored. We just need to optimize DV_1, DV_2, DV_3, DV_4 and DV_6.

2.4 Optimized Design

Optimized design is the method that to get maximum or minimum of the objective function when the design variables meet the constraints [9, 10]. The purpose of this optimization is to minimize the maximum value of angle acceleration of the thigh plate. The optimized variables are DV_1, DV_2, DV_3, DV_4, DV_6; the constraint range of design variable is its range. The curve of angle acceleration is shown in Fig.5 during optimizing. The result of optimization is shown in Table 4.

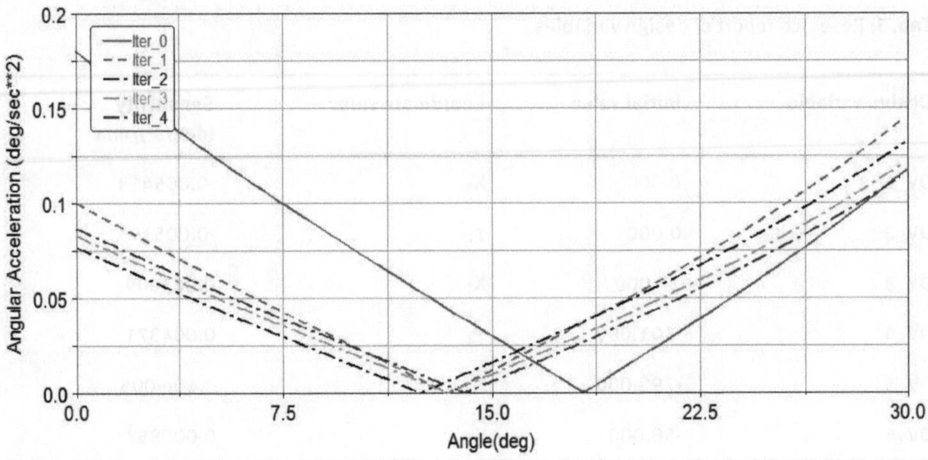

Fig. 5: Curve of angle acceleration of the thigh plate

Tab. 4: Result of optimization

Number	Maximum value of angular acceleration (deg/s^2)	DV_1	DV_2	DV_3	DV_4	DV_6
0	0.1803	0.000	0.000	27.000	-101.000	-58.000
1	0.1447	5.735	5.224	26.065	-109.820	-58.659
2	0.1319	3.086	10.000	26.494	-116.150	-58.495
3	0.1204	4.220	10.000	26.311	-116.150	-58.711
4	0.1135	6.047	10.000	26.017	-116.150	-59.048

It can be seen from table 4 that the maximum value of angular acceleration is 0.1803 (deg/s^2) when DV_1, DV_2, DV_3, DV_4 and DV_6 are taken as initial values. The maximum value of angular acceleration has been significantly reduced from the initial value 0.1803 (deg/s^2) to 0.1135 (deg/s^2) after four times optimization experiments. The pre-optimized and optimized contrast curves are shown in Fig.6 and the comparison of design variables is shown in table 5.

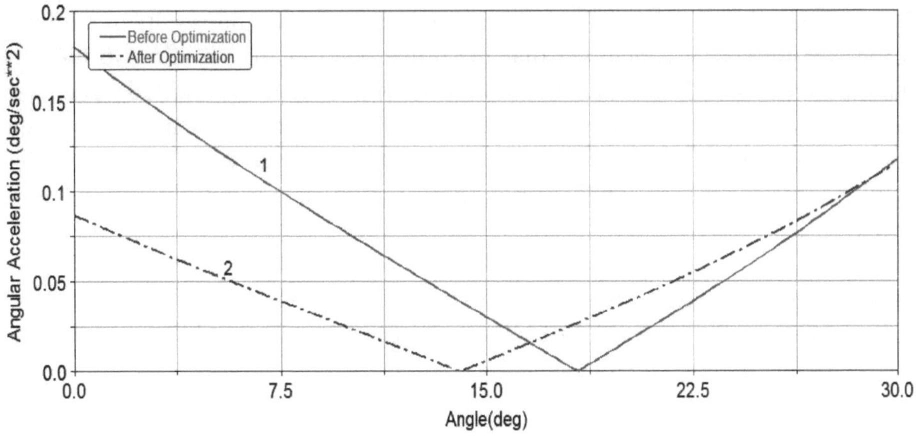

Fig. 6: Contrast curve of angle acceleration of the thigh plate

Tab. 5: Comparison of design variables

design variables	DV_1	DV_2	DV_3	DV_4	DV_6
before optimization	0.000	0.000	27.000	-101.000	-58.000
after optimization	6.047	10.000	26.017	-116.150	-59.048

In Figure 6, curve 1 is the pre-optimized curve of angle acceleration of the thigh plate and curve 2 is the optimized curve. It can be seen from the figure that the average value and maximum value of curve 2 are less than curve 1 and curve 2 is smoother than curve 1. The values of DV_1, DV_2, DV_3, DV_4 and DV_6 are 6.047, 10.000, 26.017, -116.150 and -59.048 from table 5.

3 Summary

The crank mechanism has been designed under the guidance of modern design theory and the kinematic relationship has been studied. The virtual prototype of the crank mechanism is established by using the design variables instead of the coordinates' value based on ADAMS. The influence of related design variables on the angle acceleration of thigh plate is studied. We have concluded that DV_1 ~ DV_6 has different effects on the angle acceleration of the thigh plate. The sensitivity of DV_5 is 3.60e-005 (deg/s2)/mm which is smaller than the sensitivity of other variables 2 orders of magnitude. It is shown that the abscissa of the fixed point of the electric putter has little effect on the angle acceleration of thigh plate.

DV_1, DV_2, DV_3, DV_4, DV_6 are optimized according to the results of studying on the variables. It is concluded that the hinge point O should move to the first quadrant to (6.047, 10) and the end of the L_1 extends to (26.017, -116.150). The fixed point of the electric putter moves to (-799, -59.048). At this time, the maximum angle acceleration of thigh plate is reduced from 0.1803 (deg/s^2) to 0.1135 (deg/s^2) which shows that the curve of the angle acceleration of thigh plate is more stable and more gentle.

Acknowledgement: Fund Project: Research and Development of Prospective Joint Project of Jiangsu Province (BY2015065-03); Postgraduate Research & Practice Innovation Program of Jiangsu Province.

References

[1] CHEN De-min, HUAI Chuang-feng, ZHANG Ke-tao. Proficient in Virtual Prototyping Technology of ADAMS 2005/2007 [M]. Chemical Industry Press, 2010.
[2] GONG Peng-han, HU Ren-xi, KANG Shi-yan. Virtual Prototyping from Entry to Proficient of ADAMS 2014 [M]. Mechanical Industry Press, 2016.
[3] PENG Xi-zhe, HU Zhan. China's Population Aging from the Perspective of Public Policy [J]. Chinese Social Sciences, 2011(3): 121–138.
[4] LI Guo-kang, LI Yong-wei, ZHOU Zhuo. Design of a Multifunctional Nursing Bed [J].Journal of Shenyang Ligong University, 2013, 32(3): 91–94.
[5] ZHANG Li-xun, WANG Ke-yi, ZHANG Jin-yu. Simulation of Lower Limb Rehabilitation Training of Robot [J]. Mechanical Design and Research, 2007, 23 (6): 46–50.
[6] GUO Xiao-wen, XIE Qing-sen, GUAN Pai. Research on Function Design of Nursing Bed for Lower Limbs Rehabilitation Exercise [J]. Mechanical Design, 2013, 30(9): 111–113.
[7] ZHANG Jian-guo, GUO Yan-ping, XUE Qiang. Kinematics Analysis and Simulation of Horizontal Lifting Mechanism of Multi-functional Electric Bed [J]. Modern Manufacturing Engineering, 2011(11): 72–77.
[8] HUANG Yue-hua, ZHAO Liang-liang, LI Feng. Optimization Design of the Amplitude Variation Mechanism of Supportive Arm of Power Catwalk Based on ADAMS [J]. Oil Field Equipment, 2015(7): 36–38.
[9] MA Ying, ZHU Shi-sha. Optimization of Excavator Bucket-linkage Mechanism Based on ADAMS [J]. Mechanical Engineering & Automation, 2011, (4):32–34.
[10] LI Yun. Optimization Design of Clamp Mechanism based on ADAMS [J]. New Technology & New Process, 2010, (6): 35–37.

Jie Duan, Hai-Lin Bai and Lin Cheng

Research on the Mechanism of Industry Integration between Shaanxi Advanced Manufacturing and Modern Service Industry

Abstract: With the continuous innovation of industrial development, industrial integration became a new form of industrial development. Based on the application of system theory, this paper discusesd the mechanism of industrial integration with the corresponding mathematical model. And according to the Input-Output Table of Shaanxi Province, some indicators were selected to calculate the integration of Advanced Manufacturing Industry and Modern Service Industry, the result shows that the degree of integration of the two industries tends to balance, and Advanced Manufacturing Industry capacity of the integration is significantly stronger than the Modern Service Industry, but the two industries structure is still unreasonable. The innovation of this paper lies in using two perspectives - industrial coupling and industrial integration to explain coordinate development issues, interaction relationships, operational status, and the degree of integration of the two industries.

Keywords: Advanced Manufacturing Industry, Modern Service Industry, Industry Iintegration

1 Introduction

At present, the economic development trend of China is in the transition period with both opportunity and challenge and it is in the same period in Shaanxi Province, the western industrial province. In order to actively adapt to the new economic state, Shaanxi Province pay more attention to optimize the industrial structure, accelerate the changes in the way of development, promote coordinated development of regional economic, cultivate a new industrial structure, and actively develop new service industries, which aims to the industrial structure escalating, and lay the foundation for industrial development in the thirteen five plan. In order to promote the industrial integration, in 2016, an opinion was pointed out by government of

Jie Duan, The western industrial economy research Institute of Northwestern Polytechnical University, Shaanxi Xi'an, China. Email: duanjieyong @163.com
Hai-Lin Bai, Northwestern Polytechnical University, Shaanxi xi 'an, China. Email: 18710707348@163.com
Lin Cheng, Northwestern Polytechnical University, Shaanxi xi 'an, China. Email: 598418089@qq.com.

DOI 10.1515/9783110540048-081,

Shaanxi P9783110540048rovince that they would accelerate the development of service-oriented manufacturing and promote the deep integration of both manufacturing and service industries in the "Made in China 2025" Shaanxi implementation opinion. In 2017, it was also pointed out that would follow the ideas of "stable energy, promote chemical industry, rise electronics, enhance manufacturing, expand the new industries, optimize the tradition" to promote industrial development, strengthen the manufacturing industry restructuring, and simultaneously promote the quality and efficiency of modern service industry. In this context, the promotion of Shaanxi advanced manufacturing and modern service industry integration, has become the main trend of economic development in Shaanxi Province. In this context, the integration of both the advanced manufacturing and modern service industry has become the main trend of economic development in Shaanxi Province.

2 Integration and Interaction between Advanced Manufacturing and Modern Service Industry

In this paper, the integrated industry of Advanced Manufacturing and Modern Service Industry is regarded as an industrial system. And the system is made up of two subsystems: Advanced Manufacturing Industry and Modern Service Industry. According to the existing model and theory, the C-D production function can be used to explain the mechanism of the integration. We assume that the total output of the product in this industry system is, according to the C-D function $Q_T = AL^{\alpha}K^{\beta}\eta$, $Q_T = Q_m + Q_s$. Assuming Advanced Manufacturing Industry $Q_m = \sigma Q_t$ (0<σ<1), $Q_s = (1-\sigma)Q_T$. So the Modern Service Industry and Advanced Manufacturing Industry production relationship is that: which means that. Setting up total factor production, the residual term is, is constant, we can see, industry scale pay the same or scale returns, the two industries have the necessary integration, otherwise it will produce a waste of economic resources. From this we can obtain equations:

$$\begin{cases} Q_T = A_m L_m^{\alpha\alpha_1} K_m^{\beta\beta_1} \eta_m \bullet Q_s \\ Q_s = A_s L_s^{\alpha\alpha_2} K_s^{\beta\beta_2} \eta_s = \dfrac{1-\sigma}{\sigma} A_m L_m^{\alpha\alpha_1} K_m^{\beta\beta_1} \eta_m \end{cases} \tag{1}$$

In the above formula are output elasticity coefficients of advanced manufacturing and modern service industries respectively. Therefore, the elastic coefficient of the integration industry is linked equations by yield, elaborating the relationship between the total output of industry, the output of Advanced Manufacturing and the output of Modern Service Industry. It reflects the interaction and integration between the Advanced Manufacturing Industry and Modern Service Industry.

3 Empirical Analysis

Based on the input-output method, this paper uses the Input-Output Table of Shaanxi Province in 2007 and 2012 measured the degree of integration between Advanced Manufacturing Industry and Modern Service Industry in Shaanxi Province. This paper chooses four indicators to measure the degree of integration of the two industries, namely: Advanced Manufacturing Industry input rate and demand rate, Modern Service Industry input rate and demand rate. The higher rate of investment in Advanced Manufacturing Industry, indicating that the Modern Service Industry is more reliance on the Advanced Manufacturing Industry and the stronger ability of Advanced Manufacturing Industry fuse into the Modern Service Industry; the higher demand rate of the Advanced Manufacturing Industry, indicating that the Advanced Manufacturing Industry more depend on Modern Service Industry about the intermediate consumption. And the Advanced Manufacturing Industry has stronger power fuses to the Modern Service Industry. This is also true for modern service industries.

3.1 Measurement The Degree of Integration

Intermediate input rate (F): Intermediate investment rate refers to the proportion of other industries in the middle of investment in the total investment in an industry. The formula is:

$$F_{i-j} = \frac{\sum_{i=1}^{n} x_{ij}}{\sum_{i=1}^{n} x_{ij} + N_j} (j = 1,2,\cdots,n) \tag{2}$$

Among them, $\sum_{i=1}^{n} X_{ij}$ represents the intermediate input of the j industrial sector, N_j represents the value created by the j industry sector.

(1) Intermediate demand rate (G): The intermediate demand rate refers to the proportion of other industries' demand for intermediate products of an industry in the output of this industry. The formula is:

$$G_{i-j} = \frac{\sum_{j=1}^{n} x_{ij}}{\sum_{j=1}^{n} x_{ij} + Y_i} (i = 1,2,\cdots,n) \quad (3)$$

(2) Among them. $\sum_{j=1}^{n} x_{ij}$ Represents other industries demand for intermediate products in the i industry, Y_i represents the final demand segment in the i industry sector product.
(3) Investment rate of advanced manufacturing industry (Fe-s): Fe-s=the part of investment in Modern Service Industry form the Advanced Manufacturing Industry / total investment in Modern Service Industry
(4) Demand rate of Modern Service Industry (Ge-s): Ge-s=the output of the modern service industry was consumed by the advanced manufacturing industry used in the middle part/ the total output of the modern service industry.
(5) Investment rate of Modern Service Industry(Fs-e):Fs-e=The part of investment in Advanced Manufacturing Industry form Modern Service Industry / Total Investment in Advanced Manufacturing Industry
(6) Demand rate of Advanced Manufacturing Industry (Gs-e): Gs-e=the output of the Advanced Manufacturing Industry was consumed by the Modern Service Industry used in the middle part/ the total output of the Advanced Manufacturing Industry.

Based on the above four indicators, according to the relevant data in the Input-Output Table, we calculated the corresponding indicators value and analysed the data results.

Tab. 1: Data about Advanced Manufacturing Industry and Modern Service Industry

Name of Modern Service Industry	Fe-s		Ge-s	
	2007	2012	2007	2012
Information Transmission, Computer, Services and Software	0.1483	0.1449	0.0069	0.0091
Financial Intermediation	0.0222	0.0119	0.0463	0.0662
Real Estate	0.0035	0.0029	0.0089	0.0102
Leasing and Business Services	0.0781	0.3704	0.1171	0.1732
Research and Experimental Development	0.3432	0.1270	0.0900	0.0064
Education	0.0753	0.0033	0.0055	0.0041
Culture, Sports and Entertainment	0.0129	0.0274	0.0441	0.0107
Modern Service Industry	0.0976	0.0982	0.0455	0.0400
Name of Advanced Manufacturing Industry	Fs-e		Gs-e	

	2007	2012	2007	2012
Manufacture of General Purpose and Special Machinery	0.0198	0.0346	0.0353	0.0107
Manufacture of Transport Equipment	0.0197	0.0425	0.0057	0.0016
Manufacture of Electrical Machinery and Equipment	0.0410	0.0385	0.0340	0.0620
Manufacture of Communication Equipment, Computers and Other Electronic Equipment	0.0282	0.0537	0.1852	0.6903
Manufacture of Measuring Instruments and Machinery for Cultural Activity and Office Work	0.0334	0.0433	0.5344	0.2038
Advanced Manufacturing Industry	0.0284	0.0425	0.1589	0.1937

From the perspective of investment rate, the investment rate of advanced manufacturing industry fluctuation less in 2007 and 2012. The Modern Service Industry investment rate of 2012 is 0.0425 compared with 2007 increased 1.4 percentages. But overall, the Advanced Manufacturing Investment rate is significantly higher than the Modern Service Industry. This indicating that the degree of dependence on investment Modern Service Industry on the Advanced Manufacturing Investment is significantly higher than the advanced manufacturing industry on the modern service industry. And Advanced Manufacturing Industry has stronger power fuses to Modern Service Industry. From the perspective of demand rate, the demand rate of Modern Service Industry is only about 0.04, and it decreased 0.5 percentage points in2012 compared with 2007. The Modern Service Industry has a lower dependence on the intermediate consumption of the Advanced Manufacturing Industry. The demand rate of Advanced Manufacturing Industry is high, the rate in 2007 is 0.1589 and 2012 is 0.1937. Between 5 years, there is a larger increase. That shows the Advanced Manufacturing Industry more depend on the Modern Service Industry about the intermediate consumption. And the advanced manufacturing industry has stronger power fuses to the modern service industry. The analysis of the input rate and demand rate shows that the Advanced Manufacturing Industry have stronger ability fuse into the modern service industry, and the integrate power is insufficient.

From the specific sub-industry analysis, in terms of input rate, the investment rate of the Advanced Manufacturing Industry about the two service industries including information transmission, computer service software industry and research, experiment development service industry, has decreased in 2012 compared with 2007, but it was significantly higher than other service industries, which showed that there were a higher dependency for the two industries on the Advanced Manufacturing investment. In addition, apart from the increased dependency on the Advanced Manufacturing investment in the leasing and business services and cultural, sports and entertainment industry, the investment rate of the Advanced Manufacturing Industry has declined about other Modern Service Industry. The investment rate of the Modern Service Industry about electrical machinery and equipment manufacturing industry has declined slightly in 2012 than 2007, but there become increased in other service industries, such as, it was a largest increase in communications equipment manufacturing industry, which indicated that the capacity of Modern Service Industry into the Advanced Manufacturing Industry is increasing, but the overall integration capacity is still low. In the demand rate, the top two about the demand rate of modern service industries is leasing and business service industry and research and experimental development industry in 2007, the demand rate of the leasing industry was still the highest in 2012, but the demand rate of research and experimental development industry fell sharply from 0.09 to 0.0064. In addition, the lower demand rate of information transmission and computer service software industry indicated that the demand of the manufacturing sector about the knowledge-intensive service industry is low in Shaanxi Province, the knowledge innovation ability of the manufacturing sector is insufficient and the optimization of industrial structure is difficult. The higher intermediate consumption rate of the Modern Service Industry is the high-tech manufacturing, such as communications equipment manufacturing industry and instrumentation machinery manufacturing industry, which showed that the dependency of the high-tech manufacturing industry about the intermediate consumption of the modern service industry kept a higher degree. In the advanced manufacturing industry the ability of the high-tech manufacturing industry fusing into the modern service industry is stronger.

3.2 Integration Equilibrium Degree

This paper uses the ratio of the input rate of Advanced Manufacturing and Modern Service industry to represent two industries' integration equilibrium degree of input. Using the ratio of demand ratio of Modern Service Industry and Advanced Manufacturing Industry indicates the integration equilibrium degree of the two industries in the relation of output and consumption. In theory, the higher the degree of integration, the stronger the fusion ability between the two industries. When the degree of integration equilibrium is close to 1, it shows that the two industries depend on each other to promote and develop together.

Tab. 2: Integration equilibrium degree

Year	Fe-s	Fs-e	Ge-s	Gs-e	Integration equilibrium degree(input)	Integration equilibrium degree(consume)
2007	0.0976	0.0284	0.0455	0.1589	3.44	0.29
2012	0.0982	0.0425	0.0400	0.1937	2.31	0.21

In 2007 and 2012, the integration equilibrium degree in input relationship between the Modern Service Industry and the Advanced Manufacturing Industry in Shaanxi was significantly higher than that in the consumption relationship. This is mainly due to the Advanced Manufacturing Industry have stronger ability fuse into the modern service industry. The equilibrium degree of integration of the two industries in the input relationship has declined, but the fusion value is still greater than 1, which shows that the Advanced Manufacturing Industry have stronger integration ability. However, the level of integration has declined. The consumption of the relationship of integrate equilibrium degree decreased and the value is low, this may explain the Modern Service Industry lack power fuse into the Advanced Manufacturing Industry, and the development of Advanced Manufacturing Industry depend more on Modern Service Industry in consumption, which caused integration equilibrium degree in the consumption is low. In general[1][2], the current integrate situation between the Advanced Manufacturing Industry and the Modern Service Industry in Shaanxi Province is that the Advanced Manufacturing Industry has a strong ability to integrate into the Modern Service Industry, and the driving force of the Modern Service Industry to the Advanced Manufacturing Industry needs to be further improved. Overall, the two have been in the depth of integration stage.

4 Conclusion

This article regards the industry after the amalgamation of the two industries as a system. The Advanced Manufacturing Industry and the Modern Service Industry in Shaanxi are regarded as two subsystems, and on this basis, we researched the path of the integration of the two industries.

On this basis, according to the Input-Output Table data in 2007 and 2012, we calculated the integration between two industries in Shaanxi province. And the result shows that the Advanced Manufacturing Industry has a strong ability to integrate into the Modern Service Industry. But the integration between the two is becoming more balanced. Among them, Manufacture of Communication Equipment, Computers and Other Electronic Equipment the ability to integrate into the Modern Service Industry is strongest. The intermediate demand for the Modern Service Industry in the Advanced Manufacturing Industry is declining, and the demand is higher for traditional services such as Leasing and Business Services, while the demand for knowledge intensive industries is lower. The results show that there are unreasonable industrial structure in Advanced Manufacturing Industry and Modern Service Industry in Shaanxi Province. Promoting the optimization and upgrading of the industrial structure and speeding up industrial convergence are conducive to speeding up the economic development of Shaanxi province. In fact, the present situation of Advanced Manufacturing Industry in Shaanxi province is superior to that of Modern Service Industry, which is consistent with the empirical results. So Shaanxi Province can give priority to the development of advanced manufacturing. Second, do a good job of integration and development with modern service industry. And then utilizing the high-tech content of advanced manufacturing industry to promote the upgrading and transformation of Traditional manufacturing Industry.

Acknowledgement: Projects of the Seed Foundation of Innovation and Creation for Graduate Students in Northwestern Polytechnic University (Z2017225). Project supported by the basic scientific research of Central University (3102016RW001); Projects of the National Social Science Foundation of China (No.14XJL009)

References

[1] Hudzari R M, Sapuan S M, Ssomad M A H A, et al. Simulation and analysis of innovative hand tool harvester [J]. Scientific Research & Essays, 2012(19): 1864–1871.

[2] Hudzari R M, Noorman M M, Asimi M N N, et al. Engineering Technological in Agriculture Research and Education[J]. Advanced Materials Research, 2013, 705: 493–498.

[3] Chen Xian, Huang Jianfeng. Division of Labor, Interactions and Convergence: Empirical Research on the Involvement of the Relationship between Service and Manufacturing Industries [J]. Chinese Soft Science, 2004(10):65–71. P. I.Chernousov, O. V. Golubev and A. L. Petelin, Metallurgist+, 55, 3–4, 242–250, 2011.

[4] Liu Chuan, Modern service industry and industry transformation Research on the integration degree of advanced manufacturing [J]. Jiangxi Social Sciences, 2014 (5).

[5] Ji Yahui, Cheng Bin. The Interaction and Integration of Producer Services and Advanced Manufacturing-Empirical Analysis of Input-Output Table Bused on Gansu Province [J]. Journal of Xian University of Finance and Economics, 2014(1): 20–24.

[6] Yue Hao, an Empirical Study on the Correlation between Modern Service Industry and Advanced Manufacturing Industry in Wuhan [J]. Technological Development of Enterpris, 2015(22):119-121. I. A. Krasnyanskaya, G. S. Podgorodetskii and S. N. Paderin, Metallurgist+, 59, 9-10, 887–894, 2016.

[7] Yu Hui-wen, Wang Ying, Research on the Convergence Degree of the Equipment Manufacturing Industry and the Modern Service Industry[J]. Research on Generalized Virtual Economy, 2016, 7(1): 21–27.

[8] Wang Xiao Hong, Wang Chuanrong. Integration of Manufacture and Service Industry of Industrial Transformation Condition [J]. Iindustrial Eeconomy, 2013(9): 40–47.

[9] Shi Yuan. Industrial Integration Analysis of Producer Services and Manufacturing in Anhui Province [J]. Journal of Xi an Aeronautical University, 2014(6): 30–33.

[10] Zhou Ye. The Development of Advanced Manufacturing Industry and Modern Service Industry and Enlightenment [J]. Enterprise Strategy, 2010(6): 118–121.

Yao-Zu Cao and Yi-Min Tian

Application of MATLAB in University Physics Teaching

Abstract: To make use of MATLAB as the visible sketch interface and data handling in Physics teaching, and developing the physical experiment data handling procedure and contributing to improve the interest in study and study effect of students. With the imitate of true mercury atom's spectrum, and the hydrogen atom's colour light spectrum of grating spectrum experiment, and other atoms' spectrum, elaborate MATLAB application in Optical theory and other experiment teaching.

Keywords: MATLAB application, Imitate of Mercury atom's spectrum, Hydrogen atom's spectrum

1 Introduction

Matlab is international accepted as excellent software in science and technology realm and used in most application and development. Its characteristics is to gather the science calculation, painting, system model and procedure language design together, and its function is extremely strong [1, 2]. MATLAB computes a rapid result to science and visible accurately, this help the students to turn the abstract thinking to image thinking, thus better for insight into physical meaning, concept comprehension, regulation discovering, and stir up a student to improve interest, foster the student's creative ability.

There are too many concepts and Physical regulations that's very abstract in the University Physics Optical course[3, 4], the theory teaching depend a lot on experiments, especially among them some optical phenomena and regulation lack of detailed mathematical deduce for students so it bring many difficulties for students in their Optical course. The meaning is very important to Provide our students the opportunity of observation about phenomenon in the Optical course teaching, raise their interest in study, then foster the student's thinking and innovation ability, strengthen our classroom teaching effect.

MATLAB show us the advantage in the class teaching as follows:

1) This software solves our problems simple and save time, can help us avoid a complicated operation so we can put attention to the hypostatic thinking at the physical problem.

Yao-Zu Cao, Beijing Printing Institute, Beijing, 102600, China. Email: caoyaozu@163.com
Yi-Min Tian, Beijing Printing Institute, Beijing, 102600, China. Email: tym@bigc.edu.cn

DOI 10.1515/9783110540048-082,

2) Keeps frank simulation to the physical process. Make use of the software in calculation and make diagram function, it turn very convenient in Physics process simulation.
3) MATLAB is easy to learn and easily understand, the program design language and operation is very simple and convenient, so it can raise a student's activity in the university.

Certainly, MATLAB is just the assistance of University Physics teaching, it can't replace the theory teaching. Students should insist on the foundation in control of basic physical knowledge, gradually manage about the making use of MATLAB, so as to promoting their study.

Following imitation carry out with the MATLAB7. 0 software.

To compare with monochrome spectrum simulation, the colourful spectrum of atom light interference and diffraction of computer imitating is really a blank and have never seen discussed in literatures. In the numerously experiments imitating, majority of authors all just only carry on simulation of monochrome [5, 6], colourful light spectrum imitation mostly don't show ideal [7,8]. Some successful colourful light spectrum imitation are finished by Lan haijiang etc. [9].But the colourful spectrum of the atom light imitation, such as Mercury atom spectrum, Hydrogen atom spectrum etc., don't find report in literatures.

This article tries to carry on simulation about colourful spectrum of different atoms, and synthesizes according to principle of RGB colour light and makes use of MATLAB visual program, developed to imitate grating spectrum of true mercury, hydrogen experiment and filled up the lack of colourful spectrum data in the university's Physics teaching.

2 The Colourful Spectrum Emulation of The Mercury Atom Spectrum

Suppose the light grating contains N unit, and width of each transparent slot unit as a, width of each opacity slot unit as b, the grating constant is d, then the light strength distribution of Fraunhofer diffraction as follows[10]:

$$I = I_0 (\frac{\sin \alpha}{\alpha})^2 (\frac{\sin N\beta}{\beta})^2 \quad (1)$$

In formula (1), I_0 means the light strength of center, suppose the diffraction angle is θ, λ is the wave-length of monochrome light, then

$\alpha = \pi b \sin\theta / \lambda$, $\beta = \pi d \sin\theta / \lambda$, $(\frac{\sin\alpha}{\alpha})^2$ in formula (1) is factor of single slot diffraction, $(\frac{\sin N\beta}{\beta})^2$ represents the factor of multiple light diffraction .

We combine the wave-length of the spectrum of mercury atom and with colour only to should of RGB code value, with the MATLAB tool write true procedure of imitating of a mercury spectrum as follows:

```
Clear;
Lambda= [607.3, 579.1, 577,491.6, 435.8]*1e-9;
RGB=[1,0.495,0;1,0.98,0;1,0.97,0;0,1,0.5;0,0,1;];
d=1.5e-5;
b=12e-6; N=18;
Bright=8000;
Irgb=zeros (250, 1885, 3);
IW=zeros (250, 1885, 3);
For k=1:5
Theta= (-0.015*pi: 0.00005:0.015*pi);
Phi=2*pi*d*sin (theta)/lambda (k);
Alpha=pi*b*sin (theta)/lambda (k);
If= (sinc (alpha)). ^2;
Idgs= (sin (N*phi/2). /sin (phi/2)). ^2;
I=Idf.*Idgs
For i=1:250
Iw (i,:, 1)=I*RGB(k,1) ;
Iw (i,:, 2)=I*RGB(k,2) ;
Iw(i,:,3)=I*RGB(k,3) ;
End
Irgb=Irgb+Iw;
Iw= [];
End
Br=1/max (max (max (Irgb)));
Iall=Irgb*Br*Bright;
Imshow (Iall);
clc
```

According to formula (1), combine the program showed, we can acquire different effect of colourful spectrum diagrams of the mercury atom, just as showed in figure 1. In this diagram, besides which spectral line of zero class, we can see spectrum of $\pm 1, \pm 2$ class and part of class three spectrums. Along with spectral line class turning higher, spectral line turn thinner and acuity, this is concurrent with actual experiment. Reducing grating constant, especially raise the grating constant N, we can remarkably raise the resolution ability of the grating spectrum. To acquire the thin and sharp spectrum that can satisfy to measure request, we have to consumedly increase the total number of grating, and reduce the grating constant at the same time.

Fig. 1: Mercury atom colour grating spectrum simulating diagram

3 The Colourful Spectrum Emulation of The Hydrogen Spectrum of Ballmer Series

Analysis of the hydrogen atom Ballmer series spectrum is very important in Optical Physics, during the development of Quantum Physics and modern Physics, the research of hydrogen spectrum has an incomparable function. However, because of experiment condition and funds restriction, in majority of colleges, researchers didn't develop the wave-length analysis research of hydrogen spectrum currently, so the students and teachers all could not acquire hydrogen spectrum diagrams of Ballmer series to keep a view about the spectrum, it must be an important missing of Physics experiments.

In this emulation we try to establish different light grating constant, dissimilarity penetrate width, different slot number, we can imitate different parameters of hydrogen spectrum figure. If regulate each parameter by oneself, readers can acquire arbitrary colourful hydrogen spectrum diagram, consumedly enriched people about sight-seeing and comprehension of hydrogen spectrum. Following are program of imitating about hydrogen spectrum with MATLAB. The hydrogen spectrum wave-length draws from [13].

Fig. 2: Hydrogen atom colour light grating spectra imitate grating

Figure 2 shows the four famous lines of hydrogen atom Ballmer series. Just a hard analysis about the four famous lines, opened the front door of understanding with atom world and finally caused the birth of Quantum Mechanics and modern Physics. Descend the grating constant, the appropriate increment of opacity sews width, and increment the total number of grating slots, the resolution of hydrogen spectrum consumedly raise.

4 Conclusions

Imitate a true mercury atom, hydrogen atom's colourful spectrum by MATLAB emulation, thus more clearly understand the principle, method, step and substantial etc. Combine Optical experiments and MATLAB Physics teaching, it's helpful to modern education, and contribute to innovation consciousness in trains about students, and to a student in solve of Physical concepts and regulations, stir up interests in their study and raise the teaching effect. Along with the progress of science and technology, the means of Physical teaching is continuously renewing, the computer can import a visible teaching method in our Physics teaching. Now MATLAB shows a strong graphical function in exhibiting complicated Physics formulas and regulations and thus helps a student goes more deeply into the mystery of Physical phenomenon and regulation.

References

[1] Chengsen Lin. Numerical Compute Methods [M]. Peking: Science Publisher, 2005.
[2] Zhiyong Zhang, Zuying Yang etc. MATLAB Lectures [M]. Peking: Aviation and Aerospace University Publisher, 2006.
[3] Qijun Yao. Optical Lectures (Version 4) [M]. Peking: Higher Education Publisher, 2008.
[4] Guangcan Guo, Qiang Wu. Opticss [M]. Hefei: Chinese Science and Technology University Publisher, 2003.
[5] Bo Lu, Jingjun Huang. Integrates about MATLAB and University Optical Physics Teaching Research [J].Jiangxi: Jiujiang College Journals (Natural Science version) .2010.90:109–114.
[6] Shouxin Hu, Bainian Li. Mathematical Experiments Based on MATLAB [M].Peking: Science Publisher.2004: 141.
[7] Ping Yu. Imitates about Fraunhofer Diffraction of White Light by MATLAB [J]. Normal University College Journal (Natural Science Version) in ChongQing. 2007, 24: (2) 43–45.
[8] Shixi Fan. White Light Computer Emulation about Single Slot. Peking Printing Institute Journal. 2003, 11: (4)58–60.
[9] Haijiang LAN, Xiaoming Pan, etc. MATLAB Application at Foundational Physics Teaching [J].LiuZhou College Journal 2008, 23: (3) 121–124.
[10] Shouzhu Chen, etc. Common Physics (version 6) [M].Peking: Higher Education Publisher.2006.
[11] Shenxun Ding, Kongshi Zhang. Physics Experiments Lectures [M]. Peking: QinHua University Publisher.1992: 202.
[12] Zhexue Ge. Master in MATLAB [M]. Peking: Electronics Industrial Publisher 2008: 148.
[13] Wenwei Ma, etc. Physics Lectures (Version 2) [M]. Peking: Higher Education Publisher.2006: 281.

Pimmpnarnar Roongchirarote* and Yan Zhao

The Impact of Innovation Strategy on Firm Financial Performance

Abstract: This study aims to gauge the extent to innovation strategy has diffused into the strategic plans of the Small-Medium Enterprises (SMEs) and the impact of those innovation strategies on their financial performance. Out of 700 SMEs contacted for survey purpose 409 responded ranging from manufacturing and trade to service industries across Thailand. The data was analysed using a Structural Equation Modelling (SEM) approach. The findings proved that innovation strategy had significant influence on SMEs financial performance.

Keywords: Innovation Types, Financial Performance, SMEs, SEM

1 Introduction

In high business competition, firms have to adapt themselves in term of innovation [1] so as to survive in such rapid marketplace. There are many literatures indicating that innovation drives economic growth through industrial competition [2]. According to Ref. [3], there are four types of innovation playing a vital role in business strategies, which are comprise of organizational innovation, marketing innovation, process innovation, and product innovation. Also, many studies testified relations of innovation in term of business strategy and firm performance.

Previous literatures studied relationship between business innovativeness and Small-Medium Enterprises (SMEs). Ref. [4] was interested in study of innovation conducted by small firms because SMEs become major counterparts in economy of countries in the world. Ref. [5] supported that SMEs with innovated products, services or structure models could attract niche markets and brings about the high profitability compared to competitors. In 2015, Thai SMEs were 99.7 % of total enterprises in Thailand and held 80.4 % of overall employment. Thai SMEs contained 41 % of total GDP of the country whereas expanding 5.3 % from previous year [6]. The aim of this study is to explore business strategy in term of innovation affecting to Thai SMEs performance.

***Corresponding author: Pimmpnarnar Roongchirarote**, School of Management, Shanghai University, Shanghai, P.R. China. Email: pimmp.roong@outlook.com.
Yan Zhao, Research Center for Innovation and Knowledge Management, School of Management, Shanghai University, Shanghai, P.R. China. Email: zhaoyan87@shu.edu.cn

DOI 10.1515/9783110540048-083,

2 Literature Review and Hypothesis Development

2.1 Innovation

Innovation is kind of general term which is studied from many aspects and disciplines, thus there is no a single definition and measure of innovation [7]. According to Ref. [8], innovation included any new of product, process, structure, or policy that managers perceived. Moreover, Ref. [9] mentioned that innovation is the capability to build economic value from new ideas.

2.2 Financial Performance

The firm performance is a very complicated area of scholars to determine the definition and construct its measurement. Ref. [10] explained that business performance (firm performance) included the operational and financial results. Measuring firm performance is the interpretation of the business reality outcomes to such standard symbols reflecting and relying under the same situation [11]. For SMEs, the appropriate tool to indicate firm performance is financial approaches [12].

2.3 Innovation and Financial Performance

Some researchers explored the relationship between innovation and firm performance. Ref. [13] also stated that firms with sustained innovation led to the higher profits than the ones avoiding it. Based on Ref. [3], the four types of innovation (product, process, marketing and organizational innovation) are used to develop hypothesizes of this study.

2.4 Organizational Innovation

Ref. [3] described that the implementation of new business procedures and rules to internal and external firm environments is called organizational innovation. While Ref. [14] stated that organizational innovation is the implement for business competitive advantage. Findings in previous studies indicated that organizational innovation was the fundamental component for innovative capabilities [15]. Ref. [16] pointed that both cooperation and organizational arrangement helped increasing technological innovations (product and process innovation). Ref. [17] stated that non-technological innovation (marketing and organizational innovation) stimulated product and process innovation to resulting in business achievement by affecting to firm profitability positively. Thus, the hypotheses are:

H1a: Organization innovation affects marketing innovation positively.
H1b: Organization innovation affects process innovation positively.
H1c: Organization innovation affects product innovation positively.
H1d: Organizational innovation affects financial performance positively.

2.5 Marketing Innovation

A marketing innovation is defined by Ref. [3] as practices of brand new marketing procedures in terms of significant changes in packages, designs, product pricing, promotion activities, and product placement. Responding to customer demands and market opportunities are significantly influenced by marketing innovation [18]. The interrelation between non-technological innovation and product (or service) innovation was discovered by Ref. [19] in a public organization study. Furthermore, some researchers found that the relationship between marketing innovative activities and firm or business performance were positive [20]. Thus, the hypotheses are:
H2a: Marketing innovation affects product innovation positively.
H2b: Marketing innovation affects financial performance positively.

2.6 Process Innovation

Ref. [3] gave the meaning of process innovation as new practices or improvement of the production and transportation including changes in techniques, equipment and software. Ref. [21] stated that process innovation is internal core method and driven efficiently. Process innovation is focused in manufacturing sector as core competency for competitive advantage, which is positively related to firm growth. Ref. [22] studied the relation and sequence between product and process innovation. Ref. [23] found that product and process innovation impact positively on firm performance. Hence, the hypotheses are as followed:
H3a: Process innovation affects product innovation positively.
H3b: Process innovation affects financial performance positively.

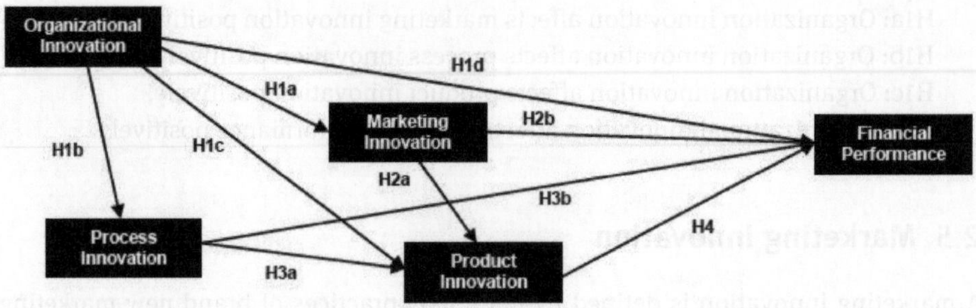

Fig. 1: The research conceptual framework

2.7 Product Innovation

Introduction of goods or services which are new or improved from the existing ones in terms of its characteristics or usage functions is called product innovation [24]. According to high competitive markets, innovation to products is mainly used in order to respond the shorter of product life cycle [25]. Ref. [20, 26] proved that product innovation was positively related to firm performance. Therefore, the hypothesis is

H4: Product innovation affects financial performance positively.

Fig. 1 presents the research hypotheses of the study.

Tab. 1: Factor and Reliability Analysis for Variables

	Factor Loading	% of Variance Explained	AVE	Cronbach's Alpha (α)
Organizational Innovation (OrgInno)				
OrgInno 1	0.739	47.788	0.845	0.972
OrgInno 2	0.848			
OrgInno 3	0.914			
OrgInno 4	0.941			
OrgInno 5	0.989			
OrgInno 6	0.963			
OrgInno 7	1.010			
Marketing Innovation (MarInno)				
MarInno 1	0.459	6.437	0.625	0.886
MarInno 2	0.792			
MarInno 3	0.882			
MarInno 4	0.858			
MarInno 5	0.880			
Process Innovation (ProcInno)				
ProcInno 1	0.870	7.292	0.676	0.921
ProcInno 2	0.805			
ProcInno 3	0.617			
ProcInno 4	0.914			
ProcInno 5	0.870			
Product Innovation (ProdInno)				
ProdInno 1	0.896	12.56	0.683	0.893
ProdInno 2	0.902			
ProdInno 3	0.726			
ProdInno 4	0.802			
ProdInno 5	0.791			
Financial Performance (FinPerf)				
FinPerf 1	0.854	4.978	0.803	0.897
FinPerf 2	0.913			
FinPerf 3	0.920			

3 Methods

3.1 Measures of The Study

This study uses questionnaire approach to explore Thai SMEs' innovation strategy in four dimensions affecting to financial performance. The Ref. [27] structured survey questionnaire (based on the concepts of Ref. [3]) is applied for the section of the four types of innovations. By following Ref. [28] survey structure, profitability, return on investments (ROI), and return on assets (ROA) are used to measure SMEs financial performance. The data collection is strictly controlled with criteria that business run by owner or manager and organization size is not more than 200 permanent employees.

In the questionnaire, the seven-point Likert scale [29] is used for exploring the relationship between the innovation strategy and firm financial performance. There are four independent variables which are the four types of innovation. They are measured by scale rating from "1=strongly disagree (SD)" to "7= strongly agree (SA)." Similarly, dependent variable of this study is financial performance. The questions of the session are measured by scale rating from "1 = much lower (ML)" to "7= much higher (MH)."

3.2 Sample and Data Collection

With the help of Department of International Trade Promotion, Thai Ministry of Commerce, the 700 questionnaires were haphazardly selected in proportion each sector from total numbers of Thai SMEs. They were distributed by mails with a telephone call for introducing the research study. A total of 409 of 700 questionnaires were usable and returned back after one follow-up. The 409 respondents or 58 % response rate is adequate for organization response rate for surveys [30].

4 Analysis and Results

In order to explore the relationship in the Fig. 1, the analysis is executed into two stages. Firstly, factor structuring analyzes the correlations of variables whether strengthening enough for grouping into a small set of variables. The study applies Exploratory Factor Analysis (EFA) by SPSS in order to decrease the larger set of variables into smaller for further analysis step. Secondly, factors relationship analysis is performed by Structural Equation Modeling (SEM) technique by AMOS. This stage includes model fit testing by using Confirmatory Factor Analysis (CFA), correlation analysis, and path analysis. The results are included in this stage.

4.1 Factor Analysis

Factor analysis is used to detect the relationships of a large numbers of variables and to summarize them into a shortened set of scores. During factor analysis procedure, all variables constructs are tested for adequacy which indicates by Kaiser-Meyer-Olkin (KMO) index. The result of KMO is 0.912 which is more than 0.7 [31]. The Sig. in the result of Bartlett's test of Sphericity is 0.000 which indicates the suitability of the data ($p<0.05$).

Tab. 2: Goodness of Fit Indices

Fit Index	Result	Reference value
χ^2 /degree of freedom	3.653	$\leq 2^a$; $\leq 5^b$
CFI (Comparative Fit Index)	0.936	$\geq 0.93^a$
NFI (Normed Fit Index)	0.914	$\geq 0.9^a$
RFI (Relative Fit Index)	0.900	$\geq 0.9^a$
IFI (Incremental Fit Index)	0.936	$\geq 0.9^a$
TLI (Tucker-Lewis Fit Index)	0.923	$\geq 0.9^a$
RMSEA (Root Mean Square Error of approximation)	0.083	$<0.08^a$; $<0.1^b$

[a] Acceptability: acceptable.

[b] Acceptability: marginal.

The four dimensions of innovation and financial performance are calculated to figure out factor loadings. Table 1 shows details of factor loadings, % of variance explained, and Cronbach's Alpha of constructs. The factor loadings of all items are satisfactory because they have the minimum value at 0.459 which is higher than the rule of thumb level of 0.4 [32]. 79.055 % of variance is total variance explained to factor analysis. It is such high scores for validity. Moreover, Average Variance Extracted (AVE) scores of the constructs ranging from 0.625 to 0.845 indicate convergent validity. Likewise, Cronbach's Alpha value of innovation types and financial performance are between 0.886 and 0.972 which are accepted at the level more than 0.70 [33]. It predicts that they have high related internal consistency or reliability.

4.2 Structural Analysis

With the complexity of the conceptual framework model carrying several mediating effects among innovations and financial performance, SEM is chosen for exploring the relationships between variables. SEM performs multilevel regression to fit the proposed model by providing in a set of weights to estimate unobserved constructs (latent variables). One of the procedures of SEM analysis is model fit testing. Table 2 indicates measurement of model fit. As shown, all the index values reach the rules of thumb. These determine that the data fit to the model.

Tab. 3: Means, Standard Deviations (S.D.), and Correlation Result

	Mean	S.D.	OrgInno	MarInno	ProcInno	ProdInno	FinPerf
OrgInno	3.26	1.33	1				
MarInno	4.65	1.22	.398**	1			
ProcInno	4.01	1.27	.633**	.385**	1		
ProdInno	4.34	1.28	.513**	.545**	.566**	1	
FinPerf	4.70	0.77	.272**	.390**	.283**	.368**	1

**. Correlation is significant at the 0.01 level (2-tailed).

Prior to proving the hypotheses, correlation analysis is applied to measure the strengths of association between variables in bivariate directions or relationships. Table 3 indicates the strong and positive relationship between the variables since there are significance ($p<0.01$). The correlation measurement identifies not only relations between the variables, but it also reveals the association of an intricate network among factors. These imply the mediation effects existing among some innovativeness and financial performance. SEM is the approach for exposing complex relationships of mediating effects shadowing in this study model in Fig. 1.

Tab. 4: Standardized Path Coefficients Result

Hypothesis			Standardized Path Estimate	P-Value	Result
H1	a	OrgInno → MarInno	0.489	<0.001	Supported
	b	OrgInno → ProcInno	0.827	< 0.001	Supported
	c	OrgInno → ProdInno	−0.045	0.435	Not supported
	d	OrgInno → FinPerf	0.165	0.040	Supported
H2	a	MarInno → ProdInno	0.414	< 0.001	Supported
	b	MarInno → FinPerf	0.335	< 0.001	Supported
H3	a	ProcInno → ProdInno	0.568	< 0.001	Supported
	b	ProcInno → FinPerf	−0.113	0.191	Not supported
H4		ProdInno → FinPerf	0.199	0.005	Supported

The outcome in Table 4 indicates that relationships among innovation types (H1a, H1b, H2a, and H3a) and between innovations and financial performance (H1d, H2b, and H4) are influenced ($p<0.05$) except the relations of OrgInno and ProdInno (H1c) and relations of ProcInno and FinPerf (H3b).

From the Table 4, Thai SMEs indicates MarInno is the most influence driver of FinPerf (standardized path estimate = 0.335, $p<0.001$). Moreover, ProdInno and OrgInno impact FinPerf of Thai SMEs with standardized path estimate = 0.199 ($p<0.05$) and standardized path estimate = 0.165 ($p<0.05$), respectively.

5 Conclusion

This study confirms two levels of relationships between innovations and financial performance. First, the interrelations among innovation types exist excluding the association between organizational innovation and product innovation. Second, organizational, marketing and product innovation impact on financial performance while process innovation may cooperate with other innovation types in order to reach firm goals. These support the literatures that innovativeness plays a vital role of firm business strategies.

In this case marketing innovation is the strongest effect to financial performance. This shows that SMEs managers focus investing on innovativeness of marketing operations rather than organizational, process, and product innovations. Investment in marketing innovation may waste if the management team is lack of experience and well-planned marketing positioning. The fallibility of wrong decision could bring about the business failure especially for SMEs owing to resources limitation.

This empirical study suggests further research studies to provide more causal explanation on the relationship between organizational innovation and product innovation and between process innovation and financial performance.

References

[1] R. F. Hurley and G. T. Hult, "Innovation, market orientation, and organizational learning: an integration and empirical examination," Journal of Marketing, 62 (3), 42–54, 1998.
[2] Zaltman, Gerald, R. Duncan, and J. Holbek, Innovations and Organizations, New York: John Wiley & Sons, 1973.
[3] OECD Oslo Manual, Guidelines for collecting and interpreting technological innovation data, OECD Publishing, 2005.
[4] Z. J. Acs and D. B. Audretsch, "Innovation in large and small firms: an empirical analysis," American Economic Review, 78 (4), 678–690, 1988.
[5] M. E. Porter, Competitive Strategy: Techniques for Analyzing Industries and Competitors, New York: Free Press, 1980.
[6] OSMEP, "Thai SMEs Annual report," 2016.

[7] R. Adams, J. Bessant, and R. Phelps, "Innovation Management Measurement: A Review," International Journal of Management Reviews, 8(1), 21–48, 2006.
[8] N. Nohria and R. Gulati, "Is Slack Good or Bad for Innovation?" Academy of Management Journal, 39(5), 1245–1264, 1996.
[9] J. L. Bercovitz and M. P. Feldman, "Fishing Upstream: Firm Innovation Strategy and University Research Alliances," Research Policy, 36, 930–948, 2007.
[10] N. Venkatraman and V. Ramanujam, "Measurement of business performance in strategy research: a comparison of approaches," Academy of Management Review, 1(4), 801–814, 1986.
[11] M. Lebas, "Performance measurement and performance management," International Journal of Production Economics, 41(1–3), 23–35, 1995.
[12] E. I. Altman, "Financial ratios, discriminant analysis and the prediction of corporate bankruptcy," The Journal of Finance, 23(4), 589–609, 1968.
[13] P. A. Geroski and S. Machin, "Do Innovating Firms Outperform Non-Innovators?" Business Strategy Review, 3(2), 79–90, 1992.
[14] Y. Y. Yeh and L. H. Chou, "Transforming a Semiconductor Company into a Learning Organisation: A Bottom-up Approach of Knowledge Management Implementation," International Journal of Technology Management, 39 (1–2), 219–234, 2007.
[15] C. Y. Lin and M. Y. Chen, "Does innovation lead to performance? An empirical study of SMEs in Taiwan," Management Research News, 30 (2), 115–132, 2007.
[16] C. Staropoli, "Cooperation in R&D in the pharmaceutical industry-the network as an organizational innovation governing technological innovation," Technovation, 18 (1), 13–23, 1998.
[17] T. Schmidt and C. Rammer, "Non-technological and technological innovation: strange bed fellows," 2007.
[18] C. R. Cano, F. A. Carrillat and F. Jaramillo, "A Meta-Analysis of the Relationship between Market Orientation and Business Performance: Evidence from Five Continents," International Journal of Research in Marketing, 21 (2), 179–200, 2004.
[19] R. M. Walker, "An empirical evaluation of innovation types and organizational and environmental characteristics: towards a configuration framework," Journal of Public Administration Research and Theory, 18 (4), 591–615, 2008.
[20] M. Varis and H. Littunen, "Types of Innovation, Sources of Information and Performance in Entrepreneurial SMEs," European Journal of Innovation Management, 13 (2), 128–154, 2010.
[21] J. M. Utterback and W. J. Abernathy, "A dynamic model of process and product innovation," Omega, 3, 639, 1975.
[22] W. J. Abernathy and J. M. Utterback, "Patterns of industrial innovation," Technology Review, June/July, 40–7, 1978.
[23] M. Ar and B. Baki, "Antecedents and Performance Impacts of Product versus Process Innovation: Empirical Evidence from SMEs Located In Turkish Science and Technology Parks," European Journal of Innovation Management, 14 (2), 172–206, 2011.
[24] OECD, Innovation Policy and Performance: A Cross-Country Comparison. Organization for Economic Co-operation and Development, OECD Publishing, 2005.
[25] G. Duranton and D. Puga, "Nursery cities: Urban diversity, process innovation, and the life cycle of products," American Economic Review, 91(5), 1454–1477, 2001.
[26] B. L. Bayus, G. Erickson and R. Jacobson, "The Financial Rewards of New Product Introductions," Management Science, 49 (2), 197–210, 2003.
[27] G. Gunday, G. Ulusoy, K. Kılıç and L. Alpkan, "Effects of innovation types on firm performance," International Journal of Production Economics, 133(2), 662–676, 2011.
[28] W. Wan Mohd Nasir, "The Relationship between Strategic Orientation and Firm Performance: Evidence from Small and Medium Enterprises in Malaysia," 2013.

[29] J. A. Krosnick and S. Presser, "Question and Questionnaire Design," in Handbook of Survey Research. 2nd Edn, P. V. Marsden and J. D. Wright, Eds., Emerald, 2010, 263–313.
[30] Y. Baruch and B. Holtom, "Survey response rate levels and trends in organizational research," Human Relations, 61(8), 1139–1160, 2008.
[31] J. F. Hair, W. C. Black, B. J. Babin and R. E. Anderson, Multivariate Data Analysis: A global perspective, NJ: Prentice Hall, Pearson Education, Inc, 2006.
[32] J. P. Stevens, Applied multivariate statistics for the social sciences (2nd edition), Hillsdale, NJ: Lawrence Erlbaum Associates, 1992.
[33] D. Iacobucci and A. Duhachek, "Advancing Alpha: Measuring Reliability with Confidence," Journal of Consumer Psychology, 478-487, 2003.

Zheng Xie, Xin-Ran Dong and Ji-An Duan*

Transverse-Load Characteristics of Long Period Fiber Gratings Fabricated by Femtosecond Laser Pulses

Abstract: The transverse-load characteristics of long period fiber gratings (LPFGs) fabricated by 800 nm femtosecond laser are demonstrated and investigated. It is observed that the resonant wavelength of LPFG is dependent on the direction of applied force. When a transverse load is applied along different orientations of the LPFG, the resonant wavelength will experience red shift or blue shift. However, the transmission attenuation is gradually decreases as the applied transverse-load increases, which is independent of the direction of applied force. This is mainly due to the asymmetrical distribution of the refractive index on the cross-section of LPFG induced by femtosecond laser pulses, the load-induced birefringence lead to the rotation of optical principal axes in the LPFG.

Keywords: transverse-load characteristics; long period fiber grating; femtosecond laser

1 Introduction

In recent year, long period fiber gratings (LPFGs) have attracted great research interests due to their unique advantages of compact size, low insertion loss, high sensitivity, and immunity to electromagnetic interference, which is widely in the sensing fields to monitor temperature, bending, refractive index [1-4]. Various methods for writing LPFG have reported, including electric discharge [5], CO_2 laser irradiation [6] , mechanical micro-bending [7] and ultraviolet laser irradiation [8] . The transverse-load characteristic is one of the most application of LPFGs. And the LPFGs obtained by different method has shown various mechanical characteristics due to different fabrication mechanisms. Liu et al. [9] and Zhang et al. [10] have studied the transverse-load characteristics of LPFGs fabricated by UV-laser in photosensitive optical fiber. They found that the resonant wavelength of LPFG split into two sub-

Zheng Xie, College of Mechanical and Electrical Engineering, Central South University, 932 South Lushan Street, Changsha, 410083, China. Email: xiezheng@csu.edu.cn
Xin-Ran Dong, College of Mechanical and Electrical Engineering, Central South University, 932 South Lushan Street, Changsha, 410083, China. Email: ziranhao2008@163.com
***Corresponding author: Ji-An Duan,** College of Mechanical and Electrical Engineering, Central South University, 932 South Lushan Street, Changsha, 410083, China. Email: duanjian@csu.edu.cn.

DOI 10.1515/9783110540048-084,

peaks as the transverse load increased and the transversal strain sensitivity is about 800 times higher than FBG sensor (~500 nm/ (kg/mm)). Wang et al [11, 12] has reported the transverse-load characteristics of CO_2 laser induced LPFGs and demonstrated that the resonant wavelength shifts have shown significantly directional correlation with the direction of applied transverse-load. Femtosecond laser present new opportunities for optical fiber sensors because of extremely short width as well as high peak power but almost without thermal affected [13, 14]. Various structures fabricated by femtosecond laser have reported and used in the sensing fields [15-17]. However, the transverse-load characteristics of long period fiber gratings (LPFGs) written by femtosecond laser have not yet seen.

In this paper, the unique transverse-load characteristics of femtosecond-laser induced LPFGs are investigated in detail. The changes of resonant wavelength and transmission attenuation with the transverse-load increase have studied. It is find that the resonant wavelength shift direction depends on the direction of applied transverse-load while the transmission attenuation of peak is independent of that. Additionally, using the related theory of optical principal axes' rotation to explain the reason of resonant wavelength shift.

2 Fabrication of LPFG and Experimental Setup

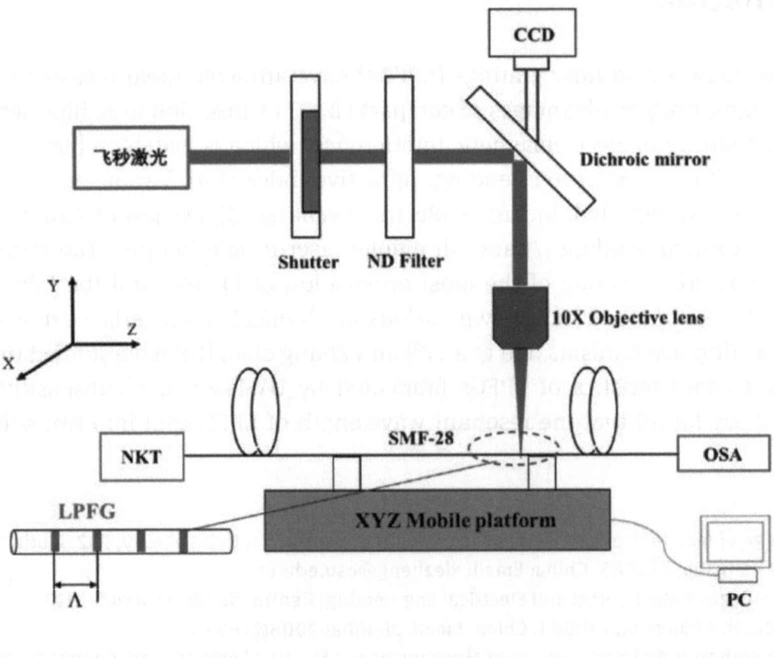

Fig. 1: The schematic diagram of femtosecond laser fabrication and detection system

Fig.1 shows the schematic diagram of femtosecond laser fabrication and detection system. An amplified 800 nm femtosecond laser (Spectra Physics, Inc.) with pulse width of 120fs and repletion rate of 1 kHz is used to fabricate LPFGs in a bare standard communication fiber (Corning SMF-28), which is mounted on a computer-controlled mobile platform with 5 nm resolution. The laser beam is tightly focused into the fiber core by a 10× microscope objective lens (NA = 0.25) and the single pulse energy is about 2 μJ measured before the objective lens. We use a broad light source (NKT) with wavelength range from 600 nm to 1700 nm and an optical spectrum analyzer (OSA, Agilent 86142B, wavelength range 600 nm~1700 nm) with a resolution of 0.06 nm to trace the transmission spectrum. In the experiment, the inscription speed, the grating period and the number of period are 20 μm/s, 490 μm and 60, respectively. The transmission spectrum of LPFG is shown in Fig.2. We can see that the LPFG exhibit three dips of ~13.5 dB, -13 dB and ~7.5 dB at 1400 nm, 1468 nm and 1608nm, respectively.

Fig. 3 illustrates the experimental setup for testing transverse-load characteristics. The two rotatable disks are used to control the grating rotation angle and the texted LPFG samples are mounted on the centers of the two disks. A force load P is applied on the surface of grating. Meanwhile, a broadband light source (NKT) and an optical spectrum analyze (OSA) are employed to trace the change of transmission spectrum on-line and in time. The length L of the loaded segment in the LPFG is 60mm, so the actual transverse-load on the unit length of LPFG is P/L. During the experiment, a series of responses of the LPFG to the transverse load from different load directions within the range of 0°~360° are recorded with an angle step of 45°.

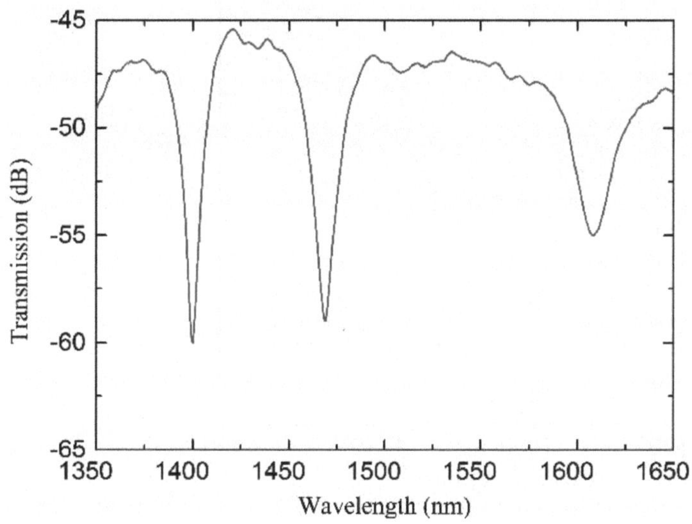

Fig. 2: Transmission spectrum of LPFG with grating period of 490μm

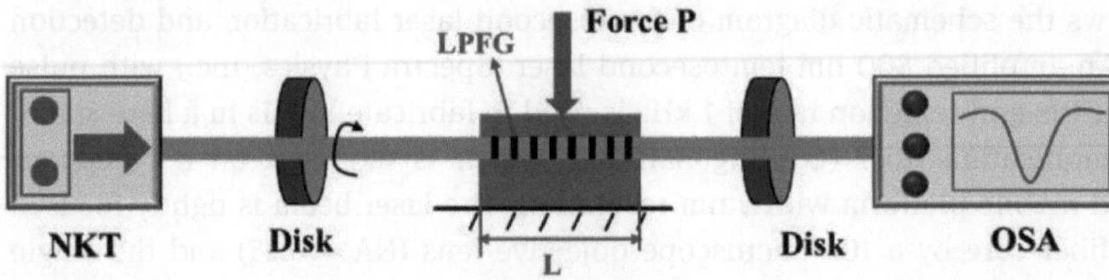

Fig. 3: Experimental setup for testing transverse-load characteristics of LPFG

3 Analysis and Discussion

We take the dip wavelength of 1400 nm as an example to study the transverse-load characteristics. Fig.4 shows the transmission spectra of LPFG with the transverse load increases along 0° load direction. It can be seen that the resonant wavelength shifts toward longer wavelength as the transverse load increases. In the range of 0~333 N•m^{-1}, the resonant wavelength changes from 1399.69 nm to 1402.55 nm, the wavelength sensitivity is about 8.59 pm/ (N•m^{-1}). The transmission attenuation decrease with the transverse load increases, changing from -60.33 dB to -54.69 dB, which is due to the decrease in the coupling efficiency between the fundamental core mode and the cladding modes.

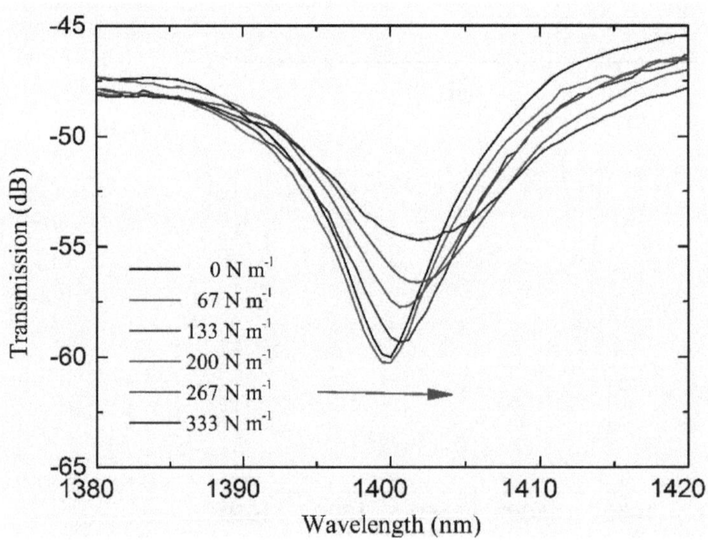

Fig. 4: Transmission spectra of LPFG with the transverse load increases along 0° load direction

When the transverse load is applied along different orientations, the resonant wavelength and peak transmission attenuation changes of LPFG written by femtosecond laser as shown in Fig.5. The resonant wavelength of LPFG is shifted toward longer wavelength when the transverse load is applied along 0°, 45°, 225°, 270°and 315° orientations. In contrast, the resonant wavelength of LPFG shifts toward shorter wavelength when the transverse load along other orientations, for example, 90°,135° and 180°. From the Fig.4 (a), it can be seen that the corresponding resonant wavelength shifts are different from different direction at the same applied transverse load. For instance, in the range of 0°~180°, for the load direction of 0°, the red-shift wavelength is 2.86 nm, for the load direction of 180°, the blue-shift wavelength is 3.73 nm. Therefore, the load sensitivities of the resonant wavelength of the LPFG written by femtosecond laser pulses strongly depend on the load directions. This is due to the asymmetrical distribution of the refractive index on the cross-section of LPFG induced by femtosecond laser pulses. The index modulation on the laser incident side of the fiber is larger than that on the laser output side of the fiber. Thus, a linear birefringence will exist in the LPFG. When the fiber is applied transverse load, the load-induced birefringence needs to overcome the original linear birefringence to rotate the optical principal axes of fiber [18, 19]. The applied transverse load in different load direction will produce different original birefringence. Therefore, the wavelength shifts depend strongly on the load directions. However, the peak transmission attenuation changes have always exhibit a tendency to decrease at the different load directions as shown in Fig.5 (b). It means that peak transmission attenuation of the LPFG written by femtosecond laser pulses is independent of the load directions. In addition, the experiment results have shown that the wavelength shift variation is almost the same in the interval of 180° loading direction as shown in Fig.5 (a). For instance, the wavelength shift variations of LPFG at the 45°and 225° load direction are 2.41 nm and 2.40 nm, respectively. For the load direction of 135° and 315°, the corresponding wavelength shift variations are -2.98 nm (negative sign presents blue -shift) and 3.6 nm, respectively.

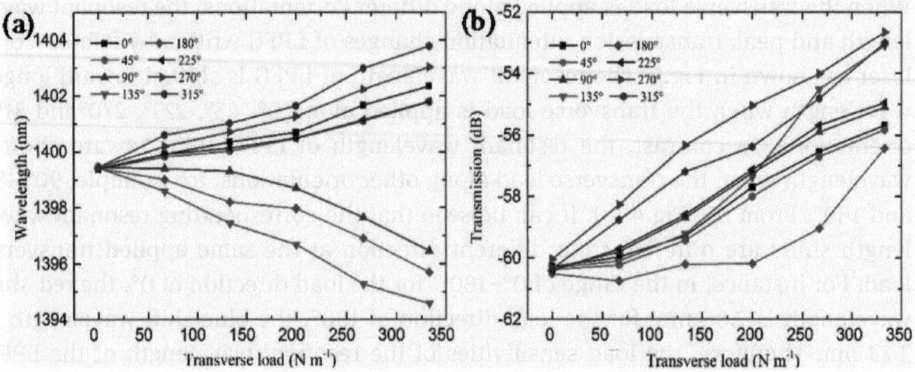

Fig. 5: (a) Resonant wavelength and (b) peak transmission attenuation of LPFG as function of the transverse load applied along different load directions

During the experiment, the transverse load of 333 N•m^{-1} is kept as same, corresponding to the different load directions, the wavelength shift variation and transmission attenuation variation are recorded as shown in Fig.6. We can see that there is one load direction that wavelength experience most sensitive red shift and that wavelength experience most sensitive blue shift in the range of 180° load directions. For example, the maximum blue-shift direction is 135° load direction and the maximum blue-shift direction is 45° load direction. In one word, the resonant wavelength shifts depend on the load directions. The function between load direction and resonant wavelength shift wavelength is approximately a cosine function of 1.25 periods as shown in Fig.6(a), which is less than that of LPFG induced by CO_2 laser (~2 periods). Moreover, the transmission attenuation variations are always negative values, which mean that it is independence of load directions. And also, the function is a cosine function of 2 periods approximately as shown in fig.6 (b). Compared with the LPFG written by CO_2 laser, the transmission attenuation changes of LPFG fabricated by femtosecond laser are more sensitive to the transverse load. This is due to the inhomogeneity of strain distribution profiles along cross section induced by femtosecond laser obviously because of its modified area are smaller [20], which leads the original birefringence in LPFG are relatively weak, thus the load-induced birefringence is significant, the load-induced the rotation angle of the optical principal axes of fiber is larger [11, 18, 19].

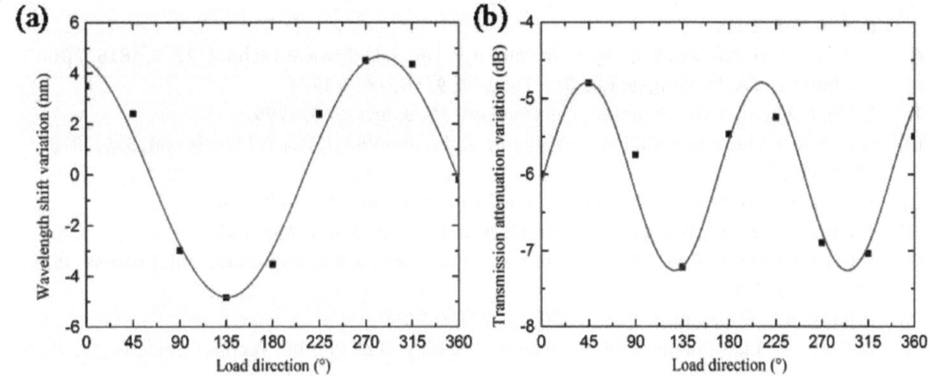

Fig. 6: (a) Wavelength shift variation and (b) peak transmission attenuation variation of LPFG as function of load direction under the applied transverse load of 333N m^{-1}

4 Conclusion

In this paper, the transverse load characteristics of LPFG written by femtosecond laser are reported. The experiment results have demonstrated that the transverse-load characteristics of the resonant wavelength are dependent on the direction of the applied transverse load, whereas that of the peak transmission attenuation are independent of the load direction. There exists one load direction of maximum sensitivity of resonant wavelength red-shift or blue-shift in the range of 180° load directions. These unique transverse-load characteristics of LPFG obtained by femtosecond laser can be used in the field of optical fiber sensors.

Acknowledgement: This research work is supported by the National Natural Science Foundation of China (NSFC) (Grant Nos. 91323301, 51505505, 51475482 and 51475481), Natural Science Foundation of Hunan Province (Grant No. 2016JJ3147), by the Fundamental Research Funds for the Central Universities of Central South University.

References

[1] L.V. Nguyen, D. Hwang, S. Moon, D.S. Moon and Y. Chung, Opt. Express, 16, 15, 11369–11375, 2008.

[2] M. Vengsarkar, P. J. Lemaire, J. B. Judkins, V. Bhatia, T. Erdogan and J. E. Sipe, J. Lightwave Technol., 14, 1, 58–65, 1996.

[3] Y.G. Han, S.H. Song, G.H. Kim, K. Lee, J.H. Lee, S.B. Lee, C.H. Jeong, C.H. Oh and H.J. Kang, Opt. Lett., 32, 15, 2245–2247, 2007.

[4] S.W. James and R.P. Tatam, Meas. Sci. Technol, 14, 5, 49–61, 2013.

[5] M. Smietana, W.J. Bock, P. Mikulic and J. Chen, Meas. Sci. Technol, 22, 1, 015201, 2010.

[6] X.Y. Zhong, Y.P. Wang, C.R. Liao, G.L. Yin, J.T. Zhou, G.J. Wang, B. Sun and J. Tang, IEEE Photon. J., 6, 4, 1–8, 2014.
[7] C.H. Lin, Q. Li, A.A. Au, Y. Jiang, E. Wu and H.P. Lee, J. Lightwave Technol. 22, 7, 1818, 2004.
[8] V. Bhatia and A. M. Vengsarkar, Opt. Lett., 21, 9, 692–694 1996.
[9] Y. Liu, L. Zhang, and I. Bennion, Electron Lett. 35, 8, 661–663, 1999.
[10] L. Zhang, Y. Liu, L. Everall, J. A. R. Williams and I. Bennion, IEEE J Sel Top Quant, 5, 5, 1373–1378, 1999.
[11] Y.P. Wang, D.N. Wang, J. Wei and Y.J. Rao, Appl. Opt., 46, 16, 3079, 2007.
[12] T. Zhu, Y.J. Rao, J.L. Wang and M. Liu, Electron Lett. 42, 8, 451–452, 2006.
[13] H.G. Limberger, C. Ban, R.P. Salathé, S.A. Slattery and D.N. Nikogosyan, Opt. Express, 15, 9, 5610–5615, 2007.
[14] L. Jiang, H.L. Tsai, J. Appl. Phys., 100, 2, 023116, 2006.
[15] D. Grobnic, S.J. Mihailov, C.W. Smelser and H. Ding, IEEE Photon. Technol. Lett., 16, 11, 2505–2507, 2004.
[16] C.R. Liao, T.Y. Hu, D. N. Wang, Opt. Express, 20, 20, 22813–22818, 2012.
[17] B. Li,L. Jiang, S. Wang, H.L. Tsai and H. Xiao, Opt. Laser Technol., 43, 8, 1420–1423, 2011.
[18] J. Calero, S. P. Wu, C. Pope, S. L. Chuang, and J. P. Murtha, J Lightwave Technol., 12, 6, 1081–1091, 1994.
[19] R. Damle and V. Rao, Smart Mater. Struct. 7, 1, 12, 1998.
[20] F. Eldiasty, Appl. Opt., 39, 19, 3197–3201, 2000.

Bei Fragen zur Produktsicherheit wenden Sie sich bitte an:
If you have any questions regarding product safety,
please contact:

Walter de Gruyter GmbH
Genthiner Straße 13
10785 Berlin
productsafety@degruyterbrill.com